CHALLENGES AND GOALS FOR

# ACCELERATORS

IN THE XXI CENTURY

# CHALLENGES AND GOALS FOR
# ACCELERATORS
# IN THE XXI CENTURY

Editors

Oliver Brüning
Stephen Myers
CERN, Switzerland

World Scientific

NEW JERSEY • LONDON • SINGAPORE • BEIJING • SHANGHAI • HONG KONG • TAIPEI • CHENNAI • TOKYO

*Published by*

World Scientific Publishing Co. Pte. Ltd.
5 Toh Tuck Link, Singapore 596224
*USA office:* 27 Warren Street, Suite 401-402, Hackensack, NJ 07601
*UK office:* 57 Shelton Street, Covent Garden, London WC2H 9HE

**Library of Congress Cataloging-in-Publication Data**
Challenges and goals for accelerators in the XXI century / editors Oliver Brüning (CERN, Switzerland), Stephen Myers (CERN, Switzerland).
pages cm
Includes bibliographical references.
ISBN 978-9814436397 (hardcover : alk. paper)
1. Particle accelerators. 2. Colliders (Nuclear physics) I. Brüning, O., editor. II. Myers, S. (Stephen), 1946– editor.
QC787.P3C47 2015
539.7'3--dc23

2014042630

**British Library Cataloguing-in-Publication Data**
A catalogue record for this book is available from the British Library.



In-house Editor: Ng Kah Fee

Typeset by Stallion Press
Email: enquiries@stallionpress.com

Printed in Singapore

# Foreword

The start of the Large Hadron Collider (LHC) operation for High Energy Physics (HEP) in 2010 and the resulting discovery of the Higgs particle in 2012 mark major achievements in the field of accelerator developments and HEP. The technology used for the LHC accelerator represents the culmination of almost 100 years of dedicated accelerator research and development. The use of accelerator technologies for understanding the building blocks of our world started in earnest in the early 1900s, when scientists started looking for ways to increase the energy of charged particles for nuclear disintegration studies and the production of exotic short lived particles. Since then the field has come a long way, starting with technologies and makeshift devices that fitted into small laboratories of individual scientists and evolving to large scale installations that require kilometres of tunnel length and collaborations of thousands of expert engineers and scientists.

In parallel with the technological developments, the field of particle accelerators diversified into a wide range of activities that extended the applications well beyond those of HEP. Accelerator technologies have found use in industrial applications (e.g. sterilization and X-ray scans), medical imaging, diagnostics and treatment (NMR and MRI imaging, X-ray diagnostics and radiation cancer treatment), material science (e.g. synchrotron light sources) and energy generation (e.g. accelerator driven spallation and nuclear fusion).

It seems therefore well justified to look back on the accelerator developments over the past 100 years and to look forward to the challenges that lie ahead for future accelerator based projects. The field is still very dynamic and leads a fast pace of developments. The pace is indeed so fast, that it was not in all cases straightforward to establish a snapshot of the technological developments in a particular area. This turned out to be particularly challenging, for example, in the developments of synchrotron light sources, high power proton drivers and radioactive isotope facilities for which we did not manage to find suitable authors with sufficient spare time at their hand for compiling a thorough overview of their field.

The book has therefore a clear bias towards the application of accelerator technologies for HEP related projects giving testimony to the background of the editors. We apologise for these shortcomings and hope the developments in these very interesting areas of accelerator technology will be covered elsewhere in the future.

The book is the result of contributions of many experts around the world. The editors would like to thank all authors of this book for finding the time to contribute to this project. It would certainly not have been possible to master the challenge without their help and support.

Dr. Oliver Brüning and Dr. Stephen Myers

# Contents

# Part 1

# Introduction

## Chapter 1

# Particle accelerators and the progress of particle physics

Michelangelo Mangano (CERN)

## 1 The Standard Model of fundamental interactions

The picture of a universe built out of a few elementary building blocks is as old as the first speculations on the nature of the physical world by the ancient philosophers. The transition from the level of philosophical speculation to scientific evidence matured over the course of centuries, as the experimental method consolidated and technological advances allowed the quantitative analysis of the phenomena that nature was making available to scientists. Starting from the last century, accelerator technology has been a key player in this intellectual enterprise, enabling, with each new technical advance, the opening of new horizons to be explored.

Since 1973, our picture of the physical laws of nature has been embodied by the so-called Standard Model (SM) of fundamental interactions.[1] A few elementary and point-like particles account for the matter we find in Nature and we create in the laboratory, and for the three fundamental forces (gravitational, strong and electroweak) that are known to act on them. The particles, all of spin 1/2, are grouped into three families. Each family contains two types (or *flavours*) of so-called quarks (each one coming in three different *colours*) and two types of so-called leptons. The interactions are mediated by the exchange of sets of spin-1 particles: 8 massless *gluons* for the strong force,

[1] Different components of the theoretical construction of the SM led to various Nobel prize awards. The unification of electromagnetic and weak interactions, with the prediction of the existence of a neutral weak interaction, gave the 1979 prize to Glashow, Salam and Weinberg. The rigorous formulation of the mathematical rules necessary for the calculation of quantum mechanical effects in the electroweak theory led to the 1999 Nobel prize for 't Hooft and Veltman. The discovery of asymptotic freedom, at the basis of the quantitative understanding of the strong interactions in the SM, led to the 2004 Nobel prize for Gross, Politzer and Wilczek. The confirmation that CKM mixing, introduced below, describes all known phenomenologies of CP violation, led to the 2008 Nobel award for Kobayashi and Maskawa.

the photon for the electromagnetic component of the electroweak force, and the massive $W^{\pm}$ and $Z^0$ for the weak interactions. The members of the first family include the *up* and *down* quarks (labeled $u$ and $d$), and the *electron* and its *neutrino* partner as leptons. The strong force binds together quarks in states known as *hadrons*: these consist of either quark triplets (*baryons*) or quark–antiquark pairs (*mesons*). In particular, triplets of up and down quarks form protons (*uud*) or neutrons (*udd*), which are then bound into nuclei in a way analogous to the binding of electrically neutral atoms into molecules. The exchange of charged $W^{\pm}$ bosons turns $u$ quarks into $d$ quarks, or electrons into neutrinos, and vice versa.[2] The two additional families of spin-1/2 particles are replicas of the first one, and only differ from it by having larger masses, which make these heavier quarks and charged leptons unstable against decay to the lightest ones. In the second family we find the *charm* and *strange* quarks $(c, s)$, and the *muon* and muonic neutrino $(\mu, \nu_\mu)$. In the third family we have the *top* and *bottom* quarks $(t, b)$, and the *tau* and tauonic neutrino $(\tau, \nu_\tau)$. The existence of these additional quarks allows for the formation of new, unstable, hadrons, made of quarks from different families.

The mathematical structure of the SM enforces all known empirical facts about particle interactions. In particular, the mass of the $W$ and $Z$ bosons, which in a generic model would break the gauge symmetry of the theory, results from the mechanism of spontaneous symmetry breaking,[3] mediated by an additional scalar field, whose associated particle is known as the Higgs boson.[4] The observed parity violation is implemented via the assignment of different weak interaction properties to the left and right helicity states of quarks and leptons.[5] Direct transitions between quarks of different families (e.g. $s \to u + W^-$ or $c \to d + W^+$) are explained by Cabibbo–Kobayashi–Maskawa (CKM) mixing, according to which the quark eigenstates of the charged weak interactions are rotated with respect to the quark mass eigenstates. The CKM rotation matrix, which allows for a

[2] In this framework, for example, nuclear $\beta$ decays are interpreted as the transition of a down quark into an up quark, transforming a neutron into a proton, via the emission of a $W^-$ boson, which then transforms into an electron and its antineutrino. The mass of the $W$ boson, much larger than the energy available in the neutron-to-proton transition, creates a potential barrier for the decay, which can be bypassed by a quantum-mechanical tunnelling phenomenon, leading to the typical very long lifetimes of nuclear $\beta$ decays.

[3] Nobel prize in 2008 for Nambu.

[4] Nobel prize in 2013 for Englert and Higgs.

[5] Right (left) helicity characterizes spin-1/2 particles whose spin points towards (opposite to) the direction of motion. Nature prescribes that only left-helicity particles, or right-helicity antiparticles, interact with $W^{\pm}$ bosons, a phenomenon known as *parity violation*.

complex phase, fully describes the known phenomenology of CP symmetry violation.[6] This symmetry violation is reflected by an intrinsic asymmetry between matter and antimatter, which is necessary to explain the dominance of matter over antimatter in the universe. The unitarity of the CKM matrix, furthermore, ensures the observed suppression of flavour-changing neutral currents (FCNC), namely weak transitions between different quark flavours of the same charge, e.g. $s \leftrightarrow d$.[7]

The few paragraphs above give only a short and incomplete summary of the key properties of the SM. A large body of experimental facts guided physicists towards formulating the theory, and later provided accurate quantitative tests of its further predictions and implications, culminating with the recent observation, at the LHC, of a new particle consistent with the Higgs boson. Accelerating and steering particles played a key role in making these experiments possible, as will be reviewed in the next few sections.

## 2 Accelerators, and the experimental path towards the standard model

Atomic spectroscopy and natural radioactivity provided the first evidence of an underlying atomic structure. The manipulation of particle beams created by natural radioactivity, and composed by what was later recognized as alpha nuclei, protons, electrons and photons, began at the turn of the 20th century,[8] allowing for the first time the direct and controlled study of the properties and interactions of these particles. The discovery of special relativity and of quantum mechanics provided the necessary kinematical and dynamical framework to describe a world made of particles whose kinetic energies were large enough to be comparable to their rest mass, but small enough to be subject to the laws of quantum phenomena. Based on the constraints of special relativity and quantum mechanics, Dirac's theory of the

---

[6] CP refers to the combined operation of charge and parity reversal, transforming e.g. a left-helicity quark into a right-helicity antiquark.

[7] FCNC processes are absent from the fundamental SM interactions, since the $Z^0$ boson only mediates couplings between same-flavour quarks. However, they could arise due to higher-order quantum mechanical effects. The family structure of the SM, and the structure of the CKM matrix, leads to large cancellations among these quantum effects, leaving only very small contributions which match the experimental results exactly.

[8] An excellent historical introduction to particle physics, covering in great detail both experimental and theoretical developments from the detection of $X$-rays (1896) to the discovery of the $W$ and $Z$ bosons (1983), is contained in A. Pais "Inward bound. Of matter and forces in the physical world", Oxford University Press, 1986. A excellent pedagogical introduction to the key ideas and results of particle physics, covering up to the end of the 90's and including a discussion of the cosmological implications of particle physics, is contained in R. M. Barnett, H. Mühry and H. R. Quinn, "The charm of strange quarks", AIP Press, Springer, 2000.

electron and of radiation (1928)[9] predicted the existence of the anti-electron, and provided the first example of what was to become a standard outcome of theoretical advances, namely the prediction of yet new particles emerging from the interplay of symmetry and dynamics. Fermi's theory of weak interactions (1933), and then Yukawa's model of strong interactions (1935),[10] consolidated the idea that the laws of the microscopic world are described mathematically by simple interactions between a few basic building blocks, the approach at the basis of modern theoretical particle physics.

Experiments with the low-, and typically fixed-energy, particle beams from natural radioactive sources helped to complete the picture of the basic components of stable matter, with the discovery of the neutron (1932),[11] the evidence for the existence of the neutrino,[12] and, much later, Madame Wu's observation of parity violation in weak decays of cobalt-60 nuclei (1957).

In parallel with the use of natural radiactive sources, the spectrum of tools for exploration was enriched by the discovery (1912)[13] of much higher energy beams driven by astrophysical accelerators, namely cosmic rays. When combined with the development of new observational instruments, such as the cloud chamber,[14] cosmic rays led to the discovery by Anderson of first the positron (1932)[15] and then the muon (1936), in this case unveiling the existence of unstable elementary particles, and of the first member of what came to become known as the "second family" of fundamental particles.

Greater experimental skills and technologies allowed physicists to exploit cosmic rays to uncover additional new particles and interactions. The weakly-interacting nature of the muon was exposed by Conversi, Pancini and Piccioni (1946), proving that it could not be the Yukawa's particle responsible for nuclear interactions. Yukawa's pion was discovered in 1947,[16] followed in the same year by Rochester's and Butler's observation of a new particle decaying into two pions, the kaon, whose exotic properties led to the proposal of the violation of parity symmetry in weak hadronic decays (1956).[17] It was

---

[9]Nobel prize in 1933 for Dirac.
[10]Nobel prize in 1949 for Yukawa.
[11]Nobel prize in 1932 for Chadwick.
[12]Indirect at first, through the continuum electron spectrum in nuclear decays, which was interpreted by Pauli in 1930 as due to the emission of a new particle; and direct, in 1956, through the observation of the conversion into positrons of antineutrinos from a nuclear reactor (Nobel prize in 1995 for Reines).
[13]Nobel prize in 1936 for Hess.
[14]Nobel prize in 1927 for Wilson.
[15]Nobel prize in 1936 for Anderson.
[16]Nobel prize in 1950 for Powell.
[17]Nobel prize in 1957 for Lee and Yang.

later recognized as a hadron containing a *strange* quark,[18] the first quark member of the second family of fundamental particles.

It was however the advent of new powerful laboratory particle accelerators (see Chapter 2 for a complete historical review), like the Berkeley synchrocyclotron (1948) and Brookhaven's Cosmotron (1952), together with the introduction of novel experimental techniques, such as the bubble chamber (1952),[19] that dramatically changed the picture. Access to higher and higher energies, the ability to modulate the beams' energy, to focus them directly toward targets, to choose among a growing range of beam particles (electrons, protons and, later, pions, kaons, muons, photons, neutrinos and respective antiparticles) and, last but not least, to trigger on the desired class of events, gave physicists a toolkit that allowed the in depth study of particles' properties with full control over the experimental conditions, allowing new discoveries and the testing of new theoretical proposals.

In the 50's and early 60's two main directions emerged. On one side the higher energies and intensities of the newly available synchrotrons (see Chapters 7 and 8) drove the explosive growth in the number of discovered unstable hadrons, stimulating and guiding the search for a suitable classification of their properties, and of its theoretical interpretation in terms of a dynamical theory of strong interactions. On the other side, the study of weak interactions started building up the experimental foundations of what was to become the theory of electroweak interactions. Neutrino beams from the decay of pion beams led to the discovery of the muon neutrino (1962),[20] providing the first hint towards the classification of fundamental particles in terms of a family structure. Kaon beams led to the discovery of CP violation (1964),[21] a cornerstone to the understanding of the matter–antimatter asymmetry in the universe, and, later, an input for the first speculation about the possible existence of a third family of quarks. Kaon beams also allowed the discovery of the $\Omega^-$ baryon (1964), giving strong experimental evidence for the hypothesis of $SU(3)$ symmetry and the quark model. The study of weak transitions of strange particles exposed the absence, or suppression, of flavour-changing neutral currents, posing a theoretical puzzle that was eventually resolved by postulating the existence of charm, the partner of the strange quark in the second family of quarks.

[18]Nobel prize in 1969 for Gell-Mann.
[19]Nobel prize in 1960 for Glaser.
[20]Nobel prize in 1988 for Ledermann, Schwartz and Steinberger.
[21]Nobel prize in 1980 for Cronin and Fitch.

In parallel, experiments with electron beams opened the way to the study of the proton's internal structure, revealing already by the mid 50's its composite nature. As the beam energies increased beyond the GeV scale, the study (1968)[22] of the electron scattering distributions off protons led Bjorken and Feynman to speculate about the existence inside the proton of point-like particles (*partons*). This was followed by a theory in which quarks are real constituents of hadrons, and the proposal in 1972 by Fritzsch, Gell-Mann and Leutwyler of Quantum Chromodynamics (QCD), the sector of the SM describing strong interactions.

In the early 60's, the first $e^+e^-$ and $e^-e^-$ colliders made their appearance (see Chapter 3): Ada in Frascati ($e^+e^-$), the Orsay linear collider ($e^+e^-$), VEP-1 in Novosibirsk ($e^-e^-$) and the Princeton–Stanford collider at SLAC ($e^-e^-$) began the exploration of QED processes at "high" energy, and the first studies of electron–positron annihilation into hadrons, paving the way for a future of major discoveries.

The overall experimental picture that was available by the early 70's was beautifully organized within the SM of fundamental interactions. By 1973, the SM included all of its theoretical ingredients: a gauge theory[23] to describe the unified electroweak and the strong interactions of two families of quarks and leptons, a consistent set of mathematical rules to perform predictive calculations of any physical observable in terms of a small set of fundamental parameters (coupling constants and masses); and a model of quark mixings that, if extended with the existence of a third family, included also a description of CP violation.

After many years in which accelerators had offered physicists more questions than answers, exposing the set of phenomena that a complete theory

[22]Nobel prize in 1990 for Tayor, Friedman and Kendall.

[23]A gauge theory is the only way to define the interactions of elementary vector particles like the photon or the bosons responsible for the weak and strong interactions that is consistent with special relativity and quantum mechanics. In short, the gauge principle requires the field strength (e.g. the electric field) to be derived from a vector potential $A_\mu(x)$, and to be invariant under local modifications of the potential induced by shifts like $A_\mu(x) \to A_\mu(x) + \partial_\mu \alpha(x)$, where $\alpha(x)$ is an arbitrary function. This is called a gauge transformation, and the invariance of the equations of motion under it is called gauge symmetry. This is automatic in the case of electromagnetism, where the field strength is defined as $F_{\mu\nu}(x) = \partial_\mu A_\nu(x) - \partial_\nu A_\mu(x)$. The gauge symmetry is needed to ensure that the time-like component of the vector field $A_\mu$, whose quantum state has a negative norm due to the Lorentz metric and is therefore an unphysical field, remains decoupled from any quantum mechanical process. The principle of gauge invariance can be applied, with some algebraic modifications, to cases where the vector potential transforms as the adjoint representation of a non-Abelian group. In the case of the SM, the full gauge group includes an $SU(3)$ component, whose vector potentials give rise to the gluon fields mediating the strong interactions, an $SU(2) \times U(1)$ component, whose vector potentials give the photon of electromagnetism, and the $W^\pm$ and $Z^0$ bosons, mediators of the weak interactions.

of particle interactions had to explain, the appearance of the SM started to change the relation between theory and experiments. The predictive power of the SM called for an extensive experimental programme to quantitatively test it, and to measure in detail the properties of the new particles and interactions that it predicted. This experimental programme has shaped the progress in accelerator technology of the last 40 years, with concrete and very ambitious challenges, which have generally been met and often surpassed.

The experimental verification of the SM relied on several different components. On one hand, the SM predicted the existence of several particles yet unobserved in the early 70's: the charm quark, a possible third family of quarks and leptons to accommodate CP violation, a $Z^0$ gauge boson giving rise to weak neutral currents, a set of gluons to mediate strong interactions, and a Higgs boson to break the electroweak gauge symmetry. On the other hand, the dynamics of the electroweak and strong interactions among quarks, leptons and gauge bosons was at the time far from being tested with any degree of precision: the inclusion in the quantitative predictions of higher-order effects, emerging from the solution of the theory's equations via a perturbative expansion in terms of the small coupling constants of the various SM interactions, and their comparison against equally accurate experimental data, was to become a crucial test of the overall mathematical consistency of the theory, and a possible probe of phenomena not accounted for by the SM.

Furthermore, the SM had barely been formulated when new theories going beyond the SM (BSM) started appearing, since the mid 70's, predicting a multitude of new possible phenomena. The key motivations for this burgeoning of theoretical work included, among others: the attempt to extend the successful unification of electromagnetic and weak interactions to a more complete unification, including strong interactions, the need to develop a quantum mechanical theory of gravitational interactions, the consideration of alternative origins of electroweak symmetry breaking and to replace the apparently ad-hoc and simplistic Higgs mechanism.

The so-called grand unified theories (GUTs) predicted the unification of electroweak and strong forces at energy scales around $10^{15}$ GeV, leading to proton decay with lifetimes in the range of $10^{31}$ years, to interesting relations between the quark and lepton masses and couplings, and suggesting the existence of neutrino masses. Supersymmetry and supergravity, candidates to bring together the SM and gravity, predicted a doubling of the spectrum, with new partners for each SM particle. The closeness of the GUT scale to the natural scale of quantum gravitational interactions, the Planck scale of

$\sim 10^{19}$ GeV, gave tantalizing hints that grand unified supergravity theories could provide the framework for a full unification including gravity, hints that were strengthened after the realization that such scenarios could naturally arise in the context of superstring theories. Compositeness models predicted a substructure of quarks, leptons and possibly weak gauge bosons. Technicolor theories proposed alternatives to the Higgs mechanism as a source of electroweak symmetry breaking and of the $W$ and $Z$ boson masses. In the last 15 years the variety of BSM models further evolved, to incorporate the most recent experimental constraints, including neutrino masses, the conclusive evidence for Dark Matter and for the inflationary phase during the Big Bang, as well as, alas, the lack of evidence of a Higgs boson and of many of the possible manifestations of the preferred early BSM theories. In particular, several new proposals appeared for alternative mechanisms to break electroweak symmetry, partly inspired by developments in string theory, including the idea of extra dimensions at the TeV scale, little Higgs, no-Higgs, and more.

While testing most of the predictions of the SM and of BSM theories called for the construction of accelerators at the highest possible energies operating in collider mode (see Chapters 4, 5, 6 for electron accelerators, Chapters 9, 10, 11, 14 and 22 for hadron accelerators), an important component of the experimental programme relied on the delivery of high intensity and diverse beams (see Chapters 7, 8, 34) to explore low-rate processes testing the flavour structure of interactions: rare meson and muon decays, CP violation in the decays of hadrons containing strange, charm and bottom quarks, and neutrino mixing. The use of these facilities, with a broad spectrum of beam types, energies and colliding modes, contributed over the course of almost 40 years to pin down with great precision the partonic structure of the nucleon, testing the predictions of perturbative QCD, measuring the strength of the QCD coupling constant $\alpha_S$,[24] and extracting the distributions of quarks and gluons (see Chapter 15) that are absolutely essential in predicting cross sections and properties of all hard processes created in today's high-energy hadronic collisions. Current and forthcoming (see Chapter 27) accelerator facilities are also dedicated to exploring the interface between perturbative and non-perturbative QCD, adopting the tools of particle physics to explore the short-distance structure of nuclei.

Some milestones of this progress are highlighted in the following subsections.

[24] This is the equivalent for strong interactions of the QED fine structure constant, $\alpha_{EM}$.

## 2.1 *The gauge bosons sector of the SM*

The SM requires the existence of a neutral weak current, describing the exchange of the $Z^0$ boson. A signature of this current are events where neutrinos do not transform into charged leptons while undergoing an interaction with a nucleon: $\nu + N \to \nu + X$. Events of this type were indeed detected in the interactions of a neutrino beam from CERN's PS (see Chapter 8) by the Gargamelle Collaboration (1973), giving the first experimental proof of the unification of electromagnetic and weak interactions. The interplay of charged and neutral weak interactions of neutrinos allowed physicists, in many subsequent neutrino-beam experiments at CERN and Fermilab, to disentangle the different contributions to the proton structure of quarks and antiquarks of different flavours, and to measure the weak mixing angle $\sin\theta_W$, a fundamental parameter of the electroweak theory.

The direct production of the carriers of both charged and neutral weak interactions, the $W^\pm$ and $Z^0$ bosons, required however accelerators capable of attaining much higher center-of-mass energies. This became possible with CERN's S$p\bar{p}$S collider (Chapter 10),[25] where the electroweak gauge bosons were observed by the UA1 and UA2 experiments in 1983. Following their discovery, the precision measurements of their properties were continued at all energy-frontier accelerators: Fermilab's Tevatron (Chapter 14), DESY's HERA (Chapter 15), SLAC's SLC (Chapter 5), CERN's LEP and LEP2 (Chapter 4) and, finally, at CERN's LHC (Chapter 22).

In analogy to the case of the electroweak gauge bosons, the direct evidence for the carrier of the strong force, the gluon, was preceded by the observation of indirect manifestations of its existence. Deep-inelastic scattering experiments at SLAC, measuring in the late 60's the momentum fraction of the proton carried by charged partons, showed that this accounts for only 50% of the total proton momentum. The rest was attributed to neutral particles responsible for the confinement of quarks. It was only in 1979 that the experiment TASSO, at DESY's PETRA collider, provided tangible direct evidence of the existence of the gluon, through the observation of 3-jet events, interpreted as the radiation of a gluon from the basic $e^+e^- \to q\bar{q}$ process.

## 2.2 *The fermionic sector of the SM*

The extension to the quark sector of the family structure of the SM exhibited by the then-known leptons, suggested in the early 70's the existence

[25]Nobel prize in 1984 for Rubbia and van der Meer.

of a fourth quark, the charm. Shortly after, the 3.1 GeV/$c^2$ $J/\psi$ meson, discovered in experiments at the SLAC $e^+e^-$ collider SPEAR and at the Brookhaven AGS (1974)[26] (see Chapter 8), was interpreted as a $c\bar{c}$ bound state, providing the first evidence for the existence of the charm quark, which was confirmed in 1976 by the discovery at SLAC of the $D^0$ meson, made of a $c\bar{u}$ pair. The quantitative studies of the spectroscopy of the $J/\psi$ and of its excited states (the charmonium system), performed at DESY and SLAC, then provided an excellent framework to test the predictions of the new theory of strong interactions, QCD.

In 1976 a new lepton, $\tau$, was observed at the SLAC $e^+e^-$ collider.[27] It was soon followed by the discovery of the 9.5 GeV/$c^2$ $\Upsilon$ meson at Fermilab (1977), interpreted as the bound state of new quark, the *bottom.* The almost simultaneous interplay of complementary accelerator technologies conspired to advance the field more than each individual discovery could have done. The appearance of both the $\tau$ and $b$ quark gave in fact immediate substance to the expectation that these were members of a new, third, family of fundamental fermions, which then required for its completion the existence of both the *tau* neutrino, and of a new quark, the *top.* While the bottom quark had been discovered with a proton beam in a fixed-target experiment (and observed shortly after (1978) at CERN's ISR $pp$ collider, see Chapter 9), its detailed properties, as in the case of the charm, would be more thoroughly explored in $e^+e^-$ collisions, starting in the 1978 with the Pluto experiment at DESY's DORIS, and from 1979 with CUSB and CLEO at Cornell's CESR. All these experiments used the resonant production of the loosely bound $b\bar{b}$ state $\Upsilon(4S)$, which dominantly decays to a pair of $B^+B^-$ or $B^0\overline{B^0}$ mesons. In 1987, the ARGUS experiment at DESY's DORIS II and the UA1 experiment, at CERN's S$p\bar{p}$S collider, observed the oscillation between $B^0$ and $\overline{B^0}$ mesons; the large oscillation frequency was indicative, in the SM, of a large value of the top quark mass, possibly larger than 100 GeV, at a time when the searches at the highest energy $e^+e^-$ and S$p\bar{p}$S colliders were sensitive to a few tens of GeV at most. The search for the top quark lasted then until its eventual observation in 1994, by the CDF and D0 experiments at Fermilab's Tevatron $p\bar{p}$ collider. The precision study of the top quark properties formed one of the keystones over the whole Tevatron programme, which witnessed over the course of the years the transition of the top from being a new, exiting and rare particle, to becoming an annoying background for the searches for yet more exotic objects, like the Higgs boson.

[26]Nobel prize in 1976 for Richter and Ting.

[27]Nobel prize in 1995 for Perl.

## 2.3 *The flavour structure of the SM*

As remarked in the introduction, the SM is characterized not only by its basic constituents (quarks, leptons) and their gauge interactions, but also by a flavour structure, which specifies the weak transitions between quarks and leptons belonging to different families. For example, the interaction of a $W^-$ boson with an up-type quark $u_i$ of the $i$th family ($u_i = u, c, t$) will generate a quark state defined by a linear superposition of down-type quarks $d_i$ of all families ($d_i = d, s, b$): $W^- + u_i \rightarrow \sum_{j=1,3} V_{ij} d_j$. The unitary matrix $V_{ij}$, known as the Cabibbo–Kobayashi–Maskawa (CKM) mixing matrix, is parameterized in terms of three real angles and one complex phase. These four parameters are fundamental inputs of the SM, like the strength of the gauge couplings or the fermion masses. Their value, together with the strength of weak interactions and the $W$ boson mass, determine the properties of all weak interactions of hadrons, including production, decay and oscillations. The complex phase, in particular, is the origin of the violation of CP invariance of the theory, and leads to matter–antimatter asymmetry. The measurement of the CKM matrix parameters created a diverse and rich field of experimental studies, developed over the last 50 years in parallel with the exploration of the SM spectrum and gauge interactions. This field has benefited from a multitude of accelerator infrastructures, necessary to deliver the large statistics necessary for the observation of extremely rare decays, and for very accurate measurements. These included experiments with kaon beams (BNL, Fermilab, CERN) and dedicated $e^+e^-$ *flavour factories* (CESR at Cornell, Doris at DESY and, more recently, PEP2 at SLAC, KEKB at KEK, and Dafne at Frascati), complemented by results obtained by the multi-purpose experiments at LEP, SLC, Tevatron and LHC. The outcome of these studies includes the verification of the unitarity of the CKM matrix to the percent level (which severely constrains the possible existence of a fourth family of quarks), the evidence that all CP violation phenomena observed in the laboratory can be explained by the CKM complex phase,[28] and a large array of strong constraints on possible extensions of the SM, which would typically lead to observables departing from the CKM predictions.

[28]The amount of CP violation predicted by the CKM matrix does not seem to be sufficient, however, to explain the cosmological abundance of matter over antimatter. This therefore suggests the possible existence of other sources of CP violation, as could arise in BSM theories, which are being actively sought in current experiments.

The discovery of a similarly rich flavour structure in the lepton sector is more recent, and follows the conclusive observation (1998) of neutrino mixings and masses,[29] using a mixture of observations driven by astrophysical neutrinos (from the Sun and from cosmic rays) and laboratory neutrinos, with beams from KEK, Fermilab, and CERN as well as neutrinos from nuclear plants. Future experiments at these and new facilities (as well as in non-accelerator laboratories) will aim to determining the absolute scale and ordering of neutrino masses (only the absolute value of differences of neutrino masses squared is available today), and at measuring a possible CP violation. While neutrino masses can be accommodated with a minor adjustment of the SM, the most compelling framework for their understanding is given by grand unified theories, in which the states required to give neutrinos a mass are part of the larger particle multiplets needed to implement the extended gauge symmetries. The embedding of leptons and quarks in the same multiplets of the grand unified symmetry leads to model-dependent correlations between the pattern of mixing in the quark, neutrino and charged-lepton sectors. For example, several models predict the decay $\mu \to e\gamma$ with branching ratios large enough ($\gtrsim 10^{-13}$) to be potentially observed in ongoing experiments at the PSI muon source, or in future $\mu \to e$ conversion experiments at Fermilab.[30] The observation of such interactions, which in the SM are suppressed, would be a major revolution for particle physics, providing a possible handle to explore the nature of fundamental interactions at the scale of grand unification.

## 2.4 *The dynamics of the SM*

The most impressive success of the SM is its ability to provide accurate quantitative predictions for the whole set of observables that have been experimentally probed. These predictions are based on first-principles calculations, using the fundamental SM parameters and the mathematical framework of quantum field theory, which prescribes the rules of the perturbative expansion and of various non-perturbative approaches such as lattice calculations. This framework also allows physicists to characterize elements of the calculations that are too hard to extract from first principles, due to the complexity of some non-perturbative problems. This is the case, for

[29] Nobel prize in 2002 for Davis and Koshiba.

[30] While not always testable with accelerators, other possible, but yet unobserved, phenomena include the neutrino-less double-$\beta$ nuclear decays and lepton-number violating processes that could trigger the dominance of matter over antimatter in the early universe.

example, of the quantitative representation of the proton in terms of fundamental constituents (quarks and gluons): while we lack a first-principles calculation of the proton structure, the SM allows us to parameterize its short-distance features in terms of so-called parton distribution functions (PDFs). These describe the density of gluons and quarks of various flavours inside the proton, as a function of the proton momentum fraction they carry. The PDFs can be connected, now via first-principles perturbative calculations, to observable quantities such as differential cross sections for various reactions involving protons in the initial state. The PDFs can be extracted from a set of benchmark measurements, and, thanks to their universality, can then be used for the prediction of any other process.

The experimental programme over the past 30 years gave great impetus to the exploration of SM dynamics. In the electroweak sector, precise measurements of rates and distributions using both leptonic and hadronic decays of $Z^0$ bosons at the LEP and SLC confirmed the SM predictions to the per-mille level. This precision allowed physicists to tightly constrain the impact of particles too heavy to be produced in $Z^0$ decays, but contributing to electroweak observables via higher-order perturbative effects. This is the case of the $W$ boson, the top quark, and the Higgs boson. The direct observation and mass measurement of these particles, at the various high-energy colliders (LEP2, Tevatron and now LHC) shows perfect consistency with the SM predictions, and allow physicists to set stringent limits on new physics processes beyond the SM.

In the strongly interacting sector, decades of measurements of the proton structure using electron, muon and neutrino beams at SLAC, Fermilab and CERN culminated in the superb accuracy achieved by the H1 and ZEUS experiments at DESY's HERA $e^{\pm}p$ collider. The extraction of PDFs from these data has now reached percent-level precision, and allows physicists to perform accurate cross section calculations for the physics programme of the LHC $pp$ collider at CERN. Hadron colliders enabled, in the last 40 years, the exploration of the energy frontier, while at the same time shedding light on the complex behaviour of hadronic interactions. CERN's ISR $pp$ collider observed for the first time the rise with energy of the total hadronic cross section, and discovered the production of particles at large transverse momentum, a confirmation of the real nature of quarks and gluons, as well as the first evidence for the production of jets, which have since become a standard tool for measurements and discoveries.

The study of masses and decay modes of hadrons, after triggering in the early 60's the modern understanding of strong interactions, has continued to challenge our ability to unveil the full details of the QCD dynamics in its non-perturbative regime.[31] Exotic hadrons, namely states made of configurations other than triplets of quarks or $q\bar{q}$ pairs, such as glueballs (made of gluon pairs), hybrids ($q\bar{q}g$ bound states), tetraquarks ($qq\bar{q}\bar{q}$ bound states) and others, are still being actively searched for at many facilities, including modern fixed-target experiments like COMPASS at CERN's SPS, and, in the case of exotic hadrons including heavy quarks such as $c$ or $b$, at the $B$ factories, the Tevatron and the LHC.

The deeper understanding of QCD in the low-energy, non-perturbative regime has also triggered rich experiental programs with low-energy and high-intensity beams, e.g. at Jefferson Lab, bridging the gap between the phenomenological models of nuclear physics and a first-principle description based on QCD.

## 3 Complementarity and synergy of different accelerator facilities

The great success in building the evidence supporting the SM is in large part due to the richness and diversity of the experimental programmes made possible by the many independent accelerator facilities available worldwide. Physicists have benefited from progress in each direction of accelerator technology: higher-energy, higher-intensity, varied beam species, operation modes, and flexibility in their use.

It is remarkable, if not surprising, that very different accelerator technologies developed in synergy and synchrony, and allowed physicists to probe in complementary, and occasionally unpredictable, ways the various facets of different problems. For example, the $J/\psi$ meson was discovered simultaneously at the Brookhaven AGS and at the SLAC $e^+e^-$ collider, the former benefiting from its sensitivity to a broad range of masses, the latter from a very precise energy resolution. The $\Upsilon$ meson was then observed, within the window of one year, in fixed-target mode first (FNAL), and in $pp$ (ISR) and $e^+e^-$ (DORIS) collisions soon after. The close temporal coincidence of the observation of the $\tau$ lepton at SLAC, and of the $\Upsilon$, opened the path to the search of the top quark. Precision measurements of $B^0-\overline{B^0}$ meson mixing suggested it might be significantly heavier than the kinematic reach

[31] The QCD property of asymptotic freedom states that at very high energy (or short distances) the interactions between quarks and gluons are weak and can be well approximated by perturbative calculations. At the low energies that are relevant for the formation of hadronic bound states, on the other hand, the strength of QCD forces becomes large, perturbative calculations are unreliable, and non-perturbative techniques, such as lattice field theory, must be used.

of $e^+e^-$ colliders. The great precision of experiments and of beam energy calibration at LEP and SLC, nevertheless allowed physicists to pin down indirectly but precisely its existence and its mass, at about the same time when the Tevatron gave evidence of direct production of $t\bar{t}$ pairs. The agreement between the indirect prediction, based on the theoretical calculation of the effects of a top quark on precision electroweak observables, and the direct measurement of its mass, was taken as the ultimate validation of the SM. Most recently, given the improved measurements of both the $W$ boson and the top quark masses, these precise comparisons have evolved to tightly constrain the mass range for the SM Higgs boson. This constraint is fulfilled by the mass of the Higgs particle, observed at the LHC in 2012 by the ATLAS and CMS experiments, yet another milestone achievement of theoretical and experimental physics, and of accelerator science.

The interplay between electron and hadron colliders has played a crucial role throughout the history of the SM, and the removal of the contributions from any single accelerator facility would leave ample gaps in the exploration. Without Tevatron or the LHC, no real top quark would have been produced as yet in the laboratory. We might still be arguing whether the indirect evidence of the top quark from LEP/SLC has a real physical counterpart, and for sure there would be theoretical models in which the top mass could be different, with other processes conspiring to reproduce the LEP results. On the other hand, if we only had direct evidence from Tevatron and the LHC, but no precision electroweak measurements from LEP/SLC, we would still be missing final evidence of the validity of the calculation of radiative corrections in the SM, and we would not have had robust predictions for a preferred mass range for the Higgs boson. The precise measurement of the total width of the $Z^0$ boson, which was done at LEP/SLC, furthermore excludes the existence of a fourth light neutrino interacting via weak interactions, and rules out the existence of a fourth family of quarks and leptons similar to the three known ones, a conclusion that could not be drawn just from the data of Tevatron and LHC, in spite of their much higher mass reach![32]

The complementarity between lepton and hadron colliders reveals also some curious twists. Major discoveries and studies related to strong interactions (the partonic structure of the proton, charmonium spectroscopy and the first quantitative tests of QCD, the discovery of the gluon, the precise measurement of the strong coupling constant $\alpha_S$, etc.) were done by

---

[32]This constraint can be evaded, however, if the neutrino of the fourth family to be too heavy to be produced in $Z^0$ decays.

experiments at electron accelerators, while the key particles of the electroweak theory, the $W$, $Z$ and Higgs bosons, were discovered at hadron colliders!

Measurements at each facility cross-fertilized the development of better tools necessary to improve the modeling and the theoretical predictive power, powering a virtuous circle: better knowledge of the proton structure from $ep$ colliders helps reduce the theoretical systematics in the measurement of the $W$ mass at the Tevatron, improving the $e^+e^-$ colliders' predictions for the Higgs mass, predictions which are in turn verified at the LHC.

As in the past, cosmic sources and accelerators of radiation remain today an essential tool for the understanding of elementary particles. It is the neutrinos produced by the Sun and by cosmic ray interactions with the Earth's atmosphere that provided the experimental evidence for neutrino masses and oscillations. And the existence of dark matter particles is still today only confirmed by astrophysical and cosmological observations, detecting the indirect effects of those particles as they were produced by the biggest accelerator of all, the primordial Big Bang. Data from accelerators, in addition to providing complementary and more direct information on these phenomena, will help improve their theoretical understanding, including advances in the modeling of the interactions of cosmic particles with the atmosphere, as it is being done with the measurements of forward phenomena with ad hoc experiments (ALFA, LHCf, TOTEM) at the LHC.

Last but not least, a multitude of dedicated accelerator facilities allow physicists to perform difficult experiments, addressing specific questions of the particle physics panorama, which can have an impact on the basic understanding, and thus on the precision of the observational tools. Among others, these include experiments to improve the measurement of parameters (such as the muon lifetime or its anomalous magnetic momentum), to test the fundamental discrete symmetries (C, P and T, or combinations thereof, including CPT), or to improve the dynamical understanding of strong interactions and of the hadronic or nuclear structure in difficult regimes such as high-density/high-temperature or low-energy regimes (leading for example, among other things, to the better modeling of the interaction of particles with different media, as needed for the precise simulation of detector responses or of neutrino beams).

## 4 The future challenges

The observation at the LHC of a new particle matching the properties of the Higgs boson closes a forty-year long campaign of exploration of the SM.

The field of particle physics has now several outstanding priorities on its horizon:

- to understand the origin of dark matter, and identify the BSM theory that explains its existence;
- to study with high precision the properties of the Higgs boson and of the electroweak symmetry breaking phenomenon, verifying up to which level they match the SM, or start showing deviations due to new physics;
- to continue testing the flavour structure of the SM, looking in particular for additional sources of violation of the CP symmetry, whether in the quark or in the neutrino and charged lepton sectors, which can explain the overall matter–antimatter asymmetry seen in the Universe;
- to keep exploring the high-energy/short-distance frontier, for direct evidence of new particles and new forces, as well as for more exotic phenomena such as the substructure of the known elementary particles or extra spatial dimensions.

It is likely that these different threads will end up pointing in the same direction, guiding us to the definition of the new "*Standard*" Model of particle physics. Which thread will give surprises first, however, we do not as yet know. Accelerator facilities and experiments exist, and are being built or being planned, to explore all these different directions.

Higher energy is the obvious and inevitable mean with which to explore phenomena at shorter and shorter distance scales, and to discover heavier and heavier new particles. Increasing the luminosity of the LHC helps probe higher energy scales, since it enables the much rarer collisions of quarks and gluons carrying the largest possible fraction of the proton momenta to take place. The factor of ten increase in integrated luminosity foreseen by the High-Luminosity LHC upgrade (see Chapter 24) would extend by over 30% the mass reach for the discovery of new particles. Going beyond this could well be necessary, however, to confirm and further explore BSM scenarios that we might have a first glimpse at the 14 TeV LHC: supersymmetry, with its large number of new particles; extra dimensions, with towers of states of ever-growing mass; and models with composite Higgs bosons, where the underlying new strong forces responsible for their binding would only fully manifest themselves at multi-TeV energies. This will require the construction of more powerful hadron colliders (see Chapter 26).

Lepton colliders (see Chapters 29, 30 and 41) cannot compete with hadron colliders in terms of absolute energy, but their discovery reach can be equally powerful. The greatest threat to hadron colliders' discovery power is the large background rate, which can totally obscure the presence of new phenomena. A standard historical example is the "missed" discovery of the $J/\psi$ meson by the ISR $pp$ collider, due to the huge background of muons that forced trigger thresholds so high as to totally deplete the potential signal. This risk remains true today at the LHC, where several searches of BSM particles are made difficult by the harsh environment. For example, the detection of chain decays of supersymmetric particles can lead to signals in which the large mass of the parent particles is dispersed in a multitude of lower-energy jets and leptons, leaving signals that either miss the trigger thresholds or are too diluted to emerge from the large SM background. The precise study of interesting Higgs production and decay modes may also suffer from the same difficulties, which will limit the ultimate precision attainable even after the High-Luminosity LHC regime is reached. High-energy lepton colliders are therefore an essential complement to the LHC. Their clean environment, with a much reduced hadronic background contamination, an excellent control over the deposited energy and ability to detect lower-energy products or signals of missing energy (such as those caused by possible dark-matter-like particles), give unequaled sensitivity to phenomena where a hadron collider could have difficulties. Furthermore, following the experience of LEP and SLC, and their ability to correctly predict the existence and mass of the top quark and of the Higgs boson on the basis of high-precision measurements, it is natural to expect that future lepton colliders will give critical input on the possible existence of new phenomena at scales higher than those accessible through direct particle production. These results would be an essential complementary addition to the high-energy programme.

The exploration of the high-energy frontier could also benefit from a higher-energy version of the HERA programme, as envisaged by the LHeC project for electron–hadron collisions in the LHC tunnel (see Chapter 28). In addition to further improving the determination of the quark and gluon distributions in the proton, leading to more precise predictions of LHC processes and probing the regime of very-high gluon density, such a facility could detect, if they exist, some BSM particles for which both hadron and lepton colliders have limited reach or limited precision. This is the case, for example, of the production of leptoquarks, namely particles carrying both lepton and baryon quantum numbers, or of supersymmetric partners of quarks, which, in certain variants of supersymmetric theories, can be resonantly produced in an electron–quark collision.

Higher energy, whether in hadronic or leptonic collider modes, is not however sufficient to complete the picture. All indications are that the fruitful complementarity of accelerator-based experiments at the highest energies, at the highest intensities, with a broad choice of beam types and in the conditions allowing for the most precise measurements, will continue to be as essential in the future as it has been in the past. The exploration of flavour in the quark, neutrino and charged-lepton sectors requires also high-intensity, low-energy accelerators: $B$-meson factories (see Chapter 20), $\tau$/charm factories (to search for possible $\tau \to \mu$ transitions, and to dig deeper in the recently observed $D^0-\overline{D^0}$ oscillations and CP violation in $D^0$ decays) and high-intensity kaon beams (to observe very rare, anticipated, but so far unseen decays, and to continue the long tradition of ground-breaking and Nobel-worthy discoveries such as P and CP violation driven by this remarkable particle), long-baseline neutrino facilities or neutrino factories (to search for leptonic CP violation), high-intensity muon sources (to search for $\mu \to e\gamma$ decays or $\mu \to e$ conversions, and for improved measurements of the muon anomalous magnetic moment). Some of the experimental facilities dedicated to these efforts, for example large underground neutrino detectors, will also contribute to other non-accelerator components of the programme, being sensitive e.g. to the proton decay predicted by grand unified theories, or improving the sensitivity to neutrinos of cosmic origin. In this connection, it is important to stress the continued and enhanced complementarity with non-accelerator based research: the persistent vitality and richness of cosmic ray measurements, using both Earth-based detectors and satellites, by now sensitive to sources of photons, neutrinos, $e^+/e^-$ and hadrons covering a huge range of energies; the growing role of neutrino beams from nuclear power stations; precision atomic and nuclear physics measurements, to explore weak interactions or fundamental electric dipole moments of electrons and neutrons; and fundamental measurements of gravity at both short and long distance scales.

Technological quantum leaps, such as the construction of dipoles with fields of several tens of Tesla, the large-scale realization of high-gradient cavities and the validation of the CLIC acceleration concept (see Chapter 29), as well as the currently far-fetched ability to store and collide high-energy $\mu^+\mu^-$ beams (see Chapter 41), or to develop plasma-driven accelerators (see Chapter 38) with intensities of interest for particle physics, will drive the long-term progress of the field. We can expect that these ambitious and far sighted R&D programmes in accelerator technology will redefine the field of high-energy physics in the XXI century, in the same way that Lawrence's invention of the cyclotron redefined, by orders of magnitude, the relation between accelerator size and delivered energy, in comparison to the previous van der Graaf technology. In the XXI century, as has been true in the XX, the progress in our understanding of the fundamental laws of nature will be inextricably linked to the progress of accelerator technology.

# Chapter 2

# Energy revolution: From static fields to cavity resonators

Oliver Brüning (CERN)

This chapter describes the origin of the development of particle accelerators as well as the main technical breakthroughs and inventions during this approximately 100-year long evolution.

## 1 Introduction

Two key physics discoveries motivated the development of particle accelerators:

- Rutherford's 1919 discovery that the bombardment of Nitrogen atoms with alpha particles from radioactive decay can disintegrate the Nitrogen nucleus [1].[1]
- Einstein's famous equation $E = mc^2$ from 1905, that states that matter and energy are related quantities and that new particles can be created in processes where sufficient energy is released [2].

Up to the end of the $19^{\text{th}}$ century matter was assumed to consist of atoms, which are indivisible building blocks for a given element (Dalton's atomic theory from 1803). The discovery by Pierre and Marie Curie in 1896 that certain atoms can decay [3, 4],[2] and the detection of a new particle, the electron, that is much smaller than the previously known atoms, by Thompson in 1896 [5, 6],[3] indicated that atoms are perhaps not fundamental building blocks of matter after all. Rather, these observations suggested that atoms are made of smaller, more fundamental building blocks. This interpretation was further supported by observations made by Rutherford together with Hans Wilhelm Geiger and Ernest Marsden in 1911 (in his

[1] The discovery earned Rutherford the Nobel Prize in chemistry in 1908.
[2] The discovery earned them and Henry Becquerel the Nobel Prize in physics in 1903.
[3] Earning Thompson the Nobel Prize in physics in 1906.

famous Gold foil scattering experiment)[7], that showed that most of the atom's mass is concentrated in a small volume inside the atom, the nucleus, and that most of the atom's volume is filled with very little matter. (Niels Bohr explained these observations in 1913 by describing an atom as a positively charged nucleus that is surrounded by electron shells in which the electrons orbit around the nucleus at a larger radius.[4]) In 1919 Rutherford showed further that the bombardment of Nitrogen atoms with alpha particles (twice positively charged He ions) from radioactive decay could disintegrate the Nitrogen nucleus, transforming the Nitrogen and alpha particle into Oxygen and a proton. This observation confirmed the suspicion that atoms are not fundamental entities and provided an interesting new tool for studying the nature and properties of the atomic nucleus: bombarding other, heavier atoms with alpha particles could lead to similar transformations of the nucleus and studying the various atomic transformations would certainly provide new insight into the composition of the atomic nucleus. Unfortunately, the kinetic energy of alpha particles from natural radioactive decay is not sufficiently high to penetrate the nuclei of elements heavier than Nitrogen. Heavier atoms have a more positively charged nucleus and the kinetic energy of alpha particles from natural radioactive decay is not high enough to overcome the Coulomb repulsion of the positively charged alpha particles and the positively charged nucleus. When Rutherford studied the potential disintegration of the Lithium nucleus with protons, he estimated that protons with a kinetic energy of 800 keV could overcome the Coulomb potential, triggering the first search for particle accelerators and starting a new research area in nuclear physics.

A second drive for the development of particle accelerators came from the search for new elementary particles. When Paul Dirac described fermions via a wave equation it was realized that Dirac's equation had additional solutions that did not correspond to yet known particles.[5] The new solutions described particles with very similar attributes to the known fermions but with opposite signs in their quantum numbers (e.g. opposite charge). These particles were later called antimatter particles and for the electron resulted in the prediction of the positron. When Carl David Anderson discovered positrons in measurements of cosmic ray collisions in 1937 [8,9],[6] the search

[4]Niels Bohr received the Nobel Prize in physics in 1922 for his contribution to the development of the atom model.

[5]The new theoretical treatment of elementary particles earned Dirac together with Erwin Schrödinger the Nobel Prize in physics in 1933.

[6]Anderson received the Nobel Prize in physics for the discovery of the positron in 1936 together with Victor Franz Hess.

for new accelerators got a second boost. Observing that new particles could be generated in collisions of sub-atomic particles provided a new tool to search for and study new elementary particles, provided particle accelerators could provide sufficiently high energies in the collisions.[7]

## 2 Particle acceleration

### 2.1 *Lorentz force*

The acceleration process of all charged particle accelerators is based on the Lorentz force, which describes the acceleration of charged particles with the help of electromagnetic fields. The Lorentz force is given by the following expression:

$$\frac{d\vec{p}}{dt} = q(\vec{E} + \vec{v} \times \vec{B}). \tag{1}$$

$\vec{p}$ is the particle momentum, $q$ and $\vec{v}$ are the charge and the velocity of the particle respectively, $\vec{E}$ stands for the electric and $\vec{B}$ for the magnetic fields that interact with the particle and $\times$ describes the rotation or cross product of two vectors. The cross product of two vectors defines a vector that points in a direction perpendicular to the two initial vectors and the Lorentz force therefore states that magnetic fields can change the direction of the particle motion but not its energy while only electric fields can change the magnitude of a particle's momentum and therefore the particle's energy.

The energy gain of a particle moving through an electric field is given by the integral of the Lorentz force along the path of motion. The particle sees an effective voltage during its passage and the product of the particle charge and the effective voltage is equal to the energy gained during the acceleration. For example, a particle of unit charge (the charge of a single proton) passing through the potential of 1 Volt receives an energy increase of 1eV. The unit 'eV' is therefore a very practical unit for measuring the energy gain in particle accelerators. Frequently used expressions for the particle energy in accelerators are keV ($10^3$ eV), MeV ($10^6$ eV), GeV ($10^9$ eV) and TeV ($10^{12}$ eV). Figure 1 shows a schematic layout for the acceleration of a beam of positively charged particles via the passage through a voltage potential.

### 2.2 *Maxwell equations and vector and scalar potentials*

Electric and magnetic field components are intimately linked to each other. This interplay can be conveniently expressed by the use of scalar and vector potentials that are functions of the spatial coordinates and time. For

[7] A very nice summary of the discovery of subatomic particles can be found in [10].

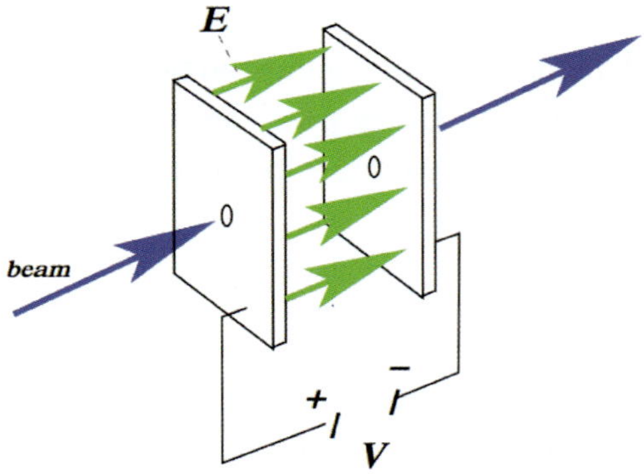


Fig. 1. Schematic illustration for the acceleration of a positively charged particle beam by the passage through a voltage potential.

example, the electric field can be written as

$$\vec{E} = \vec{\nabla}(\phi) = -\frac{1}{c}\frac{\delta}{\delta t}\vec{V}, \tag{2}$$

where $\phi$ and $\vec{V}$ are the scalar and vector potentials respectively and $c$ is the speed of light. They are also referred to as the electric and magnetic potentials respectively. The magnetic field components can be derived using Maxwell–Faraday's induction law. In a vacuum, in a region without charges and currents, Maxwell's equations lead to:

$$\frac{\partial \vec{B}}{\partial t} = -\vec{\nabla} \times \vec{E} \quad \text{and} \quad \frac{1}{c^2}\frac{\partial \vec{E}}{\partial t} = \vec{\nabla} \times \vec{B}. \tag{3}$$

Electrostatic fields can therefore occur without any magnetic field lines, but time varying electric fields must always be accompanied by time varying magnetic field lines (and vice versa).

## 3 Electrostatic accelerators

### 3.1 *High voltage acceleration columns*

The first attempts at particle acceleration were based on transformer circuits that use inductively coupled conductors, the transformer's coils and a magnetic core, to transform a low voltage high current alternating current source to a high voltage and low current alternating current output, and use a rectifier circuit to transform the alternating output voltage into a DC voltage. The concept can be described as follows: An alternating voltage drives a current through a primary coil. The current drives in turn a magnetic flux through the magnetic core and the changing flux induces via Maxwell–Faraday's induction law a current in the secondary coil. The voltage at the

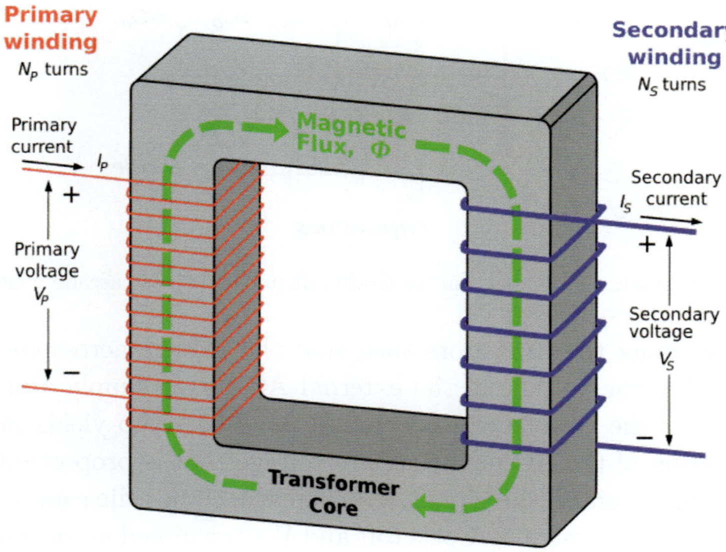


Fig. 2. Illustration for an idea magnetic core transformer [11].

output terminal of an ideal transformer is given by

$$V_s = \frac{N_S}{N_P} V_p. \tag{4}$$

$V_S$ is the output voltage, $N_S$ the number of windings of the secondary coil (that is connected to the output terminal), $V_P$ the input voltage and $N_P$ the number of windings of the primary coil (that is connected to the input voltage). Because of the required large core volume of the transformer and the difficulty in insulating the high voltage terminal from ground to prevent leakage currents, such transformer based high voltage sources are in practice limited to peak voltage outputs of approximately 100 kV. Figure 2 shows a schematic picture of an ideal magnetic core transformer.

When Rutherford proposed an experiment for the potential disintegration of the Lithium nucleus with protons, he estimated that one would need a high voltage terminal of approximately 800 kV for the acceleration of the protons. Conventional transformer power supplies were therefore not sufficient for the task and a new type of power converter was needed.

### 3.2 *Cockcroft and Walton rectifier circuit*

In 1928 J.D. Cockcroft and E.T.S Walton came up with an alternative proposal for the voltage amplification by charging a bank of capacitors with the help of an AC primary power supply [12]. The key idea of the so-called ‘cascade’ or ‘ladder’ terminal is that capacitors assure a unidirectional

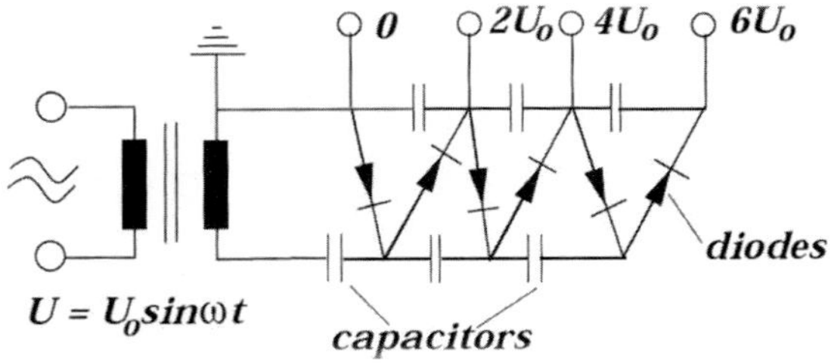


Fig. 3. Schematic circuit diagram for Cockcroft and Walton's cascade generator.

current flow along the capacitors such that the voltage across each capacitor approaches the voltage of the external AC power supply and that the voltages across the various capacitors add up. The setup yields in the end a multiplication of the AC power supply voltage that is proportional to the number of capacitors in the circuit. A schematic circuit diagram is given in Fig. 3. The original design by Cockcroft and Walton aimed at an accelerating voltage of 800 kV and was successfully used in 1932 for the acceleration of protons to 700 keV, which was sufficiently high for the disintegration of the Lithium nucleus.[8]

Cockcroft–Walton-type accelerators can be operated at megavolt levels with air insulation and locating the capacitor-diode stack in a large shielded room. Such setups have been widely used as pre-accelerators for many accelerator projects up to the late seventies. Figure 4 shows for example the Cockcroft–Walton-type accelerator at CERN that was used as the power supply for LINAC2 (a proton linear accelerator) up to the late 70s [13]. The capacitor array is seen in the foreground and the high voltage terminal in the back. The high voltage terminal contains the particle source and is connected to the LINAC2 installation through the wall in the back of the picture.

### 3.3 *Van de Graaff generator*

An alternative way to charge a high voltage terminal was developed in 1928 by Robert J. Van de Graaff at Princeton University in collaboration with Nicholas Burke [14]. The central idea of the 'Electrostatic Generator' is to transfer charges from a charge source, the positive pole of any DC power supply, to a high voltage terminal with the help of a rubber or silk conveyor belt. If the conveyor belt is connected to the inside of the high voltage terminal, the charges will flow from the conveyor belt to the outer surface of the high voltage terminal. The charges therefore accumulate on the outer

[8]This "pioneer work on the transmutation of atomic nuclei by artificially accelerated atomic particles" earned Cockcroft and Walton the Nobel Prize in physics in 1951.

Fig. 4. The Cockcroft–Walton-type accelerator for the CERN LINAC2 linear proton accelerator generated a voltage of 750 kV and was used up to the late seventies [13].

surface of the high voltage terminal and create a voltage potential between the terminal and ground. Figure 5 shows a schematic picture of a Van de Graaff generator with a setup for particle acceleration. The high voltage terminal continues to accumulate charges until the accumulation rate is equal to the rate at which charges are lost either through leakage currents or corona discharges. The larger the sphere and the better its insulation from ground, the higher the maximum voltage that can be accumulated in this process. M. A. Tuve, L. R. Hafstad and O. Dahl used such a Van de Graaff accelerator in 1935 for studying the intra-nuclear forces in proton scattering where a proton beam was shot onto various liquid or gaseous targets [15, 16]. The

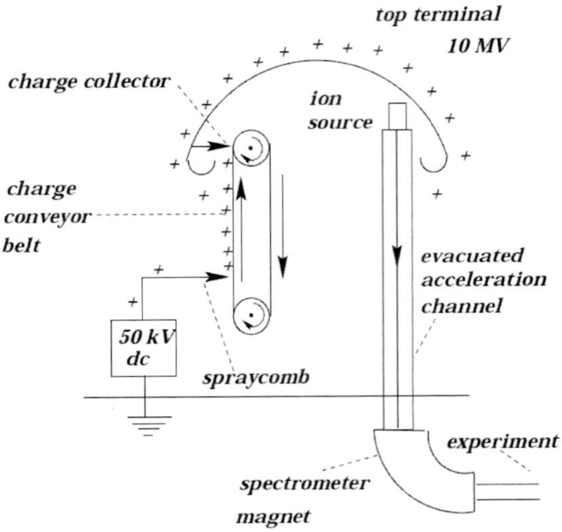


Fig. 5. Schematic picture of a Van de Graaff type voltage terminal in a setup for a particle acceleration experiment.

Fig. 6. The Million Volt Van de Graaff voltage terminal used by Tuve for proton–proton scattering experiments in 1931 [17].

Van de Graaff accelerator used by Tuve and his colleagues was capable of generating a voltage of 1.2 million volts (1.2 MV). Figure 6 shows a picture of the Van de Graaff generator used by Tuve that extended in height over two stories (similar in size to the Crockcraft–Walton generator in Fig. 4).

Placing the Van de Graaff generator into an oil tank for better insulation one can reduce the corona discharge and leakage currents from the high voltage terminal, increasing the peak voltage to the order of 10 million volts (10 MV).

### 3.4 *The Tandem Generator*

Using a negative ion source can further boost the performance of the Van de Graaff accelerator. Placing the negative ion source at the ground terminal, the ions are first accelerated towards the high voltage terminal. Placing a stripping foil in the path of the ions as they traverse the high voltage terminal can remove electrons from the ions and convert them into positively charged ions. The ions can then be accelerated again through the same voltage potential. Using, for example, a negative hydrogen ion source one can effectively double the acceleration potential of the Van de Graaff generator for protons. Figure 7 shows the schematic layout of such a Tandem Generator. Such devices have been constructed in the seventies and are still in use for particle acceleration in some places. For example, the Tandem Generator at Brookhaven National Laboratory (BNL), Laboratory in Upton New York is still in use for accelerating a wide range of ion species. The BNL Tandem has an effective voltage of two times 15 MV and is the world's highest energy Van de Graaff facility, capable of accelerating more than 50 different ion species [18]. With a maximum terminal voltage of 15 MV the capabilities of the BNL Tandem range from the acceleration of protons to 29 MeV to the acceleration of uranium ions to 385 MeV.

The BNL Tandem Van de Graaf generates the highest constant acceleration voltage for an operating particle accelerator. The Daresbury Nuclear Structure Facility generated slightly more than 20 MV accelerating voltage and provided the highest accelerating voltage. Higher constant acceleration voltages are technically difficult to realize and accelerating particles to higher

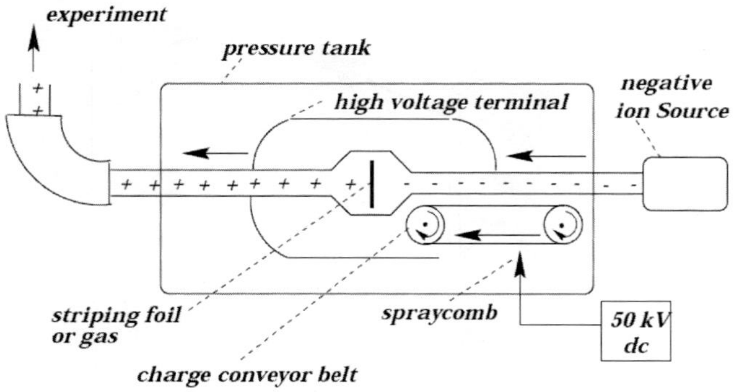


Fig. 7. Schematic picture of a Tandem Van de Graaff type voltage terminal.

energies is much more efficient by the use of time varying electric fields. This alternative acceleration concept will be discussed in the following sections.

## 4 Particle acceleration using time varying voltage generators

Parallel to the development of high voltage DC generators scientists and engineers started to look into options for using time varying electromagnetic fields for the acceleration of particles. Time varying fields allow either the combination of multiple acceleration stages or reusing one acceleration stage for multiple accelerations if the particles are bent onto a circular orbit. One prerequisite for this method to work is to divide the beam into packages such that no beam is exposed to the field lines when they have the wrong orientation for acceleration. Figure 8 illustrates the basic concept of a linear accelerator (linac) with multiple acceleration stages based on time varying fields and Fig. 9 illustrates the basic concept of a circular accelerator with a single acceleration stage. In both cases one can picture the process of acceleration by two capacitor plates generating an electric field in the appropriate direction for acceleration at the time the particles pass through the capacitor openings.

Neither of these two concepts can be pursued with constant electric fields. Staging several voltage units either implies effectively the generation of one high voltage terminal or that the particles are decelerated during the passage from one terminal to the next. Recirculating a beam through an electrostatic high voltage terminal implies likewise that the particles are decelerated on

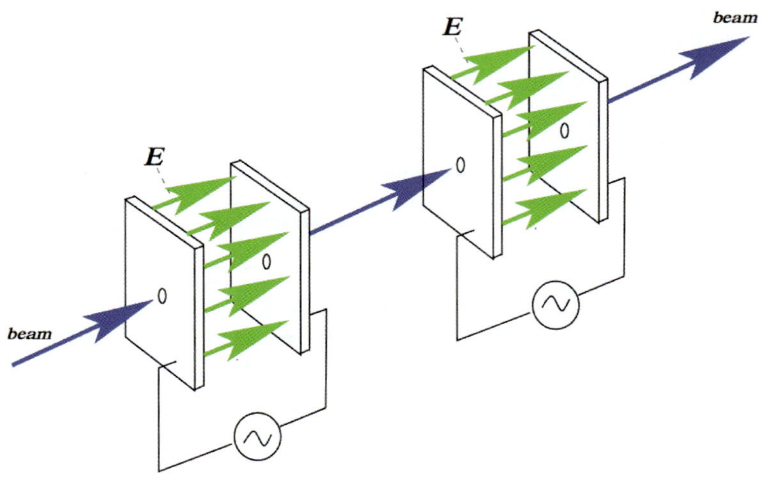


Fig. 8. Schematic picture of a linear accelerator using multiple acceleration stages.

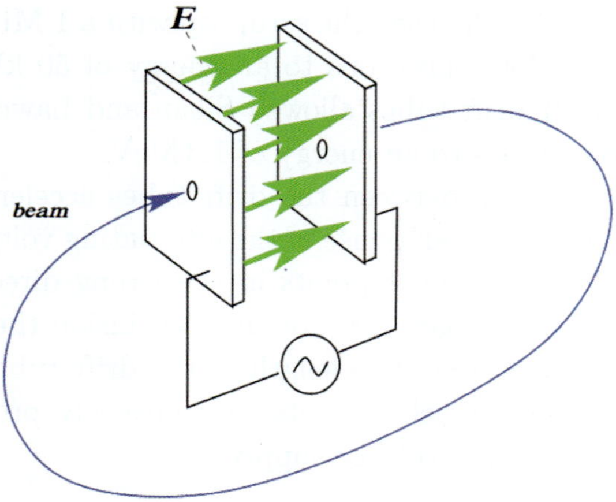


Fig. 9. Schematic picture of a circular accelerator using a single acceleration stage.

their passage back to the acceleration terminal as the particle motion is governed by a conservative Hamiltonian system where the total energy of the particle is conserved. It can merely be transformed from potential to kinetic energy and vice versa.

Using time varying fields changes this picture. The overall system is no longer conservative and the time varying power supplies can transfer energy into the system continuously.

### 4.1 *Drift tube linear accelerator*

The first design of a linear accelerator based on time varying fields goes back to G. Ising and R. Wideröe in 1924 to 1928. Ising suggested initially a linear accelerator using a travelling electromagnetic wave [19]. When Wideröe read Ising's article it inspired him to propose the concept of a drift tube linac [20, 21]; a concept that avoided the problems related to reflected waves in Ising's initial design. Figure 10 shows a schematic layout of a drift tube linac. The concept of a drift tube linac was first successfully demonstrated

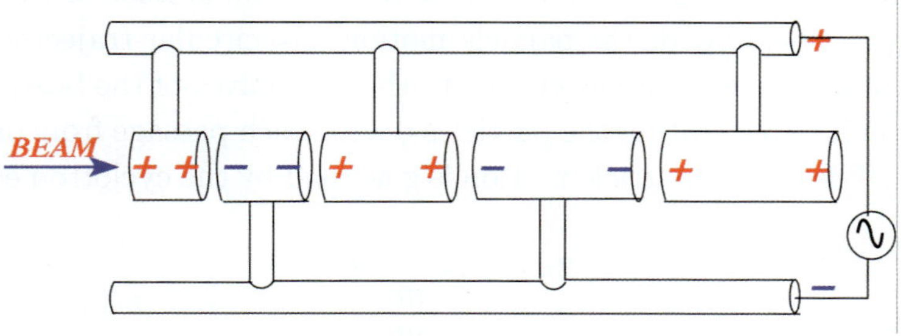


Fig. 10. Schematic picture of a drift tube linear accelerator.

by Wideröe in 1928 when he used the setup-up with a 1 MHz, 25 kV voltage supply to accelerate Potassium ions to an energy of 50 kV [20]. A similar setup consisting of 21 drift tubes allowed Sloan and Lawrence at Berkeley to accelerate Mercury ions to an energy of 1.3 MeV.

The electric field lines between the drift tubes accelerate the particles along the linac every other half phase of the alternating voltage signal. When the orientation of the field lines points in the wrong direction for acceleration along the linac, the particles are shielded inside the drift tubes and are therefore not decelerated. The length of the drift tubes must therefore be proportional to the particle velocity and inversely proportional to the frequency of the alternating voltage supply,

$$l = v_{part}/(2f_{volt}). \tag{5}$$

The drift tube linac represents an inductive load to the voltage power supply and the impedance and power consumption of the system is proportional to the frequency of the alternating voltage. In practice the inductive resistance limited the frequency of the drift tube linac to below 10 MHz before the development of high frequency and high power oscillators for radar during World War II. The limitation in the operating frequency implied in turn rather long drift tubes as the particles velocity increased. In the limiting case of a particle traveling at the speed of light, the drift tube length becomes 15 meters, which makes the concept rather impractical for the acceleration of relativistic particles.

### 4.2 *The Cyclotron*

The cyclotron exploits the option of bending particles into circular orbits and reusing repetitively the same acceleration voltage for multiple acceleration stages. Inspired by drawings in Wideröe's article [20] Ernest Lawrence designed a metallic box that is cut into two halves and is placed inside a magnetic field with field lines perpendicular to the main surface of the pill box. The magnetic field bends the particle motion into circular trajectories. Connecting an alternating voltage supply to the two halves of the box provides a mechanism for accelerating the particles during each passage from one half to the other. The overall particle motion is governed by the cyclotron equations:

$$2\pi f_{rev} = \frac{q}{m} \cdot B \tag{6}$$

$$r = \frac{m}{q} \cdot \frac{v}{B}. \tag{7}$$

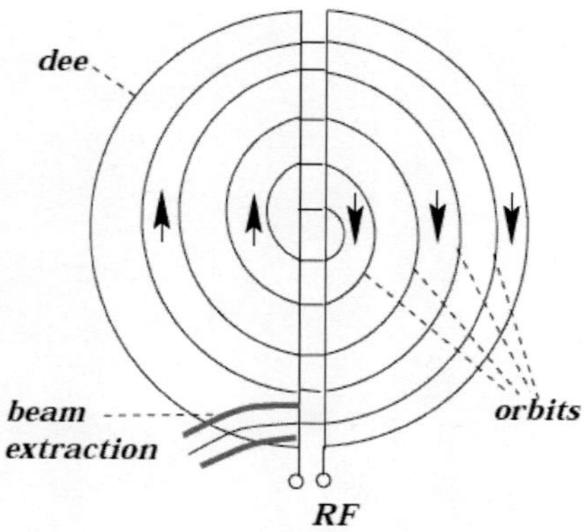


Fig. 11. Schematic picture of a Cyclotron accelerator: view from above. The particle source is placed at the center of the cyclotron. The arrows indicate the beam motion.

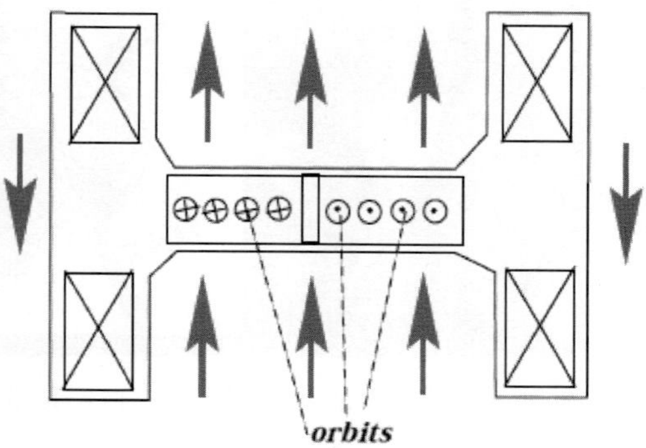


Fig. 12. Schematic picture of a Cyclotron accelerator: cross section picture from the side. The rectangles with a cross indicate the coil windings and the arrows the resulting magnetic field lines.

$m$ and $q$ are the mass and charge of the particles, $B$ is the magnetic field and $f_{rev}$ and $r$ are the revolution frequency and the revolution radius of the particle motion. As long as the particle motion is non-relativistic the mass of the particle remains approximately constant and the revolution frequency does not change during the acceleration process for a constant magnetic field. The radius of the particle motion, on the other hand, increases during the acceleration process and the particle's motion spirals outwards with increasing radius after each acceleration step. Figure 11 shows a schematic view of the setup from above and Fig. 12 shows a schematic cross section from the side where the box is placed inside a vertical magnetic field. Lawrence built the first device together with a graduate student, Stanley Livingston, in 1931. The initial device had a diameter of 5 inches and was capable of accelerating protons to 80 keV by applying an AC voltage of only 2 kV.

Fig. 13. Ernest Lawrence with the first 5 inch Cyclotron. The device accelerated $H^-$ ions to 80 keV but was compact enough to fit easily into a hand [23].

Lawrence patented the cyclotron in 1934 [22]. Laurence and the cyclotron are shown in Fig. 13. A slightly larger 12-inch cyclotron successfully accelerated protons to an energy of 1.2 MeV in 1932, an impressive achievement if one compares the size of the accelerating device (12 inches) to the size of a Van de Graaff or Cockcroft–Walton-type accelerators of comparable voltage, which have the size of a small building.[9]

Once the particle motion approaches the speed of light, the particle motion is no longer non-relativistic and the mass of the particles can no longer be assumed to be constant. The revolution frequency starts to decrease for a constant magnetic field as a consequence of the increasing particle mass. This difficulty can be overcome by either introducing stronger bending fields for the higher energetic passages at larger radii, 'Azimuthally-Varying-Field (AVF) Cyclotrons' and 'Sector-Focusing Cyclotrons', or by changing the frequency of the acceleration voltage during the acceleration process. The first

[9]The development of the cyclotron earned Lawrence the Nobel Prize in physics in 1939.

Fig. 14. The 590 MeV Cyclotron at the Paul Scherrer Institute.

variation of the Cyclotron is referred to as 'Isochronous' cyclotrons. The later variation is called a 'Synchrocyclotron', which has the key disadvantage that the device can no longer accelerate a continuous stream of particle packages but only one package per acceleration cycle due to the changing acceleration frequency.

The maximum beam energy of the Cyclotron is limited by the magnet strength and the practical size of the magnet core. Normal conducting cyclotrons are practically limited to proton beam energies of the order of 50 MeV. The use of superconducting magnets allows either an increase in the energy reach of the Cyclotron or the construction of compact cyclotrons with beam energies of the order of a few 10 MeV. Figure 14 shows as an example for a high energy Cyclotron the 590 MeV Cyclotron at the Paul Scherrer Institute in Switzerland.

Lower energy Cyclotron machines with energies of the order for 10 MeV to 20 MeV are still widely in use for medical and industrial applications (e.g. for the generation of radio-isotopes needed for Positron Emission Tomography (PET) scans).

### 4.3 *The Betatron*

The concept of the Betatron revisits the underlying mechanism for the magnetic core transformer circuit that was used for the high voltage terminals

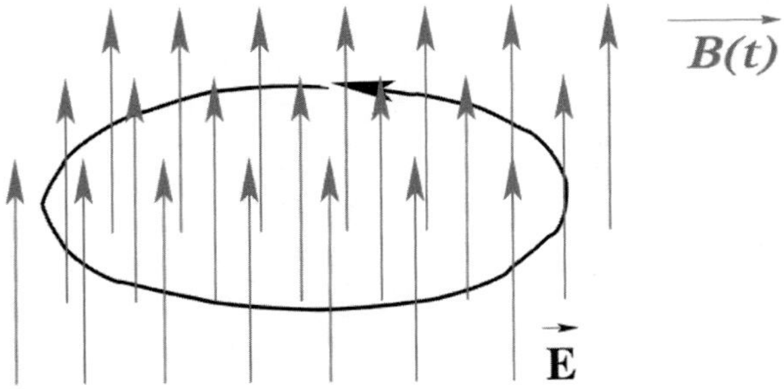


Fig. 15. Illustration of the Maxwell-Faraday induction law: a changing magnetic flux through a surface induces a rotational electric field along the boundary of the surface.

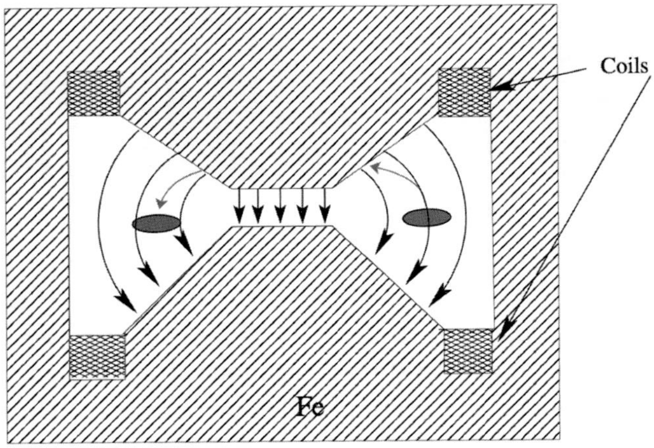


Fig. 16. Schematic side view of the Betatron induction accelerator: a changing magnetic field induces a circular electric field that accelerates the beam along its trajectory inside the Betatron.

in Section 3.1 and relies again on the Maxwell–Faraday law of induction: An alternating voltage supply drives a current through a primary coil winding. The changing current drives a time varying magnetic field inside a magnetic core and the change in the magnetic flux drives in turn a current through a secondary coil. The main difference to the transformer circuit is that for the Betatron the primary coil and the magnetic core are represented by a magnet system that also guides the beam into a circular trajectory and that the secondary coil is directly replaced by the beam. Figure 15 illustrates the Maxwell–Faraday induction law and Fig. 16 shows a schematic layout for the Betatron. Donald Kerst developed the Betatron from 1940 to 1942 [24, 25]. The basic Betatron concept can also be found in the original work of Wiederöe [20, 21]. J. Slepian patented the concept of acceleration via an induction field in 1922 [26]. However, the beam trajectory in the Betatron is not naturally stable. The beam oscillates around an ideal orbit and the beam stabilization requires a careful design and steering of the magnetic field. D. Kerst and R. Serber developed the stability condition only in 1941, long after the work by Wiederöe and Slepian [27], while D. Kerst constructed the first working Betatron machine for particle acceleration at the University of Illinois in 1941. The maximum beam energy of the Betatron is again limited, like the cyclotron, by the magnet strength and the practical size of the magnet core. The initial 2.3 MeV machine constructed by Kerst had a diameter of approximately 15 cm. The largest machine built was a 300 MeV Betatron with a beam trajectory radius of around 1.2 meters and featuring an iron magnet core with a diameter of approximately 10 meters weighing over 250 tons. It was constructed at the University of Illinois in 1950 and was operated with a repetition rate of 6 Hz. Lower energy Betatron machines with energies of the order of 40 MeV are still widely used for medical and industrial applications (e.g. as X-ray sources where the electron beam is shot onto a metal target).

## 5 Particle acceleration using electromagnetic waves

The circular accelerators in Section 4 all require a magnetic field throughout the whole area of the accelerator and the maximum attainable beam energy is therefore in practice limited by the size of the magnet system. Practical considerations limit the beam energies of these devices to a few 10 MeV for proton beams. The maximum attainable beam energy of the linear drift tube accelerator in Section 4 is essentially limited by the maximum attainable frequency and the practical diameter and length of the accelerator. Both of these limitations can be overcome by the use of electromagnetic waves and radio frequency resonators.

### 5.1 *Electromagnetic waves and waveguides*

Maxwell equations ($\frac{\partial \vec{B}}{\partial t} = -\vec{\nabla} \times \vec{E}$ and $\frac{1}{c^2}\frac{\partial \vec{E}}{\partial t} = \vec{\nabla} \times \vec{B}$) state that time varying electric and magnetic fields can exist together without local charge and current distributions. The electric and magnetic fields can form an electromagnetic wave that propagates through space. Figure 17 shows a schematic illustration of an electromagnetic wave in free space. If the electromagnetic wave is placed inside a waveguide (e.g. a tube with circular or rectangular cross section) the electromagnetic wave has to fulfill the boundary conditions that (a) magnetic field lines must run parallel to the waveguide surface and (b) the electric field lines must be normal to the waveguide surface (electric

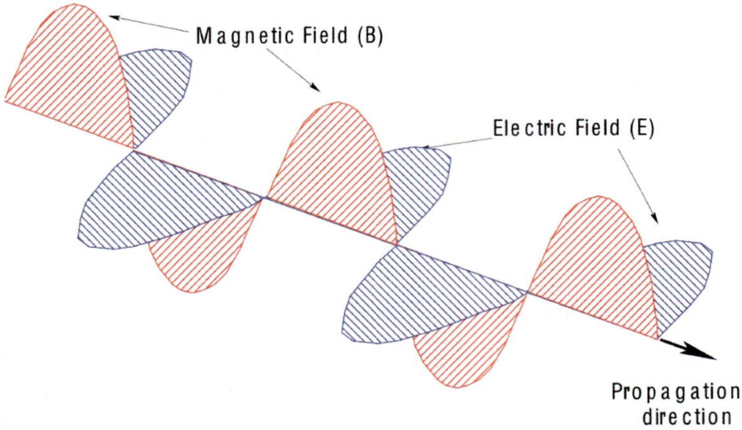


Fig. 17. Schematic illustration of an electromagnetic wave.

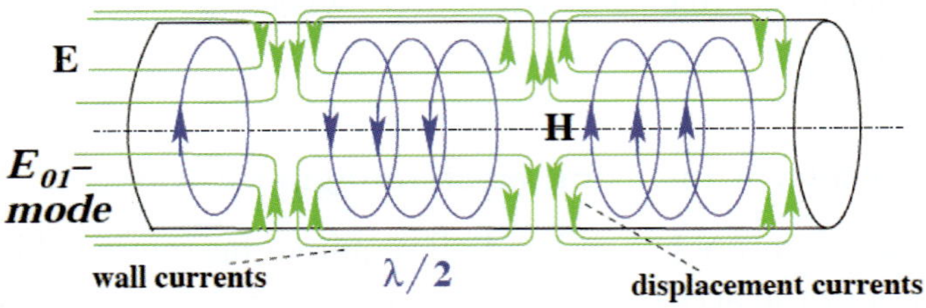


Fig. 18. Schematic illustration of a TM mode inside a waveguide with circular cross section.

field lines must end with a right angle on the waveguide surface). The electromagnetic waves that satisfy these boundary conditions can be grouped into two categories:

- Transverse Electric (TE) waves that have no longitudinal electric field component pointing in the direction of the waveguide, and
- Transverse Magnetic (TM) waves that have no longitudinal magnetic field component pointing in the direction of the waveguide.

If one wants to accelerate particles along the waveguide one is bound to choose TM waves that have a longitudinal electric field in the direction of the waveguide. Figure 18 shows a schematic illustration of a TM mode inside a waveguide with circular cross section.

### 5.2 *RF cavities*

If one terminates the waveguide with metallic plates on both sides one can trap an electromagnetic wave inside the resulting box. Figure 19 shows an

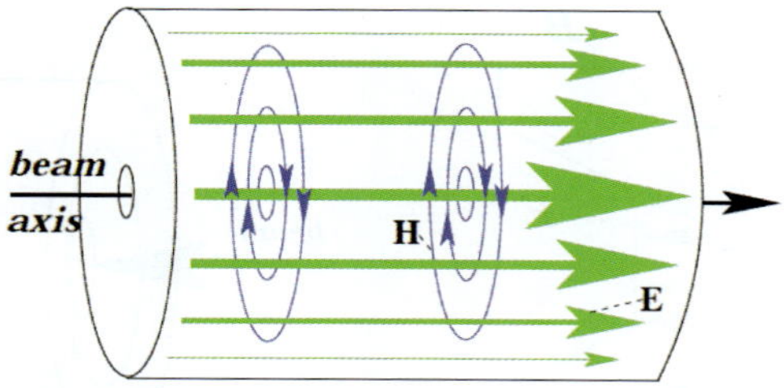


Fig. 19. Schematic illustration of a cavity resonator with a standing electromagnetic wave. The knots of the electromagnetic wave coincide with the longitudinal boundaries of the cavity.

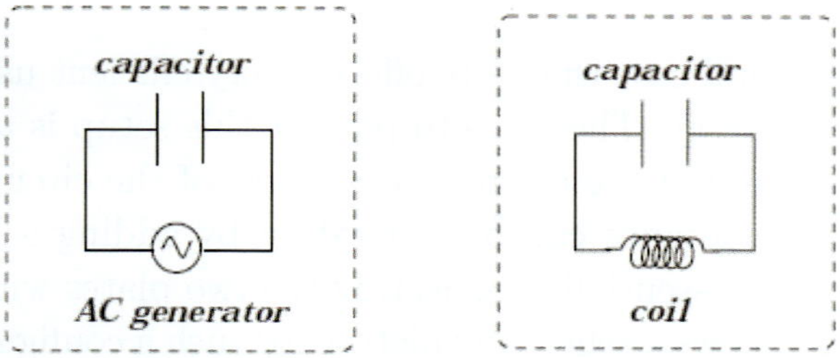


Fig. 20. Electrical circuit diagrams for capacitor plates and a resonance configuration with a parallel capacitor and inductor.

illustration of such a resonator with a standing electromagnetic TM wave. The amplitudes of the electric and magnetic fields oscillate with time and particle acceleration can only be obtained when the charged particles pass through the cavity during the appropriate phase of the field oscillation. Every other half phase of the wave oscillation the particles would be decelerated rather than accelerated and the particle beam needs to be prepared in packages such that beam is passing through the cavity only when the field orientation is correct.

An alternative illustration of the RF resonator can be obtained by revisiting again the capacitor acceleration setup in Figs. 8 and 9. A large fraction of the power of the voltage generator is required for moving charges from one side of the capacitor to the other and working against the reactive impedance of the capacitor plates. Adding an inductance to the capacitor plates (e.g. a loop current) results in a resonance circuit. Figure 20 shows the electrical diagrams for both configurations. In the first case the power converter has to perform work to move charges between the capacitor plates. In the second case the charges oscillate at the resonance frequency of the circuit and the power converter only needs to replenish the energy that is used for the particle acceleration and the resistive losses due to the surface resistance of

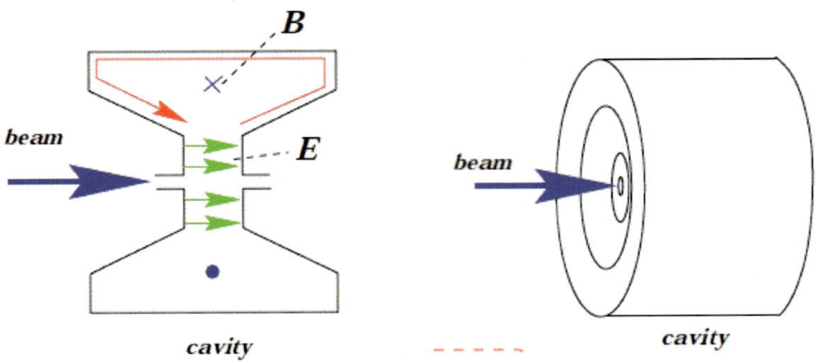


Fig. 21. Schematic picture of a 'Pill Box' cavity. Left: Transverse cross section with illustration of inductive current around the magnetic field lines. Right: Outside view.

the cavity. The second case therefore offers a very efficient usage of the electromagnetic field energy. The price to pay for this setup is a rather narrow operating regime around the resonant frequency of the circuit.

The missing inductance can be generating by adding a loop current to the capacitor plates, essentially connecting the two plates with a conducting surface. Figure 21 shows a schematic picture for such a configuration with the upper part on the left-hand side illustrating the inductive loop current. The resulting configuration resembles that of a metallic box, and the configuration is often referred to as an 'RF pill box cavity'. The argumentation via the resonance circuit recuperates the configuration of the truncated waveguide in Section 5.1. At the resonance frequency of the cavity, one obtains the standing wave configuration of Fig. 19.

### 5.3 *Alvarez tank linear accelerators*

The development of high frequency and high power oscillators for radar during World War II provided the missing ingredients for using electromagnetic waves for particle acceleration and allowed a redesign of the drifttube accelerator layout. Placing the tubes of the drifttube accelerator from Section 4.1 inside a long truncated waveguide and filling the wave guide with a resonant standing wave makes it possible to shield a passing beam from fields that point in the wrong direction for acceleration but without having to actively move charges between the tubes. Rather, the tubes are passive in such a setup, greatly reducing the reactive impedance of the system. Louis Alvarez developed the configuration in 1949 at Berkeley laboratory. The first operating accelerator was completed in 1951. It had a total length of approximately 12 meters and used a TM mode of 202.6 MHz to accelerate protons coming from a 4 MeV Van de Graaff injector to a final energy of 31.5 MeV. 200 MHz is a convenient frequency choice as this frequency corresponds to a diameter

Fig. 22. The Alvarez type structure of the old LINAC1 proton accelerator at CERN. LINAC1 used an RF frequency of 202.56 MHz and accelerated protons to an energy of 50 MeV [28].

of approximately 1 meter which is a practical transverse dimensions for a resonator tank. Louis Alvarez joined the Ernest Lawrence group at Berkeley in 1936. During the war Louis Alvarez (at M.I.T. at that time) was responsible for the development of three important radar systems. His collaboration with Lawrence at Berkeley and his experience in the development of radar placed Alvarez in a perfect situation for importing the new radar microwave technology into the field of particle accelerators.

Figure 22 shows the Alvarez type LINAC1 proton accelerator at CERN [28]. LINAC1 used an RF frequency of 202.56 MHz and was operated from 1958 until 1978 when another Alvarez type accelerator replaced it: LINAC2. Similar to the drift tube accelerator in Section 4.1 the required drift tube length needs to increase with the proton velocity and the accelerator becomes inefficient once the particles approach the speed of light. Another acceleration concept is therefore required for the acceleration of ultra-relativistic particles. However, most proton beam accelerator complexes still use an Alvarez type drift tube linear accelerator for the early acceleration stage.[10]

## 5.4 *The Synchrotron*

The development of radio frequency (RF) cavity resonators allowed a new optimization of the circular accelerators for acceleration to higher beam energies. With the goal of avoiding too large a magnet system one can use the Cyclotron Eqs. (6) and (7) to design an accelerator with a constant radius. A constant radius requires magnets for the guiding field only in the vicinity of the design orbit. For large accelerators this leads to a much more efficient accelerator design as compared to the Cyclotron or Betatron. Figure 23 shows the conceptual layout for such an accelerator with constant particle trajectory radius and a compact vacuum and magnet system.

The acceleration occurs at a dedicated insertion in the ring with the help of an RF cavity. In order to ensure that the particles experience a constant electric field at each passage through the cavity, the frequency of the electromagnetic wave inside the cavity must be an integer multiple of the revolution frequency and the beam must be prepared in individual packages such that no beam is inside the cavity when the field orientation points in the wrong direction. The resulting circular accelerator is referred to as the Synchrotron.

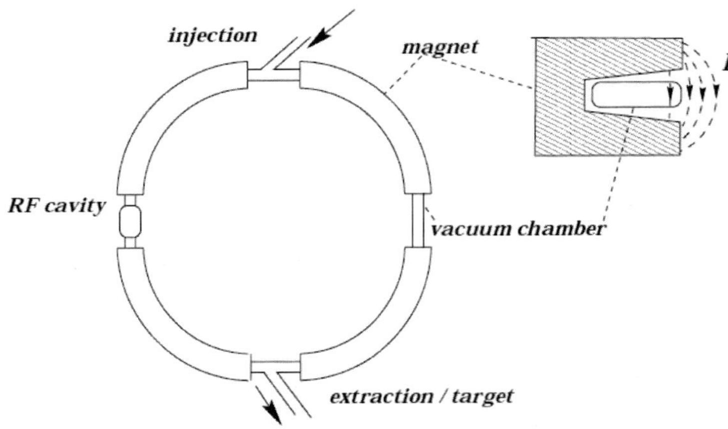


Fig. 23. Schematic illustration of a circular accelerator with constant particle trajectory radius and a compact vacuum and magnet system.

[10]Louis Avlares earned the Nobel Prize in physics in 1968 for his general contributions to elementary particle physics.

For relativistic particles it is convenient to express the particle mass as the product of the particle rest mass and the relativistic gamma factor:

$$m = m_0 \cdot \gamma \quad \text{with } \gamma = \frac{1}{\sqrt{1 - \frac{v^2}{c^2}}}. \tag{8}$$

Inserting Eq. (8) into the Cyclotron equations (6) and (7) one derives as the condition for constant radius that the magnetic field must vary proportionally to the product of the relativistic gamma factor and the particle velocity. The revolution frequency of the particles will then vary proportionally to the change in the particle velocity. The change in velocity will be negligible for ultra-relativistic particles and the operation at constant radius is thus possible with an approximately constant acceleration frequency. This allows the use of cavity resonators where the electromagnetic wave frequency is fixed to a narrow frequency range around the resonant frequency of the cavity and which is defined by the cavity geometry. The maximum beam energy of a Synchrotron depends then on its size and the available magnet technology (the attainable peak magnetic field). Solving Eq. (7) for the particle momentum ($p = m \cdot v$) one obtains for ultra-relativistic particles:

$$\frac{E}{c} = p = q \cdot r \cdot B. \tag{9}$$

In other words: the maximum attainable energy in a Synchrotron is linearly proportional to the size of the Synchrotron (or rather the radius of the beam trajectory) and the maximum magnetic field. The proportionality of the beam energy to the bending radius led over time to the development of Synchrotrons with ever-increasing size. The proportionality to the magnetic field has driven developments for high field magnet designs and led to the use of superconducting magnet technology in particle accelerators.

Vladimir Veksler at Lebedev Physical Institute [29] and Edwin McMillan at Berkeley Radiation laboratory (predecessor of the Lawrence Berkeley National laboratory) [30] independently developed the Synchrotron concept in the 1940s. McMillan devised the first operating electron synchrotron in 1945 when he worked on studies for the improvement of the Berkeley Cyclotron[30]. Similar to Alvarez, McMillan had a background in radar technology from World War II and was therefore in an ideal position to import this new technology to the field of particle accelerators. The Berkeley Synchrotron was used to create new elements, extending the periodic table of elements far beyond the 92 elements known before 1940.[11] The maximum

[11] Edwin McMillan shared the Nobel Prize in chemistry in 1951 with Glenn T. Seaborg for "discoveries in the chemistry of the transuranium elements".

attainable energy of the Synchrotron is in the end not only limited by the accelerator size and attainable peak magnetic field, as indicated by Eq. (9), but also by the energy loss due to Synchrotron radiation. The emission of electromagnetic waves by relativistic particles moving in a magnetic field was initially identified as a potential limitation to the maximum beam energy attainable in the Betatron [31]. But the radiation was first observed in a Synchrotron at the General Electric (GE) laboratory in 1947, leading to the christening of the radiation as 'Synchrotron radiation' [32]. The power emitted through Synchrotron radiation is proportional to the fourth power of the relativistic gamma factor and inversely proportional to the square of the radius of curvature in the arcs [33]

$$P_s = \frac{2r_0 c m_0 c^2 \beta^4 \gamma^4}{3\rho^2}, \tag{10}$$

where $\beta$ is the relativistic beta ($\beta = v/c$), $r_0$ the classical electron radius ($r_0 = e^2/[4\pi\varepsilon_0 m_0 c^2]$) and $\rho$ the radius of curvature of the trajectory inside the dipole field.

The radiation is therefore strongly suppressed for massive particles where the gamma factor is much smaller for a given beam energy (e.g. the effect is much smaller for protons as compared to electrons when the beam energies are comparable). This makes protons a very attractive beam species for Synchrotrons that try to push the beam energy frontier.

The Cosmotron at Brookhaven National Laboratory was the first operating proton Synchrotron. Planning for the Cosmotron started in 1948 and the machine reached its full design energy of 3.3 GeV in 1953, making it the highest energy accelerator of its time. The machine had a diameter of 23 meters featuring 288 C-shaped normal conducting iron core magnets with an aperture of 30 cm by 20 cm (width by height) and weighing in total 2000 tons. The Bevatron Synchrotron at Lawrence Berkeley National laboratory was constructed only two years later. It had approximately twice the size of the Cosmotron (a diameter of 55 meters) and achieved proton beam energies of 6 GeV (approximately twice the beam energy of the Cosmotron). This beam energy was sufficient to create the first antiprotons via pair production by shooting the proton beam onto a fixed target.[12]

The performance of Synchrotrons was greatly boosted in the 1960s by the invention of strong alternating focusing of the particle trajectories inside

[12]The discovery of antiproton in 1955 earned Emilio Segrè and Owen Chamberlain the Nobel Prize in physics in 1959.

the Synchrotron, resulting in much smaller beam sizes and thus only requiring much smaller and more compact magnet designs. Nicholas Christofilos conceived the Strong-focusing concept in 1949, but decided not to publish the idea, instead patenting it [34]. Ernest Courant, Milton S. Livingston, Hartland Snyder and J. Blewett then independently developed the strong focusing principle at Brookhaven National Laboratory in 1952 [35, 36] but acknowledged later the contribution by Christofilos [37]. The Synchrotron has become since the 60s to the present time the main workhorse for all high-energy colliders.

The Tevatron at Fermliab National Laboratory near Chicago, USA, was the first machine that used superconducting magnet technology for the main bending dipole magnets. It was operated from 1983 until 2011, had a circumference of 6.3 km, featured a superconducting dipole magnet system with a peak field of 4.2 T[13] and could eventually accelerate proton (and antiproton) beams to an energy of 980 GeV (operation started in 1983 with beam energies of 512 GeV and the energy of 980 GeV was only reached during the collider Run II which began in 2001 following extensive upgrades to the whole complex). The highest performing Synchrotron is currently the LHC at CERN, which has a circumference of 27 km and a magnet system with a peak field of 8 T. Its peak design proton beam energy is 7 TeV.

## 5.5 *Linear accelerators*

We have seen in Section 5.4 that the performance reach of circular accelerators is eventually limited by the power loss due to Synchrotron radiation. One way to circumvent this limitation is to use a linear accelerator rather than a circular one. One option for constructing a linear particle accelerator is to design a chain of RF cavities, each featuring a standing electromagnetic wave and all cavities being synchronized such that the field orientation in the cavities matches the requirements of the beam as it passes through the accelerator. Alternatively, one can come back to the concept of a travelling electromagnetic wave, as discussed in Section 5 and illustrated in Fig. 18. The challenge of using travelling electromagnetic waves for particle acceleration is that the phase velocity of the wave needs to be matched to the velocity of the accelerated particles. This can be achieved by inserting smaller aperture discs into the waveguide, as illustrated in Fig. 24. A proper design of the aperture of the irises can match the electromagnetic wave's phase velocity

[13]Approximately 100,000 times the Earth's average magnetic field. Normal conducting magnets can reach in practice a peak operating field of 2 T (when most ferromagnetic materials reach full saturation).

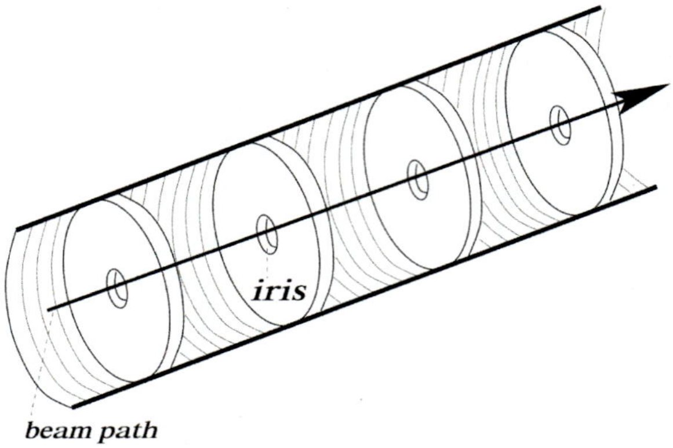


Fig. 24. Illustration of a disc-loaded accelerator structure.

Fig. 25. The LIL linear accelerator as pre-accelerator for the LEP project [38].

to that of the accelerated particles and the beam can be accelerated very much in analogy to a surfer who rides his surfboard at the crest of a wave from the sea to shore. Such linear travelling wave accelerators are commonly used as the initial acceleration phase for any electron or positron accelerator complex.

Figure 25 shows the LEP Injector Linac (LIL) of the LEP project that accelerated electrons and positron to energies of 600 MeV [38]. The Stanford Linear Collider (SLC) project at Stanford University was the first, and so far only, high-energy linear collider based on the concept of travelling wave acceleration. The SLC was a ca. 3 km long accelerator that started operation in 1988 and was capable of accelerating electron and positron beams to energies of 45 GeV. There are currently two main proposals for future high-energy linear accelerator projects under study: The Compact LInear Collider (CLIC) and the International Linear Collider (ILC) studies. Both projects target beam energies in excess of 250 GeV. The CLIC project is a linear collider based on normal-conducting RF structures and the ILC project is based on the use of superconducting RF cavities.

The search for increasing beam energies has steadily led to the development of particle accelerators of increasing size and new technologies. The evolution of particle accelerator is often illustrated by the so-called Livingston plot, Fig. 26, which shows the attainable beam energy as a function of year for the various accelerator types [39].

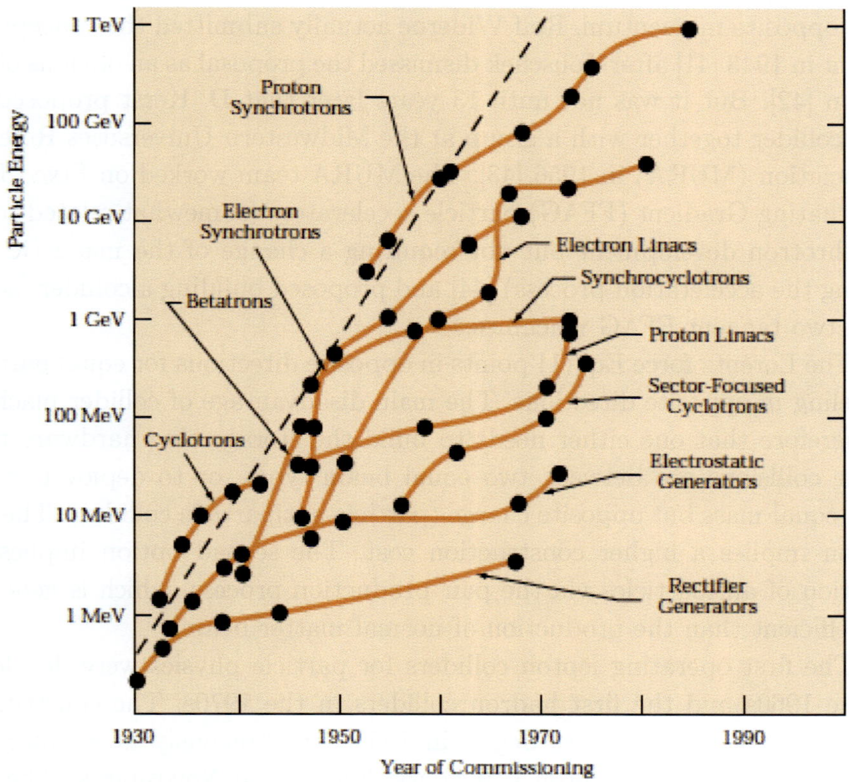


Fig. 26. Livingston Plot showing the increase in beam energies for the various types of accelerators as a function of 'year of commissioning'. Each circle indicates the start of operation of a new machine (up to 1985) [39].

## 6 Beam colliding storage rings versus fixed target colliders

The most common experimental tool for the early stage of high energy particle physics was the bubble chamber [40],[14] implying the shooting of an accelerated beam onto a fixed target. This method releases only a fraction of the total beam energy for the creation of new particles, as the overall momentum needs to be conserved in the process. The energy available for the creation of new particles increases only with the square root of the beam energy as a consequence of momentum conservation, severely limiting the maximum energy reach for studying elementary particles. Designing a particle accelerator that does not collide the beam with a fixed target but rather another beam of equal energy and opposite momentum can boost the performance of the machine significantly in terms of the maximum energy reach, as the total momentum of the initial system is zero and the energy available at the collision point is just twice the beam energy. This observation led to the development of particle colliders that accelerate not one but two beams with opposite momentum. Rolf Wideroe actually submitted the concept as a patent in 1943 [41] after Touschek dismissed the proposal as an obvious observation [42]. But it was not until 13 years later that D. Kerst proposed the first collider together with a group at the Midwestern Universities Research Association (MURA) in 1956 [43]. The MURA team worked on Fixed Field Alternating Gradient (FFAG) particle accelerators (somewhat related to the Synchrotron development but not requiring a change of the magnetic field during the acceleration process) [44] and proposed building a collider facility with two tangent FFAG machines.

The Lorentz force Eq. (1) points in opposite directions for equal particles traveling in opposite directions. The main disadvantage of collider machines is therefore that one either needs to build the storage ring hardware twice if the collisions are between two equal beam types, or to deploy particles with equal mass but opposite charge (particle-antiparticle collider). The first option implies a higher construction cost. The second option implies the creation of antiparticles via the pair production process, which is generally less efficient than the production of normal matter beams.

The first operating lepton colliders for particle physics were developed in the 1960s and the first hadron colliders in the 1970s. The construction of the first lepton colliders began in 1959 simultaneously at Stanford, in collaboration with Princeton University [39], and in Novosibirsk. The first

[14]The development earned Glaser the Nobel Prize in physics in 1960.

Fig. 27. The first collider machine at SLAC [39]. One clearly recognizes the two separate rings that house the counter-rotating electron beams. The collision point is in the center where the two beams intersect.

electron–electron collisions were detected in May 1964 in the VEP-1 (Novosibirsk) and shortly after in the Stanford machine. Figure 27 shows a picture of the Stanford–Princeton CBX collider, which featured two separate rings for collisions between two electron beams. However, electron–electron collisions mainly provide a validation of quantum electrodynamics, and are still only of limited use for the exploration of new particles. Using electron–positron beams instead provides the benefit that more energy becomes available at the collision points (the colliding particles transform into energy in the annihilation process) and that the generation of new particles starts from a pure energy state without the burden of preserving quantities defined by the initial beams.

The first electron–positron collider was the AdA ring at Frascati laboratory (a Synchrotron). The machine was proposed by Bruno Touschek in 1960 and originally conceived as a feasibility experiment to provide a sound basis for the realization of electron–positron colliders of larger center of mass energy and luminosity. AdA had a circumference of 3m and was capable of keeping particles of up to 0.25 GeV within a stainless steel vacuum vessel on a circular orbit of 65 cm radius. A radiofrequency cavity of

Fig. 28. The AdA machine in Orsay [45].

147 MHz compensated the energy loss due to Synchrotron radiation from the stored particles. The first stored beam of a few electrons was obtained at the end of May 1961, using the Frascati Electron Synchrotron as an injector. AdA was later shipped to France, to the "Laboratoire de l'Accelerateur Lineaire" of Orsay, near Paris, in order to generate a higher event rate with a high intensity linear accelerator as injector. By mid-1964, enough data had been collected to conclude observation of electron–positron collisions (the AdA detector count rates were a mixture of rest-gas and beam collision effects and data processing was required to single out the beam collision effect). Figure 28 shows the AdA machine installation when it was moved to Orsay [45]. The success of the AdA collider triggered a flood of new collider projects over the coming two decades, each machine increasing in size and moving the energy frontier for electron–positron collisions higher.

The number of events that can be detected in a collider is proportional to the beam densities at the collision point. The performance of colliding beam facilities obtained a significant boost with the invention of the so-called 'low beta' insertions that enhance the beam densities at the collision point and thus increase the collider performance by the use of dedicated focusing magnets [46, 47].

The maximum attainable energy of such circular electron–positron colliders is eventually limited by the loss due to Synchrotron radiation. Eq. (10)

shows that the power loss increases with the fourth power of the relativistic gamma factor but decreases only quadratically with the size of the storage ring. To increase the center of mass (CM) collision energy of a collider at a fixed Synchrotron radiation power loss, an increase of the collider size that is quadratic to the CM energy increase is required. This strong scaling quickly pushes electron–positron colliders to practical limits in terms of infrastructure investment. With a total circumference of 27 km and a maximum beam energy of 104.5 GeV, LEP was the largest electron–positron collider ever built.

The first hadron collider was built at CERN in 1971 and was operated until 1984. The Intersecting Storage Ring (ISR) at CERN featured two separate rings and magnet systems for two proton beams with beam energies of up to 31.4 GeV. It had normal conducting magnets in the arcs but introduced many new concepts to accelerator design (e.g. stochastic cooling [48] and superconducting magnets for the low beta focusing of the beams near the collision points). It provided center of mass energies of up to 62.8 GeV and demonstrated that a colliding beam facility is feasible for proton beams. It held the record luminosity for a hadron collider (measured by the number of events a collider can generate per second) until 2004. The development of stochastic cooling, a method for shrinking the beam size in a storage ring, opened the door to an efficient accumulation process for antiprotons and thus the proposal for a proton-antiproton collider with reasonable performance. Van de Meer invented the stochastic cooling method in the early 1970s [49,50] which led Carlo Rubbia to propose the conversion of the new CERN Super Proton Synchrotron (SPS), initially designed for fixed target beam experiments, into a proton–antiproton collider.[15] The SppS (Sp$\bar{p}$S, or simply SppS) collider had a single vacuum and magnet system for both beams. It had a circumference of 6.9 km, maximum beam energies of 450 GeV and operated as a collider from 1981 until 1984 (the SPS is still used nowadays as a fixed target synchrotron and an injector for the LHC).

The Tevatron at Fermilab National Laboratory near Chicago, USA, introduced superconducting magnet technology for the main bending dipole magnets to the collider ring concept and more than doubled the beam energies as compared to the SppS collider in a machine of comparable size and circumference. The Tevatron was operated from 1983 until 2011. The peak performance of the Tevatron was eventually limited by the rate of

[15] The invention of stochastic cooling and the proposal for a proton-antiproton collider earned Van der Meer and Rubbia the Nobel Prize in physics in 1984.

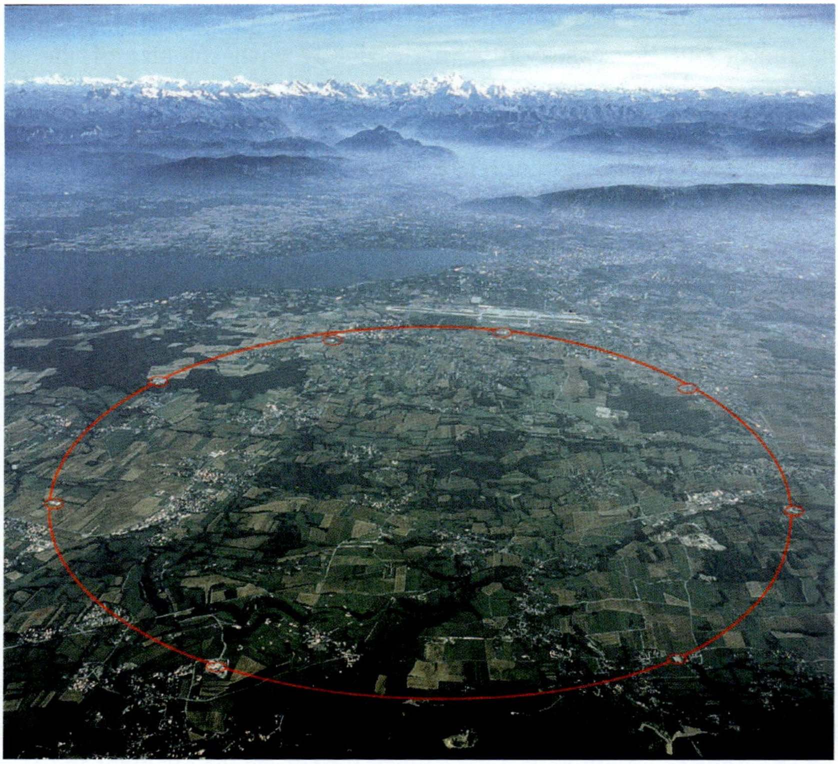

Fig. 29. The LHC collider illustrated in an aerial view of the Geneva region [53].

the antiproton production and a significant further increase in the collider performance would have either required a significant increase in the antiproton production rate or the use of two proton beams. The LHC at CERN is currently the highest energy hadron collider. Different from the SppS and the Tevatron, it went back to the two-ring collider concept of the ISR, featuring separate vacuum and magnet systems for two counter-rotating proton beams. It has a total circumference of 27 km (it was built in the old LEP tunnel) and has been designed for peak beam energies of 7 TeV. Figure 29 shows an aerial view of the LHC collider in the Geneva region next to Lake Geneva [53].

Comparing the 27 km circumference of the LHC and Lepton colliders in Fig. 29 to the size of the AdA collider in Fig. 28, featuring a circumference of 3 m, illustrates nicely the immense scalability (4 orders of magnitude) of the Synchrotron concept. The scalability is essentially only limited by practical considerations for the tunnel size, which is what makes the Synchrotron so attractive for use in high-energy beam physics projects.

A summary of the historical evolution of particle accelerators for high energy particle physics and the resulting development of the Standard Model of physics can be found in [51] and [52].

## References

[1] E. Rutherford (1919). Collision of $\alpha$ particles with light atoms, IV: An anomalous effect in nitrogen. *Philosophical Magazine* **37**, 581–587.

[2] A. Einstein (1905). Ist die Trägheit eines Körpers von seinem Energieinhalt abhängig? *Annalen der Physik* **18**(639), 1905.

[3] P. Curie and M. Curie (1898). Sur une substance nouvelle radio-active contenue dans la pechblende. *C. R. Acad. Sci. Gen.* **1898**(127), 175–178.

[4] P. Curie, M. Curie and G. Bémont (1898). Sur une nouvelle substance fortement radio-active contenue dans la pechblende. *C. R. Acad. Sci. Gen.* **1898**(127), 1215–1218.

[5] J. J. Thomson (1897). Cathode rays. *The Electrician* **39**, 104; also published in *Proceedings of the Royal Institution* April 30, 1897, 1–14.

[6] J. J. Thomson (1897). The classic measurement of the electron mass and charge. *Philosophical Magazine* **44**, 293.

[7] E. Rutherford (1911). The scattering of $\alpha$ and $\beta$ particles and the structure of the atom. *Philosophical Magazine* **21**, 669–688.

[8] C. D. Anderson (1932). The apparent existence of easily deflectable positives. *Science* **76**(1967), 238–239.

[9] C. D. Anderson (1933). The positive electron. *Physical Review* **43**(6), 491.

[10] S. Weinberg (1983). *The Discovery of Subatomic Particles*, Scientific American Library.

[11] A. Daniels (1976). *Introduction to Electrical Machines*, Macmillan Publishers.

[12] J. D. Cockcroft and E. T. S. Walton (1932). *Proc. Roy. Soc. (London) A* **136**, 619.

[13] CERN Photo Library: CERN-AC-7602012.

[14] R. J. Van de Graaff (1931). Electrostatic generator. Patent number 1991236, filed Dec 16, 1931, issued Feb 12, 1935.

[15] T. D. Cornell (1988). Merle Antony Tuve: Pioneer nuclear physicist. *Physics Today* **41**(1), 57, doi:10.1063/1.881153.

[16] J. Mehra and H. Rechenberg (2001). *The Conceptual Completion and Extensions of Quantum Mechanics 1932–1941. Epilogue: Aspects of the Further Development of Quantum Theory 1942–1999*, Volumes 1 to 6. Springer.

[17] Smithsonian Institution, Negative Number 70846. Courtesy of AIP Emilio Segre Visual Archives.

[18] "From fifty to the future: Brookhaven National Laboratory 1947 to 1997" (1997). BNL No 52538.

[19] G. Ising (1924). Prinzip einer Methode zur Herstellung von Kanalstrahlen hoher Voltzahl" (in Greman). *Arkiv för matematik o. fysik* **18**(30), 1–4.

[20] R. Wideröe (1928). Über ein neues Prinzip zur Herstellung hoher Spannungen" (in German). *Arch. F. Elektrot.* **21**, 387.

[21] M. S. Livingston and J. P. Blewett (1966). *The Development of High-Energy Accelerators*, commented reprints or translations of original papers, Dover Publish. Inc., N.Y.

[22] E. O. Lawrence (1934). Method and apparatus for the acceleration of ions. US patent 1948384, issued Feb 20, 1934.

[23] P. Preuss (2009). Accelerators and light sources of tomorrow. *Berkeley News Letter*, Dec 21, 2009, http://newscenter.lbl.gov/feature-stories/2009/12/21/accelerators-tomorrow-part1/.
[24] D. W. Kerst (1940). Acceleration of electrons by magnetic induction. *Physical Review* **58**(9), 841.
[25] D. W. Kerst (1941). The acceleration of electrons by magnetic induction. *Physical Review* **60**, 47–53.
[26] J. Slepian (1922). US patent 1645304.
[27] D. W. Kerst and R. Serber (1941). Electronic orbits in the induction accelerator. *Physical Review* **60**(1), 53–58.
[28] CERN Photo Library: CERN-AC-6808042.
[29] I. V. Veksler (1944). A new method of the acceleration of relativistic particles" (in Russian). *Doklady Akad. Nauk SSSR* **43**, 346.
[30] E. M. McMillan (1945). The synchrotron — A proposed high energy particle accelerator. *Physical Review* **68**, 143.
[31] D. Iwanenko and I. Pomeranchuk (1944). On the maximum energy attainable in a betatron. *Physical Review* **65**, 343.
[32] F. R. Elder, A. M. Gurewitsch, R. V. Langmuir and H. C. Pollock (1947). Radiation from electrons in a synchrotron. *Physical Review* **71**(11), 829–830.
[33] A. Hofmann (2004). *The Physics of Synchrotron Radiation*, Cambridge University Press.
[34] N. C. Christofilos (1950). Focusing system for ions and electrons. US patent 2736799.
[35] E. D. Courant, M. S. Livingston and H. S. Snyder (1952). The strong-focusing synchrotron — A new high energy accelerator. *Physical Review* **88**(5), 1190–1196.
[36] J. P. Blewett (1952). Radial focusing in the linear accelerator. *Physical Review* **88**(5), 1197–1199.
[37] E. D. Courant, M. S. Livingston, H. S. Snyder and J. Blewett (1953). Origin of the 'strong-focusing' principle. *Physical Review* **91**, 202–203.
[38] CERN Photo Library: CERN-AC-8607648.
[39] W. Panofsky (1997). The evolution of particle accelerators & colliders. *Beam Line* **27**(1), 36.
[40] D. A. Glaser (1952). Some effects of ionizing radiation on the formation of bubbles in liquids. *Physical Review* **87**(4), 665.
[41] R. Wideroe (1943). Anordnung zur HerbeifUuhrung von Kernreaktionen. Deutsches Patentamt, Patentschrift 876279, submitted September 1943, issued March 1953.
[42] P. Waloschek (1994). The infancy of particle accelerators: Life and work of Rolf Wiedeöe. DESY-Red-Report 94-039.
[43] D. W. Kerst, F. T. Cole, H. R. Crane, L. W. Jones *et al.* (1956). Attainment of very high energy by means of intersecting beams of particles. *Physical Review* **102**(2), 590–591.
[44] D. W. Kerst *et al.* (1956). Fixed field alternating gradient particle accelerators. In *CERN Symposium on High Energy Accelerators and Pion Physics*, Geneva, Switzerland, 1956, p. 32.
[45] J. Haissinski (1989). From AdA to ACO: Reminiscences of Bruno Touschek. In *Bruno Touschek and the birth of e+ e- physics*, Frascati, Italy, November 16, 1998, pp. 17–31.
[46] A. Hofmann *et al.* (1967). The colliding beam project at the CEA. In *Proceedings of the VI International Conference on High Energy Accelerators*, Cambridge, p. 112.
[47] A. Sessler (1993). The development of colliders. Lawrence Berkeley Laboratory, LBL-33664.

[48] S. von der Meer (1972). Stochastic damping of betatron oscillations. CERN internal report, CERN/ISR/PO/72-31.
[49] D. Mohl, G. Petrucci, L. Thorndahl and S. van der Meer (1980). *Phys. Rept.* **58**, 76.
[50] L. Bonolis (2005). Bruno Touschel vs machine builders: AdA, the first matter–antimatter collider. *Revista Del Nouvo Cimento* **28**, N11.
[51] A. D. Krisch (2000). Storage rings: Past, present and future. *AIP Conf. Proc.* **512**, 394.
[52] L. Hoddeson *et al.* (1997). *The Rise of the Standard Model: Particle Physics in the 1960s and 1970s*, Cambridge University Press.
[53] CERN Photo Library: CERN-SI-0107014.

# Part 2

# 100 Years of Accelerators

# Chapter 3

# The first colliders: AdA, VEP-1 and Princeton–Stanford

Vladimir Shiltsev (Fermilab)

The idea of exploring collisions in the center-of-mass system to fully exploit the energy of the accelerated particles had been given serious consideration by the Norwegian engineer and inventor Rolf Wideröe, who applied for a patent on the idea in 1943 (and got the patent in 1953 [1]) after considering the kinematic advantage of keeping the center of mass at rest to produce larger momentum transfers. Describing this advantage, G. K. O'Neill, one of the collider pioneers, wrote in 1956 [2]: "... as accelerators of higher and higher energy are built, their usefulness is limited by the fact that the energy available for creating new particles is measured in the center-of-mass system of the target nucleon and the bombarding particle. In the relativistic limit, this energy rises only as the square root of the accelerator energy. However, if two particles of equal energy traveling in opposite directions could be made to collide, the available energy would be twice the whole energy of one particle ..." Therefore, no kinetic energy is wasted by the motion of the center of mass of the system, and the available reaction energy $E_R = 2E_{beam}$ (while a particle with the same energy $E_{beam}$ colliding with another particle of the mass $m$ at rest produces only $E_R = (2E_{beam}m)^{1/2}$ in the extreme relativistic case). One can also add that the colliders are "cleaner" machines with respect to the fixed-target ones since the colliding beams do not interact with the target materials. The other advantage is that it is much easier to organize collisions of beams composed of matter–antimatter particles, like in electron–positron and proton–antiproton colliders.

This idea was taken seriously and three teams started working on colliding beams in the late 1950s: a Princeton–Stanford group that included William C. Barber, Bernard Gittelman, Gerry O'Neill, and Burton Richter, who in 1959, following a suggestion of Gerry O'Neill in 1956, proposed to build a couple of tangent rings to study Møller scattering; Andrei

Mikhailovich Budker initiated a somewhat similar project in Soviet Union, where electron–electron collider VEP-1 (Russian acronym for *"Встречные Электронные Пучки-1"* or "*Colliding Electron Beams-1*") was under construction in 1958; and an Italian group at Laboratori Nazionali di Frascati led by Bruno Touschek began design of the first electron–positron collider.

In the early 1960s, almost at the same time, the first colliders went into operation in the Soviet Union, Italy and USA. The Italian group built the $e^+e^-$ storage ring ADA (Anello di Accumulazione), proved the possibility of storing an accelerated beam for hours [3,4] and got enough evidence for first electron–positron collisions in mid-1964 [5,6]. The first Soviet $e^-e^-$ storage ring, VEP-1, was constructed in Moscow and moved to Novosibirsk in 1962 [7,8]. First electron–electron collisions were detected in May 1964 [9] and in 1965 VEP-1 started providing the first experimental results [10,11]. The Princeton–Stanford electron–electron collider [12] announced obtaining the first electron–electron collisions in March 1965 [13] and the first interesting results were published in 1966 [14].

## 1 AdA

On March 7, 1960, Bruno Touschek (1920–1978), a brilliant Austrian theoretician, gave a seminar in Frascati, presenting the main features of $e^+e^-$ annihilation processes and the proposal to build, as a first step, a very small ring that would use the existing 1.1 GeV electron synchrotron as an injector. Attainment of head-on collisions of electrons and positrons in flight required storing them in a magnetic device (storage ring) to allow them to collide repeatedly as they crossed at various points in their circular orbits. The beam-accumulation (*accumulazione*) problem was considered to be the most serious one, so the collider was baptized as AdA, the acronym for the Italian *Anello di Accumulazione.* In approximately one year, the ring was built and the first stored particles were obtained on February 27, 1961 (it was not clear at the time whether those were electrons or positrons [15]).

Figure 1 shows the layout and regime of AdA's operation. The electron beam from the electron synchrotron — see (1) in the left of Fig. 1 — strikes an external target (2), producing bremsstrahlung gamma rays that enter the collider ring (3) and strike the tantalum internal converter (4), producing electrons that orbit counter-clockwise and pass through the RF cavity (5). The ring is then moved laterally and rotated 180° as shown on the right. Positrons are then produced via the same procedure and orbit clockwise as shown, colliding with the oppositely circulating electrons in the ring. AdA, therefore, was the only "walking" accelerator in the history of high energy

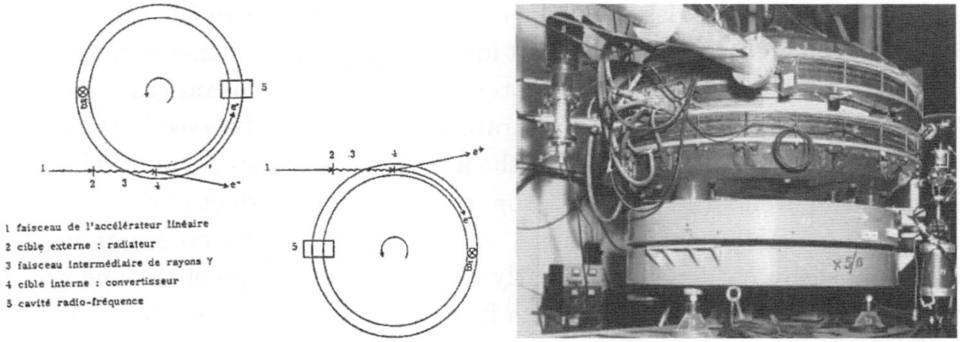


Fig. 1. (left) Layout of AdA; (right) AdA on the rotating and translating platform at Orsay. The injector beam channel is visible on the left.

Table 1. Main parameters of AdA electron–positron collider (Frascati/Orsay).

| Parameter | Value (typical operations) | Units |
|---|---|---|
| Energy, per beam | 250 (200) | MeV |
| Circumference | 4.1 | m |
| Luminosity | $\sim 10^{25}$ | $cm^{-2}s^{-1}$ |
| Beam current, per beam | $\sim$0.05 | mA |
| Injector linac beam energy | 1000 (at Orsay) | MeV |
| Max field in the rings | 1.1 | T |
| Field index $n = (dB/B)/(dR/R)$ | 0.55 | – |
| Vacuum pressure | <1 | nTorr |
| RF voltage | 7.5 (5) | kV |

physics. The main parameters of the ring are listed in Table 1. Filling of the machine with particles was quite a tedious affair, and several tricks were employed, including modulation of the amplitude of the 147 MHz RF system at injection in order to steer injected particles away from the internal converter [4]. After significant work on reduction of the vacuum pressure to below 1 nTorr, the lifetimes of feeble beams of electrons and positrons improved to many hours (up to 40). Still, the accumulation rates and stored currents were extremely low and in 1962 the AdA ring was moved from Frascati to the Laboratoire de l'Accelerateur Lineaire in Orsay (France) to employ much higher intensity of the primary electron beam from the 1000 MeV LAL linac [6].

Even with greatly improved injection rate, it took immense effort to prove that electron–positron beam–beam collisions were indeed occurring. At the end, the best method was found to be the detection of photons of the single bremsstrahlung reactions $e^+e^- \rightarrow e^+e^-\gamma$ by a lead-glass Cerenkov counter (a 150-kilogram lead-glass cylinder installed tangentially to the orbit at the interaction point). It turned out that identifying those photons was not trivial under the background of similar photons produced in the beam-gas collisions. The following method was used: the rate of gamma rays observed in the direction of beam 1 was proportional to the number of particles $N_1$ in it, while for beam–beam events the rate was proportional to the product of the numbers of particles $N_1N_2$ in both beams 1 and 2. Thus, the observed gamma-ray rate divided by $N_1$ depended linearly on $N_2$, and the slope of the line was a measure of the luminosity. The data from the last most successful AdA runs in December 1963 to April 1964 were processed by mid-1964. The data was reported on July 16, 1964 [5], providing statistically convincing evidence of the beam–beam collisions with a maximum luminosity of about $10^{25}$ $cm^{-2}s^{-1}$.

The AdA team had not only proven the basic underlying concept of the $e^+e^-$ colliders, but also tackled a number of accelerator physics issues, such as beam scattering on the electrons of the residual gas, quantum fluctuations of the synchrotron radiations and RF lifetime, coupling of vertical and horizontal betatron oscillations, and the notorious "Touschek effect". The latter manifested itself as a significant drop of the beam intensity lifetime from some 50 hours at low currents to a few hours at the highest currents ($\sim 3 \times 10^7$ particles stored). This was found to be due to intra-bunch electron–electron scattering causing transfer of energy from the transverse betatron oscillations into the longitudinal direction, and particles escape beyond the stability region of the RF bucket [16].

## 2 VEP-1 collider in Novosibirsk

Prof. Budker's team was initially formed in 1956 as the Laboratory of New Acceleration Methods at the Institute of Atomic Energy (Moscow). In 1958 the laboratory was transformed into the Institute of Nuclear Physics and moved to Novosibirsk. The work on colliding electron beams began at the end of 1956, after the Geneva conference, where the feasibility of the colliding-beam idea was first discussed. The first colliding-beam installation VEP-1 was built in Novosibirsk, assembled in Moscow and returned to Novosibirsk in 1962 for beam commissioning [17]. The first beam was captured on the VEP-1 orbit in 1963 and the first electron–electron scattering recorded on May 19, 1964. The first experimental studies on the scattering of electrons started in 1965 and $e^-e^- \to e^-e^-\gamma$ reactions, as well as the world's first $e^-e^- \to e^-e^-2\gamma$ (double bremsstrahlung) reactions were detected [18–20]. The experiments were concluded in 1967 (to open the road for the next generation $e^+e^-$ collider VEPP-2) and VEP-1 was disassembled in 1968.

The overall goal of the project was to check the limits of applicability of quantum electrodynamics at small distances by studying the angular distribution of elastic (Møller) scattering of electrons by electrons. The main purpose of the VEP-1 collider was to develop the colliding-beam method and prove that it could be used in the future for a broader class of particles and experiments. The initial proposal was to construct two installations, VEP-1 with energy $2 \times 130$ MeV and VEP-2 with energy $2 \times 500$ MeV. The VEP-1 set-up was regarded initially only as a mock-up of a "real" colliding-beam accelerator while VEP-2 was intended to be used for the QED studies. Nevertheless, after the announcement by Prof. W. Panofsky in 1958 that a similar colliding-beam facility was planned to be built at Stanford in collaboration with Princeton University, Budker abandoned the plans for construction of the 500 MeV storage rings and concentrated all the efforts on VEP-1 alone.

The main components of VEP-1 were — see Fig. 2 — a cyclic electron accelerator–injector B-2S, magnets of the storage rings, a high-vacuum system, a high-power RF system to accelerate the particles and to maintain their energy in the storage rings, a single-turn system for extraction from B-2S and injection into the rings, beam focusing and transporting system with a pulsed switching magnet to guide beam into either ring, beam diagnostics and a dedicated system of counters, and spark chambers for the QED experiments [21]. VEP-1 was the only collider with vertical orientation of the rings, one under another. The magnetic tracks of the storage ring had a radius of 43 cm. The electrons scattered at the interaction point could leave the rings through special slots in the magnet poles.

The energy of the electrons injected into the storage ring was 43 MeV (maximum, 40 MeV for routine operations). The maximum energy of the colliding beams was $2 \times 160$ MeV ($2 \times 130$ MeV in routine operation).

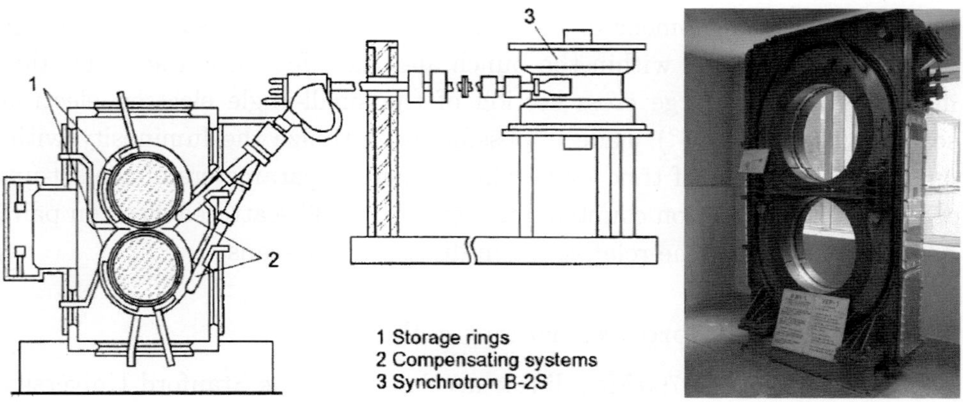


Fig. 2. Layout and photo of the VEP-1 collider.

Table 2. Main parameters of VEP-1 electron–electron collider in Novosibirsk.

| Parameter | Value (typical operations) | Units |
|---|---|---|
| Energy, per beam | 160 (130) | MeV |
| Circumference, each ring | 2.7 | m |
| Luminosity | $4 \times 10^{28}$ | $\text{cm}^{-2}\text{s}^{-1}$ |
| Injection energy | 43 (40) | MeV |
| Beam current, per beam | 100 (50) | mA |
| Max field in the rings | 1 | T |
| Field index $n = (dB/B)/(dR/R)$ | 1, 0.62 and 0 | – |
| Vacuum pressure | 30 | nTorr |
| RF voltage | 5 | kV |

The injector was an iron-free pulsed B-2S synchrotron with a spiral electron accumulation. It produced some 5 ns pulse of maximum 500 mA current of electrons (more than $3 \times 10^{10}$) with an energy spread of <0.2%. High power 100 kV pulse generators with better than nanosecond time jitter were developed and used in the extraction–injection system (1 ns rise time and some 10 ns decay). A typical collision run lasted some 10 minutes.

Several important effects were observed and studied at VEP-1 [22]. First of all, the beam–beam effects were found to be quite strong, so that, for example, the beam orbits did need to be separated during the filling time. The effects were caused by mutual influence of the electromagnetic fields produced by colliding beams on incoherent betatron oscillations of the electrons. The choice of the optimum operation regime for filling and colliding beams was one of the major operational problems. The effects of the guiding field non-linearities on the beam motion and the beam behavior near non-linear resonances were carefully investigated. The coherent phase instability of bunches was discovered and generally understood. Some other important beam dynamics phenomena in a storage ring were studied also — such as the intra-beam scattering within the bunch and the effect of ion accumulation in the beam. The large cross section of the small-angle electron–electron scattering (within 1.5°) made it possible to optimize the luminosity without appreciable loss of time by varying numerous parameters of the set-up. A system of scintillation counters registered up to 30 scattered electron pairs per second allowing the relative beam displacement scans.

## 3 Princeton–Stanford experiment collider

In the mid-1950s the 700 MeV linear accelerator at the Stanford University High Energy Physics Laboratory (HEPL) was the most powerful electron

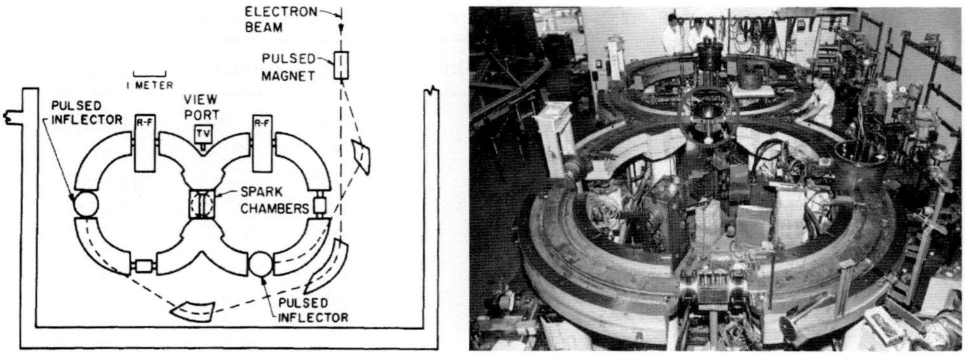


Fig. 3. Layout and photo of the Princeton–Stanford electron–electron collider.

accelerator in the world. G. K. O'Neill visited HEPL in 1957 to discuss colliding beams with W. K. H. Panofsky, then the director of the laboratory, and to seek local collaborators. As the result, the Princeton–Stanford group was set up to develop the new colliding-beam technology as well as to demonstrate it by using the new technology to test the theory of quantum electrodynamics [23]. Fast radiation damping of the high energy electrons made the injection simple, allowing accumulation of high currents. The Princeton–Stanford storage-ring experiment (CBX), being an electron–electron collision experiment, was in the form of a figure eight (Fig. 3). A detailed design of the storage ring was completed in 1958 and construction began in 1959. Initial testing at a vacuum pressure of $10^{-6}$ mm Hg ($10^{-6}$ Torr) started in October 1961. The first beam was stored on March 28, 1962; the collider became fully operational in 1965 and delivered a number of physics results testing QED. The facility was shut down in 1968.

300 MeV electrons from the Stanford Mark III accelerator were injected, on alternate beam pulses (some $10^6$ $e^-$ per pulse), into each of the 12 m circumference electron storage rings of the collider. The linac operated at 30 Hz and after a time on the order of a minute a stacked beam of up to 100 mA was built up in each ring [24, 25]. In the straight section common to the rings the circulating beam bunches can be made to pass through each other. A standard ∼30 minute physics run consisted of filling the rings to the desired current, turning off the linac and installing the spark chamber camera, switching the various beam steering magnets from their injection setting to their interacting setting, and turning on the counters and data taking for 20 minutes.

The Stanford–Princeton collider had the world's largest ultra-high vacuum system at the time (two cubic meters at 1 nTorr). This system had

Table 3. Main parameters of the Princeton–Stanford electron–electron collider.

| Parameter | Value (typical operations) | Units |
|---|---|---|
| Energy, per beam | 525 (500) | MeV |
| Circumference, each ring | 11.8 | m |
| Luminosity | $\sim 2 \times 10^{28}$ | $cm^{-2}s^{-1}$ |
| Injection energy | 300 | MeV |
| Beam current, per beam | 100 (50) | mA |
| Max field in the rings | 1.2 | T |
| Field index $n = (dB/B)/(dR/R)$ | 1.1 and 0 | – |
| Vacuum pressure | $\sim 10$ | nTorr |
| RF voltage | 20 | kV |

many windows, probes, straight sections, clearing-field electrodes, pulsed magnets, etc. Approximately one hundred gold ring gaskets were used there, two of which were 24 inches in diameter. The 25 MHz RF cavities driven by one-tube amplifiers permitted their removal during bake-out of the vacuum chamber. To inject electrons into the storage ring, an advanced delay-line inflector was used ($\sim$20 ns pulse width, including a reasonable flat top). The ring magnets were of solid iron construction. Table 3 presents the main parameters of the collider.

During the tests, commissioning and operation of the Princeton–Stanford machine, several high-beam-intensity effects were observed [14, 26, 27]. The first one was a rather violent pressure rise due to the synchrotron light. The pressure proved to be proportional to the stored beam intensity and to be strongly energy dependent (roughly a fourth-power dependence on energy.) For example, a 30 mA beam with energy of about 300 MeV radiated some 30 watts and led to the pressure rise by a factor of 300 from about 3 nTorr to 1000 nTorr. Large photoelectric currents caused by photo-desorption were found to flow to the clearing electrodes. The effect was suppressed by switching to oil-free vacuum pumps, cleaning the vacuum chambers and re-baking.

The second effect was a high-beam-density vertical instability which limited the electron beam density to $\sim 3 \times 10^{10}$ electrons per $cm^3$. The instability showed rather slow vertical growth, lasting at least many milliseconds and it could be partially stabilized by the accumulation of ions. It was believed to be due to the image-charge currents leading to a long-range-wake instability that was cured by octupole magnets to increase the tune spread (tune is the number of transverse (or longitudinal) oscillations during one revolution in the storage or collider ring). Chromatic aberrations of the magnetic system led to a head-tail instability, the effect in which the beams cannot be controlled. The aberrations were corrected and the instability suppressed.

Richter implemented sextupole and octupole magnets in the new machine before it was clearly understood why a collider or storage ring needs these non-linear corrector elements. In reply to the question "why the machine was designed with these non-linear elements", he stated "it seemed to be a good idea at the time". Another coherent coupled-beam transverse instability was fixed by separating the tunes of the two rings.

Finally, the beam–beam effects were found to be setting a very strong limitation on the luminosity. The beam–beam parameter, defined as tune shift of a particle in beam 1 colliding with beam 2:

$$\Delta Q_1 = \frac{N_2}{wh}\frac{r_e R}{\gamma Q_1}$$

(here $w$ and $h$ are width and height of beam with uniform charge distribution, $\gamma$ is relativistic factor, $Q_1$ is tune of beam 1, $r_e$ is classical electron radius, and $R$ is radius of the ring), was found to be limited, e.g., operation with $\Delta Q > 0.025$ usually resulted in significant beam degradation due to the beam–beam interaction.

## References


[1] R. Wideroe, Deutsches Patentamt, Patentschrift, Nr. 876279, klass 21g, Gruppe 36, issued on May 11, 1953.
[2] G. K. O'Neill, *Phys. Rev.* **102**, 1418 (1956).
[3] C. Bernardini, *et al.*, *Nuovo Cimento* **18**, 1293 (1960).
[4] C. Bernardini, *et al.*, *Nuovo Cimento* **23**, 202 (1962).
[5] C. Bernardini, *et al.*, *Nuovo Cimento* **34**, 1473 (1964).
[6] J. Haissinski, *Frascati Physics Series* **XIII**, 17 (1998).
[7] G. I. Budker, *et al.*, in *Proc. Workshop on Physics and Technics of Colliding Beams*, Kharkov, FTI AN USSR (1962), p. 2 (in Russian).
[8] G. I. Budker, *et al.*, in *Proc. Int'l Accelerator Conf.*, Dubna (1963), p. 274.
[9] A. P. Onuchin, *Energiya-Impuls* (Budker INP publication), No. 4, 2004, p. 3.
[10] G. I. Budker, *et al.*, *Sov. J. Nucl. Phys.* **6**, 889 (1967).
[11] P. I. Golubnichiy, *et al.*, *Atomnaya Energiya* **22**, 168 (1967).
[12] G. K. O'Neill, *et al.*, in *Proc. Int. Conf. High Energy Accelerators*, CERN (1959), p. 125.
[13] *New Scientist*, **25(435)**, 692 (1965).
[14] W. G. Barber, *et al.*, *Phys. Rev. Lett.* **16**, 1127 (1966).
[15] C. Bernardini, *Physics in Perspective* **6**, 156 (2004).
[16] C. Bernardini, *et al.*, *Phys. Rev. Lett.* **10,** 407 (1963).
[17] G. I. Budker, *Sov. Phys. Uspekhy* **9,** 534 (1967).
[18] E. A. Abramyan, *et al.*, in *Proc. IV Int. Conf. on High Energy Accelerators*, Dubna (1963), p. 274.
[19] G. I. Budker, *et al.*, in *Proc. V Int. Conf. on High Energy Accelerators*, Frascati (1965), p. 389.
[20] G. I. Budker, *et al.*, *Sov. Nucl. Phys.* **6**, 1221 (1967).
[21] A. Skrinsky, in *Proc. IEEE PAC 1995*, Dallas (1995), p. 14.

[22] A. Skrinsky, in *Proc. IIEE PAC 1973* , San Francisco (1973), p. 756.
[23] W. Barber, B. Richter, W. K. H. Panofsky, G. K. O'Neill and B. Gittelman, An Experiment on the Limits of Quantum Electrodynamics, Princeton University Internal Report HEPL-170 (June 1959).
[24] G. K. O'Neill, *Science* **141(3582)**, 679 (1963).
[25] K. O'Neill, in *Proc. 1963 Summer Study on Storage Rings, Accelerators and Experimentation at Super-High Energies*, Upton, NY, 10 June–19 July, 1963, p. 209.
[26] B. Richter, in *Proc. 3$^{rd}$ Int'l Symp. on the History of Part. Phys.*, Stanford, CA, June 24–27, 1992, SLAC-PUB-6023 (1992). [Citing from there on PBX timeline: "*The first funds were available at the end of 1958, but it was a hard job making this machine work. We had preliminary results on electron–electron scattering in July of 1963 and we published the final paper in 1965.*"]
[27] B. Gittelman, in *Proc. 1st IEEE Particle Accelerator Conference*, Washington, D.C., *IEEE Trans. Nucl. Sci.* **NS-12(3)**, 1033 (1965).

## Chapter 4

# The LEP $e^+e^-$ ring at the energy frontier of circular lepton colliders

Albert Hofmann and Jörg Wenninger (CERN)

The Large Electron Positron ring (LEP) was a circular lepton collider at CERN. It operated at beam energies around 47 GeV to produce the neutral $Z^0$ particle and above 80 GeV to create pairs of the charged $W^\pm$ bosons. At these high energies the emission of synchrotron radiation was important and demanded a very high voltage of the RF-system. It also influenced the choice of many other machine parameters. This presentation tries to show how the basic accelerator physics was used to optimize the machine and to find innovative solutions for some problems: magnets with concrete between the laminations, modulated cavities, Nb-Cu superconducting cavities, non-evaporable getter pumps, optics analysis from multi-turn data and many more.

## 1 Introduction

The Large Electron Positron ring, LEP, was a circular $e^+e^-$-collider of 27 km circumference at CERN covering beam energies between 45 and 104 GeV. It used the CERN SPS (Super Proton Synchrotron), operating in a lepton mode, as injector at 20 and 22 GeV.

Based on earlier studies, LEP was constructed between 1981 to 1989. Its exploitation as a particle physics facility went through two phases. LEP 1 was operated from 1989 to 1995 around 47 GeV per beam to produce the neutral intermediate vector boson through the reaction $e^+ + e^- \rightarrow Z^0$. It was followed by LEP 2 which was operated from 1996 to 2000 above 80 GeV to produce charged vector bosons by the reaction $e^+ + e^- \rightarrow W^+ + W^-$. These energies demanded a very high RF-voltage which represented the largest challenge. The particle interactions were measured in four experimental regions housing the detectors L3, ALEPH, OPAL and DELPHI.

LEP has been described in many publications and talks, covering its design, construction and operation as well as its components, instrumentation and physics detectors, of which only a fraction are quoted here. Detailed parameters are collected in the LEP Design Report [1]. An entertaining book by Schopper [2] gives a complete history of LEP with a balanced view of the planning, engineering, operation, components as well as the detectors and the obtained physics results. Detailed information on the design, construction and commissioning is given by Hübner [3].

The LEP ring was similar to earlier circular lepton colliders and based on the same physical principles. However, the large size and high energy led to some extreme parameters which demanded innovative solutions and optimisations. This presentation concentrates on the accelerator physics and shows how it determined the LEP parameters and guided the necessary engineering to achieve them. It emphasises innovative technical solutions which found applications in other machines.

## 2 Accelerator physics of circular lepton colliders

### 2.1 *Basic lay-out*

Most $e^+e^-$-colliders have a single ring where both beams circulate in the same vacuum chamber in opposite directions on nearly identical orbits. They consist of parts, shown in Fig. 1, which have different functions. First, there is an arc with bending magnets which determine the nominal orbit and energy and focusing magnets which keep particles with small deviations in trajectory and energy together. A dispersion suppressor, at the entrance and exit of the long straight section containing the RF-system and the interaction region, concentrates particles with energy deviations on the same trajectory to minimize their contribution to the beam size in the collision point. The RF-system replaces the energy lost by the particles due to synchrotron radiation and provides longitudinal and energy focusing to assemble them in groups, called bunches. In the interaction region strong magnetic lenses focus

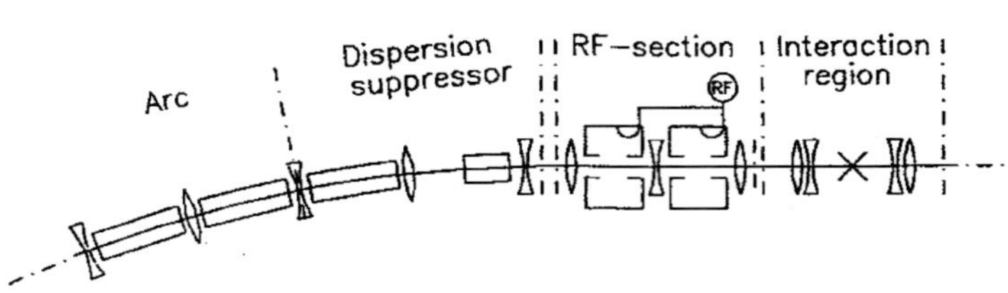


Fig. 1. Parts of a circular lepton collider.

the beam to a small cross section at the collision point inside the detector to maximise the reaction rate.

### 2.2 *Arc optics*

The arcs contain periodic cells with dipole magnets for bending and quadrupoles for focusing. Many rings use so-called FODO cells with bending magnets of homogeneous vertical field $B_y$ and alternating focusing and defocusing quadrupole lenses. The particles move with velocity $\beta c$ in the horizontal plane in longitudinal direction $s$ and get deflected by the Lorentz force into a circular trajectory of curvature $1/\rho$. This determines the momentum $p$ of the particles moving on the closed nominal orbit. Quadrupoles have a field which vanishes on the axis but has a gradient $\partial B_x/\partial B_y = \partial B_y/\partial B_x$ which deflects particles having spatial deviations $x$ or $y$ towards the central orbit in case of focusing, or away from it in case of defocusing. A single quadrupole of length $\ell$ focuses in one plane with strength $1/f$ but defocuses in the other:

$$\frac{1}{\rho} = \frac{eB_y}{p}, \quad \frac{1}{f} = \frac{\ell e}{p}\frac{\partial B_y}{\partial x}. \tag{1}$$

Arranging quadrupoles of alternating polarities in a lattice can provide overall focusing in both planes and result in a stable particle trajectories $x(s), y(s)$ where $x(s)$ and $y(s)$ stand for the horizontal and vertical deviations of the ideal reference orbit. This is illustrated in the top part of Fig. 2 for the horizontal plane (the corresponding vertical case $y(s)$ is similar and will be omitted). The figure shows the deviation $x(s)$ from the nominal orbit over many turns, called betatron motion. The envelope is large at focusing F-quadrupoles and small at defocusing D-quadrupoles. This multi-turn trajectory is described by the equation

$$x(s) = x_0\sqrt{\beta_x(s)}\cos(\phi_x(s) + \phi_0). \tag{2}$$

Both, the amplitude $\beta_x(s)$ and phase function $\phi_x(s)$, are determined by the lattice and have the same values for all particles. Of interest is the fact that $\sqrt{\beta_x(s)}$ is proportional to the envelope of the trajectories. The amplitude factor $x_0$ and the phase off-set $\phi_0$, however, are parameters of individual particles and usually determined by the emission of synchrotron radiation which establishes distributions of amplitudes and phases.

At a longitudinal location $s$, a given particle has a certain deviation in position $x(s)$ and angle $x'(s) = \mathrm{d}x/\mathrm{d}s(s)$ from the nominal orbit. Representing them as a parameter plot over many turns results in phase space ellipses which are up-right at the quadrupole centers but tilted elsewhere,

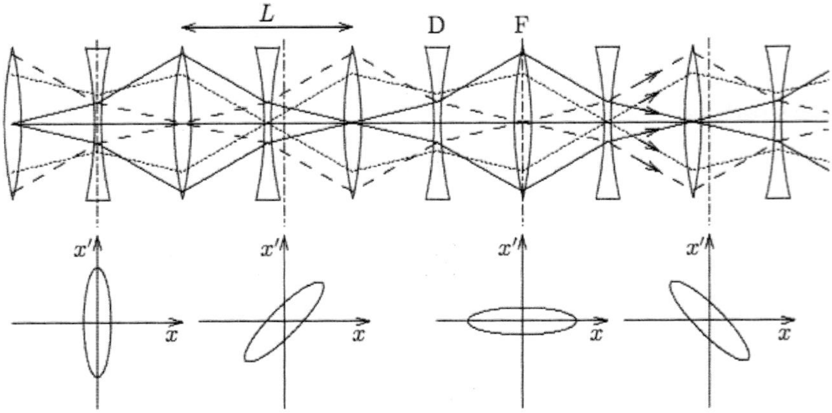


Fig. 2. Particle trajectory $x(s)$ in a FODO lattice over many turns.

as illustrated in the lower part of Fig. 2. The large displacements and small angles at the F-quadrupoles result in a flat ellipse while the reversed situation at the D-quadrupoles gives a slim ellipse. The area $A$ of the ellipse for a given particle is constant around the ring. Its normalized value $\epsilon_{ix} = A/\pi$ is called particle or trajectory emittance. For many particles of different amplitudes we have an emittance distribution with an average $\epsilon_x = \langle \epsilon_{ix} \rangle$, called beam emittance. In the centers of quadrupoles, where $\mathrm{d}\beta/\mathrm{d}s = 0$, these relations become simple:

$$\epsilon_x = \sqrt{\langle x^2 \rangle \langle x'^2 \rangle} = \sigma_x \sigma_{x'} = \frac{\sigma_x^2}{\beta_x} = \beta_x \sigma_{x'}^2, \ \beta_x = \frac{\sigma_x}{\sigma_{x'}}.$$

Strong focusing corrects spatial deviations quickly, which results in an overall small beam size and a large number $Q_x$ of oscillations per turn executed by the function $\cos(\phi_x(s))$. The phase advance $\phi_x(s)$ gives a measure of the focusing strength in the machine. $Q_x$ is the global tune and gives a measure of the overall focusing along the full machine circumference. Its value should not be an integer or a simple low-order fraction, in order to minimize the detrimental effects of magnetic field errors on the trajectory. For integer or low-order fractional tunes, the field errors add up in phase on consecutive turns resulting in a resonance excitation.

According to Eq. (2) the deviation $x$, and therefore the relative beam size, is small where $\beta_x(s)$ is small; it is a measure of the local focusing strength. For a FODO lattice this is also expressed by the phase advance $\mu_x = \Delta\phi_x(L)$ across a cell.

In Eq. (1) the curvature and the focusing strength provided by the magnetic elements are inversely proportional to momentum. As a result, local orbit, circumference $C$, revolution time $T = C/(\beta c)$ and related frequency $\omega_{\rm rev} = 2\pi/T$ as well as the tune $Q_x$ are functions of momentum deviation $\Delta p$ of the particle or the beam as a whole. This dependence is expressed by the lattice parameters: dispersion $D_x$, momentum compaction $\alpha_c$ and the chromaticity $Q'$

$$\Delta x = D_x \frac{\Delta p}{p}, \ \frac{\Delta C/C}{\Delta p/p} = \alpha_c, \ \frac{\Delta T/T}{\Delta p/p} = -\frac{\Delta\omega_{\rm rev}/\omega_{\rm rev}}{\Delta p/p} = \eta_c, \ Q' = \frac{\Delta Q}{\Delta p/p}, \quad (3)$$

where $\eta_c = \alpha_c - 1/\gamma^2$ contains a correction for the velocity dependence on energy which can be neglected for ultra-relativistic particles ($\gamma \gg 1 \rightarrow \eta_c \approx \alpha_c$). The parameters $D_x$ and $\alpha_c$ can be controlled by the linear lattice elements, dipoles and quadrupoles. Both are small and positive for strong focusing lattices. To correct the chromaticity we need an element giving extra focusing to higher momentum particles. Sextupole magnets have a focusing strength $1/f \propto x$ and can provide the desired correction if located at a finite dispersion where $x = D_x \Delta p/p$. Sextupoles are non-linear elements which have adverse effects that can be difficult to correct.

## 2.3 *RF-section*

The RF-section supplies a high frequency longitudinal field which replaces the energy $U_s$ lost by a particle due to synchrotron radiation and provides longitudinal and energy focusing. It contains cavities which are driven by a generator at a harmonic $h$ of the revolution frequency $\omega_{\rm rev}$. The total voltage seen by a particle each turn is

$$V_{RF} = \hat{V}\sin(h\omega_{\rm rev}t) = \hat{V}\sin(\omega_{RF}t),$$
$$V(t_s) = \hat{V}\sin(\omega_{RF}t_s) = \hat{V}\sin\phi_s = \frac{U_s}{e}.$$

A synchronous particle of nominal energy traverses a cavity at the synchronous time $t_s$ and gains an energy $e\hat{V}\sin\phi_s = U_s$ which replaces the radiation energy loss, as shown in Fig. 3. According to Eq. (3) a particle with excessive energy traverses a larger circumference than the reference particle and takes longer to complete one revolution if $\eta_c$ is positive (as being the case for most rings) and arrives at a later time $t_s + \tau$ at the cavity when the voltage is less than $U_s/e$ and loses overall energy. On the other hand, a particle with insufficient energy arrives early at the cavity when the voltage is higher than $U_s/e$ and gains overall energy. This corrective action leads to an oscillation around the synchronous time $t_s$, and the nominal energy $E$.

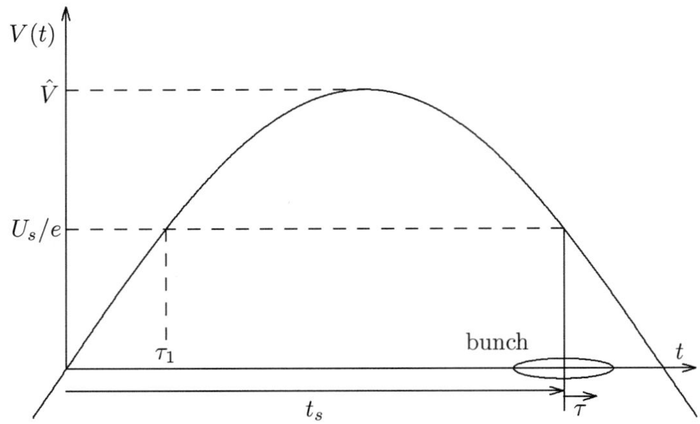


Fig. 3. Energy replacement and focusing by the RF-voltage.

By linearising the sinusoidal RF-voltage we get a harmonic oscillation with frequency $\omega_s$:

$$\tau = \hat{\tau}\sin(\omega_s t), \;\; \Delta E = \widehat{\Delta E}\cos(\omega_s t), \;\; \omega_s = \omega_{rev}\sqrt{-\frac{\alpha_c h e \hat{V}\cos\phi_s}{2\pi E_0}}. \tag{4}$$

For standard electron rings, $\eta_c > 0$, this small amplitude synchrotron frequency $\omega_s$ is real only if $\cos\phi_s < 0$, i.e. $\pi/2 < \phi_s < \pi$.

The focusing action of the RF-system assembles particles in a limited range of momentum and longitudinal position, called bunch. Synchrotron radiation establishes a Gaussian energy distribution with RMS spread $\sigma_E$. This results in a corresponding longitudinal spread $\sigma_t$ or $\sigma_s$ around $t_s$:

$$\sigma_t = \frac{\sigma_s}{\beta c} = \frac{\alpha_c}{\omega_s}\frac{\sigma_E}{E} = \frac{1}{\omega_{\rm rev}}\sqrt{\frac{2\pi E\alpha_c}{he\hat{V}\cos\phi_s}}\frac{\sigma_E}{E}. \tag{5}$$

In reality the RF-wave form is sinusoidal which limits the maximum energy deviation in a bunch. During the part of the synchrotron oscillation between $\tau = 0$ and $\tau = \tau_1$, shown in Fig. 3, a particle gains (or loses) each turn an energy $\delta E = eV(\tau) - U_s$. The total energy gained during this time is the maximum stable energy oscillation amplitude. For a given voltage $\hat{V}$ and energy loss $U_s$, this increases with the number of turns executed during this oscillation fraction which is about proportional to $1/\omega_s$, or according to Eq. (4), to $1/\sqrt{\alpha_c h}$. Therefore, to accept a certain energy spread, small values for momentum compaction $\alpha_c$ and harmonic number $h$ help to reduce the necessary RF-voltage $\hat{V}$.

The cavities are driven by a generator to induce a longitudinal field which is applied to the beam. This compensates the energy loss due to the emission of synchrotron radiation as well as the losses in the cavity itself due to its finite surface conductivity which is expressed by the cavity shunt impedance $R_s$. The total RF-power needed for two beams of current $I$ is

$$P_{RF} \approx P_{\text{radiation}} + P_{\text{cavity}} \approx 2IU_s/e + \hat{V}^2/(2R_s). \tag{6}$$

To understand possible optimisations of the system, we split the cavity impedance into two factors $R_s = Q \cdot (R_s/Q)$. The first is the quality factor given by the surface conductivity of the cavity wall. The second term depends mainly on the cavity form and its coupling to the beam, which is large for a small port. However, this also leads to beam induced fields in the form of beam loading and parasitic mode losses. For a high beam energy and not too small beam currents, one tries to make the first factor in Eq. (6) large and the second small.

## 2.4 *Synchrotron radiation*

The electrons and positrons undergo a transverse acceleration and emit synchrotron radiation. This leads to an energy loss $U_s$ per revolution for each particle which is compensated by the RF-system:

$$U_s = \frac{4\pi}{3}\frac{r_e m_0 c^2 \gamma^4}{\rho}, \ r_e = \frac{e^2}{4\pi\epsilon_0 m_0 c^2} = 2.818 \cdot 10^{-15}\text{m}, \ \omega_c = \frac{3c\gamma^3}{2\rho}, \tag{7}$$

where some fundamental constants are collected within the classical electron radius $r_e$. The radiation spectrum is wide with the critical frequency $\omega_c$ dividing it into two parts of equal power. It is emitted into a small opening angle of about $1/\gamma$. The horizontal bending of the electrons prevents the observation of the natural horizontal opening angle.

The emission of synchrotron radiation dissipates energy not only from the nominal particle motion but also from the deviations in form of synchrotron and betatron oscillations. The longitudinal RF-field, however, increases the nominal momentum only, which leads to damping of the three oscillation nodes. For the first one this is evident from the dependence of the loss on energy in Eq. (7) $\Delta U_s/U_s = 4\Delta\gamma/\gamma$. The sum of the three damping rates is constant for a given loss $U_s$:

$$\frac{1}{\tau} = \frac{1}{\tau_s} + \frac{1}{\tau_x} + \frac{1}{\tau_y} = \frac{2U_s}{ET_{\text{rev}}}(J_\epsilon + J_x + J_y), \ \sum_i J_i = 4. \tag{8}$$

For a large ring of short homogeneous dipole magnets and a beam going through the axes of all quadrupoles, the damping partitions are $J_\epsilon = 2$; $J_x =$

$J_y = 1$. By changing the RF-frequency slightly the beam will undergo a radial displacement at locations having a finite dispersion in order to keep the new orbit with changed circumference closed. Particles get some extra bending by the quadrupoles and change their momentum. The damping is now more complicated resulting in an increase of $J_\epsilon$ and decrease of $J_x$ if the beam is moved to the outside and vice versa. This RF-frequency change is a handy tool to vary damping partitions.

The photon energy is emitted in quanta $E_\gamma = \hbar\omega$ with $\hbar$ being Planck's constant divided by $2\pi$. This results in a sudden energy loss of the electron. It excites energy oscillations and, at places of finite dispersion $D_x$, also horizontal betatron oscillations. The interplay between damping and excitation determines the equilibrium energy spread and horizontal beam emittance. Since rings are designed for vanishing vertical dispersion, this dimension is little affected. However, magnet misalignments and optical errors lead to coupling and to a finite vertical emittance which is usually much smaller than the horizontal one $\epsilon_x \gg \epsilon_y$.

The emission of synchrotron radiation can lead to vertical polarisation of the electrons provided the orbit is very well corrected. This can be used for energy calibration.

### 2.5 *Interaction region and luminosity*

In the interaction region the two beams collide and the resulting reaction products are measured in a detector. Strong quadrupole lenses are used to focus the beams to a very small diameter in order to maximise the interaction rate, called "low-beta insertion". The probability for a certain interaction to occur is measured by the reaction cross-section $\sigma$, defined such that the reaction will take place if two particles collide within this area. It is usually measured in a unit called barn $= 10^{-24}\text{cm}^2$. The property which quantises the colliding beam efficiency is the luminosity, defined as the reaction event rate divided by its cross-section $L = \dot{n}/\sigma$. It is proportional to the particle number per bunch $N_b$ times the repetition rate of the collisions $f_0 = \omega_{\text{rev}}/(2\pi)$ times the number of bunches $k_b$ divided by the beam cross-section $\pi\sigma_x\sigma_y$ and a geometrical factor for the Gaussian transverse beam profile (see Eq. (9)). Besides the beam energy, the luminosity is the most important parameter of a collider. It can be optimised by increasing the number of particles per bunch $N_b$. However, instabilities due to interaction with its surroundings might limit the bunch current. If the total beam current is not limited, the number of bunches $k_b$ can be increased but a transverse beam separation is necessary to avoid collisions outside the interaction region.

A fundamental limit is given by the electromagnetic focusing of one beam on the other resulting in a so-called beam–beam tune shift $\delta Q$ (see Eq. (9)). This can have the beneficial effect of reducing the local beam size. However, this interaction has also a non-linear part resulting in a tune spread which covers resonances and imposes a limit to the maximum operationally acceptable beam–beam tune shift. This limit depends on the beam energy and synchrotron radiation damping time. For LEP the beam–beam tune shift was limited to $\delta Q \approx 0.04$. It is given below in the approximation of a flat beam $\sigma_x \gg \sigma_y$:

$$L = \frac{kN_b f_0}{4\pi\sigma_x\sigma_y}, \quad \delta Q_x = \frac{N_b r_e \beta_x^*}{2\pi\gamma\sigma_x^2} = \frac{N_b r_e}{2\pi\gamma E_x}, \quad \delta Q_y = \frac{N_b r_e \beta_y^*}{2\pi\gamma\sigma_x\sigma_y}, \tag{9}$$

where $\beta_x^*$ and $\beta_y^*$ are the beta functions at the interaction point, $N_b$ the number of particles per bunch and $k_b$ the number of bunches or particle packages per beam. Obviously a reduction of the beam cross-section increases the luminosity. It can be obtained by strong focusing in the low-beta insertion or by a low beam emittance. The first method has the advantage of leaving the beam–beam tune shift approximately unchanged. Reducing the emittance helps if the tune shift is below its limit.

### 2.6 *Optimisation and energy scaling*

For a lepton collider the beam energy is usually given by the particle physics to be studied. To optimize the ring we assume that the cost of the RF-system is about proportional to the voltage $\hat{V} \approx U_s/e \propto \gamma^4/\rho$, while the rest, including tunnel, magnet, vacuum system, etc., scales about as the bending radius $\rho$. Minimizing the total costs gives an approximate scaling [4]:

$$\rho \propto \gamma^2, \; U_s \propto \gamma^2, \; \text{cost} \propto \gamma^2, \tag{10}$$

which is used to estimate cost and size based on past experience.

## 3 LEP

### 3.1 *Energy and ring geometry*

The accelerator physics described earlier was used to optimise the LEP parameters and served as a guide for technical solutions to some problems.

As mentioned before, LEP was exploited in three steps, as LEP 1 at 47 GeV beam energy to measure the reaction $e^+ + e^- \to Z^0$, as LEP 2 at approximately 80 GeV to produce the charged vector bosons via $e^+ + e^- \to W^+ + W^-$, and at energies above 100 GeV to explore the energy frontier for $e^+ + e^-$ reactions (the maximum beam energy reached by LEP was

104.5 GeV). These three energy ranges demanded somewhat different optimisations for the RF-system and other components. The circumference of about 27 km was determined by the highest energy. By extrapolating from existing rings with the scaling laws Eq. (10) one arrives at a larger ring. However, progress in superconducting RF-technology on one side and expected difficulties to dig a tunnel into the Jura limestone led to a smaller ring with a bending radius of 3096 m and eight long straight sections of which four were equipped with RF-stations and experimental detectors.

### 3.2 *Arc optics*

The eight arcs had regular FODO cells of length $L = 97$ m each. The strength of their quadrupole lenses determined the betatron phase advances $\mu_{x/y}$ per cell which in turn influenced the momentum compaction $\alpha_c$ and the horizontal emittance $\epsilon_x$. Both get smaller for a larger phase advances up to about 137° which reduces the necessary RF-voltage and improves the luminosity in case of beam intensity limitations. As a consequence, operation of LEP at higher energy went with a higher phase advance.

The large bending radius of $\rho = 3096$ m results in bending magnets with rather low magnetic fields of $B_y = 0.05$ T at 47 GeV which does not require much iron. This led to a magnet consisting of spaced laminations with mortar in between, shown in Fig. 4, which was not only cheaper and stiffer but also less heavy than a full iron magnet [5].

This low bending field makes the use of distributed ion pumps for the vacuum system less efficient. For LEP another solution was chosen consisting of Non Evaporable Getter (NEG) pumps [6]. They worked very well and have in the mean time been used in other machines.

Sextupoles were mounted and aligned to the adjacent quadrupoles located at finite dispersion $D_x$ to correct the chromaticity. To minimize the adverse effects of these non-linear elements they were divided into families and powered at different levels.

Due to the size of LEP, optical errors were difficult to localise. Much effort was invested in the beam position measurement system to obtain not only the equilibrium orbit but also the trajectory over 1000 turns of excited coherent betatron oscillations. This allowed the measurement of relative amplitudes and phases at different locations around the ring, as shown in Fig. 5. The first one gave the ratio of the beta functions $\beta_{x1}/\beta_{x2} = (\hat{x}_1/\hat{x}_2)^2$ and the second one the betatron phase difference $\Delta\phi_x$ [7]. A local focusing error produces a phase jump which results in a beating of the beta function around the ring as shown in Fig. 6. This was used to localise and correct optical errors.

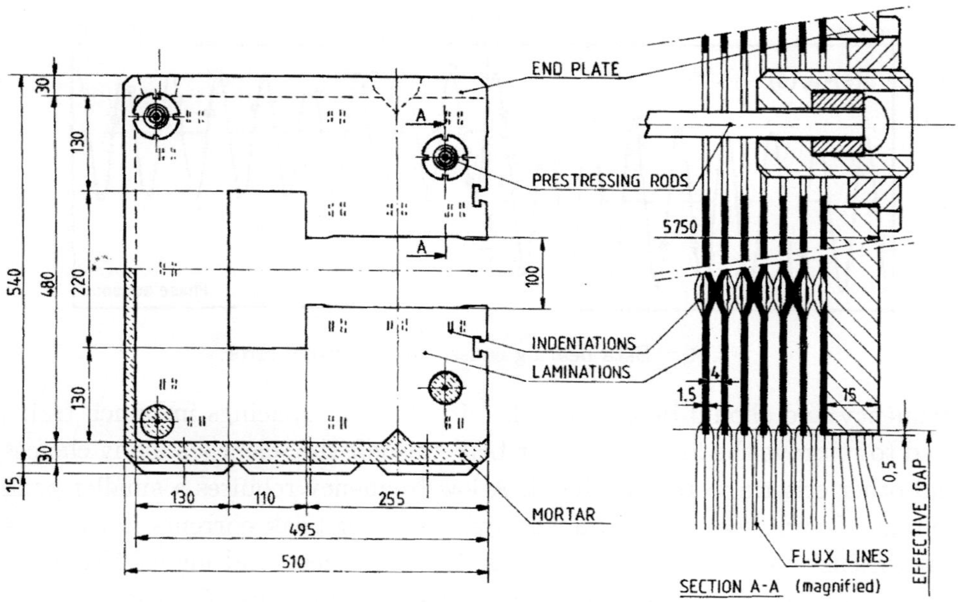


Fig. 4. Iron-concrete magnet.

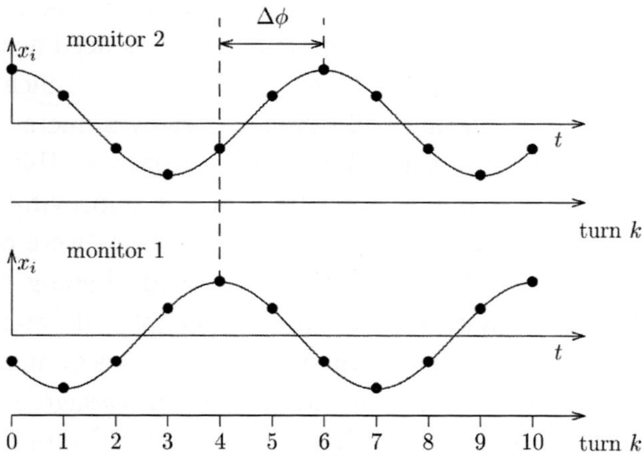


Fig. 5. Measuring betatron phase between monitors.

Measuring this phase advance as a function of RF-frequency, and thereby of the related momentum deviation, was used to check the local chromaticity correction with sextupoles.

### 3.3 *RF-system*

As a start of its optimization, the RF-frequency $\omega_{RF} = h\omega_{\text{rev}}$ was chosen. It must be a harmonic not only of the revolution frequency but also of the bunch

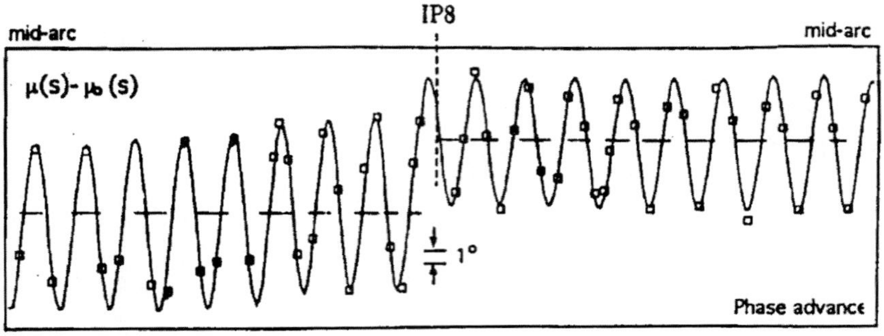


Fig. 6. Beta beating caused by a focusing error.

frequency and should be compatible with possible spacings in bunch trains used to improve the luminosity. For LEP this left the scientists many choices and other criteria were considered. A low frequency requires a smaller peak voltage $\hat{V}$ and creates long bunches with smaller peak currents having less high intensity effects and instabilities. On the other hand higher frequency super-conducting cavities reach larger voltage gradients. Considering also experiences in other laboratories and availability of power sources led to $f_{RF} = 352.21\,\text{MHz}$, similar to the one used at SLAC.

The cavities used for LEP 1 had 5 cells and a strong coupling to the beam. The bunches in LEP were far apart and experienced acceleration only during a short time fraction of a revolution. Between bunch passages the cavity fields were just oscillating and losing energy by inducing wall currents. Power could be saved by pulsing the cavities. However, this is technically difficult for the involved timing. As a solution for improving the efficiency of the cavities, the relatively lossy accelerating cavities were coupled to low loss spherical storage cavities oscillating in a mode having small fields at the wall [8]. The two cavities were tuned to slightly different frequencies and behaved like a coupled pendulum with the RF-energy moving back and forth, being in the accelerating cavity during bunch passage, see Fig. 7. This gave an improvement of about 40% in effective quality factor $Q$ and resulted in a circumferential voltage of 400 MV.

For LEP 2, superconducting cavities with four cells were used [9]. First, 16 cavities made of bulk niobium were constructed. Later, CERN developed niobium coated Cu cavities which have lower costs, higher $Q$-factor, less sensitivity to magnetic fields and improved resistance to quenches. They reached an unloaded quality factor $Q_0 = 3.2 \cdot 10^9$ and an average voltage gradient of 7.5 MV/m. With 272 cavities of Nb-Cu, together with the 16 of bulk Nb and some old Cucavities, a total circumferential voltage of 3.66 GV was achieved.

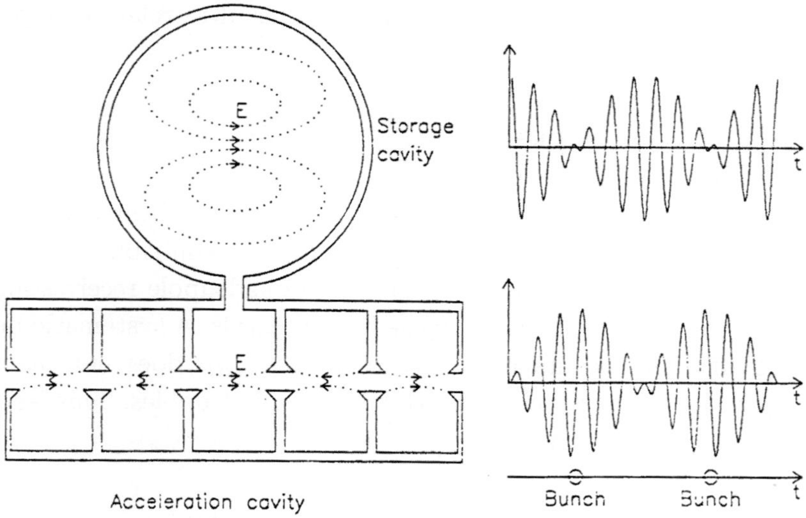


Fig. 7. Coupled acceleration and storage cavities.

### 3.4 *Energy calibration*

In the reaction $e^+ + e^- \to Z^0$ the energy of the electron and positron enters directly into the one of the $Z^0$ particle. By measuring the production rate as a function of beam energy, one obtained the energy and width of this resonance, i.e. the mass and decay time of the $Z^0$. The clean and well-defined initial state in the centre-of-mass frame of the colliding beams made LEP an ideal place to measure mass and decay width of the Z boson as well as the mass of the W bosons. Therefore, a precise knowledge of the beam energy was an important ingredient for the LEP physics program.

The determination of the beam energy depends in turn on the detailed magnetic field map around the storage ring and the total machine circumference. At a position $s$ along the ring, the local beam momentum $p_\mathrm{b}(s)$, bending field $B(s)$ and radius of curvature $\rho(s)$ are related by

$$\frac{1}{\rho(s)} = e\frac{B(s)}{p_\mathrm{b}(s)}. \tag{11}$$

The momentum $p_\mathrm{b}(s)$ may differ locally from its average value $p_\mathrm{b}$ due to energy losses through synchrotron radiation and energy gains from the RF-system. Closure of the orbit implies that the total bending angle $\theta$ integrated along the beam path must be exactly equal to $2\pi$:

$$\oint \mathrm{d}\theta = \oint \frac{\mathrm{d}s}{\rho(s)} = e\oint \frac{B(s)}{p_\mathrm{b}(s)}\mathrm{d}s = 2\pi. \tag{12}$$

To a very good approximation the mean beam for ultra-relativistic particles is defined by the integrated bending field:

$$E_{\mathrm{b}} = cp_{\mathrm{b}} = \frac{ec}{2\pi} \oint B(s)\mathrm{d}s. \tag{13}$$

The ideal nominal orbit has a circumference $C$ and the beams are centered in the quadrupoles. In that case only the dipoles contribute to the field integral. A beam passing off-center through a quadrupole receives an extra bending which has to be included in the field integrals. A systematic bending by the quadrupoles occurs if the RF-frequency $f_{RF}$ does not match the value required to center the beam orbit in the quadrupoles. This leads to a momentum change through

$$\frac{\Delta C}{C} = \alpha_c \frac{\Delta p}{p}, \quad \text{or} \quad \frac{\Delta f_{RF}}{f_{RF}} = -(\alpha_c - 1/\gamma^2)\frac{\Delta p}{p}, \tag{14}$$

where $\alpha_c$ is the momentum compaction factor that varied for LEP between $1.4 \times 10^{-4}$ and $3.9 \times 10^{-4}$. This small value made LEP very sensitive to circumference changes as will be discussed later.

An additional complexity arises for the energy in the centre-of-mass system which should ideally be exactly twice the beam energy. In practice the local energies of the two beams may differ at the collision points. Such shifts cannot be measured directly but must be evaluated using adequate models as described later.

Since tracking all magnetic fields is very difficult when relative accuracies of $10^{-5}$ are requested for the beam energy, accurate calibration methods must be developed to measure and predict the energy at any time of machine operation.

Initially the energy was determined precisely at injection energy by comparing the revolution frequency of positrons and protons. A positron beam was centered in the quadrupoles and its revolution frequency measured $f_{\mathrm{rev,e}} = \beta_e c/C \approx c/C$. Thereafter, this was repeated with a proton beam circulating on the same orbit but with a different revolution frequency $f_{\mathrm{rev,p}} = \beta_p c/C$. Since the two were sequentially on the same orbit and saw the same field, they must have had the same momentum. Comparing the two revolution frequencies gave the proton velocity $\beta_p c$ and momentum $p_p = m_p \beta_p c \gamma_p = p_e$. This method had the advantage of being independent of any magnetic measurement, but could only be done at injection energy. The interpolation of the energy to the Z resonance and to the W production threshold using magnetic field probes degraded the quality and led to large errors. This method was therefore abandoned after the advent of resonant depolarisation.

From 1991 onwards, when transverse polarization was first observed at LEP [10], resonant depolarisation became the workhorse of energy calibration at LEP. The development and results that were obtained with this technique will be described in the following sections.

### 3.5 *Transverse polarization*

The build-up of transverse polarization in $e^+e^-$ storage rings was first described by Sokolov and Ternov [11]. The emission of synchrotron radiation has a small spin-flip probability, but a large asymmetry in favor of orienting the magnetic moment along the direction of the guiding magnetic field. The maximum transverse polarisation of 92.4% is building up with a rise time

$$\tau_P = \frac{8}{5\sqrt{3}} \frac{m_e^6 c^{10} \rho^3}{\hbar r_e E_b^5}. \tag{15}$$

At 46 GeV the rise time $\tau_P$ is 310 minutes.

The motion of the spin vector $\vec{S}$ of a relativistic electron in electromagnetic fields is described by the Thomas–BMT equation [12]. In a storage ring the spin vector precesses $a\gamma$ times for one revolution, where $a$ is the magnetic moment anomaly of the electron. Its average value over all particles $\nu$, the spin tune, is directly proportional to the average beam energy $E_b$, for electrons and positrons it is given by

$$\nu = a\gamma = \frac{aE_b}{m_e c^2} = \frac{E_b \text{ (MeV)}}{440.6486(1) \text{ (MeV)}}. \tag{16}$$

In practice the maximum polarization level is reduced by resonances. Orbit and tunes must be carefully corrected and spin resonances $\nu = kQ_x + lQ_y + mQ_s + n$, where $k, l, m, n$ are integers, must be avoided. Optimum polarization levels are achieved for beam energies corresponding to a spin tune close to a half-integer $\nu = n + 1/2$. At LEP a record polarisation of 57% was observed for a beam energy of 44.72 GeV [13, 14]. At higher energies the maximum polarisation level was reduced by the large energy spread and the strong synchrotron spin resonances $\nu = n + kQ_s$. Polarisation levels of 4–5% or more were only observed up to 60.6 GeV [15]. The highest polarisation measured at LEP as a function of the beam energy is shown in Fig. 8.

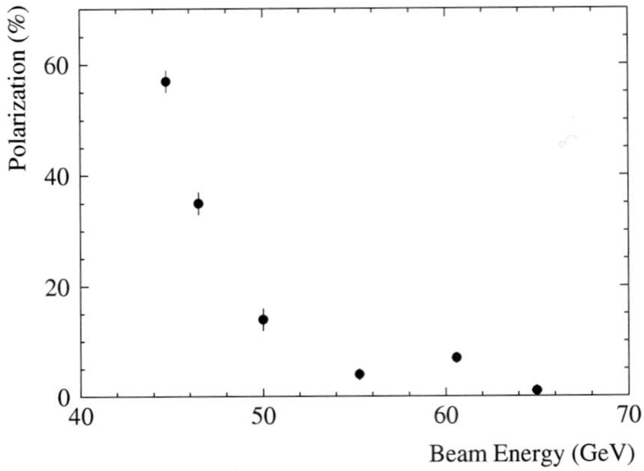


Fig. 8. Highest transverse polarisation level observed at LEP as a function of beam energy. The strong decrease with beam energy is clearly visible.

### 3.6 *Energy calibration by resonant depolarisation*

For transversely polarised beams the relation between spin tune and energy provides a means to accurately measure the average beam energy. An RF-magnet with an oscillating radial field is used to rotate the particle spins. The RF-magnet frequency $f_{\rm dep}$ is in resonance with the spin precession for

$$f_{\rm dep} = (k \pm [\nu]) f_{\rm rev}, \tag{17}$$

where $k$ is an integer, $f_{\rm rev}$ is the revolution frequency and $[\nu]$ denotes the non-integer (fractional) part of the spin tune. When the resonance condition is met, the beams may either be depolarised or the polarisation may even flip sign. The integer part of $\nu$ cannot be measured, it must be deduced from the calibration of the bending magnets. This method is referred to as energy calibration by resonant depolarisation and has been used extensively for accurate beam energy calibrations and measurements of particle masses [17]. Since the polarisation vector is an average over all spin vectors, the measured beam energy is independent of betatron and synchrotron oscillations of the individual particles and is not limited in accuracy by the beam energy spread. At LEP the measurement accuracy was typically $\pm 0.44\,$MeV [16], and a minimum polarisation level of 4–5% was required to determine the beam energy reliably.

### 3.7 *Z boson resonance scans*

For precise measurements of the Z boson parameters, the beam energy of LEP was scanned around its resonance [18, 19]. The two main scans were performed during the 1993 and 1995 LEP runs and aimed at accuracies on

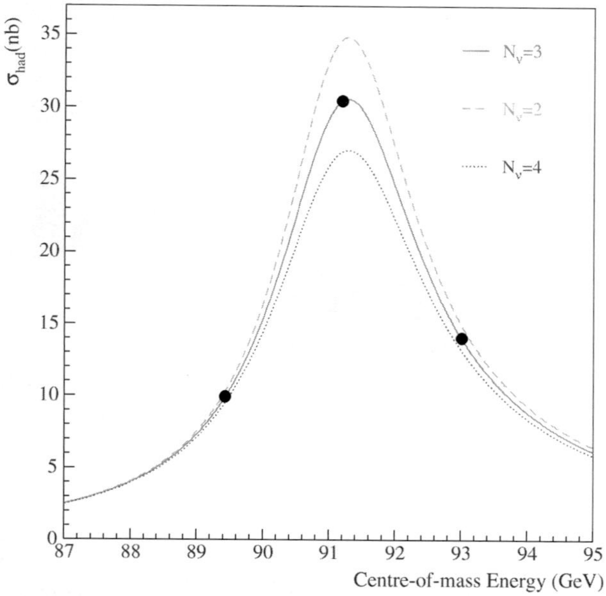


Fig. 9. Cross-section for the production of events with hadrons at LEP as a function of the centre-of-mass energy around the Z boson resonance. The maximum of the curve corresponds approximatively to the mass of the Z boson. The three points (•) correspond to energies compatible with beam polarisation which have been accurately calibrated by resonant depolarisation in 1993 and 1995. The curves indicate the expected cross-sections for 2, 3 and 4 families of light neutrinos.

the centre-of-mass energy of $\approx 1\text{–}2 \times 10^{-5}$. The ability to polarise the beam constrained the choice of beam energy, which was scanned over 3 points corresponding roughly to the peak of the Z resonance ($\nu \simeq 103.5$) and to two points located 0.9 GeV above and below the resonance peak ($\nu \simeq 101.5$ and 105.5), as shown in Fig. 9. Since luminosity operation was not compatible with beam polarisation, the energy calibrations were made regularly at the end of the physics fills with non-colliding beams and minimal changes to the machine conditions. Dedicated experiments were also performed to investigate various aspects of the energy modelling.

### 3.8 *Sources of energy variation*

The large amount of data accumulated between 1991 and 1995 during operation around the Z boson resonance unveiled many subtle effects that were influencing the LEP beam energy at the $10^{-4}$ level, more than one order of magnitude larger than the desired accuracy. The most important effects will be described below.

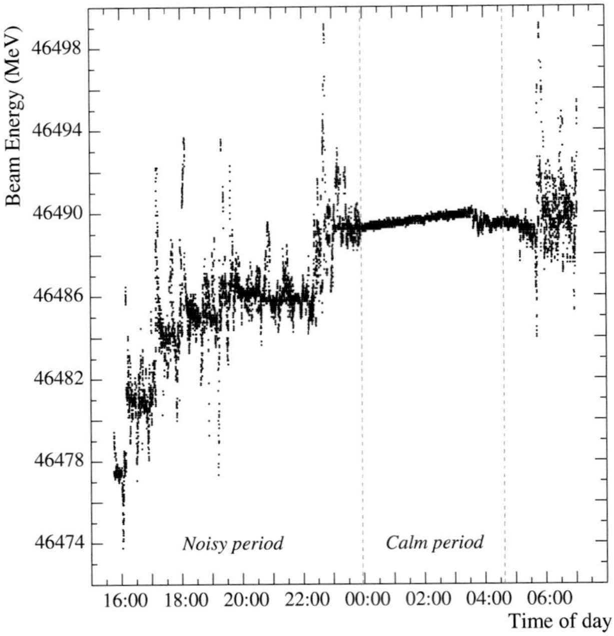


Fig. 10. Magnetic field measured in a LEP dipole by an NMR probe over 10 hours. For convenience the magnetic field has been converted to an equivalent beam energy in MeV. Large short-term fluctuations and a slow rise in field are clearly visible. Between midnight and 4:30 am the field is stable while the fluctuations disappear.

### 3.8.1 *Magnetic field stability*

In 1995 a perturbation of the dipole field was observed for the first time when Nuclear Magnetic Resonance (NMR) probes were installed in tunnel dipoles to monitor the bending field. Important short-term fluctuations and long-term field increases corresponding to $\Delta B/B = \Delta E_{\mathrm{b}}/E_{\mathrm{b}} \simeq 2 \times 10^{-4}$ were observed by the tunnel probes over 12 hours (Fig. 10). In contrast probes installed in a reference magnet located in a surface building and connected in series with the LEP dipoles registered fluctuations that were about one order of magnitude smaller. The noise showed a repetitive daily pattern with a noise-free period between midnight and about 4:30 am. The long-term upward drift of the magnetic field was driven by the short-term noise spikes.

After lengthy investigations the perturbation could be attributed to DC leakage currents flowing on the LEP vacuum chamber [20]. The currents are generated by trains circulating on a 1.5 kV DC French railway line linking the cities of Geneva and Lyon. In contrast Swiss railway lines are operated at 15 kV AC. For a given engine power, the DC currents are 10 times higher

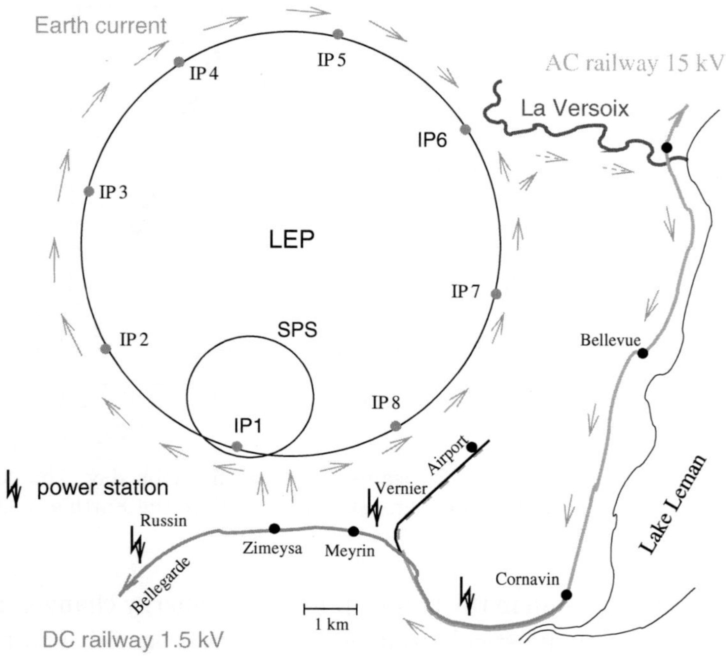


Fig. 11. Reconstructed flow of the train leakage current on the LEP vacuum chamber.

and the leakage of DC current into the ground is large, with up to 25% of the current not returning to the generator over the railway tracks. The exact amount of current seeping into the ground depends on the electrical insulation of the railway tracks. Metal conductors like pipes, wells and rivers can carry the parasitic currents. For LEP the currents entered the tunnel around IP6 and left the tunnel over the transfer lines to the SPS near IP1, see Fig. 11. The currents flowed clockwise from IP6 via IP8 to IP1, anti-clockwise via IP4. The currents themselves did not induce energy changes of the beam because the net current averaged to zero over the circumference. The dipole field however did not drop back to its initial value after a current spike but remained at a slightly higher value due to the magnet hysteresis curve. A succession of current spikes induced a slow increase of the field and the beam energy over time. The field in the magnets finally saturated due to the finite amplitude of the spikes.

### 3.8.2 *Circumference variations and tides*

In LEP the orbit length of the ultra-relativistic particles was determined by the frequency of the RF-system. Following a circumference change $\Delta C$, the beam was forced to adapt its radial position in the magnets. The additional

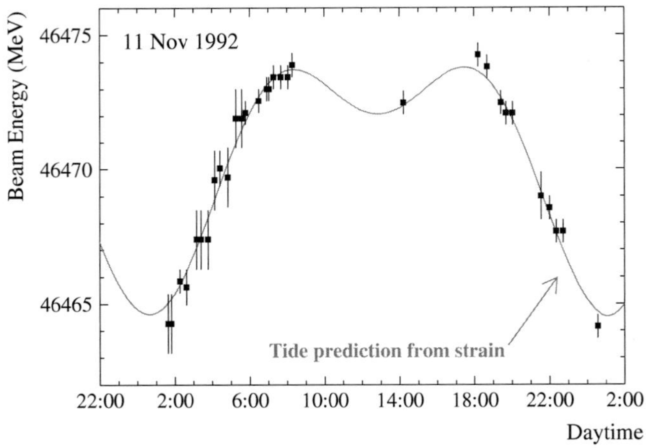


Fig. 12. Energy variation of the LEP beams during a full moon day. The curve is the energy change predicted from the horizontal strain induced by the Earth's tides.

field sampled by the beam in the quadrupoles led to energy changes according to Eq. 14. Because of the small value of $\alpha_c$ LEP was sensitive to geological movements generated by tides and hydrological strains.

Tides are caused by the $1/r^2$ dependence of the gravitational attraction of the Sun and the Moon. The presence of a celestial body leads to a quadrupolar deformation of the Earth's surface. The horizontal strain associated to the tidal forces from the Sun and the Moon induced peak-to-peak circumference changes of up to 1 mm. Figure 12 shows the result of a controlled experiment during a period of full moon. The beam energy variation followed the expected tidal distortions. The knowledge accumulated on Earth's tides allowed to model the tidal circumference and energy changes with very good accuracy [21].

Besides the periodic tidal movements LEP was also subject to much slower long-term circumference changes [22]. Such movements could be monitored by observing the radial movement of the beam relative to the quadrupoles in beam position monitors. During a typical run lasting from May to November the LEP ring experienced circumference changes of up to 2 mm as shown in Fig. 13.

The general trend was reproducible from one year to the next. The circumference increased during the summer months, some of the changes being clearly correlated to rainfall and to fluctuations in the underground water table height. Monitoring those seasonal variations of the circumference turned out to be very important to understanding the evolution of LEP beam energy since the associated energy variations reached $\Delta E_b/E_b \approx 5 \times 10^{-4}$.

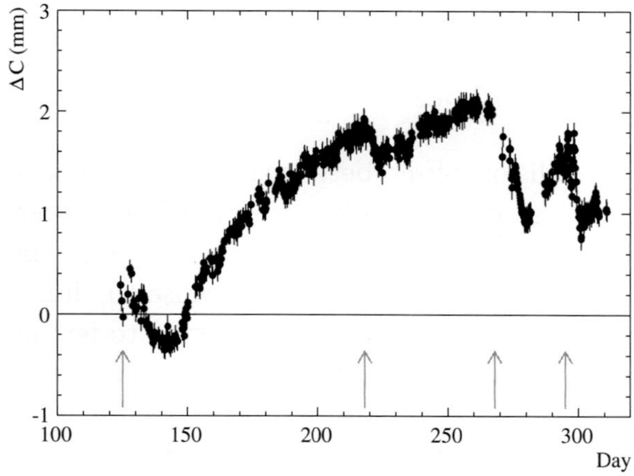


Fig. 13. Evolution of the LEP circumference (corrected for tidal changes) as a function of the day in 1999. A drift of up to 2 mm was observed during the LEP run. In the summer months the circumference increased gradually. Following periods of heavy rainfall, indicated by the arrows, the circumference shrinked for some time before expanding again.

### 3.8.3 *Centre-of-mass energies*

At the collision points of the beams the quantity relevant for all physical processes is the centre-of-mass energy, which is obtained from the average beam energy determined by resonant depolarisation after correction for local effects.

A first cause for centre-of-mass energy shifts was due to longitudinal cavity misalignments, RF phase errors between cavities as well as the actual voltage distribution. Typical corrections and fluctuations were smaller than 10 MeV, although a large systematic centre-of-mass energy shift of 16 MeV was introduced accidently at the 2 IPs where the copper RF-cavities were systematically misaligned longitudinally. A complete model of the RF-system was used to predict the correction to the centre-of-mass energy as well as the following observable quantities: the synchrotron tunes of the 2 beams, the longitudinal positions of the collision points and the horizontal orbit differences around the IPs induced by the local energy differences. Comparisons of predicted and measured observables were used to tune the model and to estimate systematic errors on the centre-of-mass energy.

Vertical dispersion was at the origin of a second correction to the centre-of-mass energy. The electrostatic separators used since 1995 to avoid parasitic encounters of the beams generated vertical dispersion of opposite sign for the $e^+$ and $e^-$ beams. When the beams collide with a vertical offset, the correlation between transverse position and energy leads to centre-of-mass energy shifts $\Delta E_{\rm CM}$ [19,23,24]:

$$\Delta E_{\rm CM} = -\frac{1}{2}\frac{\delta y^*}{\sigma_{\rm y}^2}\frac{\sigma_{\rm e}^2}{E_{\rm b}}\Delta D_{\rm y}^*, \tag{18}$$

where $\delta y^*$ is the collision offset between the two beams and $\Delta D_{\rm y}^*$ is the dispersion difference at the IP. To keep centre-of-mass energy shifts due to opposite sign dispersion under control, the collision offsets had to be carefully minimized with regular luminosity scans. Because the horizontal beam size at the IP was much larger (>200 $\mu$m as compared to few micrometers), the CM shifts were very small in that plane.

### 3.9 *Summary of LEP 1 energy calibration*

A model of the LEP beam and centre-of-mass energies was developed from the large amount of energy calibration data accumulated over the years. The models were used to extrapolate beam energies in times between energy calibrations. Best possible estimates of the energies over 15 minute intervals were given to the LEP experiments for data analysis. The final result of this work was a systematic error due to the energy calibration of the LEP beams of only 1.7 MeV on $m_{\rm Z}$ and of 1.3 MeV on $\Gamma_{\rm Z}$ [19]. The originally anticipated accuracy on $m_{\rm Z}$ was $\simeq$ 20 MeV.

### 3.10 *Energy calibration at LEP 2*

At LEP 2 a direct energy calibration by resonant depolarisation was not possible because transverse polarisation did not build up at such high energies, see Fig. 8. The beam energy had to be extrapolated from resonant depolarization measurements performed at energies between 41 and 60 GeV.

The beam energy between 80 and 104 GeV was derived from continuous measurements of the magnetic bending field by 16 NMR probes installed in LEP dipoles distributed around the ring. The relationship between the fields measured by the probes and the beam energy was calibrated against precise measurements of the beam energy between 41 and 60 GeV [25].

The validity of the NMR model was verified and corrected by three independent methods:

- A magnetic flux measurement could measure the bending field of all the dipoles of LEP.
- A spectrometer determined the energy through measurements of the deflection of the beam in a magnet of precisely known integrated field. Besides the magnet of two triplets the spectrometer consisted of BPMs installed on either side of the dipole magnet.

- The variation of the synchrotron tune with the total RF voltage was used to obtain the beam energy through the energy loss by synchrotron radiation [26].

All three methods were calibrated against resonant depolarisation between 41 and 60 GeV in a similar manner to the NMR probes. The consistency between the three methods was used to estimate the systematic errors.

To obtain the energies, corrections were applied to the NMR model to account for sources of bending field external to the dipoles (tides and other geological effects, orbit correctors, etc.), and variations in the local beam energy at each interaction point. The final relative error on the energy determination for the majority of LEP 2 running was $1.2 \times 10^{-4}$ [25].

### 3.11 *Luminosity performance*

The particle number per bunch $N_b$ in LEP was limited mainly by an instability caused by a short range interaction of the beam with the impedance of its surrounding, called Transverse Mode Coupling Instability (TMCI). It limited the bunch current to about $I_b \approx 1\,\mathrm{mA}$ as predicted. This corresponds to a peak current of about 1000 A. Due to the large spacing of the coupled bunches, instabilities between them were not important in normal operation. However, at high energies the total current according to Eq. (6) is limited by the available RF-power.

LEP 1 started at 47 GeV first with two and later with four bunches using vertical separation at the parasitic crossings. A lattice with a phase advance per cell of $\mu_{x/y} = 60°/60°$ was used at the start which was not very critical and easy to correct. This was followed by $\mu_{x/y} = 90°/60°$ which gave a smaller emittance $E_x \approx 12\,\mathrm{nm}$ and consequentially, a higher luminosity. According to Eq. (9) the bunch current was the limited by the beam–beam effect and luminosity could be further increased by increasing the number of bunches per beam $k_b$. This was achieved with an operation mode having eight equidistant bunches with horizontal separation in the arcs, called "pretzel scheme". An operation mode with four trains of two or three closely spaced bunches and local vertical separation on both sides of the interaction point were used later. The performance of the two were comparable and gave a luminosity of over $0.3 \cdot 10^{32}\,\mathrm{cm}^{-2}\mathrm{s}^{-1}$ and a total current per beam in excess of 4 mA. With these conditions the mass and width of the $Z^0$ particle were measured.

LEP 2 operated above 80 GeV to produce W pairs. For a fixed arc optics the horizontal beam emittance increases $\propto E^2$ which reduces the luminosity at a given (limited) bunch current. This could be partially counteracted with stronger focusing. Furthermore, some emittance reduction was achieved by increasing the horizontal damping partition $J_x$. Due to the sum rule $\sum J_i = 4$, this reduced the longitudinal damping and therefore increased the energy spread which demanded a higher RF-voltage. The optimum value for $J_x$ was a compromise between maximum luminosity and energy and was also adjusted to the decaying beam current. Several phase advances per cell were tried but at the end one settled for $\mu_{x/y} = 102°/90°$ which gave an emittance $E_x \approx 22$ nm for $E = 94.5$ GeV and $J_x = 1.75$. At this energy one just about reached the beam–beam limit with the available bunch current and operation with four bunches was used. This gave a maximum luminosity of $\approx 1 \cdot 10^{32}$ cm$^{-2}$s$^{-1}$ between 95 and 101 GeV. The short radiation damping time helped to get a high beam–beam tune shift of about 0.06 in both planes. Furthermore, the beam–beam focusing contributed to the small beam size at the interaction point.

At the end of the LEP operation one tried to reach the highest possible energy in search of new physics. With the circumferential voltage of 3.66 GV from all installed cavities, a beam energy of 104.5 GeV was achieved.

## Acknowledgments

This presentation reports the work done by many people at LEP. We like to thank especially D. Brandt and K. Hübner for their information about the operation of LEP during the final years.

## References

[1] LEP Design Report, CERN-LEP/84-01 (1984).
[2] H. Schopper, *LEP — The Lord of the Collider Rings at CERN 1980–2000*, Springer 2009.
[3] K. Hübner, Designing and building LEP, *Physics Reports* **403–404**, (2004) 177–188.
[4] B. Richter, Very high energy electron–positron colliding beams for the study of weak interactions, *Nucl. Instr. Methods* **136** (1976) 47.
[5] J. P. Gourber, J. Billan, N. Laeger, A. Perrot and L. Resegotti, On the Way to the Series Production of Steel-Concrete Cores for the LEP Dipole Magnets, *IEEE Trans. on Nucl. Sci.* **NS-30**(4) (1983) 3614–3616.
[6] C. Benvenuti, A new pumping approach for the Large Electron Positron collider LEP, *Nucl. Instr. Meth.* **205** (1983) 391.
[7] P. Castro, J. Borer, A. Burns, G. Morpurgo and R. Schmid, Betatron function measurement at LEP using the 1000 turns facility, in *Proceedings of the 1993 Particle Accelerator Conference,* Washington D.C., 1993.
[8] W. Schnell, Modulated Cavities, CERN ISR-LTD/76-8 (1976).
[9] A. Butterworth, P. Brown, O. Brunner, E. Ciapala, H. Frischholz, G. Geschonke, E. Peschardt and J. Sladen, The LEP2 superconducting RF system, *Nucl. Instr. Methods A* **587** (2008) 151.
[10] L. Arnaudon *et al.*, *Phys. Lett.* **B307** (1993) 187.
[11] A. A.Sokolov and I. M. Ternov, *Sov. Phys. Dokl.* **8** (1964) 1203.

[12] V. Bargmann, L. Michel and V. L. Telegdi, *Phys. Rev. Lett.* **2** (1959) 435.
[13] L. Arnaudon *el al.*, Recent achievements in transverse spin-polarisation at LEP in *Proc. of the 4th European Particle Accelerator Conf. (London),* ed. V. Suller and C. Petit-Jean-Genaz (World Sci., 1994), pp. 187–189.
[14] R. Assmann *et al.*, Deterministic harmonic spin matching in LEP, in *Proc. of the 4th European Particle Accelerator Conf. (London),* ed. V. Suller and C. Petit-Jean-Genaz (World Sci., 1994), pp. 932–934.
[15] R. Assmann *et al.*, Observation of radiative spin-polarization at 60.6 GeV, in *Proc. of the 1999 Particle Accelerator Conf. (New York),* Vol. 5, ed. A. Luccio and W. MacKay (New York: IEEE, 1999), pp. 2999–3001.
[16] L. Arnaudon *et al.*, *Z. Phys.* C **66** (1995) 45–62.
[17] V. N. Baier and Y. F. Orlov, *Sov. Phys. Dokl.* **10** (1966) 1145; A. A. Zholentz *et al.*, *Phys. Lett.* **B96** (1980) 214; A. S. Artamonov *et al.*, *Phys. Lett.* **B118** (1982) 225; Y. M. Shatunov and A. N. Skrinsky, *Particle World* **1**(2) (1989) 35; D. P. Barber *et al.*, *Phys. Lett.* **B135** (1984) 498; W. W. McKay *et al.*, *Phys. Rev.* **D29** (1984) 2483.
[18] R. Assmann *et al.*, *Z. Phys.* **C66** (1995) 567–582.
[19] The LEP Energy Working Group, *Eur. Phys. J.* **C39** (2005) 253–292.
[20] E. Bravin *et al.*, *Nucl. Instrum. Methods* **A417** (1998) 9–15.
[21] L. Arnaudon *et al.*, *Nucl. Instrum. Methods* **A357** (1995) 249–252.
[22] J. Wenninger, Observation of the radial ring deformations using closed orbits at LEP, in *Proc. of the 1999 Particle Accelerator Conf. (New York),* Vol. 5, ed. A. Luccio and W. MacKay (New York: IEEE, 1999), pp. 3014–3016.
[23] M. Böge *et al.*, Measurements of collision offsets and difference in vertical dispersion at the LEP interaction points, in *Proc. of the 5th European Particle Accelerator Conf. (Sitges),* ed. S. Myers *et al.* (Bristol: IOP, 1996), pp. 427–429.
[24] M. Bassetti and J. M. Jowett, Improving the energy resolution of LEP experiments in *Proc. of the 1987 Particle Accelerator Conf. (Washington, DC)* (New York: IEEE, 1987), pp. 115–117.
[25] A. Blondel *et al.*, *Eur. Phys. J.* **C11** (1999) 573–585.
[26] M. Müller, Precision measurements of the LEP beam energy for the determination of the W mass, PhD thesis, University of Mainz, 2000.

# Chapter 5

# SLC: The first linear collider

Nan Phinney (SLAC)

The Stanford Linear Collider (SLC) was built in the 1980s at the Stanford Linear Accelerator Center (SLAC) in California. Like LEP, it was designed to study the properties of the Z boson at a center-of-mass energy of about 91 GeV. The SLC was also a prototype for an entirely new approach to electron–positron colliders. The development of a new technology was motivated by the fact that in an electron storage ring, the electrons radiate synchrotron radiation as they are bent around the ring. To avoid excessive energy loss from this radiation, the circumference of the ring has to increase as the square of the desired energy, making very high energy rings prohibitively large and expensive. With a linear accelerator, the electrons do not need to bend and the tunnel length only grows linearly with energy.

In principle, a linear collider has two high energy linacs which accelerate electron and positron beams towards each other so that they collide at an Interaction Point (IP). Rather than build a second linac, the SLC used the 3 km long SLAC linear accelerator to accelerate both the electrons and the positrons. Present designs for linear colliders to reach 500 GeV are about 30 km long. One limitation for a linear collider is that the bunches only collide at the machine repetition rate, less than 100 times a second, compared to tens of thousands of times a second in a storage ring. For this reason, future linear colliders all have long trains of bunches. The SLC collided only a single bunch of electrons with a single bunch of positrons. A second bunch of electrons was accelerated to about 30 GeV onto a target to produce positrons.

## 1 Luminosity

In a linear collider, the luminosity is given by the formula:

$$L = \frac{n_b f_r N^2}{4\pi\sigma_x\sigma_y} H_D$$

where $n_b$ is the number of bunches, $f_r$ is the machine repetition rate, $N$ is the number of particles per bunch, $\sigma_x\sigma_y$ are the beam sizes at the IP and $H_D$ is a disruption factor. The collider luminosity depends on the total number of particles colliding per second and on the density of particles in the bunch $n_b f_r N^2$. With many fewer crossings per second, a linear collider must achieve a much smaller beam size at the IP for a given target luminosity as compared to a circular collider. The challenges for a linear collider are to produce bunches of electrons and positrons cooled to a very small phase space, or emittance; to preserve that emittance as the bunches are accelerated; and then to focus the bunches to a very small size at the interaction point. The SLC was the first linear collider ever attempted and it took many years to develop the understanding and techniques required to collide very small beams.

## 2 Layout

Because both electron and positron beams traveled down the same linac, the SLC had a shape somewhat like a tennis racket, and two circular "Arcs" transported the beams around to two final focus systems where they collided head on at the Interaction Point. A schematic of the SLC is shown in Figure 1. Geographical constraints made it impossible to build the Arcs in a plane, so they contoured up and down to avoid the tunnel of an older accelerator and other obstacles. This "terrain following" beam line made it very difficult to establish the correct optics. Eventually a technique was developed to measure beam oscillations, perform a sophisticated error analysis, and create trajectory distortions to exactly cancel the measured optical errors [1]. The technique was also capable of optimizing the orientation of the beam polarization and minimizing the polarization dilution. In spite of this success, it is unlikely that any future collider will be built in anything but a plane.

The SLC polarized electron source was a DC S-band gun with a strained GaAs photocathode [2]. The addition of a loadlock system allowed cathodes to be exchanged without venting and served as an activation chamber. The gun operated at 120 kV and produced a pair of 2 ns pulses separated by 62 ns at a repetition rate of 120 Hz. Two YAG-pumped Ti:sapphire lasers illuminated the cathode with circularly polarized light with polarization in excess of 99% and switchable each pulse. The laser intensity, timing and polarization were stabilized by feedback systems. An electron beam polarization on the order of 80% was routinely achieved.

The SLC positron source used the trailing bunch of electrons extracted from the main linac at about 30 GeV, deflected into a transport line and

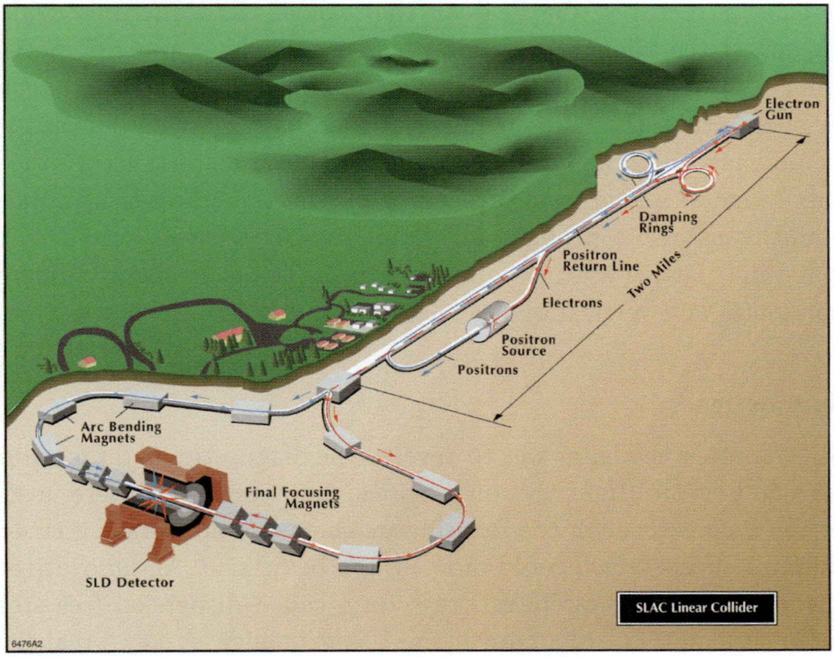


Fig. 1. A schematic of the SLC.

focused onto a target [3]. The positrons were produced in a moving target and captured in a high gradient accelerating section and solenoid field. They were accelerated to 200 MeV and transported back to the beginning of the linac through an isochronous 180° turnaround, a 2 km long transport line and a final turnaround before reinjection into the linac.

## 3 Beam dynamics

The SLC used damping rings to cool the electron and positron beams before injection into the linac. To achieve sufficient damping, they were built at an energy of 1.153 GeV and were extremely compact. The tight space constraints created numerous challenges, particularly at injection and extraction. At higher beam currents, a microwave bunch lengthening instability was discovered. While seen before in storage rings, the instability was a major problem for the SLC because the beam could be extracted at any phase of the bunch length oscillation. In the early 1990s, the damping ring vacuum chambers had to be replaced with smoother chambers to reduce the impedance seen by the beam [4, 5].

A challenge of accelerating two bunches down the same linac was controlling the wakefields. If the beams were not perfectly centered in the accelerating structures, the field on the wall of the structures could distort or deflect the bunch. A technique called "Landau damping" positioned the bunch on the crest of the RF such that the head of the bunch had higher energy than the tail. The tail was then more strongly focused by the quadrupole lattice, damping the wakefield induced distortions. In the SLC, the long-range wakefield from the leading positron bunch to the trailing electron bunch was also a problem even though they were separated by 60 ns. A "split-tune" lattice was introduced to make the optics slightly different in the horizontal and vertical planes [6, 7].

## 4 Diagnostics

For SLC, another challenge was to develop the diagnostics and procedures to measure and control the very small beams. New techniques were needed to center the beams through the focusing magnets and accelerating structures to minimize effects which could blow up the beam size. An innovative algorithm for finding the best beam trajectory, called dispersion free steering, was used first at SLC to improve the beam quality. A key breakthrough was moving micrometer-sized wires through the beams to measure the beam size non-invasively during routine operation. Feedback systems were required to stabilize the energy and position of the beams throughout the machine. Specialized feedback was used to bring the tiny beams into collision and even to optimize the final tuning of the beam size [8]. The SLC eventually had more than 40 wire scanners and 50 feedback systems. The beam size at the collision point was 1.4 micrometers horizontally and 0.7 micrometers vertically, much smaller than a human hair and about a factor of 100 smaller than in storage rings. With such small intense beams, the interaction of the two beams causes them to shrink even further in size. This effect is called "pinch enhancement" and was measured for the first time at the SLC, where it increased the luminosity by as much as a factor of two [9].

## 5 Performance

The SLC operated from 1988 through 1998, first with the Mark-II detector and later with the SLD detector. In later operation, the SLC reached bunch charges of $4.5 \times 10^{10}$ per bunch and a luminosity of $10^{30}$ [10, 11]. The SLD detector at the SLC exploited two unique features of a linear collider: the very small interaction point and a highly polarized electron beam, where the electron spins point predominantly in the same direction. These advantages allowed SLD to make measurements which were complementary to the high precision LEP measurements, including the world's most precise measurement of the critical electroweak mixing angle.

## References

[1] N. J. Walker, *et al.*, Correction of the first order beam transport of the SLC arcs, SLAC-PUB-5560, May 1991.
[2] J. E. Clendenin, *et al.*, Performance of the SLC polarized source with high polarization, SLAC-PUB-6080, Apr 1993.
[3] P. Krejcik, *et al.*, Recent improvements in the SLC positron system performance, SLAC-PUB-5786, Mar 1992.
[4] F. J. Decker, *et al.*, Longitudinal phase space setup for the SLC beams, SLAC-PUB-7892, Jul 1998.
[5] M. Minty, *et al.*, Operation and performance of bunch precompression for increased current transmission at the SLC, SLAC-PUB-7467, May 1997.
[6] C. Adolphsen, *et al.*, Flat beams in the SLC, SLAC-PUB-6118, May 1993.
[7] F. J. Decker, *et al.*, Long range wake fields and split tune lattice at the SLC, SLAC-PUB-7260, Aug 1996.
[8] L. Hendrickson, *et al.*, Luminosity optimization feedback in the SLC, SLAC-PUB-8556, Mar 1999.
[9] T. Barklow, *et al.*, Experimental evidence for beam–beam disruption at the SLC, SLAC-PUB-8043, May 1999.
[10] P. Raimondi, *et al.*, Recent luminosity improvements at the SLC, SLAC-PUB-7955, Sep 1998.
[11] N. Phinney, SLC final performance and lessons, SLAC-PUB-8556, Oct 2000.

# Chapter 6

# Asymmetric B-factories

Katsunobu Oide (KEK)

## 1 Physics motivation

The idea of asymmetric B-factories was first introduced by P. Oddone in 1987 [1] to collide $e^+e^-$ beams with different energies to measure the CP-asymmetry between the decay of $B_0$ and $\overline{B}_0$ mesons. The asymmetry of the energies of two beams boosts the generated particles longitudinally, then the difference of the decay time can be measured by the difference of the vertices, which was expected to be in about an order of 100 $\mu$m. The center-of-mass energy of the collision is set to the $\Upsilon(4S)$ resonance at 10.58 GeV. A very high luminosity around $10^{34}$ cm$^{-2}$s$^{-1}$ is required, which was more than 100 times higher than what had been achieved in colliders by that time.

## 2 Double ring collider

There may be several ways to realize the asymmetric collision. One way is to build a linear–linear or a ring–linear collider. Such a linear machine needs a very strong focusing $\beta_y^* \sim 100$ $\mu$m to achieve the luminosity, then the bunch length must be as short as $\beta_y^*$ to avoid the hour-glass effect. The bunch length itself can be obtained by bunch compressors, but the associated energy spread degrades the effective luminosity, since the width of the resonance $\Upsilon(4S)$ is only 20 MeV ($2 \times 10^{-4}$). A huge damping ring would be necessary to realize such a short bunch length and a small energy spread simultaneously. Thus linear collision schemes seemed difficult.

As for the double-ring collision, a question is the sizes of the rings. If one can collide a large high energy ring (HER), for instance at 25 GeV, with a small low energy ring (LER) at 1.2 GeV, the total cost will be saved, assuming an existing tunnel for such a high energy ring. It was pointed out [2] that the collision of rings with different circumferences has somewhat fundamental difficulty: if two rings have the ratio of circumferences $m : n$ ($m > n$), the periodicity of the system becomes very long, i.e., $\mathrm{LCM}(m, n)/m$

(Lowest Common Multiple) times the revolution period of the larger ring. Then both rings will have dense resonance lines in the tune space which reduces the operable area, especially with a large amount of the beam–beam tune shift. Thus collision of rings with different circumferences seemed difficult. Therefore only the double ring collider scheme remained.

Two projects of the asymmetric B-factories, PEP–II [3] at SLAC and KEKB [4] at KEK, were approved and the construction started by 1994. Both projects utilized the components and facilities of their previous generation colliders, PEP and TRISTAN, and built the BaBar and Belle detectors, respectively. The collision experiments for both machines started in 1999 and the operation stopped in April 2008 (PEP–II) and June 2010 (KEKB). Table 1 lists the main machine parameters corresponding to their best records [5, 6]. Both colliders achieved higher performance than their

Table 1. Progress of machine parameters of the PEP–II and KEKB B-factories. The left, center, right correspond to the highest performance of PEP–II, KEKB (no crab) and KEKB (crab), respectively. The integrated luminosities are the delivered numbers for PEP–II, and recorded for KEKB. 1/nb = $10^{33}$ cm$^{-2}$s$^{-1}$.

| | PEP–II | | KEKB (no crab) | | KEKB (crab) | | |
|---|---|---|---|---|---|---|---|
| | 8/16/2006 | | 11/15/2006 | | 6/17/2009 | | |
| | LER | HER | LER | HER | LER | HER | |
| Circumference | 2200 | | 3016 | | | | m |
| Beam energy | 3.1 | 9.0 | 3.5 | 8.0 | 3.5 | 8.0 | GeV |
| Eff. crossing angle | 0 | | 22 | | 0 (crab) | | mrad |
| Beam current | 2.90 | 1.88 | 1.65 | 1.33 | 1.64 | 1.19 | A |
| Bunches | 1722 | | 1389 | | 1584 | | |
| Bunch current | 4.02 | 1.09 | 1.19 | 0.96 | 1.03 | 0.71 | mA |
| Bunch spacing | 1.2 | | 1.8–2.4 | | 1.8 | | m |
| Hor. emittance $\varepsilon_x$ | 30 | 50 | 18 | 24 | 18 | 24 | nm |
| RF frequency | 476 | | 509 | | | | MHz |
| Bunch length $\sigma_z$ | 10 | 10 | 8 | 6 | 8 | 6 | mm |
| $\beta_x^*$ | 30 | 30 | 59 | 56 | 120 | 120 | cm |
| $\beta_y^*$ | 0.9 | 1.1 | 0.65 | 0.59 | 0.59 | 0.59 | cm |
| Hor. size at IP | 95 | 158 | 103 | 116 | 147 | 170 | $\mu$m |
| Ver. size at IP | 4.7 | 4.7 | 1.9 | 1.9 | 0.94 | 0.94 | $\mu$m |
| Beam–beam $\xi_x$ | 0.072 | 0.064 | 0.115 | 0.075 | 0.125 | 0.100 | |
| Beam–beam $\xi_y$ | 0.064 | 0.053 | 0.104 | 0.058 | 0.130 | 0.090 | |
| Luminosity | 12.1 | | 17.6 | | 21.1 | | /nb/s |
| $\int$Lum./day | 858 | | 1260 | | 1479 | | /pb |
| $\int$Lum./7 days | 5.41 | | 7.82 | | 8.43 | | /fb |
| $\int$Lum./30 days | 19.8 | | 30.2 | | 23.0 | | /fb |
| Total $\int$Lum. | 557 | | 1040 | | | | /fb |

designs, and experimentally verified the Kobayashi–Maskawa model to bring the 2008 Nobel Prize in Physics.

## 3 Luminosity

The luminosity $\mathcal{L}$ of an asymmetric ring collider can be expressed as follows:

$$\mathcal{L} = \frac{\gamma_\pm}{2er_e}\left(1 + \frac{\sigma_y^*}{\sigma_x^*}\right)\left(\frac{I\xi_y}{\beta_y^*}\right)_\pm \left(\frac{R_\mathcal{L}}{R_y}\right), \tag{1}$$

where $\gamma$, $e$, $r_e$, $\sigma_{x,y}^*$, $I$, $\beta_{x,y}^*$ are the Lorentz factor, electron charge, classical electron radius, beam sizes at the interaction point (IP), stored beam current in the ring, and the $\beta$-function at the IP, respectively. The suffix "$\pm$" denotes each beam. The expression (1) is obtained from the beam–beam tune-shift parameter

$$\xi_{\pm x,y} = \frac{r_e}{2\pi\gamma_\pm}\frac{N_\mp \beta_{\pm x,y}^*}{\sigma_{x,y}^*\left(\sigma_x^* + \sigma_y^*\right)} R_{x,y} \tag{2}$$

and the definition of luminosity

$$\mathcal{L} = \frac{N_+N_-f}{4\pi\sigma_x^*\sigma_y^*}R_\mathcal{L} \tag{3}$$

where $N$ and $f$ are the number of particles per bunch and the collision frequency ($I = Nef$), respectively, and we have assumed that the beam sizes are common in two beams. The factors $R_{\mathcal{L},x,y}$ are the geometric reduction factors due to the hour-glass effect and the crossing angle.

While a round-beam scheme may have a merit of a factor of 2 on the luminosity according to Eq. (1), a flat beam scheme has been chosen in most $e^+e^-$ colliders, as the round-beam focusing in both planes is more difficult for an extremely small $\beta^*$. For a flat beam, $\sigma_x^* \gg \sigma_y^*$, the luminosity is written as

$$\mathcal{L} \approx \frac{1}{2er_e}\left(\frac{\gamma I\xi_y}{\beta_y^*}\right)_\pm \left(\frac{R_\mathcal{L}}{R_y}\right). \tag{4}$$

Then if there is no reason to differentiate $\xi_y$ and $\beta_y^*$ in the two rings, one can set

$$\gamma_+I_+ = \gamma_-I_-. \tag{5}$$

As the ratio of beam energies gets larger, the boost at the collision becomes larger, but the low energy ring must store higher beam current. Thus the energy ratio was a compromise between the physics merit and the accelerator difficulty. PEP–II chose 3.1 GeV and 9 GeV for positrons and electrons, while

KEKB chose 3.5 GeV and 8 GeV. A larger ratio was more favored at PEP–II as it needs a magnetic separation of two beams at the IP as described later. The flavor of beams, the LER for positrons, was uniquely chosen at KEKB, where the positron acceleration for the HER was very difficult.

For the actual operation of these machines, the condition (5) was not strictly kept, as shown in Table 1. One reason was that the natural size of each beam was not equal; for instance, the LER positron beam was relatively easy to be blown up due to the electron clouds at high current. Then there was a certain limit on the positron beam current and the HER current was increased beyond Eq. (5). This tendency was stronger in KEKB than PEP–II, as the former had stronger electron cloud effects than the latter as described later.

## 4 Crossing angle

One of the design choices is the beam separation scheme near the IP. A crossing angle is a natural and easy solution for the separation, but experience at DORIS [7] casted doubts on viability of this choice. KEKB decided to apply a horizontal crossing angle $2\theta_x = 22$ mrad, relying on simulations of the beam–beam effect. The corresponding Piwinski angle ($\equiv \theta_x\sigma_z/\sigma_x^*$) was 0.86. Their conclusion at the design stage was that the effect of the crossing angle on the beam–beam interaction would not be harmful up to their design beam–beam parameter 0.05, if the operating betatron tunes were carefully chosen. Their choice was right and achieved a vertical beam–beam parameter of 0.06 in their luminosity ramp-up. Crossing angles were also successfully applied at the CESR and DAΦNE colliders in parallel with the KEKB operation. KEKB even prepared a crab-crossing scheme [8, 9] as a backup for the crossing angle scheme. The ratio of the geometric reduction factors $R_{\mathcal{L}}/R_y$ in Eq. (1) does not drastically decrease for a large crossing angle as shown in Ref. [4].

PEP–II was much more nervous about the use of a crossing angle and installed a magnetic separation scheme near the IP with permanent dipole magnets [10]. This scheme also worked, but their design around the IP had to be more complicated than with a crossing angle, and gave some limitations on the performance such as the detector background due to radiative Bhabha events [11], which was much less significant in Belle. As the space at the IP was limited, they could not install a compensation system for the detector solenoid field, which might have degraded the beam-optical performance. Another issue of the magnetic separation was the non-negligible detrimental effect due to the parasitic collisions [12], which was never observed at KEKB.

## 5 Storing high current

As described above, the luminosity is proportional to the stored current. To achieve a luminosity as high as $10^{34}$ cm$^{-2}$s$^{-1}$, a stored current of near 3 A was required, which was one order of magnitude higher than the beam current in any high energy electron storage ring at that time. The first fundamental difficulty with such high beam currents is to ensure the longitudinal stability of the beam.

The beam loading of the accelerating cavity is huge: a normal conducting cavity at the RF frequency $f_{\rm RF} = 500\,{\rm MHz}$ for the B-factories has a shunt impedance $R_s \approx 1.7$ MΩ. If the cavity is tuned to the harmonics, the 3 A beam generates 5.1 MV decelerating voltage in the cavity, which is even higher than the accelerating voltage $V_c$ of the cavity, typically 0.5 MV. Thus the detuning of the cavity is necessary and the optimal amount of the detuning frequency is given by

$$\varDelta f = -\frac{I\sin\phi_s}{2V_c}\frac{R_s}{Q}f_{\rm RF} = -\frac{P_b\tan\phi_s}{4\pi U}, \tag{6}$$

where $\phi_s$, $Q$, $P_b$, $U$ are the synchronous phase, the $Q$-value, the beam power, and the stored energy of the cavity, respectively. If the magnitude of the detuning frequency becomes higher than or comparable to the revolution frequency, the cavity impedance hits the side bands of synchrotron motion to excite strong longitudinal coupled-bunch instabilities.

This issue of the beam-loading instability was solved in the two B-factories in different ways. KEKB developed two types of cavities with large stored energy, as Eq. (6) is inversely proportional to the stored energy. Both ARES [13] and superconducting [14] cavities could store an electromagnetic energy 10 times larger than that of a conventional cavity. Then together with the cavity HOM damping mechanisms, the RF system of KEKB did not induce any beam instability up to the design current and without the help of a bunch-by-much feedback system. On the other hand, PEP–II took the alternative strategy to develop a sophisticated feedback system to reduce the effective impedance seen by the beam [15]. PEP–II applied a direct RF feedback system with newly developed sideband klystrons combing a longitudinal bunch-by-bunch feedback [16]. Both KEKB and PEP–II systems basically worked as expected nearly up to or even beyond their design currents.

Storing high beam currents caused a number of issues on the beam pipes, bellows, collimators, and even on the detectors. Direct hit of the beam of an ampere caused by beam instability or anything else easily melted down the affected components. The wakes at transitions resulted in discharge and heating. A number of models have been developed and tried for the collimators, bellows, and HOM absorbers. Also machine protection system, loss monitors, and beam abort system had to evolve as the stored beam current increased.

## 6 Electron cloud

Electron cloud was the one of the toughest issues for the asymmetric B-factories, specifically on the accumulation of the positron beam. The electron cloud had been known as a possible cause of beam instability in positively charged beams since a long time ago such as the ISR era. Its observation [21] had been made at the Photon Factory (PF) of KEK and a theoretical explanation [22] had been established well before the start of the B-factories. What was new at the B-factories was the single-bunch instability induced by electron clouds [23]. The previous instability observed at the PF had been interpreted as a coupled-bunch instability, which was supposed to be cured by a bunch-by-bunch feedback. Thus at least KEKB was not well prepared for the single-bunch phenomena which have much higher frequency contents than the available feedback. Actually, the possibility of such a single-bunch effect had been suggested [24] before the construction of the B-factories, but it had not been well recognized. The single-bunch effect was experimentally confirmed at KEKB [25] as well as at CesrTA.

The electron cloud blew up drastically the vertical beam size, and the threshold beam current was 0.4 A with 4-bucket (2.4 m) bunch spacing at KEKB. The electron cloud appeared more severely in KEKB than in PEP–II, as the former had a round Cu beam pipe while the latter an Al antechamber with TiN coating. Thus the initial startup of the luminosity at KEKB was slower than in PEP–II.

By applying a weak magnetic field at the beam pipe, the electron cloud was removed at least in the free drift space of the machine. Either permanent magnets or solenoids were installed at KEKB and PEP–II to cover almost all straight sections and inside of some magnets such as quadrupoles and weak dipoles by 2004. The mitigation worked as expected and the blowup become unnoticeable at least for 3-bucket spacing in the case of KEKB [26]. Beside the magnetic field, various other mitigation techniques have been developed and tested at the B-factories, against the formation of the electron cloud, including antechambers [17], TiN or diamond-like carbon coatings, grooved surface pipes [18], and clearing electrodes [19]. Those techniques will be effective for future super B-factories and damping rings of linear colliders. Also several measurements of the cloud density have been carried out.

Although the density of the electron cloud could be reduced below the instability threshold by the application of magnetic fields, the betatron tune shift due to the cloud still remained present in the LER at KEKB, generating a tune variation along the bunch train. A possible mitigation for the tune variation is the use of pulsed quadrupoles, as was done at KEKB [20].

## 7 Beam optics

The luminosity of a ring collider is inversely proportional to the vertical $\beta$-function at the IP as shown in Eq. (4). The B-factories have used the smallest $\beta_y^*$ in a ring collider so far. Generally speaking, a smaller $\beta_y^*$ means higher chromaticity and higher nonlinearity arisen from sextupoles for the chromaticity correction. Thus the design of the ring lattice requires special care to ensure a sufficiently large dynamic aperture. One technique applied to KEKB was the use of non-interleaved sextupole pairs separated by a $-I$ transformation that cancels the geometric nonlinearity of the sextupoles up to the second order [28]. Although the idea was very old, the application to a real ring did not occur until the B-factories, probably due to the required computing power for optimizing the sextupole settings, as the optimization requires a large number of sextupole families to extend the momentum acceptance. For instance, KEKB has 54 families of sextupole pairs. The relative betatron phase advance between the pairs became adequate by using the $2.5\pi$ cell structure in the case of KEKB arc section [29].

Another technique to enlarge the dynamic aperture was to place a special chromaticity correction section near the IP. The beam optics becomes somewhat similar to that of linear colliders in this case. KEKB designed such a section for the vertical correction, while PEP–II used a horizontal correction for their LER.

These schemes worked as expected for the B-factories, and are expected to work for future super B-factories and light sources. Once the chromaticity correction is solved, the other sources of nonlinearity are the fringe fields of the final quadrupoles and the geometric nonlinearity at the IP [30], which may be mitigated by additional octupoles placed at the final quadrupoles.

Minimizing the $x$–$y$ coupling and the residual vertical dispersion all over the ring was one of the key ingredients to achieve a high luminosity by reducing the vertical emittance. Various techniques have been applied for such optics measurements and corrections [31–33]. A counter solenoid to the detector solenoid was also effective to reduce the coupling source in the case of KEKB. This was also important in the case of crab crossing where the luminosity performance was sensitive to the chromatic $x$–$y$ coupling as described later.

## 8 Beam diagnostics and control

A number of beam diagnostic methods were developed and applied to the B-factories:

- Beam position monitors (BPMs) with a resolution better than 1 $\mu$m in the average mode. In some cases turn-by-turn or bunch-by-bunch electronics were equipped [34]. In the case of KEKB, the gain imbalance from the electrode through the electronics was calibrated using a beam-mapping technique [36]. The design of the electrodes and the electronics were carefully done for the high-current operation.
- Beam-based alignment of BPMs was regularly carried out. Displacements of BPMs near sextupoles due to the heating from the stored beam current were monitored at KEKB [35].
- Bunch-by-bunch feedback systems were installed both in PEP–II and KEKB. Only PEP–II had a longitudinal system to suppress the beam-loading instability as described above. A collaboration including the DAΦNE team has developed the system for present and future applications [43].
- Betatron tune monitor: controlling the betatron tune was extremely important to maximize the luminosity. The basic idea at KEKB was to monitor the tunes of pilot bunches in each ring that did not collide with the other beam. Tune feedback with these bunches was also applied to control them within an accuracy of $\Delta\nu \approx 10^{-4}$.
- Synchrotron radiation beam profile monitors. For the visible light, an interferometer was used specifically for the vertical size measurement [37]. Special gated cameras were also used to observe the beam size of individual bunch, in particular to diagnose the electron-cloud effects [38].
- Beam loss monitors and beam abort system: both machines were very anxious to protect the machine against accidental beam losses caused by instabilities, RF trips, wrong injection, or whatever. The most sensitive and expensive loss monitor was the BaBar and Belle detectors, which generated beam abort signals if necessary. A number of beam loss monitors such as ionization chambers and PiN diodes were distributed around the ring, especially near the collimators. The beam abort system consists of an abort kicker and a beam damp. The abort kicker had a rise time of 0.5 $\mu$s in the case of KEKB.

- The injectors had developed their own diagnostics including BPMs, wire scanners, streak cameras, etc.
- All accelerator components were controlled by computer control systems either by EPICS at KEKB [39] or a legacy system at PEP–II. An online modeling such as SAD for KEKB [32] was also important to achieve the luminosity.

## 9 Collision tuning

Starting up the colliders after a period of long shut down, the following procedures were necessary to recover the luminosity:

- Global coupling/dispersion/$\beta$-function correction all over the ring. The global orbit was then locked to the "golden" orbit that resulted from the optics correction.
- Locking the betatron tunes of the pilot bunches.
- The beam steering at the IP looking at the beam–beam deflection.
- In the case of the crossing angle at KEKB, the horizontal offset at the IP was controlled by looking at the vertical beam size measured by the interferometer [40].
- Tuning of the local coupling and dispersion at the IP by making offsets of orbits at sextupoles near the IP.
- Dithering technique was used at PEP–II to maximize the luminosity against the beam offsets [41].
- Skew sextupoles were introduced at KEKB to correct the local chromatic $x$–$y$ coupling terms at the IP.

The horizontal tunes were chosen as close to a half integer as possible, to maximize the luminosity using the dynamic-$\beta$ effect and expecting the reduction of the degree of freedom of the beam–beam interaction [42]. In the case of KEKB, the LER and the HER were operated at $\nu_x \approx 0.506$ and $\nu_x \approx 0.510$, respectively. Both the optics correction and the tune feedback were necessary to maintain a collision near the stop band.

## 10 Injector

The electron–positron injector must provide enough number of charges to the collider rings. PEP–II could fully utilize the injection system for SLC, which had more than enough performance for the PEP–II operation in terms of beam intensity, repetition rate, and beam emittance, especially with damping rings for both beams. On the other hand, the injector for KEKB was upgraded from that for TRISTAN, having only the minimum performance

Table 2. Comparison of positron injection.

| | KEKB | | PEP–II | |
|---|---|---|---|---|
| | 1999 | 2010 | | |
| Production energy | 4 | | 30 | GeV |
| Particles per pulse | 0.4 | 1.0 | 2 | $10^{10}$ |
| Repetition rate | 50 | | $\leq$120 | Hz |
| Invariant emittances H/V | $\sim$3,000/3,000 | | 3/0.3 | $\mu$m |
| $e^+/e^-$ switching time | 300 | 0.02 | 0 (simultaneous) | s |

to satisfy the requirements of the injection to KEKB as shown in Table 2. In early days, it was thought that the performances of the two machines would be eventually limited by the performance of their injector. Actually such a situation did not happen. The key was the top-up operation applied to both machines since 2004. Then the necessary strength of the injected beam became much smaller than the maximum performance even at KEKB [44]. Both machines were the first to have utilized the top-up operation for high-energy colliders, even earlier than most of the light sources. The 2-bunch/per pulse acceleration and installing a C-band section in the linac [45] also contributed to make the gap between PEP–II and KEKB smaller. KEKB has solved the conflict with the injection to the light sources by introducing a pulse-to-pulse switching of the linac.

## 11 Crab crossing

KEKB operated with crab crossing from 2007 through 2010 using superconducting crab cavities [46] and one installed cavity per ring [47]. KEKB had already achieved a luminosity of $1.76 \times 10^{34}$ cm$^{-2}$s$^{-1}$ by then with crossing angle, so the intension of the crab crossing was not simply to further increase the performance. Simulations of beam–beam effect indicated that a head-on or crab collision could increase the maximum acceptable beam–beam parameter $\xi_y$ to values even higher than 0.15 when combining with a horizontal betatron tune close to a half integer [48]. Thus the hope was to experimentally verify the possibility of operating the machine with such a high beam–beam parameter, considering super B-factories.

The crab cavities were successfully installed and operated for more than 3 years. The luminosity was actually increased as shown in Table 1. The resulted beam–beam parameter was 0.09, which was indeed higher than the value with crossing angle, but much less than the value obtained in simulation (0.15). It was not easy to single out the cause, but there were indications that remaining higher order terms of the beam optics at the IP degraded the luminosity and $\xi_y$. One example was the chromatic $x$–$y$ coupling term at the IP [49]. By installing skew sextupoles in the arc as tuning knobs, the luminosity was improved by up to 10% [6]. Then the speculation was that higher order terms at the IP could still limit the performance. It was not easy to estimate how many terms were relevant and how to correct them, as there were almost no direct beam diagnostics at the IP except the luminosity. What was verified at KEKB was that the crab crossing itself should work for any collider up to $\xi_{x,y} \lesssim 0.1$, including the Large Hadron Collider (LHC).

## References

[1] P. Oddone, *Proceedings of the UCLA Workshop: Linear Collider B-B Factory Conceptual Design*, ed. D. Stork (1987), p. 243.
[2] K. Hirata and E. Keil, *Nucl. Instr. Meth. A* **292** (1990) 156.
[3] *PEP-II: An Asymmetric B Factory. Conceptual Design Report*, SLAC-418, QCD183:S56:1993 (1993).
[4] *KEKB B–Factory Design Report*, KEK Report 95–7 (1995).
[5] J. Seeman, *Conf. Proc. C* **0806233** (2008) TUXG01.
[6] Y. Funakoshi, T. Abe, K. Akai, Y. Cai, K. Ebihara, K. Egawa, A. Enomoto, J. Flanagan, *et al.*, *Conf. Proc. C* **100523** (2010) WEOAMH02.
[7] A. Piwinski, *IEEE Trans. Nucl. Sci.* **24** (1977) 1408.
[8] R. B. Palmer, *Proc. of 1988 DPF Summer Study on High-Energy Physics in the 1990s* (Snowmass 88), Snowmass, Colorado, 27 Jun–15 Jul 1988 (1988), p. 613.
[9] K. Oide and K. Yokoya, *Phys. Rev. A* **40** (1989) 315.
[10] M. Sullivan, G. Bowden, H. DeStaebler, S. Ecklund, J. Hodgson, T. Mattison, M. E. Nordby, A. Ringwall, *et al.*, *Conf. Proc. C* **960610** (1996) 460.
[11] B. Aubert, *et al.* [BABAR Collaboration], *Nucl. Instrum. Meth. A* **479** (2002) 1.
[12] J. Seeman, M. Sullivan, M. Biagini, Y. Cai, F. J. Decker, M. Donald, S. Ecklund, A. Fisher *et al.*, *Proc. EPAC 2002*, 3–7 Jun 2002, Paris, France, pp. 434–436.
[13] Y. Yamazaki and T. Kageyama, *Part. Accel.* **44** (1994) 107.
[14] T. Furuya, *et al.*, *Proc. 7th Workshop of RF Superconductivity*, Gif-sur-Yvette, France, 17–20 Oct 1995, p. 729.
[15] H. Schwarz and R. Rimmer, *Conf. Proc. C* **940627** (1994) 1882.
[16] J. Fox, T. Mastorides, C. Rivetta, D. Van Winkle and D. Teytelman, *Phys. Rev. ST Accel. Beams* **13** (2010) 052802.
[17] Y. Suetsugu, K. Shibata, H. Hisamatsu, M. Shirai and K. Kanazawa, *Vacuum* **84** (2009) 694.
[18] M. T. F. Pivi, F. King, R. E. Kirby, TMarkiewicz, T. O. Raubenheimer, J. Seeman and L. Wang, *Conf. Proc. C* **0806233** (2008) MOPP064.
[19] Y. Suetsugu, H. Fukuma, L. Wang, M. Pivi, A. Morishige, Y. Suzuki, M. Tsukamoto and M. Tsuchiya, *Nucl. Instrum. Meth. A* **598** (2009) 372.
[20] T. Mimashi, T. Ieiri, M. Kikuchi, A. Tokuchi and K. Tsuchida, *Conf. Proc. C* **0806233** (2008) TUPD011.
[21] M. Izawa, Y. Sato and T. Toyomasu, *Phys. Rev. Lett.* **74** (1995) 5044.
[22] K. Ohmi, *Phys. Rev. Lett.* **75** (1995) 1526.
[23] K. Ohmi and F. Zimmermann, *Phys. Rev. Lett.* **85** (2000) 3821.
[24] T. O. Raubenheimer and F. Zimmermann (SLAC), *Phys. Rev. E* **52** (1995) 5487.

[25] J. W. Flanagan, K. Ohmi, H. Fukuma, S. Hiramatsu, M. Tobiyama and E. Perevedentsev, *Phys. Rev. Lett.* **94** (2005) 054801.
[26] H. Fukuma, J. Flanagan, K. Hosoyama, T. Ieiri, T. Kawamoto, T. Kubo, M. Suetake, S. Uno, *et al.*, *AIP Conf. Proc.* **642** (2003) 357.
[27] H. Jin, M. Yoon, K. Ohmi, J. W. Flanagan and M. A. Palmer, *Jap. J. Appl. Phys.* **50** (2011) 026401.
[28] At least an application for a storage ring is seen in R. Servranckx and K. L. Brown, *IEEE Trans. Nucl. Sci.* **26** (1979) 3598.
[29] K. Oide, H. Koiso and K. Ohmi, *AIP Conf. Proc.* **391** (1997) 215.
[30] K. Oide and H. Koiso, *Phys. Rev. E* **47** (1993) 2010.
[31] J. Irwin, C. X. Wang, Y. T. Yan, K. L. F. Bane, Y. Cai, F. J. Decker, M. G. Minty, G. V. Stupakov, *et al.*, *Phys. Rev. Lett.* **82** (1999) 1684.
[32] K. Akai, N. Akasaka, A. Enomoto, J. Flanagan, H. Fukuma, Y. Funakoshi, K. Furukawa, T. Furuya, *et al.*, *Nucl. Instrum. Meth. A* **499** (2003) 191.
[33] Y. T. Yan and Y. Cai, *Nucl. Instrum. Meth. A* **558** (2006) 336.
[34] T. Ieiri, K. Akai, H. Fukuma and M. Tobiyama, *Nucl. Instrum. Meth. A* **606** (2009) 248.
[35] M. Tejima, M. Arinaga, T. Ieiri, H. Ishii, H. Fukuma, M. Tobiyama and S. Hiramatsu, *Conf. Proc. C* **0505161** (2005) 3253.
[36] K. Satoh and M. Tejima, *Conf. Proc. C* **950501** (1995) 2482.
[37] T. Mitsuhashi, J. W. Flanagan and S. Hiramatsu, *Proc. Seventh EPAC*, 26–30 Jun 2000, Vienna, Austria, p. 1783–1785.
[38] J. W. Flanagan, N. Akasaka, H. Fukuma, S. Hiramatsu, T. Mitsuhashi, T. Naito, K. Ohmi, K. Oide *et al.*, *Proc. Seventh EPAC*, 26–30 Jun 2000, Vienna, Austria, p. 1119–1121.
[39] N. Akasaka, A. Akiyama, S. Araki, K. Furukawa, T. Katoh, T. Kawamoto, I. Komada, K. Kudo, *et al.*, *Nucl. Instrum. Meth. A* **499** (2003) 138.
[40] Y. Funakoshi, M. Masuzawa, K. Oide, J. Flanagan, M. Tawada, T. Ieiri, M. Tejima, M. Tobiyama *et al.*, *Phys. Rev. ST Accel. Beams* **10** (2007) 101001.
[41] L. Hendrickson, T. Gromme, P. Grossberg, T. Himel, D. Macnair, R. Sass, H. Smith, N. Spencer *et al.*, *Proc. Seventh EPAC*, 26–30 Jun 2000, Vienna, Austria, p. 1897–1899.
[42] K. Ohmi, K. Oide and E. Perevedentsev, *Conf. Proc. C* **060626** (2006) 616.
[43] A. Drago, J. D. Fox, D. Teytelman and M. Tobiyama, *Conf. Proc. C* **0806233** (2008) THPC116.
[44] Y. Ogawa, A. Enomoto, K. Furukawa, T. Kamitani, M. Satoh, T. Sugimura, T. Suwada, Y. Yano *et al.*, *Conf. Proc. C* **060626** (2006) 2700.
[45] T. Kamitani, N. Delerue, M. Ikeda, K. Kakihara, S. Ohsawa, T. Oogoe, T. Sugimura, T. Takatomi *et al.*, *Conf. Proc. C* **0505161** (2005) 1233.
[46] K. Akai, J. Kirchgessner, D. Moffat, H. Padamsee, J. Sears, T. Stowe and M. Tigner, *Proc. B Factory Workshop*, 6–10 Apr 1992, Stanford, California.
[47] K. Hosoyama, K. Akai, K. Ebihara, T. Furuya, K. Hara, T. Honma, A. Kabe, Y. Kojima *et al.*, *Conf. Proc. C* **0806233** (2008) THXM02.
[48] K. Ohmi, M. Tawada, Y. Cai, S. Kamada, K. Oide and J. Qiang, *Phys. Rev. ST Accel. Beams* **7** (2004) 104401.
[49] Y. Ohnishi, K. Ohmi, H. Koiso, M. Masuzawa, A. Morita, K. Mori, K. Oide, Y. Seimiya, *et al.*, *Phys. Rev. ST Accel. Beams* **12** (2009) 091002.

# Part 2

# 100 Years of Accelerators

## Chapter 7

# The Cosmotron and the Bevatron: The first GeV accelerators

Thomas Roser and Ernest Courant (BNL)

## 1 The development of the proton synchrotron

The idea of pulsing the magnetic field to keep the circular orbit of the accelerating particles unchanged was first proposed by M.O. Oliphant [Oliphant (1950)] in 1943. Oliphant recognized that maintaining stable orbits with changing magnetic holding field required longitudinal focusing or phase stability, as described, for example, in [Wilson (1996)]. The principle of phase stability was soon after independently discovered by V.I. Veksler [Veksler (1944)] and E.M. McMillan [McMillan (1945)] and was first applied to the synchro-cyclotron and the electron ring synchrotron. But for higher energies the cyclotron would become too massive since it requires that the entire interior of the top energy orbit be filled with magnetic field. And, at the time, the energy of an electron synchrotron was limited to about 300 MeV by synchrotron radiation. This has since been increased substantially by using large rings and massive RF power. However, the best way of getting to the highest conceivable particle energies was then, and still is, the use of the much heavier protons. In a proton synchrotron the radio frequency has to be modulated with high precision so as to track the magnetic field and at the same time keep the orbit at a constant radius, but this complication is outweighed by the lack of synchrotron radiation.

Several proposals for proton synchrotrons appeared at about the same time, first for a 1 GeV machine at Birmingham, England, [Oliphant *et al.* (1947)] in 1947 and for a 10 GeV machine at Berkeley [Brobeck (1947)]. Discussions between Leland Haworth, Director of the newly formed Brookhaven National Laboratory, Ernest Lawrence, Director of the Radiation Laboratory (now the Lawrence Berkeley National Laboratory), and the Atomic Energy Commission (now the U.S. Department of Energy) led to the decision that both Brookhaven and Berkeley, instead of competing for the 10 GeV prize,

Fig. 1. The Brookhaven Cosmotron with the injector van de Graaff in the foreground (courtesy of Brookhaven National Laboratory).

would each build a smaller proton synchrotron, one around 3 GeV and one at 6 GeV. Haworth chose the smaller size with the hope of getting it finished faster, but giving up on the possible discovery of the antiproton. In later years Haworth often said that this was the best decision he had ever made.

## 2 The first proton synchrotrons

All the first proton synchrotrons as well as the fixed field cyclotrons were "weak focusing", meaning that the magnetic bending field also provides a mainly constant focusing force on the beam in both horizontal and vertical planes [Blewett (1956); Green and Courant (1959)]. This is typically expressed in terms of the guide field index

$$n = -\frac{\rho}{B_0}\left(\frac{\partial B_y}{\partial r}\right)_{r=\rho} \tag{1}$$

Table 1. Main parameters of the first four weak focusing proton synchrotrons.

| | Birmingham | Cosmotron | Bevatron | Synchrophasotron |
|---|---|---|---|---|
| Peak energy [GeV] | 1 | 3 | 6.2 | 10 |
| Injection energy [MeV] | 0.46 | 3.6 | 9.9 | 9.0 |
| Circumference [m] | 28.27 | 69.63 | 120.16 | 208 |
| Number of straight sections | 0 | 4 | 4 | 4 |
| Bending radius [m] | 4.50 | 9.14 | 15.24 | 28 |
| Field index | 0.67 | 0.60 | 0.60 | 0.65 |
| Magnetic field range [T] | 0.02–1.26 | 0.03–1.38 | 0.03–1.54 | 0.02–1.30 |
| RF frequency range [MHz] | 0.30–9.70 | 0.36–4.18 | 0.36–2.47 | 0.18–1.44 |
| Harmonic number | 1 | 1 | 1 | 1 |
| Rise time [s] | 1 | 1 | 2 | 3.3 |
| Cycle time [s] | 10 | 5 | 6 | 12 |
| Energy gain per turn [keV] | 0.2 | 1.0 | 1.5 | 2.5 |
| Number of RF stations | 1 | 1 | 1 | 2 |
| Peak excitation current [kA] | 12.5 | 7.0 | 8.3 | 12.8 |
| Peak stored energy [MJ] | 7 | 12 | 80 | 148 |
| Magnet cross section (H×V) [m] | $2.44 \times 2.44$ | $2.38 \times 2.38$ | $6.25 \times 2.90$ | $7.5 \times 5.3$ |
| Magnet gap (H × V) [m] | $0.50 \times 0.21$ | $0.92 \times 0.24$ | $1.68 \times 0.33$ | $2.0 \times 0.4$ |
| Weight of magnet steel [tons] | 800 | 1650 | 9700 | 36000 |
| Date of completion | 1953 | 1952 | 1954 | 1957 |

where $n$ needs to satisfy $0 < n < 1$ to ensure beam stability. The betatron tunes are then equal to $\sqrt{1-n}$ and $\sqrt{n}$ for the horizontal and vertical tunes, respectively. All of the first four machines, the Brookhaven Cosmotron, the Berkeley Bevatron, the Birmingham Synchrotron and the Dubna Synchrophasotron, were all very similar in their basic design and mainly differed in their size and therefore maximum energy that could be reached. The guide field index $n$ of their pulsed magnets was about 0.6. Table 1 shows the main parameters of these four machines.

The construction of the first proton synchrotrons required a very careful shaping of the magnetic field. The weak focusing made the orbit stability a significant concern especially since these would be the first machines where particles circulate for several seconds and travel for several hundred thousand kilometers during the acceleration cycle. Even small orbit disturbances could accumulate and lead to particle loss. To evaluate the necessary size of the required magnet aperture the Berkeley team decided to construct a scale model with one quarter the linear dimension of the final Bevatron [Lofgren (1950)]. The prototype machine successfully accelerated protons, injected at 0.7 MeV, to 6.5 MeV with just 0.3% efficiency but demonstrated a functioning injection process, a sufficiently low residual gas pressure and, most importantly, that a magnet gap of 0.3 by 1.7 m is acceptable for the Bevatron.

All of the first proton synchrotrons ran with a harmonic number of one and the weak longitudinal focusing led to large radial synchrotron oscillations at the low injection energy adding to the need for a large horizontal aperture. The large aperture also complicated the injection process and made it quite inefficient. All machines injected the beam using horizontal betatron painting and then moved the beam away from the inflector by ramping the magnetic field.

The weak focusing proton synchrotrons had successfully overcome the limitations of the cyclotrons and reached multi-GeV particle energies, which led to many new discoveries, most prominently the discovery of the antiproton at the Berkeley Bevatron. However, already when the first of them, the Brookhaven Cosmotron, started operation, it became clear that they have their own limitation of ever larger and more massive magnets as the beam energy increased. A new concept, the alternating gradient focusing, was needed that allowed for more compact magnets to reach even higher beam energies. The new concept was independently invented by Nicholas Christofilos in 1949 but without publishing it, instead he decided to issue a patent on the new invention [Christofilos (1949)], and Ernest Courant, Milton S. Livingston, Hartland Snyder and J. Blewett [Courant *et al.* (1952); Courant *et al.* (1953)].

## References


Blewett, J.P., The proton synchrotron, *Rep. on Progress in Physics* **19** (1956) 37.

Brobeck, W.M., Design study for a 10-BeV magnetic accelerator, *Rev. Sci. Inst.* **19** (1948) 545.

Christofilos, N. C. (1949). Focusing System for Ions and Electrons. US Patent No. 2,736,799.

Courant, E.D., Livingston, M.S., Snyder, H.S., The strong-focusing synchrotron, *Phys. Rev.* **88** (1952) 1190.

Courant, E.D., Livingston, M.S., Snyder, H.S., Blewett J., Origin of the strong-focussing principle, *Phys. Rev.* **91** (1953) 202–203.

Green, G.K., Courant, E.D., The proton synchrotron, *Encyclopedia of Physics*, edited by Creutz, E., published by Springer Berlin Heidelberg, Volume 8/44 (1959) 218.

Lofgren, E.J., The proton synchrotron, *Science* **111** (1950) 295.

McMillan, E.M., The synchrotron — a proposed high energy accelerator, *Phys. Rev.* **68** (1945) 143.

Oliphant, M.O., The acceleration of particles to very high energies, unpublished manuscript (1943) University of Birmingham, U.K.

Oliphant, M.O., Gooden, J.S., Hide, G.S. The acceleration of charged particles to very high energies, *Proc. Phys. Soc.* **59** (1947) 666.

Veksler, V.I., A new methode of accelerating relativistic particles, *Comptes Rendus de l'Academie Sciences de L'URSS*, **43,8** (1944) 329.

Wilson, E.J.N., Fifty years of synchrotrons, EPAC96.

Chapter 8

# The PS and AGS: The first strong focusing proton synchrotrons

Kevin Brown (BNL), Massimo Giovannozzi (CERN) and Thomas Roser (BNL)

## 1 The origin of alternating-gradient accelerators

A new method of magnetic focusing for accelerators, called alternating-gradient (AG) or strong focusing, started in 1952 leading to a series of accelerators capable of much higher energies than was economically practical with earlier techniques as described in [Livingston and Blewett (1962)]. At the CERN laboratory in Geneva the first of these large accelerators was brought into operation at 28 GeV in late 1959. At Brookhaven National Laboratory (BNL) a machine of similar dimensions became operational in late 1960, with a proton energy of 33 GeV.

The principle of AG focusing originated at BNL in the summer of 1952, at a time when the Cosmotron was nearing completion [Roser and Courant (2015)]. A report of the early concepts was published, describing the possible application to a high-energy proton synchrotron and also discussing the use of magnetic quadrupole lenses in focusing linear beams of particles [Courant *et al.* (1952)].

The completion in 1952 of the Cosmotron, the first multi-GeV accelerator, had attracted to Brookhaven several scientists who were engaged in developing experimental apparatus for research studies. It was the case of a delegation of European scientists, representing the newly established CERN laboratory, with the goal of assessing the Cosmotron as a model for a 10-GeV accelerator. It was known that magnetic saturation effects limited the useful aperture of the C-shaped magnets of the Cosmotron at high fields. A possible technique would have been to retain the C-shape and also expand the useful aperture by alternating the magnets' return yoke locations from inside to outside the orbit. This would have resulted, at high fields, in a corresponding alternation in magnetic gradients from positive to negative in the successive magnets as a result of saturation. The possibility that this alternation in gradients would destroy orbital stability was considered by E. D. Courant who showed that stability was indeed improved rather than damaged. At this time H. S. Snyder joined in the design efforts and helped to develop the general principle of dynamic stability. Stability limits were identified, leading to suitable configurations of positive and negative magnet sectors and field-free straight sections between sectors around the orbit. Mechanical configurations were conceived producing the desired large magnetic gradients. The basic concept was sound, and the use of alternating gradients would allow major reductions in the transverse dimensions, as can be seen in Fig. 1, and the power requirements for magnets. Such a reduction in cost of synchrotron magnets for a given orbit radius would make it possible to design machines with much larger orbits and much higher energies.

By the time the European delegation, O. Dahl, F. Goward, and R. Wideröe, arrived at Brookhaven, the concept had been developed sufficiently to be presented as a significant improvement over the Cosmotron design. The new concept was very well received, but the CERN group was faced with the very difficult decision of going ahead on proven ground or proceeding with the new and untried idea. The CERN Council approved the latter option at its third session in October 1952 in Amsterdam.[1] The benefits of the new approach were perceived to outweigh the risks, but a vigorous plan of studies was launched to probe the critical features of the new invention. *A posteriori*, this courageous decision has been crucial to the future of CERN.

A number of studies were made in European laboratories to determine whether the orbit stability would be threatened (see, e.g., [Adams *et al.* (1953)]). Further work showed that the harmful resonances could be avoided by care in design and by suitable control of the magnetic gradients during the excitation cycle. Another difficult point was reaching and passing the so-called transition energy, but progress in the understanding of the complex physics allowed to gain confidence in the soundness of the new principle, which grew in all laboratories participating in the scientific efforts. The first relatively complete analysis of a practical design for an AG accelerator was for the 15-GeV ring at Massachusetts Institute of Technology (MIT) [MIT (1953)].

It is worth mentioning that the AG concept was developed independently elsewhere. N. C. Christofilos, an electrical engineer of American birth, educated and working in Athens, had been developing several new ideas on accelerator design in the form of private reports and patent applications. His unpublished report [Christofilos (1950)] presented the concept of AG

[1] At the same meeting, Geneva had been chosen as the location of the future laboratory.

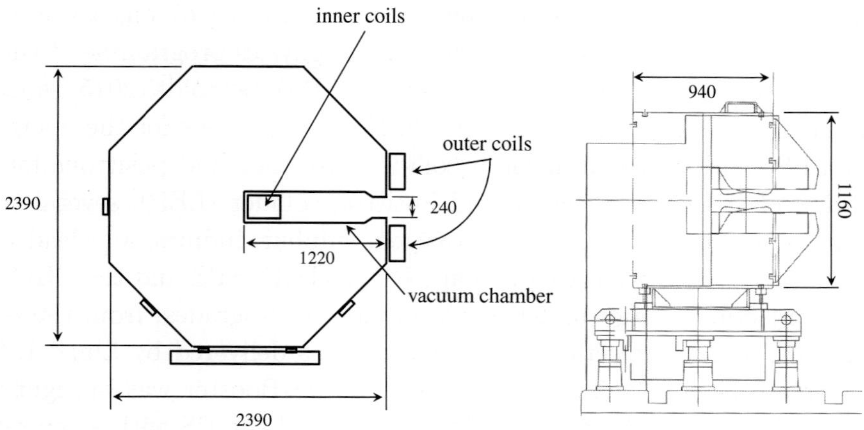


Fig. 1. Comparison between the cross section of the Cosmotron magnet (left, from [Livingston and Blewett (1962)]) and that of the PS main magnet (right). The difference in size is a consequence of the AG principle. The dimensions are in millimetres.

focusing and the possible design of an accelerator using this principle, for which he also applied for United States and European patents. After the Brookhaven publication in 1952, Christofilos came to the United States and claimed his priority, which was recognised in a brief communication [Courant *et al.* (1953)]. He was then hired by BNL and worked on a small-scale model to test experimentally whether the beam would pass transition energy. The so-called electron analogue had been built in 1954 and remained in operation until 1957. This model had a circumference of 43.1 m, accelerating electrons from 1 to 10 MeV with transition energy at 3.5 MeV. In order to reduce cost, the alternating-gradient, strong focusing was provided by electrostatic lenses and bending by electrostatic fields [Plotkin (1991)]. The test showed that transition energy could be crossed without problems, but the price was a delay in the construction of the BNL ring relative to the CERN one, which was in the end compensated by better preparation of the experimental programme as compared to the CERN machine.

## 2 The CERN Proton Synchrotron[2]

The final parameters of the CERN Proton Synchrotron (PS) were adopted on December 1954 [Regenstreif (1959)] and the completion date was set for the end of 1959. By November 1959 transition energy was successfully crossed and the circulating current already exceeded the design value. The flexibility

[2]Recently, a complete report on the PS machine has been published to celebrate its fiftieth anniversary [Gilardoni, Manglunki *et al.* (2011)].

of the AG design has been proven by the different types of ions accelerated, well beyond the original plans: protons for the physics programme at the PS or transfer to the Super Proton Synchrotron (SPS) [Schmidt (2015)] and the Large Hadron Collider (LHC) [Myers (2015)]; antiprotons for the operation of the SPS as a proton–antiproton collider; electrons and positrons for the operation of the Large Electron and Positron collider (LEP); several types of ions (deuterons, alpha particles, oxygen, sulphur, indium, and lead ions) for transfer to the Intersecting Storage Rings (ISR), SPS and the LHC.

The injection energy underwent a number of upgrades, from the original 50 MeV kinetic injection energy for protons delivered by Linac 1. The first increase to 800 MeV occurred when the PS-Booster was brought into operation in 1973, and since then it is used to fill the PS with a repetition period of 1.2 s, corresponding to about 0.8 Hz; then to 1 GeV since 1987, with the aim of better serving the antiproton production chain [Chohan and Maury (2015)]; finally 1.4 GeV came into operation in 1999 in the framework of the upgrade foreseen to generate the high-brightness beams for the LHC. A further increase to 2 GeV is planned, as proposed in [Giovannozzi *et al.* (2010)] to fully profit from the performance reach of the planned Linac 4. In any upgrade the increase of the PS injection energy was needed to mitigate space charge effects, which have become ever stronger due to the continuous increase of beam intensity.

The overall intensity evolution in the PS is visible in Fig. 2, which clearly shows how in recent years more emphasis is put on the beam brightness, as required by the LHC, than on peak intensity.

The PS lattice is made of combined function magnets, each 4.4 m long, divided into focusing and de-focusing half-units. Each unit consists of five blocks, assembled in a curved structure. The hundred main magnets are separated by a hundred straight sections, twenty of which are longer than the others and house special equipment for, e.g., injection, extraction, and radio-frequency (RF) manipulations. Four types of main magnet are present in the PS ring, depending on the focusing configuration and on whether the magnets' return yoke is oriented towards the inside or the outside of the ring. In total 35 magnets feature a D-F configuration and yoke outside, 15 a F-D configuration with yoke outside, 35 a F-D configuration, but with yoke inside, and finally 15 a D-F configuration with yoke inside. The lattice has ten super-periods and the optical functions (horizontal and vertical beta-functions and dispersion) for one super-period are shown in Fig. 3.

At low energy the tunes are controlled by means of two families of quadrupoles, the chromaticities being left at their natural values. At higher energies the control of tunes and chromaticities is achieved by means

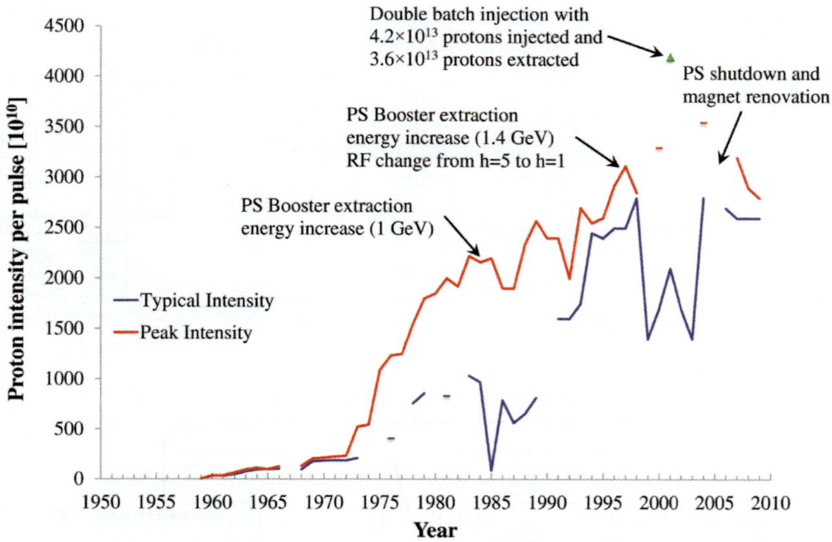


Fig. 2. Intensity evolution over the years in the PS. The marker indicates the intensity record achieved in 2001.

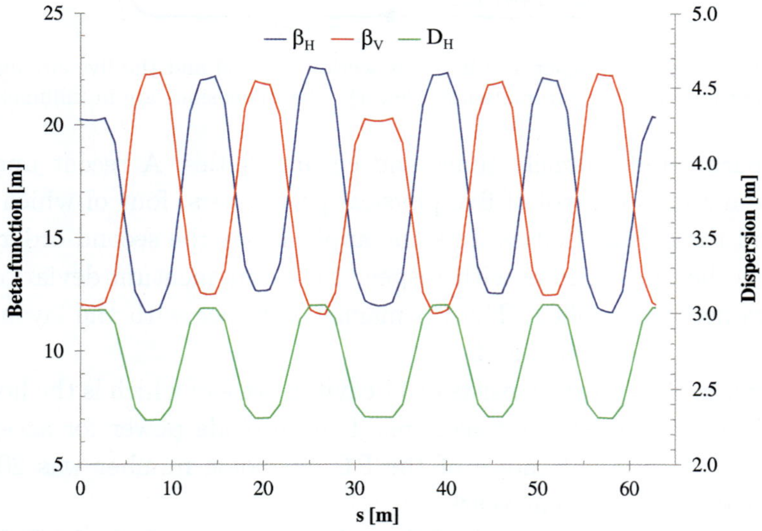


Fig. 3. Optical parameters for one super-period of the PS bare lattice. Each super-period contains two long straight sections.

of special circuits that are installed on the magnets' poles and provide a transverse variation of the magnetic gradient so as to create additional quadrupolar and sextupolar components. These circuits are the so-called pole-face windings (PFW) and figure-of-eight loop. Originally, three independent circuits were used to control three out of the four main global

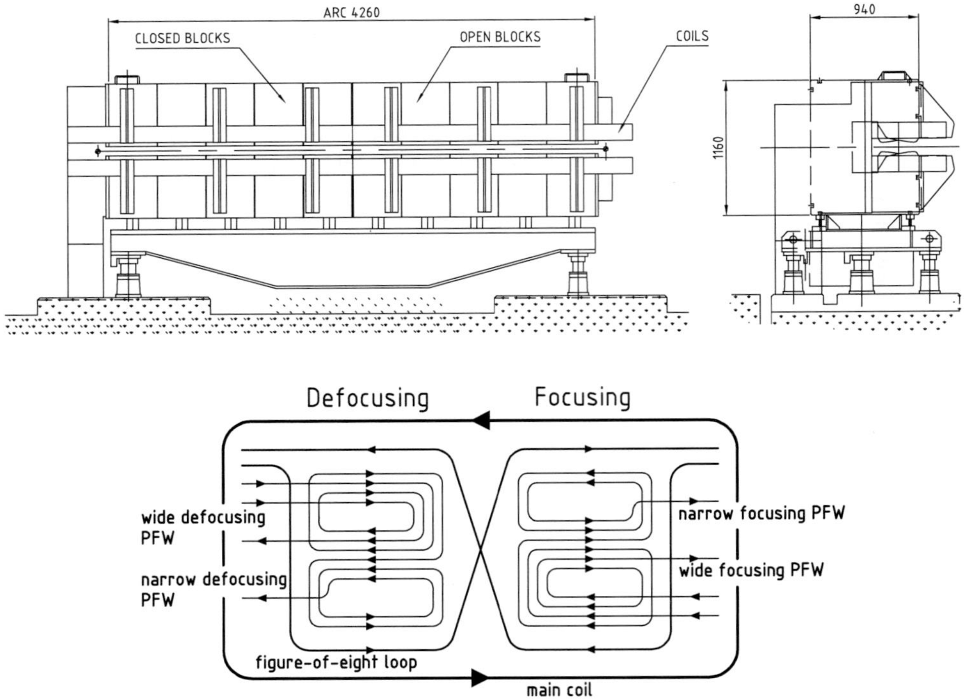


Fig. 4. The PS main magnet and its cross section (upper) and the five circuits used to control the tunes and the chromaticities (lower). The dimensions are in millimetres.

optical parameters, namely tunes and chromaticities. A recent upgrade of the system allows control of five physical parameters, four of which are the tunes and the chromaticities. In some applications the second order derivative of the horizontal tune with respect to the momentum deviation is the fifth parameter controlled. The PS magnet together with the layout of the special coils is visible in Fig. 4.

The main RF system consists of 11 cavities (one of which is the hot spare) installed in the long straight sections. They provide power for accelerating the beam. The original choice of the RF harmonic number was 20, but it was changed to 8 in recent years.

The transition energy in the PS machine corresponds to 6.1 GeV and it is crossed by applying the needed RF phase jump and a special manipulation of the beam optics, the so-called $\gamma$-jump [Hardt and Möhl (1969)], in order to faster cross transition and hence avoid or mitigate effects that might spoil the beam quality.

The evolution of the PS performance and its flexibility is reflected also in the variety of RF and extraction systems. A large number of longitudinal beam manipulations were performed [Garaby (2015)] and some are

Table 1. Basic parameters of the PS machine.

| | |
|---|---|
| Accelerated particles | $p^{\pm}$, $e^{\pm}$, and several types of ions |
| Maximum particle energy [GeV] | 26 |
| Circumference [m] | 200 $\pi$ |
| Magnetic lattice | Alternating-gradient focusing, combined-function |
| Focusing order | FOFDOD |
| Magnetic field index | $n = 282$ |
| Number of main magnets | 100 |
| Bending magnetic field [T] | 0.101 at injection (1.4 GeV), 1.24 at 26 GeV |
| Betatron oscillations/turn | 6.25 (h), 6.25 (v) |
| Transition energy [GeV] | 6.1 |
| Magnetic cycle repetition [s] | 1.2 (up to 20 GeV), 2.4 (up to 26 GeV) |
| Straight sections | number = 100, 80 of 1.6 m, 20 of 3 m |
| RF system I (tunable) | 10+1 cavities, 2.6 to 9.5 MHz, 200 kV total maximum |
| Auxiliary RF systems [MHz] | 13.3, 20, 40, 80, 200 |
| Vacuum chamber [mm$^2$] | Inconel, 150 × 80 in the bending magnets |

still routinely used to produce the various beams. Bunch merging and non-adiabatic manipulations, such as bunch rotation, are certainly well-known techniques used in the PS. However, to fulfil the requirements set by the LHC, a longitudinal bunch splitting technique had been developed [Garoby (1998)]. Each bunch injected in the PS is divided into twelve shorter and less intense bunches, but with the longitudinal emittance required for the bunch-to-bucket transfer to the SPS. This is obtained by means of a triple and two double splittings, performed by a number of ancillary RF systems working at different frequencies.

Beam extraction is also performed in several different ways, ranging from fast, single-turn, to slow, and also multi-turn extraction. The last one is performed either by slicing the beam on an electrostatic septum and transferring it to the SPS as a continuous ribbon five PS-turn long [Bovet *et al.* (1973)], or via a new technique based on transverse splitting obtained by an adiabatic crossing of a non-linear, stable betatron resonance [Cappi, Giovannozzi (2002); Gilardoni *et al.* (2006)].

The key parameters of the PS are listed in Table 1.

## 3 The BNL Alternating-Gradient Synchrotron[3]

The construction of the BNL AG (AGS) was approved in 1954 after a decision process of only four months. The construction was led by G. K. Green

[3]Recently, a complete account on the AGS has been written [Hübner (in press)].

and J. P. Blewett and beam commissioning was completed with the first protons being accelerated to 31 GeV in July 1960 [Green (1961)].

The first injector was a 50-MeV Alvarez-type proton linear accelerator. The present 200-MeV linear accelerator began operation in 1970. In 1982 $H^-$ charge exchange injection into the AGS was introduced [Barton (1983)] and in 1991 a 1.5-GeV booster synchrotron was commissioned [Ahrens (1993)]. The Booster can provide $1.5 \times 10^{13}$ protons per pulse at 1.9 GeV at the design repetition frequency of 7.5 Hz. The acceleration harmonic schemes (Booster harmonic, AGS harmonic, transfers) evolved from (3, 12, 4) to (2, 8, 4) and finally to (1, 6, 6) in pursuit of higher intensity [Brennan (1999)], thus showing the use of bunch merging in the AGS in past years prior to beam transfer.

Secondary beams from the AGS were initially provided from internal targets, which created high beam loss and activation in the accelerator not compatible with high-intensity operation. The first fast extraction was installed in the mid 60s, followed by slow extraction in 1967 which served up to six target stations and spilled out protons with repetition periods from 1.8 s to 5.8 s. To cope with the intensity increases, the AGS underwent a series of upgrades including a new main magnet power supply, addition of transverse feed-back, special magnets to provide fast crossing of the transition energy, corresponding to 8.4 GeV, and a high power RF system [Brennan (1995)]. In the early 2000s the AGS provided a slow-extracted beam of $7 \times 10^{13}$ protons per pulse at 24 GeV [Brown *et al.* (2003)], however, such a high-intensity operation was stopped at the end of 2002. The overall intensity evolution in the AGS is visible in Fig. 5. To achieve such a performance required mastering the losses induced by intensity-dependent effects. Different methods have been used, e.g., bunch flattening to mitigate space charge effects and octupoles to reduce the slow losses during the injection porch.

With the appropriate source added to the 200-MeV linear accelerator, polarised protons have been produced by the injector chains from 1985 onward for fixed target experiments. To meet injector requirements for the Relativistic Heavy Ion Collider (RHIC) [Fischer (2015)] — intensity and polarisation — the polarised source underwent a major upgrade [Zelenski *et al.* (2008)]. The polarisation transmission efficiency in the AGS has been substantially improved with the installation of two partial Siberian snakes [Derbenev and Kondratenko (1976, 1978); Roser (1988)] and a system to rapidly cross weak resonances [Schoefer *et al.* (2012)]. Polarisation at transfer to RHIC (24 GeV) is 70% (82% at 200 MeV) and with an intensity of $2 \times 10^{11}$ protons per bunch [Huang *et al.* (2009); Schoefer *et al.* (2011)].

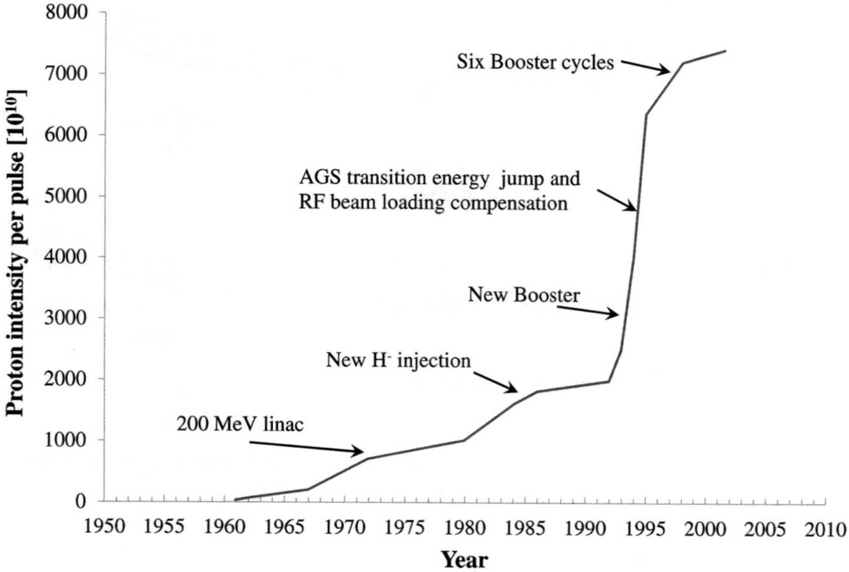


Fig. 5. Intensity evolution over the years in the AGS.

Since 1986 the Booster has also accelerated ions (deuterons to gold) using a Tandem Van de Graaff as the injector. For RHIC operation about $5 \times 10^9$ $Au^{31+}$ ions at 41.6 MeV/$n$ are injected over 60 turns into the Booster, accelerated to 101 MeV/$n$, stripped to $Au^{77+}$ and injected into the AGS. The ions are fully stripped before injection into RHIC [Gardner *et al.* (2007)]. A new pre-injector is being commissioned based on an EBIS source followed by a new linear accelerator [Pikin *et al.* (2010)]. The new system adds uranium [Alessi *et al.* (2011)] to the list of ions available to RHIC.

The AGS lattice features some similarities with the PS, but there are also some significant differences that are visible in Fig. 6.

The main magnets are not made of blocks and three variants are available, namely long open (48), long closed (96), short open (96), where the lengths are 2.29 m and 1.91 m for long and short magnets, respectively. The open magnets have a gap that flares away from the return yoke, while the gap of the closed ones flares towards the return yoke. A sketch of the AGS main magnet is shown in Fig. 7.

The total number of combinations is then six, if the two options for the orientation of the return yoke (inside or outside the ring) are also taken into account. The lattice is made of twelve super-periods, each consisting of 20 main magnets. The two groups of ten combined-function magnets feature an opposite orientation of the return yoke. The straight sections in between main magnets are of three possible lengths, namely 0.61 m (12 in

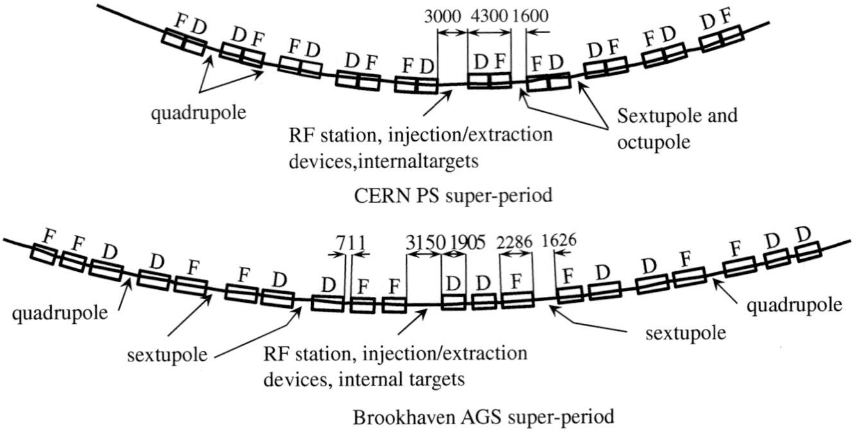


Fig. 6. Comparison between the PS (upper) and the AGS (lower) super-period from [Green and Courant (1959)]. The dimensions are in millimetres.

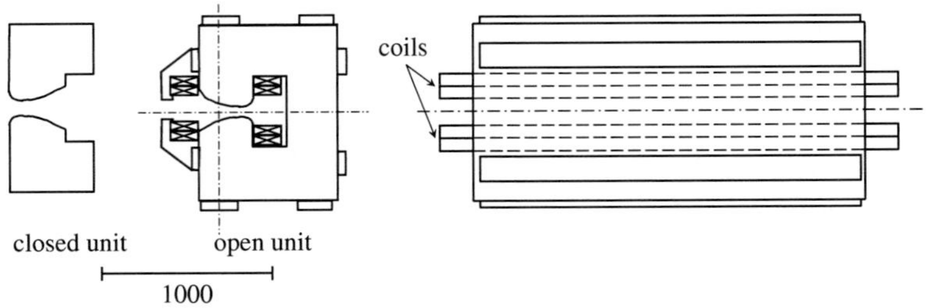


Fig. 7. The AGS long main magnet and its cross section from [Green and Courant (1959)]. The dimensions are in millimetres.

each super-period), 1.52 m (six in each super-period), and 3.05 m (two in each super-period located at the transition between different magnets' yoke orientation). Unlike the PS, where the tunes and chromaticities are controlled by means of pole-face windings, in the AGS tuning quadrupoles and chromaticity sextupoles are located in each super-period, thus enabling more convenient operation of the machine. The optical parameters (horizontal and vertical beta-functions and the horizontal dispersion) are plotted in Fig. 8 for one of the twelve super-periods.

While the minimum and maximum values of the $\beta$-functions resemble closely those of the PS, the dispersion values are clearly smaller for the AGS.

The key parameters of the AGS are listed in Table 2.

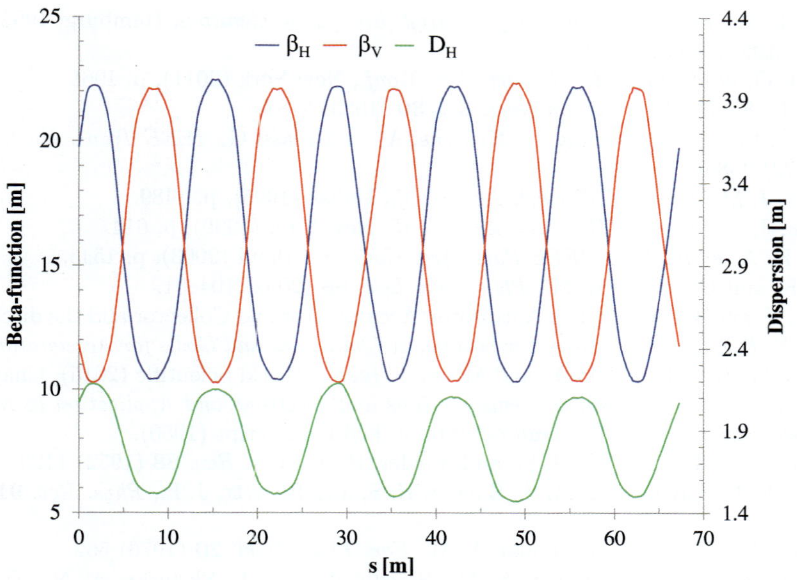


Fig. 8. Optical parameters for one super-period of the AGS bare lattice.

Table 2. Basic parameters of the AGS machine.

| | |
|---|---|
| Accelerated particles | p, polarised p and heavy ions (up to gold) |
| Particle energy [GeV] | 30, 25, and 14.5 GeV/$n$ |
| Circumference [m] | 256.9 $\pi$ |
| Magnetic lattice | Alternating-gradient focusing, combined-function |
| Focusing order | (F/2)O(F/2)(D/2)O(D/2) |
| Magnetic field index | $n = 365$ |
| Number of main magnets | 240 |
| Bending magnetic field [T] | 0.105 at injection (1.9 GeV), 1.30 at 33 GeV |
| Betatron oscillations/turn | 8.75 (h), 8.75 (v) |
| Transition energy [GeV] | 8.4 |
| Rise time/flat top time [s] | 0.6/0.5 to 2.5 |
| Straight sections | number = 240, 24 of 3.05 m, 72 of 1.52 m, 144 of 0.61 m |
| RF system I (tunable) | 10 cavities, 1.8 to 4.5 MHz, 200 kV total maximum |
| Auxiliary RF system [MHz] | 92 |
| Vacuum chamber [mm$^2$] | Inconel, 173 × 78 in the bending magnets |

## Acknowledgements

It is a pleasure to thank K. Hübner and E. McIntosh for the excellent comments on the original manuscript.

## References

Adams, J. B., Hine, M. G. N., and Lawson, J. D., *Nature* **171** (1953) 926.

Ahrens, L. A., *Proc. 15th Int. Conf. on High-Energy Accelerators*, Hamburg, 1992, World Scientific (1993), p. 109.
Alessi, J. G. *et al.*, *Proc. IEEE Part. Acc. Conf.*, New York (2011), p. 1966.
Barton, D. S., *IEEE Trans. on Nucl. Sci.* **30** (1983) 2787.
Bovet, C., Fiander, D., Henny, L., Krusche, A., and Plass, G., *IEEE Trans. Nucl. Sci.* **20** (1973) 438.
Brennan, J. M., *Proc. IEEE Part. Acc. Conf.*, Dallas (1995), p. 1489.
Brennan, J. M., *Proc. IEEE Part. Acc. Conf.*, New York (1999), p. 614.
Brown, K. A. *et al.*, *Proc. IEEE Part. Acc. Conf.*, Portland (2003), p. 1545.
Cappi, R. and Giovannozzi, M., *Phys. Rev. Lett.* **88** (2002) 104801.
Chohan, V. and Maury, S., The Antiproton Accumulator and Collector and the discovery of the W & Z intermediate vector bosons, in *Challenges and Goals for Accelerators in the XXI Century*, Brüning, O., and Myers, S. (eds.), World Scientific (2015), Chapter 11.
Christofilos, N. C., *Focusing Systems for Ions and Electrons and Application in Magnetic Resonance Particle Accelerators*, unpublished manuscript (1950).
Courant, E. D., Livingston, M. S., and Snyder, H. S, *Phys. Rev.* **88** (1952) 1190.
Courant, E. D., Livingston, M. S., Snyder, H. S, and Blewett, J. P., *Phys. Rev.* **91** (1953) 202.
Derbenev, Ya. S. and Kondratenko, A. M., *Sov. Phys. Dokl.* **20** (1976) 562.
Derbenev, Ya. S., Kondratenko, A. M., Serednyakov, S. I., Skrinsky, A. N., Tumaikin, C. M., and Shatunov, Ya. M., *Part. Accel.* **8** (1978) 115.
Fischer, W., The Relativistic Heavy Ion Collider, in *Challenges and Goals for Accelerators in the XXI Century*, Brüning, O., and Myers, S. (eds.), World Scientific (2015), Chapter 21.
Gardner, C. J. *et al.*, *Proc. IEEE Part. Acc. Conf.*, Albuquerque (2007), p. 1862.
Garoby, R., *Proc. 17th Intern. Conf. on High-Energy Accelerators*, Meshkov I. (ed.), Dubna (1998), p. 172.
Garoby, R., RF manipulations in the PSB & PS for the production of antiprotons at CERN and deceleration in the PS for LEAR, in *Challenges and Goals for Accelerators in the XXI Century*, Brüning, O., and Myers, S. (eds.), World Scientific (2015), Chapter 13.
Gilardoni, G., Giovannozzi, M., Martini, M., Métral, E., Scaramuzzi, P., Steerenberg, R. and Müller, A.-S., *Phys. Rev. ST Accel. Beams* **9** (2006) 104001.
Gilardoni, S., Manglunki, D. *et al.* (eds.), Fifty years of the CERN Proton Synchrotron, CERN-2011-004 (2011).
Giovannozzi, M. *et al.*, Possible improvements to the existing pre-injector complex in the framework of continued consolidation, in CERN-ATS-2010-026 (2010) 228.
Green, G. K., *Proc. Int. Conf. on High Energy Accelerators*, Blewett, M. H. (ed.), Brookhaven (1961), p. 39.
Green, G. K. and Courant, E. D., *Handbuch der Physik* XLIV (1959) 218.
Hardt, W., Möhl, D., Q-jump at transition, CERN ISR-300/GS/69-16 (1969).
Huang, H. *et al.*, *Proc. IEEE Part. Acc. Conf.*, Vancouver (2009), p. 4251.
Hübner, K, Brookhaven Alternating Gradient Synchrotron (AGS), in *Accelerators and Colliders*, Schopper, H., Myers, S. (eds.), Springer-Verlag, in press.
Livingston, M. S. and Blewett, J. P., *Particle Accelerators*, McGraw-Hill (1962).
MIT, Design study for a 15-BeV accelerator, MIT Research Lab. for Nuclear Sci. Tech. Rept. No. 60 (1953).
Myers, S., The Large Hadron Collider, in *Challenges and Goals for Accelerators in the XXI Century*, Brüning, O., and Myers, S. (eds.), World Scientific (2015), Chapter 22.
Pikin, A. *et al.*, *Proc. Intern. Symposium on Electron Beam Ion Sources and Traps*, Stockholm, IOP Publishing, (2010).

Plotkin, M., Brookhaven *Nat. Lab.* report-45058 (1991).
Regenstreif, E., The CERN Proton Synchrotron, CERN 59-29 (1959).
Roser, T., *Proc. 8th International Symposium High-Energy Spin Physics*, Particle and Fields Series 37, MN (1988), p. 1442.
Roser, T. and Courant, E., The Cosmotron and the Bevatron: The first GeV accelerators, in *Challenges and Goals for Accelerators in the XXI Century*, Brüning, O., and Myers, S. (eds.), World Scientific (2015), Chapter 7.
Schmidt, R., The CERN SPS proton–antiproton collider, in *Challenges and Goals for Accelerators in the XXI Century*, Brüning, O., and Myers, S. (eds.), World Scientific (2015), Chapter 10.
Schoefer, V. *et al.*, *Proc. IEEE Part. Acc. Conf.*, New York (2011), p. 2282.
Schoefer, V. *et al.*, *Proc. 3rd Intern. Part. Acc. Conf.*, New Orleans (2012), p. 1015.
Zelenski, A. *et al.*, *Proc. European Part. Acc. Conf.*, Genoa (2008), p. 1010.

# Chapter 9

# The CERN Intersecting Storage Rings

Stephen Myers (CERN)

## 1 Introduction and history

The CERN Intersecting Storage Rings (ISR) was the first proton–proton collider ever constructed. The LHC is the second. The ISR was constructed during the period 1966 to 1970 and was operated from 1971 to 1983 for physics experiments (see review article in [1]).

Following the first ideas and discussions for a proton–proton collider, studied [2] extensively by a group at the Midwestern Universities Research Association (MURA), USA, an Accelerator Research Group was set up by the CERN Council in 1956. In 1960 this group proposed a proton–proton collider attached to the CERN Proton Synchrotron (PS). In order to test the new ideas of particle stacking from MURA, a small proof-of-principle machine, the CERN Electron Storage and Accumulation Ring (CESAR), was constructed (1.9 MeV). In 1964, the principle of accumulating particles by RF stacking was proved experimentally in CESAR. CESAR was also an important test bed for the Ultra-High Vacuum (UHV) technology which was essential for the long lifetimes needed for stored beams.

In June 1965, the CERN Council decided to construct the ISR facility, and, at its meeting in December of the same year, to accept the financial plan of the project and to vote construction funds from January 1966. The plans foresaw a construction budget of 332 million CHF (1965 value) and first operation of the facility by mid-1971. Table 1 gives the basic parameters of the storage rings.

The ISR consisted of two independent storage rings intersecting at eight points with a crossing angle of 14.8 degrees! The circumference of the rings was 1.5 times that of the CPS (which supplied particles to the ISR), in order to allow space for the long straight sections in the interaction regions. The first proton–proton collisions took place in 1971 with beam momenta up to 26.5 GeV/$c$, which is the maximum momentum available from the

Table 1. Basic parameters for the ISR.

| | |
|---|---|
| Colliding particles | pp, dd, pd, $\alpha\alpha$, $\alpha$p, p$\bar{\text{p}}$ |
| Particle momentum | 3.5 to 31.4 GeV/$c$ |
| Circumference | 942.5 (300$\pi$) m |
| Number of main magnets | 132/ring |
| Magnetic dipole field | 1.33 T (max) |
| Length of main magnets | 4.88/2.44 m |
| Betatron oscillations/turn | 8.9 (h), 8.88 (v) |
| $\beta^*$ (h/v) | 21 m/12 m |
| $\beta^*$ (h/v) | 2.5 m/0.28 m in sc low-beta section |
| RF system per ring | 7 cavities, 9.5 MHz, 16 kV RF peak voltage |

CPS. The ISR operated for physics experiments from 1971 to 1983. It was decommissioned in 1984.

The combined-function magnet lattice formed two independent, interleaved rings, intersecting at eight points, five of which were used for experiments. A view of the ISR at intersection point 5 is shown in Fig. 1. The circumference of the orbits was 943 m, exactly 1.5 times the circumference of the PS.

## 2 Phase displacement and stacking

The accumulation of the very high currents in the ISR relied on a process called momentum "stacking" [2]. A "stack" was built by accumulating a few hundred PS beams across the large momentum aperture of the ISR. A single cycle involved RF capture of the PS 20 bunches at the injection momentum orbit of $-2\%$ and accelerating this beam (by changing the RF frequency) to a momentum orbit of $+2\%$. When the bunched beam reached its required momentum orbit, the RF was switched off and the beam debunched. Phase displacement occurs when an RF bucket traverses a debunched beam. The particles in the debunched beam travel around the unstable trajectories associated with the bucket (outside the separatrix, see Fig. 2). Traversing a debunched beam from low momentum to high momentum produces a decrease in the average momentum of the debunched beam by an amount equal to the phase space area of the RF buckets. A good analogy is to release droplets of mercury (RF buckets) into a cylindrical container containing some water (the debunched beam). In this analogy however the mercury droplets go from high energy to low energy and the water energy is increased. Consequently, repeatedly accelerating bunched beams in RF buckets to a fixed high momentum orbit causes the already debunched particles to be

Fig. 1. Interaction point 5 in the ISR.

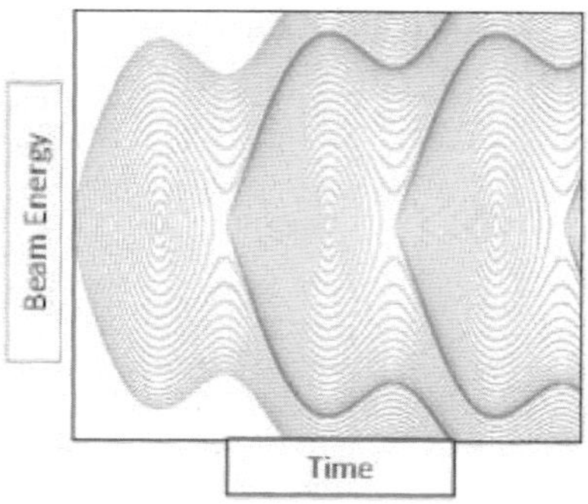


Fig. 2. Longitudinal motion of particles around the RF separatrix.

"phase displaced" to lower momentum thereby filling the momentum aperture with debunched beam.

The RF stacking technique allowed accumulation of huge intensities in the ISR; the maximum single beam current was 57 A!

## 3 Vacuum

The ISR encountered many technological challenges but one of the most important, the Ultra High Vacuum (UHV), was imperative for a long beam lifetime. The stainless steel vacuum chambers were baked *in situ* up to 300°C, and eventually all vacuum chambers were glow-discharge cleaned in order to reach the ultimate intensities.

The weakest vacuum sectors around the circumference were identified by "vacuum limit fills" where the intensity was increased to the value at which vacuum "runaway" occurred. This was an interesting challenge for the operations and RF teams as it meant that the stacking efficiency had to be maximized to produce the maximum longitudinal density in the stored "coasting" beam in order to reach the vacuum limit. The vacuum pressure and weak sectors were continuously improved during the life of the ISR (see Fig. 3). The end result was an average pressure around $3 \times 10^{-12}$ torr ($N_2$ equivalent) resulting in beam loss rates (due to beam gas collisions) of typically around one part per million per minute during physics runs (beam lifetimes of nearly 2 years!). These very long lifetimes were obtained after the beams had been centred in the aperture (see later). Beams of experiment quality could last 40 to 50 h. Maximum intensities of up to 57 A were stored per ring with 30–40 A as typical values for experiments.

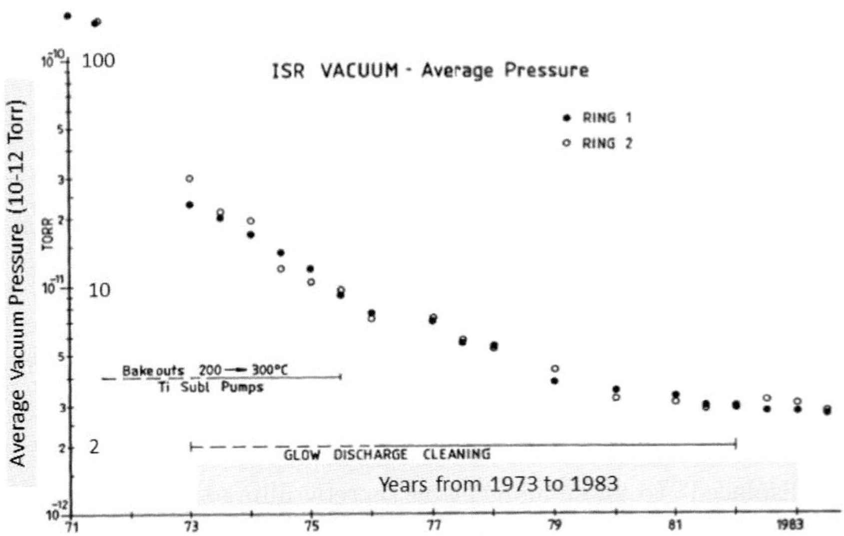


Fig. 3. Evolution of the ISR vacuum pressure.

## 4 Working lines and space charge compensation

The ISR had a working **line** not a working point. The required large tune spread resulted from the stability requirement from chromaticity and the large momentum spread needed for beam stacking. The minimum tune range of around 0.07 (see Fig. 4 [3]) created difficulties to find an area in the tune diagram which would allow the coasting beam to be free of low order non-linear resonances. The working line drawn in Fig. 4 had the stacked beam between the 3$^{rd}$ and 5$^{th}$ order resonances but necessitated traversal of the bunched beam across the family of 5$^{th}$ order resonances. It was well known that the space charge tune shift caused a "sagging" of the working line, rather like loading a beam with heavy weights. This had two effects, resonances (in Fig. 4 the beam would reach the main coupling resonance), and beam instabilities caused by the reduction of the chromaticity for the low momentum part of the stack. In order to be able to compensate the space charge effect we had (of course) to measure it. This is one of the major problems with unbunched beams: lack of diagnostics. A complicated system was developed which used beam transfer functions of empty buckets to measure the working line as a function of intensity. This system ultimately allowed measurements of the space charge tune shift which could be used for step-wise compensation [3] during stacking. The measurement system

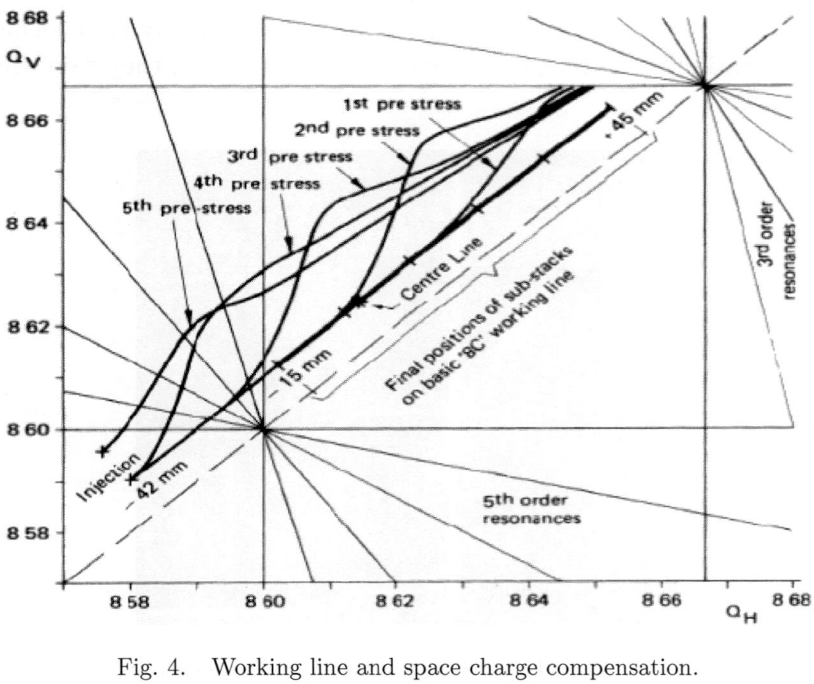


Fig. 4. Working line and space charge compensation.

was destructive to the beam (emittance) and never became robust enough to be used operationally. Figure 4 from [3] shows the procedure for space charge compensation while stacking. The working line was "pre-stressed" for currents of 3 A and after the 3 A increment of current had been stacked the next pre-stress was applied. Figure 4 shows the pre-stresses up to a total of 15 A. This space charge compensation system took advantage of the great magnetic flexibility allowed by the combined function magnets and the inclusion of 24 pole face windings. The method also greatly stressed the capabilities of the control systems of the early seventies (see later).

## 5 Schottky scans

Schottky signals result from the discrete nature of the particles in the beam. A sensitive high frequency longitudinal pick-up with some long term averaging of the signal shows a signal proportional to the longitudinal phase space density of the debunched beam. Figure 5 shows one of the first Schottky scans taken operationally in the ISR. The three scans were taken at beam currents of 10, 15 and 19.2 A. The horizontal axis is the longitudinal frequency and allows evaluation of the beam $\Delta p/p$.

Soon after discovering longitudinal Schottky scans, transverse pick-ups were used to measure the transverse Schottky scans which gave information about the tune values in the stacked beams.

The operational use of these Schottky scans completely transformed the way of operating the ISR. On the long stable beams fills, they were the only

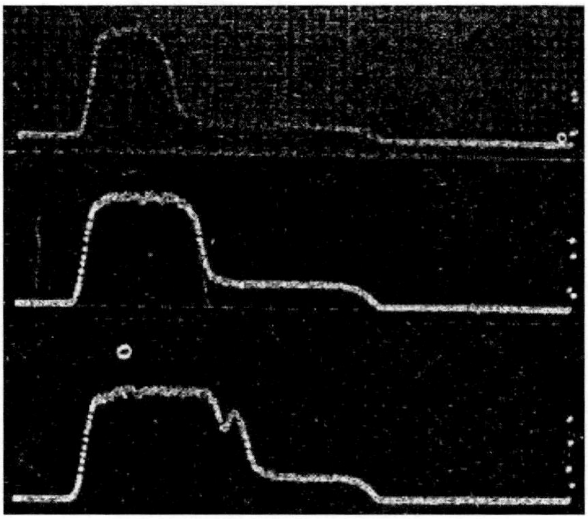

Fig. 5. The first longitudinal Schottky scans.

diagnostic available for observing the beam in a quantitative way (there was also a very useful sodium curtain [20] which allowed visual inspection of the cross section of the beam). In the longitudinal plane the longitudinal density could be evaluated as a function of $\Delta p/p$ by incorporating the value from the current meter. In addition, any "markers" on the stacks which could be identified in all planes would allow an evaluation of the location of this marker in tune space. The most usual markers for some time were the edges of the stack.

## 6 Centring the accumulated beam in the aperture

Injection to the ISR was performed by an injection kicker on a mechanically moveable girder. When the stacking process was completed the girder was moved out so that the injection kicker was outside the aperture of the machine. This of course meant that, for maximum stacking aperture, the stack was situated on the "outside" of the beam aperture (see Fig. 6, left). Consequently at the end of the stacking process, the stacked beams were not centred in the aperture. This situation remained for several years and produced high background rates in the experiments due to the diffusion of particles from the "top" of the stack to the outer chamber wall. In 1975 a technique was developed [4] for moving the beam into the centre of the aperture by increasing the main bending field appropriately. This sounds trivial but the space charge compensation for the high current stack also had to be included and synchronised with the change of the bending field. This procedure relied on the measurements made available with the Schottky scans (see Fig. 6 which shows the scans before and after centring) and the real beam position in the vacuum chamber.

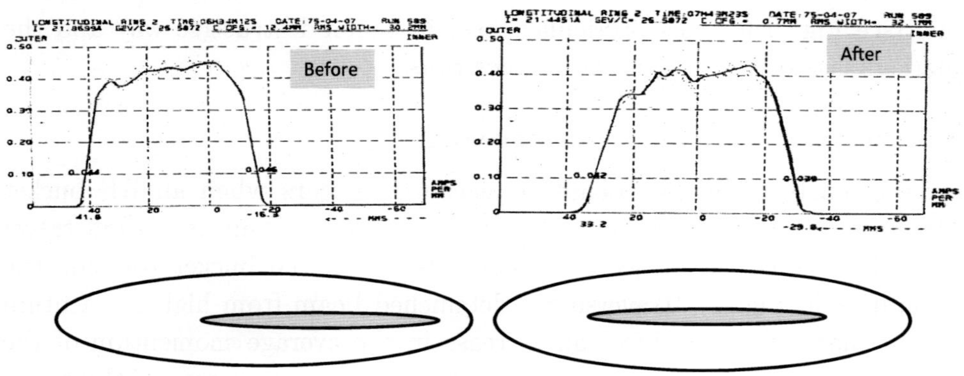


Fig. 6. Stack centring: before (left) and after (right).

Table 2. Improvements in the beam lifetime by centring the beam.

| | | | Before displacement | | After displacement | |
|---|---|---|---|---|---|---|
| Run | $I_1$ (A) | $I_2$ (A) | $\frac{di}{at}1$ (ppm/min) | $\frac{di}{at}2$ (ppm/min) | $\frac{di}{at}1$ (ppm/min) | $\frac{di}{at}2$ (ppm/min) |
| 593 | 24.0 | 24.0 | 60 | 10 | 0.8 | 0.8 |
| 594 | 24.0 | 24.0 | 10 | 10 | 0.8 | 0.8 |

This procedure produced immediate improvements in the beam lifetime (see Table 2) and in the background rates in the experiments.

## 7 Inserting markers in the stack

As previously explained, complete traversal of the stack by empty RF buckets causes a change in the average momentum of the whole beam. It is then clear that partial traversal will change the momentum of the part of the stack that has been traversed. This was a simple procedure, the RF was programmed to go from low momentum (outside the stack) to a momentum inside the stack, in this case a small reduction in the average momentum of the traversed part of the stack occurred, leaving a "marker" (lower density) at the $\Delta p/p$ where the RF traversing bucket stopped [5].

Figure 7 shows a longitudinal Schottky scan which had 4 markers inserted in this way. The markers are very clear and correspond precisely to the programmed frequency of the RF stop. Of course to be of any diagnostic use these markers must also be seen in the transverse plane. Figure 8 shows the corresponding scans for the horizontal and vertical planes. The markers are clearly visible.

The combination of these measurements allowed plotting of the working line (see Fig. 9 for this particular case) in a non-destructive way and the markers lasted throughout the physics runs.

## 8 Acceleration by phase displacement

As previously explained, phase displacement occurs when an RF bucket traverses a debunched beam. The particles in the debunched beam travel around the unstable trajectories associated with the bucket (outside the separatrix in Fig. 2). Traversing a debunched beam from high momentum to low momentum produces an increase in the average momentum of the debunched beam by an amount equal to the phase space area of the phase displacing buckets. The analogy is the release of droplets of mercury into a

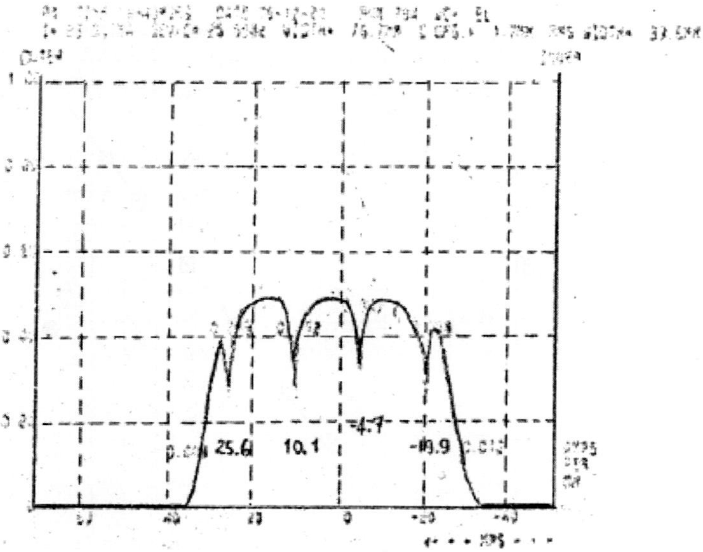


Fig. 7. Longitudinal Schottky scan with markers inserted by phase displacement.

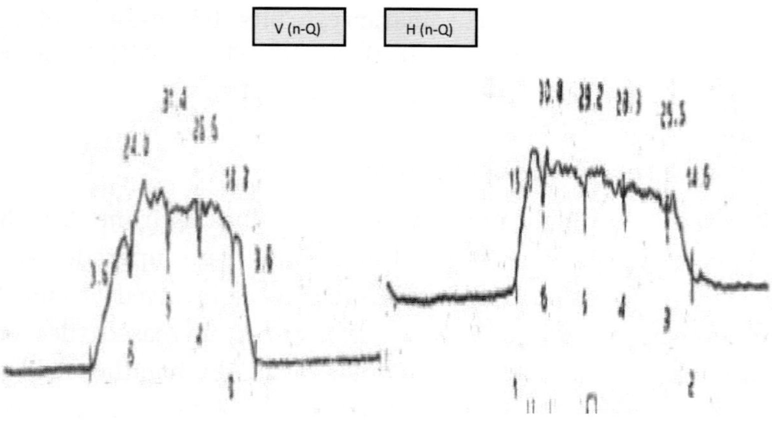


Fig. 8. Transverse Schottky scan with same markers apparent.

cylindrical container containing some water. Each mercury droplet raises the level of the water and thereby produces an increase in the potential energy.

Since the ISR circumference was larger than that of the PS, the maximum energy was also higher (31.4 compared to 26.6 GeV). In the never-ending quest for higher beam energies, it was decided to attempt to increase the energy of the accumulated beam in the ISR. However the small ISR RF system (16 kV maximum) could not capture a beam with 3% momentum spread, so it was decided to attempt to phase displace high intensity stacks of

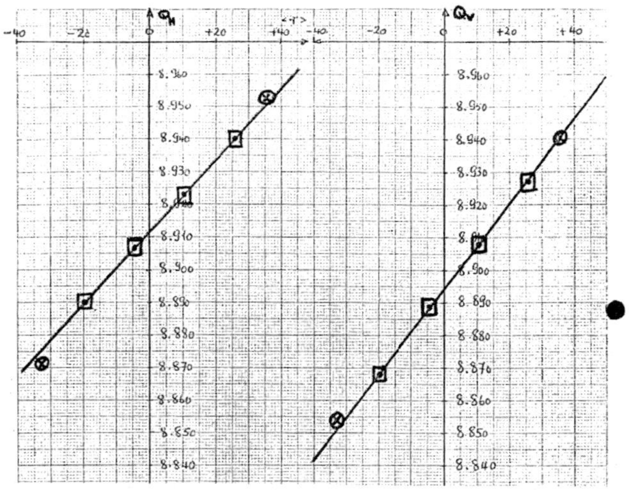

Fig. 9. Resulting measurement of the working line (betatron tunes versus beam momentum deviation).

protons. Initially the progress was slow but after some better understanding (space charge changing tunes, chromaticity, orbits, RF noise effects, absence of diagnostics...) [6] and a few break-throughs, 31.4 GeV became the preferred high luminosity operational energy of the ISR [7].

## 9 Computer control of accelerators

In order to accelerate high intensity "coasting" beams from 26 to 31 GeV, the computer control system of the ISR needed many upgrades and modifications. The acceleration by phase displacement involved around 200 RF frequency sweeps through the beam. Each sweep increased the energy by around 25 MeV and necessitated incremental of the bending field so as to keep the beam in the centre of the aperture. Due to the changing space charge effects, changes had also to be made to the tune and the chromaticity of the beams. The procedure developed [8] involved many reference "break-points" as a function of energy and the acceleration between these break-points was done by interpolation. This same procedure was subsequently employed for the acceleration of beams in the LEP and the LHC.

## 10 Working close to the integer

In the early days (lower intensity, hence lower chromaticity) the working line was situated just above the half integer (8.5). In the latter, higher-intensity days, when more tune spread was needed, we were forced to operate just

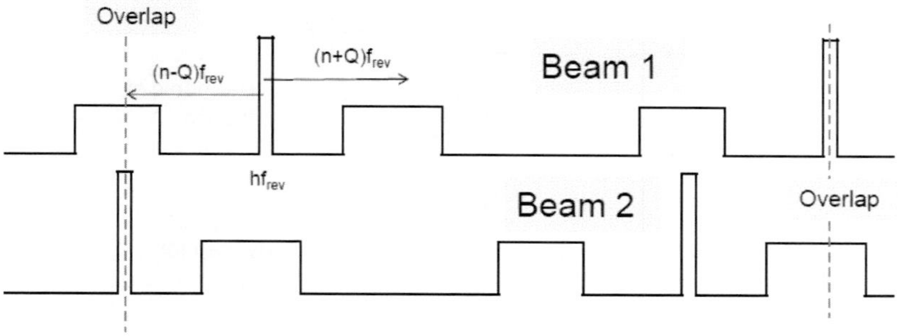


Fig. 10. Transverse frequency overlap.

below the integer resonance (9.0) since this is the most resonance free area on the tune diagram. The "top" of the stack was situated at a horizontal tune value of 8.955, just 0.045 distance from the integer. Initial operation at these new tune values was very problematic (orbit stability, transverse stability, etc.) but with time all these known problems were solved. However there was an effect unknown at the time, which caused massive emittance increase in the top portion of the stacks. The sodium curtain showed transverse cross sections of the beams which resembled lacrosse sticks. This behaviour was identified as being caused by a pulsed beam–beam effect and was given the name of "Overlap Knock-out". Several solutions were successful in reducing the effect to acceptable levels [9].

### 10.1 *Overlap knock-out*

In the ISR we had to worry about 4 beams: 2 beams per ring. The bunched beam at injection and during acceleration and the debunched already accumulated beam.

For beams that have significantly different revolution frequencies (caused by different $\Delta p/p$ or different charge/mass ratios), overlap knock-out (OLKO) was discovered to be an effect where the longitudinal harmonics of the bunch spectrum have frequency components which are equal ("overlap") to the transverse betatron frequencies and thereby, by some form of coupling, can excite the beam at its transverse resonant frequency ("RF knock-out") as shown in Fig. 11.

The OLKO resonance condition [10] Q vs $\Delta p/p$ can easily be evaluated and is depicted in Fig. 11 for the various harmonics of the bunch frequency. Clearly this condition is much more easily met at lower harmonics of the bunch frequency when the transverse tunes approach the integer.

An experiment was performed to test the strength of these newly discovered resonances. A beam of 8 A was accumulated over the tune space

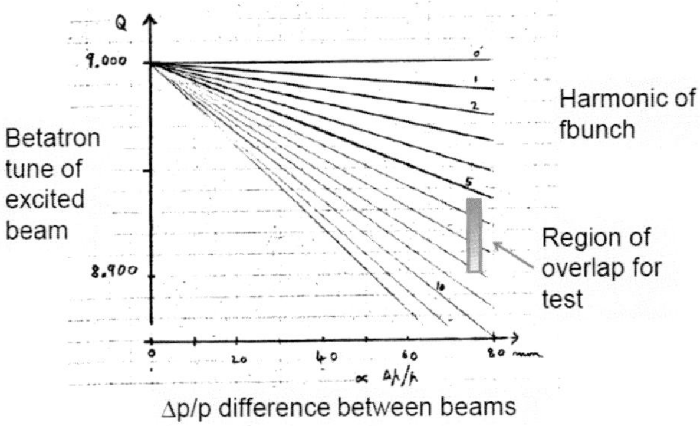


Fig. 11. The OLKO resonance conditions for the ISR.

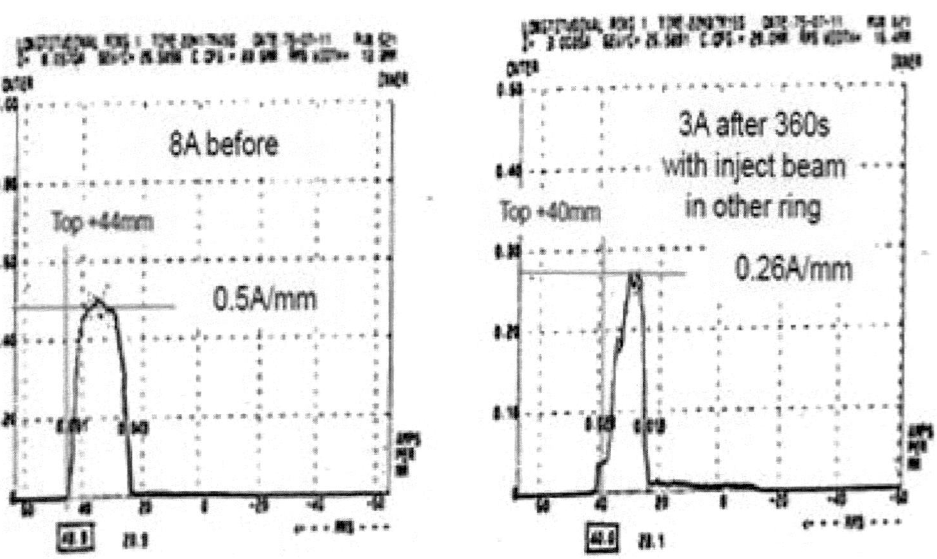


Fig. 12. Density profiles before and after OLKO.

shown in Fig. 11 and collimated by scrapers so that any emittance increase would be recorded as beam losses. A bunched beam of 80 mA was injected into the other ring and allowed to circulate for 360 seconds. Figure 12 shows the longitudinal Schottky scans before and after the 360 s presence of the injected beam. The total current was reduced from 8 to 3 A, the peak longitudinal density reduced from 0.5 to 0.26 A/mm and the whole top part of the stack had been eroded. The beam–beam tune shift exercised by the 80 mA bunched beam was of the order of $10^{-6}$!

The OLKO effect was studied extensively in the ISR and cures were found to allow operation very close to the integer. The cures used operationally

[9, 10] were:

- Reduction of the higher harmonic components of the bunch spectrum by bunch lengthening (lower RF voltage);
- Use of separations in the interaction regions so that the vector sum of beam–beam kicks over one turn is minimized.

In order to complete the study, tests were also done with bunched colliding beams with future accelerators in mind. In general, with bunched beams the resonance condition cannot be met if the RF frequencies of both beams are locked. For cases where frequency locking is not possible (e.g. resulting orbit is outside the vacuum chamber!) OLKO can be very destructive. This is particularly true for beams of different species and therefore different revolution frequencies.

## 11 Low $\beta$ insertions and luminosity

The first low $\beta$ insertions, using normal conducting insertion quadrupoles to focus the beam more strongly in the vertical plane [11], were employed in the ISR from 1974 until 1981. During the last three years of operation one low $\beta$ insertion (IP8) was upgraded with superconducting quadrupoles and produced an impressive increase in the luminosity (see Fig. 13). The proton–proton initial luminosity (design for $4 \times 10^{30}$ cm$^{-2}$s$^{-1}$) was increased from

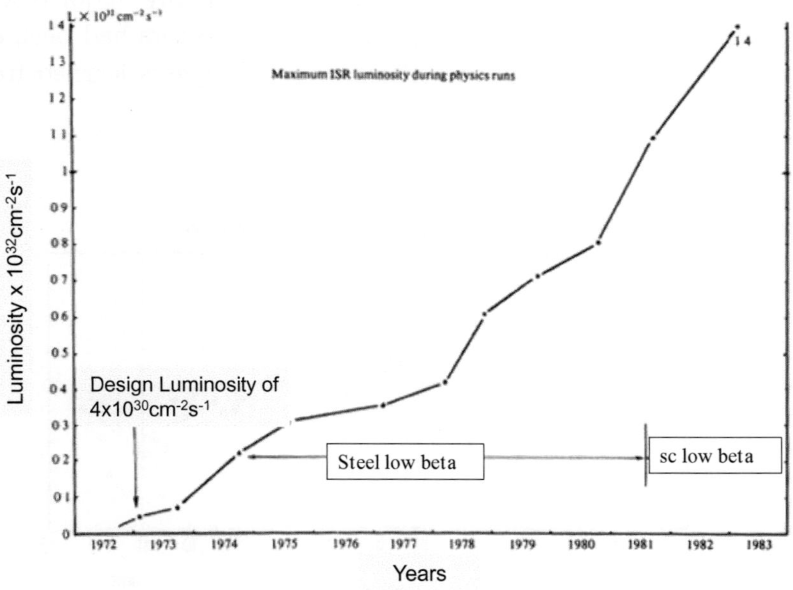


Fig. 13. Evolution of the luminosity in the ISR.

$1.6 \times 10^{30}\,\mathrm{cm^{-2}s^{-1}}$ in 1971 to $1.4 \times 10^{32}\,\mathrm{cm^{-2}s^{-1}}$ in the superconducting low-beta section installed in one of the interaction points in 1982, which stayed the world record luminosity until 1991.

## 12 Stochastic cooling

The first observations [12] of the stochastic signals in the ISR (Schottky scans) immediately turned the attention to the possibility of damping the oscillations of the particles (stochastic cooling). Significant effort in this direction was led by Wolfgang Schnell following the initial idea by Simon van der Meer [13]. A stochastic cooling test system was built as a demonstrator. The most sensitive detection of transverse beam size in the ISR was through the normalised luminosity measurement.

Figure 14 shows the results of the first conclusive observation of stochastic cooling [14]. The normalised luminosity is shown over a 13 hour period with stochastic cooling turned on and off every few hours. The effect is small but very significant: stochastic cooling worked! Very soon afterwards a similar system was designed for the Initial Cooling Experiment (ICE) with spectacular results as shown in Fig. 15.

## 13 Summary: What did ISR teach us?

The ISR was a fantastic machine for accelerator physics and was surely ahead of its time in that the detectors did not make any major discoveries even though this could have been the case if the detectors had been equally advanced. The net result was that the detector designers learned from the ISR on how to build detectors for colliders.

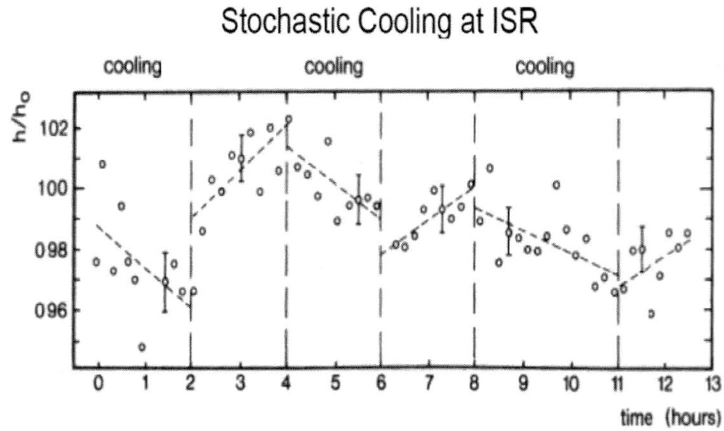


Fig. 14. First demonstration of cooling.

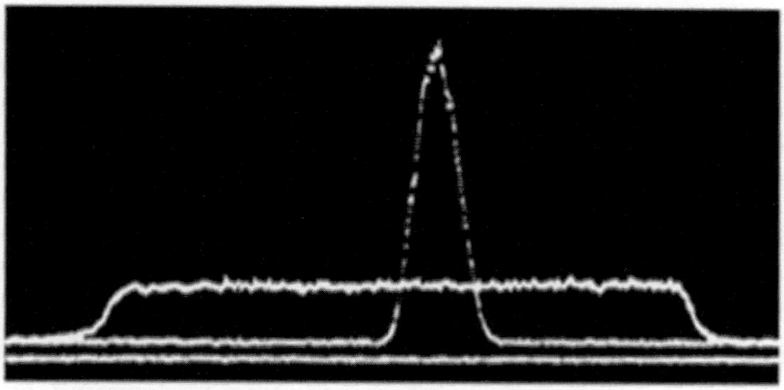

Fig. 15. Fast momentum cooling in ICE.

ISR also taught accelerator physicists how to efficiently accumulate protons (and ions) in longitudinal phase space using the principle of phase displacement. This same principle was used to accelerate ultrahigh intensities from 26 to 31 GeV per beam (the world's highest energies for many years) using a small RF system.

Ultra high vacuum is of extreme importance for all modern accelerators. In the ISR, systematic tests [15] were performed on a continuous basis in order to understand vacuum instabilities and improve the vacuum to levels never before dreamed of. Some of the more recent vacuum problems in accelerators (e.g. electron cloud) were identified in the ISR. In addition the ISR was designed with a system of clearing electrodes which were foreseen to extract the electrons produced by the beam–gas interaction. For the last few years of operation the ISR vacuum was stable at $3 \times 10^{-12}$ torr (Figure 3).

For the colliding runs of very long duration (one fill lasted nearly 2 weeks!) the precision, stability and ripple performance of the power converters were crucial. The experience gained in improving these power converters during the life of the ISR has been beneficial in the design of the power converters for more recent colliders like the LEP and the LHC.

In the never-ending quest for higher beam intensities and the ensuing beam instabilities of various kinds, it was realised that future accelerators must be designed with a low coupling impedance to the beams. During the life of the ISR, components to be installed in the machine (during routine shutdowns) were computed and measured for coupling impedance. This vigilance and approach has been extended to all accelerators built since the ISR. In LEP the LHC there were "impedance policemen" who authorised (or not) the installation of components.

The ISR was subjected to several space charge phenomena of both single beam and beam–beam origins. The compensation of the space charge tune shift was a major break-through [3] which allowed very high intensity stacks to be accumulated and the cures [9] against beam–beam effects (overlap knock-out) allowed the preservation of the transverse emittance of the "stacks" with concomitant higher luminosity.

The ISR was plagued with background in the experiments. The first use of collimators [16] in the machine produced considerable background reduction in the experiments even when the beam lifetime was not extremely low. The collimators (10 per ring) consisted of primary and secondary collimators (located at $\pi/2$ and $3\pi/2$ phase advance from the primaries). This system produced considerable improvement in reducing background and induced radioactivity in all the physics interaction points.

Another first in the ISR was the absolute calibration of luminosity using the van der Meer technique [17]. This technique required precisely known [18] relative transverse displacements of the beams as well as accurate measurement [19] of the beam current. An identical technique is now being used for the calibration of the luminosity in the LHC.

The observation of Schottky noise on the coasting beams [12] provided an essential tool to measure the beam properties in a non-destructive way. The discovery of Schottky noise led to the search for stochastic cooling of coasting beams [13], led by Wolfgang Schnell and finally tested [14] using the luminosity measurement system. This first demonstration of stochastic cooling had an enormous impact on the future of colliders and is used or proposed for almost every collider built since.

From 1976 onward deuterons were stored in the ISR so that dd and pd collisions became available. Alpha particles were stored in 1980 for $\alpha\alpha$ and $\alpha$p collisions. Initial dd luminosities reached $1.6 \times 10^{30}$ cm$^{-2}$s$^{-1}$ and were $4 \times 10^{28}$ cm$^{-2}$s$^{-1}$ in the $\alpha\alpha$ case. Antiprotons were stored as soon as the antiproton injector complex had become operational in 1981 (see p–$\bar{\text{p}}$ in the SPS, Chapter 10). This required a new transfer line from the CPS.

The ISR will be remembered for a number of break-throughs in accelerator physics and technology: UHV-technology for a large scale facility, control of intense coasting beams, discovery of Schottky scans, experimental demonstration of stochastic cooling, and absolute luminosity measurement by van der Meer scans.

## References

[1] U. Amaldi, P. Bryant, P. Darriulat and K. Hubner, 40th anniversary of the first proton–proton collisions in the CERN Intersecting Storage Rings (ISR), CERN 2012-004 (2012).

[2] K. R. Symon and A. M. Sessler, Methods of radio frequency acceleration in fixed-field accelerators with applications to high-current and intersecting beam accelerators, in *Symposium on High-Energy Accelerators and Pion Physics*, Geneva/CERN, CERN 56-26 (1956), p. 44.
[3] P. J. Bryant, D. M. Lewis, B. Nielsen and B. Zotter, On-line correction of the incoherent tune shifts due to space charge, CERN/ISR-MA-75-54 (1975).
[4] S. Myers, Stack centering using the RF and the main field, ISP Performance Report (1975).
[5] S. Myers, Non-destructive working line measurements in a stack by placing markers on the Schottky scans, ISR Performance Report (1977).
[6] S. Myers, Longitudinal behaviour of the stacked beam and the RF parameters during phase displacement to 31.4 GeV/c, CERN ISR-OP/76-44 (1976).
[7] S. Myers *et al.*, Performance of the ISR at 31.4GeV/c, *IEEE Trans. on Nucl. Sci.* **26**(3) (1979) 3155.
[8] E. Ciapala, R. Keyser and S. Myers, The CERN ISR control scheme for acceleration by phase displacement, *IEEE Trans. on Nucl. Sci.* **26**(3) (1979) 3395.
[9] H. G. Hereward and S. Myers, ELSA, overlap knock-out, ISR Performance Report (1975).
[10] J.-P. Gourber, H. G. Hereward and S. Myers, Overlap knock-out effects in the CERN intersecting rings (ISR), in *Particle Accelerator Conference*, Chicago, IL, USA, March 1977, *IEEE Trans. Nucl. Sci.* **24** (1977) 1405–1407.
[11] J.-P. Gourber, E. Keil and S. Pichler, The first high-luminosity insertion in the ISR, in *Proc. 1981 Particle Accelerator Conference*, Washington DC (1981), p. 1419.
[12] J. Borer *et al.*, Non-destructive diagnostics of coasting beams with Schottky noise, in *Proc. High-Energy Accel. Conf.*, Stanford (1974), p. 53.
[13] S. van der Meer, Stochastic damping of betatron oscillations in the ISR, CERN-ISR-PO-72-31; ISR-PO-72-31 (1972).
[14] P. Bramham, P. G. Carron, H. G. Hereward, K. Hübner, W. Schnell and L. Thorndahl, Stochastic cooling of a stored proton beam, *NIM* **125** (1975) 201–202.
[15] E. Fischer, Experience gained in the operation of the ISR vacuum system, CERN-ISR-VA/72-39 (1972).
[16] T. Risselada *et al.*, The CERN ISR collimator system, in *8th Particle Accelerator Conference*, San Francisco, CA, USA, 12–14 Mar 1979, *IEEE Trans. Nucl. Sci.* **26** (1979) 4131–4133.
[17] S. van der Meer, CERN ISR-PO/68-31 (1968).
[18] P. J. Bryant and K. M. Potter, Calibration of the beam displacements used in ISR luminosity measurements, CERN ISR-ES-BOM-82-15 (1982).
[19] K. Unser, A toroidal DC beam current transformer with high resolution, in *Proc. 1981 Particle Acc. Conf., IEEE Trans. Nucl. Sci.* **28**(3) (1981) 2344.
[20] B. Vosicki and K. Zankel, The sodium curtain beam profile monitor of the ISR, CERN ISR-VA-75-11 (1975); *IEEE Trans. Nucl. Sci.* **22** (1975) 1475–1478.

# Chapter 10

# The CERN SPS proton–antiproton collider

Rudiger Schmidt (CERN)

One of CERN's most ambitious and successful projects was the search for the intermediate bosons, W and Z [1]. The accelerator part of the project relied on a number of innovations in accelerator physics and technology. The invention of the method of stochastic cooling and the extension by many orders of magnitude beyond the initial proof of principle demonstration allowed the construction of the Antiproton Accumulator. Major modifications to the 26 GeV PS complex and the conversion of the 300 GeV SPS, which had just started up as an accelerator, to a $p\bar{p}$ collider were required. The SPS collider had to master the beam–beam effect far beyond limits reached before and had to function in a tight symbiosis with the UA1 and UA2 experiments.

## 1 Introduction

Today, the CERN Super Proton Synchrotron (SPS) is essential in the accelerator chain for delivering protons and ions to LHC and it provides beams for fixed targets experiments.

The SPS has a long and very successful history. It was designed in the beginning of the 70s [2] as a synchrotron to accelerate protons and to extract them to fixed target experiments. The accelerator has a bending radius of 1100 m and a length of 6911 m. Initially three stages of construction were planned, first for an acceleration to an energy of 200 GeV/$c$, later to 300 GeV/$c$ and finally either to 400 GeV/$c$ with normal conducting magnets, or to 700 GeV/$c$ with superconducting magnets. Later it was decided to build the accelerator in one stage with normal conducting magnets. The SPS has been delivering protons for physics experiments since January 1977. It started with a circulating beam intensity of $2 \times 10^{13}$ protons per pulse and a repetition time of 9.6 seconds.

Already during the construction period the idea came up to use the SPS for $p\bar{p}$ collisions. There was some previous experience at CERN with a hadron collider, the CERN-ISR constructed in the seventies. In the ISR two proton beams with an energy of up to 31 GeV were circulated in two separate vacuum chambers and collided with an angle, delivering a maximum luminosity of up to $1.3 \times 10^{32}\,\text{cm}^{-2}\text{s}^{-1}$.

With the experience from the ISR and the development of stochastic cooling, it was possible to accumulate dense $\bar{p}$ bunches in an accumulator ring and to transform the Super Proton Synchrotron (SPS) into a $p\bar{p}$ collider. The first collisions between protons and antiprotons in the SPS were observed in 1981 during the first phase of operation that lasted until 1986. After a substantial upgrade the machine restarted in 1987 and the last year of operation as $p\bar{p}$ collider was 1990. Table 1 shows the basic parameters of the $Sp\bar{p}S$ collider for the operation from 1988 to 1990.

The $Sp\bar{p}S$ was designed as a hadron collider operating with bunched beams. Before its commissioning it was debated if such a machine could ever work, or if beam–beam effects without the presence of damping due

Table 1. Main parameters of the $Sp\bar{p}S$ collider.

| Typical parameters | SPS $p\bar{p}$ collider |
|---|---|
| Injection momentum [GeV/$c$] | 26 |
| Top momentum [GeV/$c$] | 315 |
| Integrated luminosity in 1990 [$\text{nb}^{-1}$] | 6790 |
| Maximum initial luminosity [$\text{cm}^{-2}\text{s}^{-1}$] | $5.5 \times 10^{30}$ |
| Initial luminosity lifetime [h] | 9–12 |
| Proton bunch intensity | $12 \times 10^{10}$ |
| Antiproton bunch intensity | $5 \times 10^{10}$ |
| Number of bunches per beam | 6 |
| Number of collision points | 3 |
| Horizontal emittance $\equiv \sigma^2 \times 4/\beta$ (proton) | 11 |
| Vertical emittance $\equiv \sigma^2 \times 4/\beta$ (proton) | 11 |
| Horizontal emittance $\equiv \sigma^2 \times 4/\beta$ (antiproton) | 12 |
| Vertical emittance $\equiv \sigma^2 \times 4/\beta$ (antiproton) | 10 |
| $\beta_h$ and $\beta_v$ at IP in 1990 [m] | 0.65/0.30 |
| Linear tune shift per interaction point on protons (H, V) | 0.0037/0.0026 |
| Linear tune shift per interaction point on antiprotons (H, V) | 0.0066/0.0063 |
| Total tune shift on protons (H, V) | 0.011/0.008 |
| Total tune shift on antiprotons (H, V) | 0.020/0.014 |
| Bunch length in store $4\sigma_s$ [ns] | 2.4 |
| Operational tunes (H, V) | 26.685/27.680 |
| Bunch intensity lifetime [h] | ≥60 |
| Efficiency (time with colliding beam/scheduled time) (average 1988, 89, 90) | 51% |
| Number of stores/year (average 1988, 89, 90) | 110 |
| Number of stores/year lost due to failure (average 1988, 89, 90) | 20% |

to the emission of synchrotron radiation as in electron–positron colliders would prohibit an operation with high luminosity. Its successful operation demonstrated that beam–beam effects at high energy hadron collider can be mastered and that such machines are excellent tools for experiments in particle physics.

## 2 The SPS as a synchrotron

The SPS was finished in 1976 after five years of construction on the CERN Prevessin site. The first beam was circulating in spring 1976, and operation as a synchrotron for fixed target experiments started in January 1977.

The lattice is of separated function type and F0D0 configuration with 108 F0D0 cells. This is in contrast to the first generation of alternating synchrotrons which were of combined function type. In the separated function type the bending is provided by dipole magnets, focusing is by relatively few quadrupole magnets. With the separated function lattice the field in the dipole magnets is considerably higher than with a combined function lattice and allows to achieve higher energy with a given circumference.

## 3 From SPS to Sp$\bar{p}$S

Two concepts led to CERN's $p\bar{p}$ collider: the concept of particle–antiparticle colliders as it had been demonstrated with electrons and positrons, and beam cooling. The first realistic scheme for colliding beams was discussed in 1956 by D. W. Kerst [3]. Seven years after the first experimental confirmation of antiprotons in 1955, K. Johnson at CERN worked on the possibility of colliding protons and antiprotons in the ISR, but these studies did not result into a concrete proposal [4].

The first realistic proposal for a $p\bar{p}$ collider seems to have been made by Budker and Skrinsky at Orsay in 1966 [5]. The proposal was based on Budker's idea of electron cooling specifically for the production of antiproton beams dense enough to make $p\bar{p}$ colliders viable. In 1968 Simon van der Meer had the idea of stochastic betatron cooling but only published it in 1972 [6]. Both electron and stochastic cooling were experimentally proven in 1974 at the NAP-M storage ring in Novosibirsk. In October 1974, stochastic cooling was first observed at the ISR [7].

The discovery of neutral currents provoked Carlo Rubbia and collaborators to propose a colliding beam experiment at both CERN and Fermilab in 1976 [5] with the specific aim of producing W and Z bosons. Such machines require only a single ring, a concept already practiced at that time with electron–positron colliders. Rubbia, realizing the potential offered by the projected SPS as a 300 GeV machine proposed in 1976 to use it as a $p\bar{p}$ collider [8].

## 4 Modification of the SPS for collider operation

As soon as the project was decided in June 1978, the modifications of the SPS started [1]. The $Sp\bar{p}S$ had a six-fold symmetry with two physics experiments UA1 and UA2 at the interaction points 4 and 5.

- A new beam line was constructed, to transfer the antiprotons from the CERN Proton Synchrotron (PS) to the SPS, and a new injection system for counter-clockwise injection was added in the SPS.
- The SPS had been built for an injection energy of 14 GeV/$c$. The proton transfer line, TT10, and the injection system had to be upgraded to 26 GeV/$c$.
- A drastic improvement of the SPS beam vacuum system was needed. The design pressure of $2 \times 10^{-7}$ torr was only adequate for a synchrotron and needed to be improved by two orders of magnitude for the long beam storage time required in a 2 beam collider.
- The machine lattice had to be modified to include low-beta insertions, squeezing proton and antiproton beams to small sizes at the interaction points, for achieving design luminosity.
- The accelerating RF system with its traveling-wave structures had to undergo modifications for simultaneous acceleration of protons and antiprotons. Precise synchronization between proton and antiproton bunches had to be implemented, for collisions to occur at the centre of the detectors. The RF system was also upgraded to achieve lower noise levels which are required for long storage times.
- Beam diagnostics had to be adapted to very low beam intensities, and new devices added, such as directional couplers for independent observation of protons and antiprotons.

Further modifications to the $Sp\bar{p}S$ collider were made to operate with the ten-fold increase in the antiproton production rate in 1987.

## 5 SPS as a $\mathbf{p\bar{p}}$ collider

During 1981 to 1990, the CERN SPS was operated as a $p\bar{p}$ collider, providing high energy collisions for two major experiments located in adjacent sextants of the accelerator. This operation was almost always with three dense bunches of protons in collision with three rather weak bunches of antiprotons, with no separation of the beams at the unused crossing points.

The first proton–antiproton collisions were recorded in the summer of 1981. The first physics run took place at the end of that year when 0.2 $\text{nb}^{-1}$ of integrated luminosity was produced. Between 1982 and 1986 the peak luminosity was pushed up to $3.9 \times 10^{29}$ $\text{cm}^{-2}\text{s}^{-1}$ and more than 1.2 $\text{pb}^{-1}$ of integrated luminosity was accumulated at each of the two main experimental interactions.

Until 1983 the centre-of-mass energy was limited to 546 GeV due to resistive heating of the magnet coils. The addition of further water cooling allowed the machine energy to be pushed up to 630 GeV in 1984. The 1986 run was prematurely interrupted because of a major fault of the UA1 detector.

The main limitations to the machine performance during the initial operation phase were the scarcity of antiprotons, space charge and beam–beam effects, longitudinal stability and intra-beam scattering; all imposing constraints at different operational stages during injection, acceleration and storage.

During 1987 and early 1988, the CERN $\bar{p}$ complex upgrade was completed and provided significantly more antiprotons. During 1987 another ring, the antiproton collector (ACOL), was added to the existing antiproton accumulator in order to increase the acceptance for $\bar{p}$ of the complex.

In 1988 the available stack of antiprotons in the accumulation machine was normally between 4 and $7\times10^{11}$ particles, reaching a maximum intensity of $8.5 \times 10^{11}$ compared with the previous best stack before the upgrade of $4.5\times10^{11}$. Furthermore, the average accumulation rate was about $3.3\times10^{10}$ particles per hour, with a maximum rate of over $3.8 \times 10^{10}$, compared with $1.2 \times 10^{10}$ achieved previously.

To make the best use of this increased supply of antiprotons, the number of bunches of both protons and antiprotons injected into the collider was increased in 1988 from three to six.

- To reduce the limiting beam–beam effect, electrostatic deflectors ("separators") separated the beams at 9 of 12 collision points.
- A new RF system at 100 MHz, half the accelerating frequency of 200 MHz, increased the longitudinal acceptance at injection. The more intense bunches (up to $10^{11}$ $\bar{p}$/bunch) had a larger longitudinal emittance. By making them longer, the momentum spread was kept the same, to stay within the SPS momentum acceptance.
- The scheme for chromaticity correction was upgraded.
- In 1990, when UA2 took its last data, a super-squeezed low-beta scheme boosted the luminosity by a further factor of 2.

The yearly integrated luminosity from 1982 to 1990 is shown in Fig. 1.

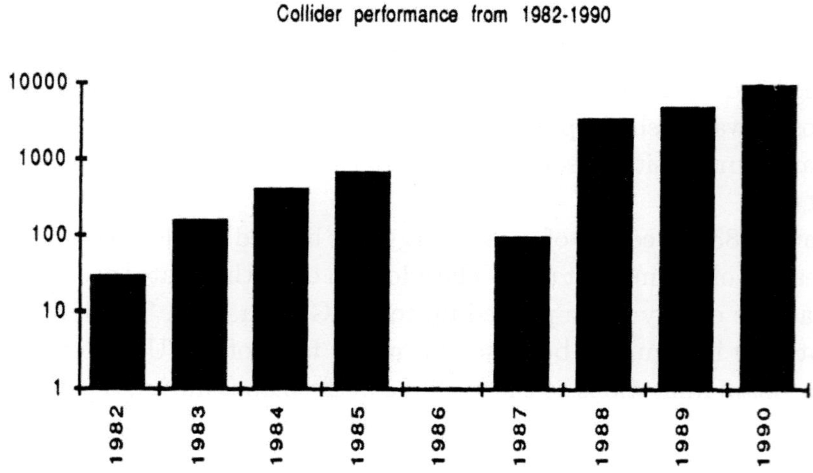


Fig. 1. Integrated annual luminosity in units of $nb^{-1}$ from 1982 to 1990 [10].

## 6 Operational cycle

After stacking antiprotons for many hours, the PS and SPS prepared for a fill. The first step was the optimisation of the injection cycle with proton and low intensity antiproton pilot bunches. The filling sequence was as follows: firstly, three (and later six) proton bunches were fabricated in the CERN PS complex and sent to the SPS at 2.4 s intervals and stored at 26 GeV. These bunches were placed symmetrically around the SPS circumference. Then a single antiproton bunch was unstacked from the $\bar{p}$ accumulator and injected into the PS at 3.5 GeV. The PS accelerated the bunch to 26 GeV, and the bunch was transferred to the SPS at an azimuthal position to within a fraction of a nanosecond for collisions in the physics experiments. This was followed by three (or six) bunches of antiprotons extracted from the antiproton stack at 2.4 s and transferred to SPS. The SPS accelerated both beams to an energy of 315 GeV.

During the first two seconds at 315 GeV, the beta functions in the two low-$\beta$ insertions for the physics detectors UA1 and UA2 were reduced down to their final values for physics operation of typically 1 m horizontally and 0.5 m vertically [11] to increase the luminosity. The collider then passed into storage, and two electrostatic separators were activated, to achieve separation at the unused crossing points away from the experiments. In the experiments, the beams were colliding head-on. Horizontal and vertical emittances were similar for both protons and antiprotons. The beams collided for many hours (see Fig. 2).

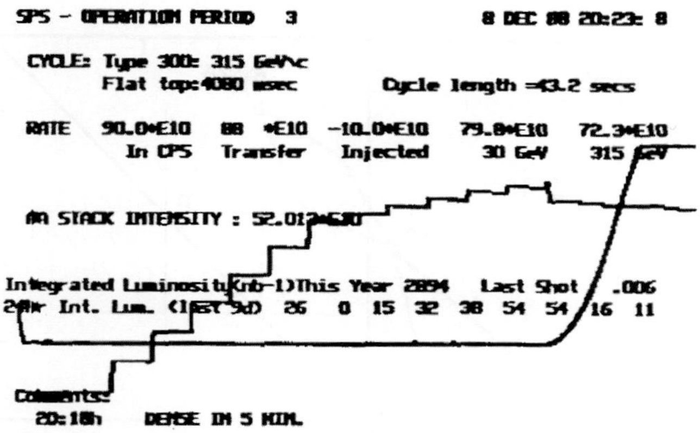


Fig. 2. The injection sequence, the energy ramp and the initial operation at 315 GeV/$c$ of the SPS collider when operating with 6 bunches per beam. The horizontal axis shows time, the step shaped curve shows the total stored beam intensity, and the solid curve shows the evolution of the main magnet current. The beam energy starts at injection values on the left and reaches collision values on the right.

The entire process of injection, acceleration and beta function squeezing was achieved in a 43.2-second cycle, and the separation required for operating with colliding beams took a further few seconds to complete. The experiments could then raise the magnetic fields in the detectors, and data taking by the experiments started within a few minutes.

## 7 Beam–beam effects

The SPS is an accelerator with normal conducting magnets with little nonlinear magnetic field perturbations. Still, resonances of second, third and fourth order turned out to be dangerous and had to be strictly avoided. The beam–beam interaction introduced higher order resonances. During the first years of operation it was understood that it is important to limit the tune spread of the particles to values that could fit between resonances of higher order [12] (see Fig. 3).

The betatron tunes of an ensemble of particles differ from the tune of a single particle because:

- The tune spread due to the beam–beam interaction is in the order of the linear tune shift. In a hadron collider the tune spread is independent of the beam energy. The tune in the SPS for a particle with small betatron amplitude is shifted upwards.
- At injection energy the Laslett space charge detuning must be considered for intense bunches, with the tune of the central particle shifted

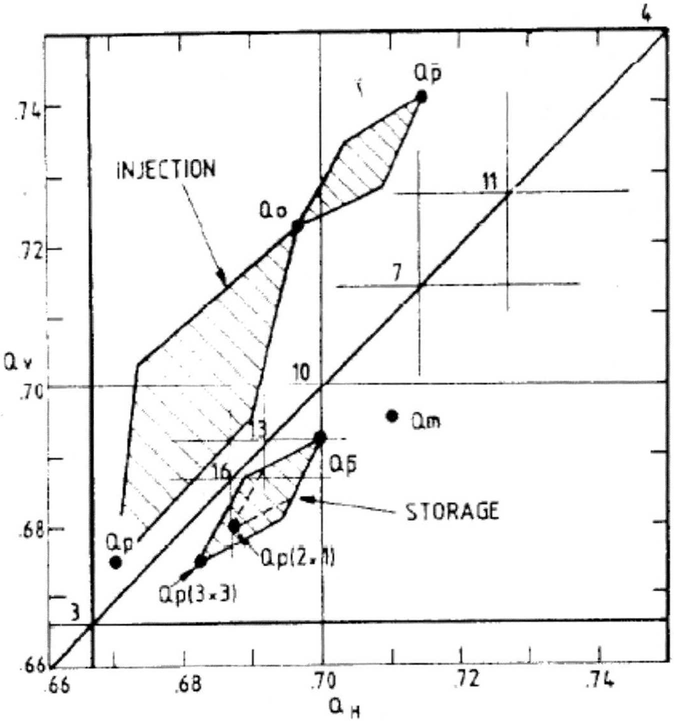


Fig. 3. Tune diagram for three-bunch operation. $Q_0$ is the single particle tune. Due to the coherent tune shift, the measured proton tune is $Q_m$. The tune shift due to Laslett detuning for small amplitude protons is $Q_p$. $Q_{\bar{p}}$ is the tune for small amplitude antiprotons, shifted due to the beam–beam effect.

downwards. The detuning decreases with $1/\gamma^2$ and is therefore negligible at collision energy.

With three proton and antiproton bunches colliding at six points, the tune shift of the antiprotons from the protons was about 0.003 per collision point. The total tune spread was about 0.018. At injection the Laslett incoherent tune spread was about $\delta Q_h = -0.03$ and $\delta Q_v = -0.05$, larger than the beam–beam tune spread but only present at injection when the beam–beam resonances were not yet excited by the beam–beam collisions. For a good transmission the particles had to be kept clear from resonances of third and fourth order during the time required for injection.

At top energy the destructive effect of 10th order resonances was observed [13]. Figure 4 shows a scan of the tune diagram with three proton bunches and a single weak antiproton bunch performed in the early days of the $Sp\bar{p}S$ collider: proton and antiproton bunch intensities recorded with a chart recorder are shown as a function of the tunes. The decay rate of

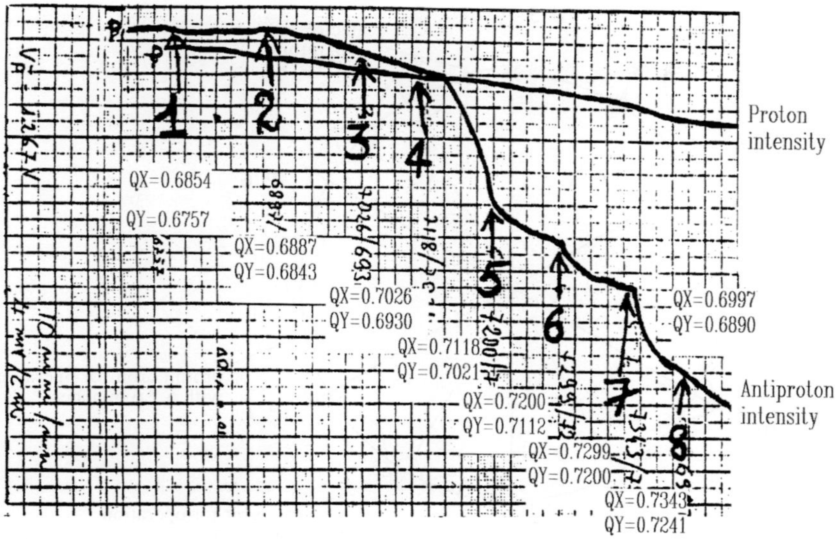


Fig. 4. Intensity of the proton and antiproton beams during a scan of the betatron tunes [14]. One can see the sharp and fast intensity losses when crossing a resonance.

the antiproton was extremely sensitive to the tunes. When they touched resonances of order 10 or lower, the lifetime dropped from some 100 hours to only 25 hours. At that time no effects from 13th and 16th order resonances were noticed.

During the first years of operation three proton bunches were colliding with three antiproton bunches. The intensity of the antiproton bunches was about 10 times less than the proton bunch intensity due to limitations in the $\bar{p}$ production.

With a substantially increased $\bar{p}$ production rate it was later possible to operate with six bunches per beam. To limit beam–beam effects the orbits of the two beams were separated using electrostatic separators at the unwanted collision points (Fig. 5). The intensity of the antiproton bunches was increased to about 60% of the proton bunch intensity and the luminosity increased by nearly one order of magnitude.

The electrostatic separators were installed close to the experiments. The beams were separated by about $6\sigma$ at the unwanted crossing points in the horizontal plane during physics operation (the separation is defined as the distance between the beams, $\sigma$ is the rms horizontal beam size). During injection and ramping one separator created an orbit deformation around the ring with opposite sign for both beams, and the separation between the beams differed for the 12 crossing points between 1.5–$6\sigma$ [15].

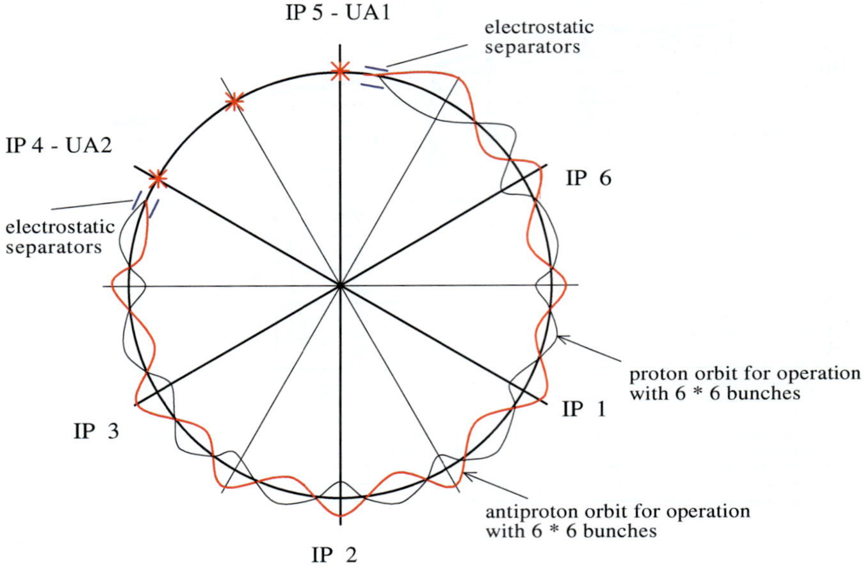


Fig. 5. Layout of the SPS proton–antiproton collider with the proton and antiproton beams separated by electrostatic deflectors.

During the commissioning of the high luminosity operation the $Sp\bar{p}S$ was operated with beams of unequal emittances. The beam–beam effect with beams of different sizes was found to be much more destructive and caused an unacceptable level of background in the physics detectors and a low lifetime for the proton beam, despite the lower intensity of the $\bar{p}$ bunches. Operation with unbalanced emittances leads to an increase of the strength of high order resonances. After the emittances of both beams were balanced, the adverse effects the of beam–beam interaction were reduced to an acceptable level.

A betatron tune modulation due to unavoidable power supply ripples was always present. With a tracking program it was demonstrated that the stability of the beam decreases due to the beam–beam interaction when such modulation was introduced in the simulation [16]. Without such modulation the particles were much more stable.

## 8 Pulsed Sp$\bar{\text{p}}$S collider

The SPS was limited to 315 GeV per beam when running as a proton–antiproton collider due to resistive heating of the coils of the main magnets. The limit can be overcome by pulsing the magnets between 100 GeV and 450 GeV. During one physics run to study $p\bar{p}$ collisions at 900 GeV c.m. energy, beams of protons and antiprotons were collided under these pulsed

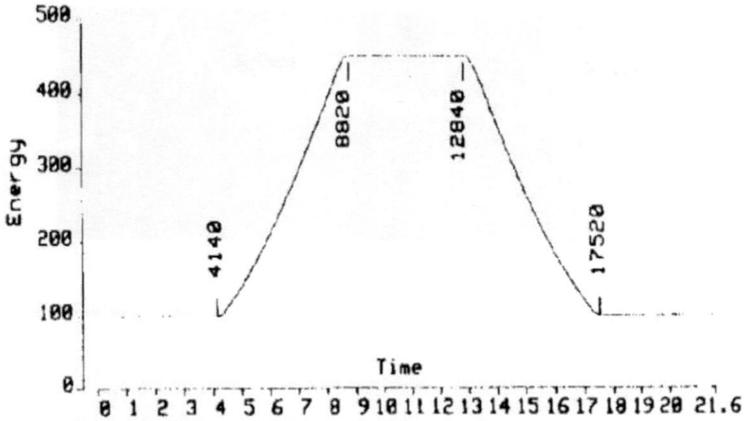


Fig. 6. The storage cycle with a length of 21.6 s of the $Sp\bar{p}S$ with two plateaus at 100 GeV/$c$ and 450 GeV/$c$.

conditions. By colliding beams of protons and antiprotons whilst ramping the SPS between 100 GeV and 450 GeV an overview of hadron physics in a very large range of energies could be obtained. Figure 6 shows the machine cycle for stored beams [17].

The first $p\bar{p}$ collisions at 900 GeV c.m. were observed in March 1985. Data taking by UA1 on colliding beams whilst ramping between 100 GeV and 450 GeV was done for a total of 95 hours. Initial luminosities were about $2 \times 10^{26}$ cm$^{-2}$s$^{-1}$ at 450 GeV beam energy and over 500000 events were recorded by the experiments.

A challenge was the control of the betatron tunes along the entire cycle. Typically the tune is adjusted in DC storage to an accuracy of about $\pm 0.001$. For the pulsed collider, hardware and software techniques were developed with the aim of adjusting the tune to a precision better than $\pm 0.005$ during the continuous ramping [18]. Two complementary methods were provided for measuring the tune. The first consisted of kicking the beam every 60 ms throughout the cycle and performing a fast Fourier transform on the position data from 256 turns following the kick. This method is robust, working well over a wide range of beam conditions and has an accuracy of 0.005. The second method was to excite the beam at a multiple of the betatron frequency and feed the response back to the excitation kicker via a phase lock loop. This has an accuracy of better than $\pm 0.001$ as shown in Fig. 7. Both results could be fed back into the power supply reference via a software loop. These techniques could also be extended to a measurement of the chromaticity throughout the cycle.

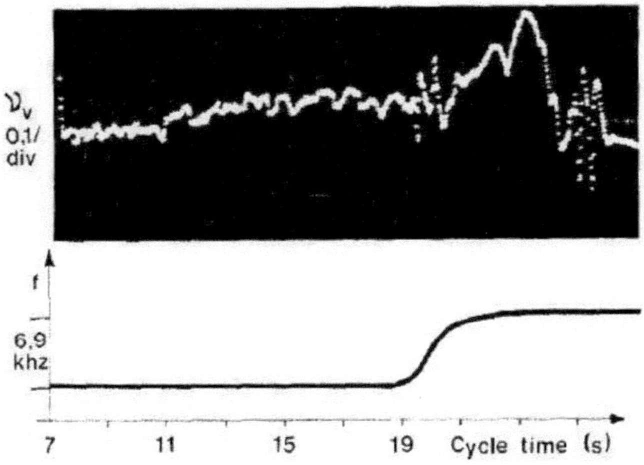


Fig. 7. Betatron tune during ramp.

## 9 Legacy of the Sp$\bar{p}$S

The SPS $p\bar{p}$ collider provided ground breaking developments for several other accelerators. In particular the LHC profited from the experience gained at the $Sp\bar{p}S$.

The $Sp\bar{p}S$ demonstrated that hadron colliders can deliver high luminosity. The lifetime for a single beam can exceed several hundred hours. The noise, in particular from the RF system, can be well controlled to avoid beam losses and emittance blowup. The beams are also stable in the presence of beam–beam effects. The effect of beams with different emittances was understood.

Operational procedures and automatic controls were vital for efficient operation. Antiprotons were scarce and it was required to reliably transfer the beam through the injector complex to the SPS, including the energy ramp and beta squeeze in SPS. Before the injection of antiprotons, a large amount of hardware must be checked for correct operation. After successful injection, various hardware systems need to switch to a different state. Data must be acquired and acceleration prepared. Once the beams are accelerated, final tuning and checks must be executed before data taking of the experiments starts. This was only possible by running automatic processes in a predefined sequence on many computers controlling the SPS. A job-control language was created to perform these tasks which also allowed real-time parallel processing across the control systems [19].

The rise in intensity over the years was in part due to hardware improvements, but a significant part came from better operational control over the beam parameters such as tune, orbit and chromaticity. Losses on the first turn after injection were eliminated by providing an automatic program to steer the injection trajectory so as to minimise the betatron oscillations on the first few turns. This program used the betatron signal from a single pickup as provided by the Q measurement system, and from this information the position and angle error at the injection point was deduced, and hence corrected.

Losses at the injection energy 26 GeV before acceleration were eliminated by reducing the closed orbit deviations at 26 GeV to less than 2 mm peak. The closed orbit was corrected by inserting appropriate beam bumps at every quadrupole around the machine. Sophisticated beam instrumentation with a large dynamic range was required to monitor beam parameters, in particular betatron tunes and emittances.

Wire scanners have been used for measuring the transverse beam size of proton and antiproton beams since the start of $p\bar{p}$ operation, by moving a wire on a circular path with a speed of 4 m/s through the beam. To increase the accuracy of the measurement a new device was developed: a thin wire was moved on a linear path through the beam with a reduced speed. This allows to measure the beam size with an error in the order of one micron [20]. Figure 8 shows the profile of a proton bunch, the horizontal scale is given in relative units. The value of sigma for this measurement is 120 micron.

## 10 Final remarks

For 35 years the SPS has been an extremely successful accelerator. The highlight was its operation as proton–antiproton collider which resulted in the discovery of the W and Z bosons. These discoveries and a new technique for cooling particles were awarded with a Nobel Prize for Carlo Rubbia and Simon van der Meer in 1984.

The SPS has been used during many years for numerous fixed-target experiments. During the operation of LEP, the SPS accelerated electrons and positrons from an energy of 3.5 GeV/$c$ to 20 GeV/$c$ for injection into LEP. The SPS has also been used by the CNGS experiment to produce a neutrino stream to be detected at the Gran Sasso laboratory in Italy, 730 km from CERN [21]. Today, the SPS operates as final injector of high-intensity proton beams and ion beams for the Large Hadron Collider (LHC), e.g. accelerating protons from 26 GeV to 450 GeV. In parallel, it continues to provide beam for fixed-target experiments and for studies on the interaction of high power beam with targets. The SPS has served as a test bench for new concepts in accelerator physics, such as an observatory for the electron

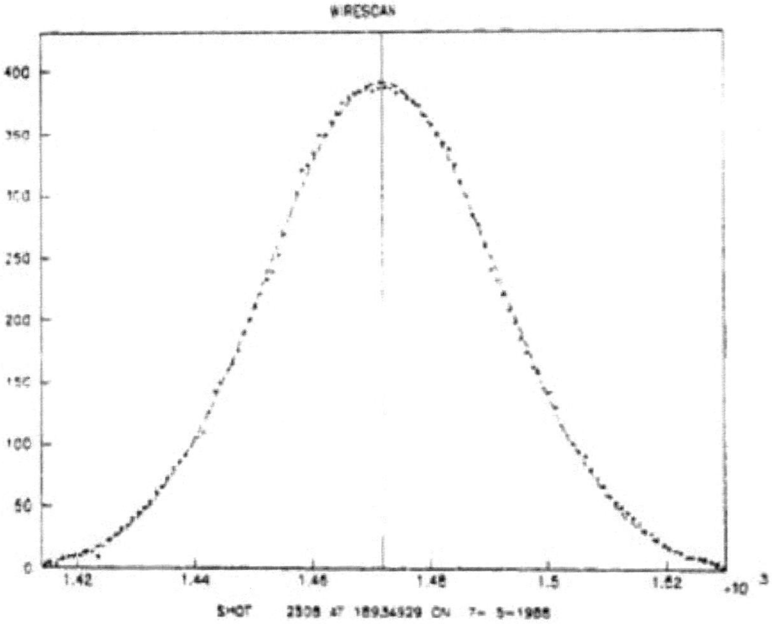


Fig. 8. Profile of a proton bunch from a measurement with a wire scanner, sigma for this measurement is 120 micron.

cloud phenomenon [22]. The experience from the SPS was essential to define the parameters for LHC. The success of the LHC Run I was to a large extent based on the performance of its injector complex (including the SPS), which delivered beams with much higher brightness.

## References

[1] H. Koziol and D. Moehl, The CERN antiproton collider programme: Accelerators and accumulation rings, *Phys. Rep.*, **403–404**, 91–106 (2004).

[2] The 300 GeV programme, CERN-1050, 1972. French version available: "Le programme 300 GeV."

[3] D. W. Kerst *et al.*, Attainment of very high energy by means of intersecting beams of particles, *Phys. Rev.*, **97**(102), 590–591 (1956).

[4] K. Johnsen, P. J. Bryant, and W. S. Newman, Opening address, in *CAS — CERN Accelerator School: Antiprotons for Colliding-Beam Facilities*, Geneva, Switzerland, 11–21 Oct 1983 (1984).

[5] C. Rubbia, P. McIntyre, and D. Cline, Producing massive neutral intermediate vector bosons with existing accelerators, *Proc. Int. Symp. on Electron and Positron Storage Rings*, Saclay, 1966 (eds. H. Zyngier and E. Crémieu-Alcan) (1967), p. 346.

[6] S. van der Meer, Stochastic damping of betatron oscillations in the ISR, Tech. Rep. CERN-ISR-PO-72-31, 1972.

[7] P. Bramham, G. Carron, H. G. Hereward, K. Hubner, W. Schnell, and L. Thorndahl, "Stochastic cooling of a stored proton beam, *Nucl. Instrum. Methods*, **125**, 201–202 (1975).

[8] H. Faissner, H. Reithler, and P. Zerwas, in *Proceedings of the International Neutrino Conference*, Aachen, June 8–12, 1976 (Vieweg, 1977).

[9] D. Boussard, RF for the CERN proton antiproton collider, Tech. Rep. CERN-SPS 85-38 (ARF), CERN, 1985.

[10] V. W. Hatton, Performance of the collider 1981–1990: Parameters related to beam losses, Tech. Rep. CERN-SL-92-20-OP, CERN, Geneva, Feb 1992.

[11] P. E. Faugeras, A. Faugier, A. Hilaire, and A. Warman, The low beta insertions of the SPS proton-antiproton collider, in *12th International Conference on High-Energy Accelerators*, Batavia, IL, USA, Aug 1983.

[12] B. de Raad, The CERN SPS proton-antiproton collider, in *11th IEEE Particle Accelerator Conference*, Vancouver, Canada, 13–16 May 1985, p. 1650.

[13] L. R. Evans, Beam effects in hadron colliders, *3rd Summer School on High-Energy Particle Accelerators: Symposium on the State of High Energy Physics*, Upton, NY, USA, 6–16 Jul 1983, pp. 243–393.

[14] L. R. Evans, The beam-beam interaction, *CAS — CERN Accelerator School: Antiprotons for Colliding-Beam Facilities*, Geneva, Switzerland, 11–21 Oct 1983, Jan 1984.

[15] L. R. Evans, A. Faugier, and R. Schmidt, Beam separation at the CERN SPS collider, *11th IEEE Particle Accelerator Conference*, Vancouver, Canada, 13–16 May 1985.

[16] M. Meddahi, *Effets faisceau-faisceau dans le collisionneur protons-antiprotons du SPS*. PhD thesis, Paris 7, Geneva, 1991.

[17] R. J. Lauckner, Pulsed operation of the CERN SPS collider, *11th IEEE Particle Accelerator Conference, pt.1*, Vancouver, Canada, 13–16 May 1985.

[18] R. Bossart, A. Chapman-Hatchett, I. Gjerpe, H. K. Kuhn, T. P. R. Linnecar, G. Paillard, C. Saltmarsh, W. Scandale, R. Schmidt, and I. Wilkie, Tune measurement and control at the CERN-SPS, in *11th IEEE Particle Accelerator Conference*, Vancouver, Canada, 13–16 May 1985.

[19] C. Saltmarsh, A multi-processor, multi-task control structure for the CERN SPS, in *Proc. Computing in Accelerator Design and Operations*, Berlin, 1983 (1985), pp. 509–517.

[20] A. Burns, J. Camas, E. D'Amico, G. Ferioli, Q. King, K. H. Kissler, J. Mann, and R. Schmidt, Wire scanner news from the CERN-SPS, in *13th IEEE Particle Accelerator Conference*, Chicago, IL, USA, 1990.

[21] E. Gschwendtner *et al.*, CNGS, CERN neutrinos to Gran Sasso, five years of running a 500 kilowatt neutrino beam facility at CERN, in *4th International Particle Accelerator Conference, IPAC2013*, Shanghai, China, May 2013.

[22] G. Arduini, K. Cornelis, W. Hoefle, G. Rumolo, and F. Zimmermann, The electron cloud instability of the LHC Beam in the CERN SPS, in *20th IEEE Particle Acclerator Conference*, Portland, Oregon, USA, May 2003.

# Chapter 11

# The Antiproton Accumulator and Collector and the discovery of the W & Z intermediate vector bosons

Vinod Chohan* and Stephan Maury (CERN)

## Preface

In the beginning of the second quarter-century of CERN's existence, one of the most successful undertakings at the laboratory was the quest for the intermediate vector bosons, $W$ and $Z$, just a couple of years after the start-up of the SPS. The Antiproton Accumulator (AA) was a key element in the overall scheme to have colliding beams of protons and antiprotons in the SPS, designed originally for fixed target physics at 300 GeV. The process of building the AA and later, the Antiproton Collector (AC) relied on a number of innovations and several novel accelerator technologies as well as major modifications to the 26 GeV PS Complex.

The AA was indeed an adventure into uncharted territory. Never before in CERN's history had a project called for such imagination, involving the whole of CERN, as for the search for the intermediate vector bosons. The proton–antiproton colliders were first discussed at CERN in 1962 [1], seven years after the antiproton discovery at the Bevatron, in Berkeley. Considerations included the CERN Intersecting Storage Rings (ISR), then in their early design stage. It was concluded in 1962 that with the expected densities of antiproton beams, luminosities were discouragingly low. However, prospects changed in 1966, with Budker's [2] and O'Neill's [1, 2] ideas of electron cooling. In [2], Budker describes the application of his invention specifically for the accumulation of antiproton beams dense enough to make proton–antiproton colliders viable. These require only a single ring, a concept already practiced at that time with electron–positron colliders. As early

*Corresponding author email:Vinod.Chohan@cern.ch.

as in 1966, Rubbia had realized the potential offered by the projected "300 GeV Machine" later built at CERN as the Super Proton Synchrotron (SPS), proposed to use it as a proton–antiproton collider [1].

The next step occurred in 1968, when Simon van der Meer invented stochastic cooling (published only in 1972 [3]). Both electron and stochastic cooling were experimentally proven in 1974, at the NAP-M storage ring in Novosibirsk [4] and at the CERN ISR [5], respectively. With these tools at hand, Rubbia *et al.* [6] revived the idea of converting the SPS into a proton–antiproton collider, with the specific aim of producing the long-awaited $W$ and $Z$ bosons. At that stage, his scheme was still based on electron cooling.

The flux of antiprotons required implied a major undertaking for the production and accumulation of these particles. Hence the need for the approval and construction of this machine; this indeed was considered as an 'experiment' because of the multitude of challenges and unknowns that had to be faced and overcome, together with the associated risks.

Two CERN working groups examined in 1976 the technical aspects of such schemes and the physics potential. Finally, CERN decided to pursue two courses of action in parallel. One was to construct rapidly a small ring (Initial Cooling Experiment — ICE) to study both electron and stochastic cooling; the other was to set up a study group to prepare a design for a $p\bar{p}$ facility using the SPS as a storage ring and collider. Initially, the study group proposed using two separate rings for collecting and cooling antiprotons, because it was clear that the electron cooling scheme would only work at low energy for the large emittance antiproton beams. Hence, the second ring was to decelerate the antiprotons. Meanwhile, the many experimental tests in the ISR, further theoretical developments and most importantly the proposal of a faster and more efficient method of longitudinal cooling (the Thorndahl filter method) gave the possibility of a solution based entirely on stochastic cooling and stacking. This then is how the AA was conceived and born — a fixed field, single DC-operated accumulator ring. The potential savings in cost and complexity over the two-ring idea were the ultimate criteria, despite the fact that it represented three orders of magnitude in extrapolation over the ISR cooling experiments.

The AA proposal called for an overall increase in antiproton density from the production target to the stack core of over $10^9$. Meanwhile the ICE ring in 1978 had given encouraging results in stochastic cooling [7], confirming cooling in all planes; however, these were at timescales (longitudinally) of the order of 10 seconds, still about 30 times slower than what was required in the AA. Nevertheless, it led to a rapid gamble to go ahead with the construction

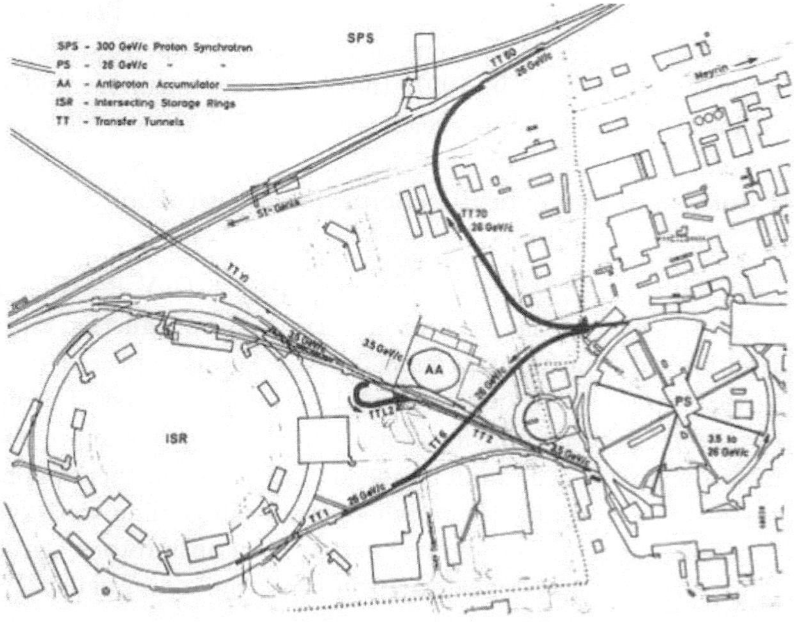


Fig. 1. The layout of the CERN accelerators, 1981, with existing and the new tunnels (bold) constructed for beam transfers to PS ISR and SPS from the AA.

of the Antiproton Accumulator (AA) based only on stochastic cooling. The antiprotons were cooled and accumulated directly at 3.5 GeV/$c$ [8], where the yield from 26 GeV (PS) protons is highest. The overall scheme (see Fig. 1), involved also some major modifications to the PS machine as well.

The stochastic stacking process, an essential feature in the accumulation scheme in the AA, could not be tested in ICE. This process involved simultaneous cooling in both transverse planes and increasing the longitudinal density by four orders of magnitude whilst moving the particles into the dense core [9]. This then was the biggest gamble in the launch of the AA because it could only be studied in detail by theoretical calculation. Fortunately, in this respect the AA performed as expected [9].

Construction of the AA began in 1979. In April 1981, the first proton–antiproton collisions occurred in the ISR, at $2 \times 26$ GeV. The SPS followed on the heels, with collisions at $2 \times 270$ GeV on 10 July 1981. The first $W$ data were taken in 1982 and the discovery of the $W$ and $Z$ was announced in 1983. In order to satisfy the ever-increasing appetite of antiproton users, the Antiproton Collector (AC) was built [10] around the AA in 1986. From 1987 onwards, after the initial running-in, it boosted the accumulation rate by an order of magnitude. After the last collider run in 1991 [11,17], the SPS returned to its mode as a fixed target physics accelerator. The Low-Energy Antiproton Ring (LEAR) [12] continued to take beam from the AC/AA complex until end of 1996. In 1997, the AA was dismantled and the AC converted into the Antiproton Decelerator (AD); LEAR was closed down.

## 1 Brief outline of the overall scheme for antiprotons of the SPS as a collider

The 50 MeV Linac, the 800 MeV Booster and the 26 GeV PS (Fig. 1) were pushed hard to deliver an intense proton beam on the production target. The burst of antiprotons emerging around 3.5 GeV/$c$ was captured in the AA (or, after 1987, in the AC). In the original AA, a fresh burst of antiprotons remained on the injection orbit for at least 2.4 s for stochastic pre-cooling of the momentum spread. The radiofrequency (RF) system then trapped and moved them to the stacking region where stack-tail cooling took over. The injection region was then free for the next burst of antiprotons, arriving 2.4 s later. This sequence was repeated during the whole accumulation period (Fig. 2).

With the addition of the Antiproton Collector ring, built encircling the AA, some of the AA's 'pre-cooling' functionalities were given to the AC ring. A powerful "bunch rotation" RF system (1.5 MV, 9.5 MHz) turned the incoming five antiproton bunches into a nearly continuous beam with reduced momentum spread in the AC. Stochastic cooling in all three planes then reduced the emittances by large factors. Another RF system (3.5 kV, 1.6 MHz) re-bunched the antiprotons, for ejection and transfer into a matched bucket on the AA injection orbit. All this and the cooling processes in the AA were repeated every 4.8 s.

In both the original and the modified AA, the antiproton stack was subjected continuously to up to six different stochastic cooling systems. Over a day, a stack with a dense core of several $10^{11}$ antiprotons was accumulated. An antiproton bunch was picked from the stack by creating an "unstacking bucket" in the core region and moved to the ejection orbit. From there it was sent through a "loop" (TTL2, Fig. 1) to the PS. In the PS, this bunch was accelerated to 26 GeV/$c$ and sent to the SPS (see Section 5). Prior to a transfer, careful checks were performed, concluded by the dispatch of a small "pilot bunch" of $\sim 10^9$ antiprotons all the way from the AA to the SPS. This assured that the big shot, containing a day's worth of accumulated antiprotons, would safely find its way. Transfers of antiprotons to LEAR were made in a different way, described in [12].

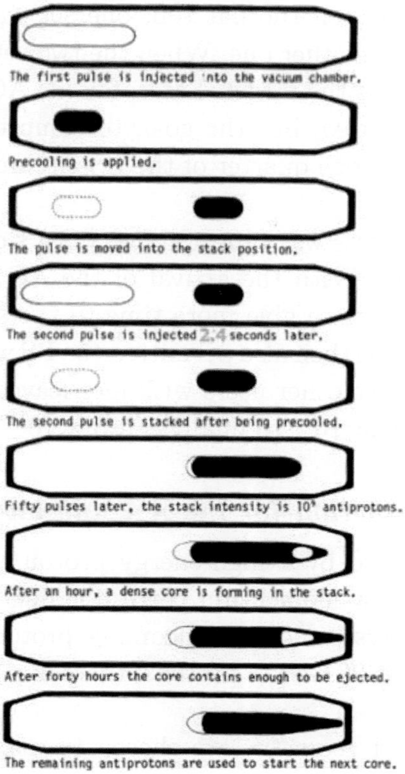


Fig. 2. Schematic sequence of cooling and accumulation in the original AA (~1981) before the advent of the additional Collector ring in 1987.

## 2 Antiproton production and accumulation

### 2.1 *High-intensity primary proton beam from the PS*

Maximizing the number of antiprotons demands a proton beam intensity as high as the target could withstand. The transverse emittances had to be small, to permit focusing to a small size over the whole length of the target. Finally, the proton burst length had to correspond to the AA circumference, one quarter that of the PS.

This meant that the PS beam, normally 20 bunches around the whole circumference, had to be merged into five bunches. The fact that the 800 MeV PS Booster consists of four superposed rings, each one quarter of the PS circumference, was helpful. The beams from two Booster rings (five bunches each) were ejected simultaneously, combined in the vertical plane, and injected into the PS. This was repeated with the other two rings, timed such that two sets of five bunches circulated on opposite sides of the PS. After acceleration to 26 GeV/$c$, on the flat top, one set was slightly accelerated and advanced towards the other one. When the two sets had fallen into step, they were ejected towards the target. Later, for the AC, different techniques of bunch merging were used, but the goal, to compress the beam into five bunches distributed over one quarter of the PS circumference, remained the same [13].

Nominally, that process was repeated every 2.4 s during the days of AA only operation till 1986. With the arrival of the AC ring, the interval was extended to 4.8 s, in order to give more time to the stochastic cooling processes in the AC ring. This then also permitted the PS with its basic repetition rate of 2.4 s cycle to serve other users with interleaved cycles. The intensity was raised from a prudent $10^{13}$ to finally $\sim 1.3 \times 10^{13}$ protons per pulse.

### 2.2 *Antiproton production for the AA and later the AC*

Antiprotons are produced by a high-energy proton beam hitting a target. The yield is highest at a momentum corresponding to production at rest in the center-of-mass system of the incoming protons and the stationary target nucleons [14]. For 26 GeV/$c$ protons, there is a flat maximum at $p\bar{p} \approx$ 3.5 GeV/$c$.

Target material and geometry are chosen to maximize proton interactions, while minimizing antiproton re-absorption. Similarly, beam optics before and after the target are vitally important [15, 16]. The result was a rod-shaped target (Fig. 3), made of heavy material, with a diameter of about 3 mm and a length, between 50 and 120 mm, of the same order as the inelastic collision length. The channel following the target has to match the "antiproton line source" (target) to the acceptance of the ring (AA, later AC). This is obtained by making the horizontal and vertical beta functions at the target about equal to its length ($\beta_{\rm h} \approx \beta_{\rm v} \approx L_{\rm t}$). The conversion rate is then (approximately) proportional to $\Delta p\sqrt{E_h E_v}$ where $\Delta p$ is the momentum acceptance and $E_h \approx E_v$ the transverse acceptances of the storage ring.

During the original AA years (1980–86), a 120 mm long copper target was used and gave a conversion rate of antiprotons to protons of the order of $0.6 \times 10^{-6}$. This was with a magnetic horn (Fig. 4) as the collector lens after the target. For the AC ring from 1987 onwards, with its larger acceptances, a 60 mm long iridium target was used. The best conversion rates obtained were $\sim 5.6 \times 10^{-6}$ with a lithium lens and $\sim 4.6 \times 10^{-6}$ with a magnetic horn [17], with incident beam of $1.5 \times 10^{13}$ protons on target.

The small-emittance primary beam was matched to the target with quadrupoles and, for a certain trial period, with an additional lithium

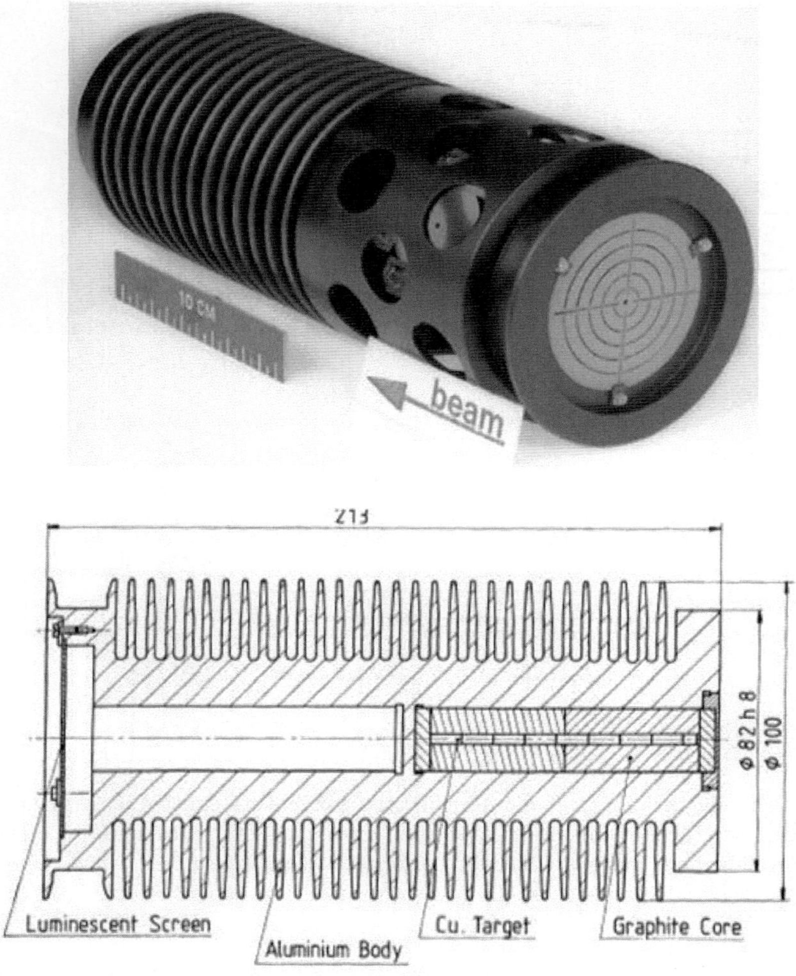


Fig. 3. An antiproton production target assembly for the AA (1980): the tungsten rod, 110 mm long, and cooling fins for forced air cooling. Correct position and size of the 26 GeV/$c$ proton beam from the PS were monitored on a scintillator screen in front of the target, with circles every 5 mm.

lens. For capturing the antiprotons from the target one used magnetic horns [18, 19] (Fig. 4), and lithium lenses [20]. Magnetic horns proved more robust and easier to replace in the highly radioactive target area. Lithium lenses gave somewhat better yield. Early operation of the AA was with a horn, then lithium lenses were used for several years, and again horns for the late AC period and for the AD.

A magnetic horn [18, 19] is a "current-sheet lens". Figure 4 pictorially shows its principle, with its inner wall "horn membrane". A current flowing

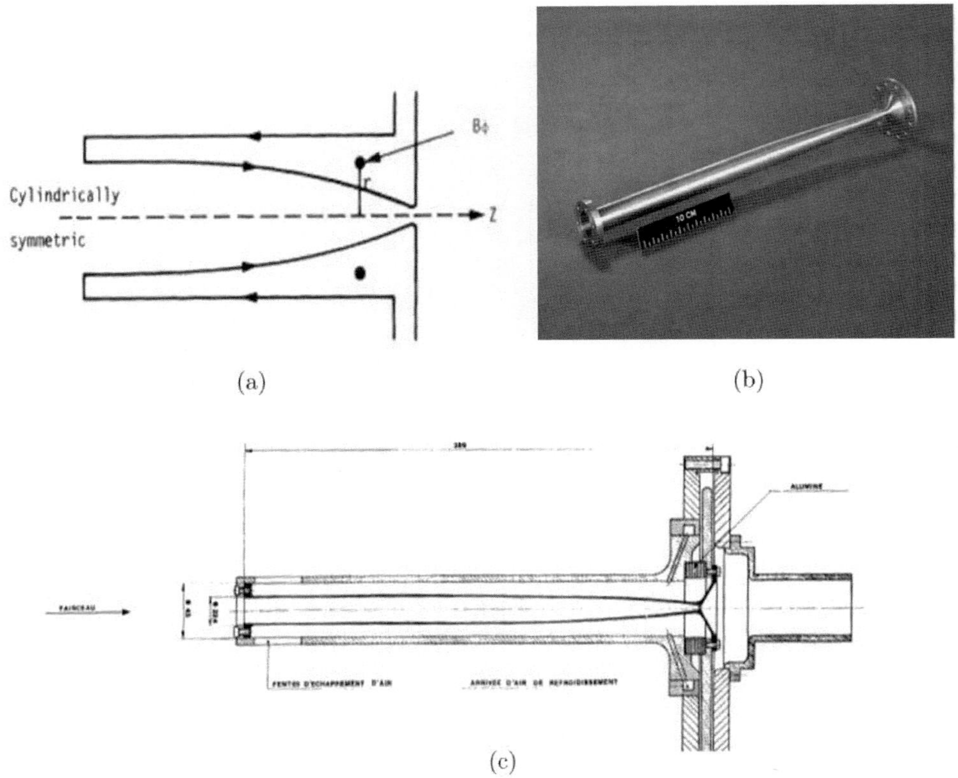


Fig. 4. The magnetic horn of the AA collected antiprotons emitted at large angles from the target: (a) the flow of electrical current in a magnetic horn and (b, c) the inner conductor. The horn was pulsed at 400 kA for 15 $\mu$s (half-sine).

through the inner wall and returning via the outer one creates an azimuthal magnetic field in the space between them, but no field inside the horn. Antiprotons produced at large angles traverse the inner wall and are bent towards the axis. The horn membrane had to be thin to avoid scattering and absorption of the penetrating particles. It was made from aluminium, about 1 mm thick. The pulse currents were of the order of several hundred kiloamperes.

Lithium lenses [20] are rods of lithium, chosen for its low interaction with protons and antiprotons, surrounded by a pulse transformer, which induces a current along the axis. The azimuthal magnetic field inside and outside the rod focuses on the particles. Rods of 34 mm diameter were used during SPS Collider operation, a 20 mm version for operation with LEAR alone. The pulse currents were of the order of 1000 kA for the 34 mm Li lens; a 20 mm Li lens was also used in routine operation for a short while. Figure 5

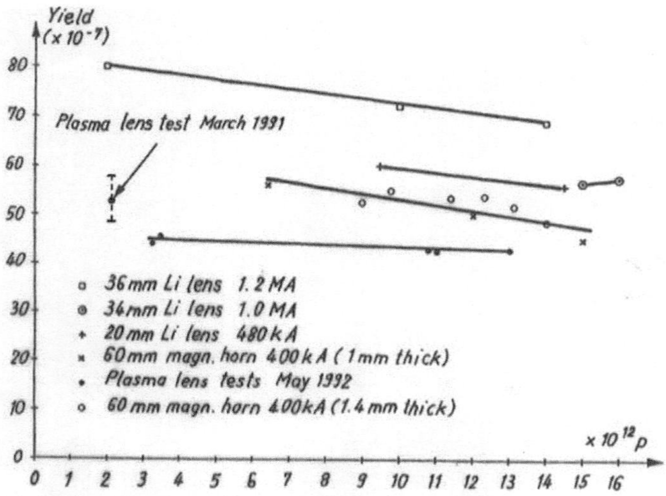


Fig. 5. Yields vs. intensity for different Collector lenses.

illustrates the yields with different Collector lenses over the years and trials to increase the antiproton yields at CERN [17].

## 3 The AA and AC storage rings

The AA [8,21] had a circumference of 157 m. Its magnetic field was constant for a beam momentum of 3.5 GeV/$c$ on central orbit. It was built in a new hall and later, covered with concrete shielding blocks (Fig. 6).

The optical properties were quite particular, dictated by the large acceptances required for antiproton collection and the needs of stochastic cooling and stacking. This lead to an impressively wide horizontal aperture, as much as 0.7 m in the regions of large dispersion; here, the orbit position depends strongest on momentum, providing spatial separation between the stored beam stack and the newly injected beam.

Quadrupoles and bending magnets (weighing 11 t and 75 t, respectively) were large in those large dispersion regions, but of conventional size in the smaller dispersion regions. Sections with vanishingly small dispersion were required for the location of kickers for momentum cooling of the stack, to avoid "heating" of the horizontal emittance. The injection kicker and the precooling devices had shutters for electromagnetic separation from the stack. Ultra-high vacuum ($10^{-11}$ torr) assured long storage times and, together with an elaborate clearing system, helped to reduce beam instability caused by ions trapped in the antiproton beam.

The AC [10,22] (Fig. 7) had a circumference of 187 m and was installed during a record short period of eleven months in 1986–87, around the AA.

Fig. 6. The AA ring in its hall (1980–81), before and after being put under heavy concrete shield.

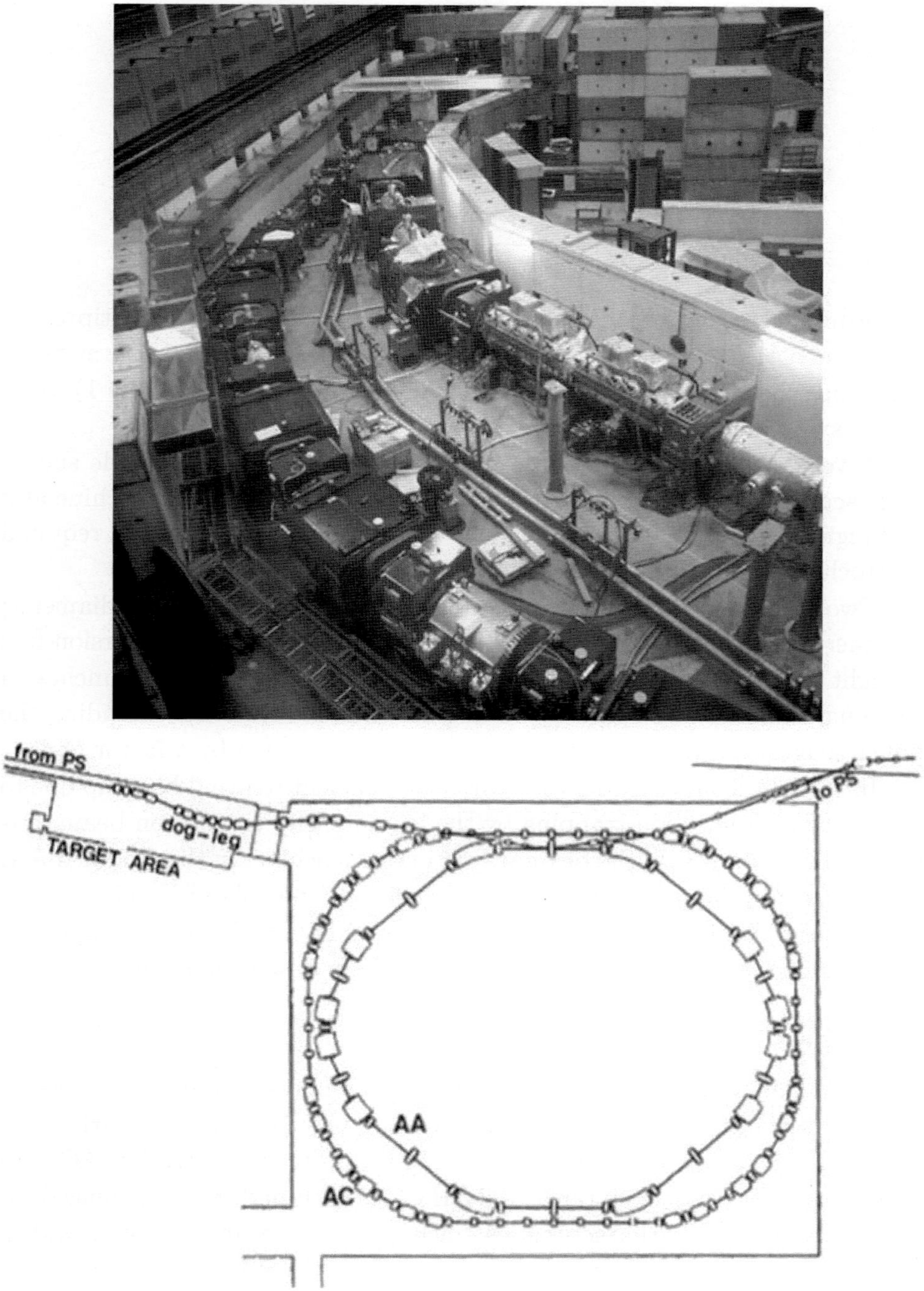


Fig. 7. In 1986/87, the AC was built around the AA and a "dog-leg" was incorporated in the antiproton injection line to diminish the flux of electrons reaching the hall. (From [10])

Table 1. Acceptances of AA and AC machines.

| | Acceptances attained simultaneously | | |
|---|---|---|---|
| Ring | Horizontal $E_h$ ($\pi$ mm mrad) | Vertical $E_v$ ($\pi$ mm mrad) | Momentum $\Delta p/p$ (%) |
| AA | 85 | 85 | ±0.75 |
| AC | 200 | 200 | ±3.00 |

Its primary task was to capture an order of magnitude more antiprotons than the AA, for which it had much larger acceptances (maximum beam emittance that can be stored in the machine without losses) (Table 1), both in transverse and in momentum planes.

A very strong focusing lattice was needed in the AC to provide such a large acceptance within an aperture not exceeding 0.35 m. The machine also had regions with strong dispersion and others with zero dispersion, required by stochastic cooling, but in a less extreme way than the AA.

Two large and powerful cavities (each 2 m long and 2.5 m in diameter; together operating at 1.5 MV, 9.5 MHz) were installed in a dispersion-free straight section. After beam injection, the cavities rotate the bunches in the longitudinal phase space by 90 degrees (bunch rotation), extending the bunch length and reducing the beam momentum spread by a factor of 4.

In the AC, a vacuum of a $10^{-8}$ torr was sufficient to avoid blow-up due to scattering as well as ion-trapping by the low-intensity antiproton beam during the 4.8 seconds that the beam stayed in this machine. In the modified AA, the higher intensity and density made an efficient control of beam instability compulsory. Additional clearing by "beam shaking" and active damping of coherent instabilities were vital to reach the new design performance.

## 4 Stochastic cooling and stacking

The AA, and later also the AC, had large numbers of stochastic cooling systems: seven in the original AA (frequency range 150 MHz to 2 GHz), five in the modified AA (frequency range 1–8 GHz), after the arrival of AC, and nine in the AC (frequency range 1–3 GHz). As each had to be optimized for a specific task, their characteristics (pickup/kicker technology, bandwidth, gain, power, etc.) differed vastly.

The pre-cooling systems, acting on newly injected beams of less than $10^8$ antiprotons with time constants of a second, required high-gain, high-power amplifiers. In the original AA, pre-cooling acted only on momentum spread, using Lars Thorndahl's notch-filter method [23]. Pickups and kickers consisted of short ferrite-frames around the beam, resembling beam current transformers. Signals were coupled out or in, via one-turn loops on one side of the frame, the opposite side being movable, acting as a shutter. The shutters were opened when the beam was moved out the precooling region. A 2 m long tank contained 100 frames, and there were two kicker tanks and two pickup tanks. The bandwidth was 150–500 MHz, and the amplifier rating 5 kW.

In the AC, pre-cooling was applied in all three phase planes. Pickups and kickers consisted of loop couplers, with electrodes left and right, or above and below the beam. The difference signal served for transverse cooling and the sum signal for momentum cooling. Two plate pairs connected in series formed a "super electrode"; 24–48 of these (depending on their frequency band) were housed in 2 m long tanks. The electrodes moved in, to follow the shrinking beam size during cooling, thus maximizing pickup sensitivity and minimizing kicker power. The low-level components (pickups, terminations, preamplifiers) were cryogenically cooled (20 K) to reduce noise. Three bands (1–1.65, 1.65–2.4 and 2.4–3 GHz) were used, with three combined horizontal/momentum and three vertical/momentum systems, six pickup and six kicker tanks. Amplifier ratings were 4.5 kW for the lower band and 2.6 kW for each of the higher bands. Thus the AA and AC cooling systems constituted as very-wide-band, high-power, low-noise, cryogenic, multi-band microwave systems, with feed-throughs and fast-moving electrodes in ultra-high vacuum.

The art of stacking by stochastic cooling and the solution adopted for the AA are described in the literature [24,25]. Stacking is done in momentum space: The pickups are located in regions of large dispersion, where particles are radially separated according to their momentum. Partial-aperture cooling systems acted over different aperture regions. The overall gain was profiled to decrease ~exponentially from the low-density stack-tail (high gain) to the high-density core, by a factor comparable to the density ratio ($10^5$). Loop couplers were used for the tail and Faltin-type slotted TEM lines [26] for the core. In the original AA, systems operating in the bands of 250–500 MHz and 1–2 GHz performed tail and core cooling, respectively.

When later pre-cooling was done in the AC, simplifications became possible in the AA. In particular, shutters on the injection kicker and on the cooling devices at the injection orbit were no longer needed. Transverse cooling was less demanding, as beams of already low emittance arrived from the AC. On the other hand, the higher intensity and density put greater demands on stack cooling. The performance was improved by further momentum pre-cooling on the AA injection orbit and by a powerful transverse stack-core cooling system using partly the difference signal from the momentum systems. All cooling systems were replaced by higher frequency ones. More details can be found in the parameter lists [21,22].

## 5 Post-acceleration of antiprotons and beams for SPS Collider

The momentum of 3.5 GeV/$c$ was considered too low for direct transfer of antiprotons to the SPS. Therefore, the antiprotons, extracted in single bunches from the AA, were sent to the PS via the newly built "loop" (TTL2, Fig. 1), for acceleration to 26 GeV/$c$. Subsequent "bunch rotation" in the PS reduced their length to about 4 ns before transfer to the SPS through the new line TT70 (Fig. 1). This process was repeated every 2.4 s, until three (and later on six) antiproton bunches were circulating on the SPS injection orbit, equidistantly spaced. To ensure fail-safe operation with 'abort' possibilities in case of need, the transfer of the same number of "cheap" proton bunches preceded that of the "precious" antiprotons, minimizing antiproton beam loss due to faulty settings and thus maximizing the efficiency for Collider operation.

## 6 Proton test beams for the AA and AC from the PS

Initial setting-up of the AA (later also of the AC) was done with protons. For this purpose, the PS produced single bunches at 3.5 GeV/$c$, which first followed the path of the 26 GeV/$c$ protons towards the target location and then, with the target removed, through the antiproton injection line to the AA (later the AC), which for that purpose had all magnets in "opposite polarity".

Precision setting-up of the AA and AC magnets had to be done in "normal polarity", and the 3.5 GeV/$c$ proton test beam from the PS came through the loop (TTL2). This was done also during routine operation, preceding antiproton transfers from the AA to the PS, in order to verify the correct settings of the AA ejection and of the beam transfer line TT2 back to the PS.

All the exercises described above demanded precise tuning of the machines to each other, in terms of magnetic field, orbit position and revolution frequency. Timing logics were intricate, timing precision had to be high, and all the settings had to be flipped from one operational mode to the other, using highly automatized procedures [27].

## 7 The *W* and *Z* discoveries and the Nobel Prize

The first SPS proton–antiproton collisions occurred in July 1981 and the first real period of physics runs took place in 1982. December 1982 saw the

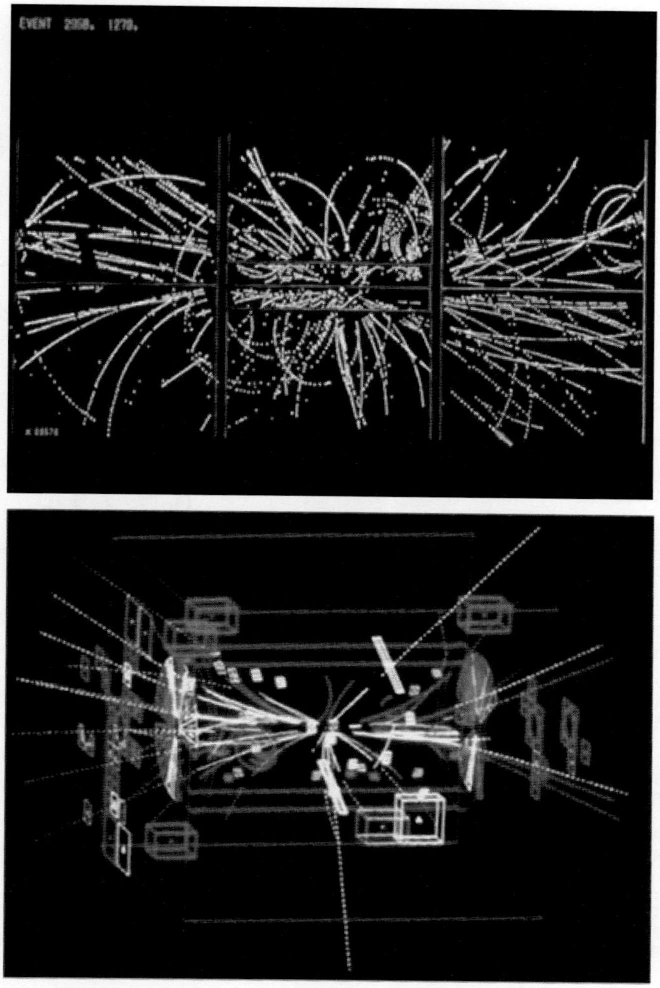

Fig. 8. First $W$ event in Dec 1982 (top picture) and the first $Z$ event in UA1 Experiment in April 1983.

Collider arriving at an integrated luminosity of 28 inverse nanobarns and Carlo Rubbia offering a champagne-only party with 28 champagne bottles! It suffices to say that the first signs of the $W$ boson were announced soon after, in January 1983. This was followed by the discovery of the $Z$ (Fig. 8), announced in May 1983. One needs not go into too many details of the $W$ and $Z$ discoveries and the Nobel Prize because these have been amply recorded elsewhere then and since then.

The Nobel citation (see Fig. 9) read that:

> *"The Nobel Prize in Physics for 1984 was awarded jointly to Simon van der Meer and Carlo Rubbia for their decisive contribution to the*

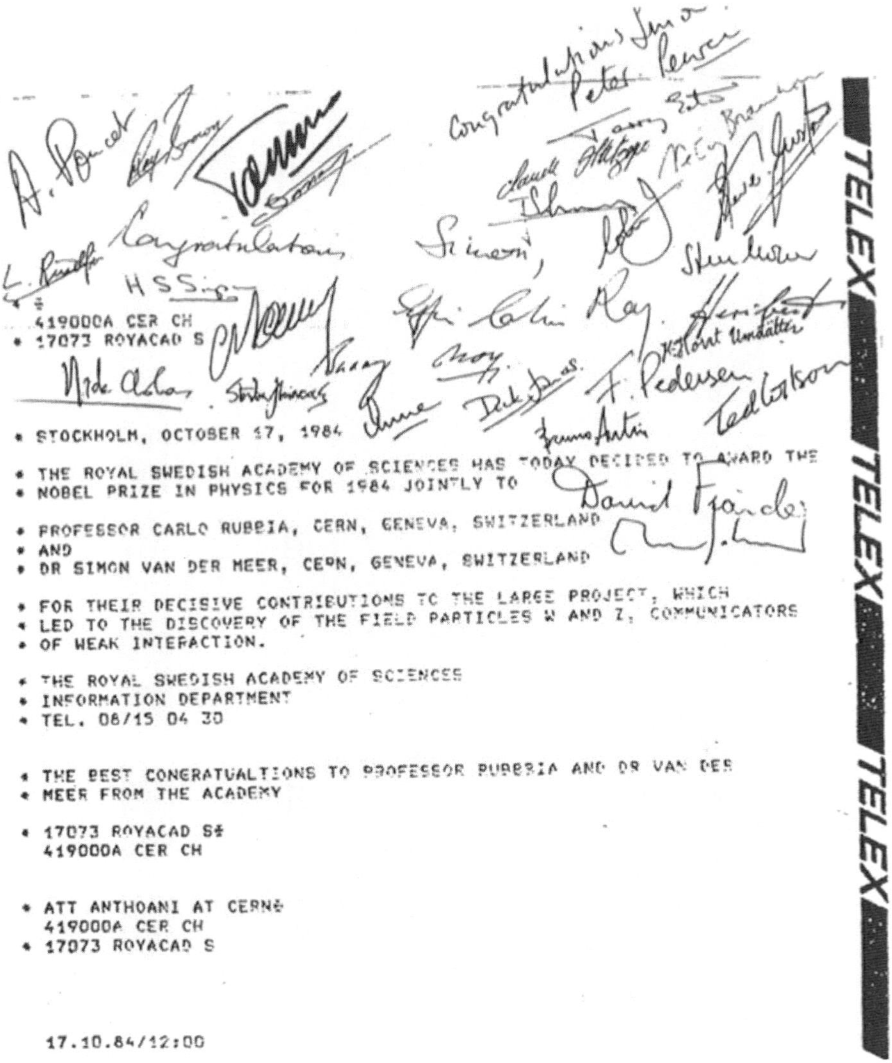
419000A CER CH
* 17073 ROYACAD S

* STOCKHOLM, OCTOBER 17, 1984

* THE ROYAL SWEDISH ACADEMY OF SCIENCES HAS TODAY DECIDED TO AWARD THE
* NOBEL PRIZE IN PHYSICS FOR 1984 JOINTLY TO

* PROFESSOR CARLO RUBBIA, CERN, GENEVA, SWITZERLAND
* AND
* DR SIMON VAN DER MEER, CERN, GENEVA, SWITZERLAND

* FOR THEIR DECISIVE CONTRIBUTIONS TO THE LARGE PROJECT, WHICH
* LED TO THE DISCOVERY OF THE FIELD PARTICLES W AND Z, COMMUNICATORS
* OF WEAK INTERACTION.

* THE ROYAL SWEDISH ACADEMY OF SCIENCES
* INFORMATION DEPARTMENT
* TEL. 08/15 04 30

* THE BEST CONERATUALTIONS TO PROFESSOR RUBBIA AND DR VAN DER
* MEER FROM THE ACADEMY

* 17073 ROYACAD S#
419000A CER CH

* ATT ANTHOANI AT CERN#
419000A CER CH
* 17073 ROYACAD S

17.10.84/12:00

Fig. 9. Copy of the Nobel Telex in the AA logbook, 17 October 1984.

*Large Project, which led to the discovery of the field particles W and Z, communicators of weak interactions."*

There was euphoria in the control rooms, particularly the local AA control room where we gloated over the telex copy, quickly stuck in the AA logbook and champagne started flowing. There were similar scenes in the UA1 cavern/control room (Fig. 10). The antiproton accelerator community certainly liked the words "large project" in the citation because it gave a collective recognition too!

Fig. 10. CERN's 1984 Nobel prize winners Carlo Rubbia (left) and Simon van der Meer, who were awarded the prize for their roles in discovering the $W^+$, $W^-$ and $Z^0$ particles.

## 8 Accumulator performance

During its initial years of operation the AA performance progressed steadily. Routine operation was attained in 1982. For AC and AA together, this was the case in 1988. Table 2 shows the peak performance obtained for the Collider operation in 1989, after the arrival of AC ring. In 1990, even further improvements were achieved to obtain a peak stacking rate of $6 \times 10^6$ antiprotons per hour [22]. Table 3 compares the AA operational performance in 1984 with operational performance of the AA+AC ensemble in 1994; the latter was for purely LEAR physics needs after the Collider operation had been terminated in 1991. It is worth noting the impressive increase in phase space density (8–9 orders of magnitude) that was consistently obtained during cooling and accumulation.

## Acknowledgements and conclusions

Much documentation and many papers exist regarding the construction of the AA and AC rings and CERN's adventure into antiprotons; citing a few references as we have hardly done justice to the work and technological ideas during the eighties; furthermore, substantial efforts made were during the $50^{th}$ birthday anniversary of CERN to summarize the work of that period and this contribution owes much to that [28]. The authors fully acknowledge the work of many and all during the period 1980–1993 that this chapter alludes to, as a brief overview of these two rings built at CERN during that epoch.

The search for the intermediate vector bosons $W$ and $Z$ was the driving force for the construction of the AA ring. The successes of that resulted

Table 2. AA+AC performance in 1988 and 1989.

| | 1988 | | 1989 | |
|---|---|---|---|---|
| | Design | Operation | Operation | Peak |
| Production beam (ppp) | $1.0 \times 10^{13}$ | $1.35 \times 10^{13}$ | $1.45 \times 10^{13}$ | |
| Repetition period | 2.4 s (1500 pulses/h) | 4.8 s (750 pulses/h) | 4.8 s (750 pulses/h) | |
| Yield (p/p) | $10.0 \times 10^{-6}$ | $5.7 \times 10^{-6}$ | $5.4 \times 10^{-6}$ | $5.8 \times 10^{-6}$ |
| $\bar{p}$ injected in AC | $10.0 \times 10^{7}$ | $7.0 \times 10^{7}$ | $7.0 \times 10^{7}$ | $8.3 \times 10^{7}$ |
| After bunch rotation | $9.0 \times 10^{7}$ | $5.6 \times 10^{7}$ | $6.2 \times 10^{7}$ | $6.8 \times 10^{7}$ |
| $\bar{p}$ after AC cooling | | $5.7 \times 10^{7}$ | $7.2 \times 10^{7}$ | $7.6 \times 10^{7}$ |
| $\bar{p}$ after transfer to AA | | $5.4 \times 10^{7}$ | $6.3 \times 10^{7}$ | $7.0 \times 10^{7}$ |
| H emittance after AC cooling (mm mrad) | | $10\pi$ | $4\pi$ | |
| V emittance after AC cooling (mm mrad) | | $10\pi$ | $4\pi$ | |
| $\bar{p}$ after AA pre-cooling | | $5.3 \times 10^{7}$ (in 4 eVs ) | $7.5 \times 10^{7}$ | $7.9 \times 10^{7}$ |
| | | $3.8 \times 10^{7}$ (in 1 eVs) | $6.2 \times 10^{7}$ | $6.2 \times 10^{7}$ |
| $\bar{p}$/pulse | $5.0 \times 10^{7}$ | $4.9 \times 10^{7}$ | $7.0 \times 10^{7}$ | $7.7 \times 10^{7}$ |
| Stacking rate ($10^{10}$/h) | 7.5 | 3.6 | 5.3 | 5.8 |
| Daily production ($10^{11}$) | 10 | 6.0 | 8.5 | 11.5 |
| Daily stacking rate ($10^{10}$/h) | | 3.3 | 4.4 | 5.16 |
| Maximum stack intensity | $1 \times 10^{12}$ | $0.85 \times 10^{12}$ | $1.03 \times 10^{12}$ | $1.3 \times 10^{12}$ |
| Transverse stack emittances | (1 to 2)$\pi$ | (2 to 3)$\pi$ | (2 to 3)$\pi$ | |
| Overall efficiency | 50% | 63% | 91% | 93% |

Table 3. Operational performance of AA (in 1984) and AA+AC (in 1994 for LEAR only).

| Characteristic | Original AA (with horn) Repetition rate 2.4 s | AC+AA (with 20 mm lithium lens) Repetition rate 4.8 s |
|---|---|---|
| 26 GeV/$c$ protons per pulse on target (Number of protons, $N_p$) | $1.2 \times 10^{13}$ | $1.4 \times 10^{13}$ |
| Antiprotons per pulse injected ($N_{\bar{p}}$) | $6.7 \times 10^{6}$ | $7.3 \times 10^{7}$ |
| Yield $=[N_{\bar{p}}/N_p]$ | $5.6 \times 10^{-7}$ | $5.3 \times 10^{-6}$ |
| Antiprotons stacked per injection | $4.7 \times 10^{6}$ | $6.0 \times 10^{7}$ |
| Best daily production [$N_{\bar{p}}$ per 24 hrs] | $1.7 \times 10^{11}$ | $1.1 \times 10^{12}$ |
| Largest stack attained | $2.8 \times 10^{11}$ (Nov 1983) | $1.3 \times 10^{12}$ (Aug 1989) |
| Corresponding stacking factor | $6.0 \times 10^{4}$ | $2.2 \times 10^{4}$ |
| Increase of phase-space density | $5.4 \times 10^{8}$ | $4.3 \times 10^{9}$ |

in the first ever Nobel Prize at CERN. Continued improvement demands and expectations of even the top quark discovery led to the construction of the AC ring; this provided an order of magnitude increase in antiproton production and subsequent gains in luminosity at the SPS Collider. While the top quark discovery eluded us due to the limited maximum beam energy of the Collider, the experience and successes at CERN in handling large physics experiments and collaborations of the AA and AC era paved the way to larger and larger international physics collaborations, whether at Tevatron in Fermilab, LEP at CERN or, now again at CERN's LHC. Furthermore, there should be little doubt that the success of the antiproton program and the SPS Collider, as another demonstration of CERN's accelerator competence, has greatly furthered the way towards the adventure of the Large Hadron Collider, now at CERN.

The AA, AC, and LEAR rings of the years 1978–1987 were all designed, constructed and operated under the responsibility of the PS division. Yet from the beginning and throughout the definition of the projects as well as the construction and operation of the rings, there was an intense and most fruitful collaboration between the accelerator experts and the experimenters across all the CERN divisions involved. The projects' team members hailed from across the CERN divisions and brought their scientific and technical effort and ingenuity for a common goal. This provided an essential ingredient to the successes of these projects (1978–87).

## References

[1] K. Johnsen, Opening remarks, in: P. Bryant, and S. Newman (Eds.), *Proceedings of the CERN Acc. School: Antiprotons for Colliding Beam Facilities,* 1983 CERN 84-15, 1984, p. 1.

[2] G. I. Budker, Status report of the work on storage rings at Novosibirsk, in: H. Zyngier and E. Cremieu-Alcan (Eds.), *Proceedings of the Symposium International sur les Anneaux de Collision,* Saclay, 1966, (Presse Universitaire de France, Paris, 1966), p. II-1-1.
[3] S. van der Meer, Stochastic damping of betatron oscillations in the ISR, CERN Int. Report ISR-PO/72-31, 1972.
[4] G. I. Budker *et al.*, Experimental study of electron cooling, *IEEE Trans. Nucl. Sci.* **22** (1975) 2093.
[5] P. Bramham *et al.*, Stochastic cooling of a stored beam, *Nucl. Instrum. Methods* **125** (1975), 201.
[6] C. Rubbia, P. McIntyre, and D. Cline, Producing massive intermediate vector bosons with existing accelerators, in: *Proceedings of the International Neutrino Conference,* Aachen, 1976 (Vieweg Verlag, Braunschweig, 1977), p. 683.
[7] G. Carron *et al.*, Stochastic cooling tests in ICE, *Phys. Lett. B* **77** (1978) 353.
[8] AA design study team, Design of a proton–antiproton colliding beam facility, CERN Int. Report PS/AA 78-3, 1978.
[9] V. Chohan and S. van der Meer, in: L. Alvarez-Gaume *et al.* (Eds.), *Infinitely CERN, 1954–2004: Memories of Fifty Years of Research* (Hurter, Geneva, 2004), p. 223. Also see: V. Chohan, Simon van der Meer (1925–2011): A modest genius of accelerator science, *Rev. Accl. Sci. Tech.* **4** (2011) 279.
[10] E. J. N. Wilson (Ed.), Design study of an Antiproton Collector for the Antiproton Accumulator, CERN 83-10, 1983.
[11] L. Evans, The proton–antiproton Collider, CERN Yellow Report CERN 88-01, http://cdsweb.cern.ch/record/186337/files/CERN-88-01.pdf.
[12] G. Plass, Design study of a facility for experiments with low energy antiprotons (LEAR), CERN-PS-DL-80-7, 1980, http://cdsweb.cern.ch/record/124681/files/198009172.pdf.
[13] R. Garoby, New RF exercise envisaged in the CERN PS for the antiprotons production beam of the ACOL Machine, CERN-PS-85-36-RF, http://cdsweb.cern.ch/record/159858/files/cer-000071189.pdf.
[14] J. Allaby, Antiproton production, in: P. Bryant and S. Newman (Eds.), *Proceedings of the CERN Acc. School: Antiprotons for Colliding Beam Facilities,* 1983 (CERN 84-15, 1984), p. 63.
[15] E. Jones, Pbar production and collection, in: P. Bryant and S. Newman (Eds.), *Proceedings of the CERN Acc. School: Antiprotons for Colliding Beam Facilities,* 1983 (CERN 84-15, 1984), p. 1.; Also see: C. Hojvat and A. van Ginneken, Calculation of antiproton yields for the Fermilab antiproton source, *Nucl. Instrum. Methods* **206** (1983) 67.
[16] M. D. Church and J. P. Marriner, The antiproton sources: Design and operation, *Ann. Rev. Nucl. Part. Sci.* **43** (1995) 253.
[17] V. Chohan *et al.*, The CERN Antiproton Accumulator Complex (AAC): Current status and operation for the nineties, in: *15th International Conference on High-Energy Accelerators,* Hamburg, Germany, 20–24 Jul 1992, pp. 106–108, http://cdsweb.cern.ch/record/240246/files/CM-P00058053.pdf.
[18] S. van der Meer, Improved collection of secondaries from a long, narrow target by a horn doublet, CERN Int. Report PS/AA 80-12, 1980.
[19] D. Boimond *et al.*, Consolidation of the 400 kA magnetic horn for AAC antiproton production, CERN Int. Report PS 94-02-AR, 1994.
[20] R. Bellone *et al.*, Performance and operational experience with CERN-lithium lenses, in: *Proceedings of the First European Particle Accelerator Conference EPAC '88,*

Rome, 1988, p. 1401; R. Bellone *et al.*, Beam tests of a 36 mm lithium lens, in: *Proceedings of the Second European Particle Accelerator Conference EPAC '90,* Nice, France, 1990, p. 1303.

[21] H. Koziol, Antiproton Accumulator (AA) parameter list, 9th edition, CERN Int. Note PS/AA/Note 80-2, 1980; 10th edition, CERN Int. Note PS/AA/Note 84-2, 1984.

[22] S. Maury and H. Koziol, Parameter list for the Antiproton Accumulator Complex (AAC), CERN Int. Report /PS 95-15 (AR/BD), 1995; S. Maury, Status of the antiproton source, in: *Workshop on Physics at SuperLEAR,* Zurich, Oct 1991.

[23] G. Carron and L. Thorndahl, Stochastic cooling of momentum spread with filter methods, CERN Int. Report /ISR-RF/ 78-12, 1978.

[24] S. van der Meer, Stochastic stacking in the Antiproton Accumulator, CERN Int. Report PS/AA 78-22, 1978.

[25] F. Caspers and D. Möhl, Stacking with stochastic cooling, in: *Proceedings of the International Workshop on Beam Cooling and Related Topics, Nucl. Instrum. Methods A* **532** (2004) 321; D. Möhl, *Stochastic Cooling of Particle Beams,* Lecture Notes in Physics, Vol. 866 (Springer, 2013).

[26] L. Faltin, Slot-type pick-up and kicker for stochastic beam cooling, *Nucl. Instrum. Methods* **148** (1978) 449.

[27] V. Chohan and S. van der Meer, Aspects of automation and applications in the CERN antiproton source, *Nucl. Instrum. Methods Phys. Res. A* **293**(1–2) (1990) 98–102; CERN-PS-89-57-AR, http://cdsweb.cern.ch/record/204708/files/cer-000116409.pdf.

[28] H. Koziol and D. Mohl, The CERN antiproton collider programme: Accelerators and accumulation rings, CERN AB 2004-098 ADM, later published as *Phys. Rep.* **403–404** (2004) 271–280, http://cdsweb.cern.ch/record/806277/files/ab-2004-097.pdf.

# Chapter 12

# Fermilab Antiproton Source, Recycler Ring and Main Injector

Sergei Nagaitsev (Fermilab)

## Introduction

The Antiproton Source for a proton–antiproton collider at Fermilab was proposed in 1976 [1]. The proposal argued that the requisite luminosity ($\sim 10^{29}\,\mathrm{cm}^{-2}\mathrm{sec}^{-1}$) could be achieved with a facility that would produce and cool approximately $10^{11}$ antiprotons per day. Funding for the Tevatron I project (to construct the Antiproton Source) was initiated in 1981 and the Tevatron ring itself was completed, as a fixed target accelerator, in the summer of 1983 and the Antiproton Source was completed in 1985. At the end of its operations in 2011, the Fermilab antiproton production complex consisted of a sophisticated target system, three 8-GeV storage rings (namely the Debuncher, Accumulator and Recycler), 25 independent multi-GHz stochastic cooling systems, the world's only relativistic electron cooling system and a team of technical experts equal to none. Sustained accumulation of antiprotons was possible at the rate of greater than $2.5 \times 10^{11}$ per hour. Record-size stacks of antiprotons in excess of $3 \times 10^{12}$ were accumulated in the Accumulator ring and $6 \times 10^{12}$ in the Recycler. In some special cases, the antiprotons were stored in rings for more than 50 days. Note that over the years, some $10^{16}$ antiprotons were produced and accumulated at Fermilab, which is about 17 nanograms and more than 90% of the world's total man-made quantity of nuclear antimatter.

The accelerator complex at Fermilab supported a broad physics program including the Tevatron Collider Run II [2], neutrino experiments using 8-GeV and 120-GeV proton beams, as well as a test beam facility and other fixed target experiments using 120-GeV primary proton beams. The following sections provide a brief description of Fermilab accelerators as they operated at the end of the Collider Run II (2011).

## 1 The Proton Source

The Proton Source consists of the Pre-Accelerator (Pre-Acc), the Linac, and the Booster. For operational redundancy, there are two independent 750-kV Pre-Acc systems which provide $H^-$ ions for acceleration through the Linac. Each Pre-Acc is a Cockcroft–Walton accelerator having its own magnetron-type $H^-$ source running at a 15-Hz repetition rate, a voltage multiplier to generate the 750-kV accelerating voltage, and a chopper to set the beam pulse length going into the Linac. The typical $H^-$ source output current is 40–60 mA.

The Linac accelerates $H^-$ ions from 750 keV to 400 MeV. Originally, the Linac was a 200 MeV machine made entirely of Alvarez-style drift tube tanks [3], but a 1991 upgrade replaced four of the drift tubes with side-coupled cavities to allow acceleration up to 400 MeV [4]. Today, the low energy section (up to 116 MeV) is made of drift tube tanks operating with 201.25-MHz RF fed from triod-based 5-MW power amplifier tubes. The high energy section (116–400 MeV) consists of seven side-coupled cavity girders powered by 805-MHz, 12-MW Klystrons providing a gradient of ≈7 MV/m. A transition section between the two Linac sections provides the optics matching and rebunching into the higher frequency RF system. The nominal beam current in the Linac is 30–35 mA.

The Booster is a 474-meter circumference, rapid-cycling synchrotron ramping from 400 MeV to 8 GeV at a 15-Hz repetition rate. (Note that while the magnets ramp at 15 Hz, beam is not present on every cycle.) Multi-turn injection is achieved by passing the incoming $H^-$ ions through 1.5-$\mu$m thick (300 $\mu$g/cm$^2$) carbon stripping foils as they merge with the circulating proton beam on a common orbit. The 96 10-foot long combined-function Booster gradient magnets are grouped into 24 identical periods in a FOFDOOD lattice [5]. The Booster RF system (harmonic number = 84) consists of 19 cavities (18 operational + 1 spare) that must sweep from 37.9–52.8 MHz as the beam velocity increases during acceleration. The ferrite tuners and power amplifiers are mounted on the cavities in the tunnel. The cavities provide a total of ≈750 kV per turn for acceleration. The Booster transition energy (4.2 GeV) occurs at 17 ms in the cycle. The Booster throughput efficiency is 85–90% for typical beam intensities of $4.5-5.0 \times 10^{12}$ protons per pulse. The majority of the proton flux through the Booster is delivered to the 8-GeV and 120-GeV neutrino production targets.

## 2 The Main Injector

The Main Injector (MI) [6] is a 3319.4-m circumference synchrotron, which can accelerate both proton and antiproton beams from 8 GeV up to 150 GeV. It has a FODO lattice using conventional, separated function dipole and quadrupole magnets. There are also trim dipole and quadrupole, skew quadrupole, sextupole and octupole magnets in the lattice. Since the Main Injector circumference is seven times the Booster circumference, beams from multiple consecutive Booster cycles, called batches, can be injected around the Main Injector. In addition, even higher beam intensity can be accelerated by injecting more than seven Booster batches through the process of slip-stacking: capturing one set of injected proton batches with one RF system, decelerating them slightly, then capturing another set of proton injections with another independent RF system, and merging them prior to acceleration. There are 18 53-MHz RF cavities, grouped into two independently controlled systems to allow slip-stacking and the flexibility for maintenance. Beam-loading compensation and active damping systems have been implemented to help maintain beam stability. For beam injections into the Tevatron, coalescing of several 53-MHz bunches of protons and antiprotons into single, high-intensity bunches also requires a 2.5-MHz system for bunch rotations and a 106-MHz cavity to flatten the potential when recapturing the beam into the single 53-MHz bunch to be injected into the Tevatron ring. A set of collimators was installed in the Main Injector to help localize beam losses to reduce widespread activation of ring components in the tunnel.

The Main Injector supports various operational modes for delivering beam across the complex. For the antiproton and neutrino production, up to 11 proton batches from the Booster are injected and slip-stacked prior to acceleration. After reaching 120 GeV, two batches were extracted to the antiproton production target while the remaining nine batches are extracted to the NuMI neutrino production target (Fig. 1). At its peak performance, the Main Injector can sustain 400-kW delivery of 120-GeV proton beam power at 2.2 sec cycle times. The Main Injector also provides 120-GeV protons in a 4-sec long slow-spill extracted to the Switchyard as a primary beam or for production of secondary and tertiary beams for the Meson Test Beam Facility and other fixed target experiments. In addition, the Main Injector serves as an “effective” transport line for 8-GeV antiprotons being transferred from the Accumulator to the Recycler for later use in the Tevatron. Protons from the Booster and antiprotons from the Recycler are accelerated

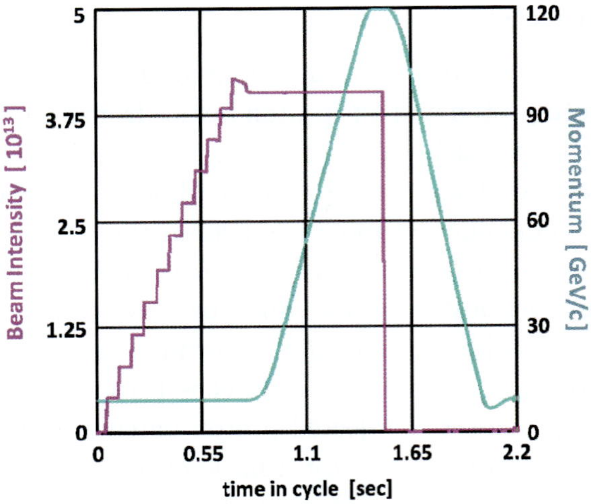


Fig. 1. The Main Injector cycle illustrating an 11-batch proton injection, acceleration and extraction.

to 150 GeV in the Main Injector and coalesced into higher intensity bunches for injection into the Tevatron for a colliding beam store.

## 3 The Antiproton Source

The Antiproton Source [7] has three main parts: the Target Station, the Debuncher, and the Accumulator. Each of these is described briefly below while outlining the steps of an antiproton production cycle. In the Target Station, batches of 120-GeV protons ($\sim 8 \times 10^{12}$ per batch), delivered from the Main Injector, strike the inconel (a nickel-iron alloy) target every 2.2 sec. The beam spot on the target can be controlled by a set of quadrupole magnets. The target is rotated between beam pulses to spread the depletion and damage uniformly around its circumference. The shower of secondary particles, emanating from the target, is focused both horizontally and vertically by a pulsed, high-current lithium lens that can provide upto 1000 T/m gradient.

Downstream of the Li lens is a pulsed dipole magnet, which steers negatively-charged particles at 8.9-GeV/$c$ momentum into the transport line toward the Debuncher. A collimator between the lens and pulsed magnet was installed to help protect the pulsed magnet from radiation damage as the incoming primary proton beam intensity increased with proton slip-stacking in Main Injector.

The Debuncher and Accumulator (Fig. 2) are both triangular-shaped rings of conventional magnets sharing the same tunnel. While the Debuncher

Fig. 2. The Antiproton Source consists of the Debuncher ring (outer circumference, left) and the Accumulator ring (inner circumference, right).

has a FODO lattice, the Accumulator lattice has particular features needed for cooling and accumulating antiprotons with stochastic cooling systems. A total of 21 independent stochastic cooling systems were implemented in the Accumulator and Debuncher [8]. Such a variety of cooling systems was possible after a series of development efforts [9, 10] allowing for more robust and less expensive pick-up arrays.

The $\sim 2 \times 10^8$ bunches of antiprotons entering the Debuncher from the transport line retain the 53-MHz bunch structure from the primary proton beam on the target. A 53-MHz RF system (harmonic number = 90) is used for the bunch rotation and debunching of the antiprotons into a continuous beam with a low momentum spread. An independent 2.4-MHz RF system provides a barrier bucket to allow a gap for extraction to the Accumulator. Stochastic cooling systems reduce the transverse emittance from 300 to 30 $\pi$ mm-mrad (rms, normalized) and momentum spread from 0.30% to <0.14% prior to injection into the Accumulator.

In the Accumulator, antiprotons are momentum-stacked and cooled by a series of RF manipulations and stochastic cooling. The incoming antiprotons are captured and decelerated by 60 MeV with a 53-MHz RF system (harmonic number = 84) to the central orbit where the beam is adiabatically debunched. Before the next pulse of antiprotons arrives (every 2.2 sec), the so-called stacktail momentum stochastic cooling system [11] decelerates the antiprotons another 150 MeV toward the core orbit where another set of

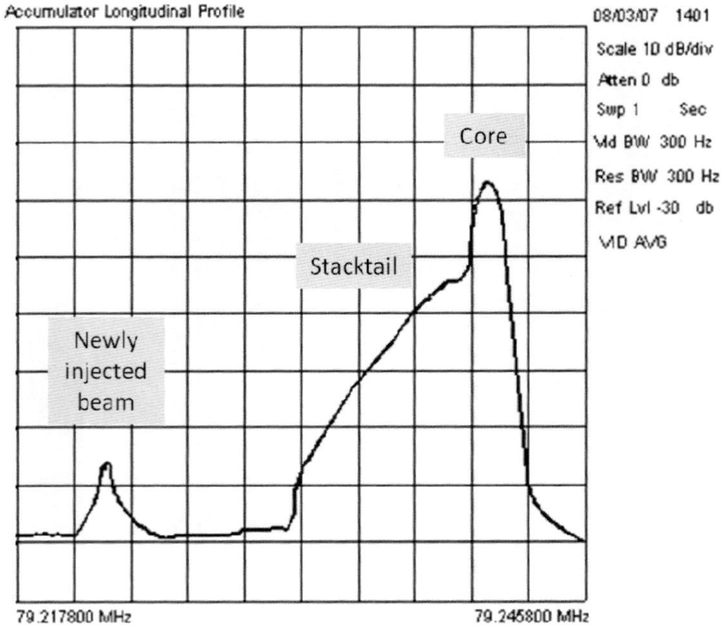


Fig. 3. The frequency (energy) distribution of antiprotons in the Accumulator highlighting incoming antiprotons (left), the stacktail beam (middle) being cooled and decelerated toward the core (right). Higher beam energy is to the left, lower energy is to the right.

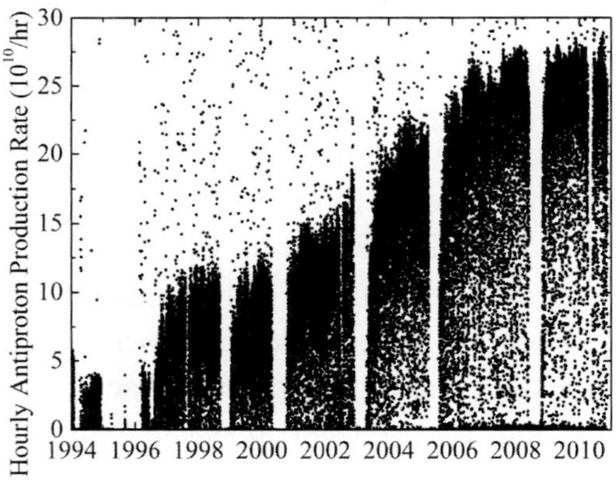


Fig. 4. The average antiproton accumulation rate since 1994 and during all of Collider Run II (including the production in the Antiproton Source and storage in the Recycler). Note that some data points at the highest rates, in particular in the early years, are merely artifacts of the data acquisition and logging system.

independent betatron and momentum stochastic cooling systems provides additional cooling while building a "stack" of antiprotons. Figure 3 illustrates the frequency (energy) distribution of antiprotons in the Accumulator. Figure 4 shows the average antiproton accumulation rates since 1994; typical values for recent Run II operations are in the range 24–26 $\times 10^{10}$/hr, with a maximum recorder rate of $28.5 \times 10^{10}$/hr.

## 4 The Recycler

The Recycler [12] is a permanent-magnet, fixed momentum (8.9 GeV/$c$) storage ring located in the Main Injector tunnel (Fig. 5). As conceived, the Recycler would provide storage for very large numbers of antiprotons (up to $6 \times 10^{12}$) and would increase the effective production rate by recapturing unused antiprotons at the end of collider stores (hence the name Recycler). Recycling of antiprotons was determined to be ineffective and was never implemented. However, the Recycler was used as a final antiproton cooling and storage ring for accumulating significantly larger stashes (so called to differentiate from Accumulator "stacks") of antiprotons that can be accumulated in the Antiproton Accumulator. The main Recycler magnets are combined-function strontium ferrite permanent magnets arranged in a FODO lattice. Powered trim magnets are used to provide orbit and lattice corrections.

An important feature of the Recycler is an electron cooling system [13]. It augments the Recycler's cooling capability and complements the stochastic cooling system and its inherent limitations. The Pelletron (an electrostatic

Fig. 5. The Recycler (top) and the Main Injector (bottom) rings installed in a common tunnel.

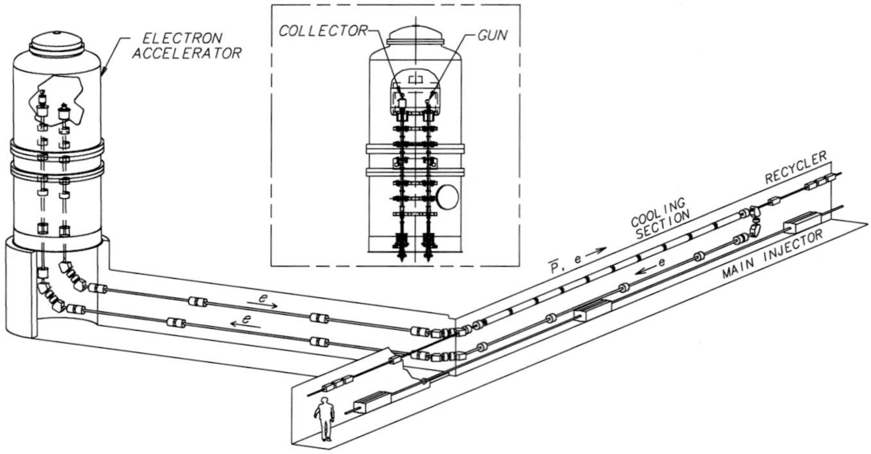


Fig. 6. Schematic layout of the Recycler electron cooling system and accelerator cross section (inset).

accelerator manufactured by the National Electrostatics Corp.) provides a 4.3-MeV electron beam (up to 500 mA) which overlaps with the 8-GeV antiprotons, circulating in the Recycler, in a 20-m long section and cools the antiprotons both transversely and longitudinally. Figure 6 shows the schematic layout of the Fermilab electron cooling system. The DC electron beam is generated by a thermionic gun, located in the high-voltage terminal of the electrostatic accelerator. This accelerator is incapable of sustaining DC beam currents to ground in excess of about 100 $\mu$A. Hence, to attain the electron DC current of 500 mA, a recirculation scheme is employed. A typical relative beam current loss in such a process is $2 \times 10^{-5}$.

The Fermilab system employed a unique beam transport scheme [14]. The electron gun is immersed in a solenoidal magnetic field, which creates a beam with large angular momentum. After the beam is extracted from the magnetic field and accelerated to 4.3 MeV, it is transported to the 20-m long cooling section solenoid using conventional focusing elements (as opposed to low-energy electron coolers where the beam remains immersed in a strong magnetic field at all times). The cooling section solenoid removes this angular momentum and the beam is made round and parallel such that the beam radius, $a$, produces the same magnetic flux, $Ba^2$, as at the cathode. The magnetic field in the cooling section is quite weak (100 G), therefore the kinetics of the electron–antiproton scattering is weakly affected by the magnetic field.

After becoming operational in September 2005, electron cooling in the Recycler allowed significant improvements in Tevatron luminosity by providing higher intensity antiprotons with smaller emittances. With electron cooling, the Recycler has been able to store up to $6 \times 10^{12}$ antiprotons. In routine operations, the Recycler accumulated $3.5-4.0 \times 10^{12}$ antiprotons with a ~200-hr lifetime for injection to the Tevatron [15].

Among other unique features of the Recycler was the so-called barrier-bucket RF system [16] which allowed for crucial longitudinal beam manipulations of the accumulated antiproton beam.

## 5 Antiproton flow and Tevatron shot-setup

As mentioned previously, stacks of freshly produced antiprotons were stored temporarily in the Accumulator. The Accumulator antiproton stack was periodically transferred to the Recycler where electron cooling allowed for a much larger antiproton intensity to be accumulated with smaller emittances. Typically $22-25 \times 10^{10}$ antiprotons were transferred to the Recycler every ~60 minutes. Prior to electron cooling in the Recycler, antiprotons destined for the Tevatron were extracted from the Accumulator only. Since late 2005, all Tevatron antiprotons were extracted from the Recycler only. Figure 7 illustrates the flow of antiprotons between the Accumulator, Recycler and Tevatron over a 1-week period.

A typical Tevatron collider fill cycle is shown in Fig. 8 [17]. First, proton bunches are injected two at a time on the central orbit. Then, electrostatic

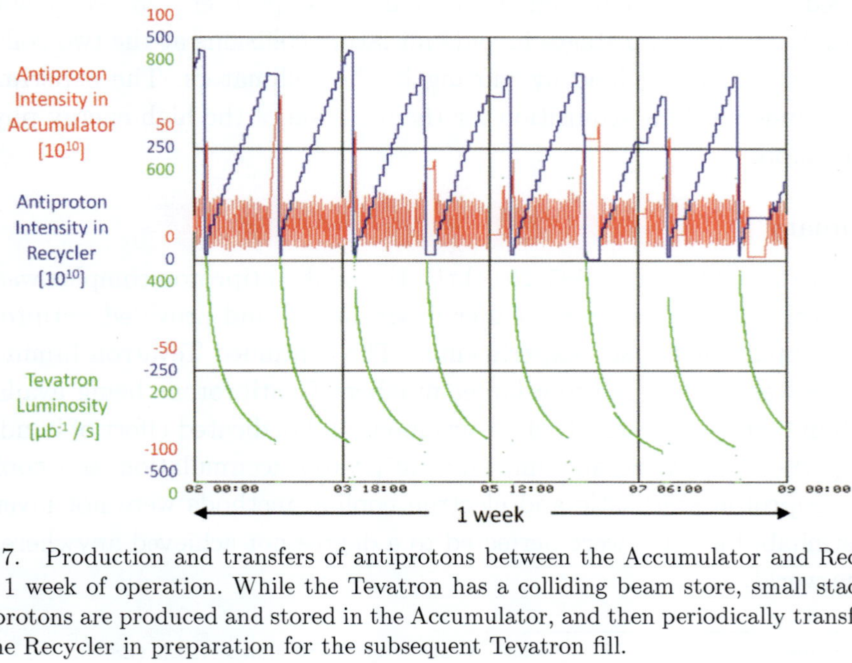


Fig. 7. Production and transfers of antiprotons between the Accumulator and Recycler over 1 week of operation. While the Tevatron has a colliding beam store, small stacks of antiprotons are produced and stored in the Accumulator, and then periodically transferred to the Recycler in preparation for the subsequent Tevatron fill.

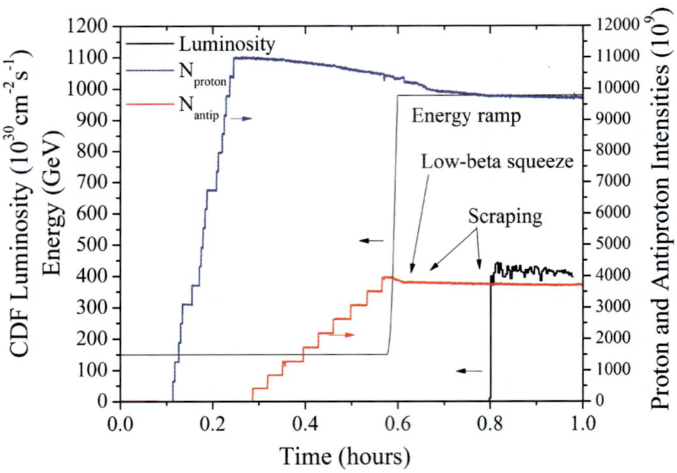


Fig. 8. The collider fill cycle for store #8709 (May 2011).

separators are powered to put the protons onto a helical orbit. Antiproton bunches are then injected four bunches at a time into gaps between the three proton bunch trains. After each group of three antiproton transfers, the gaps are cleared for the subsequent set of transfers by "cogging" the antiprotons — changing the antiproton RF cavity frequency to let them slip longitudinally relative to the protons. Once the beam loading is complete, the beams are accelerated to the top energy (in 86 seconds) and the machine optics is changed to the collision configuration in 25 steps over 125 sec (low-beta squeeze). The last two stages include initiating collisions at the two collision points and removing halo by moving in the collimators. The experiments then commence data acquisition for the duration of the high-energy physics (HEP) store.

## Summary

For more than 25 years (1985–2011) the Fermilab antiproton complex was the centerpiece of the Tevatron collider program [18] and provided antiprotons for other particle physics experiments.[1] The continued Tevatron luminosity increase was mainly due to a larger number of antiprotons being available, which in turn was the result of a continuous and dedicated effort of hundreds of experts to optimize and improve antiproton accumulation and cooling. The antiproton stochastic and electron cooling methods were not invented at Fermilab, but they were perfected to a degree not achieved anywhere else in the world.

[1]For example: Charmonium (E760/E835), Antihydrogen (E862), and APEX (E868).

## References

[1] D. Cline *et al.*, Proposal to construct an antiproton source for the Fermilab accelerators, proposal 492, in *Proceedings of 1976 NAL Summer Study on Utilization of the Energy Doubler/Saver*, Fermilab, Batavia U.S.A. (1976), p. 309.

[2] S. D. Holmes (Ed.), *Tevatron Collider Run II Handbook*, Fermilab Preprint TM-2484 (1998).

[3] C. D. Curtis *et al.*, Linac $H^-$ beam operations and uses at Fermilab, *IEEE Trans. Nucl. Sci.* 26(3) (1979) 3760–3762.

[4] C. W. Schmidt, The Fermilab 400-MeV Linac upgrade, in *Proceedings of 1993 Particle Accelerator Conf.*, Washington, DC, 1993, pp. 1655–1659.

[5] E. Hubbard *et al.*, *Booster Synchrotron,* Fermilab Technical Memo TM-405 (1973).

[6] S. D. Holmes, Design criteria and performance goals for the Fermilab main injector, *Particle Accelerators* **58** (1997) 39.

[7] M. Church and J. Marriner, The antiproton sources: Design and operation, *Ann. Rev. Nucl. Par. Sci.* **43** (1993) 253–295.

[8] R. J. Pasquinelli, Implementation of stochastic cooling hardware at Fermilab's Tevatron collider, *JINST* **6** (2011) T08002.

[9] D. McGinnis, Theory and design of microwave planar electrodes for stochastic cooling of particle beams, *Microw. Opt. Technol. Lett.* **4** (1991) 439.

[10] D. McGinnis, Slotted waveguide slow-wave stochastic cooling arrays, in *Proceedings of 1999 PAC, IEEE Conference*, New York, U.S.A. (1999), p. 1713.

[11] V. Lebedev, Improvements to the stacktail and Debuncher momentum cooling systems, in *Proceedings of COOL 09, MOA1MCCO02*, Lanzhou, China (2009).

[12] G. Jackson (Ed.), *Fermilab Recycler Ring Technical Design Report*, Fermilab Preprint TM-1991 (1997).

[13] S. Nagaitsev *et al.*, Experimental demonstration of relativistic electron cooling, *Phys. Rev. Lett.* **96** (2006) 044801.

[14] A. Burov *et al.*, Optical principles of beam transport for relativistic electron cooling, *Phys. Rev. Spec. Top. Accel. Beams* **3** (2000) 094002.

[15] A. Shemyakin and L. R. Prost, Ultimate performance of relativistic electron cooling at Fermilab, in *Proceedings of COOL11, THIOA01*, Alushta, Ukraine (2011).

[16] C. M. Bhat, Longitudinal momentum mining of beam particles in a storage ring, *Phys. Lett. A* **330** (2004) 481.

[17] C. Gattuso *et al.*, Optimization of integrated luminosity of the Tevatron, in *Proceedings of PAC 2009*, Vancouver, B. C., Canada.

[18] S. Holmes *et al.*, Overview of the Tevatron collider complex: Goals, operations and performance, *JINST* **6** (2011) T08001.

# Chapter 13

# RF manipulations in the PSB & PS for the production of antiprotons at CERN and deceleration in the PS for LEAR

Roland Garoby (CERN)

## 1 The antiproton production beam

Antiproton production required maximizing the number of protons sent by the proton synchrotron (PS) on the antiproton production target in a burst of $\frac{1}{4}$ of a revolution period (ratio of circumferences between Antiproton Accumulator (AA) and PS). The Proton Synchrotron Booster (PSB) being made up of four rings of $\frac{1}{4}$ of the PS size, using the beam from a single ring was a potential solution, but with a modest intensity. Multiple techniques were therefore tried to make use of multiple rings.

Transverse (vertical) recombination at injection of the beams from two different PSB rings was first implemented, but was quickly abandoned because of the excessive beam loss due to the large transverse emittance of the recombined beam. Longitudinal recombination techniques were more successful.

The first scheme involved "slip stacking" of the beams from two PSB rings with slightly different energies [1]. The principle is sketched in Figure 1. Two different RF frequencies $f_1$ and $f_2$, corresponding to the energies of the PSB beams, are simultaneously applied into the PS, splitting the C10 RF cavities into two groups of 5. With a frequency difference $f_2 - f_1 \sim 5f_s$, where $f_s$ is the synchrotron frequency in the centre of an unperturbed bucket of one family, two families of buckets coexist and slip past each other because of their frequency difference. Therefore the separation between the two sets of bunches decreases until they are superimposed in azimuth. If the RF is then suddenly switched to the average frequency $(f_1 + f_2)/2$ on both groups and the voltage increased by a large enough factor, pairs of bunches are captured in the corresponding large buckets.

Feed-forward beam loading compensation was essential for a correct RF operation at such a high intensity. However this process also suffered from

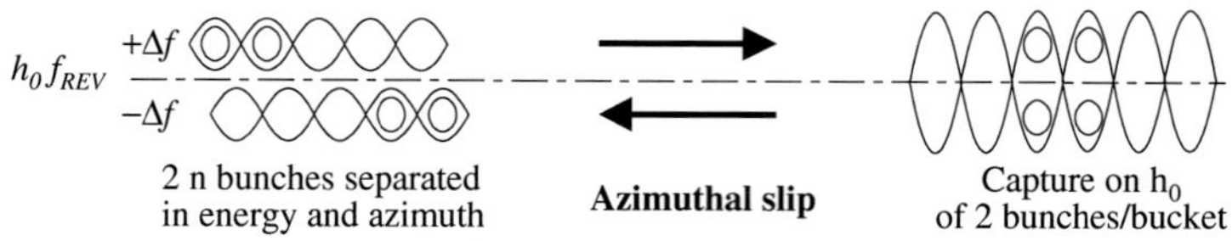


Fig. 1. Slip stacking.

excessive beam loss, probably due to the presence of a large proportion of particles with very large emittances after filamentation and to limited acceptances at low energy. Hence a similar but much more sophisticated process was implemented to slip stack the beams at high energy. Bunches from two PSB rings were injected in diametrically opposite locations and accelerated to 25 GeV. Then the RF cavities were again split into two groups. Combining four cavities operating at the same peak voltage $\hat{V}$, with two cavities on $h = 20$, one on $h = 19$ and one on $h = 21$, each of these groups generates a 100% amplitude modulation at the revolution frequency $\omega/2\pi$ of a carrier on $h = 20$. Indeed:

$$V_{total} = \hat{V}\cos(19\omega t) + 2\hat{V}\cos(20\omega t) + \hat{V}\cos(21\omega t)$$
$$\Leftrightarrow V_{total} = 2\hat{V}\cos(20\omega t) \cdot [1 + \cos(\omega t)]$$

↑ Carrier on $h = 20$ ↑ Peak amplitude modulation at the revolution frequency

With a proper phasing of the different harmonics with respect to the beam, the maximum amplitude of a group can be centered onto a set of five bunches so that each set "sees" a maximum voltage from his RF group and a minimum from the other one. Separate beam controls can hence handle separately both sets, giving them a difference in energy and "slip stacking" begins. This process taking place immediately before ejection, there is no need to recapture pairs of bunches in the same bucket: the beam is simply ejected onto the target when both beam sets are azimuthally superimposed. The mountain range display in Figure 2 shows a typical result at $10^{13}$ p/p.

The practical implementation of this process was rather complex. It included repetitive closing of gap short-circuit relays to reduce the impedance of cavities being re-tuned to $h = 19$ and $h = 21$. This technique was operationally used for many years, until the end of 1989. During a few years it was combined with another strongly non-adiabatic technique in an attempt to exploit all PSB rings and maximize the intensity. With a specially designed RF deflecting dipole in the transfer line between PSB and PS, bunches of two PSB rings were interlaced (Figure 3) so that each PS bucket captured two of them [2]. This "funneling" process is intrinsically imperfect, both in

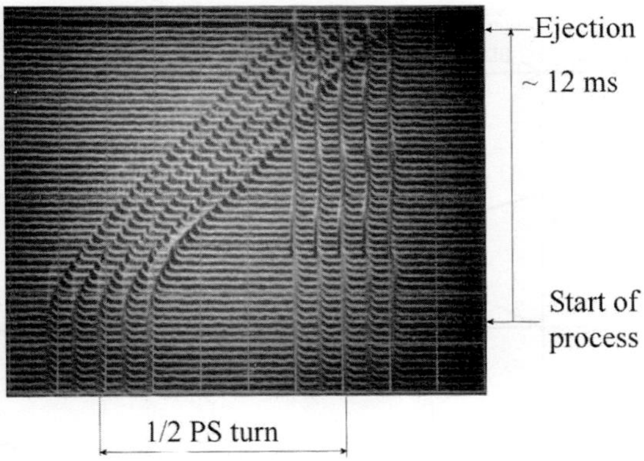


Fig. 2. Mountain range display of longitudinal recombination at 25 GeV.

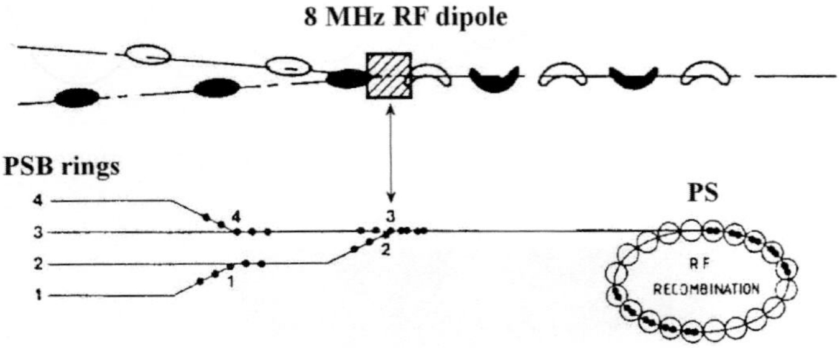


Fig. 3. "Funneling" with an RF dipole in the PSB–PS transfer line.

transverse phase planes, because all particles in a bunch do not get the same deflection, and in the longitudinal phase plane because capture is strongly non-adiabatic, generating a large blow-up after filamentation. An acceptable operational compromise was nevertheless found, adjusting the distance between bunches to optimize the overall result and minimize beam loss.

The upgrade programme launched in 1983 and based on the Antiproton Collector (ACOL) added the requirement that proton bunches should be short for bunch rotation in the ACOL to be efficient. A two-step quasi-adiabatic process was therefore proposed, using bunch merging and batch compression [3].

The bunches from two PSB rings are injected in adjacent locations in the PS, filling $\frac{1}{2}$ of the circumference, and accelerated to an intermediate energy of 3.57 GeV. Merging pairs of bunches is carried out at this energy by "smoothly" changing the RF voltages simultaneously applied on $h = 20$ to $h = 10$, with the $h = 10$ stable phase centered in the middle of the $h = 20$ buckets (Figure 4). If the rate of voltage changes is slow enough, the time

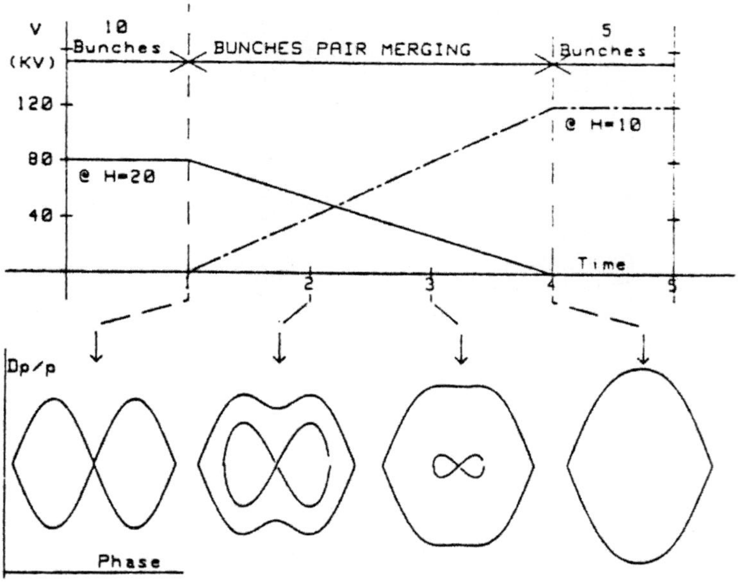


Fig. 4. RF voltages and buckets during bunch pair merging.

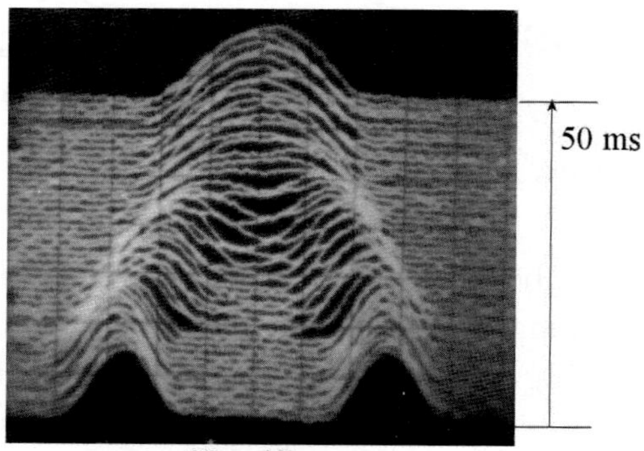


Fig. 5. Mountain range display of bunches merging at 3.57 GeV/$c$.

structure on $h = 20$ disappears progressively, being replaced by an $h = 10$ periodicity. The process is then quasi-adiabatic and longitudinal emittance is preserved, as illustrated in Figure 5.

After acceleration on $h = 10$ up to the 25 GeV flat top, the five bunches resulting from merging still occupy one half of the PS circumference. Batch compression is then applied to concentrate them in one quarter of the circumference. This is obtained by increasing the RF harmonic "seen" by the beam in five steps of two units, to cover the range from $h = 10$ to $h = 20$. At each step, two RF groups are active, one on $h_0$ and the other on $h_0 + 2$.

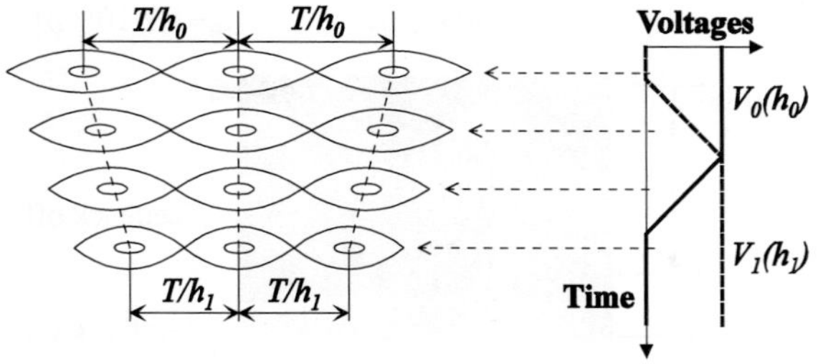


Fig. 6. Principle of batch compression (single harmonic step).

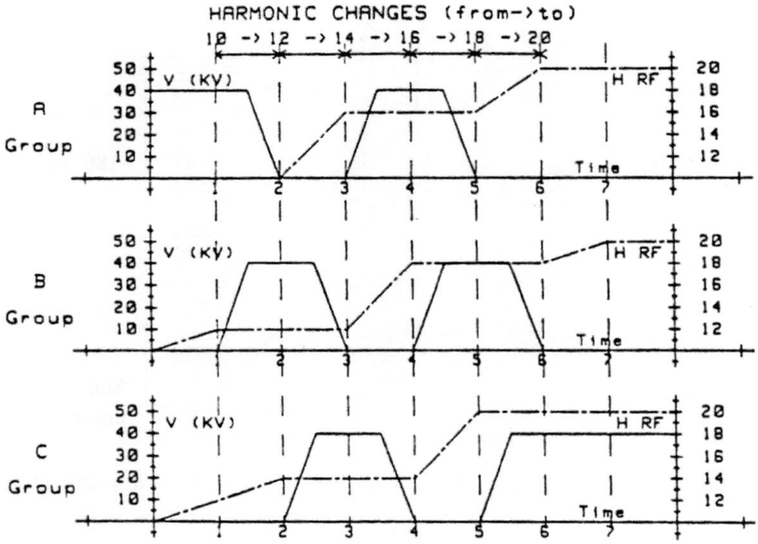


Fig. 7. Voltage and harmonics of the 3 groups of cavities during batch compression ($h = 10 \rightarrow h = 20$),

The voltage on $h_0$ is slowly ramped down to 0, while the voltage on $h_0 + 2$ is increased to maximum. Phasing is such that bunches converge towards the centre of the batch. The principle of such a step is shown in Figure 6, and the complexity of the practical implementation of the complete process is illustrated in Figure 7.

Cavities are split into four groups, operating at three different frequencies and the low level RF has to be able to provide all these harmonics with the correct phasing. Fast RF feedback was combined with gap short circuit relays to minimize the imperfections due to the beam-induced voltage when cavities are at low voltage, so-called transient beam loading. Beam performance in these conditions was much better than with the RF dipole.

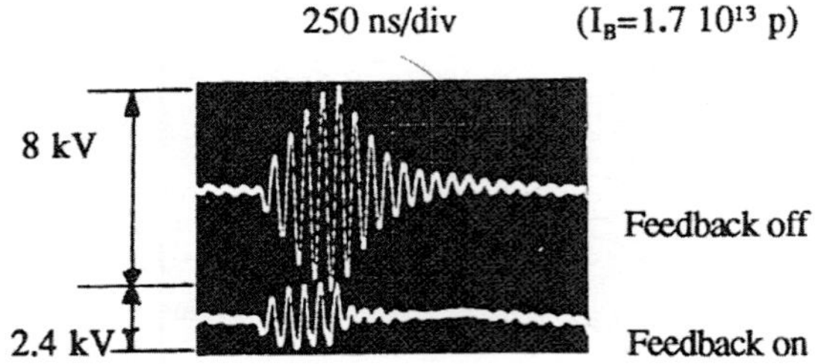


Fig. 8. Voltage induced in a cavity by five bunches without (bottom)/with (top) one-turn delay feedback.

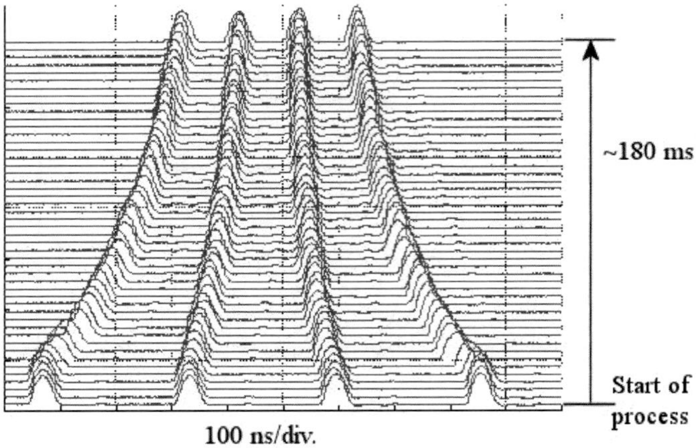


Fig. 9. Mountain range display of batch compression ($h = 8 \rightarrow h = 20$) at 25 GeV for the AD.

Transient beam loading remained however a source of imperfection which was addressed with a one-turn delay feedback [4] reducing the impedance by an additional factor of ~4 in a limited bandwidth of ±3 revolution frequency harmonics around the cavity tune. A typical example of the effect on the voltage induced in a cavity by a beam of five bunches on $h = 20$ (end of the batch compression process) is shown in Figure 8.

All these measures and renewed low level electronics are nowadays combined to provide the performance regularly achieved for the Antiproton Decelerator (AD) (Figure 9).

## 2 Antiproton acceleration

The main difficulty with the acceleration of antiprotons in the PS resulted from the difficulty to provide short enough bunches which could be captured

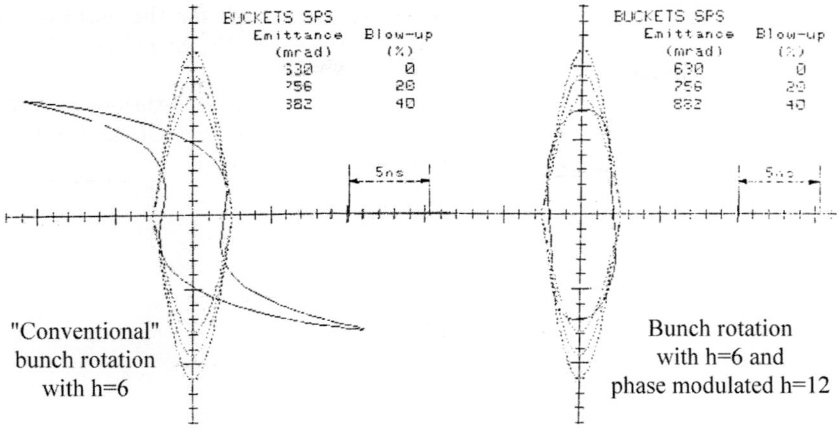


Fig. 10. Bunches after bunch rotation on $h = 6$ (left)/$h = 6 + 12$ (right) at 25 GeV.

into a single 200 MHz RF bucket in the SPS. Although the lowest possible harmonic compatible with the tuning range of the ferrite cavities was used ($h = 6$), the quality of a "conventional" non-adiabatic bunch rotation at 25 GeV remained insufficient (Figure 10, left). That was solved with a more involved gymnastics combining RF on $h = 6$ and $h = 12$ and varying amplitudes and phase during bunch rotation to linearize the focusing voltage only over the instantaneous length of the bunch (Figure 10, right). Synchronization with the SPS revolution frequency to $\pm 0.5$ ns remained a demanding challenge because of the large voltage and phase transients due to the voltage and phase manipulations required by the bunch rotation gymnastics.

## 3 Antiproton deceleration

For LEAR, the PS had to decelerate for the first time beam from 3.57 GeV/$c$ down to 600 MeV/$c$. To stay within the tuning range of the ferrite cavities, the RF system was operating on $h = 10$ instead of $h = 6$. To deal with very low bunch intensities, the beam control was based on an accurate frequency programme and an AC-coupled phase loop, without radial loop.

## References

[1] D. Boussard and Y. Mizumachi, Production of beams with high line density by azimuthal combination of bunches in a synchrotron, CERN-SPS-ARF-79-11, http://cdsweb.cern.ch/record/1020585/files/CM-P00063699.pdf

[2] G. Nassibian and K. Schindl, RF beam recombination ("funnelling") at the CERN PSB by means of an 8 MHz dipole magnet, CERN-PS-85-28-BR, http://accelconf.web.cern.ch/accelconf/p85/PDF/PAC1985_2760.PDF

[3] R. Garoby, New RF exercise envisaged in the CERN PS for the antiprotons production beam of the ACOL machine, CERN-PS-85-36-RF, http://cdsweb.cern.ch/record/159858/files/cer-000071189.pdf
[4] F. Blas and R. Garoby, Design and operational results of a "One-turn-delay feedback" for beam loading compensation of the CERN PS ferrite cavities, CERN-PS-91-16-RF, http://cdsweb.cern.ch/record/220107/files/CM-P00059458.pdf

## Chapter 14

# Tevatron: The world's first fully superconducting collider and the discovery of the top and bottom quarks

Vladimir Shiltsev (Fermilab)

## 1 Synopsis

For almost a quarter of a century, the Tevatron proton–antiproton collider was the centerpiece of the world's high energy physics program — beginning operation in December 1985 until it was overtaken by the LHC in 2011. The aim of this unique scientific instrument was to explore elementary particle physics reactions with center-of-mass collision energies of up to 1.96 TeV. The initial design luminosity of the Tevatron was $10^{30}\,\mathrm{cm}^{-2}\mathrm{s}^{-1}$, however as a result of two decades of upgrades, the accelerator has been able to deliver 430 times higher luminosities to each of two high luminosity experiments, CDF and D0. The Tevatron has been shut off since September 30, 2011. The collider was arguably one of the most complex research instruments ever to reach the operation stage and is widely recognized for many technological breakthroughs and numerous physics discoveries.

## 2 History and performance

The Tevatron was conceived by Bob Wilson [1] to double the energy of the Fermilab complex from 500 GeV to 1000 GeV. The original name, the "Energy Saver/Doubler", reflected this mission and the accrued benefit of reduced power utilization through the use of superconducting magnets. The introduction of superconducting magnets in a large scale application allowed the (now named) Tevatron (see Fig. 1) to be constructed with the same circumference of 6.3 km, and to be installed in the same tunnel as the original Main Ring proton synchrotron which would serve as its injector (at 150 GeV). Development of superconducting magnets was initiated in the early 1970's

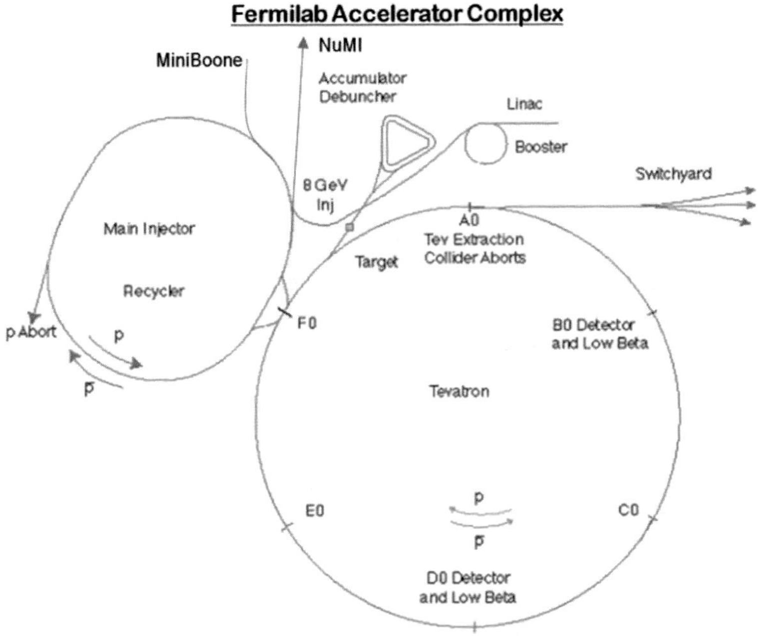


Fig. 1. Layout of Fermilab's accelerator complex during Tevatron Run II.

and ultimately produced successful magnets, leading to commissioning of the Tevatron in July 1983.

In 1976 D. Cline *et al.*, proposed a proton–antiproton collider with luminosities of about $10^{29}$ cm$^{-2}$sec$^{-1}$ at Fermilab [2] or at CERN, based on the conversion of an existing accelerator into a storage ring and construction of a new facility for the accumulation and cooling of approximately $10^{11}$ antiprotons per day. The motivation was to discover the intermediate vector bosons. The first antiproton accumulation facility was constructed at CERN and supported collisions at 630 GeV (center-of-mass) in the modified SPS synchrotron, where the W and Z particles were discovered in 1983. Meanwhile, in 1978 Fermilab decided that proton–antiproton collisions would be supported in the Tevatron, at a center-of-mass energy of 1800 GeV and that an Antiproton Source facility would be constructed to supply the flux of antiprotons needed for a design luminosity of $1 \times 10^{30}$ cm$^{-2}$sec$^{-1}$.

The Tevatron as a fixed target accelerator was completed in 1983 [3]. It comprised nearly 1000 high-quality superconducting magnets. The Tevatron cryogenic plant was the largest cryo-factory in the world, delivering about 23 kW of cooling power at the liquid helium temperature of 5 Kelvin and was later recognized as an International Mechanical Engineering landmark. The Antiproton Source [4] was completed in 1985 and the first collisions were observed in the Tevatron using some operational elements of the CDF detector (then under construction) in October 1985. Initial operation of the Collider for data taking took place during a period from February through May of 1987. A more extensive run took place between June 1988 and June 1989, representing the first sustained operation at the design luminosity. In this period of operation a total of 5 $pb^{-1}$ were delivered to CDF at 1800 GeV (center-of-mass) and the first western hemisphere W's and Z's were observed. The initial operational goal of $1 \times 10^{30}$ $cm^{-2}sec^{-1}$ luminosity was exceeded during this run. Table 1 summarizes the actual performance achieved in the 1988–89 run. (Short runs at $\sqrt{s} = 630$ GeV and $\sqrt{s} = 1020$ GeV also occurred in 1989.)

In the early to mid-1990's a number of improvements were implemented to prepare for operation of Collider Run I (August of 1992 through February 1996):

Electrostatic separators aimed at mitigating the beam–beam limitations by placing protons and antiprotons on separate helical orbits, thus allowing an increase in the number of bunches and proton intensity: twenty-two 3 m long electrostatic separators operating at up to ±300 kV across a 5 cm gap had been installed in the Tevatron by 1992. During Run II (2001–2011), four additional separators were installed to improve separation at the nearest parasitic crossings.

Low beta systems which ultimately allowed operations with $\beta*$ less than 30 cm: The 1988–89 Run did not have a matched insertion for the interaction region at B0 (where CDF was situated). Two sets of high performance quadrupoles were developed and installed at B0 and D0 (which came online for Run I in 1992).

Cryogenic cold compressors lowered the operating Helium temperature by about 0.5 K, thereby allowing the beam energy to be increased to 1000 GeV, in theory. In operational practice 980 GeV was achieved.

Antiproton Source improvements: A number of improvements were made to the stochastic cooling systems in the Antiproton Source in order to accommodate higher antiproton fluxes generated by continuously increasing the proton intensity on the antiproton production target. Improvements included the introduction of transverse stochastic cooling into the Debuncher and upgrades to the bandwidth of the core cooling system. These improvements supported an accumulation rate of $7 \times 10^{10}$ antiprotons per hour.

Run I consisted of two distinct phases: Run Ia which ended in May 1993, and Run Ib which was initiated in December 1993. The 400 MeV linac upgrade (from the initial 200 MeV) was implemented between Run Ia and Run Ib with the goal of reducing space-charge effects at injection energy in the Booster and provide higher beam brightness at 8 GeV. As a result, the total intensity delivered from the Booster increased from roughly $3 \times 10^{12}$ per pulse to about $5 \times 10^{12}$. This resulted in more protons being transmitted to the antiproton production target and, ultimately, more protons and antiprotons in collision in the Tevatron.

Run I ultimately delivered a total integrated luminosity of 180 $pb^{-1}$ to both CDF and D0 experiments at $\sqrt{s} = 1800$ GeV. By the end of the run the typical luminosity at the beginning of a store was about $1.6 \times 10^{31}$ $cm^{-2}sec^{-1}$, a 60% increase over the Run I goal. (A brief colliding run at $\sqrt{s} = 630$ GeV also occurred in Run I.)

In preparation for the next Collider run, construction of the Main Injector synchrotron and Recycler storage ring was initiated and completed in the spring of 1999 with the Main Injector initially utilized in the last Tevatron fixed target run.

The Main Injector was designed to significantly improve antiproton performance by replacing the Main Ring with a larger aperture, faster cycling machine [5]. The goal was a factor of three increase in the antiproton accumulation rate (to $2 \times 10^{11}$ per hour), accompanied by the ability to obtain 80% transmission from the Antiproton Source to the Tevatron from antiproton intensities up to $2 \times 10^{12}$. An antiproton accumulation rate of $2.5 \times 10^{11}$ per hour was achieved in Collider Run II, and transmission efficiencies beyond 80% for high antiproton intensities were routine.

The Recycler was added to the Main Injector Project midway through the project (utilizing funds generated from an anticipated cost under-run.) As conceived, the Recycler would provide storage for very large numbers of antiprotons (up to $6 \times 10^{12}$) and would increase the effective production rate by recapturing unused antiprotons at the end of collider stores [6]. The Recycler was designed with stochastic cooling systems but R&D in electron cooling was initiated in anticipation of providing improved performance. Antiproton intensities above $5 \times 10^{12}$ were ultimately achieved although routine operation was eventually optimized around $4 \times 10^{12}$ antiprotons. Recycling of antiprotons was never implemented, as the most efficient use of the new machine was operationally found to be as an additional storage ring to accumulate and cool antiprotons from the Antiproton Source and to optimally reformat the beam for injection to Tevatron.

The Main Injector (MI) and Recycler (RR) completed the development of the Fermilab accelerator complex — see the ultimate scheme of operational

Table 1. Achieved performance parameters for Collider Runs I and II (typical values at the beginning of a store).

| | 1988–89 Run | Run Ib | Run II | |
|---|---|---|---|---|
| Energy (center-of-mass) | 1800 | 1800 | 1960 | GeV |
| Protons per bunch | 7.0 | 23 | 29 | $\times 10^{10}$ |
| Antiprotons per bunch | 2.9 | 5.5 | 8.1 | $\times 10^{10}$ |
| Bunches in each beam | 6 | 6 | 36 | – |
| Total antiprotons | 17 | 33 | 290 | $\times 10^{10}$ |
| Proton emittance (rms, norm.) | 4.2 | 3.8 | 3.0 | $\pi\mu$m |
| Antiproton emittance (rms, norm.) | 3 | 2.1 | 1.5 | $\pi\mu$m |
| $\beta*$ at the IPs | 55 | 35 | 28 | cm |
| Luminosity (typical, start of store) | 1.6 | 16 | 350 | $\times 10^{30}$ cm$^{-2}$s$^{-1}$ |
| Luminosity integral | $5 \times 10^{-3}$ | 0.18 | 11.9 | fb$^{-1}$ |

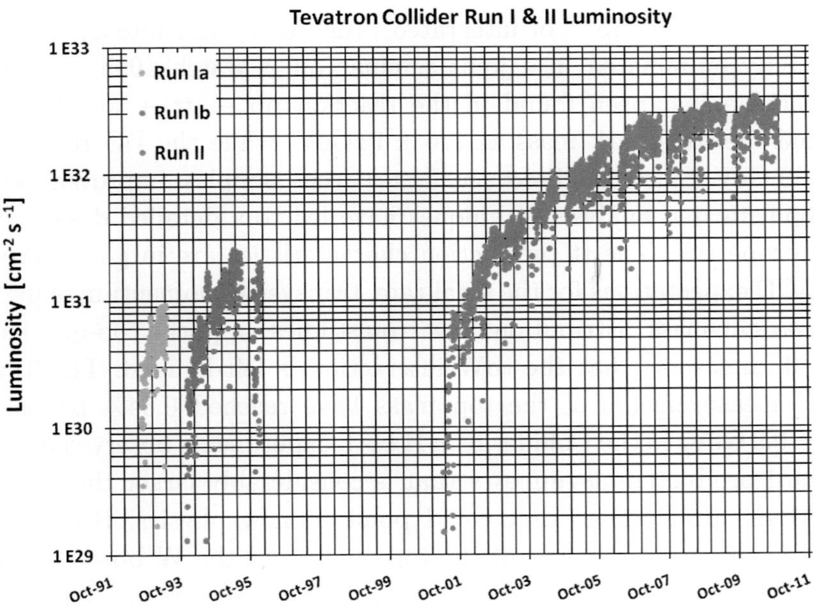


Fig. 2. Initial luminosity for all Collider stores.

accelerators in Fig. 1 — and constituted the improvements associated with Collider Run II [7]. The luminosity goal of Run II was $8 \times 10^{31}$ cm$^{-2}$sec$^{-1}$, a factor of five beyond Run I. However, incorporation of the RR into the Main Injector Project was projected to provide up to an additional factor of 2.5.

Run II was initiated in March 2001 and continued through September 2011. A number of difficulties were experienced in the initial years of operation. These were ultimately overcome through experience accumulated in the course of operation and the organization and execution of a “Run II Upgrade Plan”. At the end of the Run II, typical Tevatron luminosities were well in excess of $3.4 \times 10^{32}$ cm$^{-2}$ sec$^{-1}$, with record stores exceeding $4.3 \times 10^{32}$ cm$^{-2}$ sec$^{-1}$ — see the achieved performance parameters in Table 1.

## 3 Particle physics discoveries at the Tevatron: The top and bottom quarks

The legacy of the Tevatron experiments includes many results for which the high energy of a hadron collider was decisive [8]. Chief among these is the discovery of the top quark in 1995, which for 15 years could be studied only at the Tevatron. The hunt took off with the growing data-sets available to both CDF and DØ in 1992–1993. A growing body of observations that probed quantum corrections to the electroweak theory pointed to a top-quark mass in the range 150–200 GeV. Eighteen months of the Collider running (which delivered some 0.15 fb$^{-1}$ of integrated luminosity) and intense experimental activity culminated in a joint seminar on 2 March 1995, demonstrating that top had been found in the reaction pp → tt + anything. CDF reported the top-quark mass at $176 \pm 13$ GeV, while the DØ result was at $199 \pm 30$ GeV. Since the discovery, larger event samples, improved detectors and sophisticated analysis techniques have led to a detailed dossier of top-quark properties. The tt production characteristics are in good agreement with QCD expectations for the total rate, transverse-momentum dependence and invariant-mass distribution. Tevatron studies support a top-quark charge of +2/3, and show that the tbW interaction is left-handed. The Tevatron Run II measurements of the top mass have reached 0.54% precision, at $173.2 \pm 0.9$ GeV. In the Collider Run II both experiments observed a forward-backward production asymmetry that is considerably larger than the Standard Model predictions, that could point to new physics. High precision measurements of the masses of the top quark and the W boson and of the frequency of $B_s$ oscillations showed that hadron colliders can be quite precise instruments. Robust multivariate analysis techniques that enhance the sensitivity of searches in the face of challenging backgrounds and detector innovations such as the first hadron–collider silicon vertex detector and secondary vertex trigger, and multilevel triggering, are now part of the standard particle detector toolkit.

## 4 Major accelerator physics and technology achievements

The Tevatron program impressively propelled the field of accelerator technology and beam physics [9, 10]. The most notable examples are given below.

### 4.1 *Tevatron superconducting magnets*

Superconducting magnets define the Tevatron, the first synchrotron built with this technology [11, 12]. The Tevatron magnets employed NbTi supercoducting cable operating at 4.5 K and had warm iron yoke — see Fig. 3. The 4.5 T dipole magnets had a beam aperture of 76 mm and were about 6 m long. Issues that had to be addressed included conductor strand and cable fabrication, coil geometry and fabrication, mechanical constraint and support of the coils, cooling and insulation, and protection during quenches. Magnetic field quality — one of the most demanding parameters — is given by the multipole coefficients in the expansion:

$$B_x + i \cdot B_y = B_0 \sum_{n=0} (b_n + ia_n) \left[\frac{x + iy}{R_0}\right]^n, \qquad (1)$$

where $R_0$ is the reference radius (1 inch in the Tevatron), the pole number is $2(n+1)$ and $b_n(a_n)$ are the normal (skew) multipole coefficients, and $b_0$ is unity. The multipoles allowed by dipole symmetry, $b_2$ $b_4$ $b_6$ ... are designed to be small and would be zero for a pure $\cos\theta$ winding. The precise coil placement, and hence good magnetic field uniformity at the relative level of few $10^{-4}$, had the biggest effect on the accelerator performance. The magnets, designed in the 70's, performed beautifully over the years, though

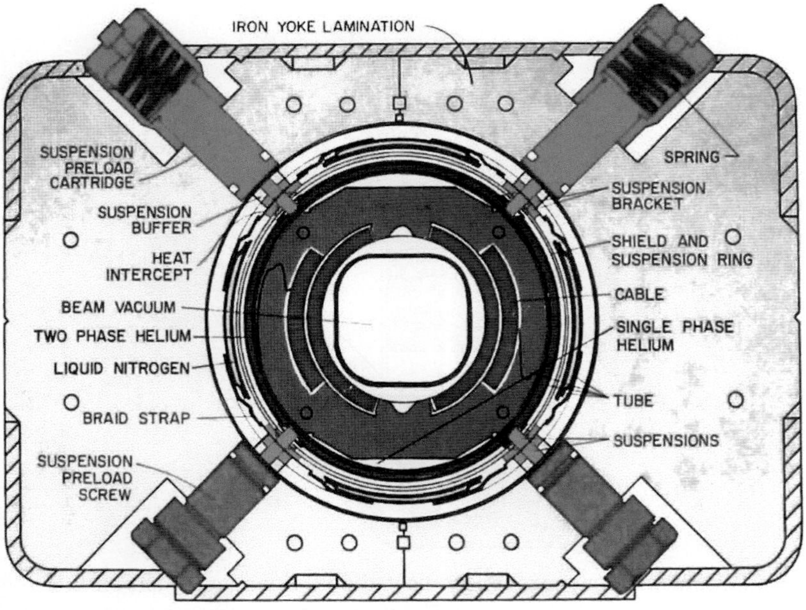


Fig. 3. Cross section of the Tevatron superconducting dipole magnet.

they also offered us a number of puzzles to resolve for optimal operation, like the "chromaticity snap-back" effect [13] and coupling due to cold-mass sagging [14]. The Tevatron experience paved the way for other high energy hadron colliders, like HERA (9 m long, 75 mm bore, 5.3 T dipoles made of NbTi conductor, with Al collar and cold iron), RHIC (simple and economical design of 9 m long, 80 mm bore, 3.5 T dipoles) and LHC (15 m long, 56 mm double bore dipoles operating at 2 K super fluid He temperatures with design field of 8.3 T).

### 4.2 *Recycler permanent magnets*

The Recycler was the first high energy accelerator ever built with permanent magnets — see Fig. 4. It was also arguably the cheapest accelerator built (per GeV) and it employed a 362 gradient dipole and 109 quadrupole magnets made of SrFe (peak field of about 1.4T) [15]. The biggest challenge was to compensate for the intrinsic temperature coefficient of the ferrite field of $-0.2\%$ per °C. This was canceled down to the required $0.01\%/°C$ by interspersing a thin NiFe "compensator alloy" strip between the ferrite bricks the pole tips. The magnetic field drifted (logarithmically slowly) by a minuscule 0.04% over many years of operation [16].

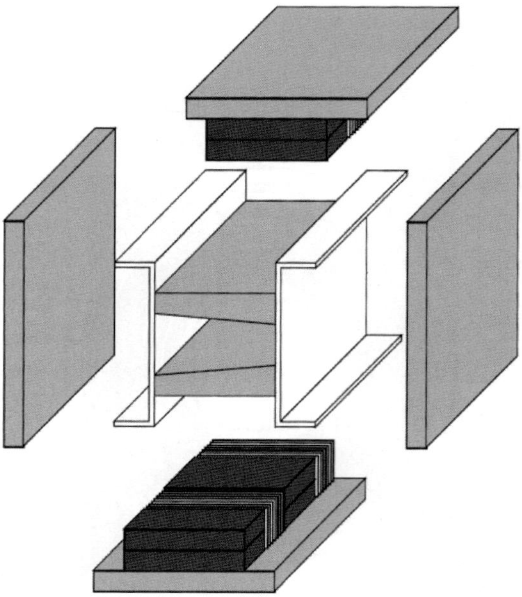

Fig. 4. Recycler permanent magnet gradient dipole components shown in an exploded view. For every $4''$ wide brick there is a $0.5''$ interval of temperature compensator material composed of ten strips.

### 4.3 *Production, stochastic cooling and electron cooling of antiprotons*

The Tevatron antiproton production and cooling complex is described in detail in Chapter 12. Here we can just mention that the combination of high proton intensity on the target, better targetry, and electron cooling in the Recycler resulted in world record antiprotron stacking rates of about $28 \times 10^{10}$ per hour; and also record high stacks in excess of $300 \times 10^{10}$ and $600 \times 10^{10}$ were accumulated in the Accumulator ring and the Recycler respectively.

### 4.4 *Slip-stacking and barrier-bucket RF manipulations*

Two innovative methods of longitudinal beam manipulation were developed and implemented in operation and were crucial for the success of the Tevatron Run II: (a) multi-batch slip stacking [17] that allowed to approximately double the 120-GeV proton bunch intensity for antiproton production; (b) the RF barrier-bucket system with rectangular 2kV RF voltage pulses [18] allowed for a whole new range of antiproton beam manipulation in the Recycler including operational "momentum mining" of antiprotons for the Tevatron shots [19] — see Figs. 5 and 6.

### 4.5 *Electron lenses for beam–beam compensation*

Electron lenses [20, 21] are a novel accelerator technology used for compensation of the long-range beam–beam effects in the Tevatron [22, 23], operational DC beam removal out of the Tevatron abort gaps [24], and, recently, for hollow electron beam collimation demonstration [25]. Two electron lenses were built and installed at the A11 and F48 locations of the Tevatron ring —

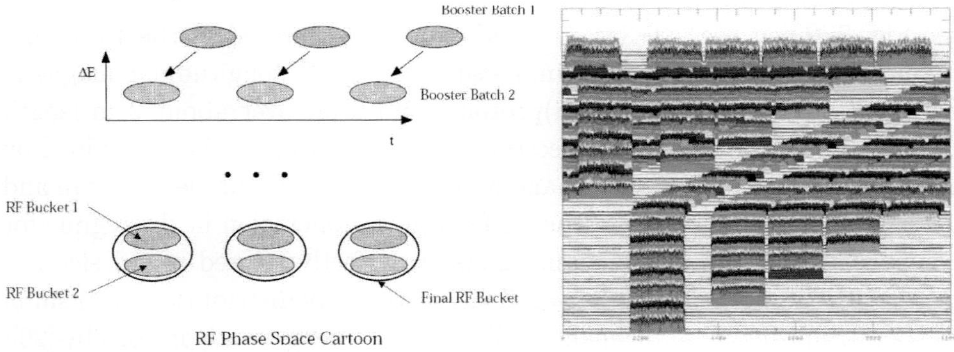


Fig. 5. (left) The principle of the "slip-stacking" method for doubling the proton bunch intensity; (right) "mountain range" plot showing 11 batch slip stacking process in the FNAL Main Injector. Horizontal scale is 10 $\mu$sec.

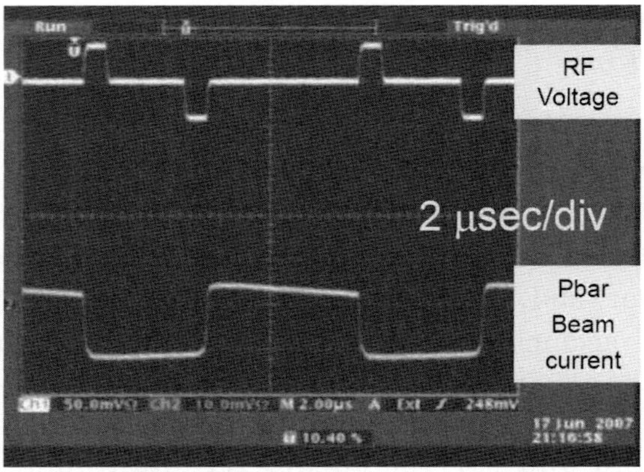


Fig. 6. Beam manipulation using the RF "barrier bucket": scope traces of the ±2 kV RF barrier buckets in the Recycler (top) and antiproton beam current profile at one of the stages of operation prior to so-called "momentum mining" (bottom).

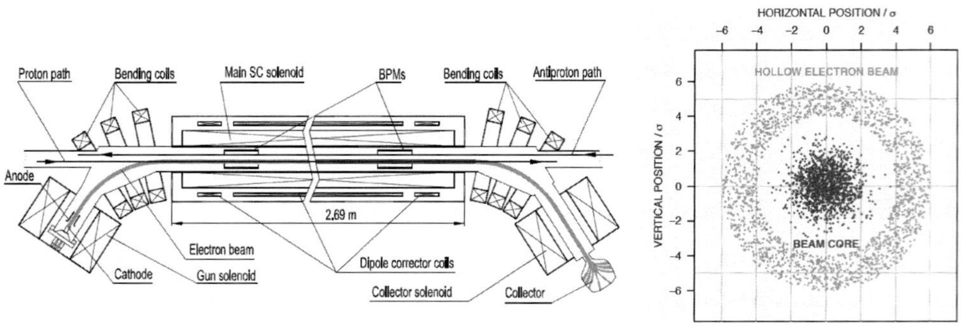


Fig. 7. (left) General layout of the Tevatron Electron Lens; (right) the method of the Hollow Electron Beam Collimation (HEBC) — low energy magnetized electron beam current surrounds the high energy beam of (anti)protons in the Tevatron.

see Fig. 7. They use a 1–3 A, 6–10 kV e-beam generated at the 10–15 mm diameter thermionic cathodes immersed in a 0.3 T longitudinal magnetic field and aligned onto the (anti)proton beam orbit over about 2 m length inside a 6 T SC solenoid. The electron beam current profile is determined by the configuration of the cathode and anode electrodes of the electron gun and determines the application of the TELs, e.g., for operation in the regime for compensation of the beam–beam effects, a perfectly steered round electron beam with "smoothed-edge-flat-top" current density distribution $j_e(r)$ shifts the betatron tunes $Q_{x,y}$ of small amplitude high-energy (anti)protons by [20]:

$$dQ_{x,y} = \pm \frac{\beta_{x,y} L_e r_p}{2\gamma e c} \cdot j_e \cdot \left( \frac{1 \mp \beta_e}{\beta_e} \right), \tag{2}$$

where the sign reflects focusing for protons and defocusing for antiprotons, $\beta_e = v_e/c$ is the electron beam velocity, $\beta_{x,y}$ are the beta-functions at the location of the lens, $L_e$ denotes the effective interaction length between the electron beam and the protons or antiprotons, $r_p = e^2/mc^2 = 1.53 \cdot 10^{-18}$ m is the classical proton radius, and $\gamma_p = 1044$ the relativistic Lorentz factor for 980-GeV protons. If the electron beam is much wider than the (anti)proton beam, then all of the high-energy particles acquire the same $dQ_{x,y}$. The factor $1 \pm \beta_e$ reflects the fact that the contribution of the magnetic force is $\beta_e$ times the electric force contribution, and its sign depends on the direction of the electron beam; both TELs direct the beam against the antiproton flow. In another configuration for Hollow Electron Beam Collimation (HEBC), a hollow electron beam which has no electric or magnetic fields inside — see Fig. 7 (right) — affects only halo (anti)protons causing quick diffusion to collimators [25].

We should emphasize that the Tevatron Collider Runs not only delivered excellent performance (integrated luminosity), but also greatly advanced the whole accelerator field by studies of beam–beam effects [26, 27], crystal collimation [28], electron cloud [29] and beam emittance growth mechanisms [30–32], new theories of beam optics [33, 34], intra-beam scattering [35] and instabilities [36], sophisticated beam–beam and luminosity modeling [37] and more efficient beam instrumentation [38–45]. Comprehensive review of the Tevatron Collider can be found in [46–48].

## References

[1] R. R. Wilson, *Physics Today* (October 1977), p. 23.

[2] D. Cline *et al.*, Proposal 492, in *Proc. 1976 NAL Summer Study on Energy Doubler/Saver*, Fermilab (1976), p. 309.

[3] H. Edwards, *Ann. Rev. Nucl. Part. Sci.* **35**, 605 (1985).

[4] M. Church, J. Marriner, *Ann. Rev. Nucl. Part. Sci.* **43**, 253 (1993).

[5] S. D. Holmes, *Particle Accelerators* **58**, 39 (1997).

[6] G. Jackson (ed.), *Fermilab Recycler Ring Technical Design Report*, Fermilab Preprint TM-1991 (1997).

[7] S. D. Holmes (ed.), *Tevatron Collider Run II Handbook*, Fermilab Preprint TM-2484 (1998).

[8] C. Quigg, *CERN Courier* **51**(8), 20 (2011).

[9] Tevatron Accelerators, special issue, *JINST* **7** (2012).

[10] V. Shiltsev, *Mod. Phys. Lett. A* **27**(1), 1230001 (2012).

[11] A. V. Tollestrup, FermilabUPC086, Batavia (1979).

[12] R. Palmer and A. V. Tollestrup, *Ann. Rev. Nucl. Part. Sci.* **34**, 247 (1984).

[13] R. Hanft *et al.*, Fermilab Preprint TM-1542 (1988).

[14] M. Syphers and D. Harding, in *Proc. IEEE PAC2005* (2005), p. 1967.

[15] G. W. Foster, in *Proc. IEEE PAC95* (1995), p. 1298.

[16] J. Volk, *JINST* **6**, T08003 (2011).

[17] K. Seyia *et al.*, in *Proc. IEEE PAC07* (2007), p. 742; I. Kourbanis *et al.*, FERMILAB-CONF-09-151 (2009).
[18] C. M. Bhat, FERMILAB-FN-0916 (2011).
[19] C. M. Bhat, *Phys. Lett. A* **330**, 481 (2004).
[20] V. Shiltsev *et al.*, *Phys. Rev. ST-AB* **2**, 071001 (1999).
[21] V. Shiltsev *et al.*, *Phys. Rev. ST-AB* **11**, 103501 (2008).
[22] V. Shiltsev *et al.*, *Phys. Rev. Lett.* **99**, 244801 (2007).
[23] V. Shiltsev *et al.*, *New Journ. Phys.* **10,** 043042 (2008).
[24] X. L. Zhang *et al.*, *Phys. Rev. ST-AB* **11**, 051002 (2008).
[25] G. Stancari *et al.*, *Phys. Rev. Lett.* **107**, 084802 (2011).
[26] V. Shiltsev *et al.*, *Phys. Rev. ST-AB* **8**, 101001 (2005).
[27] A. Valishev *et al.*, arXiv:0906.0386 (2009).
[28] N. Mokhov *et al.*, *JINST* **6**, T08005 (2011).
[29] R. Zwaska, in *Proc. IEEE PAC11*, MOOBS4.
[30] V. Shiltsev and A. Tollestrup, *JINST* **6**, P08001 (2011).
[31] T. Sen, *JINST* **6**, P10017 (2011).
[32] V. Shiltsev, G. Stancari and A. Valishev, *JINST* **6**, P08002 (2011).
[33] V. Lebedev and A. Bogacz, *JINST* **5**, P10010 (2010); V. Lebedev, in *Proc. ICAP2004*, p. 299.
[34] Y. Alexahin and E. Gianfelice-Wendt, *JINST* **6**, P10006 (2011).
[35] S. Nagaitsev, *Phys. Rev. ST-AB* **8**, 064403 (2007).
[36] A. Burov and V. Lebedev, *Phys. Rev. ST-AB* **10**, 054202 (2007); *Phys. Rev. ST-AB* **12**, 034201 (2009).
[37] A. Valishev *et al.*, arXiv:0906.0386 (2009); A. Valishev, in *Proc. RUPAC2006* (2006), p. 7.
[38] H. Pfeffer and G. Saewert, *JINST* **6**, P11003 (2011).
[39] C. Y. Tan and J. Steimel, *JINST* **6**, T09002 (2011).
[40] R. Moore, A. Jansson and V. Shiltsev, *JINST* **4**, P12018 (2009), and references therein.
[41] J. Crisp and B. Fellenz, *JINST* **6**, T11001 (2011).
[42] R. Thurman-Keup *et al.*, *JINST* **6**, T10004 (2011).
[43] J. Crisp *et al.*, *JINST* **6**, T10001 (2011).
[44] R. Thurman-Keup *et al.*, *JINST* **6**, T09003 (2011).
[45] G. Annala *et al.*, *JINST* **6**, T11005 (2011).
[46] V. Shiltsev, in *Proc. International PAC2011* (2011), p. 903.
[47] S. Holmes and V. Shiltsev, *Ann. Rev. Nucl. Part. Sci.* **63**, 413 (2013).
[48] V. Lebedev and V. Shiltsev, *Accelerator Physics at the Tevatron Collider* (Springer, 2014).

# Part 2

# 100 Years of Accelerators

## Chapter 15

# The HERA lepton–proton collider

Ferdinand Willeke (BNL)

## 1 Introduction and overview

The lepton–proton collider HERA (Hadron–Elektron Ring Anlage) was built during the years 1984 to 1991 in Hamburg, Germany and was operated by the DESY laboratory between 1992 and 2007. Scattering experiments of high energy electrons on protons (deep inelastic scattering [1]) carried out in the nineteen-sixties and seventies have discovered quarks as the smallest constituents of matter and contributed to the foundations of quantum chromodynamics as the theory of strong, quark–gluon interactions. However, precise comparison of theory and experiments was limited by the accelerators available at that time. Leptons were accelerated by linear accelerators or synchrotrons and the high energy beam was directed on a fixed target. The center-of-mass energy in such collisions is limited resulting in a limited spatial resolution. Alternatively, high energy muons or neutrinos were produced by collision of high energy hadron beams with a fixed target, and the collisions of these secondary beams with a target were studied. This technique provided access to higher center-of-mass energies but the data rate and kinematic range obtained were still rather limited. These shortcomings motivated the proposal to construct a circular collider which accelerates counter rotating electron and proton beams to high energy in two separate synchrotrons in the same accelerator tunnel, thereby providing simultaneously high center-of-mass energy and a high collision rate in a so far unexplored kinematic range.

Under the leadership of Bjoern Wiik, a team at DESY in Hamburg developed the design and performed superconducting magnet R&D for the HERA lepton–proton collider. The design was based on an 820 GeV, superconducting proton synchrotron of 6 km circumference and a 30 GeV normal conducting electron ring of the same size and promised a center-of-mass energy of 318 GeV and a luminosity of $L = 10^{31}\,\mathrm{cm^{-2}s^{-1}}$. This design was published in the HERA Proposal in 1981 [2]. Three years later, the construction of the collider was approved and started immediately in 1984. An important element in the approval of HERA was the fact that a significant part of the components was contributed by collaborating foreign laboratories with funding from outside Germany. The major contributions consisted of half of the superconducting magnet production from Italy and half of the superconducting quadrupole construction from France. Together with contributions from a total of ten nations, this constituted about one third of the HERA construction cost (excluding civil construction). This so-called "HERA Model" for the construction of large accelerator systems in collaboration turned out to be very successful and has since been applied to other large accelerator facilities.

The construction of HERA went as planned. The accelerator tunnel and the four underground halls became available for installation of accelerator components in 1987. The electron ring was commissioned in 1988–1989 and the superconducting magnet production in industry, the greatest technical challenge, was successful. The construction was completed within schedule and budget with proton ring and collider commissioning in 1991.

In the summer of 1992, HERA started colliding beam operations, providing collisions of its 820 GeV proton beam with the 27.5 GeV electron beam for the experiments H1 and ZEUS located in the straight sections North and South respectively. In 1995, a 3rd experiment, HERMES, for the detection of electron scattering on an internal gas target was added in the straight section East and finally in 2003, the 4th experiment, HERA-B, started to take data from collisions of high energy protons on an internal wire target.

During its life cycle, HERA underwent some major upgrades: The interaction region in the straight section EAST was modified to accommodate the HERMES experiment (1994/5). This upgrade included two spin rotator assemblies for the production of longitudinally polarized lepton beams at the HERMES collision point HERA-East. The accommodation of the 20 m long HERA-B (1996) in the utility straight section HERA-West was a major upgrade which required rebuilding a 300 m long straight section for both electron and proton beams. In order to enable HERA's full potential, a comprehensive upgrade of the vacuum system, an energy upgrade to 920 GeV, additional RF systems and many other improvements were performed in 1997 resulting in efficient 27.5–920 GeV operation and the possibility to run with electron beams (initially HERA was operated with positrons and protons to mitigate the impact of dust particles which degraded the lifetime of electron beams). HERA reached its design luminosity in 1997 and run in this configuration until the summer of 2000. Finally, the colliding beam interaction regions were re-optimized and rebuild in 2000/2001 which resulted in a luminosity increase of a factor three which boosted the peak luminosity above $5 \cdot 10^{31}\,\text{cm}^{-2}\text{s}^{-1}$. During this upgrade, spin rotator assemblies were installed around the colliding beam experiments ZEUS and H1, which enabled the collisions of longitudinally polarized leptons with high energy protons. The high flexibility built in the new HERA interaction regions allows varying the center-of-mass energy by changing the proton beam energy. This provided access to fundamental measurements of the proton structure.

HERA concluded its very successful physics program in 2007. The collider was continuously improved and reached the peak of its performance in the last year of operation. HERA provided a total integrated luminosity of 800 $\text{pb}^{-1}$ of high energy lepton–proton collisions.

## 2 Luminosity considerations

The luminosity goal formulated in the HERA proposal was $L = 1 \cdot 10^{31}\,\text{cm}^{-2}\text{s}^{-1}$ at 318 GeV center-of-mass energy. Implicitly it was assumed that HERA would provide collisions during 3000 hours per year and thus would produce 1 $\text{fb}^{-1}$ in ten years of operation.

The luminosity of a collider is given by the product of number of particles per bunch in each beam times the collision frequency divided by the effective cross section of the colliding beams in the plane perpendicular to the beam axis. For HERA we express the number of leptons per bunch times the collision frequency $N_e \cdot f_c$ as the total lepton beam current $I_e/e$. The number of protons per bunch divided by the beam cross section $\sigma^*_{xp}\sigma^*_{yp}$ becomes $\frac{1}{2}N_p\gamma_p\varepsilon_N^{-1} \cdot (\beta_{xp} \cdot \beta_{yp})^{-1/2}$ with $\gamma_p$ the proton relativistic factor, normalized proton beam emittance $\varepsilon_N$ and the proton beam envelop functions $\beta^*_{xp}$, $\beta^*_{yp}$ at the collision point. In summary, we have

$$L = \frac{\gamma_p I_e N_p}{4 \cdot \pi e \varepsilon_N \sqrt{\beta^*_{xp}\beta^*_{yp}}} \tag{1}$$

The lepton beam current is ultimately limited by the available RF power. This limit arises primarily from cost constraints. At 27.5 GeV, each electron radiates 86 MeV of its energy in the form of synchrotron radiation in the dipole magnets with a bending radius of 608 m. Replacement of this energy requires an RF voltage of approximately 120 MV and the power delivered to the beam for 50 mA of beam current amounts to 4.3 MW. The amount of power needed to build up the accelerating fields in the 86 normal conducting 5 and 7-cell RF cavities and 16 superconducting cavities is about 3 MW. Thus a total RF output power of about 8 MW is required. The total power consumed by the RF system exceeds 15 MW under these circumstances. Thus a beam current in the order of 50 mA constitutes a practical, though soft limit for the electron beam intensity and corresponds to what has been achieved.

The factor $N_p/\varepsilon_N$ is called proton beam brightness. It is ultimately limited in the low energy part of the acceleration chain. The limiting parameter is the space charge tune shift experienced during accumulation of proton beam at injection energy of 50 MeV (kinetic, $p = 0.29\,\mathrm{GeV}/c$) of the proton synchrotron DESY-III ($C = 318\,\mathrm{m}$). The space charge tune shift for a bunch of $10^{11}$ protons of a normalized emittance of $\varepsilon_N = 5 \cdot 10^{-6}$ rad·m is $\Delta Q = 0.5$ which constitutes a very respectable upper value. Typical values of the beam brightness achieved in HERA operations amounted to $N_p/\varepsilon_N = 0.75 \cdot 10^{11}/3.5 \cdot 10^{-6}$ rad·m.

The beta functions of the proton beam are limited by apertures in the focusing magnets, by the available space in the straight section, and by limitations arising from the chromaticity associated with the final focusing. There are two constraints which limit proton focusing in HERA: The large difference in beam energy of leptons and protons requires that the beams must be completely separated before the large focusing forces for the proton beam occur. Furthermore, since the lepton emittance is naturally larger than the proton emittance, strong lepton focusing in the horizontal plane is required to match the proton beam size for optimum collision operation. In order to keep the chromaticity of the lepton beam within tolerable limits, it is required that the electron beam must be fully focused before full beam separation is achieved. The other early design decision which limits the beta functions of the protons is to focus the protons with normal conducting magnets. The original decision was based on schedule considerations in view of the large amount of time needed to develop superconducting focusing magnets in presence of strong synchrotron radiation from the lepton arising from the magnetic beam separation. In order to provide sufficient aperture for the protons, the beta functions were originally limited to $\beta*_{xp} = 10\,\mathrm{m}$ and $\beta*_{yp} = 1\,\mathrm{m}$. After the reconstructions of the interaction region which included superconducting separation and focusing magnets for the lepton beam inside the colliding beam detectors, the proton beta function could be reduced to $\beta*_{xp} = 2.45\,\mathrm{m}$ and $\beta*_{yp} = 18\,\mathrm{cm}$. These values are close to the lower limit given by the minimum bunch length of the protons of 25 cm. The minimum bunch length is closely related to the proton beam brightness (short bunches limit the brightness by a linear relationship) and in a more soft way by the amount of RF focusing. The HERA proton ring has four 208 MHz and two 52 MHz cavities providing a voltage of approximately 1 MV. This amount of RF focusing provides a good compromise between short bunches and small energy spread required for beam–beam stability. The proton bunch length at collisions is 25 cm and this bunch length reduces the luminosity by 5%. With these parameters, the maximum HERA luminosity is calculated as $L_{\max} = 5.3 \cdot 10^{31}\,\text{cm}^{-2}\text{s}^{-1}$ which is also the maximum recorded value.

Writing the luminosity in the above form implies that a number of effects can potentially introduce further constraints. They need to be taken into account appropriately in the design of the accelerator rings which is discussed below.

Leptons and protons experience different beam–beam forces. The experience with HERA suggests that this configuration does not cause any problems as long as the maximum tolerable beam–beam tune shift values do not exceed values achieved in $e^+e^-$ and $p\bar{p}$ collisions respectively. The different values of energy spread and bunch length do not impose additional limitations. Typical limitations for the proton beam–beam tune shift are $\Delta Q = 0.003$ [3]. While such values have been achieved during machine studies in HERA, the values obtained in routine operations were smaller due to limitations in the electron beam current. The beam–beam effect experienced by the proton beam is further controlled by distributing the total electron beam current over a maximum number of bunches. The corresponding low level of beam–beam force allows for stable, halo-free collision operations. The HERA bunch distance is 96 ns (due to the 10 MHz RF in DESY-III) which, given the constraint of the injection system, allows to fill 180 proton and lepton bunches. The bunch current for the leptons is only 260 $\mu$A or $3.4 \cdot 10^{10}$ $e$/turn and the corresponding proton beam–beam tune shift is below $10^{-3}$.

Furthermore, the number of bunches needs to be compatible with single bunch intensity limitations for the lepton beam. The maximum single bunch lepton current in HERA (>1 mA) exceeds the operating values (0.3 mA) by at least a factor of three.

The distance between bunches needs to be sufficiently large to avoid detrimental long range beam–beam forces between protons and leptons. The first parasitic collision in HERA occurs after a distance of $(96\text{ ns}/2)\cdot c =$ 14.4 m from the interaction point (IP) which is well beyond the point where the beams are fully separated (they travel in different beam pipes beyond 11 m from the IP).

The lepton beam cross section is chosen to be equal to the proton beam cross section. HERA operation confirmed earlier findings at the SPS collider that this is necessary to minimize the nonlinear effect of the beam–beam interaction on the proton beam and assure a stable and halo-free proton beam.

Emittance and beta functions at the IP for leptons have to be chosen such that the beam–beam forces for the lepton beam result in a tolerable beam–beam tune shift for stable operations. This is a constraint with major consequences for the lattice design of the lepton beam. For given proton beam parameters, the choice of the lepton beta functions at the IP determines the beam–beam tune shift of the lepton beam. For stable beam–beam operations, the lepton beam tune shift was required to stay below $\Delta Q_y = 0.050$. For a proton bunch charge of $10^{11}e$, the beta values of the leptons at the IP must not exceed $\beta^*_{ex} = 0.62$ m and $\beta^*_{ey} = 0.26$ m. With these values, the chromaticity contributions from the IP focusing are significant and an elaborate chromaticity correction is needed to avoid performance limitations.

Proton and lepton beam sizes are then matched with a horizontal lepton beam emittance of 20 nm. Limiting the emittance to 20 nm rad requires a fairly strong focusing in the FODO arcs of the lepton ring. A phase advance of 60 degrees per FODO cell is superior with regard to dynamic aperture and chromaticity correction but the corresponding emittance is almost twice as large as required. The chosen phase advance is 72 degrees per cell. In addition to this fairly strong focusing, a slight RF frequency detuning of 250 Hz is required which forces the electron beam to travel 1 mm off-axis (0.1% off-energy) through the arcs. This provides the desired horizontal lepton emittance of 20 nm. (Measured values resulted in a slightly larger value of 25 nm.) Dynamic aperture is an important consideration in the 72-degree lattice. Orbit control is very critical to avoid enlarged nonlinear fields and excitation of synchro-betatron resonances [4]. The vertical emittance of the lepton beam required to match the proton beam size is 2.5 nm of 12.5% of the horizontal emittance. This is achieved fairly easily. Control of this value, however, is required to maintain halo-free proton beams in collisions. Emittance tuning bumps which generate vertical dispersion are used to keep proton and electron vertical beam sizes well matched.

Only head on collisions will provide the required stability for the proton beam. These conclusions have been achieved in beam–beam collision simulations in 1984 and in 1996. While the HERA ring geometry did explicitly include the possibility of a crossing angle geometry, the arguments in favor of colliding the beams under a finite angle were never strong enough to seriously

Table 1. HERA main parameters (as achieved in routine operations).

| Parameter | Unit | Protons | Electrons |
|---|---|---|---|
| Beam energy | GeV | 920 | 27.5 |
| Beam current | mA | 100 | 45 |
| Circumference | m | 6336 | |
| Number of colliding bunches | | 174 | |
| Number of non-colliding bunches | | 6 | |
| Bunch charge | $10^{10}e$ | 7.3 | 3.3 |
| Horizontal emittance | nm | 4 | 20 |
| Vertical emittance | nm | 4 | 3 |
| Beta $x$ at IP | m | 2.45 | 0.62 |
| Beta $y$ at IP | m | 0.18 | 0.26 |
| Luminosity | $10^{31}\,\text{cm}^{-2}\text{s}^{-1}$ | 5.3 | |
| Luminosity life time/run time | hr. | 9 | |
| Luminosity per run | $\text{pb}^{-1}$ | 0.5 | |
| Bunch length | cm | 16 | 0.9 |
| Hour glass reduction of luminosity | % | 94.3 | |
| Number of interaction points | | 2 | |
| Horizontal beam–beam tune shift/IP | | 0.0023 | 0.03 |
| Vertical beam–beam tune shift/IP | | 0.0007 | 0.03 |

consider this option. The results of the above discussion are summarized by the following Table 1 of main HERA parameters.

## 3 Lattice design and IR layout

The overall size of the collider was determined by two constraints: The facility should be built near the DESY site in Hamburg and the collider tunnel should be located on public land. This limited the circumference of the accelerator rings to 6 km and required a deep accelerator tunnel 20 m below the surface.

In order to limit the cost of the large accelerator complex, the construction teams under G.-A. Voss and B. Wiik made use of the existing accelerator complex as HERA injectors: The former 23 GeV PETRA electron–positron collider provided 12 GeV electrons and positrons and 40 GeV protons for injection into HERA. The complex included the 450 MeV electron/positron LINAC and two fast cycling booster synchrotrons. Together with a new 50 MeV drift tube LINAC these accelerators formed the HERA injector complex. This choice however implied low proton injection energy and a fairly slow injection rate (up to one every 5 min). The mitigation of these disadvantages was one of the challenges in building and operating the HERA collider. Another compromise made to control the HERA cost was to carry over 86 normal conducting high impedance 5-cell and 7-cell RF cavities from PETRA to HERA which together with 16 superconducting cavities provide the accelerating fields of >120 MV for the lepton beam.

Besides luminosity, the most important parameter for HERA (and any other collider) is the center-of-mass energy $E_{\mathrm{cm}} = 2\cdot(E_p\cdot E_e)^{1/2}$ given by the product of proton $E_p$ and electron $E_e$ beam energy. The maximum energy of the proton beam is given by the maximum technically feasible dipole field $B$, the total circumference of the accelerator ring and the fill factor, the ratio of circumference and total length of the bending magnets. Thus superconducting magnets are mandatory to reach high beam energy. For the maximum available circumference, the electron energy is limited by the fill factor and the available RF power which can be delivered to the beam to balance synchrotron radiation losses. All these factors are ultimately limited by the cost for construction and to operate the facility.

The most effective focusing, thus the highest fill factor is provided by a FODO lattice. The choice of the basic FODO cell parameters are a compromise of large fill factor and magnet cost: The required magnet aperture scales with the square root of the length of the FODO cell $L$ and the ratio of length of focusing and dipole magnets scales with $L^{-1/2}$. Thus while a longer cell length provides a better fill factor, the cost of the dipole magnets and the needed transverse space in the accelerator tunnel increases quadratically with $L$. Optimization leads to a 47 m long food cell (protons) with four 10 m long dipole magnets and two 1.8-m long quadrupole magnets. The magnet aperture is 54 mm. The state of superconducting magnet technology provide a 5 T bend field and a 75 T/m quadrupole field which focus the beam with betatron phase advance of 90 degree/FODO cell. The maximum proton beam energy is 920 GeV. The four 1270 m long arcs have 416 dipole and 212 quadrupole magnets and are separated by four 314 m long straight sections. Each half cell is equipped with a horizontal or vertical corrector magnet. A specialty of HERA is that the 6 m long sextupole and quadrupole correctors are wound on the beam pipe inside each dipole magnet (extending of 2/3 of the dipole length at the upper and lower end of the two dipoles respectively in a half cell). These provide not only for tune and chromatic corrections but also compensate quasi-locally the substantial sextupole component of the dipole field (relative field error is 0.3% at a radius of 25 mm) due to superconducting eddy currents. These maximize near the relatively low injection energy of 40 GeV, the maximum achievable energy of the PETRA injector ring.

For the electron ring, the optimization of the FODO cell is driven by the fact the beam emittance scales with the 3rd power of the number of FODO cells, which suggests smaller FODO cell lengths to avoid unreasonably large magnet apertures. There is an advantage to keep the two FODO structures for electrons and protons in phase, which leads to a 23.5 m long FODO cell with two 9.2 m long dipoles ($B = 0.3$ T) and two 0.75 m long quadrupoles. The dipole gap allows a 20 mm vertical half aperture. The horizontal half aperture is limited in the quadrupoles to 40 mm. With this lattice maximum beam energy of 27.5 GeV has been achieved. The synchrotron radiation power loss of the beam is 86 MeV per electron and turn. As most of the RF amplitude is generated in the normal conducting cavities carried over from PETRA, the total RF power required to reach this beam energy is about 8 MW, which includes 4.3 MW delivered to the beam of 45 mA.

The vacuum chambers in the proton ring consist of copper coated stainless steel and the electron ring chamber consists of extruded copper with integrated pumping.

The relatively long straight sections (360 m) of HERA serve many purposes: They accommodate the low beta interaction region (IR) for the four experiments H1 (North straight section), ZEUS (colliding beam, South), HERMES (e-beam gas target, East), HERA-B (p-beam wire target, West). There are seven 50 m long RF sections with 12 normal conducting 500 MHz 5-cell or 7-cell cavities, one on each side of the IRs, and a 60 m long superconducting cavity section in HERA West-Right providing an RF accelerating voltage of 130 MV. The straights further accommodate three pairs of 60 m long spin rotator magnets around the interaction regions North, South and East, the proton injection elements, the proton beam dump, the two proton 52 MHz cavities and the four 208 MHz cavities, as well as active damper systems and diagnostic equipment. The overall layout of the accelerator is shown in Fig. 1. Electron injection can be fitted in the NW arc using special strong dipole magnets to provide space for the three kicker magnets and the pulsed septum.

The layout of the interaction region is dominated by the decision to only use normal conducting magnets for the final focus of both proton and lepton beams. This concept that was originally based on cost and schedule constraints was not altered in the later upgrade of the IRs for the same reasons, except for weak superconducting air-coil magnets inside the field of the ZEUS and H1 detector solenoids. Because of the large energy difference, the proton and electron beams need to be separated sufficiently before focusing elements for the protons can be placed. The corresponding rapid magnetic separation is achieved by the superconducting air-coil combined function bending and

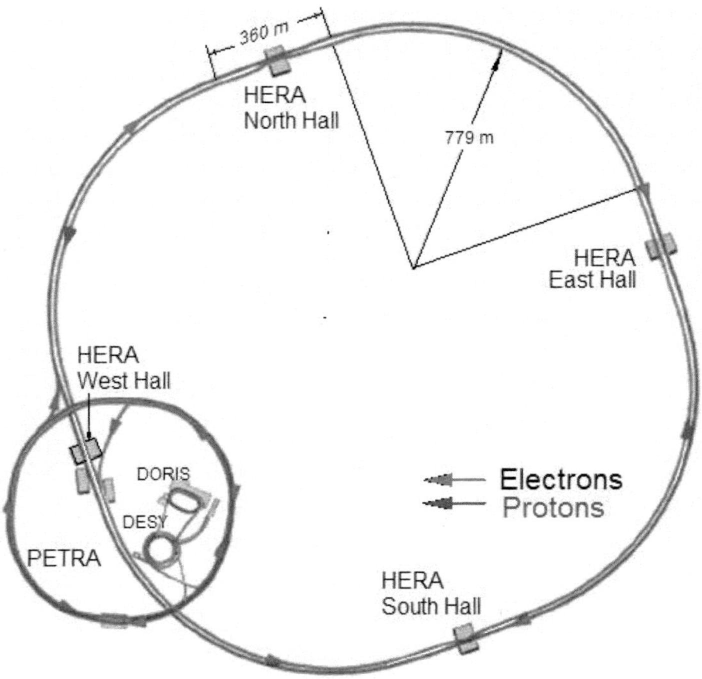


Fig. 1. HERA *ep* collider footprint.

low beta focusing elements for the electron beam which was integrated in the colliding beam detectors of H1 and ZEUS. The elements of the low beta triplet outside the detector (iron yoke quadruples) are placed off center on the design orbit contributing that way to the beam separation. This provides a beam separation of 60 mm at 11 m from the IP which is sufficient spacing to place the first focusing element for the protons, a special low beta half-quadrupole magnets with a nickel vertical mirror plate, the field of which is not seen by the electron beam. This is the first element of the 50 m long proton double doublet final focusing. The double doublet structure balances the chromaticity and maximum beta in both planes and allows fairly small values of the beta at the IP with $\beta_x* = 2.45$ m and $\beta_y* = 0.18$ m (near the bunch length limit). Strong synchrotron radiation ($\sim$30 kW/IR) is generated inside the beam separation magnets before and after the collision point. The radiation passes through the detector and through key-hole shaped vacuum chambers inside the downstream low beta quadrupole magnets. It is absorbed sufficiently far away from the detector to keep backscattering photons at a level tolerable for the sensitive near-beam detectors needed for *ep* colliders. Figure 2 provides a top view (note the different horizontal and vertical scale) of the interaction region. Details are described in Ref. [5].

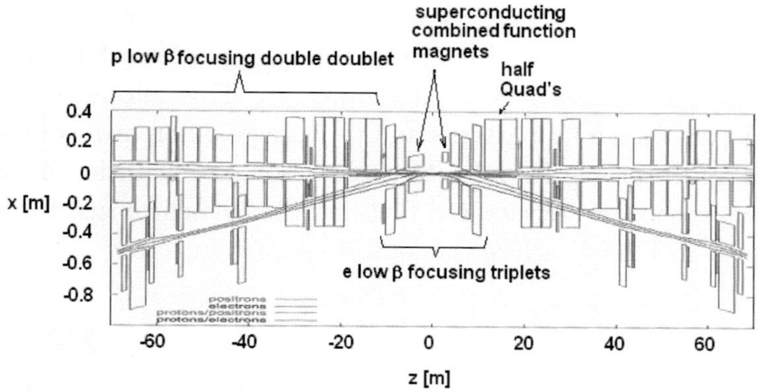


Fig. 2. Top view of the HERA interaction region (note the distorted scales).

## 4 Injectors

The lepton accelerator chain consists of a 450 MeV s-band linear accelerator. Positrons are created by focusing a 200 MeV electron beam on a tungsten target. Positrons are collected at 450 MeV in the PIA collector ring over a period of 80 ms, the cycle time of the DESY-II 12 GeV, 12.5 Hz separated function injector synchrotron. It accelerates the lepton beam to 7 GeV prior to transfer to the PETRA ring which accelerates the leptons to 12 GeV, the HERA injection energy. The hadron beam starts from a magnetron type $H^-$ source. An RFQ accelerates the $H^-$ beam to 180 keV, the injection energy of the drift tube linac which accelerates the ions to 50 MeV kinetic energy. The next element in the acceleration chain is the DESY-III proton synchrotron. The $H^-$ ions undergo charge transfer upon injection through an aluminum oxide foil and protons are being accumulated over ten turns, thereby achieving a beam current of 50 mA to 150 mA (the revolution time reduces by a factor of three during acceleration). Acceleration to the top energy of 8 GeV (7.5 GeV/$c$ momentum) occurs without crossing the transition energy which is well above the extraction energy due to DESY-III's low alpha lattice. The PETRA synchrotron also serves as the proton pre-accelerator for HERA. Its magnets can be excited to reach the required HERA injection energy of 40 GeV. The focusing is intentionally weak in the horizontal plane which yields a large dispersion and low transition energy below the injection energy. A bypass for protons is added to the PETRA ring to avoid the high impedance RF region in the South straight section. Due to path length differences, revolution time differs between lepton and proton at HERA injection. The synchrotrons in the HERA acceleration chains are operated with different RF frequencies (the corresponding 500 MHz reference frequencies differ by as much as 10 kHz) which keeps the machines in continuous "cogging" mode. An intricate timing and trigger system triggers transfer whenever the desired start and target buckets are lining up. This allows for maximum flexibility in the bunch fill patterns and to manage different revolution times for leptons and protons. In order to avoid resonant excitation of one beam by another, protons are accelerated first and then synchronized to the lepton RF frequency in HERA before the lepton beam is injected.

## 5 Superconducting magnets

The required large proton beam energy of $E > 820$ GeV together with the space constraints for the collider as well as electrical power cost considerations mandated superconducting magnets for the HERA proton ring. HERA is the second large hadron accelerator which uses superconducting magnets. The HERA dipole magnet has a two layer cosine-theta type superconducting coil wound with a Rutherford type of cable. The geometric tolerance of the conductor is 20 microns which limits the relative magnetic field error to the required $10^{-4}$. The coil aperture is 72 mm leaving space for the one-phase He supply flow and the cold beam pipe with 54 mm aperture. The beam pipe carries correction windings for trim quadrupoles, sextupole corrections as well as corrector for a substantial decapole component of the main field at low excitation. The coil is secured by an aluminum alloy collar (Al-Mg-Mn) surrounded by the return yoke made of steel. This forms the inertia mass inside the liquid He vessel. The magnet is cooled by supercritical He of a pressure of 18 bar. This mixed phase coolant penetrates the superconducting coil and provides excellent cooling of the superconducting cable. The He vessel is surrounded by a heat shield which is cooled with 60 K cold He gas. The temperature of the supercritical He is 4.5 K for a magnet excitation of 4.1 tesla (at 820 GeV). When HERA was upgraded to 920 GeV operations, the temperature was lowered to 4.2 K in order to maintain operational safety margin. Figure 3 shows a schematic cross section of the HERA dipole magnet.

The magnet performance is excellent. The dipole magnets did not show any substantial "training" and full performance was achieved and maintained at the first or 2nd excitation cycle. With a coolant temperature of 4.5 K, all magnets reached comfortably $> 5$ T. None of production units of the HERA experienced spontaneous quenches. The HERA quench protection system consists of quench heaters to distribute quenches safely over a large coil area. Cold diodes are switched in parallel to each half coil to bypass the magnet currents around the quenches magnet. A measurement bridge

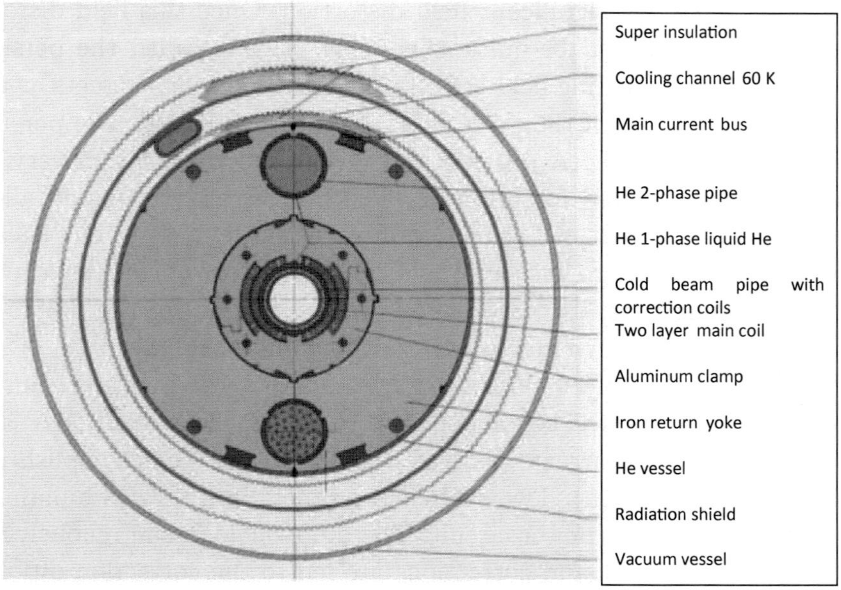


Fig. 3. Schematic cross section of the HERA dipole magnet.

monitors any difference in voltage across the two half coils, in the case of which it triggers the quench heaters and magnet current bypass switches in each ring-octant which sends the current to quench resistors to dump the magnetic energy. The same method is used when comparing the voltage of four magnets to protect against symmetric quenches of the two coils. This very robust quench protection system never failed in the 17 years of HERA operation. Quenches only occurred as a consequence of substantial beam loss in the coils. No magnet was ever damaged because of the excellent quench performance of the magnet. Because of sufficient operational margin, the magnet system did not show any quench propagation.

While the field quality is excellent for the HERA magnets at high excitation due to mechanical tolerance of the conductor position in the coil, at low excitation, the magnet is subject to a systematical effect of high field magnets made of type-II superconductors. Magnetic flux lines can penetrate the conductors due to so-called pinning centers, impurities in the lattice of the superconductor. While this is helpful to build a robust superconducting high field magnets, it produces an issue at low fields. Field lines penetrate the superconducting filament of the cosine-theta coil. If the magnet field is increased, eddy currents around the field lines counteract the increase in flux generated by the coil. This effect weakens and distorts the fields generated by the coil. It leads not only to a dipole field reduction but, what is more important, it causes a significant field distortion. Once this field distortion is excited, it will persist for quite some time. This is called the persistent current. The corresponding field error in the HERA dipole magnets at injection field (roughly 5% of the peak field) is quite significant. The persistent current sextupole in HERA peaks at 0.3% at a radius of 25 mm very close to the set point required for injection. The field error contains all allowed components of the dipole coil geometry, in particular a fairly strong decapole component in addition to the sextupole. The persistent current sextupole is sufficiently strong that, if compensated only at the end of the dipole by a lumped corrector, the dynamic aperture would suffer significantly.

What makes this effect more detrimental is that the persistent currents are not really persistent. They are subject to a slow decay. In HERA, about 10% of the persistent currents decayed during the time to accomplish injection into HERA (20 min). Two reference magnets in a service building are part of the main dipole power circuit. The dipole field is continuously monitored by a NMR probe and corrections are fed to the correction dipoles to maintain the injection field to a precision of $10^{-4}$. Rotating coils measure the sextupole and provide input to the sextupole correction circuits. This helps to control the chromaticity to within one unit, even during the start of the energy ramp when all decayed persistent currents are re-induced quasi-instantaneously.

The main quadrupole magnets of the HERA proton ring consisted of 1.8 m long superconducting magnets with a field gradient of 75 T/m. The quadrupole design has features similar to those of the dipole. Besides the quadrupole, sextupole, and decapole correctors wound on the dipole beam pipe, superferric dipole correctors and quadrupole trims completed the suite of correction elements. A fairly recent, concise description of the HERA superconducting magnet system can be found in Ref. [6].

## 6 Operational challenges

One of the largest difficulties in operating HERA is the low injection energy of protons which implies strong and time dependent sextupole components of the dipole field due to slowly decaying superconducting eddy currents and their quick restoration at the start of acceleration. At low energy, the beam size is very large and strong field distortions are present. If the chromaticity is chosen too large, the dynamic aperture would be poor and a significant fraction of the beam would be lost during injection and early acceleration. If the chromaticity is too close to zero, the fast changing persistent current sextupole which depends on magnet history could drive the chromaticity below zero in which case the beam would become unstable due to a special form of head-tail instability [7]. The instability can be controlled by a broad band damper system but the dampers increase the proton beam emittance and reduce luminosity. While this effect is well studied and well understood, it requires a lot of attention and skilled operators.

Early operation with electrons suffered from a sudden breakdown of beam lifetime with a low intensity threshold. The beam lifetime would not recover even for the smallest beam currents but would be restored after a beam dump and new injection. This was attributed to dust in the vacuum system. The explanation was that dust particles on the positively charged plates of the integrated getter pumps would get ionized by synchrotron radiation and then accelerated and captured inside the beam leading to massive beam loss due to bremsstrahlung. This problem was resolved by running with positrons until 1998. The vacuum system was cleaned in 1997 and the ion pumps were replaced by NEG pumps. After 1998 operation with electrons was possible, though occasional dust trappings were observed in later years as well.

An important feature of HERA was longitudinal polarization of the lepton beams at the interaction regions North, South and East. The lepton beam is polarized in the vertical direction by the Sokolov–Ternov effect. To achieve good polarization it is important that all horizontal magnetic fields from imperfections must be kept small and non-resonant in order to maintain the equilibrium polarization direction (called "N-axis") in the arcs parallel to the dipole field with a tolerance in the order of mrad. Otherwise the distribution of spin precession frequencies in the beam would lead to a reduction of the polarization vector by precession phase dilution. Strong distortions of the N-axis lead to complete depolarization of the lepton spin. A system of harmonic bumps was designed and used successfully to keep the N-axis angle within tolerances. Beam polarizations of up to 70% were achieved that way [8]. Only longitudinal polarization is useful for experiments. Sixty-meter long spin rotators generated vertical orbit bumps and an interleaved s-shaped horizontal orbit chicane that rotate the spin from the vertical into the longitudinal direction and vice versa [9]. In order to restore the vertical spin orientation for all particles it is important that the horizontal fields in quadrupoles between the rotators do not cause a net change in spin orientation independent of betatron amplitude and phase. This implies special beam optical conditions which are known as spin matching. HERA spin matching in the three interaction regions was very successful. A polarization of close to 60% with three spin rotator pairs was achieved routinely in HERA operations [10].

Large lepton storage rings are subject to strong buildup of spurious dispersion due to closed orbit distortions. The distorted orbit stays in phase with the dispersion it causes by corresponding dipole components in the quadrupoles. A dispersion wave caused by closed orbit can drive a strong contribution to the synchro-betatron resonance $Q_x + 2Q_s$ = integer. This causes a problem for high polarization beam operations since high polarization requires a low horizontal tune in the vicinity of such resonances. An additional interesting effect made it initially impossible to run the machine at the tune needed for high polarization: The 72-degree lattice seemed to perform best with only two sextupole families. This caused a strong chromatic sideband of the half integer resonance $2Q_x + Q_s$ = integer, far away from the operating tune. However, the interference of the two resonances $Q_x + 2Q_s$ = integer and $2Q_x + Q_s$ = integer, generated a strong resonance $Q_x + 3Q_s$ = integer, and the combination of all these effects made it impossible to operate HERA with the desired low horizontal betatron tune resulting in poor beam spin polarization. Once understood, the cure was straightforward: A quarter integer phase advance in both planes between the North and South interaction region, reduced the chromatic sideband $2Q_x + Q_s$ = integer of the half integer resonance. In addition, continuous orbit feedback avoided any enhancement of the $Q_x + 2Q_s$ = integer resonance by orbit driven dispersion waves. These two measures allowed running HERA at low tune, high polarization, high luminosity and good lifetime.

## 7 Operational results and conclusion

In the first four years of HERA operation, the progress was slow but in 1997, the design luminosity was reached and a respectable 35 pb$^{-1}$ per year could be accumulated. During a large shutdown for installing the NEG pumps in 1997–1998, a number of measures were implemented which drastically improved the operational reliability: An 8th RF section was installed increasing the total RF power to nominally 12 MW thereby providing sufficient operational margin. The coupler of the superconducting cavities was improved by a DC bias voltage, the proton injection line optics was improved, and the power supply system was overhauled and the original inadequate control system was replaced by up-to-date controls This increased the reliability of the accelerator complex by a factor of ~1.5 and the accelerator availability reached 80%. The accumulated luminosity in the middle of 2000 amounted 200 pb$^{-1}$.

After the luminosity upgrade in 2000/2002, HERA had a slow start due to an unfortunate combination of circumstances. Despite the fact that high

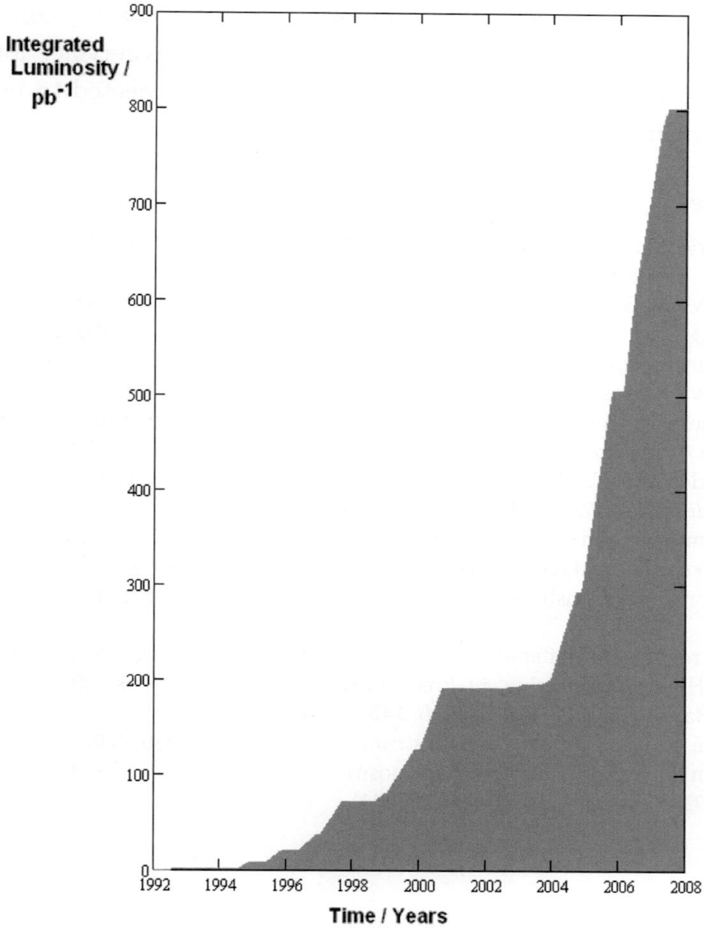


Fig. 4. HERA integrated lifetime luminosity

luminosity could be demonstrated on a few bunches within a few months after turn-on, it took until 2004 to be able to operate HERA with full performance. In 2005 and 2006, a yearly luminosity of 200 $pb^{-1}$ was accumulated. In its last year of operation 2007, HERA was operated at lower proton energies in the range between 450 GeV and 920 GeV. When HERA operations were terminated in June 2007, HERA had provided an integrated luminosity of 800 $pb^{-1}$ (see Fig. 4).

HERA and the detectors ZEUS, H1 and HERMES have produced a wealth of high precision data on deep inelastic scattering, which provided for experimental verification of quantum chromodynamics, electroweak physics and related areas as the fathers of HERA had envisioned. While the hope for new physics and new discoveries with HERA have not come true, the impact of HERA on our understanding of hadronic matter constitutes an important landmark of physics. HERA physics and experience form a solid base for the understanding of the physics which is expected to be observed in the Large Hadron Collider.

## References


[1] J. I. Friedman *et al.*, in *Proc. 14th Int. Conf. High-Energy Physics*, Vienna, 1968 (CERN, Geneva, 1968).

[2] G.-A. Voss, B. H. Wiik *et al.*, HERA — A proposal for a large electron–proton colliding beam facility at DESY, DESY report HERA 81/10 (1981).

[3] K. Ohmi, Beam beam limit in a hadron collider (THYB01), in *Proceedings of IPAC-12*, New Orleans, Louisiana, USA (2012).

[4] F. Willeke, Overcoming performance limitations due to synchro-betatron resonances in HERA, in *Proceedings of EPAC04*, Lausanne, Switzerland (2004).

[5] M. Seidel and F. Willeke, Layout of the upgraded HERA interaction regions, in *Proceedings of EPAC2000*, Vienna (2000).

[6] P. Schmueser and F. Willeke, The electron-proton collider HERA, in *Elementary Particles*, eds. S. Myers and H. Schopper, Landolt-Boernstein — Numerical Data and Functional Relationships in Science and Technology: New Series, Vol. 21C (Springer, 2013).

[7] E. Metral, G. H. Hoffstaetter and F. Willeke, Destabilizing effect of linear coupling in the HERA proton ring, in *Proceedings of EPAC2002*, Paris (2000).

[8] D. P. Barber *et al.*, *Phys. Lett. B* **343** (1995) 436.

[9] J. Buon and K. Steffen, *Nucl. Instrum. Meth. A* **245** (1986) 248.

[10] E. Gianfelice *et al.*, Longitudinal positron polarization in HERA-II, *Proceedings of EPAC04*, Lausanne, Switzerland (2004).

# Part 2

# 100 Years of Accelerators

# Chapter 16

# A vision unfulfilled: The hopeful birth and painful death of the Superconducting Super Collider*

Michael Riordan
*University of California, Santa Cruz (emeritus)*

As Europe began to approach parity with the United States in high energy physics around 1980, US physicists were becoming increasingly concerned about the future of their national program. While Europeans were pushing ahead with ambitious plans for two advanced particle colliders, LEP at CERN and HERA at DESY, the United States found itself faced by knotty problems in manufacturing the 4-T superconducting magnets required for the new Isabelle proton–proton collider at Brookhaven [1]. Its ballooning costs, which had swelled from an initial $275 million to an estimated $600 million (and counting), threatened to drain funds from the rest of the high energy physics program, including at Fermilab and SLAC. And by mid-1982, it seemed likely that physicists doing research at CERN's new proton–antiproton collider would soon discover the massive W and Z particles predicted by the Standard Model, among Isabelle's primary targets.

Thus physicists led by Fermilab Director Leon Lederman began calling for a bold new US initiative at a summer meeting hosted in Snowmass, CO, by the APS Division of Particles and Fields [2]. The Snowmass workshop was an unusual gathering in several respects. It brought together representatives from the *entire* US high energy physics community — theorists, experimenters and machine builders alike — to deliberate and make projections collectively. Previously, new facilities had usually arisen in response

*This article is based in part on a scholarly book-length history of the Super Collider published in 2015 by the University of Chicago Press, titled *Tunnel Visions: The Rise and Fall of the Superconducting Super Collider*, by Michael Riordan, Lillian Hoddeson and Adrienne W. Kolb. The second section has been derived from Chapters 2 and 3 of that manuscript, written principally by Hoddeson and Kolb.

to proposals from various universities and national laboratories, not the community as a whole. Even Fermilab had begun as a proposal for a 200 GeV proton accelerator to be built as an addition to the Lawrence Berkeley Laboratory [3]. Leading voices in US high energy physics started suggesting that this was not the best way to meet the daunting European challenge.

What emerged from Snowmass were visions of a proton collider dubbed the Desertron — partly because the required real estate might be available only in the American Southwest. This enormous machine would stretch 100 to 200 km in circumference, have a collision energy of 20 to 40 TeV, and cost up to $3 billion according to rough estimates. In some ways, this collider resembled the Very Big Accelerator that had been envisioned previously by Lederman, Robert R. Wilson and others as an international project [4]. But it soon came to be championed by the US high energy physics community (or at least its proton-smashing contingent) as the best way to reinvigorate the national program and restore US leadership in the discipline.

In early 1983, as CERN announced the long-expected W boson discovery [5], the US Department of Energy (DOE) empanelled a group of physicists to follow up on the Snowmass proposals and make long-range recommendations for the future of the US high energy physics program. Chaired by Stanford University physicist Stanley Wojcicki, it met for the next five months, considering proposals and letters from throughout the community. By then Brookhaven had resolved its superconducting magnet problems and redesigned Isabelle, renaming it the Colliding Beam Accelerator.[1] Fermilab chimed in with a new proposal to build a 4–5 TeV proton–antiproton collider known as the Dedicated Collider that could extend to the limits of its existing site [6]. But, encouraged by President Reagan's Science Adviser George Keyworth to "think big," the Wojcicki Panel had much more adventurous projects in mind [19].

Meeting in Woods Hole, MA, and then at Columbia University's Nevis Laboratory in June and July, the panel recommended that Isabelle be terminated and that the United States proceed with the design and eventual construction of the Desertron, renamed the Superconducting Super Collider, or SSC. It would stretch nearly 100 km in circumference, have a collision energy of 20–40 TeV and luminosity of $10^{33}\,\mathrm{cm}^{-2}\mathrm{s}^{-1}$, and cost about $2 billion — not including experiments, R&D, contingency, and miscellaneous start-up costs [6]. The other alternatives were deemed insufficient to address the most important physics research (now that the W and Z bosons had been

[1]See Ref. 1, Crease, "Quenched!" part 2, 441–443.

discovered), expected to occur at TeV energies.[2] The US High Energy Physics Advisory Panel (HEPAP) met in Washington July 11–12 and unanimously endorsed the Wojcicki Panel report. In an accompanying letter, HEPAP Chair Jack Sandweiss of Yale stressed that the SSC "would be the forefront high energy facility of the world and is essential for a strong and highly creative United States high energy physics program into the next century." [8]

## 1 Establishing the SSC project, 1983–1988

The crucial work of formulating a conceptual design for the SSC was addressed by a large team of physicists from all across the country organized by Cornell accelerator physicist Maury Tigner and headquartered at LBL [9]. Major design decisions to be made were the superconducting magnet field strength, which affected the machine circumference, and the magnet aperture. By mid-1985 the SSC Central Design Group had selected the "high-field" option, with fields up to 6.5 T, and a dipole aperture of 4 cm; these choices allowed a circumference of only 83 km for a 20 TeV (per beam) accelerator and would help to hold down total construction costs.[3] In March 1986 the group issued the SSC Conceptual Design Report including these design choices and parameters [10]. It estimated the SSC construction costs at $3.01 billion (in 1986 dollars), including a $529 million contingency but not such costs as R&D, computers, commissioning, detectors, and inflation. After an extensive DOE review added just $5 million to the project cost, the report was widely hailed as "a model of scientific clarity" [11]. But a subcommittee of the DOE review panel expressed concerns that the 4 cm aperture might be too risky and suggested that the option of increasing it to 5 cm be kept open as a future possibility. For some reason, however, DOE officials did not add anything to the project contingency to account for this major uncertainty.

By the time the SSC proposal reached the White House in late 1986, DOE had included most of the additional costs and now estimated the SSC price tag at $4.5 billion (or $6 billion in as-spent dollars over a ten-year

---

[2]In 1983 theoretical research suggested that the Higgs boson might exist at masses up to 1 TeV and that other mass-generating particles or phenomena could occur at up to 1.8 TeV. To generate enough collisions between quarks and gluons to create such massive particles or phenomena, the proton collision energy had to be at least 10 TeV and preferably more.

[3]A "two-in-one" magnet design, in which the two proton beams could be enclosed within a single cryostat, was considered among the possibilities but rejected as too adventurous. Later, European accelerator physicists criticized this decision as too conservative, driving costs up. But it is questionable whether the state of the art in superconducting magnet design had advanced by the mid-1980s to the point where such a daring design, which was later employed in the CERN Large Hadron Collider (LHC), could be seriously considered for the SSC.

construction period, including inflation.) This sum became too much to stomach for scientists in other disciplines, who argued that a US commitment to build the Super Collider would inevitably press upon their own hoped-for funding. Voices of opposition to the SSC got louder and louder as its total estimated costs grew [12]. Meeting in late 1986 and early 1987 to decide the matter, Reagan's Domestic Policy Council was split evenly on whether or not to proceed. But at a 29 January 1987 Council meeting with Reagan present, he personally made the crucial decision to go ahead with it, saying the nation should "throw deep" [13]. In a press conference the following day, Secretary of Energy John Herrington told reporters that the SSC would be "an American project [with] American leadership," but that the DOE also intended "to seek maximum cost-sharing funding from other countries" [14].

That decision set in motion a site-selection process lasting well over a year. At an estimated cost of some $5 billion, the SSC project was much too large to be sited at an existing laboratory like Fermilab, as most high energy physicists would have preferred. The centrifugal forces of the US political process, in which Representatives and Senators compete to bring federal funds back to their home district or state, militated against this approach. Thus DOE officials figured that they had to throw the site-selection process open to all states, with the decision to be made based on objective criteria such as site geology and accessibility. It was also considered a good approach to generate broad national recognition of, and support for, the project. By the 2 September 1987 deadline, DOE had received 43 proposals from half the states. The list was winnowed by late 1987 to a group of eight "best-qualified" sites, which included one south of Dallas, Texas, and an Illinois site next to Fermilab, using its existing accelerator as the proton injector.

Back at Central Design Group headquarters, things were not going well. Spread over four laboratories and coordinated from Berkeley, the magnet R&D program had experienced numerous failures testing its superconducting dipoles, only two of which had attained the design current by September 1988 [15]. With the Isabelle fiasco still fresh in their memories, anxious DOE officials began pressing Tigner for more successes, exacerbating their deteriorating relationship. And a brief attempt to reallocate $8 million from his budget to cover unexpected costs of the site-selection process worsened the growing tensions. By the summer of 1988, DOE officials had effectively excluded Tigner as a possible director of the SSC Lab, whose management structure was then increasingly under discussion.

In August the DOE issued a request for proposals to manage and operate the new laboratory, seeking competitive bids rather than award a "sole-source" contract to the Universities Research Association (URA), which then managed Fermilab and the Central Design Group [9]. The approach was partly a reaction to US political reality, as sole-sourcing such a large and important contract would have unleashed a storm of protest among other interested parties. URA leaders decided that they had to "team" with industrial partners in order to address the stated DOE requirements; they eventually chose EG&G, Inc. and the Sverdrup Corporation, both of which had extensive experience doing large-project military and industrial contracting but essentially no familiarity with the US high energy physics community. And a group of high-level URA consultants, including many respected leaders of this community, hastily recommended that physicist Roy Schwitters of Harvard University be designated SSC lab director in its proposal.[4] Tigner was also included in the proposal as deputy director, but it was becoming increasingly clear to Schwitters that he was not enthusiastic about it [16].

As no other proposals came in by the 4 November 1988 deadline, DOE began negotiating with URA on a contract to manage and operate the SSC Lab. And on November 10, two days after the election of George H. W. Bush to be the next US President, DOE officials announced that Waxahachie, TX, would be the site of the SSC Laboratory, not Fermilab as most US high energy physicists had expected [17]. To many it was a stunning surprise, eliciting charges of undue Texas political influence from the losing states. Indeed, with an adopted Texan headed for the White House and the powerful Texas delegation prepared to support the project in Congress, DOE officials had seemingly made an astute political choice. But DOE insiders pointed to the excellent characteristics of the Waxahachie site as the basis of the selection [18]. Another important factor was the public opposition to the project in Illinois. After much political outcry, the choice was upheld.

## 2 Settling uneasily in Texas, 1989–1991

On 18 January 1989, days before Reagan stepped down as US President and after resolving some serious disagreements, DOE and URA officials finally signed the management and operations contract for the SSC Laboratory. That same day Tigner resigned as deputy director, saying that he would return to Cornell. He and Schwitters had had long discussions about his assuming the role of SSC project manager, but they could not come to agreement on the authority Tigner would have in this position [16]. After his departure, many Central Design Group physicists, particularly its leadership,

[4]See Ref. 7, p. 265.

followed his example and returned to their home institutions. This was a serious loss for the fledgling laboratory, for these individuals carried with them much of the group's institutional memory and *espirit-de-corps*; they might instead have formed the foundation of the new laboratory organization.[5]

To flesh out his core management team, Schwitters increasingly looked beyond the high energy physics community. While Bruce Chrisman and Helen Edwards of Fermilab were already on board as business manager and technical director, he named Livermore physicist Richard Briggs, who had only limited accelerator-building experience, to serve as deputy director and project manager. And to direct the Magnet Division responsible for developing and procuring the thousands of superconducting magnets needed, he named engineering physicist Tom Bush, whose major professional experience had occurred in manufacturing nuclear-tipped missiles for the US Navy's Trident submarine fleet.

As physicists and engineers headed for Texas in the spring of 1989, to work mostly at a leased office park south of Dallas, Schwitters commissioned a thorough review of the SSC dipole magnet R&D program. By June it had come in, acknowledging that the program was behind schedule, still needing to show that 10 percent operating margins above the design current could be consistently attained in full-length test magnets [20]. The report also took up the question of the magnet aperture, observing that a 5 cm aperture would be much less risky but that, if pursued, a costly two-year delay could occur in the onset of the magnet industrialization program, delaying the project by a similar period. It also raised the possibility of increasing the proton injection energy to avoid beam losses at low energies due to "persistent currents" in the superconducting dipole magnets.

By that June, the nucleus of the laboratory's technical staff led by Edwards had begun developing a "site-specific" Super Collider design. This task involved adapting the generic collider design developed by Tigner's team to the peculiar characteristics of the Waxahachie site — for example, its geology — which would ordinarily have involved only minor changes. But prodded by Edwards, who followed a more conservative design philosophy than Tigner, this group built in added safety margins by increasing the dipole aperture to 5 cm and the injection energy from 1 to 2 TeV [21]. Computer simulations suggested that proton losses at injection could be reduced by such measures, maintaining the design luminosity. Other more conservative parameter choices were included, such as increasing the number

[5]Wojcicki makes a similar point in Refs. 7 and 19, esp. p. 292.

of superconducting dipole magnets from 8,600 to 8,800 and the main ring circumference to 86 km. All told, the changes promised to boost the SSC cost by at least a billion dollars.

That September the US Congress had officially given the project a green light, overwhelmingly approving a construction start based on a DOE estimated total cost of $5.9 billion — including R&D, detectors, contingency and inflation. When the new Energy Secretary, Admiral James D. Watkins, visited the Dallas SSC headquarters on November 1, Schwitters advised him that several design changes were being considered. Watkins replied that these changes should not push the total cost above the $5.9 billion figure approved by Congress. But the injunction seemed to fall on deaf ears. Later that month, word leaked from DOE headquarters in Washington that major, costly changes in the SSC design were underway. According to a front-page *Washington Post* article, the cost overruns could total as much as $2 billion [22].

The fleeting honeymoon between DOE and the SSC physicists was over. When he learned of the design changes and associated cost increases, Watkins "went non-linear," according to an unnamed source in a *Physics Today* article [23],[6] insisting that the SSC be built for $5.9 billion. But in early December, the SSC Machine Advisory Committee, which included several of the world's foremost accelerator builders, endorsed the design changes proposed by Edwards, saying they were required to reduce commissioning and operating risks. When a URA committee overseeing the SSC project met to consider the looming crisis, its chair Wolfgang Panofsky, a former SLAC director as well as esteemed accelerator builder, suggested that the proton beam energy could be lowered to 17 TeV to reduce costs, but nobody else seemed interested [24].[7] Instead, DOE tasked a blue-ribbon panel of eminent physicists (including five Nobel laureates) to assess the proposed SSC design changes and whether they were necessary to address the project's physics goals — including whether or not 17 TeV proton beams would be sufficient. Chaired by SLAC physicist Sidney Drell, the panel supported the design changes and opposed to reducing the beam energy below 20 TeV. HEPAP endorsed the panel's report, stating that "any substantial reduction in the

[6] Other accounts confirm this reaction in different terms.

[7] The energy reduction would still not reduce the total project cost down to $5.9 billion, but Panfosky astutely argued that it was politically important to show a sense of restraint, given that many other political constituencies (including other sciences) were already under severe budgetary pressures.

energy of the SSC would compromise our ability to elucidate the nature of electroweak symmetry breaking," the uppermost goal of the project [25].[8]

Watkins accepted the HEPAP recommendations, and the DOE began a year-long process of "rebaselining" the SSC project according to the site-specific design [26]. Meanwhile, the magnet R&D teams at Brookhaven, Fermilab and LBL (which did almost all the superconducting magnet work) began retooling to address the new specifications. But chastened by the cost overrun, Watkins had lost any faith he may have had in the ability of academic physicists to manage such a complicated multibillion-dollar project. Instead, he deliberately began inserting into the SSC organization individuals from the US military-industrial complex, in whom he had much more confidence [27]. The first of these was Joseph Cipriano, a career civilian manager at the Pentagon, then serving as the director of weapons and combat systems in the US Naval Sea Systems Command. In May 1990 he assumed the position as head of the DOE site office in Texas, with the telling title of "project director" and a direct reporting line to Watkins. Cipriano soon started hiring a large staff that included many former employees of the Pentagon, other government agencies and associated industries, few of whom had any experience working with the high energy physics community.

By March 1990, Schwitters had come around to the recognition that his initial choice for SSC project manager, Richard Briggs, was not working out as hoped and began a search for a replacement. After considering good candidates from Bechtel and NASA (who decided against the job), the search narrowed in August to high energy physicist Paul Reardon, who had had extensive project management experience at Brookhaven, Fermilab, and Princeton.[9] But when the DOE was asked to approve his appointment, Watkins put forward his personal choice, Edward Siskin, an accomplished engineer who had served in the nuclear navy and more recently as executive director of Stone and Webster Engineering Corporation.[10] URA

[8]Left unspoken in the panel report was consideration that a 17 TeV beam energy would narrow the gap between the SSC and a proposed lower-energy European LHC project at CERN, thus enhancing arguments made by SSC opponents that it would be much more cost effective to collaborate on that project. At the time, theoretical arguments held that the Higgs boson mass might come in as high as 1 TeV and that other symmetry-breaking phenomena could occur up to 2 TeV. The likelihood of a relatively light Higgs boson, with a mass near 100 GeV, came to the fore only during the 1990s.

[9]Reardon had served as Associate Director for High Energy Physics Facilities at Brookhaven and as a business manager on the Fermilab Energy Doubler; he had also managed the construction of the Tokamak Fusion Test Reactor at Princeton Plasma Physics Laboratory.

[10]Stone and Webster had a long history of large-project construction management stretching back at least to the Manhattan Project, in which it managed construction of the Oak Ridge Laboratory.

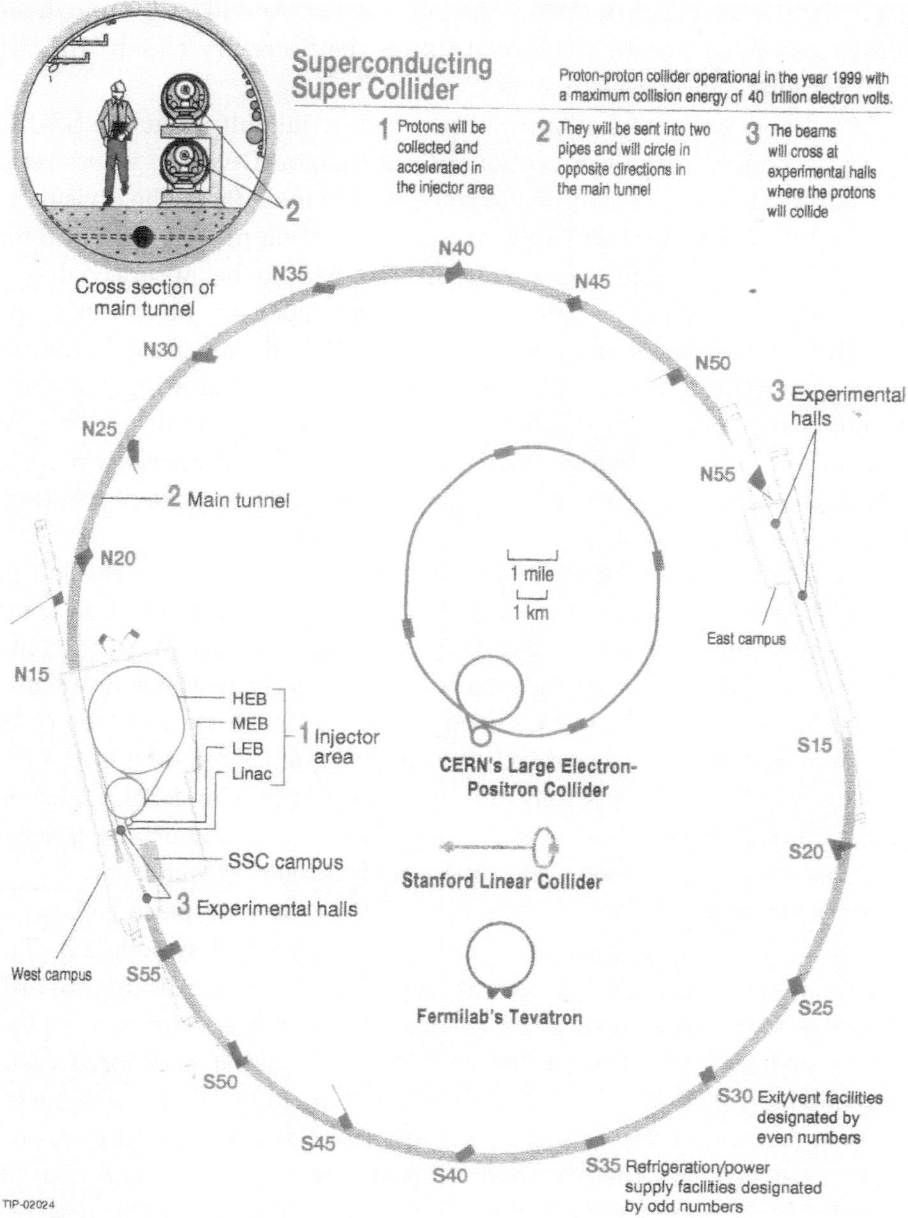


Fig. 1. Schematic diagram of the SSC, compared with the other large particle colliders in operation during the early 1990s. Credit: Fermilab Archives.

officials dealt with this proposal by including Siskin in the SSC organization in a newly created position as general manager, reporting to Schwitters. Siskin brought in much-needed management expertise, but it added yet another level of bureaucracy to an already complicated organization, further distancing the technical staff from the lab director. And like Cipriano, he had a direct reporting line to Watkins, to keep the Secretary closely appraised of lab activities and developments.

The deepening presence of a military-industrial culture at the SSC Lab could also be felt in the Conventional Construction Division, where retired Navy Admiral John Ives stepped in to head it in mid-1990. The division had gotten off to a slow start when Sverdrup Corporation proved unable to meet its contractual obligations during 1989 and had to be abruptly dropped from the SSC leadership team. Although Ives filled an important gap in the SSC management and helped to assure DOE officials that construction activities were in capable hands, his presence reinforced growing perceptions by high energy physicists that an alien culture was taking over at the new laboratory.[11] This perception made recruiting good physicists even more difficult than it already was due to other undesirable features of working in the Dallas area.

To be fair, there were serious management deficiencies at the lab, not least of which was the absence of a functioning cost-and-schedule control system (also called a project-management control system). With the failure of Sverdrup and other seemingly more-pressing early teething problems at the new lab, this difficult but necessary task had fallen through the cracks. A few managers had tried to develop such a system, as required in the DOE contract, but the lack of strong buy-in from Edwards and Schwitters had delayed implementation for over a year. This was an important task that Siskin began to address soon after stepping in as general manager.

But he clashed with some of the physicist managers, especially Edwards, who could not seem to grasp the value of such a control system [29]. Their conflict came to a head in the early spring of 1991, when Edwards, Reardon and others recommended that the quadrupole magnet apertures should also be increased to 5 cm like the dipoles, at an estimated additional cost of about $50 million. When Siskin and Cipriano (backed by Watkins) denied the request unless a better case could be made based on core SSC physics goals, Edwards decided that the time had come for her to return to Fermilab [30].

In February 1991 the DOE had at long last established the new baseline total project cost at $8.25 billion [31, 32], a growth of $2.35 billion. This small additional change threatened to increase the SSC price tag even further just as the House of Representatives was about to take up the DOE request for the project for the next fiscal year. While the proposed magnet change made good sense technically, its timing was terrible from a

[11]For more on this topic, see Refs. 11 and 28.

political perspective. Yet another such cost overrun, however small, would have provided SSC opponents valuable ammunition and further undermined Congressional backing. As it was, House support dropped substantially to 251–165 when an amendment to kill the project was offered that May. When the Senate voted on a similar amendment later in 1991, it supported the SSC by a more comfortable 62 to 37 margin.

In contrast with collider construction activities, SSC physicists controlled the design and construction of the experimental detectors. These projects began to take shape in 1991, as the laboratory reviewed three letters of intent for large, general-purpose detectors from the high energy physics community [33]. One was accepted in May, from the Solenoidal Detector Collaboration (SDC) led by LBL physicist George Trilling, which was given approval and funding to prepare a full proposal. Two others were turned down, including the L* proposal from a major international collaboration led by MIT physicist Samuel Ting; its rejection implied that a large number of foreign physicists interested in doing SSC research would not be involved in its experimental program.[12] Remnants of the unsuccessful collaborations then teamed with other physicists to propose the GEM (for "Gammas, Electrons and Muons") detector [34]. Led by Barry Barish of Caltech and William Willis of Columbia, this collaboration received a green light to proceed to a full proposal. The successful collaborations were large, extended, hierarchical bureaucracies with high energy physicists in control of all the major decisions. Here the engineers involved reported ultimately to physicists.

To serve as the research director in charge of the SSC Lab's experimental program, Schwitters appointed theorist Fred Gilman, whom he knew from the years they had worked at SLAC during the 1970s. Gilman would be overseeing the expenditure of close to a billion dollars on the two detectors, then estimated to cost about $500 million apiece — about the entire cost (in constant dollars) of building SLAC in the early 1960s. Of the $8.25 billion total estimated SSC cost, DOE had allocated $760 million (in 1991 dollars) for the experimental program, of which $550 million was designated for the two general-purpose detectors. It meant that almost another half a billion dollars had to come from "non-federal" sources, mainly science budgets of foreign governments, to cover contributions of their physicists involved in the detector-building process.

The culture of the SSC Laboratory was thus bifurcating into two distinct subcultures that coexisted uneasily. Engineers with backgrounds in the armed forces or industry dominated the collider building process. But

[12] For details and discussion, see Ref. 19, pp. 274–275.

authority over its detector projects remained in the hands of high energy physicists, who reported to the director through a chain of command that included only other physicists.

## 3 Washington and the wider world, 1987–1992

The 1983 HEPAP recommendation to proceed with the SSC had not been very well received in Europe. For the centerpiece of the CERN long-range plan was eventually to add superconducting magnets to the 27 km LEP tunnel, just then beginning construction, and thereby convert this facility into a proton collider known as the Large Hadron Collider (LHC) capable of energies up to 18 TeV [35].[13] The tunnel had been deliberately designed large enough to accommodate such magnets. The LHC would be the natural next step in a gradual, cost-effective machine-building process that had been occurring there for decades. Ever since the 1950s, CERN had built one accelerator or collider after another as extensions of its existing facilities while growing other laboratory infrastructure — both physical and human — accordingly.

High-ranking physicists from other nations, especially Japan, expressed concern at 1983 and 1984 meetings of the International Committee on Future Accelerators that the "nationalistic" US approach would subvert the dreams of a World Accelerator, which had for years been envisioned as a truly international partnership [4, 36]. The Reagan Administration's public rhetoric that the SSC would be an American project intended to restore US hegemony in high energy physics only reinforced these adverse perceptions. So when DOE officials visited counterparts in Canada, Europe and Japan in 1987 to discuss possible SSC contributions, they received polite but cool receptions — and no formal commitments.

The model of foreign collaboration that had long been followed in high energy physics entailed in-kind contributions to the experimental detectors, not to the accelerators or colliders feeding them. In the construction of its HERA collider, however, West Germany's DESY lab was pioneering a new approach in which other nations — in this case, France and Italy — made major contributions (of superconducting magnets) to the collider, too. This so-called HERA model of international collaboration was increasingly envisioned as the ideal way for other nations to become partners in the SSC

[13]The LHC energy was later reduced to 14 TeV after magnet R&D had shown the higher figure was unattainable.

project, for foreign contributions to the SSC detectors alone would likely total less than 10 percent of its overall costs.

But this hoped-for approach ignored the realities of European science funding, which was increasingly impacted by the CERN budget putting undue pressure on national science budgets and threatening to crowd out other worthy science. These pressures grew increasingly onerous during the 1980s. In the mid-1980s, Great Britain's Kendrew Commission recommended cutting CERN's budget by 25 percent to reduce these pressures. It sent shock waves through the CERN management, which began to moderate its financial demands on Member nations; in 1987 the Abragam panel recommended major cuts in CERN staffing levels to achieve that moderation [37,38]. And European nations' foremost priority in high energy physics had to remain the success of the research program at CERN, widely regarded as a shining beacon of European unity [39]. Thus any US hopes for major European contributions to building the Super Collider, not just its detectors, could only be regarded as wishful thinking.

Japan was another story, however. After commissioning the TRISTAN electron-positron collider in 1986, Japanese high energy physicists were about to begin looking for their next major effort, most likely by joining a proton-collider project in Europe or the United States. And the booming Japanese economy of the late 1980s, combined with the fact that the country was relatively laggard among OECD nations in its support of basic research, made it a prime target for US officials and physicists seeking major contributions to the SSC construction. High-level delegations began visiting Japan in 1984, and Japanese high energy physicists became closely involved in SSC superconducting magnet R&D [40].

But Japan had another hopeful high energy physics suitor: CERN. During the late 1980s and early 1990s, Italian physicist and Nobel laureate Carlo Rubbia, who had stepped in as director-general, began vigorously promoting the LHC as a viable, much more cost-effective alternative to the SSC. To cover its anticipated costs in CERN's now-constrained funding environment, he began to look beyond its Member states for other nations to join the LHC project as partners [38]. Among them were many of the same countries then being approached by DOE officials, primarily Japan.

These officials were trying to meet requirements established by the Office of Management and Budget (OMB), the powerful executive agency that develops the President's budget every fiscal year and tries to oversee its implementation. Never great fans of the SSC project, OMB officials had established a rough goal that one third of its funding had to come from "non-federal sources" — meaning other nations or state governments.[14] When the estimated project cost was $5.9 billion, that is, just under $2 billion had to come from non-federal sources; when it rose to $8.25 billion, over $2.7 billion was required. After the Waxahachie site decision in late 1988, a Texas commitment to contribute $1 billion from state coffers became public, reducing the need for foreign contributions to about $1 billion. But the subsequent cost overruns and rebaselining of the project in 1990–1991 increased this target amount to just over $1.7 billion.

In May 1990 the House of Representatives passed the Superconducting Super Collider Authorization Act capping the federal contribution to the project at $5 billion and requiring that at least 20 percent of the costs come from foreign sources. At the $8.25 billion level soon to be estimated, that would have meant over $2 billion in foreign funding. But the Senate never took up the House bill, letting it die. Still, House Members continued to view this bill as "the will of the House" that SSC advocates could ignore only at their peril [41].

Congress was not of one mind on this question of foreign contributions. For the SSC had been touted as a means to restore US industrial competitiveness; thus letting other nations contribute high-tech items, especially superconducting magnets, made little sense. But at the billion-dollar levels that were becoming necessary, there was no other option. Other nations would never agree to send cash. A few powerful figures in Congress, such as Louisiana Senator J. Bennett Johnston, maintained that the United States should build the SSC itself and pay all the costs of doing so. As its price increased in the 1990s and the nation began to address its burgeoning deficits, however, many politicians and government officials recognized that major foreign participation had become mandatory.

In November 1990 Congress passed the Omnibus Budget Enforcement Act, which would have largely unforeseen but ominous consequences for the SSC. This bill enacted a "firewall" between domestic discretionary and defense spending, so that cutbacks in the latter during the waning Cold War could not be used to fund increases in the former. There would be no "peace dividend," that is. And domestic spending could increase only at the rate of inflation. Thus the SSC project, with its annual budget expected to grow beyond a billion dollars in the mid-1990s,[15] would inevitably exert

[14]This standard seems to have emerged in 1987, based in part on what DOE officials were then claiming should be possible for foreign contributions, but we could not identify its actual source.
[15]See Ref. 31, chart of SSC annual budgets.

tremendous pressure on other energy and water projects, unless foreign contributions somehow arrived to relieve it.

The lack of major foreign contributions to the SSC was increasingly cited among the arguments against the project when the annual appropriations bills came to the floors of the House and Senate in the early 1990s. One could grasp the growing pressure on other high energy physics projects such as Fermilab's Main Injector, which several Representatives tried to "zero out" in 1991 but was restored by others, at markedly reduced funding levels [42]. After that skirmish, the already shaky SSC support of the Illinois Congressional delegation withered, as did that of other midwestern states, and soon effectively died. In July 1991 the House and Senate compromised on an SSC budget of $484 million for fiscal 1992 — $50 million less than the Bush Administration had requested earlier that year.

Administration officials increasingly recognized that in order to obtain $1.7 billion in foreign contributions, at least $1 billion would be required from Japan — the only nation then thought able to shoulder such a financial burden.[16] But there were only about 100 Japanese physicists who might become involved in the SSC research, most of them on the SDC experiment. That might justify an SSC contribution of about $100 million, a sum its education and science ministry, Monbusho, might have been able to offer on its own. Making a contribution ten times as large would necessarily be a *political* choice, based on political trade-offs and requiring approvals at the highest levels of the Japanese government.

Thus in the fall of 1991, a succession of high-level US delegations headed for Tokyo, ahead of a summit meeting between President Bush and the Japanese Premier Kiichi Miyazawa, to invite their counterparts to join the United States in a "partnership" to build the SSC [43, 44]. That October Bush Science Advisor D. Allan Bromley flew there with high DOE officials to make a formal pitch to the Science Council of Japan, among others. After Secretary of State James Baker visited that November, Watkins followed up in December, talking to various agency heads about joining the project. Everything seemed set for an early January summit.

But major changes were afoot in the core White House staff, as Bush's poll numbers plummeted that fall in advance of the 1992 elections, largely because of the nagging recession in the US economy. What had originally been touted as a celebration of trans-Pacific friendship and cooperation

[16] Few recognized at the time, however, that the booming Japanese economy had already peaked and entered a decade of decline.

became an effort to extract concessions from the Japanese auto industry to accept quotas on imports of US auto parts.[17] This did not rest well with the summit hosts. Add to the fact that Bush fell ill at a state banquet, and the hoped-for Presidential request for Japan to join the United States in an SSC partnership did not happen. But aides agreed behind the scenes to establish a bilateral working group to address unresolved issues. To many in Congress, this was another indication of Japanese waffling and delay tactics. But to those familiar with its culture, this was an important signal that its government was indeed seriously interested in pursuing an SSC partnership.[18]

In both nations, one of the principal arguments against the Super Collider continued to be its likely impact on the support of other sciences. Japan had long under-funded basic research, favoring applied research benefitting its export-led economy. Shipping a billion dollars' worth of high-tech products overseas to help establish a massive new US physics facility was hardly going to ameliorate this indigenous scientific problem.

In the United States, the impact of the SSC on other sciences remained a cogent part of the political arguments being made against it, especially since the Budget Enforcement Act of 1990 meant that its annual increases thereafter would press upon other domestic spending. "There is growing concern that valuable 'small science' activities will be severely curtailed by a few capital-intensive 'big science' programs," stated influential Michigan Congressman Howard Wolpe in May 1991 [48]. In the House SSC debate later that month, impact on other sciences was among the primary concerns expressed, exceeded only by the project's cost growth. And it did not help that the SSC was often grouped with NASA's Space Station, whose ever-increasing annual budget was beginning to crowd $2 billion.

Vocal SSC opponents also included other scientists, principally materials scientists like Rustum Roy of Penn State and condensed-matter physicists such as Princeton Nobel laureate Philip Anderson. "Big science has gone berserk," Roy told the *New York Times*. "Good minds and a lot of money are going into areas that are not relevant to American competitiveness... or even [to] the balanced development of American science." [49] For almost two decades, Anderson had been waging a campaign against big science in general

[17]See Ref. 43, p. 213. Indeed, Bromley was not included in the contingent accompanying Bush, while executives from US automakers were added at the last minute by the new White House Chief of Staff, Samuel Skinner [45].

[18]To have embarked on such a process and later declined would have caused serious Japanese loss of face [46]. Other sources agree that Japan was interested in partnering with the United States on the SSC but question whether it was prepared to make a billion-dollar commitment to do so.

and high energy physics in particular;[19] his words carried special weight in Congress. Open questioning of the SSC by him and other reputable scientists gave Congressional foes powerful ammunition to use in their increasing efforts to kill the project.[20] Leaders of the high energy physics community, among them Nobel laureates Leon Lederman and Steven Weinberg, tried to rebut these arguments in their own testimony. The two belatedly published *The God Particle* and *Dreams of a Final Theory* to remind readers of the exciting physics questions the SSC was being built to resolve.

## 4 The demise of the Super Collider, 1992–1993

As 1992 dawned the Soviet Union no longer existed, and a revolutionary CERN digital-communications platform called the World Wide Web was finding its way into North America via SLAC and Fermilab. By then the SSC Lab had its fifth project manager (including acting managers) in less than four years. SLAC accelerator builder John Rees stepped up in December to replace Reardon, who bristled working under general manager Siskin and instead moved over into the SSC detector-building arena.[21] Rees had had extensive experience in managing three SLAC collider projects; he also established a trusting relationship with Siskin, whose time was then consumed by efforts to get the cost-and-schedule control system working properly. Together they began trying to bring the project under control.

Under Rees's leadership the SSC Lab awarded contracts early that year to build its magnet test laboratory and to begin tunneling the main collider ring.[22] Superconducting dipole magnets with 5 cm apertures were beginning to roll off the Fermilab assembly line, fabricated by workers from General Dynamics and Westinghouse, which had won these magnet contracts but were awaiting the completion of magnet-manufacturing facilities in Louisiana and Texas. In May construction began on the linear accelerator to feed low-energy protons into a series of booster rings that would accelerate them to the 2 GeV injection energy.

[19]For an early, influential essay, see Ref. 50.

[20]Congress is generally favorable to science, said Boehlert aide Goldston, so its members do not want to be perceived as against science, per se. It is much easier for them to be seen as setting priorities between conflicting disciplines or projects [41].

[21]In typical fashion, Rees had joined the project on a consulting basis for two years, remaining on the SLAC payroll with the SSC Lab reimbursing his salary and benefits until he stepped up as project manager. He initially supervised the chain of room-temperature accelerators supplying lower-energy protons to the superconducting magnet rings [51].

[22]The first tunneling contract went to Dillingham/Obayashi, a US/Japanese consortium, but it could not begin tunneling until after a successful accelerator-systems string test later that year.

Meanwhile, back in Washington, Bush's fiscal 1993 budget called for $650 million for the SSC — an increase of 34 percent above the level enacted in 1992 [52]. After much behind-the-scenes wrangling, it also requested $631 million for the high energy physics "base budget," including $30 million for the Fermilab Main Injector project. But the House subcommittee on energy and water projects pared the SSC budget back to its 1992 level, $484 million, and shaved $15 million off the Main Injector. In this severely constricted funding environment, there was just not enough money in its FY1993 allocation to meet Administration requests.

On 17 June 1992, with little warning, lightning suddenly struck the SSC [53]. An amendment to kill the project succeeded on the House floor by 232 to 181, a margin of 51 votes, reversing the comfortable 86-vote margin on a similar 1991 amendment.[23] As this was essentially the same Congress, with only a few new members, that meant nearly 70 Representatives had changed their minds about the project. The proximate cause was a divisive, rancorous vote six days before, in which a balanced-budget amendment narrowly failed to achieve the necessary two-thirds majority [54]. "Now we've got to roll up our sleeves and get to work," Budget Committee Chairman Leon Panetta told reporters after this battle, "so that we truly exercise the discipline that we have to if we're serious about getting the deficit under control."[24] In this worsening fiscal climate, a poorly understood, multibillion-dollar science project perceived to be mismanaged and experiencing continuing cost overruns was an easy political target for budget-cutters. Or as Congressman Joe Barton from the SSC Lab's Texas district put it, House members "wanted a budget scalp they could take home and wave in front of voters" [55].[25]

SSC advocates in the Administration and Senate quickly mobilized to rescue the project. Bush met with Schwitters at the White House and personally urged wavering Senators to support the SSC in the forthcoming vote, then flew to Dallas for a meeting and photo-op at the lab. Bennett Johnston led the Senate fight, staging public hearings that touted the SSC as critical to US science and twisting his colleagues' arms behind the scenes [56]. When the Senate vote occurred on August 3, the margin was 62–32 to continue the

[23]Compared to 1991, the project lost 14 votes in California, 6 in Illinois and 5 in New York, for a combined loss of 25 votes in states with existing high energy physics labs. The Illinois vote was particularly noteworthy, 18 to 2 against the SSC [53].

[24]Panetta reversed his prior SSC vote on this occasion, voting against it, see Ref. 54, p. 1688. He would later become OMB Director and then White House Chief of Staff in the Clinton Administration, as well as CIA Director and then Secretary of Defense in the Obama Administration.

[25]Knowledgeable Washington observers claimed that the anti-SSC vote was partly a rebuke to Barton for his House leadership on the balanced-budget amendment.

Fig. 2. SSC Director Roy Schwitters shows President George H. W. Bush several of the superconducting dipole magnets in July 1992, just before the successful accelerator systems string test. Accompanying them at far left is Presidential Science Adviser D. Allan Bromley and (next to him) local Texas Congressman Joe Barton. Credit: Fermilab Archives.

project, even better than the previous year. After House and Senate Appropriations Committee leaders met in September to resolve the differences, the SSC Lab came away with $517 million for fiscal 1993, a small increase. It had survived for another year but emerged from the 1992 Congressional battle badly wounded and vulnerable.

SSC construction began anew as the nation headed into the pivotal 1992 elections, which resulted in a new President and 127 new faces in Congress.[26] After a successful superconducting magnet string test that August, tunneling of the main ring could start; this came in January 1993. By April 1993 four boring machines were in operation, and four miles of the tunnel had been completed [58]. There was extreme urgency to get as much work as possible finished before the inevitable next round of Congressional budget battles. "As scientists, technicians and workers rush from one task to the next, there is a sense that a desperate campaign is under way, to beat the clock and win

[26]There were 114 new Representatives and 13 new Senators, a turnover not seen since 1974, many elected on promises to bring budget deficits under control [57].

Fig. 3. View along the SSC tunnel and up one of its vertical shafts, as seen in early 1993. Credit: Fermilab Archives.

against tough odds," observed a *New York Times* reporter who visited the SSC site in March [59].

The SDC experiment received DOE approval to proceed with construction of its gargantuan detector, now expected to cost over $600 million. As only $275 million had been budgeted, however, more than half its cost now had to come from elusive foreign sources. With about 100 physicists on the SDC experiment, Japan might have come up with $100 million once the issue of its "partnership" in the SSC project was resolved. But what about the other $225 million?

The new Clinton Administration, still in formation, was debating what to do about a Republican project in a state Democrats had lost badly. The new OMB Director Panetta favored jettisoning it, while Texan Lloyd Bentsen defended it. Out of that debate arose a plan to stretch out SSC construction by three to four years, but that would add over $1 billion to its total cost. According to Clinton's Science Advisor Jack Gibbons, an SSC stretch-out was advocated partly to allow more time to resolve magnet problems and pin down foreign contributions [60].

SSC advocates still hoped Japan would finally agree to become a partner in the laboratory at a mid-April summit meeting between Clinton and Premier Miyazawa. Most of the thornier issues had been resolved in 1992 and Miyazawa was prepared to make such a commitment in 1993, needing only to be asked by the new President [61].[27] But at the April 16 meeting, trade concerns were uppermost on Clinton's agenda, particularly obtaining concessions on Japanese auto-parts imports.[28] The question of partnership in the SSC project never came up, and what had become known as the crucial "golden handshake" between the two heads of state did not occur.

With a quarter of the House seats now filled by new members, most of them seeking big-ticket items to delete from the 1994 budget before them, the SSC had little chance of surviving the appropriations battle in that body without a major Japanese commitment. Indeed, the House overwhelmingly rejected the $620 million allocated by the Appropriations Committee for the project in a June 24 vote, by 280 to 150 — just a day after the Space Station had narrowly survived a similar amendment by 216 to 215 [62].

Hopes for an easy repeat of the 1992 reversal dimmed, however, when feared Congressman John Dingell of Detroit called a televised June 30 hearing before his committee. Cipriano, Siskin, Schwitters, URA President John Toll, and the new Energy Secretary Hazel O'Leary testified under oath. Most damaging were charges of continuing SSC cost overruns and the use of parallel accounting systems [63]. O'Leary agreed to investigate these charges and sent a team of 70 DOE staffers to Waxahachie to quiz lab employees and pore through its records. When the team issued its report in mid-August,

[27]But knowledgeable insiders say that political trade-offs would necessarily have been required, implicitly or explicitly, to reach such an agreement.

[28]This conclusion was reached from a variety of sources, including Ref. 19, p. 280. As Richard Barth, who had served on the National Security Council under Bush, observed in a phone conversation, "You're talking about something a few thousand scientists had a strong interest in versus something that a few hundred thousand autoworkers were interested in."

the estimated total cost of the project had risen to $10 billion, due largely to probable magnet cost increases and the recognition that sufficient foreign contributions were not likely to come in for the detectors [64]. And that figure did not include the additional costs to accrue due to the three-year schedule stretch-out, which would add at least another $1 billion.

Johnston called an August 4 Senate hearing featuring Gibbons, O'Leary, Schwitters, and other SSC supporters. O'Leary revealed early findings of the forthcoming report, saying she was going to shake up the SSC Lab management and replace URA by an industrial contractor that had "world-class experience in managing large construction projects".[29] In press accounts the SSC cost was now regularly being reported as $10–11 billion.[30] On September 30 Johnston worked his usual Senate magic, delivering a narrower 57–42 vote against an amendment to kill the project [67]. But ensuing celebrations at the SSC Lab proved premature.

In October leaders of the House and Senate Appropriations Committees met in a closed-door session, reporting out a "compromise" bill that included the full $640 million Clinton request for the SSC and ignoring the House vote from the previous June. Outraged by this egregious slight, SSC opponents mobilized for an October 19 counterattack on the House floor, entering a motion to send the entire energy and water bill back to committee with instructions to delete SSC funding. "Are we to continue to be suckers or not?" asked Kansas Democrat Jim Slattery, the author of the motion, which succeeded by a resounding 282 to 143 margin [68].[31] Johnston and other supporters caved in the following day, agreeing to kill the SSC, provided the $640 million could be used to close down the project in an orderly fashion. Both House and Senate overwhelmingly rubber-stamped this compromise the following week, and Clinton signed the bill into law. The SSC was officially dead, with its burial to be completed by late 1994.

## 5 Summary and conclusions

Immediate reactions to the SSC termination ranged from stunned disbelief and dismay that Congress could do such a terrible thing to smug satisfaction on the part of opponents. Uppermost on the list of reasons given were the

[29] In her statement, O'Leary conveniently ignored the fact that URA did not really have full control of the project.

[30] One of the House staffers who worked behind the scenes to defeat the project said two years later that he doubted SSC opponents could have succeeded had its estimated costs been kept convincingly below $10 billion [66].

[31] According to Boehlert legislative aide Goldston, this was a "precedent-shattering event" — the first time the House had ever overturned an Appropriations Committee conference report [69].

project's ever-growing costs and the lack of major foreign participation. Before the SSC construction began in 1990, wrote New York Congressman Sherwood Boehlert, the House had approved a bill that "capped Federal spending at $5 billion and required foreign contributions to cover 20 percent of the costs." The DOE failure "to take those requirements seriously sealed the project's fate" [70].[32]

If the United States was to remain involved in physics research at the high energy frontier, it now had to join the LHC project at CERN. "The only good thing to come out of this will be more international collaboration," said SLAC Deputy Director Drell. "It's important that we move forward on this frontier of research, and the LHC is the one machine we know how to build." [71] Chaired by Drell, another HEPAP subpanel recommended that US physicists do just that, along with completing the Fermilab Main Injector and SLAC B Factory, to maintain the health of US high energy physics research [72]. It took nearly four more years, however, before US officials finally signed a formal agreement to join the LHC project and make over $500 million worth of in-kind contributions to the collider and its detectors.[33] This 1997 agreement established a completely new way of doing business for US high energy physics; the nation was not the leader but a partner among equals on a major accelerator project.

While the LHC project also experienced serious growth problems and cost overruns, increasing from an estimated 2.74 billion Swiss francs (CHF) in 1996 to over 5 billion CHF in 2009,[34] it managed to survive; experiments began in early 2010, several years later than originally targeted. The visions and steadfastness of LHC builders were rewarded in mid-2012 by the discovery of the long-sought Higgs boson near 125 GeV — a relatively low mass compared to the theoretical predictions of the 1980s that had driven SSC design. This achievement in the face of similar teething problems raises the

[32]In 2001 Boehlert stepped up as Chair of the House Science Committee and remained in that position through 2006; David Goldston served as his Chief of Staff.

[33]The DOE agreed to spend $450 million on the collider and detectors and NSF to add another $81 million on the detectors [73, 74].

[34]Of the 2.74 billion figure, 2.53 billion CHF was for the collider and experimental areas but did not include the detectors or computers, which were less well understood in 1996; it grew by 48 percent to 3.756 billion CHF in 2009, including inflation — and probably should include another 2 to 4 percent more to repair damages that occurred during commissioning on 19 September 2008. Similar cost growth occurred on the detectors and computers. These figures do not include CERN labor costs or overheads, which would raise the total cost of the LHC to nearly 10 billion CHF, including outside contributions, or close to US$10 billion in 2009 dollars. Nor do they include the value of the existing LEP tunnel and other CERN infrastructure used in construction of the LHC [75].

natural question: why did CERN (and its non-Member partners) succeed where the United States had failed?

From the SSC's early days, many observers thought that it should have been sited at or close to Fermilab, to take advantage of its existing physical and human infrastructure built up over 20 years.[35] CERN had followed this path for decades, building one machine after another as extensions of existing facilities and reusing parts of the older machines in new projects, thereby keeping costs down. Perhaps as important, this lab had also gathered and developed some of the world's most experienced accelerator physicists and engineers, who worked together smoothly as a team in achieving an extremely complex project like the LHC. Fermilab possessed equally capable machine builders as well as substantial physical infrastructure that would not have had to be built anew in Texas. They could also have turned to other productive work when the inevitable funding shortfalls occurred during the Congressional appropriations process. And the difficult cultural clashes between high energy physicists and military-industrial engineers that characterized the SSC project in its early years would not have erupted. While the state of Texas had offered to spend $1 billion to support the project, knowledgeable observers (often with substantial hindsight) estimated the actual value of Fermilab infrastructure at two to three times that figure.[36]

These arguments however ignore the realities of the American political process. A lucrative new project costing over $5 billion and promising more than 2,000 jobs cannot be "sole-sourced" to an existing US laboratory, no matter how powerful its Congressional delegation may be. As politically savvy DOE leaders recognized, the SSC project had to be offered up to all states able to provide a suitable site, with the decision based (at least publicly) on objective criteria. A smaller project costing perhaps $1 billion and billed as an upgrade of its existing facilities *might* have been sole-sourced to Fermilab, given the political climate of the mid-1980s, but not one as huge and expensive as the SSC. It unfortunately *had* to be placed on the US auction block, and Texas made the best bid according to the official judgment criteria.

Unlike the SSC, the LHC project had strong project management from start to finish by a single individual, physicist Lyn Evans, who brought decades of experience working on hadron colliders at CERN [76]. This was undoubtedly an important factor in its success. He enjoyed strong support

[35]See, for example, Ref. 19, pp. 291–292.
[36]For example, C. Llewellyn-Smith suggested a figure of $2 billion at the May 2009 APS meeting.

of top management as well as from a deeply experienced cadre of accelerator physicists and engineers who worked together without the cultural clashes that occurred at the SSC Lab. And on the LHC project, engineers reported ultimately to physicists, the users of the machine best able to make the required tradeoffs when things did not occur as originally hoped. The LHC project encountered daunting difficulties, serious delays and major cost overruns, too, but its core management team led by Evans held together and worked through these problems. They also shared a common technological culture — and understood and supported the project's principal scientific goals. Similar observations cannot be made regarding the military-industrial engineers who came to dominate SSC construction.

CERN also enjoys an internal structure, governed by its Council, that largely — but not entirely — insulates its leaders and machine builders from the political vagaries and machinations of individual Member states.[37] Unlike in the United States, the lab director or project manager could not be hauled before a parliamentary investigations subcommittee and required to testify under oath about LHC management problems or cost overruns. Nor did the project face annual appropriations battles and threats of termination. Serious problems that arose, such as a 20 percent cost overrun in 2001, had to be addressed in Council, which represents science ministries from Member states and generally operates by consensus on major decisions. This governing structure ultimately helped to keep project control in the hands of the scientists involved, instead of politicians or other government officials.

Because the Council also has to address the wider interests of national science ministries, CERN leaders must be sensitive to the pressures its annual budget, new projects and cost overruns could exert on other European science. The mid-1980s recommendations of the Kendrew and Abragam panels had a chastening effect upon CERN management [37, 78].[38] In this manner, European small science had a valuable voice that was heard within the CERN governing process. The LHC project was therefore tailored to address such concerns before Council could grant its final approval [74]. No similar mechanism existed within US science, other than for disgruntled physicists to complain publicly in prominent guest editorials and Congressional hearings

[37]There were certainly political problems that had to be addressed, such as when Germany and Great Britain initially refused to support the proposed LHC project in 1994 [74]. Another political problem that had to be addressed was claims that the LHC might produce black holes that could swallow and destroy the Earth [77].

[38]Discussions with Llewellyn–Smith and CERN historian John Krige were especially helpful in understanding the Council's crucial role in representing the interests of other European scientists within the CERN governing structure.

after SSC costs got out of hand in 1989–91. The consequent polarization of the US physics community helped to undermine what had originally been fairly strong Congressional support for the project, aiding SSC opponents in that forum.[39]

And because of these pressures within Council, CERN had to effectively internationalize the LHC project — gaining major commitments from such non-Member states as Canada, China, India, Japan, Russia and the United States — *before* going ahead with it. These contributions allowed Evans and his team to proceed with a collider design able to reach the full 14 TeV rather than the initial phase of a descoped machine that might have been buildable with less funding [74, 76]. When the full LHC project gained final Council approval in 1996, it was a truly international scientific project with financial backing from over 20 nations.

In the final analysis, too, the LHC was (somewhat fortuitously) much more appropriately sized to its primary scientific goal: the discovery of the Higgs boson. The likelihood that this elusive quarry could come in at such a low mass as 125 GeV was not well appreciated until 1990, when supersymmetry theories began to suggest that such a light Higgs boson might occur in Nature [79].[40] But by then the SSC dice had been cast — in favor of a gargantuan 40 TeV collider able to reveal the origins of spontaneous symmetry breaking even if such phenomena were to occur at masses as high as 2 TeV. After that fateful decision, which was endorsed unanimously by a HEPAP subpanel but added billions to the SSC cost, the US high energy physics community committed itself to a project that became increasingly difficult to sustain politically amidst the constricting budget climate of the early 1990s. With the end of the Cold War and the absence of a hoped-for "peace dividend," plus an ensuing recession, the United States entered a period of austerity not unlike what is happening today in much of the West. In such a difficult fiscal climate, a poorly understood scientific project experiencing huge cost overruns and lacking any major foreign contributions was an easy political target for Congressional budget-cutters to "sacrifice."

A 20 TeV proton collider — or perhaps just a billion-dollar extension of Fermilab facilities such as its proposed Dedicated Collider — would likely have survived the budget axe and discovered this light symmetry-breaking

[39]Had such a voice been present in DOE deliberations, project advocates might have taken more seriously Panofksy's proposal to reduce or descope the SSC project when it faced cost increases of some $2 billion due to design changes in late 1989 and early 1990.

[40]By the end of the 1990s, the Fermilab discovery of the top quark with a mass close to 175 GeV and precision measurements of Standard Model parameters suggested that the Higgs boson would most likely come in at less than 200 GeV. See, for example, Ref. 80.

boson long ago.[41] Such smaller projects would surely have been more defensible during the economic contractions of the early 1990s, as they accorded better with the US high energy physics community's diminished "political capital" in Washington. Their construction would also have proved much easier for physicists to manage and control without having to involve military-industrial engineers.

Unlike historians gazing into the past, however, accelerator physicists do not enjoy the benefit of hindsight when designing a new machine. Guided partly by the dominant theoretical paradigm, they operate with a cloudy crystal ball through which they can only guess at likely discoveries to be made, and must plan accordingly.[42] And few of them can foresee with clarity what will occur in the economic or political realms that might affect an enormous project requiring a decade to complete and costing billions of dollars, euro, or Swiss francs — or, relevantly, a trillion yen. This climate of uncertainty argues for erring on the side of fiscal conservatism, trying to reduce expenses by building a new machine at an existing laboratory and recycling its infrastructure, both physical and human. Such an incremental approach has been followed successfully at CERN for six decades now, and to a lesser extent at the other high energy physics labs.

But US physicists chose to depart from this well-worn path in the case of the Superconducting Super Collider. It took a huge leap of faith to imagine that they could construct an enormous new collider with over 20 times the energy of any machine they had previously achieved, at a green-field site where everything had to be assembled anew from scratch — including its management team — and defend the project before Congress in trying fiscal times. When the SSC cost grew beyond \$10 billion, its fate was sealed. A more modest project sited instead at Brookhaven or Fermilab would likely have survived and still be producing good physics today. As one prominent high energy physicist observed, the SSC was probably "a bridge too far" for this once-influential scientific community.

[41]While the proposed machine, a proton–antiproton collider, might not have achieved sufficient luminosity to discover the 125 GeV Higgs boson, a subsequent proton–proton Dedicated Collider suggested by L. Teng would likely have had enough. This design, using superconducting dipole magnets developed for the SSC, could have generated a collision energy of 6 TeV at a luminosity of $10^{33}\ \mathrm{cm^{-2}s^{-1}}$ [81]. I mention this hypothetical machine mainly to illustrate the wisdom of pursuing a conservative, gradualist approach to accelerator design in a constrained fiscal environment.

[42]As a prominent CERN director-general (who will remain anonymous) once told me in this regard, "Do not trust theorists too much, only a little!"

## Acknowledgments

I thank Ezra Heitowit, Lillian Hoddeson, Chris Llewellyn-Smith, Steve Myers, Burton Richter and Herwig Schopper for helpful comments on and corrections to earlier drafts of this article. Any residual errors of fact or opinion are solely the responsibility of the author.

Research for this article and *Tunnel Visions* has been supported by the National Science Foundation under award numbers SBR-94-11671 and SES-0823296, as well as by the US Department of Energy, the John D. Guggenheim Memorial Foundation, and the Richard Lounsbery Foundation.

## References

[1] J. R. Sanford, "Isabelle, a Proton–Proton Colliding Beam Facility at Brookhaven," *IEEE Transactions on Nuclear Science*, Vol. NS–24, No. 3 (June 1977) 1845–48. For a history of this aborted project, see R. Crease, "Quenched! The ISABELLE Saga," parts I and II, *Physics in Perspective*, Vol. 7, No. 3 (September 2005) 330–76, and Vol. 7, No. 4 (December 2005) 404–52.

[2] B. Levi, "A Look at the Future of Particle Physics," *Physics Today* (January 1983) 19–21. The published proceedings of this workshop are available as R. Donaldson *et al.*, *Proceedings of the 1982 DPF Summer Study on Elementary Particles and Future Facilities*, June 28–July 16, 1982 (Fermi National Accelerator Laboratory, 1983).

[3] L. Hoddeson *et al.*, *Fermilab: Physics, the Frontier, and Megascience* (Chicago: The University of Chicago Press, 2008) pp. 25–27.

[4] A. Kolb and L. Hoddeson, "The Mirage of the 'World Accelerator for World Peace' and the Origins of the SSC, 1953–1983," *Historical Studies in the Physical Sciences*, Vol. 24, No. 1 (1993) 101–124.

[5] John Krige, "Distrust and Discovery: The Case of the Heavy Bosons at CERN," *Isis*, Vol. 92, No. 3 (2001) 517–40.

[6] US Department of Energy, "Report of the 1983 HEPAP Subpanel on New Facilities for the U.S. High Energy Physics Program," Report No. DOE/ER-0169 (July 1983), pp. 33–35.

[7] S. Wojcicki, "The Supercollider: The Pre-Texas Days," *Reviews of Accelerator Science and Technology*, Vol. 1 (2008) 259–302. See pp. 263–264 on the Panel and Keyworth's "think big."

[8] J. Sandweiss, letter to A. Trivelpiece, 12 July 1983, in US Department of Energy, "Report of the 1983 HEPAP Subpanel," i-iii.

[9] L. Hoddeson and A. Kolb, "The Superconducting Super Collider's Frontier Outpost, 1983–1988," *Minerva*, Vol. 38 (2000) 271–310.

[10] SSC Central Design Group, "Conceptual Design of the Superconducting Super Collider," Report No. SSC-SR-2020 (March 1986). Copy in Fermilab Archives.

[11] D. Ritson, "Demise of the Texas Supercollider," *Nature* (16 December 1993) 607–610.

[12] I. Goodwin, "Amazing Race: the SSC Contest Generates Disorder and Discord," *Physics Today* (May 1988) 69–74.

[13] M. Crawford, "Reagan Okays the Supercollider," *Science* (6 February 1987) 625; I. Goodwin, "Reagan Endorses the SSC Project, A Colossus Among Colliders," *Physics Today* (March 1987) 47.

[14] US Department of Energy, press release, "DOE Requests Funding to Build the Superconducting Super Collider," 30 January 1987. See also transcript of press conference by Ace Federal Reporters, Inc., copy in Irwin Goodwin Collection, Niels Bohr Library and Archives, College Park, MD.
[15] B. G. Levi and B. Schwarzschild, "Super Collider Magnet Program Pushes Toward Prototype," *Physics Today* (April 1988) 17-21; W. J. Broad, "Giant Atom Smasher Encounters Vexing Technical Obstacles," *New York Times* (13 December 1988) D23.
[16] R. Schwitters, interviews by M. Riordan.
[17] I. Goodwin, "DOE Picks Texas for 'Gippertron,'" *Physics Today* (February 1989) 95–98.
[18] US Department of Energy, "SSC Site Evaluations: A Report by the SSC Task Force," Report No. DOE/ER–0392 (September 1988); R. Diebold, interview by M. Riordan.
[19] S. Wojcicki, "The Supercollider: The Texas Days," *Reviews of Accelerator Science and Technology*, Vol. 2 (2009) 265–301.
[20] G. Ross (*sic*) and T. Kirk, "Report of the SSC Collider Dipole Review Panel," SSCL Report No. SSC-SR-1040. The name of co-author Gus Voss of DESY was misspelled and the error not caught.
[21] A. Chao, interview by M. Riordan.
[22] R. J. Smith, "Supercollider Could Face Cutbacks," *Washington Post* (19 November 1989) A1.
[23] I. Goodwin, "Trying Times: Cost of Remodeling SSC Causes Texans to Circle Their Wagons," *Physics Today* (January 1989) 45–47.
[24] W. K. H. Panofsky, "A Personal View on SSC Project Cost Growth and Control," 13 December 1989, Panofsky Collection, SLAC Archives; W. K. H. Panofsky, interview by M. Riordan and S. Weiss.
[25] F. E. Low to J. M. Decker, "Report of the 1990 HEPAP Subpanel on SSC Physics," DOE Report No. DOE/ER–434 (January 1989); I. Goodwin, "HEPAP and its Subpanel Approve Redesign and Higher Cost of the SSC," *Physics Today* (February 1990) 67–68.
[26] J. R. Sanford and D. M. Mathews, eds, "Site-Specific Conceptual Design of the Superconducting Super Collider," Report No. SSCL-SR-1056 (July 1990).
[27] Admiral J. D. Watkins, interview by M. Riordan and S. Weiss.
[28] M. Riordan, "A Tale of Two Cultures: Building the Superconducting Super Collider Laboratory, 1988–1993," *Historical Studies in the Physical and Biological Sciences*, Vol. 32, No. 1 (Fall 2001) 125–144.
[29] E. Siskin, interview by M. Riordan.
[30] K. A. McDonald, "Supercollider Scientists Are Embroiled in Dispute; Outcome Could Raise Project's $8 Billion Cost," *Chronicle of Higher Education* (10 April 1991) A19.
[31] D. P. Hamilton, "The SSC Gets Its (Official) Price Tag: $8.3 billion," *Science* (15 February 1991) 741.
[32] Office of Energy Research, "Report on the SSC Cost and Schedule Baseline," Report No. DOE/ER–0468P (January 1991).
[33] R. Crease, "Choosing Detectors for the SSC," *Science* (21 December 1990) 1648–1650.
[34] D. P. Hamilton, "Ad Hoc Team Revives SSC Competition," *Science* (21 June 1991) 252.
[35] H. Schopper, "LEP and Future Options," in F. T. Cole and R. Donaldson, *Proceedings of the 12th International Conference on High-Energy Accelerators*, 11–16 August 1983 (Batavia, IL: Fermilab, 1983) pp. 658–663; H. Schopper, interview by M. Riordan.
[36] B. Richter, memo to R. DeVries and W. Kornack, 1 September 1983, copies in Richter Collection, SLAC Archives, and also in Fermilab Archives.

[37] C. Llewellyn-Smith, interview by M. Riordan.
[38] W. Sweet, "Abragam and Rubbia Reports Chart Future for CERN," *Physics Today* (September 1987) 71–75.
[39] John Krige, private communications.
[40] M. Riordan and K. Takahashi, "Cooperation in High-Energy Physics Between the United States and Japan," *SLAC Beam Line* (Spring 1992) 1–9.
[41] D. Goldston, private communication with M. Riordan. During the 1990s Goldston was a key legislative aide to NY Congressman Sherwood Boehlert, a leading House SSC opponent.
[42] "House Vote Revives Fermilab Injector," *Science* (7 June 1991) 1373.
[43] D. A. Bromley, *The President's Scientists: Reminiscences of a White House Science Advisor* (New Haven: Yale University Press, 1994) 213.
[44] D. E. Sanger, "Japan Wary as U.S. Science Comes Begging," *New York Times* (27 October 1971) E13.
[45] M. Wines, "Bush in Japan: Bush Reaches Trade Pact with Japan But Automakers Denounce It," *New York Times* (10 January 1992) A1.
[46] K. Takahashi and I. Itabashi, private conversations with M. Riordan.
[47] B. Richter, memo to J. Gibbons, "Japan and the SSC," 14 January 1993, Richter Collection, SLAC Archives.
[48] H. Wolpe, "Status of the Superconducting Super Collider Program," Report of the House Subcommittee on Investigations and Oversight, Committee on Science, Space & Technology (9 May 1991) 1.
[49] W. J. Broad, "Vast Sums for New Discoveries Pose a Threat to Basic Science," *New York Times* (27 May 1990) 1.
[50] P. W. Anderson, "More is Different: Broken Symmetry and the Nature of the Hierarchical Structure of Science," *Science* (4 August 1972) 393–396.
[51] J. Rees, interviews by M. Riordan.
[52] I. Goodwin, "Cheers for Bush's 1993 R&D Budget Cut Short by Problems and Pessimism," *Physics Today* (June 1992) 55–61, see especially the table of DOE funding on pp. 56–57.
[53] D. P. Hamilton, "Lightning Strikes the SSC," *Science* (26 June 1992) 1752–1753. For a deeper analysis of Congressional dynamics, see H. Idelson, "House Denies Atom Smasher Its 1993 Expense Account," *Congressional Quarterly* (20 June 1992) 1773–1785.
[54] J. R. Crawford, "Defeat of Budget Amendment Fans Anti-Deficit Flames," *Congressional Quarterly* (13 June 1992) 1683–1689.
[55] I. Goodwin, "Victim of Budget Balancing War, SSC Now Faces Uncertain Future in Senate," *Physics Today* (July 1992) 53–54.
[56] I. Goodwin, "What's Gone Wrong with the SSC? It's Political, Not Technological," *Physics Today* (August 1992) 58–60; I. Goodwin and B. Schwarzschild, "Good News for the SSC," *Physics Today* (September 1992) 55.
[57] I. Goodwin, "Despite Retirements and Defeats, Congress Retains Friends of Science," *Physics Today* (February 1993) 71–72.
[58] B. Schwarzschild, "Tunnel Boring Begins at Superconducting Super Collider," *Physics Today* (March 1993) 19; *SSC News* (May 1993).
[59] M. W. Browne, "Building a Behemoth Against Great Odds," *New York Times* (23 March 1993) C1. This was the article in which Schwitters was quoted making his infamous comment about "the revenge of the C students."
[60] J. Gibbons, interview by S. Weiss; I. Goodwin, "Clinton's Hands-on Economic Plan: Technology Gains, Big Science Loses," *Physics Today* (April 1993) 43; S. Begley, "Sparing Those Sacred Cows," *Newsweek* (1 March 1993) 17.

[61] J. Kondo, private conversation with M. Riordan.
[62] C. McIlwain, "SSC Falls Victim to Congressional Austerity," *Nature* (1 July 1993) 6.
[63] I. Goodwin, "Collision Course"; C. Anderson, "University Consortium Faulted on Management, Accounting," *Science* (9 July 1993) 157; C. McIlwain, "Barrage of Mud Fails to Stick to Super Collider," *Nature* (8 July 1993) 92.
[64] US Department of Energy Office of Field Management, "Report of the DOE Review Committee on the Baseline Validation of the Superconducting Super Collider," Report No. DOE/ER-0459P (August 1993). This was often called the "Scango Report" after team leader John Scango.
[65] I. Goodwin, "As SSC Faces 'Live or Die' Vote, O'Leary Shakes up Management," *Physics Today* (September 1993) 52–53; C. Anderson, "DOE Pulls Plug on SSC Contractor," *Science* (13 August 1993) 822–823. Goodwin called the SSC a $10 billion collider while Anderson put the cost at $11 billion.
[66] D. Pearson, interview by M. Riordan and S. Weiss.
[67] I. Goodwin, "Senate Vote Rescues SSC, But Final Act Awaits Conference," *Physics Today* (October 1993) 110–111.
[68] E. Pianin, "House Deals a Big Defeat to Atom Smasher in Texas," *Washington Post* (20 October 1993) A1; E. Pianin and S. LaFraniere, "Proponets of Collider Give Up," *Washington Post* (21 October 1993) A1, quote on p. A12.
[69] D. Goldston, private conversations with M. Riordan and S. Weiss.
[70] S. Boehlert, "Collider Project Doomed Itself by Excess," letter to editor, *New York Times* (4 November 1993).
[71] C. McIlwain, "SSC Decision Ends Post-war Era of Science-Government Partnership," *Nature* (28 October 1993) 773–774.
[72] F. Flam, "Panel Presents Vision for Physics after the Supercollider," *Science* (3 June 1994) 1397; US Department of Energy, "HEPAP Subpanel on Vision for the Future of High-Energy Physics," Report No. DOE/ER-0614P (May 1994).
[73] I. Goodwin, "Four Years after SSC's Demise, US Reaches Agreement on 'Unprecedented' Collaboration in CERN's LHC," *Physics Today* (January 1998) 43–44.
[74] C. Llewellyn-Smith, "How the LHC Came to Be," *Nature* (19 July 2009).
[75] J. Gillies, email communications with M. Riordan; L. Evans and C. Llewellyn-Smith, private communications with M. Riordan; A. Unnervik, "The Construction of the LHC," in L. Evans, ed., *The Large Hadron Collider: A Marvel of Technology* (Lausanne: EPFL Press, 2009), pp. 38–55, esp. Table 1 on p. 40.
[76] G. Brumfield, "The Machine Maker," *Nature* (18 December 2008) 862–868.
[77] R. Dieter-Heuer, interview by M. Riordan.
[78] R. Herman, "High-energy physics splits the scientists," *New Scientist* (20 June 1985) 10.
[79] J. Ellis, G. Ridolfi, and F. Zwirner, "Radiative Corrections to the Masses of Supersymmetric Higgs Bosons," *Physics Letters,* Vol. 257 (1990) 83–88; H. Haber and R. Hempfling, "Can the Mass of the Higgs Boson of the Minimal Supersymmetric Model be Larger than $m_Z$?" *Physical Review Letters,* Vol. 66 (1991) 1815–1818.
[80] M. Riordan, P. C. Rowson, and S. L. Wu, "The Search for the Higgs Boson," *Science,* Vol. 291 (12 January 2001) 259–160.
[81] L. Teng, "A 3 TeV on 3 TeV Proton–Proton Dedicated Collider for Fermilab," Fermilab Technical Report No. TM-1516 (30 March 1988).

# Part 2

# 100 Years of Accelerators

# Chapter 17

# The GSI heavy ion facility

Norbert Angert and Bernhard Franzke (GSI)

## 1 Introduction

Interest in experiments with fast heavy ions increased considerably in the mid-1950s. This was caused by the conjecture that certain atomic nuclei with masses up to about 300 u may possess measurable lifetimes. Extrapolations of the nuclear shell model suggested the existence of a stability island at $Z = 120 \pm 6$ and $N = 180 \pm 15$ ($Z$ = number of protons, $N$ = number of neutrons). In order to reach those exotic proton and neutron numbers by the fusion of existing nuclei, it was necessary to overcome their repulsive Coulomb force by high enough kinetic energy. Prompted by this interest in accelerated heavy ions, a study group at Heidelberg University, Germany, headed by Ch. Schmelzer, started in 1963 a design study for a universal heavy ion linear accelerator. The aim was to work out an accelerator concept which would make it possible to accelerate any ion up to uranium to specific energies well above the nuclear reaction threshold (Coulomb barrier) of 5–6 MeV/u for any target nucleus. In addition, the accelerator should include the possibility of varying the energy continuously over the widest possible range up to the maximum energy of about 10 MeV/u without affecting the beam quality [Schmelzer 1968].

When designing a linac for heavy ions, special issues must be considered which are not relevant for proton or light ion machines [Böhne *et al.* 1969]. The Heidelberg study group performed both fundamental research and design studies for optimizing the layout together with technological developments.

Important prerequisites for the envisaged type of accelerator are ion sources which can generate highly charged ions for the full spectrum of elements [Krupp 1970, Illgen *et al.* 1972, Wolf 1972]. Knowledge of the equilibrium ion charge after passage of targets (strippers) in order to increase the ion charge during the acceleration process and of the yield in the charge states after stripping is important when optimizing a heavy ion accelerator layout [Betz *et al.* 1966, Franzke *et al.* 1967]. Knowledge of charge-changing cross sections of partially stripped highly charged ions is important for determining vacuum conditions [Betz *et al.* 1967, Angert 1968, Möller 1968]. A review of the research on these fundamental issues for heavy ion acceleration performed by the Heidelberg study group was given just before the start of the Unilac construction [Schmelzer 1969].

In parallel to the research, the technical optimization of the machine layout included the selection of suitable RF accelerating structures and focusing concepts for the different energy ranges. The choice of radio frequencies and RF-sources was accompanied by particle dynamics calculations and prototype studies of accelerator components [Böhne 1969]. By the end of the 1960s, the initial linac concept, based on single-gap accelerating cavities only, had been developed to a technical proposal for a heavy ion linac ready for construction. It consisted of Wideröe-type structures in the low energy range before stripping of ions at 1.4 MeV/u and Alvarez-type drift-tube and single-gap cavities in the post-stripper sections.

In the mid-1960s, nuclear physicists at universities (Darmstadt, Frankfurt, and Marburg) in the State of Hesse, Germany, started discussing different options for a joint research laboratory for future nuclear physics research. A facility for research with energetic heavy ions was considered finally to offer the most interesting perspectives, and received also the political support. The Gesellschaft für SchwerIonenforschung (GSI, Association for Heavy Ion Research) was founded jointly by the Federal Republic of Germany and the State of Hesse at the end of 1969. The construction of this national heavy ion research laboratory was started north of Darmstadt at the end of 1971, with the universal heavy ion linac (Unilac), as proposed by the Heidelberg study group, as the central facility. It was built in the years 1972 to 1975. Experiments started in early 1976. Uranium beams have been accelerated since spring 1976. An expansion of the GSI heavy ion facility by an 18 Tm synchrotron SIS18 and a 10 Tm Experimental Storage Ring (ESR) started in 1986. Commissioning of SIS18–ESR took place in 1989/1990. The layout of the GSI heavy ion facility is shown in Fig. 1. Key features as well as the upgrades of the three accelerators (Unilac, SIS18, and ESR) are described in the following sections.

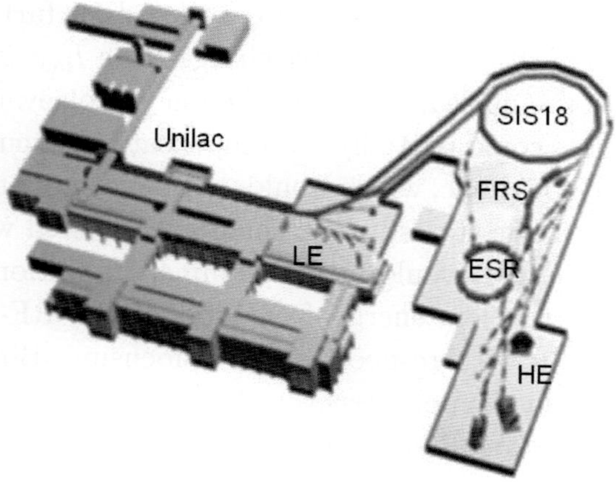


Fig. 1. Layout of the GSI heavy ion accelerator facility with the Unilac, the heavy ion synchrotron SIS18 and the Experimental cooler-Storage Ring ESR. Experiments for low energy beams from the Unilac and for high energy beams from SIS18–ESR are installed in the areas LE and HE, respectively. FRS indicates the area for the production and separation of in-flight-produced radioactive beams.

## 2 The Unilac, a universal heavy ion linear accelerator

### 2.1 *Special issues of the Unilac design*

When the Unilac was designed in the 1960s, charge states which could be obtained for heavy ions with sufficient beam current from the available — and at that time commonly used — heated-cathode Penning ion sources were at maximum 10+ to 12+ [Pasyuk *et al.* 1967]. Therefore, the charge-to-mass ratio $q/A$ from the ion source was rather low for heavy ions ($q/A \leq 0.05$). The lifetime of the ion sources in operation with very heavy ions was not very long (10–20 h). Therefore, the DC pre-accelerator voltage for the ion source platform should allow an easy access for maintenance or source replacement and should not exceed a few hundred kV. As a consequence, the initial velocities $\beta = v/c$ of the ions after DC pre-acceleration were rather low ($\beta \leq 0.5\%$). A linear accelerator for a specific energy $\geq 5$ MeV/u for ions of such low charge-to-mass ratios would be rather long and expensive, both in construction and operation. Therefore, stripping-off electrons by sending the beam through thin foil or gas targets at intermediate energy in order to increase the ion charge state was recommended. However, the intensity of the uniformly charged beam is split in the stripping process into a Gaussian shaped charge state distribution, from which one charge state, with only a fraction of the primary intensity, is usually selected for further acceleration. The intensity yield of the most probable charge state $I_{\max}$ is proportional to $1/Z^{1/2}$, where $Z$ is the nuclear charge of the ion [Nikolaev 1965, Betz 1972].

The intensity loss could partially be compensated by multi-charge operation in the post-stripper linac, which would, however, affect the beam quality. Using lower initial charge states, which could be delivered with higher beam intensity from ion sources, would both recommend additional stripping and make the design of the low energy front end of the RF-linac even more challenging, especially with respect to radial focusing [Böhne *et al.* 1969]. The radiofrequency quadrupole structure (RFQ) had been invented at that time, but was not yet available in the western hemisphere.

### 2.2 *The accelerator*

Taking these special issues for heavy ions into account, two identical ion source platforms and 320 kV DC pre-accelerators, housed in Faraday cages, were installed. During the first years of Unilac operation, Duoplasmatron ion sources have been used for light ions and Penning ion sources for metals and heavy elements [Böhne 1977]. In later years, only the more versatile heated-cathode Penning ion sources were used for routine operation. This made ion source maintenance and Unilac operation easier. Charge and mass separations of the pre-accelerated 11.7 keV/u DC-beams are performed in the low energy beam transport line to the RF-linac.

For RF-acceleration in the low energy range from 11.7 keV/u to 1.4 MeV/u a 27 MHz Wideröe-type drift-tube structure was chosen. It was designed for a duty cycle of 25% (5 ms/50 Hz) at a maximum mass-to-charge ratio of $A/q = 24$. For reasons of mechanical stability a coaxial-line structure has been chosen (see Fig. 4). Magnetic quadrupole lenses were installed only in every second drift-tube [Böhne *et al.* 1969]. In order to achieve radial focusing at the very low initial velocities a $3\pi/\pi$ drift-tube structure (instead of $\pi/\pi$) had to be used in the first of the four Wideröe-tanks for the acceleration to 187 keV/u. Magnetic quadrupoles were installed in the $3\pi$ drift-tubes [Böhne 1972]. Figure 2 shows the Faraday cages housing the ion source platforms at a potential of up to 320 kV, the charge and isotope separating beam transport system, and the Wideröe-type pre-stripper RF-linac.

At a specific energy of 1.4 MeV/u the beam passes a stripper with an areal density of 1–40 $\mu g/cm^2$, which consists either of an $N_2$ gas jet or a carbon foil. The initial ion charge state is increased there by a factor of 3 to 8, depending on the ion species. For uranium the charge state 10+ is increased to an average charge state 28+ in the gas jet or to 40+ in a carbon

Fig. 2. View onto the Unilac during the installation phase with the two Faraday cages for the ion source platforms (left and right), the beam transport lines to the RF-linac, where charge and isotope separations are performed, and the Wideröe-type pre-stripper accelerator (yellow) in the foreground. The Alvarez-type post-stripper accelerator (violet) is visible in the background. The stripper and the following charge analyzing system are installed in between. (All photographs in this chapter are by Zschau, GSI)

foil. Usually, the most abundant charge state (e.g. $U^{28+}$ after the $N_2$ jet) of the resulting charge distribution is sorted out for further acceleration by an achromatic magnetic charge state separator. Up to the end of the 1980s a low energy experiment area on the left hand side of the linac tunnel had been supplied with a parasitic 1.4 MeV/u beam, which had been peeled off from the charge state distribution in the separator for experiments in atomic physics, bio-physics, and materials research [Böhne 1977].

Initially, the first part of the post-stripper linac consisted of only two 108.48 MHz Alvarez-type tanks, each about 10 m long. The first one accelerated to fixed specific beam energy of 3.6 MeV/u, the second subdivided tank to 4.7 or 5.9 MeV/u. The following 20 single-gap cavities increased the beam energy for $U^{28+}$ to 8.5 MeV/u and for $U^{40+}$ to 10.2 MeV/u. The 20 single-gap cavities could be used both in acceleration or deceleration mode to obtain any desired energy between 2.5 MeV/u and the maximum energy.

### 2.3 *Unilac upgrades*

Increase of beam energy: In order to reach higher injection energies into a planned future synchrotron two additional Alvarez tanks with output energies of 8.5 and 11.4 MeV/u were installed in 1981 [Angert 1983]. Figure 3

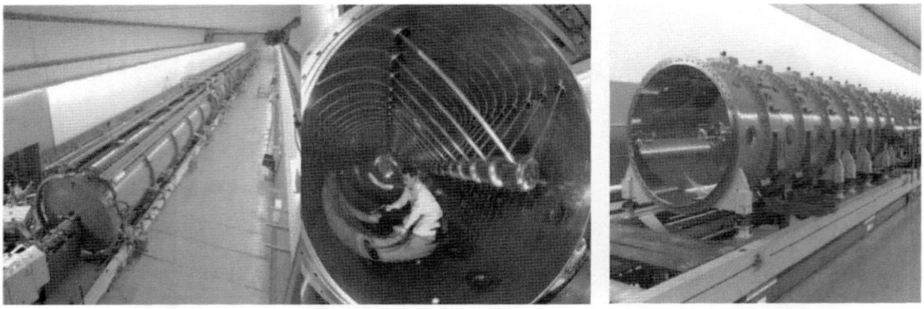

Fig. 3. Left: Picture of the upgraded Alvarez post-stripper linac with four Alvarez tanks; Center: View into Alvarez tank 1; Right: Single-gap cavities.

shows the four tanks of the expanded 108 MHz Alvarez post-stripper linac, the accelerating structure of the Alvarez tank 1, and the 108 MHz single-gap cavity resonators.

Second 1.4 MeV/u linac and dual ion operation: The low energy experiment hall for parasitic 1.4 MeV/u beams was shut down in 1988 and a second 1.4 MeV/u RF-linac was installed there, using an Electron Cyclotron Resonance (ECR) ion source, which could deliver charge states high enough (for example $U^{28+}$) for the acceleration in the post-stripper linac [Angert *et al.* 1992]. After separation of charges and isotopes in a magnet spectrometer the beam is accelerated by a 108 MHz 4-rod-RFQ from 2.5 keV/u to 0.3 MeV/u. A novel 108 MHz IH-type drift-tube structure, with quadrupole-free drift-tube sections and only two integrated magnet triplets, accelerates ions to 1.4 MeV/u. This new high charge state RF-linac (Hoch–Ladungs–Injektor, HLI) delivers a second ion species to the post-stripper accelerator via a fast pulsed switching magnet (see Fig. 4, top right). The RF and focusing systems of the post-stripper accelerator had been modified in order to allow fast switching (up to 50 Hz) between the acceleration of ions with different charge-to-mass ratios to different output energies [Glatz 1986]. With the new injector and the fast switching post-stripper linac, two different ion species could be delivered with different beam energies and pulse repetition rates to the three experiment areas of the Unilac (see Fig. 5).

Intensity upgrade and multi-beam operation: After the commissioning of SIS18–ESR the Unilac could deliver sufficient beam current for light ions (e.g. Ne) to fill the SIS18 up to the space charge limit [Baer 1998]. For heavy ions, however, the beam currents from the Wideröe pre-stripper linac were too low by more than two orders of magnitude. Therefore, at the end of 1990s, the Wideröe drift-tube structure (Fig. 4) was replaced by a new high current pre-stripper linac for 1.4 MeV/u (Hoch–Strom–Injektor, HSI). It consists of

Fig. 4. Top left: The 27 MHz Wideröe $\pi/\pi$ or $\beta\lambda/2$-drift-tube structure (tank 2 to 4), which was used until 1998 for the acceleration of ions with $A/q \leq 24$ to an energy of 1.4 MeV/u. Top right: Existing high current IH pre-stripper linac (HSI) for the acceleration of ions with $A/q \leq 65$ (commissioned in 1999), with the stripper and charge analyzing system behind. The beam transport line coming from the high charge state linac (HLI) is visible on the right. Bottom left: IH-type RFQ of the new high current pre-stripper linac HSI with mini-vanes. Bottom right: IH drift-tube structure of the HSI; see also text.

a 9.4 m long 36 MHz IH-type RFQ for the acceleration from 2.2 keV/u up to 120 keV/u and two tanks with a 36 MHz IH drift-tube structure, each tank about 10 m long, for the acceleration to 1.4 MeV/u (Fig. 4) [Ratzinger 1996]. For the HSI the IH drift-tube concept was designed for the first time for high beam currents at a large tune depression of ~0.70. The peak intensity for short beam pulses of very heavy ions could be increased by two orders of magnitude.

The HSI can be operated at a high duty cycle of up to 25% (5 ms/50 Hz) with charge states from the Penning ion source ($U^{10+}$). In low duty cycle operation (≤0.3 ms/3 Hz), as required for the injection into SIS18, much higher average accelerating gradients (up to 4.3 MV/m) are possible. That

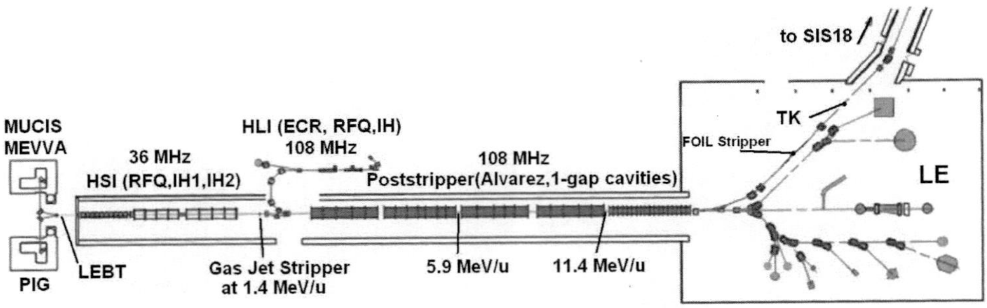


Fig. 5. Schematic plan view of the upgraded Unilac. From left to right: Ion source terminals, high current pre-stripper accelerator (HSI, 36 MHz RFQ and IH), high charge state accelerator (HLI, 108 MHz), Alvarez-type post-stripper accelerator (with energy steps of 3.6, 4.7, 5.9, 8.5, and 11.4 MeV/u), and single-gap cavity section for acceleration, deceleration, or fine tuning of the beam energy. Right: low energy (LE) experiment area and beam transport (TK) line to the SIS18 with foil stripper and charge separator behind.

allows the acceleration of ions with a mass-to-charge ratio of up to $A/q = 65$ (with a total voltage of 91 MV instead of 34 MV). For instance, $U^{4+}$ ions delivered from a MEVVA-type ion source with beam pulses of more than 30 mA [Brown 1994, Spädtke *et al.* 1998] can be accelerated to 1.4 MeV/u this way. Commissioning of the HSI with beam was performed in 1999 [Barth 2002]. Figure 5 shows the plan view of the upgraded Unilac in 1999.

Fast switching between the acceleration of ions with different $A/q$-values is also possible in the HSI. From the two source platforms, beams of two different ion species can be accelerated with different duty cycles and repetition rates on request from the experiments. Together with the HLI, the alternating acceleration of three different ion species is possible now in the post-stripper linac [Dahl 2009]. With this flexibility, the Unilac has delivered for example a low energy lead beam with high repetition rate (5 ms, 38 Hz) from the PIG ion source to one experiment and a low energy carbon beam with low repetition rate (5 ms/10 Hz, from the ECR ion source) to another one, both with the desired energy. In between, short beam pulses (0.3 ms) of carbon and uranium ions (from a MEVVA ion source) have been delivered during the same run for injection into the SIS18 every few seconds. In order to meet the intensity requirements for the future operation of Unilac/SIS18 as injector for the FAIR facility, additional upgrade measures at the Unilac have been and will be performed [FAIR 2001]. However, matching between ion source and RFQ, the control of beam brightness through the HSI and the post-stripper linac, and the matching to the SIS18 acceptance is still a challenge in high current operation [Dahl *et al.* 2012].

## 3 The synchrotron/storage-ring facility SIS18–ESR

Considerations for an expansion of the GSI accelerator facility by a synchrotron, with the aim to reach relativistic energies, started immediately after the commissioning of the Unilac in 1976 [Blasche *et al.* 1977]. In 1979 a 14 GeV/u facility for relativistic heavy ion beams was proposed. Two concepts for this facility have been studied in the following years: A one-stage concept, with one 100 Tm synchrotron (SIS100 with $dB/dt = 2\,T/s$ ramp-rate), and a two-stage concept with an 18 Tm fast cycling booster synchrotron (SIS18, $dB/dt \leq 10\,T/s$), and a 100 Tm high energy synchrotron (SIS100, $dB/dt \leq 0.5\,T/s$), with either normal or superconducting magnets [Blasche *et al.* 1981].

There was great interest by part of the nuclear physics community to get relativistic heavy ions as soon as possible. A GSI–LBL–Heidelberg–Warsaw collaboration made a proposal for the acceleration of oxygen ions up to 13 GeV/u in the CERN PS-complex [Stock *et al.* 1982] that was accepted by CERN in 1983. A new ion injector for the CERN Linac1 was built by a CENG–CERN–GSI–LBL collaboration. Oxygen and sulfur beams were accelerated in the CERN PS-SPS up to 200 GeV/u in 1986 and 1987, respectively [Angert *et al.* 1988].

Discussions about the concepts for the expansion of the GSI facility went on in parallel. None of the proposals received the unanimous support of the science community. Finally, the way to the extension of GSI was paved by the decision to build first only the booster synchrotron of the two-stage concept, together with a storage ring for experiments with cooled heavy ion beams. This step into the new field of cooling highly charged heavy ion beams was strongly stimulated by successful demonstrations of electron cooling method at BINP, Novosibirsk [Budker *et al.* 1976] and both the electron and the stochastic beam cooling at ICE and at the Low-Energy Antiproton Ring (LEAR) of CERN [Carron *et al.* 1979, Lefèvre *et al.* 1980]. The SIS18–ESR proposal — consisting of the 18 Tm heavy ion synchrotron SIS18, the 10 Tm ESR, the 18 Tm magnetic FRagment Separator (FRS), and several experiment stations (see Fig. 1) — was submitted to the responsible ministries of the Federal Republic of Germany and the state of Hesse in autumn 1984 [Kienle 1985, Blasche *et al.* 1985].

The SIS18–ESR project was approved in spring 1985. The construction of the facility started at the end of 1986 and the first beam was injected into the SIS18 on the occasion of the 80th birthday of Ch. Schmelzer on November 17th, 1988. A Ne beam was accelerated to the full energy of 2 GeV/u in mid-1989 [Böhne *et al.* 1990]. In spring 1990, a beam of fully stripped argon ions

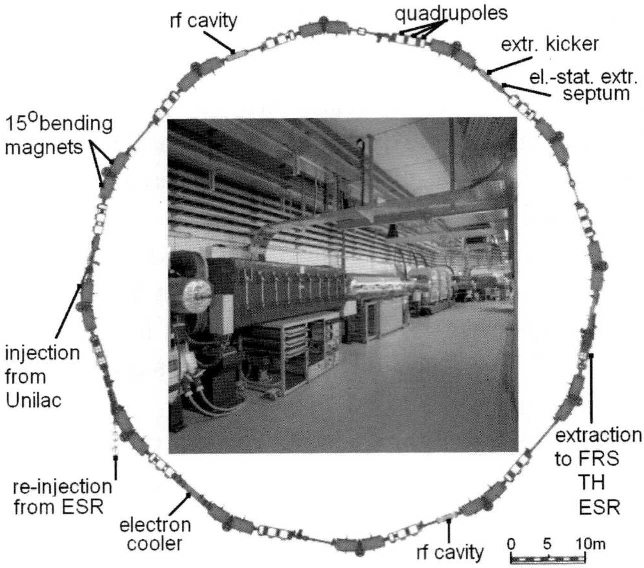


Fig. 6. Lattice of the SIS18 and main components of the ring equipment. Center: Picture of a ring section with bending magnet (red), quadrupole lenses (yellow), and one of the two RF-accelerating sections in between.

could be stored and electron-cooled in the ESR for the first time [Franzke *et al.* 1990].

### 3.1 *The heavy ion synchrotron SIS18*

A strictly periodic lattice with 12 identical cells was chosen for the SIS18 [Blasche *et al.* 1992]. Figure 6 displays the lattice structure of the synchrotron ring and shows a picture of the synchrotron tunnel.

The circumference of SIS18 is 216 m. Each of the 12 machine periods contains two bending magnets and a triplet group with two long doublet lenses and one additional triplet lens. The main reason for the strict periodicity (no super-periods) was the wish for sufficient working space, free from systematic resonances up to the 5th order. $Q_{\mathrm{H}} > Q_{\mathrm{V}}$ and FD rather than DF doublets have been chosen in order to shift $\gamma_t$ to high energy. The focusing changes during acceleration from the triplet to a doublet structure. The field strength of the quadrupoles is thereby changed such that the horizontal and vertical tune values are nearly constant $Q_{\mathrm{H}} = 4.2$ and $Q_{\mathrm{V}} = 3.3$. The triplet focusing at injection leads to a fairly large transverse machine acceptance of $A_{\mathrm{H}}/A_{\mathrm{V}} = 200/50$ mm mrad. That is required because of the rather low injection energy of 11.4 MeV/u and the relatively weak ion currents for heavy ions, which make multi-turn injection necessary. The doublet focusing at high energies requires less quadrupole strength and makes beam extraction and chromaticity control easier. The main magnet power supplies can be operated either for fast field ramping of 10 T/s up to 1.2 T dipole field or for a reduced ramping rate of 4 T/s up to 1.8 T. The higher ramping rate should allow operation up to 3 Hz for higher average intensities at lower and medium energies.

A novel scheme for the control of the magnet power supplies was applied, which made it possible to include all necessary corrections for the effects of iron saturation and eddy currents once and for all during commissioning. That together with a new computer control system, allowed the operation for a time-shared use of up to 16 different machine settings, with various combinations of ion mass and charge, energy, ramping rate, optics, type of extraction (fast or slow), and destination, in a super-cycle.

After passing a foil stripper in front of the synchrotron at 11.4 MeV/u, which increases the average charge state for heavy elements by almost a factor of three, maximum beam energies up to 1 GeV/u can be reached for the heaviest elements ($U^{73+}$) and 2 GeV/u for light nuclei, with a charge to mass ratio of $q/A = 1/2$ ($He^{2+}$ to $Ne^{10+}$). These beams can be delivered (see Fig. 1) either directly to several fixed target experiments in the high energy target area (HE) or to the ESR through a beam line with stripper and charge separator. A third possibility is to transport the beam onto a thick target where fast radioactive beams are produced by fragmentation or fission (in-flight method). With the FRagment Separator (FRS) pure beams of exotic nuclei can be prepared [Geissel 1992]. The analyzed isotopes can be either used for investigations of nuclei with lifetimes as short as microseconds in fixed target set-ups or transferred to the ESR.

### 3.2 *SIS18 upgrades*

Upgrade of the computer control system for cancer therapy: Patient treatment at GSI with light ion beams — preferably carbon — was proposed jointly by the Heidelberg University Radiology, the German Cancer Research Centre, Heidelberg, and GSI in 1993 [Kraft *et al.* 1993]. In order to fulfill the requirements for the proposed treatment method in an optimum way, the accelerator control system was modified in such a way that settings with up to 255 energy steps (in the range from 80 MeV/u to 430 MeV/u) in combination with intensity variations over two orders of magnitude (from $10^6$ to $10^8$ ion/s) and up to seven beam diameter steps at the patient irradiation station could be pre-programmed for a patient treatment run, according to the treatment planning of the radiologists [Eickhoff 1996]. The high charge state pre-stripper linac HLI has been routinely used as ion source for the therapy program, because of its superior long term stability with carbon beams. Treatment of patients started at the end of 1997.

SIS18 electron cooler: Within an intensity upgrade program for heavier ions [Blasche *et al.* 1994], which started in mid-1990s, an electron cooler system was supplemented to the synchrotron. It opened up the possibility to obtain higher intensities in the SIS18 by cooler-assisted fast accumulation of heavy ions at the injection energy. The electron cooler was designed and manufactured in collaboration between GSI and BINP, Novosibirsk. As the typical synchrotron cycles with slow extraction are a few seconds long and the cooling times for the stripped heavy ions at injection energy are of the order of 100 ms ($\propto A/q^2$), the intensity gain by the cooler was up to one order of magnitude for the very heavy ions [Steck *et al.* 1999]. After the installation of the HSI at the Unilac, the electron cooler is available for the preparation of high quality beams and the accumulation of beams of rare isotopes from the HLI, utilizing the low material consumption of the ECR ion source of only a few mg/h.

SIS18 upgrades with respect to the booster operation for FAIR: The future Facility for Antiproton and Ion Research (FAIR) at GSI [FAIR 2001] will use the SIS18 as booster synchrotron. The intensities planned for FAIR can only be achieved by avoiding the intensity loss due to stripping in front of the SIS18. For uranium the acceleration of the intermediate charge state 28+ instead of 73+ raises the incoherent space charge limit ($\propto A/q^2$) in the SIS18 considerably. A multifaceted upgrade and improvement program had to be launched for the SIS18 after the approval of the FAIR proposal in 2003. That included, among other measures, a new injection system for a safe inflection of the intermediate charge states at 11.4 MeV/u, longitudinal and transverse feedback systems, the installation of a new $h = 2$ acceleration cavity and upgrades of the main magnet power supplies in order to reach higher cycling frequencies, and also a new connection to the local power station [Spiller *et al.* 2004, Dahl *et al.* 2012].

A major challenge has been the dynamic pressure increase in operation with the multi-electron heavy ions ($U^{28+}$) at beam intensities far below the calculated incoherent space charge limit of the SIS18. This phenomenon, which has been observed also at other facilities (LHC, RHIC), is caused by beam ions which are lost either due to charge exchange or during the injection processes. That results in ion-induced desorption of a large number of heavy gas atoms or molecules per incident heavy ion [Mahner 2008]. Besides the new injection system, the NEC coating of all dipole and quadrupole vacuum chambers has been performed to increase considerably the distributed

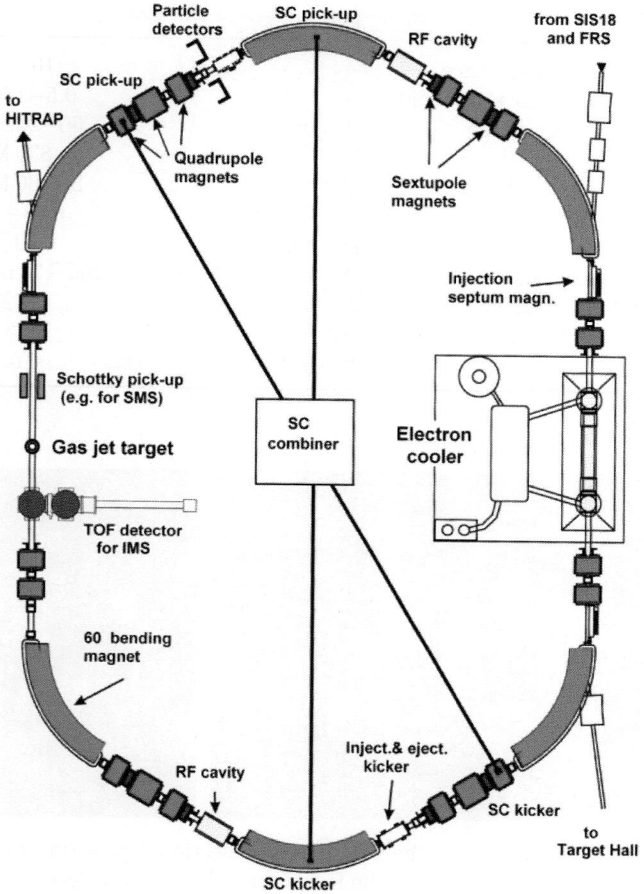


Fig. 7. Schematic plot of the ESR including some information about installations for electron and stochastic (SC) beam cooling and in-ring experiments.

pumping speed. Catchers for lost beam ions have been installed at dedicated ring positions in combination with additional NEC pumps [Spiller *et al.* 2010]. Optimization with respect to beam losses at the injection and during the acceleration process is still an issue. The intensities for accelerated and extracted $U^{28+}$ ions have been increased by almost two orders of magnitude up to several $10^{10}$ ions per SIS18 cycle this way.

### 3.3 *The Experimental Storage Ring* (*ESR*)

The scheme of the ESR layout is shown in Fig. 7, and some basic parameters are listed in Table 1 [Franzke 1987]. The ring geometry is a hexagon defined by six 60° dipole magnets with a maximum bending power of 10 Tm. In one of the magnet-free straight sections of about 10 m length an electron cooler is installed. The other contains an internal gas jet target and other

Table 1. Basic parameters of the ESR.

| | | |
|---|---|---|
| Central orbit circumference | – | 108.36 m |
| Magnetic rigidity ($B \times \rho$) range | – | 0.5–10.0 Tm |
| Specific ion energy range | Proton | 30–2200 MeV |
| | $Ne^{10+}$ | 3–830 MeV/u |
| | $U^{92+}$ | 3–560 MeV/u |
| Bending magnet | field | 0.08–1.6 T |
| | ramping rate | $\leq 1$ T/s |
| Transverse acceptances | $A_H/A_V(\delta p/p = 0)$ | 400/170 mm mrad |
| Momentum acceptance | $\Delta p/p_{max}(\varepsilon_H, \varepsilon_V = 0)$ | ±2% |
| Dispersion amplitude $D_H$ | arc | 6.3 m |
| | straight section | $\leq 0.5$ m |

Fig. 8. Left: The picture shows the long straight section for in-ring experiments with one of the six 60° bending magnets in the foreground. Right: ESR electron cooler (yellow) between compensation solenoids (violet) counterbalancing the horizontal–vertical coupling by the cooler solenoid field.

experimental equipment (see photographs in Fig. 8). The focusing structure consists of four quadrupole triplets in the arcs and two quadrupole duplets in each long straight section.

The ESR orbit circumference of 108.36 m corresponds to one half of that of SIS18. This ratio seemed to be the best choice considering the bunch-to-bucket beam transfer from SIS18, electron cooling speed, and luminosity for in-ring experiments. Due to rather large magnet apertures the ion optical design provides comfortable transverse and momentum acceptances (see Table 1) allowing to adapt the beam optics to the different tasks of the ESR: accumulation of stable and radioactive ions up to 560 MeV/u for $U^{92+}$ and 830 MeV/u for $Ne^{10+}$, stochastic and electron cooling, and simultaneous storage of several charge states of the heaviest ions or even storage of more than 100 nuclear species with different mass-to-charge ratio $A/q$. Until 2005, fast ejection of fully stripped heaviest ions back to the SIS18 enabled further acceleration of fully stripped heaviest ions [Franzke *et al.* 1993].

Deceleration is possible as well as fast and slow extraction of cooled beams to external experiments. The ion optical design allows to keep the betatron tune constant for the different operation modes at $Q_H = 2.2$ and $Q_V = 2.4$. Two ferrite loaded RF-cavities with maximum voltage of 5 kV$_{pp}$ and tunable in the range from 0.85–5 MHz are used for beam de-bunching after injection, classical RF-stacking, and energy variation in the synchrotron mode. Typical ion energies at injection are about 400 MeV/u. Because of the lower limit of the tuning range the RF harmonic number $h$ has to be changed from $h = 2$ to $h = 4$ during deceleration to specific energies below 30 MeV/u. Deceleration of $U^{92+}$ beams down to 3 MeV/u is possible this way [Steck *et al.* 2004a], though electron cooling (see next section) of RF-bunches at the initial and an intermediate energy of 30 MeV/u is necessary. At 4 MeV/u, the cooled beams can be ejected to the IH/RFQ decelerator of the HITRAP facility [HITRAP 2003].

The lifetime of beams in the ESR is determined by the charge changing cross sections for collisions between the highly charged ions and residual gas atoms. The most critical process at lower energies is the electron capture from heavier components of the residual gas (mainly $N_2$, $O_2$, CO, $CO_2$, $H_2O$ and Ar). The required $N_2$ equivalent pressure is in the low $10^{-11}$ mbar range (UHV). For $U^{92+}$ beams at 4 MeV/u, lifetimes between 10 s and 20 s have been measured under these UHV conditions.

## 4 Ion beam cooling

### 4.1 *Electron cooler and cooling results*

Basic design parameters of the ESR electron cooler are given in Table 2; a photograph is shown on the right side of Fig. 8 [Angert *et al.* 1990]. The maximum electron energy used for stable cooling of ion beams is about 230 keV. The guiding field for the electron beam is operated in the range 0.01–0.1 T; electron currents range from 0.01 to 2.0 A. Electron cooling is applied also to decelerated beams. At 3 MeV/u the electron energy is as low as 1.53 keV, the electron current is typically 0.1 A and the solenoid field 0.015 T. Low electron currents are applied to avoid beam loss due to high radiative recombination (RR) rates in the cooler.

Primary beams of heavy ions from Li to U, mostly fully stripped or having only few electrons left, are stored and cooled in the ESR. As the electron cooling rate scales with $q^2/A$, the cooling times range from seconds down to ten milliseconds, depending on initial transverse emittances $\varepsilon_H, \varepsilon_V$

Table 2. Design parameters of the ESR electron cooler. In parentheses: operational values.

| | | |
|---|---|---|
| Electron | energy range | 16.5–310 (1.53–250) keV |
| | beam current | $\leq$10 ($\leq$2) A |
| | beam diameter | 50 mm |
| Specific ion | energy range | 30–560 (3–490) MeV/u |
| Cathode | diameter | 50 mm |
| | temperature | 1100 K |
| Solenoid | field | 0.01–0.25 T |
| Length | installation | 6.0 m |
| | effective | 2.5 m |

and momentum spread $\delta p/p$ of primary beams. The number of stored ions determines the beam parameters in the equilibrium between electron cooling ($\mathrm{EC} \propto q^2/A$) and intra-beam scattering ($\mathrm{IBS} \propto q^4/A^2$) rates.

Figure 9 shows the equilibrium values of horizontal emittance and momentum spread of electron-cooled 400 MeV/u $U^{92+}$, $Zn^{30+}$, and $C^{6+}$ beams as a function of the number of stored ions. Below thresholds between 500 to 1000 ions the equilibrium values are no longer defined by the balance between electron cooling (EC) and intra-beam scattering (IBS). In fact, IBS is suppressed and the ions are ordered in a longitudinal string, where ions cannot pass each other longitudinally. In this situation, the measured $\delta p/p$ below $10^{-6}$ is determined mainly by the stability of the power supplies of the ring magnets rather than by the very low longitudinal electron temperature in the order of 1 meV.

### 4.2 *Stochastic pre-cooling*

Beams of radioactive isotopes produced by projectile fragmentation or fission of heavy primary nuclei in the thick FRS-target emerge with large momentum spread in all phase planes, but with comparable average velocities. Many isotopes with about the same magnetic rigidity may be injected into the ESR. However, for these "hot" radioactive fragment beams with large emittance and momentum spread, the cooling time with electron cooling is increased to tens of seconds. Therefore, the stochastic cooling (SC) method invented [van der Meer 1972] and applied for the first time at CERN [Carron *et al.* 1977] has to be applied as the first cooling step.

The design of the SC-system at the ESR aimed at a cooling time of a few seconds for the reduction of $\delta p/p$ from $\pm 0.35\%$ to $\pm 0.1\%$ and of $\varepsilon_{\mathrm{H,V}}$ from $2\times 10^{-5}$ rad m to $2\times 10^{-6}$ rad m. The electrode systems are optimized for an ion velocity of $\beta = v/c \approx 0.75$. The RF-network allows the matching of the

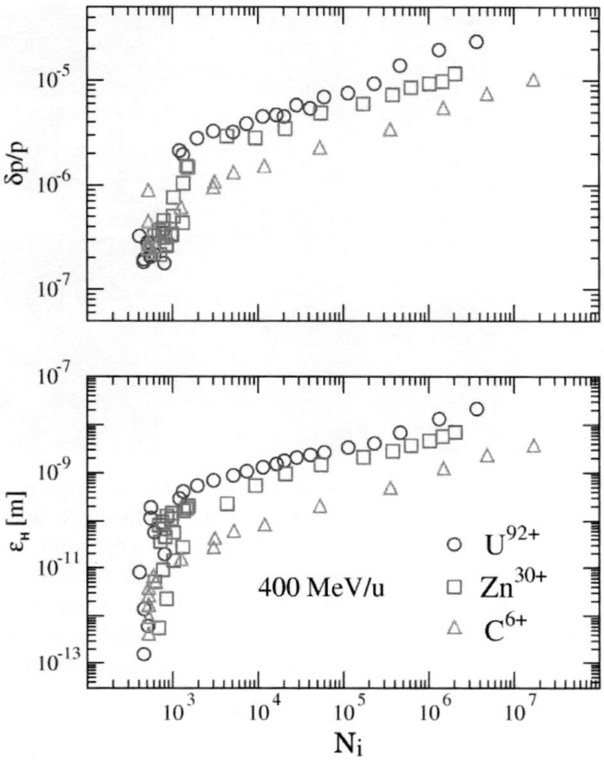


Fig. 9. Momentum spread $\delta p/p$, deduced from spectral analysis of signals from beam noise pick-up [Schaaf 1991] (top), and horizontal emittance $\varepsilon_H$, measured by means of an ionization beam profile monitor (bottom), of various beams at the equilibrium between electron cooling and intra-beam scattering. It is seen that the number of stored ions $N_i$ determines momentum spread and transverse emittances (here the horizontal emittance is plotted) of the circulating beam [Steck *et al.* 1996, Steck *et al.* 2004b].

signal delay between pick-ups and kickers for ions in the energy range 400–550 MeV/u. A total RF-power of 2 kW is available in the frequency range 0.9 to 1.7 GHz. The installation of the stochastic cooling system at the ESR started in 1996. Due to limitations of free installation space, the pick-up and kicker electrodes (see Fig. 10) are embedded in the vacuum chambers of main bending and quadrupole magnets — as shown in Fig. 7 — at locations with suitable dispersion and $\beta$-functions.

First successful tests were performed in 1997 [Nolden *et al.* 1997]. The system was improved step by step for routine operation within the subsequent years [Nolden *et al.* 2000]. The combination of SC with electron cooling of an artificially heated $U^{92+}$ beam is illustrated by the waterfall diagram of Fig. 11. Meanwhile, SC has been applied successfully also to mixtures

Fig. 10. View through the aperture of stochastic cooling pick-ups installed in the gap of the dipole magnet in the center of the northern arc of the ESR, where the dispersion amplitude has a maximum value of 6.3 m. The correction electrodes are installed in the center of the southern arc. The hot beam on the injection orbit (left side) is passing through the electrodes where it can be pre-cooled before it is decelerated by means of RF to the permanently electron-cooled stack (right side) for beam accumulation (RF-stacking). For mass measurements, electron cooling is applied at injection energy immediately after stochastic pre-cooling (see Fig. 11).

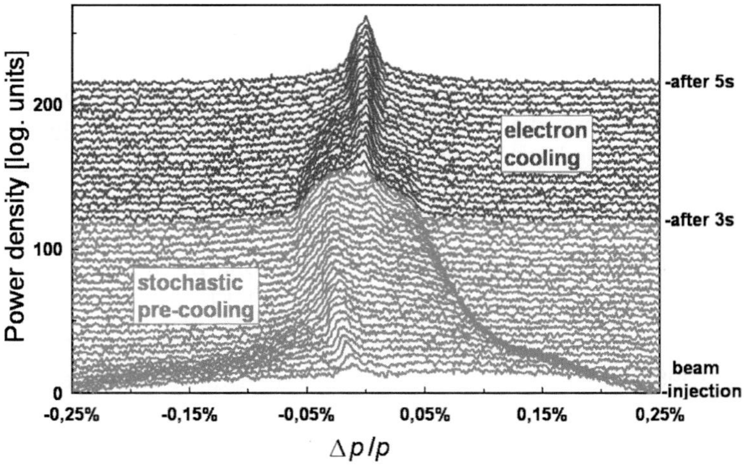


Fig. 11. Waterfall diagram of the Schottky power density from a $U^{92+}$ beam at 400 MeV/u vs. relative momentum deviation at different times after beam injection. The primary beam from SIS18 was heated by passing through a thick target in front of the FRS [Nolden 2003].

of nuclear fragments. The total cooling time is reduced from several 10 s to 5–10 s this way.

### 4.3 *Mass spectrometry at the ESR*

Electron-cooled beams of highly charged heavy ions enabled novel experimental conditions in atomic and nuclear physics. One example is the test of quantum electrodynamics (QED) in the strong electric field of hydrogen-like uranium or gold ions [Stöhlker *et al.* 2003]. Another highlight is $e^-$-decay of nuclei into bound states mentioned in the next section. However, the combination of FRS and ESR opened up unique possibilities for direct mass determination of unstable nuclear fragments as described briefly in the following.

The frequency spectrum of the beam noise (Schottky spectrum) from an electron-cooled, multi-component beam in the ESR consists of more or less well separated lines. The mean frequency differences between the lines are determined by the mass-to-charge ratio $A/q$. The relative difference $\Delta f/f$ in the revolution frequency $f$ between ions of different $A/q$-ratio is given by

$$\frac{\Delta f}{f} = -\frac{1}{\gamma_t^2}\frac{\Delta(A/q)}{A/q} + \left(\frac{1}{\gamma^2} - \frac{1}{\gamma_t^2}\right)\gamma^2\frac{\delta\beta}{\beta}, \tag{1}$$

where $\gamma_t$ is the value of $\gamma = (1-\beta^2)^{-1/2}$ at the transition point and $\delta\beta/\beta$ is the relative velocity spread in the ion beam. To achieve the highest resolution in the frequency spectrum the second term of Eq. (1) should be as small as possible. This can be attained in two ways: For the *Schottky Mass Spectrometry* (SMS) extremely small values of the velocity spread $\Delta\beta/\beta \to 0$ are achieved by means of electron cooling of less than 1000 stored ions as mentioned above. For the *Isochronous Mass Spectrometry* (IMS) the ESR is operated at $\gamma = \gamma_t$, hence $(\gamma^{-2} - \gamma_t^{-2}) = 0$ and the beam does not need to be cooled. This method is applied to unstable nuclei with lifetimes much shorter than the cooling time to equilibrium. It is obvious that in both cases a circulating beam of fragments with different $A/q$ generates narrow lines in the frequency or time-of-flight (TOF) spectrum.

Extensive investigations on the SMS method [Schlitt *et al.* 1997, Franzke *et al.* 1995] were followed by a series of successful experiments in the following ten years. An impressive demonstration of the potential of SMS is given by Fig. 12, where about 100 fragment species with different $A/q$ are recorded in a spectrum of 320 kHz band-width. To date, more than 350 previously unknown atomic masses and about the same number of poorly known ones have been determined by means of SMS at the ESR with mass resolution values $A/\Delta A$ higher than $10^6$.

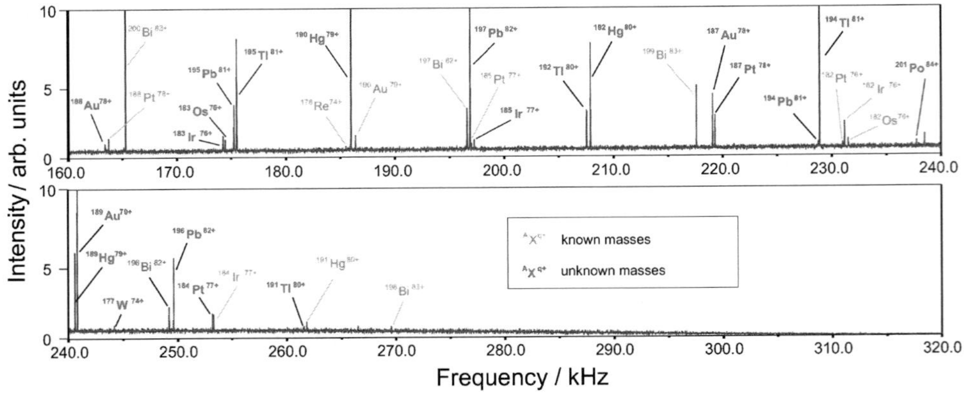


Fig. 12. Upper half of a Schottky frequency spectrum of electron-cooled nuclear fragments of primary uranium ions measured with 320 kHz band-width. The complete spectrum represents about 100 known (green) and previously unknown or poorly known (red) atomic masses [Litvinov 2003].

Systematic IMS measurements were started after extensive studies of the isochronous mode of the ESR at a strongly decreased transition point from $\gamma_t = 2.2$ to $\gamma_t \approx 1.4$ at nearly constant horizontal betatron tune $Q_H \approx 2.27$. The latter tune value is necessary for the adequate phase advance between injection septum and injection kicker magnet. Another topic of the machine investigations was the correction of the horizontal chromaticity $\delta Q_{\mathrm{H}}/(\delta p/p)$ in order to increase the fraction of the aperture, where the necessary condition $\gamma \approx \gamma_t$ is fulfilled [Hausmann *et al.* 2000]. The primary signals used for IMS come from secondary electrons generated by a single or very few circulating ions passing through the thin foil of the TOF detector (see Fig. 7). The ions may circulate over up to 1000 turns, because — due to the isochronous mode of the lattice — the energy loss in the foil should not influence the circulation period. Differences in the period are caused only by differences in the $A/q$-ratio (see Fig. 13).

An impressive illustration of the SMS and IMS results over about ten years is given in Fig. 14. The intensity and energy increase envisaged at the future FAIR facility will extend the experimental access mainly towards the neutron drip line (see overview by [Franzke, Geissel and Münzenberg 2008]).

## 5 Highlights in research and application

The GSI heavy ion accelerator facility has provided over many years unique beams, which have been the basis for a very successful scientific program in the fields of nuclear and atomic physics as well as for the application of energetic ions in materials and biophysics research. Figure 15 shows landmark

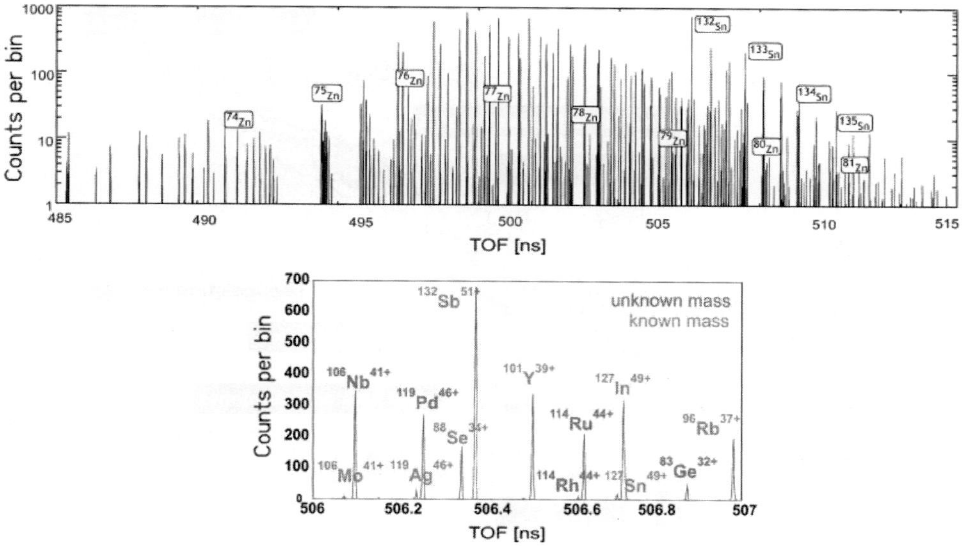

Fig. 13. Top: Revolution time spectrum for uranium fission fragments. The prominent peaks are labeled with the identification of nuclides. Bottom: The measured mass-over-charge range covers about 10% [Matos 2004].

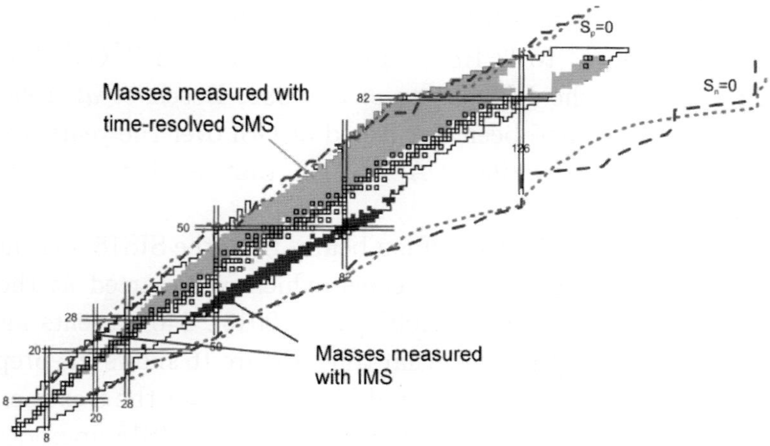


Fig. 14. The gray areas in the chart of nuclides indicate about 1100 isotopes whose masses were measured at the FRS–ESR facility. More than 350 of these are new [Litvinov *et al.* 2007]. The dark borderlines enclose the presently known ground state masses. The dashed (blue) and dotted (red) lines mark proton ($s_p = 0$) and neutron ($s_n = 0$) drip lines predicted by two different theories.

results in the field of nuclear physics from experiments at the Unilac, at the SIS18–FRS facility and at the ESR. Those include for example the discovery of in-flight produced super heavy nuclei from element number 107, Bohrium, to 112, Copernicium, at the Unilac [Hofmann 1998, 2009], pionic atoms

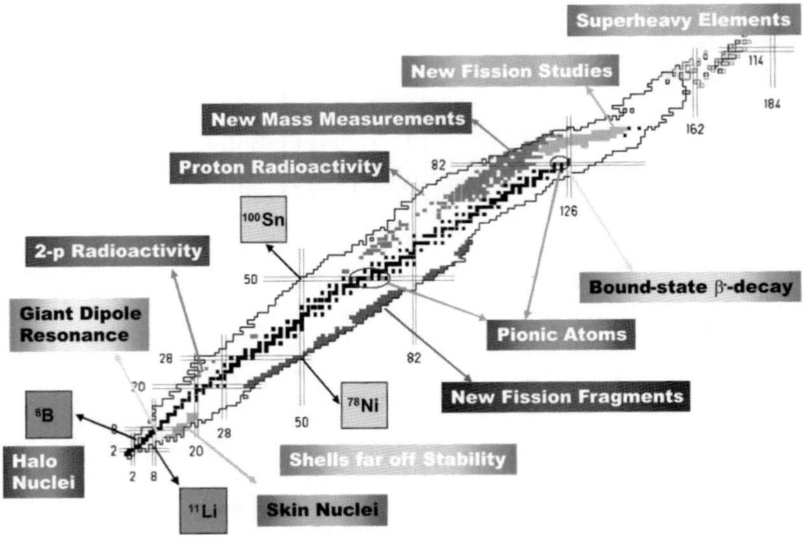


Fig. 15. Landmark results from experiments with in-flight produced super heavy elements (SHE), with exotic beams at the Synchrotron–FRagment-Separator facility (SIS–FRS), and at the Experimental Storage Ring (ESR) at GSI Darmstadt. [Geissel 2011].

produced with the SIS18–FRS [Geissel *et al.* 2002], and the bound state beta decay of $^{163}Dy^{66+}$ to $^{163}Ho^{66+}$ and of $^{187}Re^{75+}$ to $^{187}Os^{75+}$, observed for the first time at the ESR [Jung *et al.* 1992, Bosch *et al.* 1996]. More than 370 new isotopes had been discovered at GSI over the years up to 2011 [Thoenessen 2012]. Meanwhile, in April 2013, that number has increased to 432 [GSI 2013].

A very successful application of ion beams from the SIS18 was the cancer therapy with high energy carbon beams, which was started at the end of 1997 and performed at GSI until 2008 [Kraft 2013]. 440 patients have been successfully treated at GSI during that time. Figure 16 shows the preparation of a patient (left) by the radiologist and his assistant for the treatment in the irradiation cave at GSI (right). Due to the Unilac and SIS18 upgrades during the 1990s as described in Secs. 2.3 and 3.2, respectively, patient treatment could be performed within three therapy blocks each year of four weeks each, parallel to the continuing scientific experiments at GSI.

In 1998, in parallel to the start of the Darmstadt pilot project, a proposal for the Heidelberg Ion-beam Therapy (HIT) was submitted (by Radiologische Universitätsklinik Heidelberg, DKFZ Heidelberg and GSI) to the government and was approved in 2003 [Debus *et al.* 1998]. Major aspects of HIT were influenced by the experiences at GSI. Requirements, however, went beyond those of the pilot project at GSI. The facility has been constructed

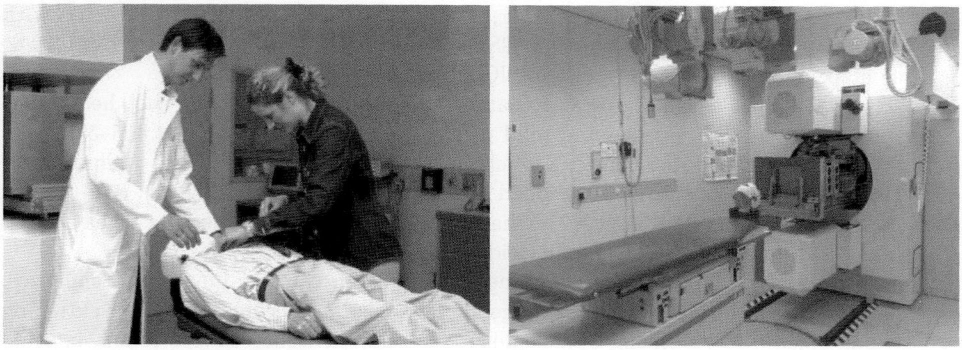

Fig. 16. Patient preparation (left) for treatment in the cancer irradiation cave (right) at GSI.

with strong support from GSI [Eickhoff *et al.* 2000, Eickhoff *et al.* 2004]. HIT started operation in 2009 with both proton and carbon beams.

The success of that rather challenging multiple ion beam operation and the co-existence of science activities and beam application with the requirements for therapy was also an important proof that the existing accelerator facility is capable to perform the time-shared multi-beam operation requested for the FAIR project.

## References


Angert, N. (1983). Status report on the upgraded Unilac, *IEEE Trans. Nucl. Sci.* **30** (4), 2980–2982.

Angert, N., Bossler, J., Dahl, L., Glatz, J., Gutowski, W., Hutter, G., Klabunde, J., Kube, G., Nicklas, A., Ratzinger, U., Schäfer, D., Schwedhelm, R., Spädtke, P., Vinzenz, W., Wolf, B., Deitinghoff, H., Friedrich, J., Klein, H., and Schempp, A. (1992). Commissioning and first operation experience of the new heavy ion injector of the UNILAC, in *Proc. EPAC92*, see http://accelconf.web.cern.ch/AccelConf/e92/PDF/EPAC1992_0167.PDF.

Angert, N., Bourgeois, W., Emig, H., Franzke, B., Langenbeck, B., Leible, K.D., Odenweller, T., Poth, H., Schulte, H., Spädtke, P., and Wolf, B.H. (1990). The 320 keV ESR-Electron Cooler, in *Proc. EPAC90*, see http://accelconf.web.cern.ch/AccelConf/e90/PDF/EPAC1990_1374.PDF.

Angert, N., Brouzet, E., Garoby, R., Hancock, S., Haseroth, H., Hill, C., Schindl, K., and Tetu, P. (1988). Accelerating and separating mixed beams of ions with similar charge-to-mass ratio in the CERN PS complex, in *Proc. EPAC88,* see http:// accelconf.web.cern.ch/AccelConf/e88/PDF/EPAC1988_1367.PDF.

Angert, N., Franzke, B., Möller, A., and Schmelzer, Ch. (1968). Der Elektroneneinfang von Jod-Ionen zwischen 19 und 55 MeV, *Phys. Lett. A* **27**, 28–29.

Baer, R., Blasche, K., Dolinskii, A., Dymnikov, A., Eickhoff, H., Franczak, B., Hofmann, I., Moritz, P. (1998). SIS operation at high beam intensities, in *Proc. EPAC98*, see http://accelconf.web.cern.ch/AccelConf/e98/PAPERS/WEP01A.PDF.

Barth, W., Dahl, L., Glatz, J., and Richter, S. (2002). Experiences during operation with high current $U^{4+}$ beams in the new GSI high current injector, in *Proc. LINAC2002*, see http://accelconf.web.cern.ch/AccelConf/l02/PAPERS/TU420.PDF.

Betz, H.-D. (1972). Charge states and charge-changing cross sections of fast heavy ions penetrating through gaseous and solid media, *Rev. Mod. Phys.* **44** (3), 465–539.

Betz, H.-D., Hortig, G., Leischner, E., Schmelzer, Ch., Stadler, B., and Weihrauch J. (1966). The average charge of stripped heavy ions, *Phys. Letters* **22** (5), 643–644.

Betz, H.-D., and Schmelzer Ch. (1967). Charge exchange cross sections of fast heavy ions, UNILAC-Report 1-67, Universität Heidelberg.

Blasche, K., Bock, R., Franzke, B., Greiner, W., Gutbrod, H.H., Povh, B., Schmelzer, Ch., and Stock, R. (1977). Überlegungen zur Physik der Kernmaterie unter extremen Bedingungen und zu einem Beschleuniger für relativistische schwere Ionen, internal GSI-Report GSI-P-2-77, 39 pages (in German).

Blasche, K., Böhne, D., Franczak, B., Franzke, B., and Riedel, C. (1981). SIS — a two-stage synchrotron facility for heavy ions, *IEEE Trans. Nucl. Sci.* **28** (3), 2013–2015.

Blasche, K., Böhne, D., Franzke, B., and Prange, H. (1985). The SIS heavy ion synchrotron project, *IEEE Trans. Nucl. Sci.* **32** (5), 2657–2661.

Blasche, K., and Franczak, B. (1992). The heavy ion synchrotron SIS, in *Proc. EPAC92*, see http://accelconf.web.cern.ch/AccelConf/e92/PDF/EPAC1992_0009.PDF.

Blasche, K., and Franzke, B. (1994). Status report on SIS-ESR, in *Proc. EPAC94*, see http://accelconf.web.cern.ch/AccelConf/e94/PDF/EPAC1994_0133.PDF.

Boehne, D. (1969). Linac cavity technology, *IEEE Trans. Nucl. Sci.* **16** (3), 390–394.

Boehne, D. (1972). Status report on the Unilac-project, in *Proc. 1972 Prot. Lin. Acc. Conf.*, Los Alamos, *Nucl. Instr. Meth.* **103**, 25–32.

Boehne, D. (1977). Status report on the Unilac, *IEEE Trans. Nucl. Sci.* **24** (3), 1070–1075.

Boehne, D., Blasche, K., Franczak, B., Franzke, B., and Steiner, R. (1990). Performance of the SIS and developments at GSI, in *Proc. EPAC90*, see http://accelconf.web.cern.ch/AccelConf/e90/PDF/EPAC1990_0018.PDF.

Boehne, D., and Schmelzer, Ch. (1969). Existing and proposed heavy ion accelerators, in *Linear Accelerators*, eds. Lapostolle, P.M., and Septier, A.L. (North-Holland Publ. Co., Amsterdam, 1970), pp. 1047–1070.

Bosch, F., Faestermann, T., Friese, J., Hein, F., Keene, P., Wafers, E., Zeitelhack, K., Beckert, K., Franzke, B., Klepper, O., Kozhuharov, C., Menzel, G., Moshammer, R., Nolden, F., Reich, H., Schlitt, B., Steck, M., Stöhlker, T., Winkler, T., and Takahashi, K. (1996). Observation of bound-state $\beta^-$-decay of fully ionized $^{187}$Re: $^{187}$Re–$^{187}$Os cosmochronometry, *Phys. Rev. Lett.* **77**, 5190.

Brown, I.G. (1994). Vacuum arc ion sources, *Rev. Sci. Instrum.* **65** (10), 3061–3081.

Budker, G.I., Dikansky, N.S., Kudelainen, V.I., Meshkov, I.N., Parkhomchuk, V.V., Pestrikov, D.V., Skrinsky, A.N., and Sukhina, B.N. (1976). Experimental studies of electron cooling, *Part. Acc.* **7**, 197–211.

Carron, G., Faltin, L., Schnell, W., and Thorndahl, L. (1977). Experiments with stochastic cooling in the ISR, *IEEE Transact. Nucl. Sci.* **24** (3), 1402–1404. see http://accelconf.web.cern.ch/AccelConf/p77/PDF/PAC1977_1402.PDF.

Dahl, L. (2009). The GSI Unilac upgrade program to meet FAIR requirements, in *Proc. HIAT09,* pp. 193–197, see http://accelconf.web.cern.ch/AccelConf/HIAT2009/papers/fr-01.pdf.

Dahl, L.A., Barth, W.A., Bellachioma, M.C., Groening, L., Kester, O.K., Kirk, M.M., Ondreka, D., Pyka, N., Spiller, P.J., Stadlmann, J., Vormann, H., Yaramyshev, S.G., Bozyk, L.H.J., El-Hayek, Y., and Xiao, C. (2012). Development of the intensity and quality of heavy ion beams at GSI, in *Proc. HIAT12,* see http://accelconf.web.cern.ch/AccelConf/HIAT2012/papers/thb04.pdf.

Debus, J., Wannenmacher, M., Ertner, D., Fuss, M., Heeg, P., Wenz, F., zur Hausen, H., Bendel, R., Bortfeld, T., Hartmann, G., Jäkel, O., Karger, C.P., Kriessbach, A., Lappe, C., Massimino, M., Ölfke, U., Schlegel, W., Specht, H.J., Angert, N., Badura, E., Becher, W., Böhne, D., Brand, H., Brusasco, C., Dolinskii, A., Eickhoff, H., Essel, H.G., Franzcak, B., Geiss, O., Haberer, Th., Hoffmann, J., Krämer, M., Kraft, G., Kraft-Weyrather, W., Krause, U., Kurz, N., Langenbeck, B., Ott, W., Pavlovic, M., Poppensieker, K., Richter, M., Schardt, D., Schempp, A., Schlitt, B., Scholz, M., Spiller, P., Steiner, R., Stelzer, H., Voss, B., Weber, U., Pobell, F., Enghardt, W., Hasch, B., Hinz, R., Lauckner, K., Pawelke, J., and Sobiella, M. (1998). Proposal for a dedicated ion beam facility for cancer therapy, eds. Gross, K.D. and Pavlovic, M., pp. 1–98, submitted in September 1998, see also: Proposal for the Construction of a Clinical Therapy Facility for Cancer Treatment with Ion Beams, ed. Gross, K.D. pp. 1–13 gsi.de/documents/DOC-2004-Feb-120-1.pdf.

Eickhoff, H., Boehne, D., Haberer, Th., Schlitt, B., Spiller, P., Debus, J., and Dolinskii, A. (2000). The proposed dedicated ion beam facility for cancer therapy at the clinic in Heidelberg, in *Proc. EPAC00*, see http://accelconf.web.cern.ch/AccelConf/e00/PAPERS/WEP3A13.pdf.

Eickhoff, H., Franczak, B., Krause, U., Riedel, C., and Steiner, R. (1996). Accelerator aspects of the cancer therapy project at the GSI Darmstadt, in *Proc. EPAC96*, see http://accelconf.web.cern.ch/AccelConf/e96/PAPERS/THPG/THP052G.PDF.

Eickhoff, H., Haberer, T., Schlitt, B., and Weinrich, U. (2004). HICAT — The German Hospital Based Light Ion Cancer Therapy Project, in *Proc. EPAC04*, see http://accelconf.web.cern.ch/AccelConf/e04/PAPERS/FRYACH01.PDF.

FAIR (2001), FAIR CDR — An international accelerator facility for beams of ions and antiprotons, Conceptual Design Report; FAIR (2006), FAIR Baseline Technical Report, see http://www.fair-center.eu/for-users/publications/fair-publications.html.

Franzke, B. (1987). The heavy ion storage and cooler ring project at GSI, *Nucl. Instr. Meth. B* **24/25**, 18.

Franzke, B., Angert, N., Möller, A., and Schmelzer, Ch. (1967). Umladung von Jod-Ionen der Energie 24 bis 74 MeV in einem Überschall-Quecksilberdampfstrahl, *Phys. Lett. A* **25**, 769–770.

Franzke, B., Beckert, K., Bosch, F., Eickhoff, H., Franczak, B., Gruber, A., Klepper, O., Nolden, F., Raabe, P., Reich, H., Spädtke, P., and Steck, M. (1993). Heavy ion beam accumulation, cooling, and experiments at the ESR in, *Proc. 1993 Part. Acc. Conf.*, see http://accelconf.web.cern.ch/AccelConf/p93/PDF/PAC1993_1645.PDF.

Franzke, B., Beckert, K., Eickhoff, H., Franczak, B., Nolden, F., Poth, H., Schaaf, U., Schulte, H., and Spädtke, P. (1990). Commissioning of the heavy ion storage ring ESR, in *Proc. EPAC90,* see http://accelconf.web.cern.ch/AccelConf/e90/PDF/EPAC1990_0046.PDF.

Franzke, B., Beckert, K., Eickhoff, H., Nolden, F., Reich, H., Schaaf, U., Schlitt, B., Schwinn, A., Steck, M., and Winkler, T. (1995). Schottky mass spectrometry at the experimental storage ring ESR, *Physica Scripta* **T59**, 176.

Franzke, B., Geissel, H., and Münzenberg, G. (2008). Mass and lifetime measurements of exotic nuclei in storage rings, *Mass Spectrometry Rev.* **27**, 428.

Geissel, H. (2011). Private communication.

Geissel, H., Armbruster, P., Behr, KH., Brünle, A., Burkard, K., Chen, M., Folger, H., Franczak, B., Keller, H., Klepper O., Langenbeck B., Nickel, F., Pfeng, E., Pfützner, M., Roeckl, E., Rykaczewski, K., Schall, I., Schardt, D., Scheidenberger, C., Schmidt,

K.H., Schröter, A., Schwab, T., Sümmerer, K., Weber, M., Münzenberg, G., Brohm, T., Clerc, H.-G., Fauerbach, M., Gaimard, J.-J., Grewe, A., Hanelt, E., Knödler, B., Steiner, M., Voss, B., Weckenmann, J., Ziegler, C., Magel, A., Wollnik, H., Dufour, J.P., Fujita, Y., Vieira, D.J., and Sherrill, B. (1992). The GSI projectile fragment separator (FRS): a versatile magnetic system for relativistic heavy ions, *Nucl. Instr. Meth. B* **70**, 286.

Geissel, H., Gilg, H., Gillitzer, A., Hayano, R.S., Hirenzaki, S., Itahashi, K., Iwasaki, M., Kienle, P., Münch, M., Münzenberg, G., Schott, W., Suzuki, K., Tomono, D., Weick, H., Yamazaki, T., and Yoneyama, T. (2002). Deeply bound 1s and 2p pionic states in $^{205}$Pb and determination of the s-wave part of the pion-nucleus interaction, *Phys. Rev. Lett.* **88**, 122301.

Glatz, J. (1986). The Unilac as a fast-switching variable ion and energy accelerator, in *Proc. 1986 Lin. Acc. Conf*, SLAC-Report 303, pp. 302–305.

GSI (2013), GSI-Magazine Target No. 9, April 2013 (in German only), see https://www.gsi.de.

Hausmann, M., Attallah, F., Beckert, K., Bosch, F., Dolinskiy, A., Eickhoff, H., Falch, M., Franczak, B., Franzke, B., Geissel, H., Kerscher, Th., Klepper, O., Kluge, H.-J., Kozhuharov, C., Löbner, K.E.G., Münzenberg, G., Nolden, F., Novikov, Yu.N., Radon, T., Schatz, H., Scheidenberger, C., Stadlmann, J., Steck, M., Winkler, T., and Wollnik, H. (2000). First isochronous mass spectrometry at the experimental storage ring ESR, *Nucl. Instr. Meth. A* **446**, 569.

HITRAP Technical Design Report (2003). GSI-Report 2003-69-2, eds. Beier, Th., Dahl, L., Kluge, H.-J., Kozhuharov, C., and Quint, W.

Hofmann, S. (1998). New elements — approaching Z = 114, *Rep. Prog. Phys.* **61**, 639–689.

Hofmann, S. (2009). Superheavy elements, *The Euroschool Lectures on Physics with Exotic Beams*, Vol. III, Lecture Notes in Physics Vol. 764 (Springer Berlin-Heidelberg, 2009), p. 203.

ICE: Carron, G., Herr, H., Lebee, G., Koziol, H., Krienen, F., Möhl, D., Petrucci, G., Rubbia, C., Sacherer, F., Sadoulet, B., Stefanini, G., Thorndahl, L., van der Meer, S., and Wikberg, T. (1979). Experiments on stochastic cooling in ICE (Initial Cooling Experiment), *IEEE Trans. Nucl. Sci.* **26** (3), 3456–3461, see http://accelconf.web.cern.ch/AccelConf/p79/PDF/PAC1979_3456.PDF.

Illgen, J., Kirchner, R., Schulte in den Bäumen, J. (1972). Duoplasmatron ion sources, in *Proc. ICOMCHISAAS 1971, IEEE Trans. Nucl. Sci.* **19** (2), 35–47.

Jung, M., Bosch, F., Beckert, K., Eickhoff, H., Folger, H., Franzke, B., Gruber, A., Kienle, P., Klepper, O., Koenig, W., Kozhuharov, C., Mann, R., Moshammer, R., Nolden, F., Schaaf, U., Soff, G., Spädtke, P., Steck, M., Stöhlker, Th., and Sümmerer, K. (1992). First observation of bound-state $\beta^-$-decay, *Phys. Rev. Lett.* 692164.

Kienle, P. (1985). The SIS/ESR-Project of GSI, GSI Report 85–16.

Kraft, G. (2013). History of the Heavy Ion Therapy at GSI, pp. 1–17 see http://three.usra.edu/articles/Krafts_GSI.pdf.

Wannenmacher, M., Specht, H.J., zur Hausen, H., Gademann, G., Kraft, G., Böhne, D., Lorenz, W.J., and Hartmann, G. (1993). Einrichtung einer experimentellen Strahlentherapie bei der Gesellschaft für Schwerionenforschung Darmstadt, GSI Report 93-23, eds. Kraft, G. and Gademann, G., ISSN 01714546.

Krupp, H. (1970). Production of multiply charged ions in gas ion sources, *Nucl. Instr. Meth.* **90**, 167–172.

Lefèvre, P., Möhl, D., and Plass, G. (1980). The CERN low energy antiproton ring (LEAR) project, in *Proc. of 11th Internatl. Conf. on High-energy Accel.*, Geneva, Switzerland, 1980, *Experientia Suppl.* **40**, 819–823.

Litvinov, Yu.A. (2003). Basic nuclear properties of neutron-deficient nuclei investigated via high precision mass measurements in the element range of $36 \leq Z \leq 92$, PhD thesis, Justus-Liebig-Universität-Giessen, GSI-Report Diss. 2004-05.

Litvinov, Yu.A., Bosch, F., Geissel, H., Kurcewicz, J., Patyk, Z.,Winckler, N., Batist, L., Beckert, K., Boutin, D., Brandau, C., Chen, L., Dimopoulou, Ch., Fabian, B., Faestermann, T., Fragner, A., Grigorenko, L., Haettner, E., Hess, S., Kienle, P., Knöbel, R., Kozhuharov, C., Litvinov, S.A., Maier, L., Mazzocco, M., Montes, F., Münzenberg, G., Musumarra, A., Nociforo, C., Nolden, F., Pfützner, M., Plass, W.R., Prochazka, A., Reda, R., Reuschl, R., Scheidenberger, C., Steck, M., Stöhlker, Th., Torilov, S., Trassinelli, M., Sun, B., Weick, H., and Winkler, M. (2007). Measurement of the $\beta^+$ and orbital electron-capture decay rates in fully-ionized, hydrogen-like, and helium-like $^{140}$Pr ions, *Phys. Rev. Lett.* **99** (26), 262501.

Mahner, E. (2008). Review of heavy-ion induced desorption studies for particle accelerators, *Phys. Rev. Spec. Top. Acc. Beams* **11**, 104810.

Matos, M. (2004). Isochronous mass measurements of short-lived neutron-rich nuclides at the FRS-ESR facilities, PhD thesis, Justus-Liebig-Universität Giessen.

Möller, A., Angert, N., Franzke, B., and Schmelzer, Ch. (1968). Der Elektronenverlust von Jod-Ionen zwischen 6 und 68 MeV, *Phys. Lett. A* **27**, 621–622.

Nikolaev, V.S. (1965). Electron capture and loss by fast ions in collisions, *Sov. Phys. Uspekhi* **8**, 269.

Nolden, F. (2003), Experience and prospects of stochastic cooling of RI beams at GSI. Invited talk at *COOL'03 Workshop,* Tokyo.

Nolden, F., Beckert, K., Caspers, F., Franczak, B., Franzke, B., Menges, R., Schwinn, A., and Steck, M. (2000). Stochastic cooling at the ESR, *Nucl. Instr. Meth. A* **441**, 219.

Nolden, F., Böhne, D., Bourgeois, W., Franzke, B., Steck, M., and Schwinn, A. (1997). ESR stochastic precooling, *Nucl. Phys. A* **626**, 491.

Pasyuk, A.S., Tretiakov, Y.P., and Gorbachev, S.K. (1967). The production of multiply charged argon-, krypton-, xenon-, and tungsten-ions in an experimental arc source, Dubna-Report 3370.

Ratzinger, U. (1996). The new GSI pre-stripper linac for high current heavy ion beams, in *Proc. LINAC1996,* see http://accelconf.web.cern.ch/AccelConf/l96/PAPERS/TU 202.PDF.

Schaaf, U. (1991). Schottky-Diagnose und BTF-Messungen an gekühlten Strahlen im Schwerionen-Speicherring ESR, PhD thesis, Universität Frankfurt, GSI Report GSI-91-22, ISSN 0171–4546 (in German).

Schlitt, B., Beckert, K., Bosch, F., Eickhoff, H., Franzke, B., Fujita, Y., Geissel, H., Hausmann, M., Irnich, H., Klepper, O., Kluge, H.-J., Kozhuharov, C., Kraus, G., Münzenberg, G., Nickel, F., Nolden, F., Patyk, Z., Radon, T., Reich, H., Scheidenberger, C., Schwab, W., Steck, M., Sümmerer, K., Winkler, Th., Beha, T., Ferch, M., Kerscher, Th., Löbner, K.E.G., Jung, H.C., Wollnik, H., and Novikov, Y. (1997). Schottky mass spectrometry at the ESR: a novel tool for precise direct mass measurements of exotic nuclei, *Nucl. Phys. A* **626**, 315c.

Schmelzer, Ch. (1968). Study of a variable energy heavy-ion linear accelerator, in *Proc. 1968 Prot. Lin. Acc. Conf.*, BNL Report 50170, pp. 735–738.

Schmelzer, Ch. (1969). Special problems in heavy ion acceleration, in *Linear Accelerators*, eds. Lapostolle, P. M., and Septier, A. L., (North-Holland Publ. Co., Amsterdam, 1970), pp. 1029–1044.

Spädtke, P., Emig, H., Leible, K.D., Mühle, C., and Wolf, B.H. (1998). Ion sources for the new high current injector at GSI, in *Proc. LINAC98,* see http://accelconf.web.cern.ch/AccelConf/l98/PAPERS/TH4052.PDF.

Spiller, P.J., Blasche, K., Hülsmann, P., Krämer, A., Ramakers, H., and Reich-Sprenger, H. (2004). High intensity uranium operation in SIS18, in *Proc. EPAC04*, see http://accelconf.web.cern.ch/AccelConf/e04/PAPERS/TUPLT020.PDF.

Spiller, P.J., Bozyk, L., Eickhoff, H., Kollmus, H., Puppel, P., and Reich-Sprenger, H., (2010). Acceleration of intermediate charge state heavy ions in SIS18, in *Proc. IPAC2010,* see http://accelconf.web.cern.ch/AccelConf/IPAC10/papers/mopd002.pdf.

Steck, M., Beckert, K., Beller, P., Franczak, B., Franzke, B., and Nolden, F. (2004a). Improved performance of the heavy ion storage ring ESR, in *Proc. EPAC2004*, see http://accelconf.web.cern.ch/AccelConf/e04/PAPERS/TUPLT016.PDF.

Steck, M., Beckert, K., Beller, P., Franzke, B., and Nolden, F., (2004b). Observation of ultra-cold heavy ion beams with micrometer size by scraping, in *Proc. EPAC 2004*, see http://accelconf.web.cern.ch/AccelConf/e04/PAPERS/WEPLT055.PDF.

Steck, M., Beckert, K., Eickhoff, H., Franzke, B., Nolden, F., Reich, H., Schlitt, B., and Winkler, T. (1996). Anomalous temperature reduction of electron-cooled heavy ion beams in the storage ring ESR, *Phys. Rev. Lett.* **77**, 3803.

Steck, M., Groening, L., Blasche, K., Eickhoff, H., Franczak, B., Franzke, B., Winkler, T., and Parkhomchuk, V.V. (1999). Electron cooling assisted beam accumulation in the heavy ion synchrotron SIS by repeated multi-turn injection, in *Proc. 1999 PAC,* see http://accelconf.web.cern.ch/AccelConf/p99/PAPERS/TUP97.PDF.

Angert, N., Bailkowska, H., Bock, R., Gutbrod, H.H., Harris, H., Maier, M.R., Poskanzer, A.M., Pugh, H.G., Pühlhofer, F., Renford, R.E., Ritter, H.G., Sandoval, A., Schröder, L.S., Skrzypczak, E., Stock, R., Ströbele, H., Szwed, R., Warwick, A., Weik, F., Wiemann, H., and Wolf, K.L. (1982). Study of relativistic nucleus–nucleus reactions induced by $^{16}$O-beams of 9–13 GeV/u at the CERN PS, in *Proc. the Bielefeld Workshop on Quark Matter Formation and Heavy Ion Collisions*, May 1982, Bielefeld, eds. M. Jacob, M. and Satz, H., pp. 557–582.

Stöhlker, T., Beier, T., Beyer, H.F., Kühl, T., and Quint, W. (2003). Test of strong-field QED, in *The Physics of Multiply and Highly Charged Ions*, Vol. 1, Part III, (Springer, Netherlands), pp. 351–386.

Thoenessen, M., (2012). The discovery of the nuclides, *Nucl. Phys. News* **22** (3), 19–22.

van der Meer, S. (1972). Stochastic damping of betatron oscillations in the ISR, CERN Report CERN/ISR-PO/72-31.

Wolf, B. (1972). Characteristics of a cold cathode type penning source and low energy isotope separation of high charge states, in *Proc. ICOMCHISAAS 1971, IEEE Trans. Nucl. Sci.* **19** (2), 74–79.

# Chapter 18

# ISOLDE and REX: A rare isotope facility

Richard Catherall (CERN)

## 1 Introduction

The chart of nuclides maps all known isotopes of the elements that make up the periodic table. An element is characterized by the number of protons in its nucleus; however, the stability of an isotope against radioactive decay is determined by the number of associated neutrons. The ISOLDE facility is dedicated to the production of radioactive nuclei in the form of low-energy ion beams, providing an abundance of isotopes for a variety of research experiments. The first isotope separation on-line (ISOL) experiment at CERN took place at the 600 MeV Synchro-Cyclotron (SC) in 1967 [1–3]. After a complete reconstruction in 1973, the ISOLDE-2 facility continued to operate until the shut-down of the SC in 1990. With the completion of a new facility in 1992, ISOLDE continues to develop and provide radioactive ion beams to an expanding scientific community and has become the reference in a world-wide network of facilities adopting the ISOL technique.

Situated on the border between France and Switzerland on CERN's Meyrin site, the ISOLDE facility is part of CERN's Proton Synchrotron (PS) accelerator complex where it receives protons directly from the PS Booster. The success of ISOLDE is mainly due to the large variety of available radioactive ion beams. With over 1000 isotopes from more than 70 elements already available, ISOLDE continues to develop new radioactive ion beams with emphasis on purity and quality. The Radioactive EXperiment at ISOLDE (REX-ISOLDE) can be considered as a second facility dedicated to the post-acceleration of radioactive ion beams delivered by ISOLDE and is considered as one of the recent major improvements of the facility. The layout of ISOLDE is shown in Fig. 1.

There are over 500 world-wide users of the ISOLDE facility. They are represented by the ISOLDE Collaboration, with representatives from 15 member states, who meet regularly to define the user's requirements and to

Fig. 1. The ISOLDE machine showing the proton beam lines in the foreground, the two mass separators GPS (left) and HRS (right) and the secondary beam line distribution. On the platform is the REXEBIS preceding the REX Linac.

shape the future physics program [4–6]. The main line of research includes: nuclear structure physics, nuclear astrophysics, atomic physics, solid state physics, life sciences and fundamental reactions. Production rates between 1 and $10^{10}$ ions $s^{-1}$, measured half-lives as little as 4.35 ms ($^{14}Be$) and versatile ion manipulation at an energy as low as $10^{-6}$ eV have all contributed to an exciting program in modern nuclear structure physics. And with REX-ISOLDE, post-acceleration up to 3 MeV/u of isotopes as light as $^{8}Li$ and as heavy as $^{224}Ra$ has been used for Coulomb excitation, few-nucleon transfer reaction or fusion evaporation studies primarily using the Miniball detector and segmented silicon detector arrays.

## 2 Operation

At ISOLDE, radioactive nuclei are produced by the fragmentation, spallation or fission process induced by the interaction of high-energy proton beam from the PS Booster with thick target materials. The radioactive nuclei are then transferred to a dedicated ion source for ionization to the $1^{+}$ state before being accelerated up to 60 keV to produce radioactive ion beams composed of multiple elements. The radioactive ion beam is then mass separated using an electro-magnet before being distributed to the experiments situated at the end of numerous beam lines distributed throughout the ISOLDE experimental hall.

The ISOLDE facility has two mass separators: the General Purpose Separator (GPS) and the High Resolution Separator (HRS). The availability of two separators enables the efficient use of both the proton beam and the time for experimental physics. While one separator is being used for the production of radioactive ion beams, the other can be prepared for the next physics experiment.

### 2.1 *The PS Booster*

The proton driver beam is delivered by the PS Booster, a stack of four superimposed synchrotrons accelerating protons, delivered by a Linac at 50 MeV/u, to 1.4 GeV before further injection into CERN's accelerator complex. The PS Booster delivers 1 pulse of up to $3.5 \times 10^{13}$ protons every 1.2 seconds. Up to 50% of these pulses are typically delivered to ISOLDE, providing a DC-equivalent beam intensity of $2\,\mu$A. Although the proton beam energy is 1.4 GeV, this can be lowered to 1 GeV upon request. The high-intensity, low-repetition rate of the proton beam essentially results in the delivery of pulsed radioactive ion beams to the ISOLDE experiments. There are two advantages of receiving the proton beam with a pulsed structure; first, the high intensity of the pulsed beam enhances the release time of radioactive nuclei from the target material and secondly, the background due to secondary neutrons is considerably reduced as they are no longer an issue for neighboring experiments ~100 ms after proton beam impact. The proton beam can be steered and focused and the pulse length modified in order to optimize the radioactive nuclei production rate of a specific target.

## 3 Targets, ion sources and selectivity

### 3.1 *The target unit*

At the heart of ISOLDE operations is the target unit: a compact vacuum vessel containing the target material, the transfer line and the ion source. At approximately 25 kg, the water-cooled aluminium chamber also has connections for other auxiliary devices making it a versatile ensemble for both isotope production and target and ion source development. Up to 30 target units per year are produced and operated at ISOLDE in order to provide a large variety of radioactive ion beams for a demanding physics program. The combination of target and ion source and other options depends on the isotope of interest and its required purity. With a choice of approximately

25 target materials and up to 10 different ionization techniques, each target unit is tailor-made to the experiment's requirements.

### 3.2 *Target materials*

The radioactive nuclei production rate partially depends on the target material thickness and its properties. Over the years, different target materials have been investigated and characterized at ISOLDE providing an optimal choice of target material for a specific element [7]. The materials are diverse in thickness and form, ranging from light calcium oxide at 4 $\text{gcm}^{-2}$ to molten lead at 227 $\text{gcm}^{-2}$. Probably the most versatile and most requested target material is uranium carbide produced at ISOLDE from uranium oxide and graphite. The target material is placed inside a tantalum target container, 20 mm in diameter and 200 mm long (see Fig. 2), which is heated up to 2000°C to enhance the diffusion of the radioactive nuclei from the target material. One of the main challenges at ISOLDE is to improve the release time of radioactive nuclei from the target and ion source system to produce and measure isotopes with very short half-lives. In terms of target material, one approach has been to investigate the use of nano-materials. Successful tests have been carried out on silicon carbide, yttrium oxide, and calcium

Fig. 2. The interior of a target unit showing the suspended tantalum oven containing the target material.

oxide, and further tests, notably on sub-micron size grains of uranium carbide, are on-going [8–10].

### 3.3 *Ion sources*

The high production rate of radioactive nuclei also depends on high ionization efficiency. One of the more recent developments in this field is the Versatile Arc Discharge Ion Source (VADIS) [11]. The efficiency of the VADIS for the ionization of noble gases has been recorded between a factor of 5 and 20 times higher than the previously used Forced Electron Beam Induced Arc Discharge (FEBIAD) ion sources. For radon, the ionization efficiency has been recorded at 60% and has contributed to $^{229}$Rn being measured for the first time in 2009 [12]. The VADIS has become the standard plasma ion source used at ISOLDE and continues to provide excellent results in the ionization of most chemical elements.

Elements with a low ionization potential can be positively ionized to the $1^+$ state by interaction with a hot surface of a metal with a high work function. At ISOLDE, the surface ionization technique often takes the form of a 30 mm long tungsten tube with an inner diameter of 3 mm, though other materials such as tantalum, rhenium and lanthanum hexaboride are also available for specific elements. Elements with a low ionization potential can be ionized by surface ionization to a high efficiency; however, the incidental production of isobars can be a problem when striving for the production of pure radioactive ion beams of a selected isotope. One of the greatest challenges at ISOLDE is the production of pure radioactive ion beams and selective target and ion source systems play a key role in all future developments.

### 3.4 *Selectivity*

Although mass separation is done in the magnet spectrometer, beam purification starts in the target and ion source system. This begins with the choice and preparation of the target material to its purest form, and includes the cleanliness of all target components and a vacuum pressure down to $10^{-7}$ mbar. The use of a cold transfer line between the target container and ion source where, with the exception of noble gases and molecules, the radioactive nuclei are condensed is an example of beam purification. Further developments of this technique include a temperature controlled transfer line and the inclusion of a quartz tube; the latter being known for its ability to absorb alkalis at a given temperature [13]. However, one of the more versatile purification techniques comes from ionization by lasers tuned for the ionization of specific elements.

### 3.5 *The resonance ionization laser ion source*

The thermal nature of the more standard hot surface and plasma discharge ion sources in many cases render these ionization processes non-selective. The resonance ionization laser ion source (RILIS) is based on laser step-wise resonance ionization of atoms in a hot cavity [14, 15]. Within a specific ionization scheme, each laser wavelength is precisely tuned so that the photon energies match the successive transition energies of a given element. In practical terms, the lasers are guided through the windows of each separator magnet to the atomic vapor generated in the cavity of a hot surface ion source. Selectivity is obtained by reducing the cavity temperature to minimize the overall ionization of most elements while tuning the laser wavelength to optimize the ionization efficiency of a specific element. The combination of mass separation with this high degree of selectivity has resulted in RILIS being used for the production of pure radioactive ion beams for above 50% of the yearly ISOLDE schedule.

The RILIS laser set-up is capable of generating a tunable laser radiation in the range of 210–950 nm. It consists of two Nd:YAG lasers, three dye lasers, three solid-state titanium:sapphire (Ti:Sa) lasers and devices for generation of higher harmonics [16]. The dye and Ti:Sa laser systems can be combined or operated individually to provide more flexibility when changing from one ionization scheme to another and to establish most efficient ionization schemes that require the use of dye and/or Ti:Sa lasers.

### 3.6 *Laser ion source and trap*

Another recent development to provide a pure isotopic beam while optimizing the time structure and the transversal emittance of the produced ion beam is the Laser Ion Source and Trap (LIST) [17]. In this design, the performance of the conventional laser ion source is improved through the installation of a repeller electrode and a linear Paul trap/ion guide structure in the target unit. Through the rejection of surface ionized contaminants coming from the hot cavity of the ion source and resonant laser ionization inside the trap, suppression factors up to three orders of magnitude can be achieved for certain isobaric contaminants, while keeping the losses in laser ionization efficiency below a factor of 30 compared to normal RILIS operation. Recent examples include the resonant laser ionization of Mg and Po while suppressing Na, Al and Fr isobars.

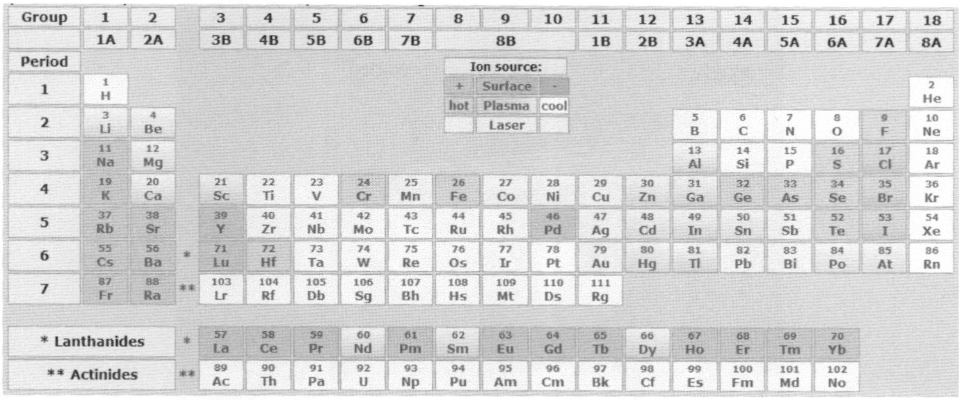


Fig. 3. Elements that have been produced as ion beams at ISOLDE.

### 3.7 *Neutron-induced fission*

Neutron-induced fission in actinide target materials can also be considered as enhancing selectivity within a target and ion source system. It is a two-step process where the primary proton beam is diverted on to a tungsten rod adjacent to the actinide target material. The secondary neutrons favor the production of neutron-rich isotopes within the fission process thus increasing the ratio of neutron-rich isotopes to neutron-deficient isobars. One drawback of the present neutron converter/target geometry is the production of n-def isotopes due to the divergence of protons upon impact with the neutron converter. Developments addressing this problem are on-going with the goal of producing a concentric target geometry surrounding the neutron converter [18]. An added advantage of the two-step process is the reduction of damage to the target material due to the energy deposition from the proton beam directed on to the target material.

## 4 The ISOLDE machine

### 4.1 *The ISOLDE frontends*

The target station, more commonly known as the Frontend, is the starting point for all radioactive beams. The Frontend can be broken down into two parts separated by a ceramic insulator; the first part, or the target coupling table, has all the connections for the operation of the target unit: vacuum coupling, electrical and hydraulic connections and mechanical moving parts for the clamping of the target unit and the opening of its vacuum valve. When in operation, the coupling table is at an electrical potential of up to 60 kV and consequently, all connections, cabling, tubing and the power

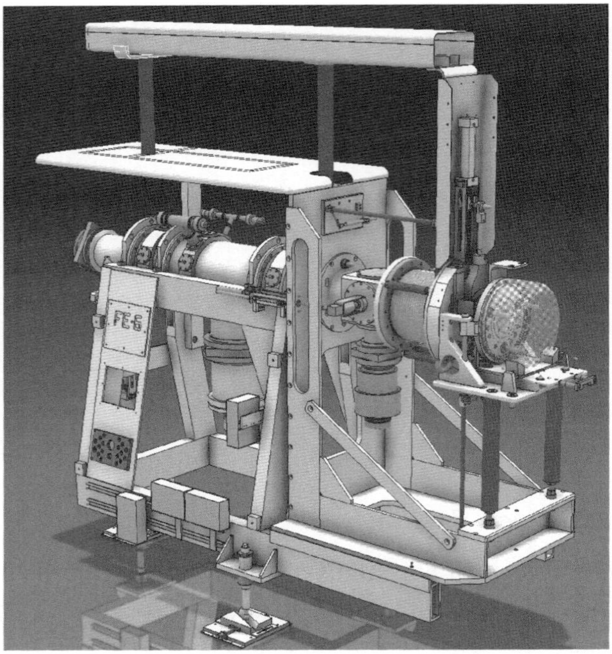


Fig. 4. The ISOLDE Frontend.

supplies are at the same potential. The second part of the Frontend is at ground potential and consists of a vacuum chamber, turbomolecular pumps, electrostatic quadrupoles, beam diagnostics and a movable extraction electrode. The latter is moved into the target unit to a specific distance from the ion source for the optimization of beam tuning. It is retracted to its home position to enable the closure of the vacuum valve on the target unit and the sealing of the Frontend during a target change. Each Frontend is identical so that a target unit can be used on either separator. However, the GPS does have an extra RF connector for the testing of prototype target units.

Both Frontends are situated in sealed Faraday cages at ground potential. The cages are supplied with a continuous stream of dried air for the prevention of sparking between the coupling table and ground, especially since the surrounding air is incidentally ionized by the secondary particles produced when the proton beam impinges on the target.

### 4.2 *The acceleration voltage*

One of the main challenges of the high voltage system used for the acceleration of radioactive ion beams comes from the timing structure of the PS Booster proton beam. Failure of the high voltage power supply may result from the load generated by the intense ionization of the air inside the

Frontend Faraday cage due to the highly charged secondary particles. The instantaneous beam current is close to 2 A. To overcome this problem, a dedicated power supply with a modulator was designed to reduce the high voltage from 60 kV to 0 V 35 $\mu$s before proton beam impact and to restore it to its nominal value ($\pm 1$ V) within 10 ms. Consequently, the change in energy of the radioactive beam implies that, when passed through the fixed magnet field of the mass separator magnets, the full spectrum of masses are distributed to the different beam lines. To overcome the consequential problem, a dedicated beam gate or deflector is initiated during the ramping sequence of the high voltage system.

### 4.3 *The mass separators*

The General Purpose Separator (GPS) magnet is an H-magnet type with a bending radius of 1.5 m and a bending angle of 70 degrees. The resolving power $dm/m$ is approximately 800. Both the magnet and its vacuum chamber have a window, providing a direct view to the target and ion source, which is used for the transport of laser light to the ion source from the external laser barrack.

The High Resolution Separator has two magnets: a 90-degree magnet and a 60-degree magnet. Both are C-magnet types and both have a bending radius of 1 meter. The overall resolution of both magnets has been measured at approximately 6000. Similar to the GPS magnet, the 90-degree magnet has a window for the transport of laser light to the target on the HRS Frontend.

Through a combination of beam instrumentation and mechanical apertures, the separator magnets are calibrated as a function of mass for a given acceleration voltage. The calibration is done using ion beams of stable isotopes purposely added to the target and ion source system. The majority of radioactive ion beams cannot be measured using the standard beam instrumentation and the experimentalist relies upon the correct calibration of the separator magnets on a stable beam before switching to the mass of a radioactive beam.

### 4.4 *The radio frequency quadrupole cooler and buncher*

After mass separation, typical emittances of the ion beam are 25–35 $\pi$ mm mrad with an energy spread of 5 eV at 60 keV. The phase space associated with this emittance contributes to beam losses both during transport and at experiments with a small acceptance. By bunching the radioactive ion beam, many decay measurement experiments benefit from an increase in the signal-to-noise ratio provided the space charge limit is not exceeded. With

Fig. 5. The Radio Frequency Quadrupole Cooler and Buncher installed at the exit of the HRS separator.

these criteria in mind, the Radio Frequency Quadrupole Cooler and Buncher (RFQCB) was developed and installed at the exit of the HRS separator in 2009 [19–21]. The ISCOOL (ISolde COOLer), as it is more commonly known, is defined as a general purpose cooler and buncher of radioactive ion beams. The incoming ions lose their energy through collisions with a neutral buffer gas while the oscillating quadrupole field of the linear RF trap focuses the ions axially. A combination of electrostatic elements provides an electric field gradient and allows the beam to be bunched before extraction. The time of flight through the ISCOOL is 300 $\mu$s with a bunch width measured at less than 5 $\mu$s. With a signal-to-noise ratio of a few orders of magnitude for certain elements and an emittance measured at $3\pi$ mm mrad, experiments at ISOLDE have largely benefited from the beam quality provided by ISCOOL [22].

### 4.5 *Beam lines*

The beam lines coming from both separators serve the central beam line in the experimental hall which in turn distributes the beam to various experiments. The low energy of both stable and radioactive beams, <60 keV, permits the use of electrostatic quadrupoles and deflectors operating with a tension <6 kV for both focusing and steering. The different beam lines are connected by switchyards containing fixed electrostatic plates into which the beams are deflected. The GPS separator switchyard is different in that it contains movable electrostatic deflector plates allowing the deflection of neighboring beams within ±15% of central mass range to be distributed to the GPS low and high mass beam lines while the central mass is distributed to the beam line array in the experimental hall.

### 4.6 *The tape station*

Optimization of a radioactive ion beam is done with the use of a dedicated tape station. At a specific time after proton beam impact, a sample of the radioactive beam is made on an aluminium tape which is then moved in front of $4\pi$ beta detectors or a gamma detector to measure the count rate of the collected sample. In the first instance, this technique is used to scan the proton beam across the target container in order to optimize its position in terms of radioactive nuclei production. Further sampling is done for the optimization of the target temperature and ion source tuning. Modifying the delay time for each sample taken enables the characterization of the target unit in terms of yield production and the release time of a specific element.

## 5 REX-ISOLDE

Experiments at ISOLDE make use of radioactive ion beams with an energy of up to 60 keV. A natural extension of such experiments was to include reaction studies such as Coulomb excitation, capture reactions and transfer reactions, all of which require a higher energy. To accomplish such experiments and to take full advantage of the large variety of radioactive ion beams (RIB) available at ISOLDE, the implementation of a universal post-acceleration scheme was required. In 1994, the concept of REX-ISOLDE, an acronym of Radioactive beam EXperiments at ISOLDE, emerged where the post-accelerated beams would achieve an energy of a few million electron volts per atomic mass unit (MeV/u) [23, 24]. The post-acceleration process starts with the preparation of the radioactive $1^+$ ions from ISOLDE. They are accumulated and phase-space cooled in a buffer-gas-filled Penning trap before being bunched and transported to the electron-beam ion source (EBIS) [25]. Charge breeding, where the ions are converted from a $1^+$ state to a $q^+$ state, takes place in the EBIS by bombardment with a dense, energetic electron beam. The highly charged ions with a low mass to charge ratio ($A/q < 4.5$) are then extracted and mass separated before being accelerated in a room temperature linear accelerator (Linac). Originally designed for the post acceleration of neutron-rich Na and K isotopes with $A/q < 50$, REX has extended its range of masses from light $^{6}$He to heavy $^{224}$Ra. To maintain

Table 1. REXTRAP data.

| |
|---|
| Buffer gas = argon, neon |
| Buffer gas pressure $<10^{-3}$ mbar |
| Magnetic field $B = 3$ T |
| Trap length = 0.9 m |
| Cooling time = typically 20 ms |
| Trap capacity $<10^{8}$ ions/bunch |
| Trans. emittance $\sim 10\pi$ mm mrad at 30 keV |
| Efficiency 40–60% |

Table 2. REXEBIS information.

| |
|---|
| Solenoidal magnetic field = 2 T |
| $E_{\text{electron}} = 3-6\,\text{keV}$ |
| $\tau_{\text{confinement}} = 3$ to $>200$ ms |
| Beam current $I_e < 0.4$ A |
| Current density $j_e = 100-150$ A/cm$^2$ |
| Trap length $L_{\text{trap}} = 0.8$ m |
| Warm bore |
| $A/q < 4.5$ |

a mass-to-charge ratio of <4.5 as required by the Linac, a charge state $>50^+$ for the heavier masses is achieved by increasing the breeding time in the REXEBIS [26, 27]. The time required for cooling and charge breeding is of little importance for radioactive elements with long half-lives; however, this is not the case for radioactive isotopes with short half-lives where decay losses inevitably lead to a reduction in isotope production for the experiment. By optimizing the cooling and breeding times, or even excluding the use of REXTRAP by injecting the ISOLDE beams directly into REXEBIS, REX-ISOLDE has been able to successfully post-accelerate radioactive isotopes with half-lives as short as 8.5 milliseconds ($^{11}$Li).

The first acceleration stage of the REX-ISOLDE Linac, from 5 keV/u to 300 keV/u, is provided by a 4-rod Radio Frequency Quadrupole (RFQ). The beam then passes through a combination of two magnetic quadrupole triplet lenses and a rebuncher in order to match the acceptance of the Interdigital H-type (IH) structure which accelerates the beam to 1.2 MeV/u. The beam is extracted from the IH structure into the 7-gap cavities to accelerate the beam to 2.2 MeV/u before finally passing through the 9-gap IH structure to boost the beam energy to 3 MeV/u. Through stepped activation of the six accelerating cavities, the energy of the ion beam can be varied from 300 keV/u (the energy of the RFQ cavity) up to the maximum energy.

Once again, purity is a key factor for the production of radioactive ion beams. With a production rate often in the region of a few thousand ions per second, there is a clear necessity to suppress any contaminating components within the beam of interest. This is achieved partially by the excellent vacuum pressure of $10^{-11}$ mbar but this does not completely eliminate residual gases such as C, N, O, Ar and Ne, the latter being the buffer gas used in REXTRAP. However, by adjusting the breeding time to tune the $A/q$ value of a specific element so that it does not coincide with the $A/q$ value of other elements, a pure beam without contamination from the residual gases can be produced.

Another advantage of REX-ISOLDE is its ability to produce radioactive ion beams that are not readily available from ISOLDE. This is done by the "in-trap decay" method where a specific isotope that is readily produced at ISOLDE is allowed to decay to its daughter isotope in REXEBIS before being ionized and post-accelerated.

Further purification techniques include using REXEBIS to break up molecular beams intentionally produced at ISOLDE for improved mass separation and by using the inherent mass selectivity of the Penning trap to extract only the ions of interest. Using the latter technique, a mass resolution in the order of 30000 has been achieved for ions with mass number $A$ between 30 and 40.

Fig. 6. A schematic drawing of the low energy part of the REX-ISOLDE facility showing the REXEBIS on the platform above REXTRAP and the mass separator.

# 6 Outlook

## 6.1 *The HIE-ISOLDE project*

ISOLDE continues to expand and the HIE-ISOLDE project (High Intensity and Energy for ISOLDE) will be a major improvement of the ISOLDE facility [28]. Approved by CERN management in 2010, the project is essentially divided into two parts: the high energy upgrade where post-accelerated radioactive ion beams at REX-ISOLDE will attain an energy of 10 MeV/$u$ after the installation of a superconducting Linac and the Design Study [29] to evaluate the consequences of an increase in proton beam intensity and to improve on the beam quality for experiments.

The superconducting Linac will be installed in three phases. The first phase, including the installation of two high-beta cryomodules each containing five quarter wave resonators (QWR) and one solenoid, will provide post-accelerated RIB with an energy up to 5.5 MeV/$u$ by the end of 2015, while the second phase includes the installation of a further two high-beta cryomodules to attain 10 MeV/$u$. Finally, to attain energy variation and more flexibility, the existing variable energy section will be replaced by two low-beta cryomodules, and a multi-harmonic buncher, upstream of the REX RFQ and chopper system, will be installed to increase the bunch spacing from 10 ns to approximately 100 ns.

The high intensity upgrade will result from the commissioning of the new Linac 4 [30] at CERN which will be able to provide a proton driver beam of $1 \times 10^{14}$ protons per pulse or a DC-equivalent intensity of 5.6 $\mu$A. The Design Study will be addressing issues such as radiation protection, radiation resistant materials, operation and intervention. By building on past experience, new designs will be tested, validated and implemented during the second scheduled long shutdown of the CERN accelerator complex. Also within the Design Study, beam quality will be addressed providing designs and solutions for a reduced transversal emittance, pre-mass separation, higher mass resolution and an improved electron beam ion source for REX-ISOLDE.

## 6.2 *Booster energy upgrade*

In order to provide beams with improved luminosity for CERN's accelerator complex, there are plans to upgrade the proton beam energy of the PS Booster from 1.4 GeV to 2 GeV [31]. Should this upgrade be made available to the ISOLDE facility, a several-fold increase in the yields of radioactive nuclei will be possible [32]. Based on the extrapolation of previous measurements of isotope production at 600 MeV, 1 GeV and 1.4 GeV and on cross-section calculations, an average gain of 40% for fission products, a factor of

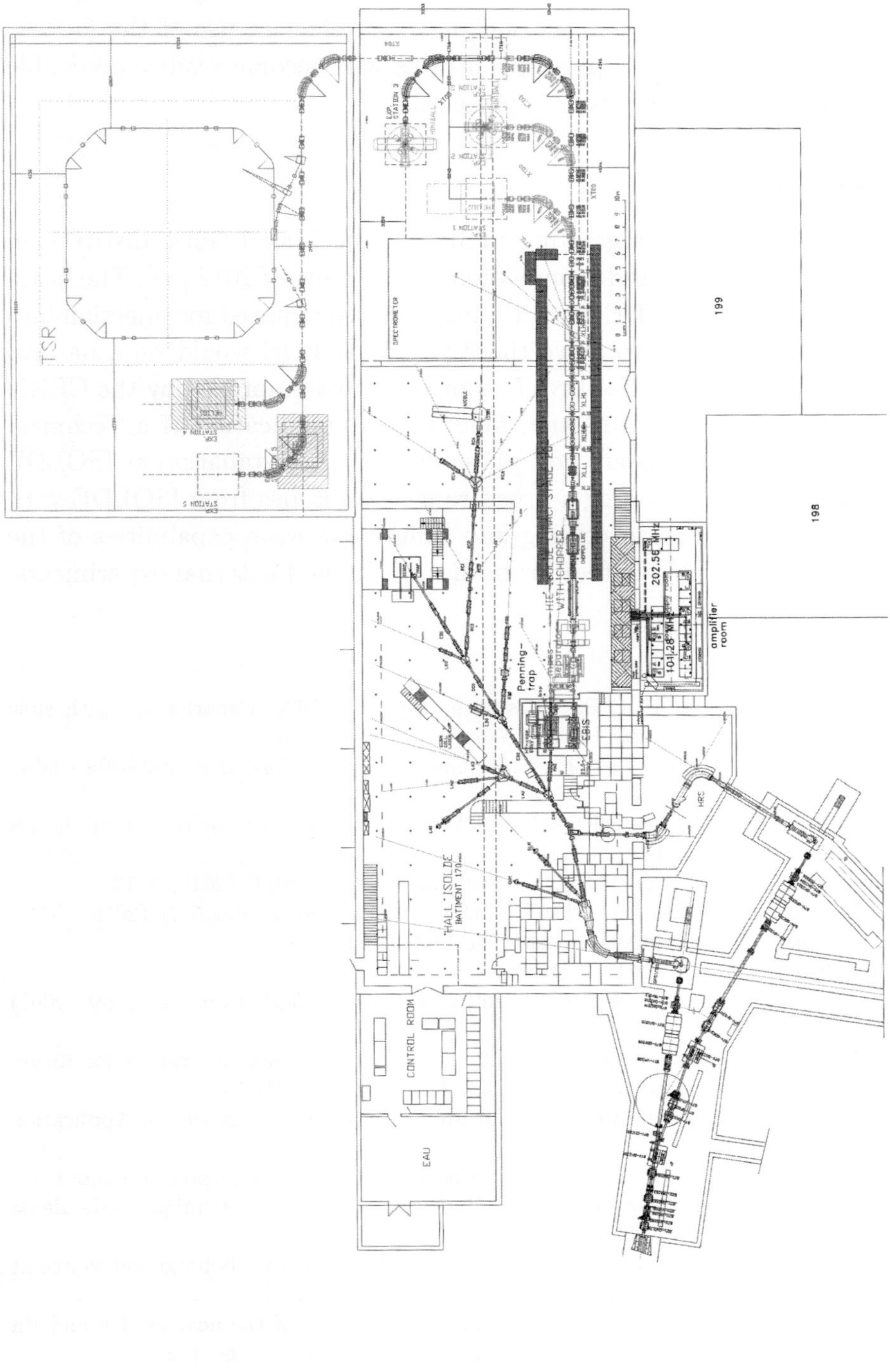


Fig. 7. The layout of the ISOLDE machine showing the superconducting Linac and the proposed layout of the TSR.

$\times 2$ to $\times 5$ gain for fragmentation products and an increase by a factor of six for exotic spallation products can be expected. Obvious advantages of higher production rates mean that the physics program can be completed in a shorter period of time, thus making more efficient use of the facility, and exotic nuclei with low production rates will become readily accessible for present and future experiment proposals.

### 6.3 *TSR@ISOLDE*

The heavy-ion Test Storage Ring (TSR) at the Max Planck Institute in Heidelberg, Germany, was decommissioned at the end of 2012 [33]. The beam parameters of HIE-ISOLDE perfectly match those required for injection into the TSR and therefore, to move the TSR to ISOLDE would be a natural extension of the ISOLDE facility. This proposal was approved by the CERN Research Board in May 2012 and, following the publication of a Technical Design Report [34], plans are now under way for its installation at ISOLDE as an experiment. Combining a broad range of isotopes from ISOLDE with the energy range, electron cooling and continuous wave capabilities of the TSR will provide a unique tool for both in-beam and external experiments.

## References


[1] B. Jonson *et al.*, ISOLDE PS Booster facility at CERN: Experiments with slow radioactive beams, *Nuclear Physics News* **3**(2) (1993) 5–16.

[2] J. Aysto, CERN's longest serving experimental facility, *Phys. Rep.* **403–404** (2004) 459–469.

[3] E. Kugler *et al.*, The new CERN-ISOLDE on-line mass-separator facility at the PS Booster, *NIMB* **70**(1–4) (1992) 41–49.

[4] A. Herlert, The ISOLDE facility, *Nuclear Physics News* **20**(4) (2010) 5–12.

[5] B. Jonson and K. Rissager, The ISOLDE facility, *Scholarpedia* **5**(7) (2010), 9742, http://www.scholarpedia.org/article/The_ISOLDE_facility.

[6] http://isolde.web.cern.ch/ISOLDE/

[7] U. Koster, ISOLDE target and ion source chemistry, *Radiochim. Acta* **89** (2001) 77777–77785.

[8] T. Stora *et al.*, Oxide target designs for high primary beam intensities for future radioactive ion beam facilities, *AIP Conf. Proc.* **1099** (2009) 764.

[9] T. Stora *et al.*, Nanostructured target for isotope production, U.S. Patent Application 13/120, 709.

[10] S. Fernandes, Submicro- and nanostructured porous materials for production of high-intensity exotic radioactive ion beams, PhD thesis, Ecole Polytechnique Federale de Lausanne, Switzerland (2010).

[11] L. Penescu *et al.*, Development of high efficiency versatile arc discharge ion source at CERN ISOLDE, *Rev. Sci. Instrum.* **81**(2) (2010) 02A906.

[12] D. Neidherr *et al.*, Discovery of $^{229}$Rn and the structure of the heaviest Rn and Ra isotopes from Penning-trap mass measurements, *PRL* **102** (2009) 112501.

[13] E. Bouquerel *et al.*, Purification of a Zn radioactive ion beam by alkali suppression in a quartz line target prototype, *Eur. Phys. J. Special Topics* **150** (2007) 277–280.
[14] V. I. Mishin *et al.*, Chemically selective laser ion-source for the CERN-ISOLDE on-line mass separator facility, *NIMB* **73** (1993) 550–560.
[15] V. N. Fedoseyev *et al.*, The ISOLDE laser ion source for exotic nuclei, *Hyperfine Interactions* **127** (2000) 409–416.
[16] V. N. Fedosseev *et al.*, Upgrade of the resonance ionization laser ion source at ISOLDE on-line isotope separation facility: New lasers and new ion beams, *Rev. Sci. Instrum.* **83** (2012) 02A903, doi:10.1063/1.3662206.
[17] D. Fink *et al.*, First application for the Laser Ion Source and Trap (LIST) for on-line experiments at ISOLDE, in *EMIS 2012 Proc.*, *NIMB* **317** (2013) 417–421.
[18] R. Luis *et al.*, Optimization studies of the CERN-ISOLDE neutron converter and fission target system, *Eur. Phys. J. A* **48** (2012) 90.
[19] A. Jokinen *et al.*, RFQ-cooler for low-energy radioactive ions at ISOLDE, *NIMB* **204** (2003) 86–89.
[20] I. Podadera *et al.*, Preparation of cooled and bunched ion beams at ISOLDE-CERN, *Eur. Phys. J. A* **25**(s01) (2005) 743–744.
[21] H. Franberg *et al.*, Off-line commissioning of the ISOLDE cooler, *NIMB* **266** (2008) 4502–4504.
[22] E. Mané *et al.*, An ion cooler–buncher for high sensitivity collinear laser spectroscopy at ISOLDE, *Eur. Phys. J. A* **42** (2009) 503–507.
[23] F. Wenander and K. Riisager, Isotope toolbox turns 10 (2012), http://cerncourier.com/cws/article/cern/48345.
[24] O. Kester *et al.*, *Hyperfine Interactions* **129**(1–4) (2000) 43–66.
[25] F. Wenander, Charge breeding of radioactive ions with EBIS and EBIT, *J. Instrum.* **5** (2010) C10004.
[26] P. Van Duppen and K. Riisager, Physics with REX-ISOLDE, *J. Phys. G: Nucl. Part. Phys.* **38** (2011) 024005.
[27] D. Voulot *et al.*, Radioactive beams at REX-ISOLDE: Present status and latest developments, *Nucl. Instrum. Meth. B* **266** (2008) 4103–4107.
[28] The HIE-ISOLDE Project (2012), http://hie-isolde.web.cern.ch/hie-isolde/.
[29] R. Catherall *et al.*, An overview of the HIE-ISOLDE design study, in *EMIS 2012 Proc.*, *NIMB* **317** (2013) 204–207.
[30] R. Garoby *et al.*, Plans for the upgrade of the LHC injectors, in *IPAC 2012 Proc.* (2012), p. 1010.
[31] G. Arduini *et al.*, Possible improvements to the existing pre-injector complex in the framework of continued consolidation, in *Proc. Chamonix 2010 Workshop on LHC Performance*, CERN-ATS-2010-026 (2010), p. 228.
[32] M. Borge *et al.*, Motivations to receive a 2 GeV proton beam at ISOLDE/HIE-ISOLDE: Impact on radioisotope beam availability and physics program, CERN-INTC-2012-069/INTC-O-016.
[33] The ion storage ring TSR for REX-ISOLDE at CERN (TSR@ISOLDE), http://www.mpi-hd.mpg.de/blaum/storage-rings/tsr/index.en.html
[34] M. Grieser *et al.*, *Eur. Phys. J. Special Topics* **207**(1) (2012) 1–117.

# Chapter 19

# LEAR and AD antiproton facilities: Production of antihydrogen

Pavel Belochitskii, Horst Breuker, Tommy Eriksson, Stephan Maury, Walter Oelert, Gerard Tranquille (CERN)

## 1 Preface

There is the justified question: Why does antimatter physics very often hit the headlines of scientific and public interest throughout its historical development? On the scientific side it proves the enormous potential for discoveries in fundamental and basic physics. The public attention is driven to some part at least by the enjoyable science fiction literature and adventures leading to unrealistic speculations of unlimited energy and infinite speed of spaceships started with the episodes of Star Trek in 1966. This latter part is not the topic of this article. It would be beyond the aim of this contribution to identify all the thoughts of opposite subjects to matter discussed in the antic history and we only will mention here the amusing, speculative and whimsical contribution of Sir Arthur Schuster [1] from 1898 introducing the term of "negative atoms" by arguing: "If there is negative electricity, why not negative gold, ..."

In 1928 the modern theory of antimatter was initiated by Paul Dirac [2], see Fig. 1, when he realized that his relativistic version of the Schrödinger equation predicted electrons with negative energy, the anti-electrons. In fact these particles — named positrons — were first observed by Carl D. Anderson in 1932 [3].

Antimatter consists of antiparticles in the same way as normal matter is assembled of particles. For example, a positron (the antiparticle of the electron or $e^+$) and an antiproton ($\bar{p}$) can form an antihydrogen atom in the same way that an electron and a proton form a "normal matter" hydrogen atom. The existence of the antiproton was experimentally confirmed in 1955 by O. Chamberlain, E. Segrè, C. Wiegand and T. Ypsilantis [4].

$$\left(i\partial\!\!\!/ - m\right)\psi(x) = 0$$

Fig. 1. A well known photo of Dirac and one representation of his famous equation.

The excitement of antimatter discoveries and its physics entrapped Werner Heisenberg in 1972 to the statement [5]: "I think that the discovery of antimatter was perhaps the biggest jump of all the big jumps in physics in our century."

Forty years after the discovery of the antiproton, in 1995, eleven antihydrogen atoms were first detected [6] at the Low Energy Antiproton Ring (LEAR) of CERN [7] and confirmed at Fermilab [8] by following a concept from Munger *et al.* [9]. However, these antihydrogen atoms were extremely energetic and therefore not suitable for precision studies of the properties of antihydrogen in comparison to normal matter hydrogen. For this trapped and cold antimatter atoms at rest are essential.

One of the most fundamental open problems in physics concerns the asymmetry between particles: Why is the observable and visible universe apparently composed almost entirely of matter and not of antimatter. Still, antimatter galaxies may exist and if so they are supposed to have the same features and properties as those of normal matter. Beside other objectives the AMS experiment [10] is aiming to investigate the existence of such exotic galaxies.

To produce cold antihydrogen the antiprotons used have to be cooled by (i) decelerating them with the Antiproton Decelerator (AD), (ii) passing them through a thin foil, and (iii) finally capturing them in a Penning–Malmberg trap. The overall cooling process works, but is highly inefficient: approximately $3 \times 10^7$ antiprotons leave the AD and only up to 150000 are ideally trapped, an efficiency of less than 0.5%.

These days antihydrogen atoms are produced frequently by the three collaborations at the AD: ATRAP [11], ALPHA [12], and ASACUSA [13] with essentially similar methods, mostly developed at ATRAP. Whereas ATRAP and ALPHA aim to produce antihydrogen at rest, ASACUSA intends to have a flux of these atoms for hyperfine transition studies.

The antiprotons are still hot when initially trapped. To cool them further, they are mixed with an electron plasma which cools itself via cyclotron radiation, and then sympathetically cool the antiprotons via Coulomb collisions. Eventually, the electrons are removed, leaving the antiprotons with energies less than 100 MeV. In parallel a cloud of positrons is captured from radioactive sodium and trapped near the antiprotons. Finally the antiprotons are moved gently into the positron plasma, where some form antihydrogens. In 2002 both groups, first the ATHENA collaboration [14] and shortly thereafter the ATRAP group [15], announced the creation of the first "cold" antihydrogen. Still, since the neutral antihydrogen atom is unaffected by the electric fields used to trap its charged components, the antihydrogen hits the trap walls and annihilates.

Any high-precision tests of properties of antihydrogen can only be performed if the antihydrogen atoms are cold enough that they can be held in place for a relatively long time. The spins of the components of the antihydrogen atoms have a magnetic moment which can interact with an inhomogeneous magnetic field; some of the antihydrogen atoms can be attracted to a magnetic minimum. In fall 2010 the ALPHA collaboration reported first the success of 38 trapped antihydrogen atoms [16]. A year later lifetimes of more than 15 minutes of the antihydrogen atoms were observed by both collaborations ALPHA [17] and ATRAP [18]. Finally ALPHA reported lately the very first spectroscopy of an antimatter atom [19], demonstrating resonant quantum transitions in antihydrogen by manipulating the internal spin state.

It is worthwhile to add that new additional experiments at the AD, such as AEGIS [20] and GBAR [21], are presently (status of 2012) preparing for precise measurements of the gravitational interaction between matter and antimatter and that the new decelerator ELENA [22] is supposed to accelerate the progress drastically. There is a brilliant future in physics to be expected here.

## 2 Observation of first antihydrogen atoms at LEAR

The first observation of antihydrogen atoms, a bound system consisting of positrons (known since 1932) and antiprotons (known since 1955), was reported only in 1995 in the PS210 experiment [6] at the Low Energy Antiproton Ring (LEAR) of CERN. A method was used [9] where the antihydrogen atoms were formed in the collision of antiprotons with a heavy element as an internal cluster target.

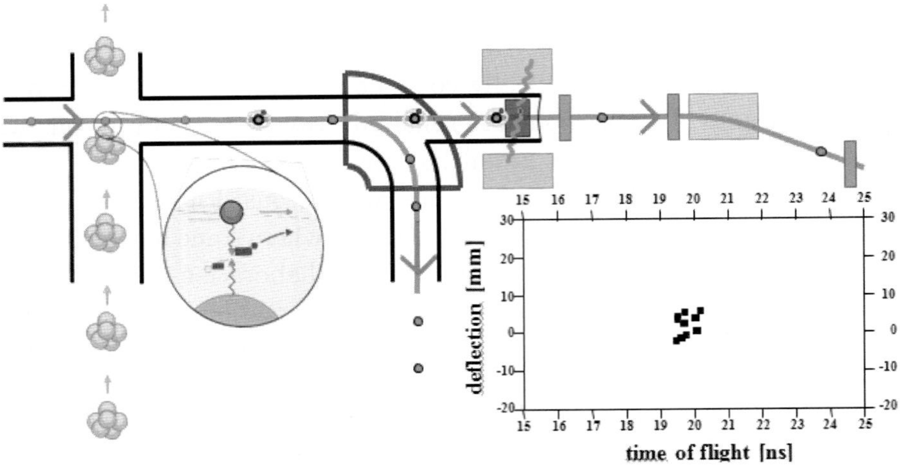


Fig. 2. Principle layout of the PS210 experiment and its final results observing the first production of eleven antihydrogen atoms. For details see Ref. [6].

The experiment was a real challenge both for the experimenters and for the LEAR operation team, but they succeeded in producing and observing the first ever seen antihydrogen atoms.

As shown in Fig. 2, antiprotons circulating in LEAR were passing through the Coulomb field of a nucleus (here Xe) and might create an $e^+e^-$ pair. Occasionally the antiproton will capture the positron from the produced pair and together form a fast moving antihydrogen atom, which — as a neutral object — will exit tangentially in the accelerator bending following its creation point. The antihydrogen atom was then stripped in a silicon detector. Two 511 keV photons from the $e^+$ annihilation were detected back-to-back in a cylindrical NaI counter. The antiproton resulting from the stripped antihydrogen continues through a set of detectors and an analyzing magnet, such that the time of flight versus the deflection could be measured. This way the signal of eleven observed productions of antihydrogen atoms could be clearly determined.

Since the antimatter atoms produced were traveling nearly at the speed of light they could not be used for further investigations into the properties of antimatter physics. In 1996 LEAR was closed for physics with antiprotons and was converted to a heavy ion accumulator for the use as one of the injectors to the LHC for ion operation. Still the successful production of antihydrogen atoms combined with the strong scientific and public interest in antimatter physics resulted in the conclusion that a delivery of low energy antiprotons was and is absolutely essential. And thus the construction of AD with its present extension of ELENA was conceived as a logical continuation of the antimatter program.

## 3 Low Energy Antiproton Ring (LEAR)

### 3.1 *History of LEAR*

The history of LEAR began some time in 1976 when, over a coffee in the CERN Main Building cafeteria, Kurt Kilian and Dieter Möhl discussed the possibility of making experiments with low energy antiprotons at the CERN PS. The dream of high fluxes of low energy antiprotons which, in addition, would be free of contamination by other particles and cold (i.e. extremely well "collimated" and mono-energetic) was becoming a reality. Stochastic and electron cooling had been demonstrated to work a few years earlier, and a paper of the Novosibirsk Group on "Electron Cooling and New Possibilities in Elementary Particle Physics" presented by A. Skrinsky at the 1976 International Conference on High Energy Physics held in Tibilisi discussed a wide range of ideas from antihydrogen production to a high energy proton–antiproton collider.

Around the same time at CERN, a working group spearheaded by Carlo Rubbia had started to look into the feasibility of antiproton accumulation for proton–antiproton collisions in the SPS. The proposed scheme was based on electron cooling and consisted of a ring to bring the antiprotons down from 4 GeV, where they are copiously produced at the PS, to about 50 MeV where they can be efficiently cooled by electrons. The antiprotons are then re-accelerated to 4 GeV before being transferred to the PS. The developments for the SPS lead to a natural question: Why not use a similar small deceleration and cooling ring to provide dense and clean beams of antiprotons for low energy experiments?

At the 1977 International Conference on High Energy Accelerators held in Serpukov, Dieter, Kurt and Ugo Gastaldi published a paper promoting the use of "Rubbia's antiprotons" at low energies. LEAF (Low Energy Antiproton Facility), as it was known, would accomplish the antiproton collection and deceleration very much like what the AD (Antiproton Decelerator) has been doing since 2000. Over the following two years, the project evolved into an "add-on" to the antiproton source baptised as the APR (Anti Proton Ring) and later renamed LEAR: Low Energy Antiproton Ring. This new ring would use an antiproton beam decelerated in the PS and after cooling and further deceleration extract it over a long period (15 minutes) such that the PS and AA could continue to supply the ISR and the SPS with minimum disturbance. The LEAR idea found full support from the PS Division and in

May 1980 the final design report was published. A month later the Research and Management Boards approved the project under very strict conditions:

- Construction of the facility should not interfere with the other duties of the PS Division.
- Once running, LEAR should only consume a small fraction (6%) of the AA production, it would have low priority compared to the ISR and the SPS proton–antiproton programme, and it should not interfere with the "normal" operation of the PS.

A call for Letters of Intent launched at the end of 1977 had an unexpectedly large echo which led to a first-phase physics programme with 16 experiments involving more than 240 physicists from 44 institutions.

### 3.2 *The LEAR machine*

LEAR was almost square in shape (Fig. 3) with a circumference of 78 m (1/4 of the PS). Its four-period lattice with compact 90° bending magnets and eight quadrupole doublets provided four long straight sections, each of 8 m

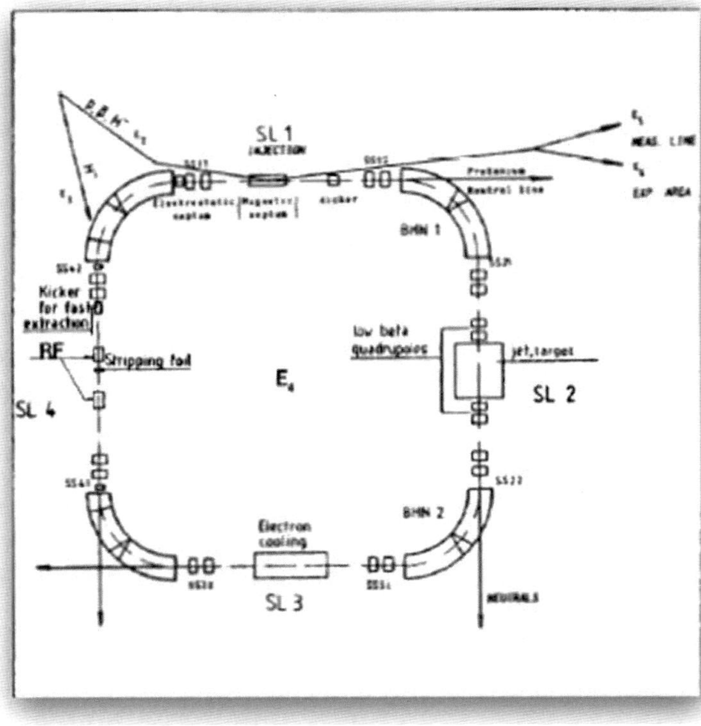


Fig. 3. Schematic view of LEAR.

Table 1. LEAR basic parameters.

| | |
|---|---|
| Momentum (energy) range | 0.1–2 GeV/c (5.3 MeV–1.3 GeV) |
| Circumference | 78.54 m |
| Focussing structure | 4 superperiods, separated function BoDFOFDoB |
| Betatron wave numbers | $Q_h = 2.3$, $Q_v = 2.7$ |
| Momentum compaction | $\gamma_{tr}^{-2} = -0.048$ |
| Maximum acceptances | $\varepsilon_h = 240\ \pi$ mm mrad, $\varepsilon_v = 48\ \pi$ mm mrad |
| RF system frequency range | 0.4–3.5 MHz |
| Design pressure | $10^{-11}$–$10^{-12}$ Torr |

free length. These served for the installation of large equipment, in particular the electron cooler and the internal gas jet target experiments. Eight short straight sections, each 1 m long, accommodated less bulky equipment. The C-type magnets were open to the outside of the ring. This simplified the injection, the ejection, and the design of "exit lines" for neutral states formed in flight in the straight sections ($H^0$, antineutrons, proton–antiproton bound states). These exit lines also greatly eased the detection of antihydrogen atoms formed by antiproton interaction with an internal gas target.

A particularity of the optics was the very strong focusing: a phase advance of ~250°/period yielded an "imaginary transition energy" (decrease of orbit length C with momentum, i.e. $(dC/C)/(dp/p) \equiv \alpha_p = \gamma_{tr}^{-2} < 0$). This leads to a large dispersion of the revolution frequencies $(df/f)/(dp/p) \equiv \eta = \gamma^{-2} - \gamma_{tr}^{-2}$, beneficial for cooling to small momentum spreads and for control of instabilities. The basic parameters are shown in Table 1.

### 3.2.1 *LEAR operation modes*

A single bunch, of usually a few $10^9$ antiprotons, was skimmed off the AA stack at intervals ranging from 15 min to several hours. The average consumption, $10^6\ \bar{p}$/s, was less than 10% of the maximum accumulation rate of the AA. The bunch was decelerated in the PS to 609 MeV/c and transferred to LEAR, where it could either be decelerated to as low as 100 MeV/c (5.3 MeV kinetic energy), or accelerated up to nominally 2000 MeV/c (1270 MeV).

In the "beam stretcher mode", used for most of the experiments, ultra-slow extraction provided a continuous spill until the next fill. In the "internal target" mode for the JETSET experiment, a beam with an initial intensity of as much as $5 \times 10^{10}\ \bar{p}$ was kept circulating for many hours, even days, until most particles had been consumed by interaction with the gas jet target. An online procedure for creating a new ejection momentum by interpolation between existing ejection settings was also developed. Within two hours the machine and experimental lines could be set up for any "non-standard" momentum. For trap experiments one or several bunches were extracted by a fast kicker placed a quarter of a betatron wavelength upstream of the magnetic septum.

### 3.2.2 *Beam cooling*

In LEAR both stochastic and electron cooling were used to counteract the adiabatic emittance growth during deceleration and beam blow-up due to residual gas scattering. Stochastic cooling was optimised for several strategic momenta on the low energy cycle and at 1, 1.5 and 1.94 GeV/c on the high energy cycle when the internal gas jet was operated. Electron cooling was applied at 310 MeV/c and below and contributed to enhancing the machine's duty cycle by reducing the deceleration time to 100 MeV/c by a factor of 2.

The transverse stochastic cooling system consisted of a number of pick-ups and kickers with coaxial relays commuting between different cable-delays used to compensate the changing particle velocities over the momentum range. Pre-adjusted delays were used at the fixed "flat-tops" to provide fast cooling and a second set of delays and cables provided slower transverse cooling at any intermediate momenta. This slow cooling was also used to maintain high beam quality during the long extraction process. The longitudinal system also consisted of two parts using pick-ups and kickers. A line filter of fixed electrical length was used for cooling at the injection momentum of 609 MeV/c and another line filter of variable length was used at all momenta below 609 MeV/c.

The electron cooling device used at LEAR was inherited from the ICE (Initial Cooling Experiment) ring which had been set up to determine which of the two cooling methods (stochastic or electron) would be more appropriate for the cooling and accumulation of high energy antiprotons. In order to fit in one of the eight metre-long straight sections of the machine, the interaction length of the cooler had to be reduced by half. The high voltage and the control systems of the device were also completely refurbished and a dedicated equipment building was erected close to the LEAR ring. The actual installation of the cooler took place during the summer of 1987 followed by the conditioning of the cathode and further tests to monitor the evolution of the LEAR vacuum in the presence of the electron beam. By the autumn of 1987 the cooler was ready to cool its first beam. The first cooling tests were made on a 50 MeV proton beam injected directly from the Linac 1 and the initial results confirmed all expectations from this device. As of 1988 electron cooling was used routinely at all the plateaus of the low energy antiproton cycle.

Apart from being the first cooler to be used routinely for accelerator operations, the apparatus was also the first to demonstrate the cooling and stacking of ions. In 1989 a machine experiment was devoted to studies on $O^{6+}$ and $O^{8+}$ ions coming from the Linac 1. An increase by a factor of 20 in intensity was achieved by applying electron cooling during the longitudinal stacking process. Later these ions were accelerated to an energy of 408 MeV/u and extracted to an experiment measuring the depth dose distribution in tissue equivalent types of plastic.

### 3.2.3 *Ultra-slow extraction*

In the stretcher mode, a spill as constant as possible of some $10^6$ $\bar{p}$/s was required by the users. The filling sequence was determined by the smallest intensity that the PS was able to handle. The limit was pushed down to $10^9$ $\bar{p}$, some four orders of magnitude below its usual value for protons. Even so, the spill length had to be at least 15 min, a formidable challenge, as so far extraction times of only a few seconds had been achieved. It meant beam stretching ("extraction time/revolution time") by nine orders of magnitude with, on average, less than one particle extracted per turn. A novel ultra-slow extraction technique was devised for this purpose. It is based on (yet another!) idea by van der Meer, which was brought to maturity by R. Cappi and W. Hardt.

Conventional slow extraction uses a programmed tune change, driving the beam towards a resonance, which eats into the tune distribution of the beam. The time structure of the spill is very sensitive to all sorts of tune ripple and exhibits spikes and holes (detrimental to the experiments) when the sweep is slow.

Ultra-slow ("stochastic") extraction uses RF noise to diffuse the particles within an appropriate range of dp/p, thereby producing a very-low-density tail on the momentum distribution. The chromaticity, (dQ/Q)/(dp/p), adjusted with sextupole lenses, leads to a corresponding tail in the Q-distribution and the extraction resonance is placed at a Q-value inside that tail. This largely reduces the influence of Q-ripple, as the density near the resonance is low and particles perform a random walk around it. The spill rate is controlled by the level of the noise transporting particles from the stack into the tail.

This concept worked admirably and permitted smooth 15-min spills in the first runs in 1983. Very soon, 1-h spills became common. At the end of the LEAR era, the number of transfers per day was minimized by taking from the AA batches of the highest intensity compatible with safe operation

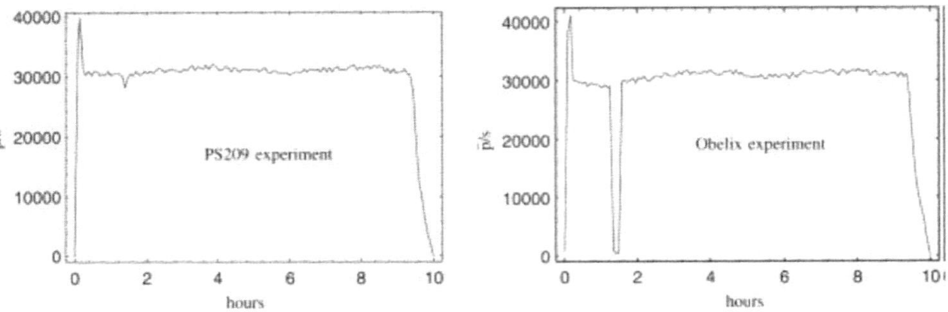


Fig. 4. A typical LEAR ultra-slow extraction lasting 10 hours. The beam is shared by two experiments and each curve represents the rate recorded by the experiment averaged over 10 seconds.

for the experiments. Figure 4 illustrates a 10-h spill; the record spill length was 14 h.

### 3.3 *Conversion to LEIR*

Towards the end of the antiproton program LEAR was already preparing its conversion to LEIR (Low Energy Ion Ring) which would cool and accumulate lead ions for CERN's flagship accelerator, the LHC. A series of machine experiments using Pb ions with various charge states (52+ to 55+) not only demonstrated the feasibility of the proposed scheme, but also brought to light an anomalously high recombination rate between the cooling electrons and the $Pb^{53+}$ ions (which had initially been the proposed charge state) leading to lifetimes that were too short for cooling and stacking in LEAR. Consequently it was decided to use $Pb^{54+}$ ions instead as they are produced in equal quantities to the 53+ charge state.

### 3.4 *Conclusion*

The low energy antiproton programme had been conceived as an "add-on" to the SPS collider, at little extra cost and consuming only a small fraction of the antiprotons production. But the results obtained with LEAR soon made it an important and very visible part of CERN's activities. The interest had enable LEAR to continue for five years beyond the end of the collider.

From the outset, the low energy antiproton programme presented a major challenge to the accelerator community. LEAR was an unconventional enterprise, conceptually and technically. The success of the AA and LEAR has made popular the novel extraction and cooling techniques, and set the example for a dozen ion cooling rings ("king LEAR's daughters") built in Europe and the USA.

## 4 Antiproton decelerator ring (AD)

### 4.1 *Introduction*

The antiproton program at CERN started around 1980 when a first production and storage facility was brought into operation. This was called the Antiproton Accumulator (AA) which was a single ring with about 150 m circumference and an associated production target area. Antiprotons produced at the target were transferred to the ring where they were collected at a momentum of 3.57 GeV/c. Stochastic cooling was then applied using several systems in order to reduce beam emittances and make accumulation of several pulses possible. A major upgrade to the facility took place in 1986 when a second ring with increased beam acceptances was added; this was to be called the Antiproton Collector (AC) and permitted a 10-fold increase in accumulation rates. The facility, now called the Antiproton Accumulation Complex (AAC) also included upgrades of the target area and its transfer lines. Up to $6 \times 10^{10}$ antiprotons could be accumulated per hour with a repetition rate of once every 4.8 seconds and stored antiproton "stacks" of up to $10^{12}$ particles were achieved. AAC was used until 1996 to supply batches of up to a few $10^{10}$ antiprotons per pulse to the SPS proton–antiproton collider program and also to the low energy antiproton physics program at LEAR (Low Energy Antiproton Ring).

By 1996 it had been decided to reduce the number of physics programs at CERN and to simplify the accelerator complex in order to free up resources for the LHC project. During 1997 and 1998, the AAC was thus converted from a two-ring fixed-energy accumulator complex to a simpler single-ring collector/decelerator including a new experimental area housed within the same building. Supplying antiprotons suitable for low energy experiments was now greatly simplified with one less storage ring to operate and with production, deceleration and transfer to the experiments done within the same facility, thus avoiding transfers to and from the PS ring where deceleration was made previously [23]. Commissioning of the new machine, baptised the Antiproton Decelerator (AD) took place in 1999 and physics could start in 2000 with three experiments, ATHENA, ATRAP and ASACUSA receiving single batches of 100 MeV/c (5.3 MeV kinetic energy) antiprotons (Fig. 5).

### 4.2 *From AC to AD*

The AD ring is based on the AC ring of the AAC which was modified in order to accommodate an electron cooler. The four-fold symmetry layout gave place to a two-fold symmetry with a six-metre long insertion in one of

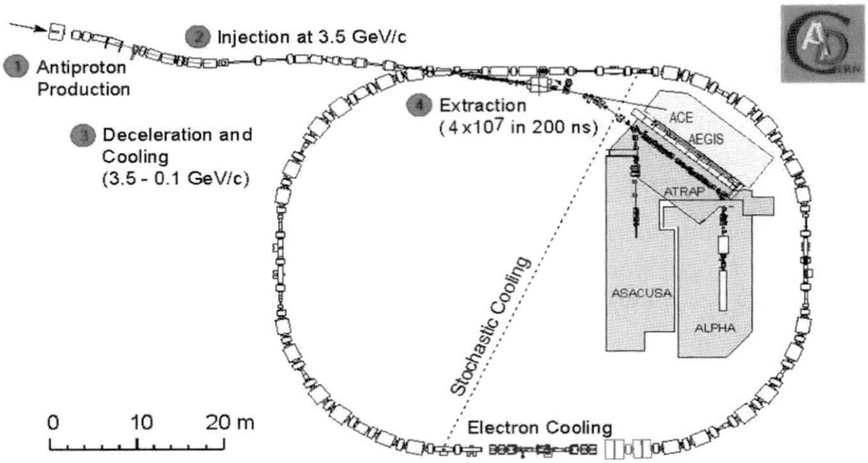


Fig. 5. Layout of AD and the experimental areas.

the long straight sections accommodating the cooler and its associated orbit correctors. Removal of several stochastic cooling pickup and kicker tanks was also necessary, keeping only 1/3 of the system bandwidth. This resulted in a modest increase in the duration of the deceleration cycle. To be able to control orbit excursions during deceleration, additional horizontal and vertical orbit corrector dipoles were installed wherever possible. Since not enough space was available for the required number of correctors, individual trim power supplies were installed on 12 of the 24 main bending magnets so that these could act also as horizontal orbit correctors. The vacuum system needed improvements due to much lower beam energies. By adding around 100 sublimation pumps around the ring, average ring pressures of a few $10^{-10}$ Torr could be achieved during beam operation. New ejection lines were installed, joining the previous AC to AA transfer line and leading to the experimental zone of the machine hall which was installed in the central part of the AD hall. Finally, modifications to the control system and the ring power converters made it possible to run the ring with a deceleration cycle.

### 4.3 *AD overview*

For each AD cycle (Fig. 6), about $1.5 \times 10^{13}$ protons at 26 GeV/c are transferred from the Proton Synchrotron (PS) to the AD target area where they impinge on a water-cooled iridium target situated in a separate shielded building. After focussing using a Magnetic Horn, secondary particles are being led through a dog-leg shaped transfer line before entering the AD hall. Around $5 \times 10^7$ antiprotons are then injected into the ring at 3.57 GeV/c where due to the large spread in energy and production angles

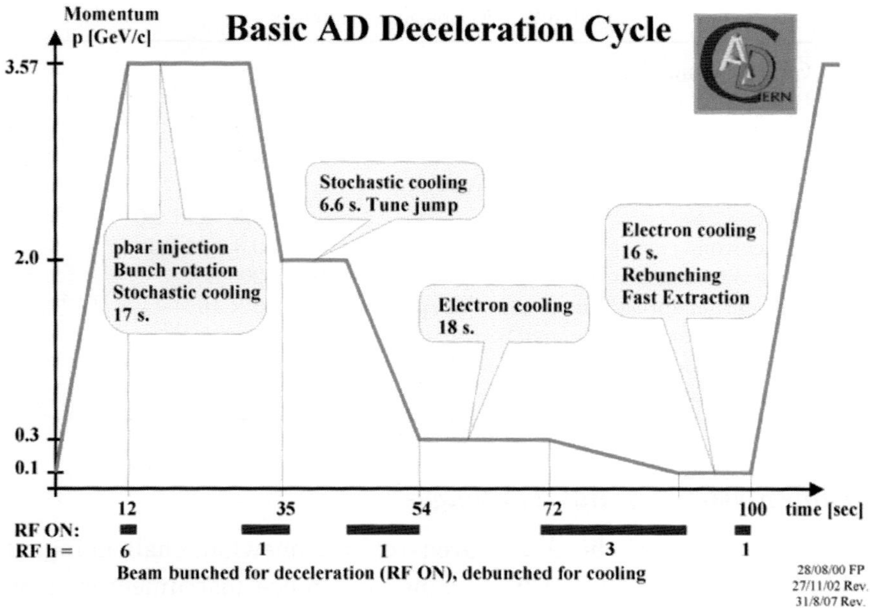


Fig. 6. AD cycle.

they completely fill the ring acceptances. Before any deceleration can be done, beam size and energy spread must be reduced. This is done using stochastic cooling. Taking advantage of the short bunches, 30 ns as generated in the PS, a 9.5 MHz 1.2 MV RF-system performs a 90-degree rotation in longitudinal phase-space (so-called Bunch Rotation) thereby stretching the bunches while reducing their energy spread from ±3% to ±0.75% which corresponds to the maximum acceptance of the stochastic cooling system. Cooling can now take place with good efficiency which is a pre-requisite for loss-less deceleration. To counteract adiabatic blow-up and to increase the density of the extracted beam, further beam cooling is applied at 2 GeV/c which is the lowest level for the stochastic cooling system and then at both 300 and 100 MeV/c, now using the electron cooler. Total cycle duration is around 100 seconds, mainly limited by beam cooling duration and limitations in the deceleration rates which is due to the ring magnet design being optimised for fixed field operation. Before ejection at 100 MeV/c, the cooled beam will be bunched and rotated again in longitudinal phase-space in order to create one short bunch, <200 ns, which is necessary for efficient trapping in the experiments. Alternatively, multi-bunch ejection of up to six bunches with 2.4 seconds interval can be done as required. At higher beam energies, single bunches can be extracted at momenta up to 500 MeV/c into the ACE/AEGIS experimental area. The basic parameters are shown in Table 2.

Table 2. AD basic parameters.

| | |
|---|---|
| Circumference [m] | 182 |
| Production beam [protons/cycle] | $1.5 \times 10^{13}$ |
| Injected beam [pbars/cycle] | $5 \times 10^{7}$ |
| Beam momenta, max.–min. [GeV/c] | 3.57–0.1 |
| Momenta for beam cooling [GeV/c] | |
| Stochastic | 3.57 and 2.0 |
| Electron | 0.3 and 0.1 |
| Transverse emittances, max.–min. [$\pi$ mm mrad] | 200–1 |
| Momentum spread, max.–min. [dp/p] | $6 \times 10^{-2}$–$1 \times 10^{-4}$ |
| Vacuum pressure, average [Torr] | $4 \times 10^{-10}$ |
| Cycle length [s] | 100 |
| Deceleration efficiency [%] | 85 |

### 4.4 *Commissioning and challenges*

The commissioning of the AD proved to be somewhat challenging, using up the better part of the 1999 machine run. Since instrumentation was a bit marginal for beams of a few $10^7$ particles, the initial setting-up of the deceleration cycle was done using proton beams of a few $10^{10}$ particles, either in the reverse direction using the old AA to PS transfer line or in the normal direction where beam cooling could be applied, now with the machine in reversed polarity and the production target moved out of the way. Due to unpredictably large orbit and Betatron tune swings, the cycle had to be built up by decelerating in small steps stopping at several levels for corrections. An additional complication was ring magnet Eddy current effects when arriving at the different flat parts of the cycle which increased the foreseen length of each ramp. Several improvements had to be made to the different systems to reduce the observed orbit fluctuations at 100 MeV/c which not only reduce the beam cooling performance but also had proven to be detrimental to the operation and efficiency of the experiments. At the end of the setting-up period, most of the design specifications had been reached except for the cycle length which was around 100 instead of 60 seconds. After several machine study sessions during the initial physics runs, the AD performance was significantly improved, especially regarding deceleration efficiency and intensity and density of the ejected beam. Table 3 summarises the basic parameters of the extracted beams.

### 4.5 *Experimental areas for the AD complex*

The experimental areas were designed to house at least four experiments [24]. All experimental beam lines are installed within the AD circumference.

Table 3. Extracted beam parameters.

| Parameters (at extraction) | Operational 2012 | | | |
|---|---|---|---|---|
| Beam momentum [MeV/c] | 100 | 300 | 500 | 100, multiej. |
| Transverse emittances H/V [95%] [$\mu$m] | $<1\pi$ | $<2\pi$ | $8\pi$ | $<1\pi$ |
| Total energy spread [$4\sigma$] [$10^{-3}$] | 0.8–0.4 | 0.5 | 2 | >1 |
| Bunch length [ns] | 150 | 300 | 500 | 50 |
| Number of antiprotons [$10^7$] | 3.6 | 4.0 | 4.0 | $0.5 \times 6$ |
| Repetition rate [s] | 95 | 90 | 85 | 112 |

The limited space inside the AD machine imposed tight constraints on the topology of the beam lines. In order to achieve the space needed, the old AA was dismantled and the inner shielding of the AD machine was reduced to a minmum. Additional space was made available by installing two experiments (ATRAP 1 and 2) vertically. The horizontally installed experiments are ALPHA, ASACUSA and AEGIS. The beam transfer lines are roughly perpendicular to the major axis of the AD machine and 1.2 m above the ground level. The optics of the experimental lines were designed to have minimum beam sizes and zero dispersion functions at the focus locations. The overall effective emittance at the focus is less than $1\pi$ mm mrad. The lines were kept as short as possible and their geometries were carefully chosen in order to increase the distance between experimental lines, and thus reducing the electro-magnetic influence between neighbouring lines. Pulse to pulse switching of the beam among the experiments is prohibited by the severe constraints of the field stability of the beam line magnets. Today three experiments receive the beam with a switching time of eight hours with the fourth experiment on hold for one week. The experimental areas were designed complying with the CERN safety rules. Physicists work in experimental areas, counting rooms, laser huts and workshops during the functioning of the machine. Easy access and evacuation if needed is provided by at least two different paths per zone. Radiation protection is provided by concrete walls separating the experimental zones with a thickness 0.4 m and a layer of 0.8 m of concrete in the injection region over a length of 18 m.

### 4.6 *Conclusion*

The AD is presently (status 2012) running 4500 hours per year with an availability of around 90%. Cooled antiproton beams are being delivered to several different experiments working around the clock. In view of the ELENA project [22], a lifespan of at least 20 more years can be expected which has prompted CERN to initiate a major consolidation project for ELENA. Not only the machine itself is being addressed, but also its infrastructure, safety issues and the user facilities.

## References


[1] A. Schuster, *Nature* (18 August 1898) 367.
[2] P.A.M. Dirac, *Proc. Roy. Soc. A* **133** (1931), 60.
[3] C.D. Anderson, *Phys. Rev.* **43** (1933) 491.
[4] O. Chamberlain, E. Segrè, C. Wiegand and T. Ypsilantis, *Phys. Rev.* **100** (1955) 947.
[5] W. Heisenberg, Development of Concepts in the History of Quantum Theory, in J. Mehra (ed.), *The Physicist's Concept of Nature* (Springer, Netherlands, 1973), p. 271.
[6] G. Baur *et al.*, *Phys. Lett. B* **368** (1996) 251.
[7] R. Garoby, RF manipulations in the PSB & PS for the production of antiprotons at CERN and deceleration in the PS for LEAR, in O. Brüning and S. Myers (eds.), *Challenges and Goals for Accelerators in the XXI Century* (World Scientific, Singapore, 2015), Chapter 13.
[8] G. Blanford *et al.*, *Phys. Rev. Lett.* **80** (1998) 3037.
[9] C.T. Munger *et al.*, *Hyperfine Interactions* **76** (1993) 175.
[10] See e.g.: http://ams.cern.ch.
[11] http://gabrielse.physics.harvard.edu/gabrielse/overviews/Antihydrogen/Antihydrogen.html.
[12] http://alpha-new.web.cern.ch/.
[13] http://asacusa.web.cern.ch/ASACUSA/.
[14] M. Amoretti *et al.*, *Nature* **419** (2002) 456.
[15] G. Gabrielse *et al.*, *Phys. Rev. Lett.* **89** (2002) 213401.
[16] G.B. Andresen, *Nature* **468** (2010) 673.
[17] G.B. Andresen, *Nature Physics* **7** (2011) 558.
[18] G. Gabrielse *et al.*, *Phys. Rev. Lett.* **108** (2012) 113002.
[19] C. Amole *et al.*, *Nature* **483** (2012) 439.
[20] http://aegis.web.cern.ch/aegis/home.html.
[21] http://irfu.cea.fr/en/Phocea/Vie_des_labos/Ast/ast_technique.php?id_ast=2095.
[22] AD User Community, ELENA: An Upgrade to the Antiproton Decelerator, CERN-SPSC-2009-026/SPSC-P-338 (2009).
[23] S. Maury, The antiproton decelerator: AD, *Hyperfine Interactions* **109** (1997) 43–52; S. Maury, Status of the antiproton decelerator: AD, *Nuclear Physics B (Proc. Suppl.)* **56 A** (1997) 349–357; S. Maury and D. Möhl, Simplified antiproton schemes for antihydrogen production in traps, PS/AR Note 95-17.
[24] M. Giovannozzi, J.-Y. Hemery, C. Metzger and U. Mikkelsen, Experimental area of the CERN antiproton decelerator, CERN-PS/98-047 (CA) (1998).

# Part 3

# Particle Accelerators for the XXI Century

# Chapter 20

# Super B factories

Katsunobu Oide (KEK)

## 1 Introduction

The B-meson factories in the previous generation, KEKB and PEP–II, have achieved peak luminosities, 2.1 and 1.2 × $10^{34}\,\text{cm}^{-2}\text{s}^{-1}$, respectively. Thus a Super B factory as the next step should realize a luminosity around $10^{36}\,\text{cm}^{-2}\text{s}^{-1}$. The luminosity $\mathcal{L}$ of a ring collider, including asymmetric ones, can be expressed by the following expression:

$$\mathcal{L} = \frac{\gamma_\pm}{2er_e}\left(1+\frac{\sigma_y^*}{\sigma_x^*}\right)\left(\frac{I\xi_y}{\beta_y^*}\right)_\pm\left(\frac{R_\mathcal{L}}{R_y}\right), \tag{1}$$

where $\gamma$, $e$, $r_e$, $\sigma^*_{x,y}$, $I$, $\beta^*_{x,y}$ are the Lorentz factor, electron charge, classical electron radius, beam sizes at the interaction point (IP), stored beam current in the ring, and the $\beta$-function at the IP, respectively. The suffix $\pm$ denotes each beam. The expression (1) is obtained from the beam–beam tune-shift parameter

$$\xi_{\pm x,y} = \frac{r_e}{2\pi\gamma_\pm}\frac{N_\mp\beta^*_{\pm x,y}}{\sigma^*_{x,y}\left(\sigma^*_x+\sigma^*_y\right)}R_{x,y} \tag{2}$$

and the definition of luminosity

$$\mathcal{L} = \frac{N_+N_-f}{4\pi\sigma_x^*\sigma_y^*}R_\mathcal{L} \tag{3}$$

where $N$ and $f$ are the number of particles per bunch and the collision frequency ($I = Nef$), respectively, and we have assumed the beam sizes are common in two beams. The factors $R_{\mathcal{L},x,y}$ are the geometric reduction factors due to the hour-glass effect and the crossing angle.

To boost the luminosity, the most essential parameters are in the ratio $I, \xi_y$, and $\beta^*_y$ in Eq. (1). One design, once considered in KEK, was to go for an ultimate of a head-on collision scheme, hoping to achieve a very high $\xi_y \sim 0.3$

with a horizontal tune close to a half-integer. Such a scenario assumed gains of $\times 5$ in $\xi_y$, $\times 4$ in $I, \times 2$ in $1/\beta_y^*$ to achieve $\times 40$ in the luminosity. The actual numbers were around $\xi_y = 0.3, I = 9$ A, $\beta_y = 3\,\mathrm{mm}$, and $\mathcal{L} \sim 8 \times 10^{35}\,\mathrm{cm}^{-2}\mathrm{s}^{-1}$. Such a head-on scheme had a number of difficulties:

- The assumed very high $\xi_y$, predicted in simulations, was not verified in the crabcrossing experiment for three years in KEKB. The observed $\xi_y \sim 0.09$ was significantly smaller than the prediction $\sim 0.15$ in the case of KEKB, even after application of sophisticated tuning of the machine including chromatic $x{-}y$ coupling correction.
- The operation near a half-integer horizontal tune with a high beam–beam parameter induces large dynamic $\beta$ and dynamic emittance, which limit the aperture at the final quadrupoles. They also cause heavy synchrotron radiation at these quadrupoles which increase the detector background and heating. While they could be mitigated by a larger crossing angle, that would make the crabcrossing difficult.
- It was necessary to reduce the bunch length down to $\sigma_z \lesssim \beta_y^*$ in the head-on scheme to avid an hour-glass effect. Such a short bunch length can generate a significant amount of coherent synchrotron radiation with the very high beam current, causing severe longitudinal instability that makes such a short bunch impossible.

These reasons were enough to abrogate the head-on scheme eventually. Fortunately an alternative idea, *nano-beam scheme*, developed by P. Raimondi saved the next generation B-factories as described below.

## 2 Nano-beam scheme

The idea by Raimondi consists of:

- A large crossing angle. in terms of *Piwinski angle*, $\theta_x\sigma_z/\sigma_x \gg 1$, where $\theta_x$ is the half horizontal crossing angle.
- A very short $\beta_y \ll \sigma_z$.
- Small horizontal/vertical emittances.

The formula for the luminosity, Eq. (1) is still valid in the nano-beam scheme. The major gain of the luminosity comes from a very short $\beta_y^*$. Unlike a head-on scheme, the bunches intersect one another only at the central part close to the IP in the nano-beam scheme due to the large Piwinski angle as shown in Fig. 1.

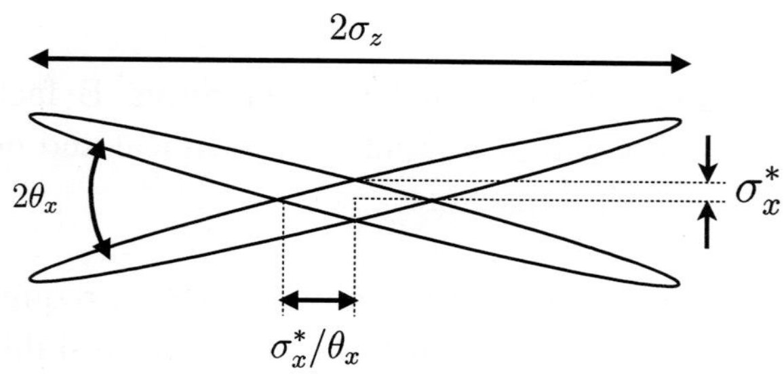


Fig. 1. Beam crossing in the nano-beam scheme.

As the length of the intersecting region is $\sigma_x^* \ll \sigma_z$, the condition to avoid an hour-glass effect becomes $\beta_y^* \lesssim \sigma_x^* \ll \sigma_z$. In the case of SuperKEKB, it is possible to choose $\beta_y^* \sim 0.3\,\mathrm{mm}$, which gives $\times 20$ gain compared to KEKB. Then the bunch length $\sigma_z$ is not necessary to be very short, thus it is possible to avoid unfavorable effects such as the coherent synchrotron radiation. This scheme does not require a very high $\xi_y$ ($\xi_y \sim 0.09$) which has been achieved at KEKB, and is assumed at SuperKEKB. The crossing angle itself can be larger than the previous machines, since it does not need crab crossing anymore. As it does not require an operation near a half-integer tune, and also the horizontal beam–beam parameter is very small, the dynamic emittance effect is not an issue. The situation of the nano-beam crossing is similar to a collision with many micro-bunches which have a short bunch length $\sigma_x^*/\theta_x \ll \sigma_z$.

Actually Raimondi has proposed another idea, *crabbed waist scheme*, on top of the nanobeam scheme. The crabbed waist aligns the vertical waist along the center line of the other beam to reduce the dependence of the beam–beam effect on the horizontal displacement of a particle. This scheme should improve $\xi_y$ by reducing the synchrotron–betatron coupling caused by the crossing angle. Although the crabbed waist scheme has merits on the collision itself, its realization needs further study. A simple way to introduce the crabbed waist is to install a pair of sextupoles in both sides of the IP. Although the unnecessary nonlinearities of these sextupoles can be canceled by $-I$ or $I$ transformation between the pair, the inevitable nonlinearity around the IP interferes the cancellation to reduce the dynamic aperture drastically. Such nonlinearities include the fringe field of the final quadrupoles, geometric nonlinearities at the IP, and nonlinear fields in the quadrupoles and solenoids. As these terms increase for smaller $\beta_y^*$, the solution may be non-trivial and not yet found at least at SuperKEKB. Thus the crabbed waist scheme is not considered as the base line at SuperKEKB.

## 3 Issues

Although no fatal issues have been found for Super B factories with the nano-beam scheme, there will be a number of technical and beam dynamical issues exist to realize such a machine:

- The small spot size at the IP, especially in vertical, requires the stability of the orbit and beam size during the collision. Stability at a few nm at the IP will be necessary to avoid beam blowup due to the collision offset. Also some sensitive magnets such as final focus quadrupoles must be stable at a few $\mu$m level to keep the vertical emittance.
- Although the horizontal and vertical emittances of Super B factories are not smaller than some advanced light sources, it will not be easier with the existence of the beam–beam effects and strong nonlinearities around the IP. The intrabeam scattering and space charge effects will add more difficulties and complications.
- The beam–beam effect with the large crossing angle is less trivial. Although a strong–weak model has been carried out, a strong–strong simulation is not yet complete as it requires high number of slices and particles to reach reliable results. Also combined effects with lattice non-linearities and space–charge effects will be important.
- As the $\beta_y^*$ is so small as a linear collider, the resulting dynamic aperture is small even with dedicated chromaticity correction system. It will reduce the Touschek lifetime especially in the low energy ring and also the injection efficiency.
- Electron clouds in the positron ring may need special attentions considering the orders smaller vertical emittance than the previous B factories.
- Injection to the collider rings need a low-emittance beam at the injector. A damping ring is necessary at least for the positrons.
- All power supplies, beam diagnostics, feedbacks need higher precision stability, and quicker response.

## 4 SuperKEKB

As of the year 2013, SuperKEKB is the only Super B factory under construction. The construction has been going on since 2010 by inheriting and modifying many components of previous KEKB. It will start the beam operation in early 2015 to accumulate luminosity of 50 ab$^{-1}$ in around 10 years. The design parameters are listed in Table 1.

Table 1. Design parameters of SuperKEKB compared to KEKB with/without crab crossing. 1/nb = $10^{33}$ cm$^{-2}$s$^{-1}$.

| | SuperKEKB design | | KEKB (no crab) 11/15/2006 | | KEKB (crab) 6/17/2009 | | |
|---|---|---|---|---|---|---|---|
| Parameters | LER | HER | LER | HER | LER | HER | Unit |
| Circumference | | | 3016 | | | | m |
| Beam energy | 4.0 | 7.0 | 3.5 | 8.0 | 3.5 | 8.0 | GeV |
| Eff. crossing angle | 83 | | 22 | | 0 (crab) | | mrad |
| Beam current | 3.8 | 2.6 | 1.65 | 1.33 | 1.64 | 1.19 | A |
| Bunches | 2500 | | 1389 | | 1584 | | |
| Bunch current | 1.5 | 1.0 | 1.19 | 0.96 | 1.03 | 0.71 | mA |
| Bunch spacing | 1.2 | | 1.8–2.4 | | 1.8 | | m |
| Hor. emittance $\varepsilon_x$ | 3.2 | 4.6 | 18 | 24 | 18 | 24 | nm |
| RF frequency | | | 509 | | | | MHz |
| Bunch length $\sigma_z$ | 6 | 5 | 8 | 6 | 8 | 6 | mm |
| $\beta_x^*$ | 3.2 | 2.5 | 59 | 56 | 120 | 120 | cm |
| $\beta_y^*$ | 0.27 | 0.3 | 0.65 | 0.59 | 0.59 | 0.59 | cm |
| Hor. size @ IP | 10 | 11 | 103 | 116 | 147 | 170 | $\mu$m |
| Ver. size @ IP | 0.048 | 0.048 | 1.9 | 1.9 | 0.94 | 0.94 | $\mu$m |
| Beam–beam $\xi_x$ | 0.0028 | 0.0012 | 0.115 | 0.075 | 0.125 | 0.100 | |
| Beam–beam $\xi_y$ | 0.088 | 0.081 | 0.104 | 0.058 | 0.130 | 0.090 | |
| Luminosity | 800 | | 17.6 | | 21.1 | | /nb/s |
| $\int$Lum./day | — | | 1260 | | 1479 | | /pb |
| $\int$Lum./7 days | — | | 7.82 | | 8.43 | | /fb |
| $\int$Lum./30 days | — | | 30.2 | | 23.0 | | /fb |
| Total $\int$Lum. | 50,000 (goal) | | 1,040 | | | | /fb |

## References


[1] P. Oddone, *Proceedings of the UCLA Workshop: Linear Collider BB Factory Conceptual Design*, ed., D. Stork, p. 243 (1987).

[2] K. Hirata and E. Keil, *Nucl. Instr. Meth. A* **292** (1990) 156.

[3] PEP-II: An asymmetric B factory. Conceptual design report, SLAC–418, QCD183:S56:1993 (1993).

[4] KEKB B–Factory design report, KEK Report **95–7** (1995).

[5] J. Seeman, *Conf. Proc. C* **0806233** (2008) TUXG01.

[6] Y. Funakoshi, T. Abe, K. Akai, Y. Cai, K. Ebihara, K. Egawa, A. Enomoto, J. Flanagan *et al.*, *Conf. Proc. C* **100523** (2010) WEOAMH02.

[7] A. Piwinski, *IEEE Trans. Nucl. Sci.* **24** (1977) 1408.

[8] R. B. Palmer, *Proc. of 1988 DPF Summer Study on High-Energy Physics in the 1990s* (Snowmass 88), Snowmass, Colorado, 27 Jun–15 Jul 1988 (1988), p. 613.

[9] K. Oide and K. Yokoya, *Phys. Rev. A* **40** (1989) 315.

[10] M. Sullivan, G. Bowden, H. DeStaebler, S. Ecklund, J. Hodgson, T. Mattison, M. E. Nordby, A. Ringwall *et al.*, *Conf. Proc. C* **960610** (1996) 460.

[11] B. Aubert *et al.* [BABAR Collaboration], *Nucl. Instrum. Meth. A* **479** (2002) 1.

[12] J. Seeman, M. Sullivan, M. Biagini, Y. Cai, F. J. Decker, M. Donald, S. Ecklund, A. Fisher *et al.*, *Proc. EPAC 2002*, 3–7 Jun 2002, Paris, France p. 434–436.
[13] Y. Yamazaki and T. Kageyama, *Part. Accel.* **44** (1994) 107.
[14] T. Furuya *et al.*, *Gif-sur-Yvette 1995, RF superconductivity*, Vol. **2** (1995), p. 729.
[15] H. Schwarz and R. Rimmer, *Conf. Proc. C* **940627** (1994) 1882.
[16] J. Fox, T. Mastorides, C. Rivetta, D. Van Winkle and D. Teytelman, *Phys. Rev. ST Accel. Beams* **13** (2010) 052802.
[17] Y. Suetsugu, K. Shibata, H. Hisamatsu, M. Shirai and K. Kanazawa, *Vacuum* **84** (2009) 694.
[18] M. T. F. Pivi, F. King, R. E. Kirby, T. Markiewicz, T. O. Raubenheimer, J. Seeman and L. Wang, *Conf. Proc. C* **0806233** (2008) MOPP064.
[19] Y. Suetsugu, H. Fukuma, L. Wang, M. Pivi, A. Morishige, Y. Suzuki, M. Tsukamoto and M. Tsuchiya, *Nucl. Instrum. Meth. A* **598** (2009) 372.
[20] T. Mimashi, T. Ieiri, M. Kikuchi, A. Tokuchi and K. Tsuchida, *Conf. Proc. C* **0806233** (2008) TUPD011.
[21] M. Izawa, Y. Sato, T. Toyomasu, *Phys. Rev. Lett.* **74** (1995) 5044.
[22] K. Ohmi, *Phys. Rev. Lett.* **75** (1995) 1526.
[23] K. Ohmi and F. Zimmermann, *Phys. Rev. Lett.* **85** (2000) 3821.
[24] T. O. Raubenheimer and F. Zimmermann (SLAC), *Phys. Rev. E* **52** (1995) 5487.
[25] J. W. Flanagan, K. Ohmi, H. Fukuma, S. Hiramatsu, M. Tobiyama and E. Perevedentsev, *Phys. Rev. Lett.* **94** (2005) 054801.
[26] H. Fukuma, J. Flanagan, K. Hosoyama, T. Ieiri, T. Kawamoto, T. Kubo, M. Suetake and S. Uno *et al.*, *AIP Conf. Proc.* **642** (2003) 357.
[27] H. Jin, M. Yoon, K. Ohmi, J. W. Flanagan and M. A. Palmer, *Jap. J. Appl. Phys.* **50** (2011) 026401.
[28] At least an application for a storage ring is seen in R. Servranckx and K. L. Brown, *IEEE Trans. Nucl. Sci.* **26** (1979) 3598.
[29] K. Oide, H. Koiso and K. Ohmi, *AIP Conf. Proc.* **391** (1997) 215.
[30] K. Oide and H. Koiso, *Phys. Rev. E* **47** (1993) 2010.
[31] J. Irwin, C. X. Wang, Y. T. Yan, K. L. F. Bane, Y. Cai, F. J. Decker, M. G. Minty and G. V. Stupakov *et al.*, *Phys. Rev. Lett.* **82** (1999) 1684.
[32] K. Akai, N. Akasaka, A. Enomoto, J. Flanagan, H. Fukuma, Y. Funakoshi, K. Furukawa and T. Furuya *et al.*, *Nucl. Instrum. Meth. A* **499** (2003) 191.
[33] Y. T. Yan and Y. Cai, *Nucl. Instrum. Meth. A* **558** (2006) 336.
[34] T. Ieiri, K. Akai, H. Fukuma and M. Tobiyama, *Nucl. Instrum. Meth. A* **606** (2009) 248.
[35] M. Tejima, M. Arinaga, T. Ieiri, H. Ishii, H. Fukuma, M. Tobiyama and S. Hiramatsu, *Conf. Proc. C* **0505161** (2005) 3253.
[36] K. Satoh and M. Tejima, *Conf. Proc. C* **950501** (1995) 2482.
[37] T. Mitsuhashi, J. W. Flanagan and S. Hiramatsu, *Proc. Seventh EPAC2000*, 26–30 Jun 2000, Vienna, Austria, p. 1783–1785.
[38] J. W. Flanagan, N. Akasaka, H. Fukuma, S. Hiramatsu, T. Mitsuhashi, T. Naito, K. Ohmi, K. Oide *et al.*, *Proc. Seventh EPAC2000*, 26–30 Jun 2000, Vienna, Austria, p. 1119–1121.
[39] N. Akasaka, A. Akiyama, S. Araki, K. Furukawa, T. Katoh, T. Kawamoto, I. Komada, K. Kudo *et al.*, *Nucl. Instrum. Meth. A* **499** (2003) 138.
[40] Y. Funakoshi, M. Masuzawa, K. Oide, J. Flanagan, M. Tawada, T. Ieiri, M. Tejima, M. Tobiyama *et al.*, *Phys. Rev. ST Accel. Beams* **10** (2007) 101001.

[41] L. Hendrickson, T. Gromme, P. Grossberg, T. Himel, D. Macnair, R. Sass, H. Smith, N. Spencer *et al.*, *Proc. Seventh EPAC2000*, 26–30 Jun 2000, Vienna, Austria p.1897–1899.
[42] K. Ohmi, K. Oide and E. Perevedentsev, *Conf. Proc. C* **060626** (2006) 616.
[43] A. Drago, J. D. Fox, D. Teytelman and M. Tobiyama, *Conf. Proc. C* **0806233** (2008) THPC116.
[44] Y. Ogawa, A. Enomoto, K. Furukawa, T. Kamitani, M. Satoh, T. Sugimura, T. Suwada, Y. Yano *et al.*, *Conf. Proc. C* **060626** (2006) 2700.
[45] T. Kamitani, N. Delerue, M. Ikeda, K. Kakihara, S. Ohsawa, T. Oogoe, T. Sugimura, T. Takatomi *et al.*, *Conf. Proc. C* **0505161** (2005) 1233.
[46] K. Akai, J. Kirchgessner, D. Moffat, H. Padamsee, J. Sears, T. Stowe and M. Tigner, *Proc. B Factory Workshop*, 6–10 Apr 1992, Stanford, California.
[47] K. Hosoyama, K. Akai, K. Ebihara, T. Furuya, K. Hara, T. Honma, A. Kabe, Y. Kojima *et al.*, *Conf. Proc. C* **0806233** (2008) THXM02.
[48] K. Ohmi, M. Tawada, Y. Cai, S. Kamada, K. Oide and J. Qiang, *Phys. Rev. ST Accel. Beams* **7** (2004) 104401.
[49] Y. Ohnishi, K. Ohmi, H. Koiso, M. Masuzawa, A. Morita, K. Mori, K. Oide, Y. Seimiya *et al.*, *Phys. Rev. ST Accel. Beams* **12** (2009) 091002.

# Chapter 21

# The Relativistic Heavy Ion Collider

Wolfram Fischer (BNL)

The Relativistic Heavy Ion Collider (RHIC), shown in Fig. 1, was build to study the interactions of quarks and gluons at high energies [Harrison, Ludlam and Ozaki (2003)]. The theory of Quantum Chromodynamics (QCD) describes these interactions. One of the main goals for the RHIC experiments was the creation and study of the Quark–Gluon Plasma (QGP), which was expected to be formed after the collision of heavy ions at a temperature of approximately 2 trillion kelvin (or equivalently an energy of 150 MeV). The QGP is the substance which existed only a few microseconds after the Big Bang. The QGP was anticipated to be weakly interacting like a gas but turned out to be strongly interacting and more like a liquid. Among its unusual properties is its extremely low viscosity [Auerbach and Schlomo (2009)], which makes the QGP the substance closest to a perfect liquid known to date. The QGP is opaque to moderate energy quarks and gluons leading to a phenomenon called jet quenching, where of a jet and its recoil jet only one is observable and the other suppressed after traversing and interacting with the QGP [Jacak and Müller (2012)].

After the discovery of the QGP its characterization required collider operation over a wide parameter range (see Table 1) for the exploration of the QCD phase diagram, i.e. the study of QCD matter as a function of baryon density and temperature — similar to the study of water as a function of pressure and temperature. Unlike colliders for high-energy physics that run usually with a single particle combination at a specific energy, RHIC has operated at many energies and with a number of ions, including different ions in both rings. This flexibility was an important design criterion, and continues to guide the upgrades.

Discoveries that were only possible with large integrated luminosities are the observation of the heaviest anti-matter nuclei to date in gold–gold

Fig. 1. Aerial view of RHIC with the two experiments STAR and PHENIX, and the injector complex consisting of the LINAC (protons), Electron Beam Ion Source (EBIS) (heavy ions), Booster, and AGS. NSRL is the NASA Space Radiation Laboratory, a facility to irradiate materials and organisms with beams from the Booster.

Table 1. RHIC species combinations and ion energies from 2000 to 2012.

| Species | Ion energy [GeV/nucleon] |
|---|---|
| $^{238}U^{92+}$ — $^{238}U^{92+}$ | 96.4 |
| $^{197}Au^{79+}$ — $^{197}Au^{79+}$ | 3.85, 4.6, 5.75, 9.8, 13.5, 19.5, 27.9, 31.2, 65.2, 100.0 |
| $^{63}Cu^{29+}$ — $^{197}Au^{79+}$ | 99.9/100.0 |
| $^{63}Cu^{29+}$ — $^{63}Cu^{29+}$ | 11.2, 31.2, 100.0 |
| d — $^{197}Au^{79+}$ | 100.7/100.0 |
| p↑ — p↑ | 31.2, 100.2, 204.9, 249.9, 254.9 |

collisions, such as the anti-hypertriton ($^{3}_{\Lambda}\overline{\rm H}$) [STAR (2010)] and anti-helium-4 nucleus ($^{4}\overline{\rm He}$) [STAR (2011)], a record that is poised to stand for a long time.[1] Recently the Large Hadron Collider (LHC) has also started to collide lead ions [Jowett *et al.* (2011)], as well as protons with heavy ions. Heavy ion experiments at the LHC and RHIC are complimentary. The LHC energy is

[1] The probability to obtain anti-nuclei is reduced by a factor of more than 100 for every anti-nucleon added to the anti-nucleus [STAR (2011)], and there are no stable matter nuclei with 5 nucleons. One would then expect the need for 10000 times more luminosity to produce anti-lithium-6 ($^{6}\overline{\rm Li}$) at the same energy. Although anti-matter production rates at the LHC are higher due to the higher energy this may still not be sufficient.

up to 27 times higher than RHIC's thereby probing the QCD matter at a higher temperature, further away from the phase transition to the QGP.

RHIC is also the only collider of spin-polarized protons [Alekseev *et al.* (2003)]. Even after 20 years of experimental efforts, the question of individual parton (quarks and gluons) contributions to the spin of the nucleon is not yet solved. Several deep inelastic scattering experiments showed that the spin of the nucleon cannot be explained by the contribution of the quarks alone. The newest results from world data affirm that the contribution from quarks in the currently measured region (parton momentum fraction $x > 0.001$) gives only 30% of the total spin of the proton. The remainder of the proton spin must come from the gluons and the orbital angular momentum of the quarks and gluons. As of today no clear indication of the size of the orbital angular momentum of quarks and gluons is available. The polarized proton beams provide unique opportunities for the study of the polarized quark and gluon spin structure of the proton and QCD dynamics at a high energy scale and is therefore complementary to existing deep inelastic scattering experiments. Recent data from RHIC have for the first time shown a non-zero contribution of the gluons to the proton spin of 10% in the currently still limited momentum range $0.05 < x < 0.2$ [Aschenauer (2012)]. Despite this exciting new result the mystery of the proton spin is not yet solved and needs further studies. These require an increase in the center-of-mass energy in the collisions or different particle beams, like polarized neutrons. Neutrons cannot be accelerated alone but must be attached to charged nuclei in the form of, for example, polarized deuterons or polarized $^3$He.

One of the RHIC hadron rings can also be the basis for a new collider, the electron–ion collider eRHIC [Ptitsyn *et al.* (2011a)]. Although at a lower center-of-mass energy than the previously existing electron–proton collider HERA, eRHIC would go beyond HERA in a number of ways: (i) the targeted luminosity is about two orders of magnitude higher; (ii) the collision energy is variable over a wide range; (iii) the hadron beam species can range from protons to uranium; and (iv) polarized light ions like protons and $^3$He are available. eRHIC is to explore the parameter regime of strong gluon interactions. As the name already suggests, the gluons bind matter into strongly interacting particles, and the binding energy of the hadrons accounts for almost all of their mass.

With substantial upgrades completed, and further upgrades under way, RHIC has another decade of productive investigations of the Quark–Gluon Plasma and spin physics ahead [Vigdor (2012)], and can then be the basis for the election–ion collider eRHIC [Aschenauer *et al.* (2012)].

## 1 Machine overview

The major design considerations for RHIC were (i) the energy reach and range; (ii) the luminosity; (iii) flexibility in both ion species and collision energies; and (iv) spin polarization for proton operation.

*Energy.* RHIC was designed to fit into the existing ISABELLE tunnel [Sanford (1977)]. With a circumference of 3834 m and superconducting magnets with a maximum dipole field strength of 3.5 T, an energy of 100 GeV/nucleon can be reached for fully stripped gold ions. The equivalent proton energy is 250 GeV. Previous heavy ion collision experiments were done with fixed targets and gold beams with an energy of 10 GeV/nucleon at the AGS (e.g. [Bloomer, Love and Waters (1991)]), and lead beams with an energy of 158 GeV/nucleon at the CERN SPS (e.g. [Blaizot, Dinh and Ollitrault (2000)]). The center-of-mass energy is approximately $\sqrt{2mc^2E}$ for fixed target experiments (with $m$ and $E$ the rest mass and total energy of a particle, and $c$ the speed of light) but $2E$ for colliders, giving colliders a large advantage in the energy reach. With RHIC the center-of-mass energy was increased by more than an order of magnitude compared to the SPS. The LHC increased the center-of-mass energy again by up to a factor of 27, so that a very wide energy range is now available to study QCD matter. The energy range accessible through RHIC spans the transition energy between cold nuclear matter and the QGP, making the collider the tool of choice for studying the phase transitions between these different forms of matter.

*Luminosity.* While the energy determines what physical process can be observed in principle, the luminosity $\mathcal{L}$ determines how many events are created, i.e. the likelihood that a particle in one beam interacts with a particle in the other beam. The event rate is given by $\mathcal{L}\sigma_i$ where $\sigma_i$ is the cross section for the process under study. A collider has a large advantage in energy reach compared to a fixed target experiments, but it has a disadvantage when it comes to luminosity since beams are much less dense than fixed targets (for the RHIC gold beams by about a factor of $10^{13}$ compared to the solid material). Luminosity considerations are therefore central to design, operation and upgrades of colliders. The luminosity is only determined by the geometric properties of the colliding beams at the interaction point (IP) and can be expressed as

$$\mathcal{L}(t) = \frac{f_{rev} N N_b^2(t)}{4\pi\sigma_t^2(t)} h(\sigma_t(t), \sigma_s(t), \theta) \qquad (1)$$

where $f_{rev}$ is the revolution frequency of the particles, $N$ is the number of bunches, $N_b$ the number of ions per bunch, $\sigma_t$ the transverse rms beam size (assumed to be the same for the horizontal and vertical plane, and for both beams), and $\sigma_s$ the longitudinal rms bunch length. The factor $h(\sigma_t, \sigma_s, \theta)$ is smaller than and of order 1, and describes the luminosity reduction due to the so-called hourglass effect (long bunches have a longitudinal hourglass shape at the interaction point), and a finite crossing angle $\theta$. From Eq. (1) one can deduce that for high luminosity it is desirable to have a large number of bunches $N$ that contain a large number of ions $N_b$, and are tightly focused at the interaction point so that $\sigma_t$ is small. The time-dependence of the luminosity $\mathcal{L}(t)$ comes from the fact that throughout a store ions are lost from the bunches, and typically the beams increase in transverse and longitudinal size.

RHIC was designed for $N = 60$ with an upgrade option to double the number of bunches. The number of ions per bunch $N_b$ can be limited by many effects. For RHIC the two most important ones are the injector chain and an instability during acceleration in RHIC when the so-called transition energy is crossed and, for a short moment, bunches become very short and lose longitudinal focusing.

The injector chain was prepared over a number of years to produce intense bunches of heavy ions ($N_B$ in Eq. (1)) at a source and accelerate them through the Tandem [Steski *et al.* (2001)], AGS Booster [Weng *et al.* (1991)], and AGS [Ahrens *et al.* (1994)]. During the several acceleration stages the charge state $Z$ of the ions has to be carefully selected to reduce both space charge effects (therefore favoring small $Z$ at low energies) and charge–change processes (electron stripping and capturing), which lead to beam loss and vacuum pressure increases (see e.g. [Roser, Ahrens and Hseuh (1994)]). Recently a significant upgrade was made to the ion source (see below), increasing its flexibility and ion species range, and reducing its operating cost.

The transverse beam size $\sigma_t$ at the interaction point is determined by the beam emittance $\epsilon_n$ and the focusing parameter $\beta^*$ so that the beam size is given as $\sigma_t = \sqrt{\epsilon_n \beta^* / (\beta\gamma)}$, where $(\beta\gamma)$ are the relativistic factors. Focusing is achieved through a triplet of focusing quadrupoles near the interaction point, and the focusing strength is limited by the achievable gradient in the quadrupoles [Anerella *et al.* (2003)].

In Eq. (1) the bunch intensity $N_b(t)$ and the transverse and longitudinal beam dimensions $\sigma_{t,s}(t)$ are written as time-dependent quantities. There are many effects that can change $N_b$ and $\sigma_{t,s}$. Of course $N_b$ is reduced by the collision of particles in one beam with the particles in the other beam (often referred to as "burn-off"), an effect for which the collider was built. But there are other effects that lead to beam size growth and eventual particle loss due to the dynamic and physical apertures. For heavy ions the most important beam growth effect is intrabeam scattering (IBS) [Chao and Tigner (1998)]. IBS is a process in which ions in the beam scatter off each other due to their electric charge, leading to beam size growth. Growth rates scale with $Z^4/A^2$ (charge state $Z$ and mass number $A$) and are particularly severe for the highly charged fully stripped heavy ions ($Z = 79$ for gold). In RHIC the IBS growth rates for gold are an order of magnitude larger than those for protons, and are of order one hour. Because of IBS, strong focusing is desirable in all dimensions. Strong transverse focusing was achieved with relatively short FODO cells [Chao and Tigner (1998)], in which quadrupoles are only about 10 m apart. For enhanced longitudinal focusing a second radio frequency system (with frequency 197 MHz) was installed in addition to the 28 MHz acceleration system. IBS leads to rapidly decreasing luminosities in stores as the curve for un-cooled beams in Fig. 2 illustrates. The primary constraint from IBS has guided the main upgrade with stochastic cooling (see below). For protons the main effect for beam size growth and loss is the beam–beam effect in which the long-term stability of the particle motion is affected by the periodic exposure to the nonlinear electromagnetic field generated by the other beam [Chao and Tigner (1998)].

*Flexibility.* Two Tandem accelerators allowed for various ion species including two different ones. RHIC was designed with two independent rings to allow for maximum flexibility in the selection of the species. Only two dipoles are shared between the rings in each of the six interaction regions so that the beams can be joined for collisions, and separated again.

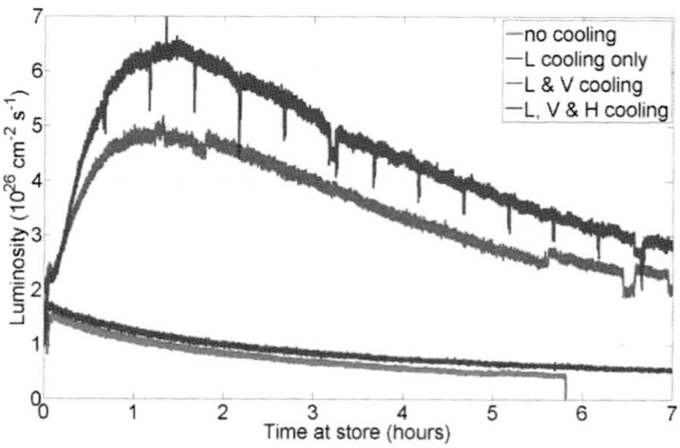


Fig. 2. Luminosity (collision rate) for stores of colliding uranium beams in RHIC without cooling; with longitudinal cooling only; with longitudinal and vertical cooling; and with cooling in all planes.

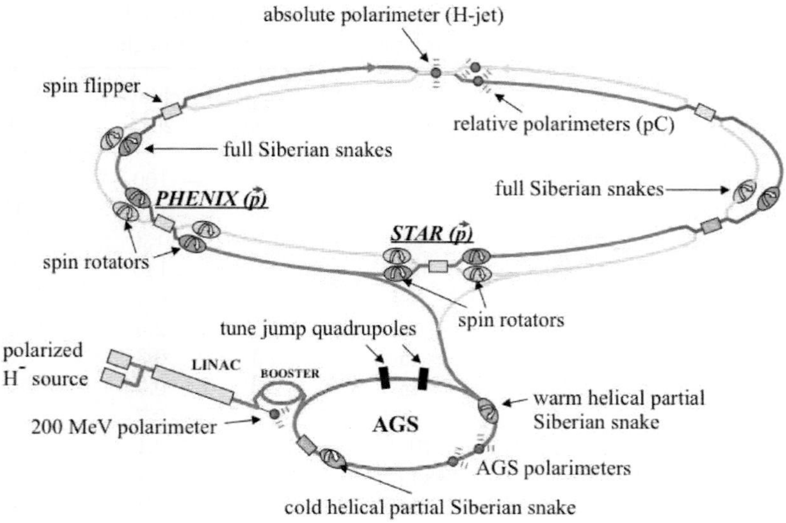


Fig. 3. Overview of special devices for polarized proton operation: source, relative and absolute polarimeters, partial and full snakes, tune jump quadrupoles, spin rotators, and spin flipper.

*Polarized protons.* RHIC is also the only collider of spin-polarized protons [Alekseev *et al.* (2003)]. Protons do not self-polarize when stored due to the lack of synchrotron radiation. They must be created in a polarized source [Zelenski (2010)], and polarization must be maintained throughout the acceleration and storage during collisions [Hoffstaedter (2006); Lee (1997)]. This requires special spin-manipulating magnets, so-called Siberian snakes in the AGS and RHIC (see Fig. 3). With these snakes the harmful effects of depolarizing resonances can be overcome. In the AGS there are also two fast quadrupoles installed to jump over 80 relatively weak depolarizing resonances [Schoefer *et al.* (2012)]. The RHIC snakes are installed in opposite locations in both rings, and flip the spin from up to down, or down to up thereby canceling depolarizing effects that build up over many revolutions. Spin rotators near the experiments rotate the spin from the stable vertical direction in the arc into the longitudinal direction in the experiments, and back. Polarized proton operation also requires polarization measurements at various energies [Alekseev *et al.* (2003)].

Initially RHIC provided collisions for two large experiments (STAR and PHENIX) and two smaller experiments (PHOBOS and BRAHMS). The two small experiments have completed their data taking period in 2005 and 2006 respectively. The two large experiments were upgraded several times, with further upgrades planned. The parameters for gold and polarized proton beams are listed in Table 2 and Table 3 respectively.

Table 2. RHIC gold beam parameters with design, achieved, and upgrade values. Unless noted otherwise parameters are given for the beginning of store.

| Parameter | Unit | Design 2000 | Achieved 2011 | Upgraded ≥2014 |
|---|---|---|---|---|
| ion energy $E$ | GeV/nucleon | | — 100 — | |
| no. of bunches $N$ | ... | 60 | 111 | 111 |
| ions per bunch $N_b$ | $10^9$ | 1.0 | 1.3 | 1.6 |
| average beam current/ring | mA | 59 | 146 | 176 |
| stored beam energy | MJ | 0.12 | 0.47 | 0.56 |
| norm. rms emittance $\epsilon_n$ | $\mu$m | 2.5 | 2.5 | 4.2 |
| rms bunch length $\sigma_s$ | m | 0.3 | 0.3 | 0.3 |
| IP envelope function $\beta^*$ | m | 2.0 | 0.75 | 0.7 |
| hourglass factor $h$ | ... | 1.0 | 0.7 | 0.9 |
| beam–beam parameter $\xi$/IP | $10^{-3}$ | 2.3 | 2.1 | 1.5 |
| initial luminosity $L_{peak}$ | $10^{26}$ cm$^{-2}$s$^{-1}$ | 8 | 50 | 60 |
| average luminosity $L_{avg}$ | $10^{26}$ cm$^{-2}$s$^{-1}$ | 2 | 30 | 60 |
| calendar time in store | % | 60 | 59 | 60 |
| integrated $L$ | nb$^{-1}$/week | 0.07 | 1.0 | 2.2 |

## 2 Heavy ion operation and upgrades

RHIC commissioning with gold beam began in 2000. After operation at full energy was established in 2001, the design luminosity was also demonstrated in that year (Table 2) [Fischer (2013)]. In 2003 the first asymmetric operation followed (d–Au), and in 2004 the energy scan for Au–Au collisions started. A full energy scan, extending to energies as low as 40% of the nominal injection energy, followed in 2010 for the determination of the critical point in the QCD phase diagram. All species combinations and energies are listed in Table 1. Table 2 shows the gold beam parameters as of 2011, showing an increase in the average store luminosity by a factor of 15 over the design value.

Figure 4 displays the integrated luminosity $L_{NN} = A_1 A_2 \int \mathcal{L} dt$ for all the ion runs. By multiplying the luminosity with the mass numbers $A_{1,2}$ of species in the two beams, one can better compare the performance with different ion species. Note that the operating periods have become shorter over time, which requires greater efficiency in setting up a particular mode. For the Au–Au runs there was an increase in the slope by a factor 14 from 2001 to 2004, and about a factor of two when comparing the Au–Au runs in 2004 to 2007, 2007 to 2010, and 2010 to 2011. These dramatic increases in the luminosity are based on a series of upgrades.

The first upgrade was in the reduction of $\beta^*$ from the design value of 2 m to 1 m and later to 0.75 m (Table 2). Reductions in $\beta^*$ require a good

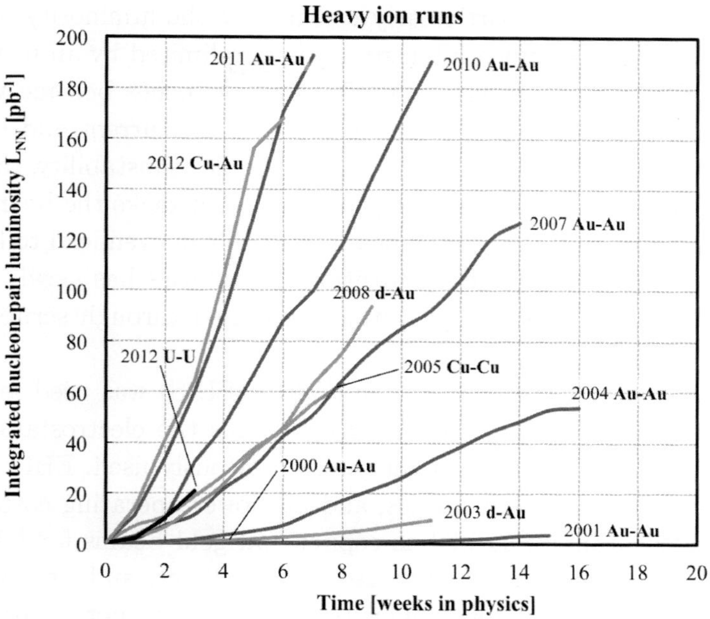


Fig. 4. Integrated luminosity as a function of operating time for the RHIC heavy ion runs.

control over all beam parameters as the beam size increase in the final focus triples, and beam losses become more likely.

In 2003 the first attempt was made to double the number of bunches. This led to an increase in the vacuum pressure in large parts of the ring, making operation impossible. After launching an investigation it was determined that the pressure increases were caused by electron clouds [Fischer *et al.* (2008)]. The positively charged beams can accelerate electrons in the beam pipe, generated for example through residual gas ionization, and when these electrons hit the beam pipe they can generate new electrons in turn. With a Secondary Electron Yield (SEY) greater than one (SEY $> 1$) one has electron multiplication and the formation of an electron cloud. Detrimental effects of the electron clouds include the release of gas molecules from the beam pipe wall, and direct interactions with the beam leading to instabilities and beam size growth. An upgrade program for the vacuum system was implemented which included baking of all warm sections, replacing warm beam pipes with pipes that have a Non-Evaporable Getter (NEG) coating with a low SEY and and large pumping speed, and pre-pumping the cold arcs before the cool-down so that less than a mono-layer of gas is adsorbed.

Over the years the bunch intensity was steadily increased in the injector chain, exceeding the design value by more than 50% in 2012. An increase in the bunch intensity is particularly useful since the luminosity scales with $N_b^2$ (see Eq. (1)). The bunch intensity $N_b$ is also limited by an instability at the so-called transition energy in RHIC, when bunches become very short and lose longitudinal focusing. The increased peak current and the loss of longitudinal focusing both make a beam vulnerable to instability. In addition, the short bunches create dense electron clouds that make the beams further unstable. Instabilities increase the beam size and can even lead to beam loss. The intensity threshold for the transition instabilities has slowly increased as the quality of the beam pipe surface increased through scrubbing with the generated electrons.

In 2012 a new Electron Beam Ion Source (EBIS) was used for the first time [Alessi *et al.* (2011)]. This source replaces the electrostatic Tandem accelerators [Thieberger (1984)] that were previously used. EBIS allows for greater flexibility in the ion species, and has lower operating costs. In 2012 EBIS delivered uranium as well as copper and gold beams for RHIC. Uranium is the heaviest element ever used in a collider, and uranium beams with intensities sufficient for collider operations could not be produced at the Tandem. Uranium ions are of interest to the experiments because of their prolate shape, which can create a QGP with even higher density than the more spherical gold nuclei.

The most impressive luminosity upgrade for heavy ions was the implementation of stochastic cooling during store [Blaskiewicz, Brennan and Mernick (2010)]. Stochastic cooling had been used in lower energy machines previously. This was the first implementation in a collider. In uranium operation with cooling in the longitudinal and both transverse planes the peak luminosity was increased by a factor of three and the average luminosity by a factor of five (Fig. 2). With cooling the lowest emittances in a hadron collider were achieved, and at the time of maximum luminosity in Fig. 2, 97% of the uranium ions are lost due to burn-off, also a record number for a hadron collider.

But even with longitudinal cooling the bunch length still increases due to IBS and yet another radio frequency system is scheduled to be installed to provide even more longitudinal focusing. This is the first superconducting radio frequency system in RHIC (56 MHz frequency). The voltage in the cavity, which is common to both beams, is generated by the beam itself [Ben-Zvi (2009)].

To reduce the setup time, and increase the reproducibility, beam-based tune and coupling feedback was developed for the energy ramps [Cameron *et al.* (2006)]. Later, orbit feedback and chromaticity feedback were

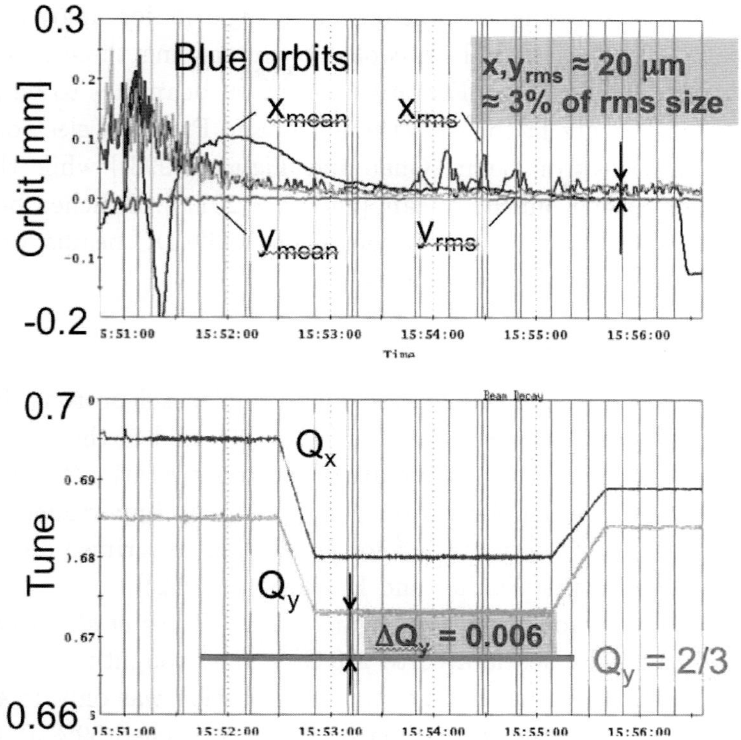


Fig. 5. Orbit mean and rms values in the RHIC Blue ring for the horizontal and vertical plane (top), and horizontal and vertical tunes (bottom) for a polarized proton energy ramp in RHIC. The tune excursion on the ramp increases the polarization transmission.

added [Minty *et al.* (2011)]. Figure 5 shows an example for a polarized proton energy ramp, which has even tighter constraints than the heavy ion ramps. Every energy ramp now has orbit, tune, and coupling feedback, which greatly reduced the number of lost ramps. Chromaticity feedback is only used for setup as the chromaticity is stable as long as the orbit is stable. Only with the reduced setup time due to the feedbacks was it possible to operate three different species combinations at four energies for physics production in 2012.

For the future, heavy ion upgrades concentrate on further increases in the bunch intensity $N_b$, low-energy cooling, and possibly new ion species combinations. While the bunch intensity $N_b$ was previously limited by transition instabilities that increased the beam size (see above), stochastic cooling allows for beam size increases as long as no ions are lost. The beams can be cooled down to a smaller size again.

The energy scan with Au–Au in 2010 led to strong hints for a critical point in the QCD phase diagram, but an order of magnitude more luminosity at energies below the nominal injection energy is necessary to have a better signal. Low-energy cooling is therefore one of the major upgrades for the coming years. The beams will be cooled by immersing them in a bath of electrons of low temperature. For this an electron beam has to be prepared with the same Lorentz factor $\gamma$ as the ion beam. Previous electron coolers have used a DC electron beam [Chao and Tigner (1998)] while the RHIC cooler will likely use a bunched beam [Fedotov (2012)]. Bunched beams can be prepared with radio frequency acceleration, while DC beams are accelerated through electrostatic accelerators.

## 3 Polarized proton operation and upgrades

RHIC holds the record for the highest energy polarized protons [Bai *et al.* (2006); Fischer and Bazilevsky (2012)] (Table 3), and is the only existing collider of polarized hadrons. In polarized proton operation both luminosity $\mathcal{L}$ (see Eq. (1)) and polarization $P$ are of great importance since the experimental figure of merit is $\mathcal{L}P^2$ for single-spin experiments or $\mathcal{L}P^4$ for double-spin experiments [Fischer and Bazilevsky (2012)].

The polarized proton operation began in 2002 [Roser *et al.* (2011)], when polarized protons were accelerated to 100 GeV and brought into collision for the first time, although the average store polarization was only 15% for this run. Since then the average store polarization has reached 59% at 100 GeV, and beams were accelerated to 255 GeV with an average store polarization of 52%. The design and achieved parameters as of 2012 are listed in Table 3, and the integrated luminosities per run are shown in Fig. 6.

Polarized protons are created in several steps [Zelenski (2010)]. First, polarized electrons are obtained. The valence electrons of alkali metals (rubidium in the case of the RHIC source) are polarized through optical pumping. In this process the spin-dependent electron energy levels are separated in a magnetic field (Zeeman effect) and a laser is used to lift the electrons from the energy ground state into an excited state. From there the electrons will decay into a specific lower state with the desired polarization, given by quantum mechanical selection rules, and accumulate in this state. Through the choice of the laser light wavelength and polarization one can produce electrons in both polarization states. A low-energy proton passing through the alkali vapor picks up the polarized electron (alkali metals readily loose their outermost electron), and the electron polarization is transferred to the proton in a so-called Sona transition [Sona (1967)] when the longitudinal magnetic field is reversed. The then neutral hydrogen atom picks up another electron in a sodium jet to form a negatively charged $H^-$ ion. This is done for easier accumulation in the first ring accelerator, the AGS Booster (Fig. 1).

Table 3. RHIC polarized proton parameters with design, achieved, and upgrade values. Unless noted otherwise parameters are given for the beginning of store.

| Parameter | Unit | Design 2004 | Achieved 2012 | Upgraded ≥2014 |
|---|---|---|---|---|
| proton energy $E$ | GeV | 100/250 | 100/255 | 100/255 |
| no. of bunches $N$ | ... | 111 | 107 | 107 |
| protons per bunch $N_b$ | $10^{11}$ | 2.0/2.0 | 1.6/1.7 | 2.0/2.5 |
| average beam current/ring | mA | 280 | 214/230 | 268/328 |
| stored beam energy | MJ | 0.36/0.87 | 0.27/0.75 | 0.34/1.1 |
| norm. rms emittance $\epsilon_n$ | $\mu$m | 2.5 | 3.3/3.5 | 3.0 |
| rms bunch length $\sigma_s$ | m | 0.87/0.55 | 0.63/0.55 | 0.5/0.4 |
| IP envelope function $\beta^*$ | m | 1.0/1.0 | 0.85/0.65 | 0.85/0.60 |
| hourglass factor $h$ | ... | 0.8/0.9 | 0.7/0.8 | 0.9/0.85 |
| beam–beam parameter $\xi$/IP | $10^{-3}$ | 9.6 | 5.9/6.1 | 8.2/9.8 |
| initial luminosity $L_{peak}$ | $10^{30}$ cm$^{-2}$s$^{-1}$ | 90/250 | 46/165 | 100/500 |
| average luminosity $L_{avg}$ | $10^{30}$ cm$^{-2}$s$^{-1}$ | 60/150 | 33/105 | 60/300 |
| average polarization $P$ | % | 70 | 59/52 | 70 |
| calendar time in store | % | 60 | 59/54 | 60 |
| integrated $L$ | pb$^{-1}$/week | 22/55 | 9.3/32 | 22/100 |

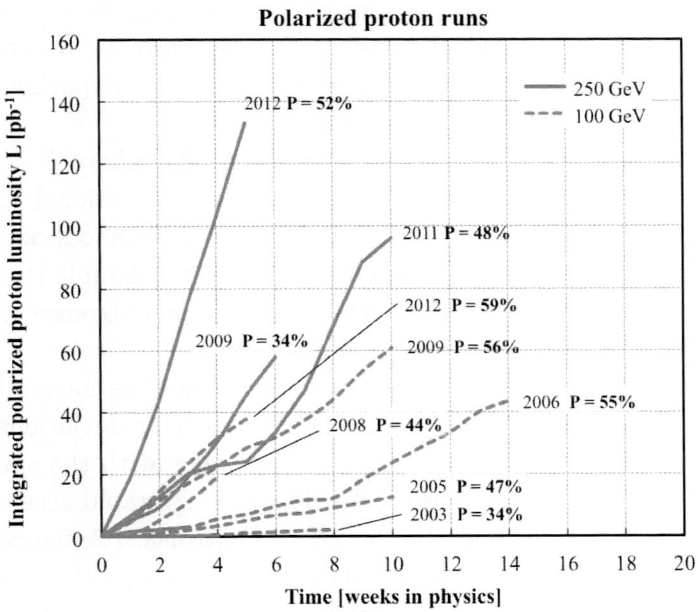


Fig. 6. Integrated luminosity as a function of operating time for the RHIC polarized proton runs.

The polarized source went through several upgrades leading to both higher intensity and polarization. The latest upgrade is expected to increase the source intensity again, this time by an order of magnitude [Zelenski *et al.* (2011)]. The high intensity will be used to retain the well-polarized core of the beam, and still allow for high bunch intensities $N_b$ in RHIC to increase the luminosity (see Eq. (1) and Table 3).

In flat rings the vertical spin direction is in principle stable as the spin precesses around the vertical magnetic fields that bend the proton trajectories. The number of precessions per turn is the spin tune $Q_{sp} = G\gamma$, where $G = 1.7928$ is the anomalous gyromagnetic ratio of the proton. Protons, however, will experience horizontal magnetic fields when they travel with a vertical offset through a quadrupole. And the horizontal fields tilt the spin axis out of the stable vertical direction.

If the effects of the horizontal fields add up coherently turn after turn, a so-called resonance condition, depolarization can occur. There are two main types of spin-depolarizing resonances: imperfection resonances and intrinsic resonances. In imperfection resonances the protons are exposed to horizontal magnetic fields in quadrupoles because of vertical orbit errors. Imperfection resonances occur when $G\gamma = k$ where $k$ is an integer, i.e. with every increase of the Lorentz $\gamma$ by 0.558 a resonance is encountered. In the acceleration to 255 GeV, a few hundred of these resonances have to be crossed. The resonance strength also increases proportionally with $\gamma$ requiring better and better orbit control as the proton energy increases.

Intrinsic resonances occur when $G\gamma = kP \pm Q_v$, where $P$ is the super-symmetry of the ring (i.e. the number of identical sections) and $Q_v$ the number of vertical oscillations per turn. Intrinsic resonances occur because the beam has a finite size and therefore particles will sample the horizontal fields in the quadrupoles. These resonances also become stronger with energy, but only proportionally to $\sqrt{\gamma}$ [Lee (1997)].

There are a number of techniques to minimize the harmful effects of depolarizing resonances. For intrinsic resonances orbit deviations are minimized in general, and orbit harmonics in resonance with the spin tune $G\gamma$ in particular. Resonances can be crossed fast through rapid acceleration, i.e. a large $d\gamma/dt$, or tune jumps in the case of intrinsic resonances (a rapid change in $Q_v$). Resonances can also be crossed slowly so that the spin will adiabatically flip sign during the resonance crossing. These techniques are all used, but at high enough energies (AGS and RHIC) they are not sufficient to maintain the polarization.

For these high energies Siberian snakes are used, which are devices to deliberately tilt the spin axis by a large amount [Derbenev and Kondratenko

(1989); Lee (1997), Hoffstaedter (2006)]. With a strong enough snake one can force a spin flip at all imperfection resonances (AGS), or change the spin tune so as to avoid the resonance condition for intrinsic resonances (AGS and RHIC).

The AGS started with a solenoid partial snake (which tilts the spin axis by less than 180 degrees). In 2004 a new helical warm snake was installed, which increased the polarization by 20%, and in 2005 an even stronger superconducting helical snake. In RHIC there are 2 full snakes in each ring, which create an energy-independent spin tune of 1/2, thereby eliminating all low-order resonance conditions.

The orbit, tune and coupling feedback mentioned above [Minty *et al.* (2011)] (Fig. 5) was particularly useful for the polarized proton operation, since it allowed to control the orbit deviations within a few percent of the beam size. It also allowed to accelerate the beam close to a low-order orbit resonance, where the polarization transmission is highest.

An important aspect of the polarized proton program is polarimetry. An absolute measurement of the polarization in RHIC is provided by a polarized atomic hydrogen jet [Bazilevski *et al.* (2008)]. These measurements are long (at least one store length of several hours). Therefore, there is also a fast polarimeter based on an ultra-thin carbon target [Huang *et al.* (2002)], which needs to be calibrated with the hydrogen jet. Since polarization measurements are slow and have a relatively large error, it is not possible to tune the machine based on polarization measurements. Rather machine setup and tuning must rely on measured beam properties such as orbit deviations and tunes.

While injector limits on the bunch intensity $N_b$ will be overcome by the new polarized source [Zelenski *et al.* (2011)], the bunch intensity is also limited by the beam–beam effect [Keil (1995)]. When the beams are colliding, the particles in one beam experience the electromagnetic field of the other beam. The change in the tune of small-amplitude particles due to these fields is the beam–beam parameter $\xi$. This field is unlike any field created by magnets: the field strength increases with the transverse amplitude of the test particle, and then decreases again. The nonlinear nature of this lens sets a limit for $\xi \propto N_b/\epsilon$. To overcome this limit an electron lens is installed in each of the RHIC rings so that the beam–beam effect is partially reversed allowing for an increase in $N_b/\epsilon$ [Fischer *et al.* (2012); Luo *et al.* (2012)]. It is expected that the electron lenses will allow for a doubling of the luminosity.

Another upgrade under study is polarized $^3$He. In addition to polarized protons, experimenters would also like to study polarized neutrons but these can be accelerated only when attached to a charged nucleon. The lightest combination of a neutron with another charged nucleon is the deuteron, but the deuteron has a small anomalous gyromagnetic ratio $G$, i.e. its spin axis can not easily be affected by magnetic fields. While this is helpful at low energies in accelerating deuteron beams, it also makes Siberian snakes and spin rotators inefficient, and prevents acceleration to high energies. A polarized $^3$He source is under development at MIT [Epstein (2011); Milner (2010)], and unpolarized $^3$He was accelerated in the Booster and AGS. An upgrade to the polarimeter is also necessary, and six instead of only two snakes may be required to preserve the polarization of $^3$He in RHIC (with $G = -4.19$, twice as many imperfection resonances have to be crossed than for protons). Polarized $^3$He is also an important species for the electron–ion collider eRHIC.

## 4 eRHIC

The design requirements for the electron–ion collider eRHIC were listed in the introduction above. The luminosity of eRHIC is also limited by the beam–beam effect, in this case by the effect that the ion beam has on the electron beam (which has a much lower rigidity). One way to overcome this

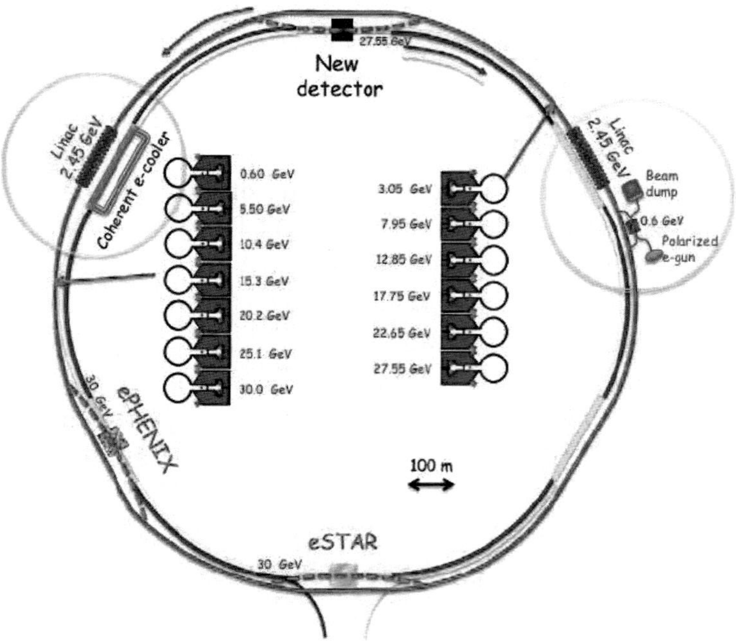


Fig. 7. Possible layout of the electron–ion collider eRHIC. A multi-turn energy recovery linac is installed in the exisitng RHIC tunnel, accelerating electrons up to 30 GeV energy.

limit is to have only a single beam–beam interaction, i.e. accelerate the electrons, collide them, and decelerate them again to extract their considerable energy [Ptitsyn *et al.* (2011b)]. A layout of such a machine is shown in Fig. 7. To limit the number of acceleration stations, the electrons are accelerated through several arcs with increasing energy.

Stochastic cooling as implemented for ions in RHIC (see above) is not effective for protons, which have two orders of magnitude larger number of particles per bunch. To boost the luminosity in electron–proton collisions a new form of cooling is needed [Litvinenko and Derbenev (2009)].

## References

Ahrens, L.A. *et al.* (1994) Heavy ion acceleration strategies in the AGS accelerator complex — 1994 status report, in *Proceedings of the European Particle Accelerator Conference,* London, UK, pp. 378–380.

Alekseev, I. *et al.* (2003), Polarized proton collider at RHIC, *Nucl. Instrum. and Methods A* **499**, 392–414.

Alessi, J. *et al.* (2011), Performance of the new EBIS preinjector, in *Proceedings of the Particle Accelerator Conference,* New York, NY, USA, pp. 1966–1968.

Anerella, M. *et al.* (2003), The RHIC magnet system, *Nucl. Instrum. and Methods A* **499**, 280–315.

Aschenauer, E. (2012), The RHIC spin program — achievements and future opportunities, in *Fall Meeting of the Division of Nuclear Physics of the American Physical Society,* Newport Beach.

Aschenauer E. *et al.* (2012), Electron Ion Collider: The next QCD frontier, understanding the glue that binds us all, White Paper of the nuclear physics community.

Auerbach, N. and Schlomo, S. (2009), $\eta/S$ ratio in finite nuclei, *Phys. Rev. Lett.* **103**, 172501.

Bai, M. *et al.* (2006), Polarized proton collisions at 205 GeV at RHIC, *Phys. Rev. Lett.* **96**, 174801.

Bazilevski, A. *et al.* (2008), Precision measurements of the proton beam absolute polarization with RHIC polarimeters, *AIP Conf. Proc.* **1149** 723–726.

Ben-Zvi, I. (2009), Superconducting storage cavity for RHIC, BNL-81965-2009-IR, C-A/AP/#337.

Blaizot, J.-P., Dinh, P.M., and Ollitrault, J.-Y. (2000), *Phys. Rev. Lett.* **85** (19), 4012–4015.

Blaskiewicz, M., Brennan, J.M. and Mernick, K. (2010), Three-dimensional stochastic cooling in the Relativistic Heavy Ion Collider, *Phys. Rev. Lett.* **105**, 094801.

Bloomer, M.A., Love, W. and Waters, L. (1991), New results from AGS heavy-ion experiments, *Nuclear Physics A* **527**, 595–600.

Cameron, P. *et al.* (2006), Simultaneous tune and coupling feedback in the Relativistic Heavy Ion Collider, and possible implications for the Large Hadron Collider commissioning, *Phys. Rev. ST Accel. Beams* **9**, 122801.

Chao, A.W. and Tigner, M. (editors) (1998), *Handbook of Accelerator Physics and Engineering,* World Scientific.

Derbenev, Y. and Kondratenko, A., (1989), Criteria for preservation of polarization in accelerators with Siberian snakes, *Sov. Phys. Tech. Phys.* **34** 1152–1155.

Epstein, C. (2011), Development of a polarized $^3$He ion source for RHIC, *AIP Conf. Proc.* **1441**, 643.

Fedotov, A. (2012), Towards demonstration of electron cooling with bunched electron beam, BNL-96824-2012-IR, C-A/AP/#445.

Fischer, W. (2013), Run overview of the Relativistic Heavy Ion Collider, http://www.rhichome.bnl.gov/RHIC/Runs.

Fischer, W. *et al.* (2008), Electron cloud observations and cures in the Relativistic Heavy Ion Collider, *Phys. Rev. ST Accel. Beams* **11**, 041002.

Fischer, W. *et al.* (2012), Construction progress of the RHIC electron lenses, in *Proceedings of the International Particle Accelerator Conference,* New Orleans, Louisiana, USA, pp. 2125–2127.

Fischer, W. and Bazilevsky, A. (2012), Impact of three-dimensional polarization profiles on spin-dependent measurements in colliding beam experiments, *Phys. Rev. ST Accel. Beams* **15**, 041001.

Harrison, M., Ludlam, T. and Ozaki, S. (2003), RHIC project overview, *Nucl. Instrum. and Methods A* **499**, 235–244.

Hoffstaedter, G.H. (2006), *High-Energy Polarized Proton Beams,* Springer Tracts in Modern Physics, Volume 218.

Huang, H. *et al.* (2002), Commissioning CNI proton polarimeters in RHIC, in *Proceedings of the European Particle Accelerator Conference,* Paris, France, pp. 338–340.

Jacak, B.V. and Müller, B. (2012), The exploration of hot nuclear matter, *Science* **337**, 310.

Jowett, J.M. *et al.* (2011), First run of the LHC as a heavy-ion collider, in *Proceedings of the International Particle Accelerator Conference,* San Sebastian, pp. Spain, 1837–1839.

Keil, E. (1995), Beam–beam dynamics, CERN 95-06, Vol. 1, 539–555.

Lee, S.Y. (1997), *Spin Dynamics and Snakes in Synchrotrons,* World Scientific, Singapore.

Litvinenko, V. and Derbenev, Y. (2009), Coherent electron cooling, *Phys. Rev. Lett.* **102**, 114801.

Luo, Y. *et al.* (2012), Six-dimensional weak-strong simulation of head-on beam–beam compensation in the Relativistic Heavy Ion Collider, *Phys. Rev. ST Accel. Beams* **15**, 051004.

Milner, R. (2010), Development of a polarized $^3$He ion source for RHIC, White paper, http://cepstein.scripts.mit.edu/he3/wp-content/uploads/2011/06/He3_WhitePaper.pdf.

Minty, M. *et al.* (2011), Simultaneous orbit, tune, coupling, and chromaticity feedback at RHIC, in *Proceedings of the Particle Accelerator Conference,* New York, NY, USA, pp. 1394–1398.

Ptitsyn, V. *et al.*, (2011a), High-luminosity electron–hadron collider eRHIC, in *Proceedings of the International Particle Accelerator Conference,* San Sebastian, Spain, pp. 3726–3728.

Ptitsyn, V. *et al.* (2011b), High-luminosity electron–hadron collider eRHIC, in *Proceedings of the Particle Accelerator Conference,* New York, NY, USA, pp. 693–695.

Roser, T., Ahrens, L.A., and Hseuh, H.C. (1994), Charge exchange studies with gold ions at the Brookhaven Booster and AGS, in *Proceedings of the European Particle Accelerator Conference,* London, UK, pp. 2441–2443.

Roser, T. *et al.* (2011) First polarized proton collisions at RHIC, in *Workshop on Increasing the AGS Polarization,* University of Michigan, 2002.

Sanford, J.R. (1977), ISABELLE, a proton–proton colliding beam facility at Brookhaven, *Proceedings of the Particle Accelerator Conference,* Chicago, Illinois, USA, *IEEE Transactions on Nuclear Science,* **24** (3) 1845–1848.

Schoefer, V. *et al.* (2012), Increasing the AGS beam polarization with 80 tune jumps, in *Proceedings of the International Particle Accelerator Conference,* New Orleans, LA, USA, pp. 1015–1019.

Sona, P.G. (1967), A new method proposed to increase polarization in polarized ion sources of $H^-$ and $D^-$, *Energia Nucleare,* **14** (5), 295–299.

Steski, D.B., Alessi, J., Benjamin, J., Carlson, C., Manni, M., Thieberger, P., and Wiplich, M. (2001), Upgrade and operation of the BNL Tandems for RHIC injection, in *Proceedings of the Particle Accelerator Conference,* Chicago, Illinois, USA, pp. 2545–2547.

The STAR Collaboration (2010), Observation of an antimatter hypernucleus, *Science* **337**, 58–62.

The STAR Collaboration (2011), Observation of the antimatter helium-4 nucleus, *Nature* **473**, 353–356.

Thieberger, P. (1984), The Brookhaven double MP facility: recent developments and plans for the future, *Nucl. Instrum. and Methods* **220** 45–53.

Vigdor, S. (2012), The case for continuing RHIC operations, White paper with extensive feedback from the RHIC user and support community.

Weng, W.T., Ahrens, L., Damm, R., and McNerney, A.J. (1991), Construction and early commissioning results of the AGS Booster, in *Proceedings of the Particle Accelerator Conference,* San Francisco, California, USA, pp. 52–56.

Zelenski, A. (2010), Review of polarized ion sources (invited), *Review of Scientific Instruments* **81** 02B308.

Zelenski, A. *et al.* (2011), High-intensity, high-brightness polarized and unpolarized beam production in charge-exchange collisions, in *Proceedings of the Particle Accelerator Conference,* New York, NY, USA, pp. 2555–2557.

# Chapter 22

# The Large Hadron Collider

Stephen Myers (CERN)

The Large Hadron Collider (LHC) was first suggested (in a documented way) in 1983 [1] as a possible future hadron collider to be installed in the 27 km "LEP" tunnel. More than thirty years later the collider has been operated successfully with beam for three years with spectacular performance and has discovered the long-sought-after Higgs boson. The LHC is the world's largest and most energetic particle collider. It took many years to plan and build this large complex machine which promises exciting, new physics results for many years to come.

I describe the LHC design objectives, review some of the more relevant beam effects, define the major accelerator components and parameters, and finally give an overview of the commissioning and operational performance since the initial turn on of the collider. The latter will include the major accident which took place in September 2008 and the subsequent repair and redesign of the faulty components.

The first attempts to circulate beam in the LHC in September 2008 were initially very successful. However after only nine days of preliminary beam commissioning, on September 19, disaster struck: the last octant was being ramped up in preparation for high energy operation when a magnet interconnect failed and the enormous energy stored in the superconducting magnets was released in an uncontrolled way and damaged around 600 metres of the LHC installed equipment. The next 14 months were crucial for the machine. A crash programme for the repair of the damaged sector was initiated as well as in depth studies to understand the cause of the failure and make design changes which would ensure that such an accident could never reoccur in the future.

In this paper, the story of the four years of the intensive activity since the accident is described starting with the repair of the damaged area and followed by the three very successful years of beam operation.

## 1 Introduction and short history of the project

In his book describing the construction of the LEP Collider [2], Herwig Schopper emphasized that the only reason for choosing (in 1981) the 27 km for the LEP tunnel was the LHC, since even for LEP 200 a circumference of about 23 km would have been sufficient. He wrote

> "In view of the long-term future of CERN and of a later project, a proton collider in the LEP tunnel, a possibility which had already been considered at that time, I thought it would be decisive to have as large a tunnel as possible."

The first documented performance estimates for a proton–proton collider within the LEP tunnel were already considered at the very beginning of the LEP project in April 1983 [1], long before the first electron–positron collisions in LEP in 1989. This paper pointed out many of the fundamental performance limitations in the future machine and made first evaluations of possible performance.

One year later marked the beginning of the detailed design study needed for such a machine and detectors with a workshop in Lausanne [3] organized by G. Brianti. Brianti led the LHC design work from 1984 through 1993 and during this time prepared the project technically for approval by the CERN Council. The years 1987 until 1993 were very difficult ones for the CERN project. In the United States the approval of the Superconducting Super Collider (SSC) in 1987 severely jeopardized hopes for the approval of the less powerful LHC collider. However following a series of technical and cost reviews which pushed the SSC estimated cost from 4.4 billion to more than 11 billion dollars, the US congress cancelled the project in 1993. Following very difficult negotiations in 1994, the CERN Council at its December meeting (1994) approved the LHC project for two-stage construction with a further budget review in 1997 when the amount of the additional contributions from non-member states would be known. Chris Llewellyn-Smith (DG) and Lyn Evans (project leader) spent the next two years maximizing the extra contributions from the US, Japan, Russia, India, and Canada. Thanks to their untiring efforts the machine was approved for single-stage construction in December 1996, with the budget deficit to be financed by loans.

The early years of the project were of course not easy, since from 1994 until 2000, CERN was simultaneously operating LEP for physics, upgrading LEP to higher energy (LEP2) as well as preparing the LHC project. However in 2000 the CERN Council decided to end LEP2 operation, allowing massive redeployment of skilled and experienced CERN staff from LEP2 to the LHC design and construction. With this new impetus, the design of the LHC gathered real momentum with L. Evans leading the overall project and the three accelerator department heads leading the technical design of the components: P. Lebrun (magnets, cryogenics and vacuum), P. Ciriani (infrastructure and technical services), S. Myers (radio frequency, accelerator physics, beam diagnostics, controls, injection, extraction and beam dump, machine protection, and power supplies). It is important to emphasise and continue to highlight the enormous contributions of the CERN technical groups and group leaders in the design, fabrication, and testing of these very complex systems. The strength of CERN is imbedded in its technical groups.

In June 2008 the LHC was declared operational and no longer a project. In October 2008, the CERN Council nominated S. Myers as Director of Accelerators and Technology (for the years 2009 through 2013), responsible for all CERN accelerators with special emphasis on the LHC repair, re-commissioning, and operational performance. The LHC performance was nothing short of sensational during the three years of "Run 1" (2010–2012) exceeding all expectations and culminating in the discovery of the "Higgs" boson in 2012.

It is imperative to note that during the long years of the LHC design, construction, commissioning and beam operation, each of the successive CERN Director Generals made enormous personal contributions to the project: H. Schopper (1981–1988), C. Rubbia (1989–1993), C. Llewellyn-Smith (1994–1998), L. Miaini (1999–2003), R. Aymar (2004–2008) and R. Heuer (2009–2015). Without the full support of each and every one of these DGs, the project would undoubtedly have failed. It is equally important to highlight the enormous role of Lyn Evans as head of the LHC project for 14 years from 1994 until 2008. He was the thread of continuity throughout the difficult and arduous years of approval and construction. He led the project through many major crises.

Figure 1 shows a schematic layout of all CERN accelerators and also indicates the year of first operation of each accelerator. Figure 2 shows an artist's view of the LHC installed in the 26.7 km underground tunnel on the Franco-Swiss border near Geneva, Switzerland.

## 2 The LHC design and objectives

The key objectives of the LHC were the exploration of the Standard Model in the TeV energy range, and the search for the Higgs boson and potential new physics beyond the Standard Model.

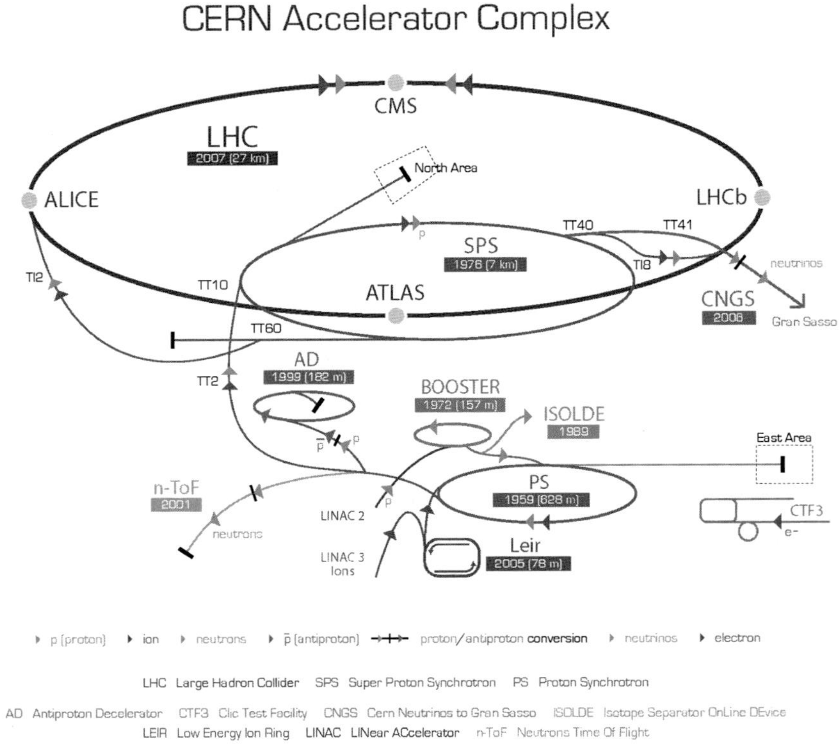


Fig. 1. Schematic layout of CERN accelerators.

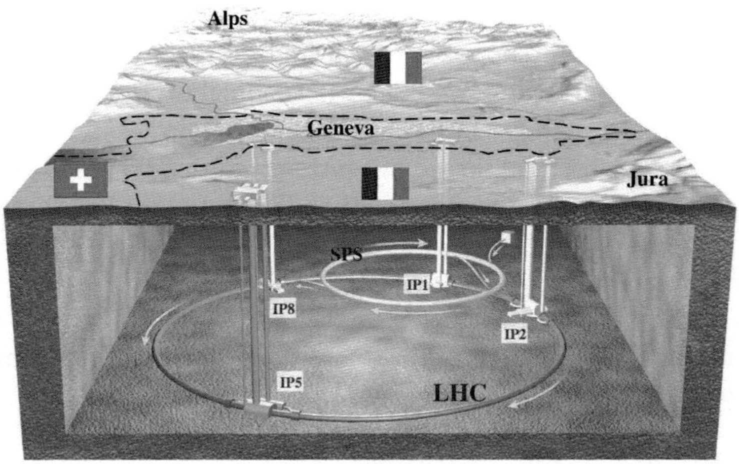


Fig. 2. Artist's view of the LHC.

Hadron colliders are ideal instruments for the exploration of new physics as the collisions naturally cover a wide energy range. The very weak synchrotron radiation (see later) in a circular hadron collider is another argument for using hadron rather than lepton beams. A hadron collider therefore offers the potential of much greater collision energies.

There are two different types of particle colliders.

1. Using a single magnetic channel;
2. Using two (virtually) independent magnetic channels.

In the former case (1), bunches of particles circulating in one direction collide with their anti-particle counterparts circulating in the opposite direction. Without complicated separation schemes, the number of collision point is simply twice the number of bunches. This fact severely limits the luminosity achievable in a single-ring collider. Recent examples of this type of collider are LEP (with electrons colliding with positrons), the Sp$\bar{\text{p}}$S and the Tevatron collider, both of which collided protons with anti-protons.

In the latter case (2), particles of the same or different types circulate in separate magnetic channels and collide in pre-defined intersecting points around the circumference. Examples of this type of collider are the ISR (the first ever proton collider), HERA and RHIC. In this case, with bunched beams (note that the ISR operated with coasting or un-bunched beams), the revolution frequencies of the different rings must be synchronized to ensure stable collisions in the intersection points.

The high luminosity needed for the LHC objectives imposed a two-ring design with separate magnet and vacuum systems for the two counter-rotating hadron beams. Designing the LHC with two separate magnet channels and vacuum systems for the two counter-rotating beams allows in addition the operation with more complex hadron species (e.g. ion beams). In order to minimize the cost, the LHC followed a novel 2-in-1 magnet design where the two apertures for the counter-rotating hadron beams have separate magnet coils (with opposite field polarities) but share the same cryostat and powering infrastructure.

During the course of the project, the LHC design parameters were significantly changed due, to a certain extent, to the continuous comparison and competition with the SSC [4] (Carlo Rubbia insisted that, from a discovery point of view, the factor of ten higher luminosity in the LHC more than compensated for the lower energy of the CERN machine). This resulted in a factor of ten increase in the LHC luminosity from the first design projections to the design study [5] and the nominal LHC performance specification [6].

The LHC has been designed with four experimental insertions: two insertions for high luminosity experiments (ATLAS and CMS) using proton–proton collisions (with luminosities $\geq 10^{34}\,\text{cm}^{-2}\text{s}^{-1}$), one B-meson experiment (LHCb) requiring medium luminosities for proton–proton collisions ($10^{32}\,\text{cm}^{-2}\text{s}^{-1}$), and one dedicated experiment for ion collisions (ALICE) requiring low luminosities ($10^{29}\,\text{cm}^{-2}\text{s}^{-1}$) for the operation with proton beams. For the high luminosity detectors a major design issue was related to the number of events per bunch–bunch crossing (event pile up). The detectors were originally designed for a maximum number of pile up events of around 25 with 25 ns bunch spacing.

## 3 Accelerator beam effects

### 3.1 *Synchrotron radiation*

When charged particles traverse perpendicular magnetic fields, they radiate photons and lose energy due to "synchrotron radiation". The energy loss is proportional to the 4th power of the relativistic energy ($\gamma^4$). At the same absolute beam energy (GeV), protons are much less relativistic (since the rest mass of protons is almost 2000 times higher than that of electrons). Hence, for the same primary beam energy, the synchrotron radiation losses of protons are much lower ($\sim 8 \times 10^{-14}$) than that of electrons.

Synchrotron radiation has both advantages and disadvantages for particle beams. In fact hundreds of dedicated synchrotron "light sources" have been built all over the planet in order to use the properties of the high energy photons generated by bending beams of electrons. These secondary beams of photons are used for a multitude of disciplines from life sciences to semi-conductor developments.

For particle colliders, the synchrotron radiation can be used (with appropriate design) to damp the longitudinal and transverse oscillations of the primary beam particles. For electron–positron colliders, where the radiation is significant with respect to the beam energy, this damping can be very fast and can produce very low emittance beams (e.g. in LEP at the highest beam energies, more than 3% of the beam energy was lost on each turn of the machine, 11,200 times per second). For the present energies of hadron colliders the damping rates are still rather low. The LHC at its design energy has a transverse emittance damping time of around 26 hours.

On the negative side synchrotron radiation causes heating of the components surrounding the primary beam. For the case of LEP this was an enormous problem with around 20 MW of synchrotron radiation at high energy. In the LHC, at the energies operated at until now, this heating effect is negligible in comparison to the heating produced by the electro-magnetic fields associated with the bunched beams. At design energy the synchrotron radiation power reaches just under 4 kW per beam (~0.2 W/m) and is evacuated by the beam screen's cooling circuit which is maintained between 5 and 20 K while the cold bore is operating at 1.9 K.

For energy upgrades of the LHC the synchrotron radiation from the protons becomes increasingly more important, both positively as well as negatively. A more critical aspect of the synchrotron radiation at even low LHC energies is the production of photo-electrons which can enhance the electron cloud effect (see later).

### 3.2 *Collective effects*

The motion of the charged particles is governed by the Lorentz force. However there are additional perturbing electro-magnetic fields associated with the charged particle beams themselves. These perturbing fields interact with their environment and can produce many undesirable effects on the beams.

Depending on the beam intensity and on the bunch filling pattern, they can cause

- Beam instabilities;
- Parasitic losses;
- Degradation of the beam quality by emittance growth and poor lifetime of all or some specific bunches;
- "Radio frequency" heating of the components surrounding the beam.

Collective effects include incoherent phenomena, concerning the behaviour of a single particle in the electro-magnetic field produced by all the others, and coherent interactions of the beam with its surroundings. The charged particles interact with themselves (leading to space charge effects) and with their environment, inducing charges and currents in the surrounding structures, which create electro-magnetic fields often referred to as wake fields. Furthermore, the charged particles can interact with other charged particles present in the accelerator (e.g. the electron cloud effect mentioned previously) and with the counter-rotating beam in a collider (the beam–beam effect). The motion of the particles has to be calculated using the **total** electro-magnetic fields, which are the sum of the external and perturbation fields.

Single-bunch collective effects are associated with the broad-band impedance of low-Q structures, while multi-bunch or coupled-bunch effects are dominated by the narrow-band impedance of high-Q resonators.

The coupling of the beam to its environment is characterized by the evaluation of the "coupling impedance". A complete impedance evaluation of all structures installed around the LHC beams has been carried out in order to calculate the ultimate beam performance of the collider. In addition many design features have been introduced in order to reduce this impedance and allow higher intensities without beam instabilities. An example of this is the cryogenic part of the LHC beam pipe (mainly beam screen) which is fabricated from copper cladded stainless steel to keep the resistance as low as possible both for the reduction of instabilities and ohmic heating.

Collective effects are the ultimate limitation to the performance of the LHC. The existing cures against instabilities include feedback systems (see later), chromaticity control, and Landau damping of the coherent beam oscillation modes.

Landau damping in general causes the frequency of oscillation of the particles to become dependent on the amplitude of the oscillation. In this way resonant build-up of oscillations can be eliminated or at least reduced. In the momentum plane Landau damping can be introduced by non-linearities in the RF acceleration fields (usually by the addition of a second higher harmonic frequency cavity system). In the transverse plane (horizontal or vertical) Landau damping can also be produced by the addition of non-linear fields. This can be accomplished in a controlled way by powering of octupole magnets. The highly non-linear electro-magnetic field experienced by colliding beams (beam–beam effect) is a strong producer of Landau damping. In the LHC the maximum amount of Landau damping results from collisions of the beams and can be a source of stabilization.

Chromaticity is defined as the dependence of the transverse (betatron) oscillation frequency of the particles on their momentum amplitude. Chromaticity can be introduced and/or corrected in the optics by powering sextupole magnets at locations of momentum dispersion in the lattice.

The modes of operating the LHC for the highest intensities have been evaluated by computer simulations [7] with variations of the parameters, feedback ("damper") gain, chromaticity (Q'), and the strength in the octupole magnets. Figure 3 shows the results of such a simulation in a three dimensional plot with two cases: single beams (SB), and coupled beams (CB). The present strategy (2012) is to operate with high chromaticity, high damper gain, and high current in the octupole.

### 3.2.1 *Beam induced heating*

As described above, the electro-magnetic fields associated with the charged particle beams interact with their environment and produce "RF" beam heating of the components "seen" by the beam. As previously stated, the

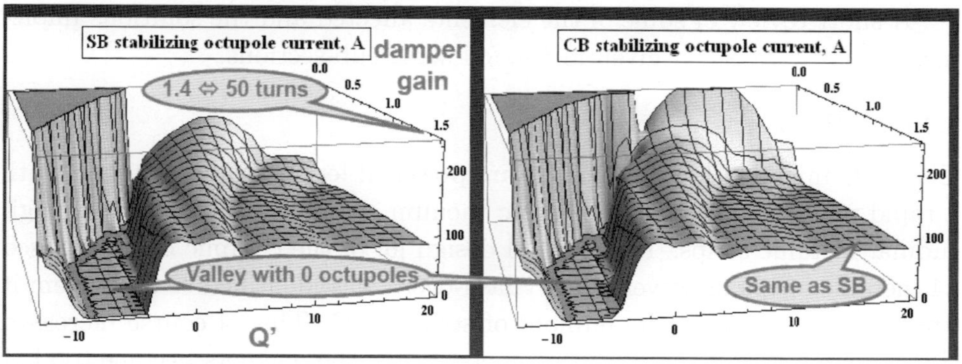


Fig. 3. Comparison of single beams (SB) and coupled beams (CB) thresholds for instability. Vertical axis is octupole current, horizontal chromaticity, and into page is gain of the damper. (Courtesy of Metral and Burov)

coupling impedance (which produces the beam heating) for all installed components has been evaluated by numerical techniques and experimentation. A likely source of error is always the "RF fingers" which ensure electrical continuity between devices as the components expand and contract with temperature variations. This can be particularly problematic during warm-up from superconducting temperatures. Excess heating can be produced if the electrical contacts are not perfect.

Beam heating is a well-known issue in all high intensity accelerators and colliders. In the LHC, the first signs of significant heating were observed during operation in 2011. With the continuous increase in the beam intensity, beam heating became an important operational issue during 2012. Figure 4 shows the temperature increase produced in some of the LHC components as a function of the beam intensity.

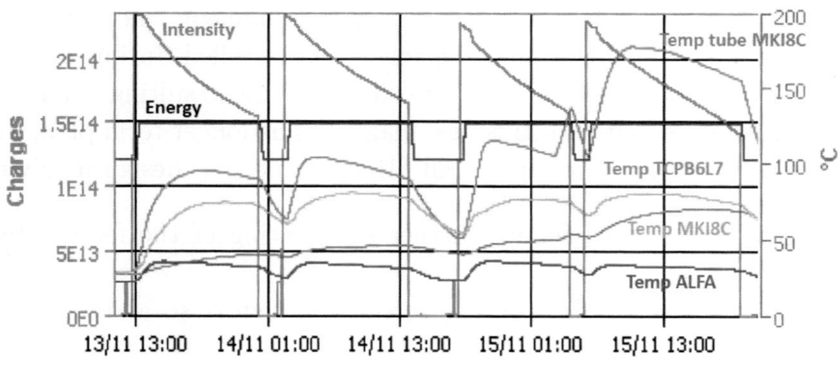


Fig. 4. Plots of beam intensity (charges), beam energy, and temperature increase for kicker (MKI8C), collimator (TCPB6L7), detector (ALFA) during 4 LHC fills in mid-Nov 2012. (Courtesy of Barnes, Goddard, and Mertens)

A short review of some of the elements affected and the solutions implemented and foreseen is given in the following sub-sections.

#### 3.2.1.1 LHC injection kickers (MKI)

The LHC injection kickers (MKIs) are powerful ferrite loaded magnets with a rapid rise time aided by a ceramic vacuum chamber equipped with longitudinal metallic strips. The original design [6] of the ceramic chamber used 24 metallic strips, however when "flash-over" was identified as a problem it was decided to reduce the number of strips to 15. This of course increased the beam heating of the ferrite magnets. When the magnets become too hot and reach their "Curie temperature" the resulting magnetic fields are less than with cool ferrites and efficient injection is no longer possible. During 2012, heating of the MKIs during long physics data-taking periods increased the temperatures of the magnets towards their limit value. Consequently, following the beam abort at the end of the data-taking, the magnets were too hot to allow injection. Sometimes several hours of cool down time were needed before the injections for the next physics fill could be started. This of course reduced the overall availability of the collider. The only short term solution to mitigate (slightly) the situation was to remove selected bake out jackets to allow better cooling. This reduced the measured maximum temperatures slightly (3 to 7°C depending on the magnet) and reduced the waiting time after long physics fills. However it is preferred in the future to keep the jackets on the kicker tanks during operation.

For the longer term, several paths are being followed. One of them is to increase the number of metallic strips so as to reduce the heating of the ferrites. This of course implies finding solutions to the previously encountered flash-over problem. As this issue had been foreseen, a new MKI design (using 19 metallic strips) had been under development since the start of LHC operation. This development magnet was installed during a technical stop (TS3) in September 2012. Figure 5 shows the resulting temperature improvement measured with this new magnet. The longer term plan, during long shutdown (LS1), is to replace all MKIs with a new design involving

- 24 metallic strips to reduce further the heating of the ferrites by the beam;
- Reduce the electric field on the screen conductors so as to reduce the probability of flash-over;
- Improve heat radiation from the ferrite by increasing the emissivity of the tanks.

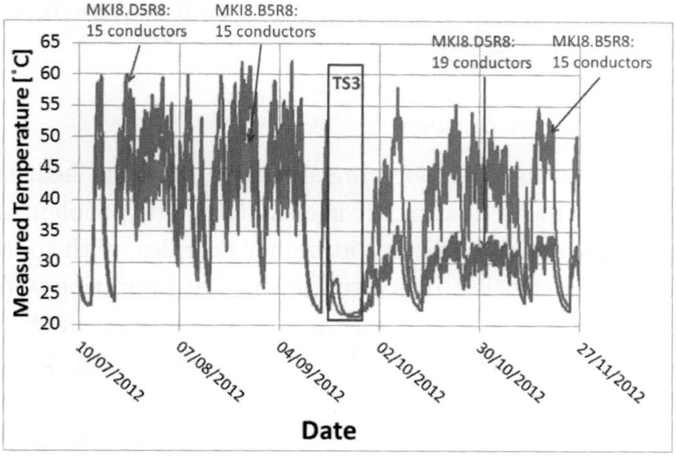


Fig. 5. Temperature impact produced by the new MKI design.

### 3.2.1.2 Synchrotron radiation monitor (BSRT)

The synchrotron radiation monitors (BSRT) [6] are critical diagnostic devices for LHC operation: they allow non-destructive measurement of the beam sizes (emittances) and allow monitoring of the particles in the abort gap. During high intensity running in the latter part of 2012, the beam 2 BSRT mirror and its copper mirror support suffered damage due to significant heating. The cause of this damage has not yet been completely understood in spite of the combined efforts of many of the accelerator groups. Detailed simulations and bench measurements are presently on-going to check the impedance of the device. A new robust design will be needed for the start-up after LS1 implying a great effort in the early part of the shutdown to allow the installation before the end of the shutdown.

### 3.2.1.3 ALFA "roman pot"

Roman pots are devices which contain experimental forward detectors. These pots are moved very close to the beam so as to observe particles at very low angles. Unlike the TOTEM roman pots, ALFA was designed without active cooling of the detector and subsequently suffered larger temperature rises. Studies have indicated that the temperature increases are consistent with the impedance heating of the ferrite damper ring which is doing its job in preventing even more heating.

In the short term the ALFA team decided (like for the MKIs) to remove the bake out jackets and add some air cooling fans. In the longer term they will implement a new design with reduced longitudinal impedance and active cooling circuits.

### 3.2.2 *Summary on beam heating*

In 2012 the LHC saw the first worrying signs of equipment heating and damage due to beam induced heating. It is clear that this problem will increase with the future operating conditions at higher energies, with shorter bunches and higher bunch intensities. It is also known that operation with 25 ns will produce slightly more beam heating due to the increase in the total beam current as well as the change in the frequency spectrum of the 25 ns trains. From a technical point of view, many improvements are foreseen for implementation during the long shutdown 1. Other measures are presently being investigated in case the already foreseen measures are inadequate. The most obvious measure would be to modify the frequency spectrum of the bunches so as to reduce the coupling to the frequency spectrum of the impedances. The frequency spectrum of the bunches can be improved by making longer bunches, however this can imply a loss of luminosity. Alternatively, the addition of a second, higher harmonic RF system (Landau cavities) has the possibility of changing the higher frequency spectrum of the bunches without the concomitant reduction in luminosity. This possibility is being pursued as a longer term measure. In addition more detailed, more accurate temperature monitoring on critical devices is being pursued.

### 3.2.3 *Beam–Beam*

The highest luminosity of the LHC requires a large number of bunches (~2800) and to avoid unwanted collisions, a crossing angle is needed to separate the two beams upstream and downstream of the wanted collision point. The amplitude of the crossing angle is limited by the available aperture in the final quadrupole triplet.

Strong sources of non-linearities are the electro-magnetic fields generated by the particles themselves. These non-linearities can cause significant perturbations to the opposing beam (beam–beam effect) in the case of a colliding beam facility like the LHC.

These perturbations include single particle (incoherent) effects, such as betatron tune spreads (produced by the non-linear head-on and long-range collisions) as well as a reduction in dynamic aperture in the case of insufficient beam separation at the unwanted encounters. There are also coherent effects including closed orbit distortions, betatron tunes and chromaticity changes related to the different bunches.

It is crucial to take into account both the head-on and the long-range interactions for the total evaluation of the beam–beam effect. The impact of the head-on collisions is usually defined by the linear beam–beam strength parameter, which is approximately equivalent to the linear tune shift of small amplitude particles.

Because of the non-linearity of the beam–beam forces, the particles exhibit betatron tune dependence on their betatron amplitudes. The resulting tune spread must fit in the tune diagram without crossing dangerous low order non-linear resonance lines. Thus design of the interaction regions must take into account the total spread originating from head-on and long-range beam–beam interactions and must be kept as small as possible. Experience from the SPS and the Tevatron shows that the total tune spread should not exceed a value of around 0.015. Figure 6 shows the tune distribution (tune footprint) for the LHC nominal proton parameters with three head-on collisions, one offset collision (IP2) and the full number of long-range interactions. The overall spread is slightly above 0.010.

Another important beam–beam effect is the reduction of the beam lifetime due to the diffusion of large amplitude particles beyond the dynamic aperture limit. The combined effects of the triplet errors and the beam–beam interaction are significant for collisions as well as for injection. In both cases the beam–beam interaction produces a sizeable reduction in the dynamic aperture.

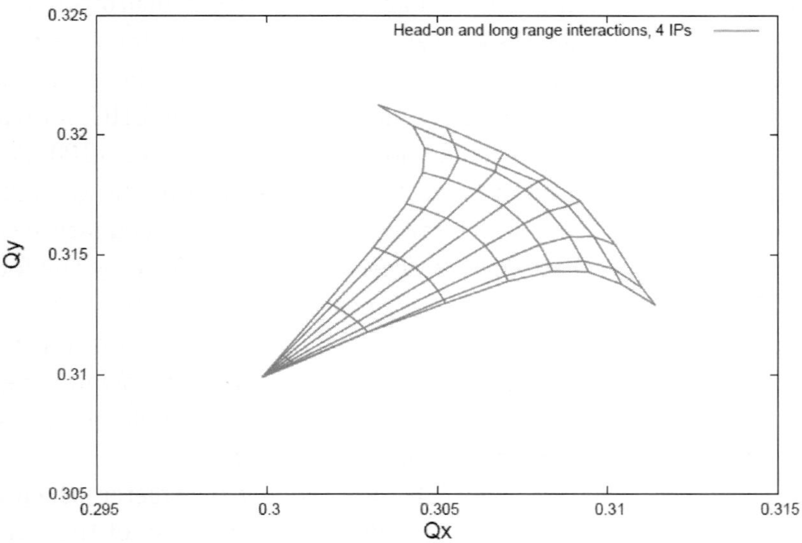


Fig. 6. Tune footprint (tune spreads) produced by the beam–beam effect, Qx and Qy are horizontal and vertical tunes respectively. (Courtesy of W. Herr)

For technical reasons (such as the injectors, and the necessity to allow for the rise time of injection and the beam abort kickers), the bunches in the LHC do not form a continuous train of equidistant bunches spaced by 25 ns, but are composed of trains of bunches. Due to the symmetric layout, all bunches collide head-on in the diametrically opposed interaction points 1 and 5 (ATLAS and CMS). However, the bunch position in a train determines the beam–beam forces seen by that bunch. For example, the bunches at the beginning and end of a train experience only half the long-range interactions at each interaction point. This implies that each bunch in the LHC behaves as an "individual" and exhibits different orbits, tunes and chromaticities due to the different beam–beam forces. Since it is technically very difficult to control the properties of each individual bunch, it is important to minimize these effects. In the LHC these effects have been minimized by a careful design of the crossing schemes and bunch filling patterns (trains) involving partial compensation in one interaction point by another one [6].

### 3.3 *Electron cloud*

One of the most crucial performance limitations to the LHC, foreseen at the design stage, was the "electron cloud" effect [8], see also http://ab-abp-rlc.web.cern.ch/ab-abp-rlc-ecloud/. The mechanism for the build-up of an electron cloud is as follows. Seed electrons are generated by ionization of the residual gas in the vacuum chamber at injection energy, or by photo-electrons liberated by the large number of hard ultra-violet synchrotron radiation photons at 7 TeV. When these electrons are in the vicinity of the beam potential, they experience the electro-magnetic force of the beam, and are accelerated towards the vacuum pipe inner surface. In the LHC, for example, the proton bunch intensity is large enough to give a kick of ~100 eV to the stray electrons. On interaction with the inner surface of the vacuum pipe, these electrons stimulate gas desorption and produce secondary electrons. The secondary electrons are further accelerated to ~100 eV by the following bunch, 25 ns apart, leading to an avalanche in the production of electrons. This avalanche mechanism leads to the fast build-up of an electron cloud, which increases the heat load deposited on the LHC arc beam screens [6]. In addition it degrades the vacuum pressure, induces single or multi-bunch instabilities, and causes long-term emittance growth.

The secondary electron yield (SEY) is the key parameter which defines the vacuum level in a beam tube. It is defined as the ratio of the number of produced electrons to the number of incident electrons. Typical SEY values for metallic surfaces are ~2.

In the LHC design [6], a number of counter-measures are implemented, which aim either at suppressing electron cloud build-up or at reducing its effect. These measures include

- Having a sawtooth chamber in the arcs, which reduces the photon reflectivity;
- Shielding the pumping holes inside the arc beam screen so as to prevent multipacting electrons from reaching the cold bore of the dipole magnets;
- Coating the warm regions by a special getter material, with low secondary emission yield; and
- Conditioning the arc chamber surface by the cloud itself (beam scrubbing), which ultimately provides a lower secondary emission yield. During commissioning the bunch spacing and/or the beam energy can be reduced to process the chamber while staying within the available cooling capacity.

Since the operation with beam commenced, electron cloud effects have been measured and have limited range of parameter space for performance improvements. In particular, operation at 25 ns has been delayed due to the detrimental effect of the electron cloud effect at this bunch spacing. Operation at half the number of bunches (with 50 ns bunch spacing) has provided excellent luminosity but with the down side for the experiments that the event pile up is significantly above the design value for the ATLAS and CMS detectors. During 2011 and 2012 the most effective measure to condition the vacuum surfaces has been "beam cleaning" (or scrubbing). This process involves provoking electron cloud by injection of the maximum number of high intensity bunches for relatively long periods of time (many days). With time the secondary electron yield (SEY) of the vacuum surface decreases as do the number of electrons involved. Eventually the SEY decreases below the threshold needed for the electron cloud instability. An example of beam cleaning is shown in Figure 7 where "scrubbing" was carried out with 25 ns bunch spacing for 3.5 days.

### 3.4 *Radiation to electronics*

During the design phase of the LHC, the assumed "safe" fluence (for $\geq$20 MeV hadrons) for LHC was $10^9$ n cm$^{-2}$. However, during initial operation with another CERN project (CNGS) in 2007, it was found that many electronics components failed due to single event errors (SEEs) forcing the run to be prematurely abandoned. These failures occurred at much lower values (factor of 100) than expected of the radiation dose. Using these values as

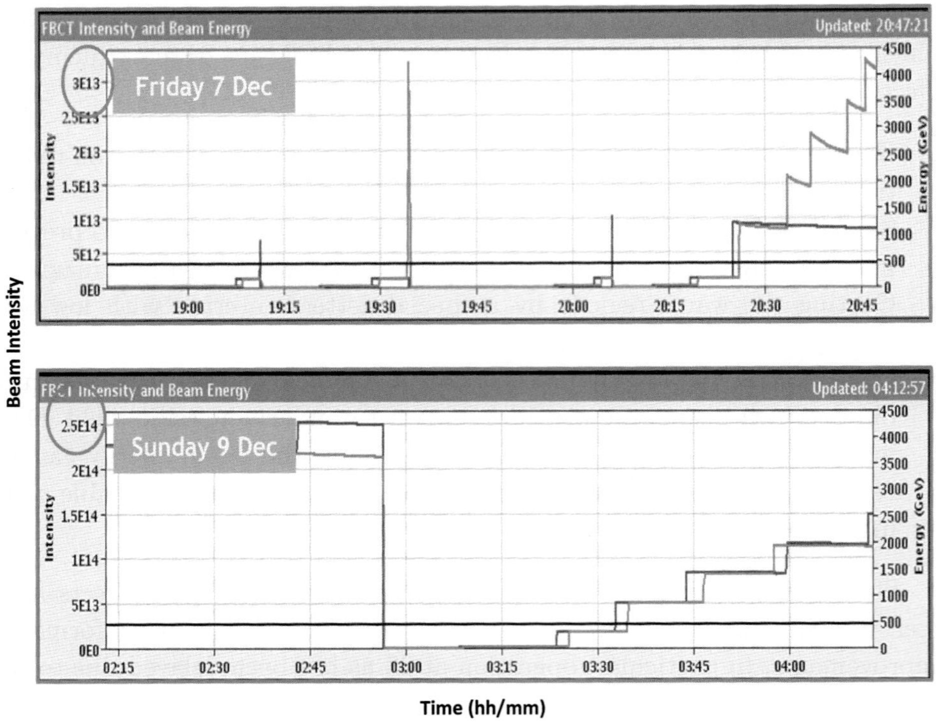


Fig. 7. An example of beam cleaning in 2012, beam intensity versus time. Upper trace shows poor lifetime at low intensities and lower trace shows greatly increased lifetime with an order of magnitude more intensity after two days of beam cleaning. (Courtesy of J.-M. Jimenez)

benchmarks, a detailed campaign was re-launched to evaluate the radiation levels and tunnel locations where such SEEs could affect the performance of the LHC. The initial findings of this study are shown in Table 1 with numbers given for the equivalent dose (for 200 days of nominal operation) and in column 4 related to the failures in the CNGS. It was clear that there was a serious problem waiting to occur as the intensity and luminosity were increased. Consequently a crash project was set up to:

- Evaluate the most critical areas in terms of radiation dose following assumptions on the increase of the performance of the LHC;
- Identify the most susceptible electronics installed in these areas (usually "off the shelf" electronics and controllers);
- Make a prioritized list of components which would fail first;
- Prepare mitigation for these electronics (shielding, removal to a safer area, separation of susceptible components, redesign of the most critical); and

Table 1. 20 MeV fluence comparison with CNGS as unity.

| Location | Assumption | 20 MeV fluence | Multiple of CNGS equivalent |
|---|---|---|---|
| Sea Level | Annual | $1 \times 10^4$ | 0.001 |
| Airline Altitude | Annual | $1 \times 10^6$ | 0.1 |
| CNGS Failure | Annual | $1 \times 10^7$ | 1 |
| UA67 | Annual | $5 \times 10^8$ | 50 |
| RR13 | Annual | $7 \times 10^8$ | 70 |
| UJ67 | Annual | $3 \times 10^9$ | 300 |
| LSS1 and 5 | Annual | $1 \times 10^{10}$ | 1000 |
| DS 1 and 5 | Annual | $5 \times 10^9$ | 500 |
| ARC | Annual | $1 \times 10^9$ | 100 |

- Estimate levels, test electronic components, work on new designs (separation), shielding.

A comprehensive scheduled programme was set up to implement these measures in the most vulnerable areas during all short technical stops and shutdowns. In addition very close monitoring and verification of the models used were carried out on an almost continuous basis.

The critical parameter for LHC's efficient operation is the number of beam aborts (dumps) cause by SEEs. It should be noted however that it is not always immediately obvious if a beam dump has resulted from a SEE. The overwhelming source of the radiation producing SEEs come from the collisions in the high luminosity insertions around ATLAS and CMS (LSS1 and LSS5). Hence a good figure of merit is the number of beam dumps per unit of integrated luminosity. Operation with beam in 2011 produced a figure of merit of 12 beam dumps per inverse femtobarn of accumulated luminosity per high luminosity experiment. Figure 8 shows the evolution of the number of SEE failures plotted as a function of the integrated luminosity. A linear fit is made to the "SEU Induced LHC Dumps" (dots) showing for 2012 a figure of merit of 3 dumps per $fb^{-1}$, a factor of 4 less than in 2011 (also shown on the plot). Since the integrated luminosity in 2012 was approximately a factor of 4 above that of 2011, this means that the total number of beam dumps was approximately the same. Clearly future increases in the LHC luminosity will require further reductions in this figure of merit. The aim for the period after long shutdown 1 (LS1), from 2015 until 2018, is a further gain of about another factor of 6 (0.5 beam dumps per $fb^{-1}$). This requires a continued aggressive campaign of work during LS1.

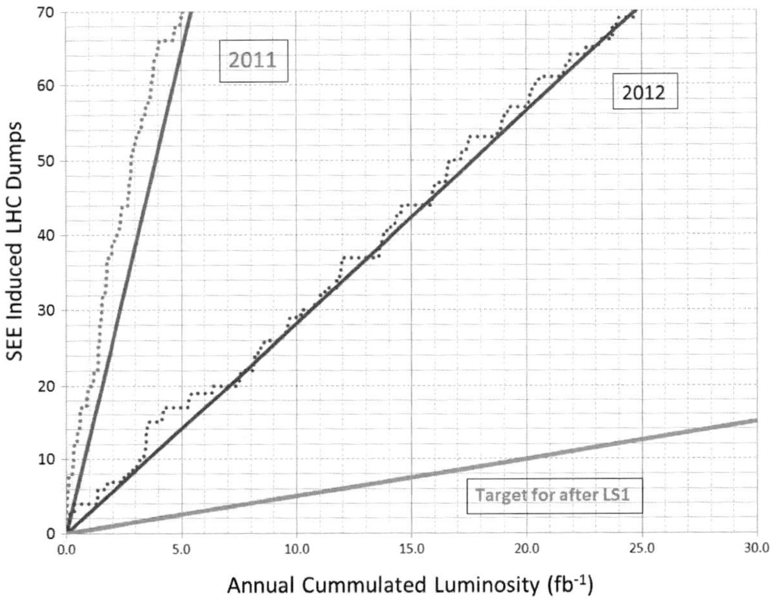


Fig. 8. Evolution of the number of SEE failures plotted as a function of the integrated luminosity. (Courtesy of M. Brugger)

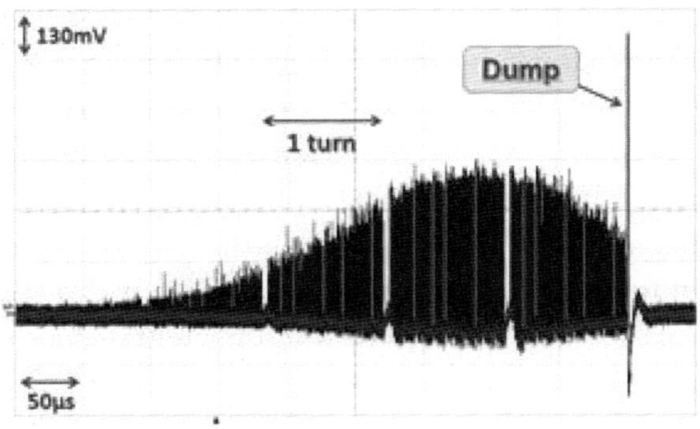


Fig. 9. Beam loss monitor signal produced by a UFO. (Courtesy of T. Baer)

## 3.5 *Unidentified falling objects (UFOs)*

During early operation in 2010, several spurious beam dumps were triggered by beam losses exceeding the set thresholds on the beam loss monitors. Analysis of the beam loss monitor signals showed that the losses were of very short duration (1 ms, see Figure 9) and it was also noticed that the signals were similar to those produced by flying wire scans. For many decades accelerator scientists have been pre-occupied by the idea of dust-like particles inside the vacuum chamber. It was quickly assumed that these losses were produced by micrometer sized dust-like particles “falling” into the beam and causing scattering of the protons which was picked up by the beam loss monitors. Since the nature of these dust particles was unidentified, it was decided to name the effect UFOs standing for Unidentified Falling Objects. These UFO events are observed around the whole machine, for both beams, and for proton as well as lead ions.

Since July 2010, UFOs have caused around 40 spurious beam dumps of LHC fills and were a major source of inefficiency of operation [9]. However starting in the middle of 2011, it was decided that the machine protection could be maintained at a safe level even if the thresholds for the beam loss monitors were carefully and gradually increased. The resulting increases in the BLM thresholds have considerably reduced the number of beam dumps attributed to UFOs.

Although most of the UFO events are far below the threshold of the BLMs, each UFO is monitored and recorded. In 2011, more than 16,000 candidate UFO events have been detected. The arc UFO rates in 2011 and 2012 are shown in Fig. 10. Even though the beam intensity was increased from 228 to 1380 bunches, the arc UFO rate decreased in 2011 by about a factor of 5.

It can also be seen from Fig. 10 that following the Christmas stop in 2011/2012 the UFO rate increase by about a factor of 2.5. Although many events are identified in the regions of the LHC injection kickers, studies of the spatial distribution of UFOs have indicated that they occur all around the LHC circumference.

On several occasions “storms” of UFOs have been observed at the LHC injection kickers (MKI). Because of their importance for efficient beam

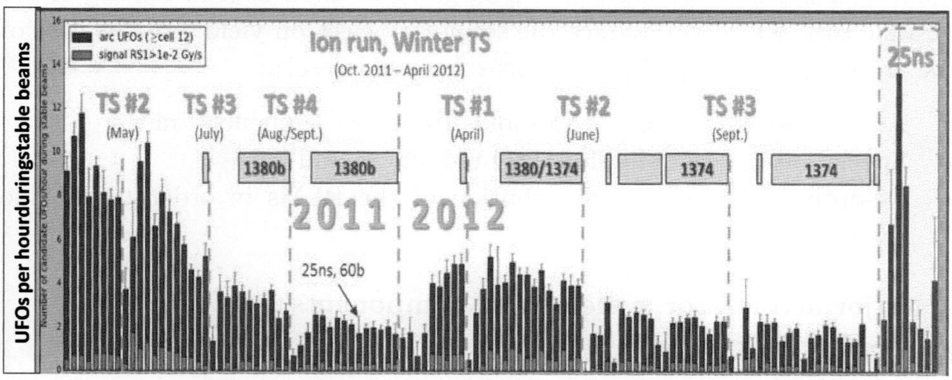


Fig. 10. UFO rates in the LHC arcs. (Courtesy of T. Baer)

operation, extensive studies have been carried out on these UFOs. These studies included improvements in instrumentation, theoretical studies, laboratory experiments, dedicated machine studies in the LHC, as well as extensive computer studies using the FLUKA code. The preliminary conclusion of these studies is that the MKI UFOs are macro-particles of around 100 $\mu$m diameter and originate from the ceramic beam screen.

Based on FLUKA predictions the energy deposition due to an arc UFO increases with the beam energy (factor of 3–4), meaning that the expected amplitude of the BLM signal with 7 TeV operation will be significantly higher than that at 3.5 TeV. It can also be seen from Fig. 10 that the tests performed at the end of 2012 indicate a significant rise in the UFO rate when operating with 25 ns bunch spacing. These two observations coupled with the fact that the quench limit for the superconducting magnets decreases (factor of 4 to 5) with increasing beam energy cause some concern for the number of beam dumps while operating at higher energies. However on the positive side, first measurements of the quench limits of the superconducting magnets indicate that the values calculated (many years ago) may be pessimistic. Consequently it is important to measure the parameters associated with the production of UFOs very early in the beam operation after long shutdown 1.

During LS1, several improvements are foreseen in order to mitigate against beam dumps due to UFOs when operating at high energy with 25 ns bunch spacing. For the MKIs the situation with electron cloud in these areas will be improved by NEG coating of interconnects, by-pass tubes and beam instrumentation equipment. The cleaning procedure for the tanks will be improved by reducing the dust contamination and additional cleaning of the metallic strips. As mentioned previously, the number of metallic strips will eventually be increased to 24 which will reduce the electric field during the injection pulse. Studies are also on-going to allow special coating of the ceramic tube which will reduce the secondary electron yield. For the case of the arc UFOs several mitigation measures are foreseen:

- An increase in the BLM thresholds towards the quench limit;
- Special beam scrubbing for UFO reduction; and
- Re-optimisation of the distribution of the BLMs in order to provide improved protection.

## 4 Major accelerator systems and components

### 4.1 *Civil engineering*

The 26.7 km "LHC" tunnel was actually constructed for the previous CERN project: the Large Electron–Positron Collider (LEP). Figure 11 shows the

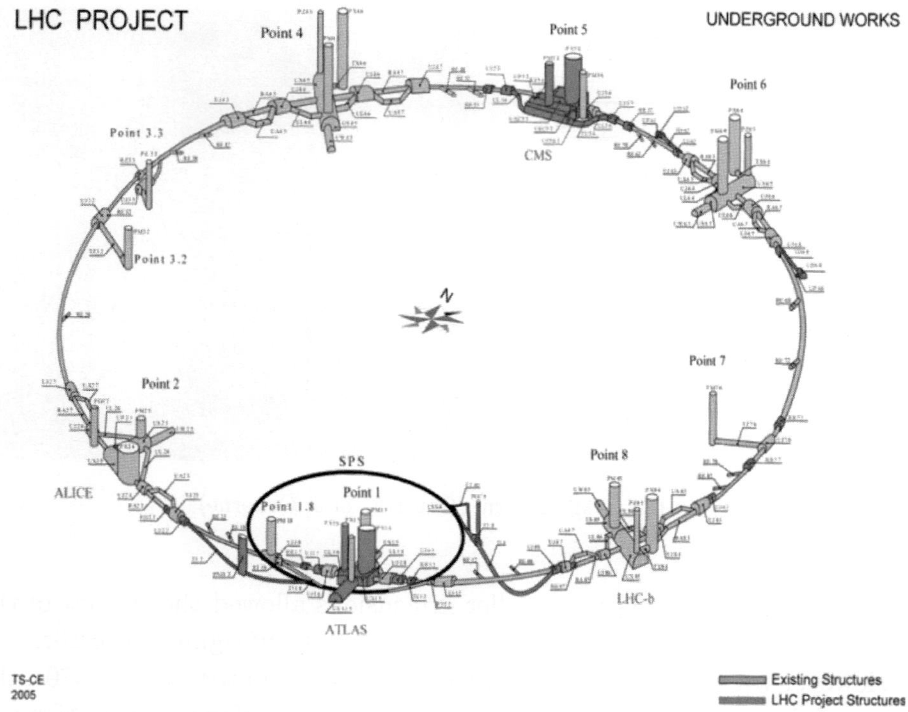


Fig. 11. Additional civil engineering for the LHC. (Courtesy of J. Osborne)

additional civil engineering work needed to convert the existing infrastructure to the LHC needs. A major part of this work was the construction of the caverns for ATLAS and CMS and the provision of the long proton transfer lines between the SPS and the LHC. Figure 12 shows, as a typical civil example, engineering drawings and a final photograph of the CMS tunnel in point 5.

## 4.2 *Survey, geodesy*

Particle accelerators impose very tight tolerances in absolute and relative accuracy for the positioning of their installed components. These tolerances result from beam dynamics and mechanical (geometrical) issues both of which influence the aperture available for the circulating beam. The task of the surveyors is to measure and align the position, orientation, shape and the size of all major accelerator components as well as the particle detectors to accuracies not needed in any other domain. In order to position a component, a reference system (frame) must be defined [10]. A co-ordinate system which defines the position of the object is attached to this frame. The reference system at CERN has evolved with the increase of the size of

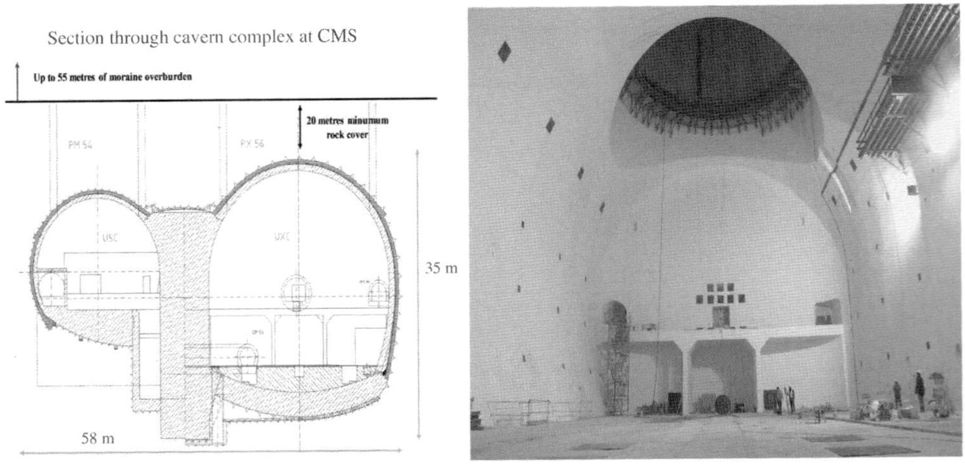


Fig. 12. CMS cavern. (Courtesy of J. Osborne)

its installations. Initially, the smaller dimensions allowed the surface of the earth to be considered as a plane without introducing significant errors.

However with the advent of the much larger accelerators of the 1970s, the surface of the earth could no longer be considered as a plane. A new reference surface was adopted using a sphere and the average sea level throughout the continents. A new co-ordinate H, the altitude, was defined as the distance measured with respect to this surface.

As the circumference of the accelerators increased to the size of LEP (27 km), it was necessary to consider an ellipsoid of revolution as reference surface of the earth. In addition, the gravitational effects of the nearby Jura Mountains and Lake Geneva had to be taken into account. An equipotential surface of gravity, called the geoid, to which the force of gravity is perpendicular everywhere has been defined by means of zenithal camera and gravimetric measurements. The measurements taken with survey instruments are therefore linked to this geoid.

Presently the CERN reference system is a local ellipsoid which fits the earth in the CERN area. The deviation of vertical between this ellipsoid and the local "plum line" vertical has been calculated and taken into account in the new geoid.

The absolute positioning of accelerator components, known in an XYZ co-ordinate system, is ensured by means of a geodetic network. The first level of this geodetic network is a surface network which is comprised of monuments solidly anchored to the earth, forming a very well defined basic framework and from which the links to national and international reference system can be established. The determination of the co-ordinates of these monuments is done by very accurate triangulation, trilateration, levelling measurements and more recently by GPS measurements. The accuracy of these network points has to be in the range of one mm.

This surface network, once measured, is transferred to the underground network of the accelerator. The underground network is comprised of tripods regularly spaced along the accelerator tunnel. The topographical traverse linking one access shaft to the adjacent one is realized by gyro-theodolites, theodolites and very accurate electronic distance measuring devices. Offset distances with respect to a nylon stretched wire are also frequently measured in order to improve the "smoothness" of the network.

### 4.3 *The magnets system*

The LHC magnets have three main functions:

1. To bend the particles on quasi-circular paths around the 27 km vacuum channels (dipole magnets);
2. To focus the particles into tight beams which fit well within the vacuum enclosure (quadrupole magnets);
3. To provide small dipolar corrections to steer the bunches with minimum deviation from a perfect "closed orbit" and to provide focussing corrections to avoid nonlinear resonances and instabilities (correction elements)
   The magnets guide the particle beam by virtue of the Lorentz force

$$\mathbf{F} = q \cdot \mathbf{v} \times \mathbf{B}$$

where $q$ is the electrical charge of the particle, $\mathbf{v}$ its velocity, and $\mathbf{B}$ the magnetic field. Hence a uniform magnetic field with a single component perpendicular to the particle velocity tends to bend the particle on a circular trajectory. Equating the Lorentz force to the centripetal force (depicted in Fig. 13) gives a simple relationship (for protons) between the strength of the field $B$, the radius of the circumference $\rho$ and the particle momentum $p$:

$$B \cdot \rho\,(\mathrm{Tm}) = 3.36 \cdot p \left(\frac{\mathrm{GeV}}{\mathrm{c}}\right)$$

where the momentum is expressed in practical units of (GeV/c), and $B\rho$ is known as magnetic rigidity. This simple equation shows clearly the trade-off between the bending magnetic field $B$ and the size of the machine (related to $\rho$). In the case of the LHC, with the existing tunnel providing the value of the bending radius ($\rho$), the maximum beam momentum (loosely referred to as energy) is clearly proportional to the maximum field strength which

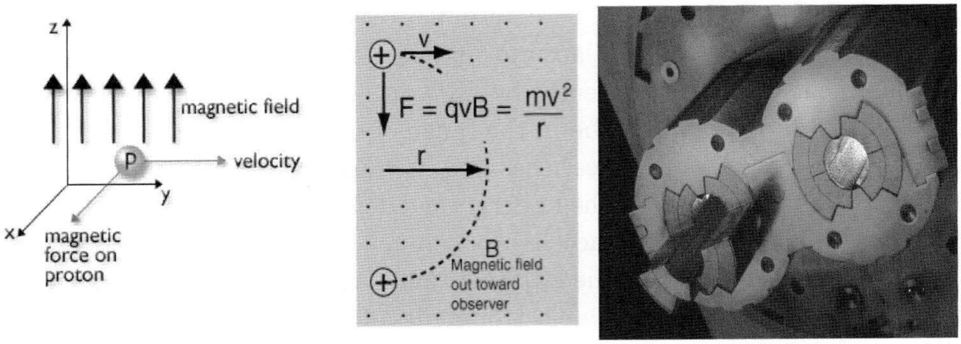


Fig. 13. Particle motion in a bending field (left, centre) and cross-section of an LHC dipole (right).

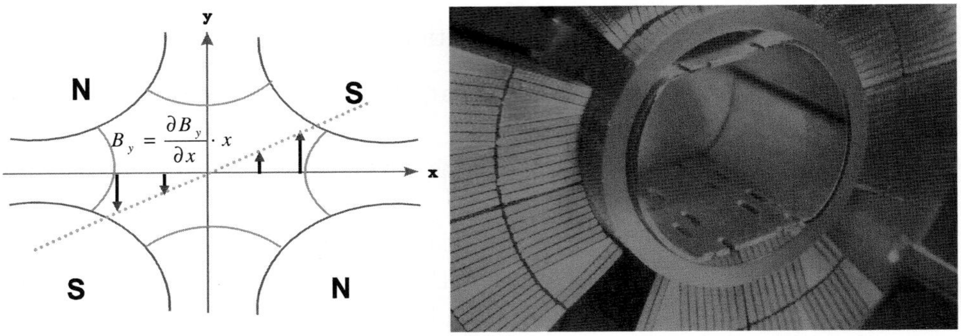


Fig. 14. Quadrupolar field (left) and cross-section (right) of an LHC quadrupole.

can reliably be produced in the bending field of the dipoles. A maximum field strength of 8.3 T in the LHC allows for a maximum energy per beam of 7 TeV. Figure 13 also shows a cross-section of the LHC superconducting "two-in-one" dipole magnet.

Magnets having fields which vary linearly with displacement from the magnetic centre (quadrupoles) are employed to focus the beams of particles (see Fig. 14), in analogy to the use of optical focusing lenses to focus a beam of light. Figure 14 (right) also shows a cross-section of an LHC quadrupole.

The design of magnets to be used in an accelerator has many additional requirements. A summary list of these requirements is given below.

- Physical constraints (space, transport, weight, ...);
- Magnetic field strength;
- "Good field" region;
- Field quality under different working conditions;

- Physical aperture;
- Powering constraints;
- Cooling by water or by cryogenic system;
- Radiation resistance of the coils;
- Ability to align and re-align during complete lifetime;
- High reliability;
- Protection against the uncontrolled release of the stored magnetic energy.

Field quality requirements are crucial in order to minimize non-linear particle motion driven by the non-linearities (higher order multi-poles) inherent in all magnetic fields.

The magnetic field may be expressed by a Taylor series expansion. Higher order terms in this expansion can drive non-linear motion of the particles and ultimately cause beam loss. In superconducting magnets this is one of the most crucial requirements and the non-linear magnetic terms usually determine the "dynamic aperture" of the machine. The dynamic aperture is defined by the largest transverse amplitude trajectories that particles can have and still be stable.

The dynamic aperture is derived by computer "tracking of particles" in a simulation for millions of turns in full six dimensional phase space, and with the particles subjected to the measured higher order field components in the magnets. As a simple example of the impact of non-linearities, Fig. 15 shows a one dimensional phase space plot of particle motion close to a 5th order non-linear resonance. The left plot is for a completely linear machine and the right plot results from the addition of a single non-linear element (a sextupole in this case), showing clearly the distortion of the phase space caused

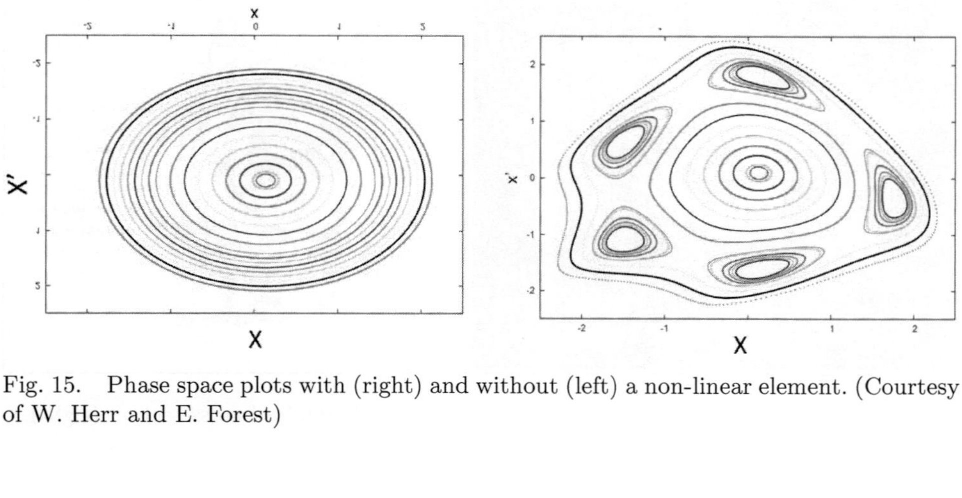


Fig. 15. Phase space plots with (right) and without (left) a non-linear element. (Courtesy of W. Herr and E. Forest)

by the non-linearity as well as the chaotic motion between the resonant islands.

### 4.3.1 *Superconducting magnets*

The magnetic field in a superconducting magnet is generated by the distribution of the current carrying coils around the beam aperture whereas the fields in normal conducting magnets are derived from the magnetization of the iron yoke. Consequently superconducting magnet technology is highly dependent on the ability to produce superconducting materials in the form of high current cables. Figure 16 shows the superconducting cable used in the LHC magnets (Rutherford cable). In total over 7000 km of this cable was needed in the construction of the LHC magnets. Figure 17 shows an artist's view of the LHC superconducting magnet (left) as well as a mechanical engineering cross-section (right). Figure 18 shows a coil cross-section of the

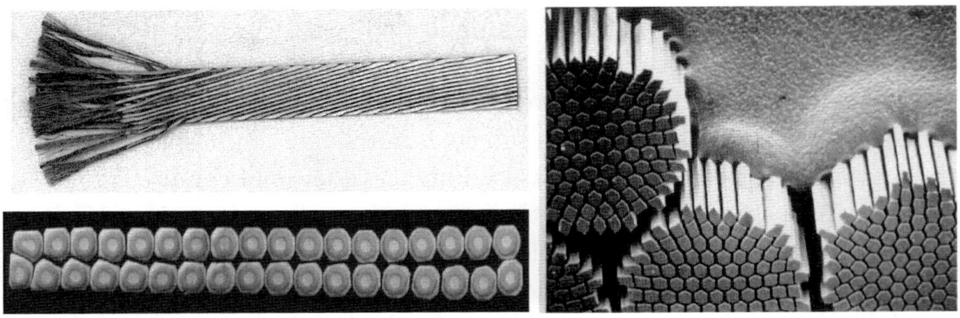

Fig. 16. 7000 km of superconducting cable. (Courtesy of L. Rossi)

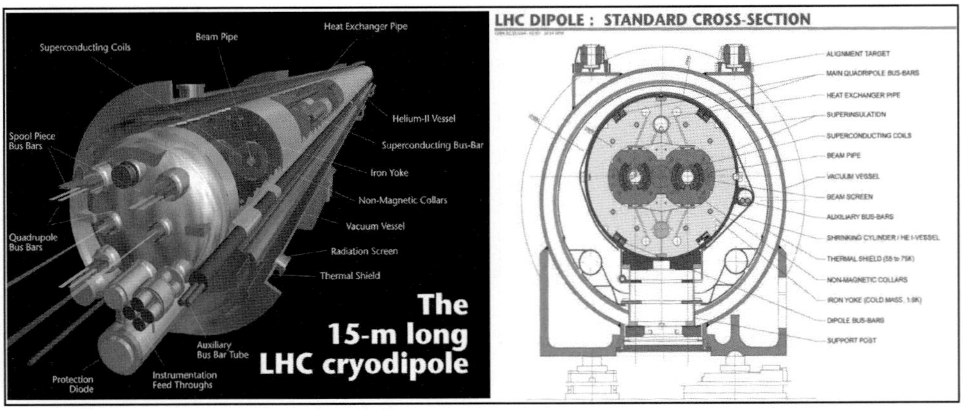


Fig. 17. Dipole magnet: artist's impression (left) and standard cross-section (right).

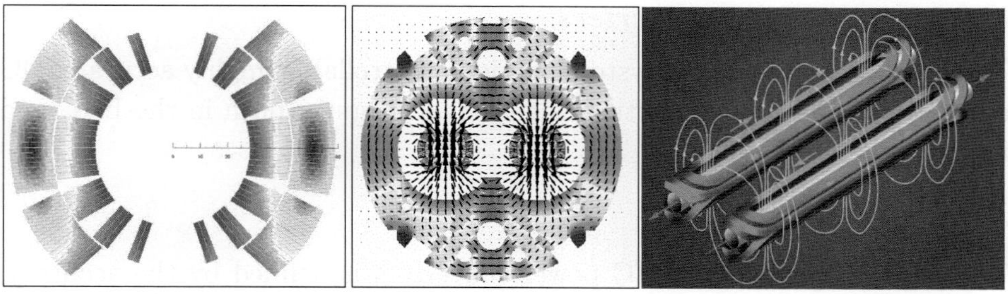

Fig. 18. Coil distribution (left) and field lines (centre, right).

LHC dipole, the field distributions in the "two-in-one" magnet as well as a pictorial view of the field lines.

#### 4.3.1.1 Magnet (quench) protection

The transition of a superconductor to a "normal" conducting state is called a "quench" and can be induced by any sudden temperature increase resulting from (for example) internal mechanical energy release or by the loss of some of the high energy beam particles. During the initial stages of the quench, the resistance of the superconducting cable increases and therefore produces resistive heating ($i^2R$). This heating can result in thermal runaway and ultimately damages the coils.

It is not possible to completely exclude all possible causes of quenches in superconducting magnets operated in the hostile environment of high energy beams. Consequently the magnets must be actively protected against damage resulting from their own stored magnetic energy. The stored energy in a SC magnet is simply

$$\text{Stored Energy} = \frac{1}{2}i^2L$$

where $i$ is the current flowing (amps) and $L$ is the inductance of the magnet.

Each LHC dipole has an inductance of 98.7 mHenrys. Consequently at ultimate field (9 T) the total stored energy in all 1232 dipoles is more than 10 GJ (enough energy to heat up and evaporate more than 13,000 kg of a metal such as copper).

The protection system measures the resistance in the magnet coils and, in the event of detecting an irreversible resistive transition, triggers a system to discharge the stored magnetic energy into a safe place. Such incidences are of course to be avoided as they must be preceded by an abort of the circulating beam, reducing the operational efficiency. In addition quenches may reduce the useful life of the magnet.

### 4.4 *Magnet power converters*

The LHC power converter system provides controllable, highly accurate, DC current for the superconducting and normal magnets both in the LHC and in the detectors.

#### 4.4.1 *Performance*

The performance of the LHC power converters is defined by the accuracy, reproducibility, stability, and resolution. All performance requirements are given in ppm (parts per million) of the maximum (nominal) current.

The accuracy defines the long-term uncertainty (period of one year) in the setting, taking into consideration the full range of permissible changes in operating and environmental conditions. The accuracy of the feedback control is determined primarily by the current measurement transducer and, in the case of digitally controlled power converters, the control algorithm and the ADC employed in the feedback loop.

The reproducibility is defined as the uncertainty (period of one day) in returning to a set of previous working values from one cycle to the next, and without any adjustment of the calibrated parts (e.g. DCCT, ADC).

The stability is defined as the maximum deviation over a period of half an hour, with no changes in operating or environmental conditions other than main network.

The resolution is defined as the smallest increment in current that can be induced or discerned, and is directly determined by the analogue to digital conversion system.

Table 2 [6] gives a summary of the performance of the LHC power converters.

The accelerator physics requirements for the main bending and quadrupole magnets presented a major challenge for these power converters. In order to achieve the very difficult specifications, an enormous development effort was needed in the following domains: Analogue-to-Digital Conversion techniques, high-current DC current measurements, digital control technologies and new soft-switching power converter topologies. These challenges have been met and the stability of the beams in the LHC is testimony to the performance of the power converters.

#### 4.4.2 *Tracking*

The LHC machine is divided into eight sectors, each powered by a separate dipole power converter. Again for beam dynamics reasons, very precise tracking is required between each of the segmented eight main dipole converters

Table 2. Performance specifications for the LHC power converters.

| Circuit type | Nominal current (A) | Current polarity | One year accuracy (±ppm of nominal) | One day reproducibility (±ppm of nominal) | 1/2 hour stability (±ppm of nominal) | Resolution (±ppm of nominal) |
|---|---|---|---|---|---|---|
| Main bends, quads | 13000 | Unipolar | 50 and 20 with calibration | 5 | 3 | 1 |
| Inner triplet | 8000/6000 | Unipolar | 100 and 20 with calibration | 20 | 10 | 15 |
| Dispersion suppressors | 5000/6000 | Unipolar | 70 | 10 | 5 | 15 |
| Insertion quadruples | 4000/5000 and 6000 | Unipolar | 70 | 10 | 5 | 15 |
| Separators (D1,D2,D3,D4) | 5000/7000 | Unipolar | 70 | 10 | 5 | 15 |
| Trim quads | 600 | Bipolar | 200 | 50 | 10 | 30 |
| SSS correctors | 600 | Bipolar | 200 | 50 | 10 | 30 |
| Spool pieces | 600 | Bipolar | 200 | 50 | 10 | 30 |
| Orbit correctors | 120/60 | Bipolar | 1000 | 100 | 50 | 30 |
| Warm magnets | 850/1000 | Unipolar | 200 | 50 | 10 | 15 |

as well as tracking the dipoles with the main quadrupoles. For this reason the power converters have been designed to track a given reference function. For static reference functions, the performance is ensured by the accuracy and reproducibility of the converters, however for the dynamic part, additional sources of errors come from timing errors or lagging errors in the controls regulation.

The digital current loop of the power converters is designed with no lagging error; i.e. independence of the load time constant. The lagging error between the current and the field must therefore be known and is corrected by a timing shift at the start of the ramp. The different current–field lagging errors depend mainly on the vacuum chamber and on the beam screen.

There are a total of 1720 power converters with a total steady-state input power of more than 60 MW and a peak power of more than 80 MW. In total current, they provide about 1.8 MA characterised by the requirements of high current and low voltage from the SC magnets. The performance of the power converters is dictated by beam dynamics considerations related to the control of the betatron tune. This is exacerbated by the complex relationship between the magnetic field and the coil-currents in a SC magnet. The high current requirement dictated (at the time) that the power converters be installed underground near the loads in order to avoid enormous cabling costs. The large "RF" galleries (excavated at the even points for the LEP RF complex) were used for the installation of the high current power converters. At the odd points special tunnel enlargements (RRs) were used for the installation of the auxiliary power supplies (dispersion suppressors, insertion quadrupoles, insertion orbit correction, with current range from 6 kA to 120 A converters). Warm magnets are powered from the surface using the existing surface buildings and cable ducts from LEP.

Additional requirements, resulting from the re-use of the LEP infrastructure, resulted in low volume and high efficiency for the power converters design. Also, the modular design approach allows fast replacement of faulty components in the restricted access of the underground areas. In order to minimize heat losses to air in the tunnel and the concomitant ventilation costs, all converters (with the exception of the low power orbit correctors) are water cooled. In order to avoid interference with other crucial equipment, electro-magnetic compatibility was and is an important criterion in the design of the converters.

### 4.5 *Vacuum systems*

In particle accelerators, the beam pipes must be at ultra-high vacuum in order to reduce the beam–gas interactions, i.e. the scattering of beam particles resulting from collisions with the molecules of the residual gas. These interactions reduce the beam lifetime (nuclear scattering) and the luminosity (multiple coulomb scattering). They can also cause intensity limitations provoked by pressure instabilities and by electron induced instabilities (for positive beams only).

Beam–gas scattering can also increase the background in the detectors and the radiation dose rates in the accelerator tunnels. The latter leads to material activation, increased radiation dose rates to intervention crews, premature degradation of tunnel infrastructures like cables and electronics and finally higher probability of electronic single events errors (SEEs) produced by neutrons. SEEs are of great concern for the electronics in the tunnel as well as in the service galleries (see earlier).

The design of an accelerator vacuum system faces severe additional constraints which must be taken into account at the design stage. Amongst these constraints, the coupling "impedance" seen by the beam must be minimized in order to preserve beam stability, the generation of radio frequency higher-order-modes (HOM) must be minimized so as to avoid local heating by the beam, and the beam aperture in the magnets must be maintained sufficient to allow good beam lifetime.

The LHC has the particularity of having three vacuum systems [6, 11]: insulation vacuum for cryo-magnets, insulation vacuum for helium distribution line (QRL) and beam vacuum. There are of course two almost independent beam vacuum systems, one for each circulating beam. Clearly the levels of vacuum are vastly different for each system: the requirements for the beam vacuum (driven by the required beam lifetime and the background in the experimental areas) are the most stringent.

Vacuum sectorisation is essential to transform each vacuum system into manageable lengths. Vacuum barriers define the sector extremities for the insulation vacuum and sector valves for the beam vacuum. For the QRL and the magnet insulation vacuum, the sector lengths are 428 m and 214 m. The vacuum sector lengths are variable in most cases except the distance between two cryo-magnets. Since there are no sector valves in the cold arc, the length of these sectors is approximately 2900 m.

The design of the LHC vacuum system had to take into account several phenomena associated with particle beams. Synchrotron radiation (described previously) illuminates the vacuum chambers particularly in the bending arcs, and electron cloud build-up (described previously) endangers particle stability, vacuum pressure as well as increasing the cryogenic heat load. In spite of the enormous effort to minimize these effects at the design stage, it is nonetheless necessary to condition the vacuum with beam in order to reach design performance.

The requirements on the design of the LHC beam vacuum system are much more complex than those of a classical vacuum system. Low vacuum pressures are needed to allow beam lifetimes compatible with long (tens of hours) data taking runs in a cryogenic environment where heat input to the 1.9 K helium circuit must be minimised and where significant quantities of gas can be condensed on the vacuum chamber. The main sources of heat load coming from the beam are

- Synchrotron radiation by the high energy circulating beam;
- Energy loss by nuclear scattering;
- Electro-magnetically induced image currents (coupling impedance);
- Energy dissipated during the development of electron clouds.

Reducing the heat input to the cryogenic system introduced constraints on the vacuum system design (e.g. the necessity of a beam screen), on the materials (e.g. the introduction of a copper layer) and on the gas density to be achieved in the LHC vacuum system.

### 4.5.1 *The arc beam screens*

The beam screen has been designed with a race-track shape in order to maximise the aperture and leave space for the cooling channels (see Fig. 19). The flat parts of the beam screen are perforated with slots, covering about 4% of the total area so as to allow condensation of gases on surfaces which are

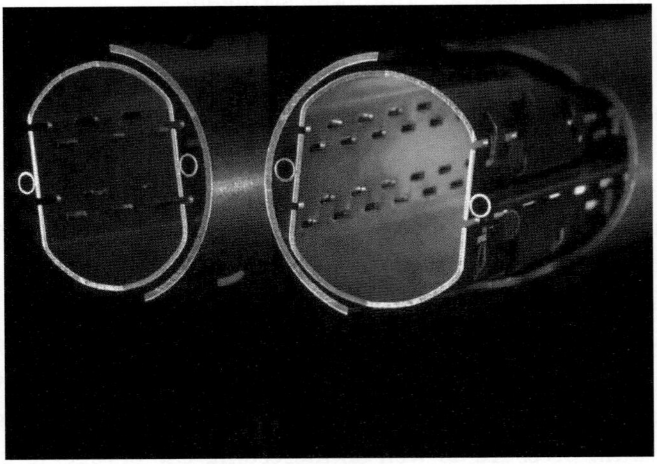

Fig. 19. The LHC beam screen with pumping slots and cooling channels.

protected from the direct impact of energetic particles (ions, electrons and photons). The locations of the slots have been chosen to minimise the coupling impedance of the beam and their size designed to minimise RF losses through the holes. A thin copper layer on the inner surface of the austenitic stainless steel beam screen provides a low resistance path for the image current of the beam, and a rolled-in saw tooth structure allows interception of photons at normal incidence thereby reducing the number of reflected photons.

Two stainless steel tubes cool the beam screen with helium which is temperature regulated (20 K) at the output of each cooling circuit at each half-cell. This produces a temperature of the cooling tubes between 5 K and 20 K for nominal cryogenic conditions.

### 4.5.2 *Cold interconnects*

The continuity of the beam vacuum envelope, the helium flow in the cooling tubes, as well as a smooth transition between beam screens are all ensured by the beam vacuum interconnects. The beam envelope has been designed to present a low resistance to the beam image currents and minimise coupled-bunch instabilities. It must also have a low inductance to minimise the longitudinal single-bunch instability. In order to meet these difficult requirements, complex interconnect modules have been designed and built. These interconnects provide shielded bellows with sliding contacts (made from gold-plated copper-beryllium slides on a rhodium-coated copper tube) , also called "plug-in modules", and allow for thermal expansion as well as compensation of mechanical and alignment tolerances. Figure 20 shows a photograph of a cut-through plug-in module to illustrate the design.

### 4.5.3 *Vacuum pumping*

In particle accelerators, discrete pumping is the most commonly used solution and is often obtained by ion pumps combined with sublimation or non-evaporable getter (NEG) cartridge pumps, cryogenic and turbo-molecular pumps.

Ion pumps are widely used since they are very reliable and provide a high pumping speed. Ion pumps have also the advantage of pumping all gas species and once baked-out at 250–300°C, the ultimate pressure is in the low $10^{-10}$ Pa.

Sublimation pumps are often used to speed up the pump down to the UHV pressure range or as a complement to ion pumps at very low pressures.

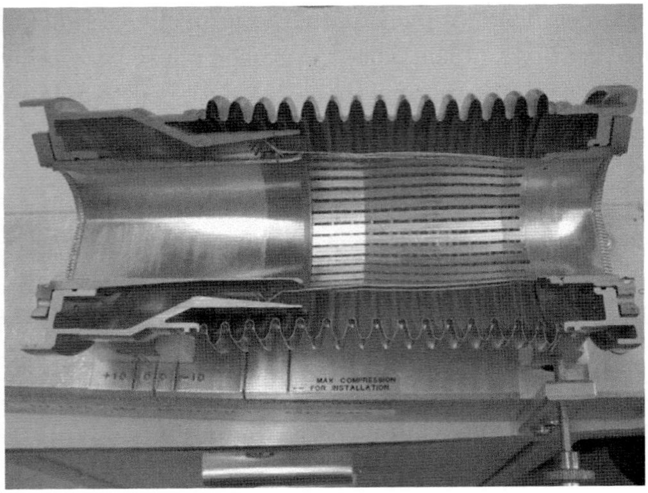


Fig. 20. Photograph of a cut-through plug-in module.

Similar to NEG pumps, the major limitation of the sublimation pumps is related to the pumping speed of noble gases and methane.

## 4.6 *Machine protection and interlock systems*

### 4.6.1 *Beam stored energy*

The stored energy in a particle beam is simply

$$E_b = k_b \cdot n_b \cdot e \cdot E \text{ (eV)} = I_{\text{total}} \cdot E \cdot t_{\text{rev}}$$

where $k_b$ and $n_b$ are the number of bunches and the number of particles per bunch respectively, $e$ is the proton charge, and $E$ the particle energy in eV. $I_{\text{total}}$ is the total beam current and $t_{\text{rev}}$ is the revolution time.

For the LHC nominal (design) parameters the stored energy at 7 TeV is 362 MJ per beam. This quantity of stored energy is capable of heating and melting about 500 kg of a typical metal such as copper. More importantly, since the cross-section of the beam is extremely small, the energy density of the beam is enormously destructive. For this reason a very comprehensive machine protection system has been designed and tested to protect machine components against the uncontrolled loss of all or part of the stored beam energy.

The stored energy in the beams is compared in Fig. 21 for several accelerator facilities. It is seen that modern accelerators operate or are designed for beam momenta between a few GeV/c to a few TeV/c. The stored beam energies are in the range of 10 KJ to 500 MJ.

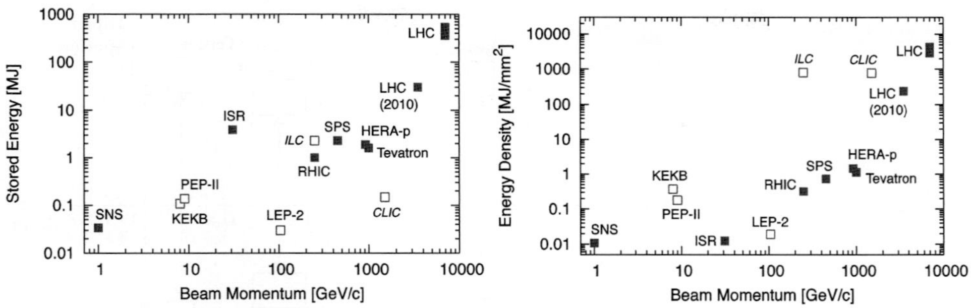


Fig. 21. The stored energy (left) and the density of stored energy (right) are shown versus the beam momentum for various accelerators. Filled symbols refer to proton, while open symbols indicate electron/positron accelerators. ILC and CLIC are design studies. (Courtesy of R. Assmann)

### 4.6.2 *Collimation system*

Nominal luminosity of the LHC requires the storage and collision of 7 TeV beams each with more than 360 MJ of stored energy. Beam losses, partial or total are unavoidable; consequently the LHC must be capable of handling these losses. For partial losses of 1% of the total beam intensity in a 10 s period, a peak power load of 500 kW is produced at 7 TeV. These losses occur in an environment of the superconducting (SC) magnet which cannot be exposed to an energy deposition of more than ~8.5 W/m, without quenching.

The collimation system has been designed and optimized [12, 13] to prevent beam-induced quenches of the SC magnets and for protection of beam-line components against large radioactive doses. This is accomplished by ensuring that all losses occur at the collimators and not elsewhere in the machine. This implies that the collimators must under all conditions define the bottle-neck of the aperture of the machine. By so doing, all particles which grow to large amplitudes are intercepted by the collimators, thus protecting the rest of the machine, in particular the SC magnets.

The three-stage collimation system is installed in the dedicated cleaning insertions in IR3 and IR7, and has been designed to ensure that only a small fraction of lost protons escape from these cleaning insertions. Figure 22 shows the hierarchical settings of the various collimator settings with respect to the centre of the beam.

The main types of collimators are

- Primary collimators (TCP) with robust carbon–carbon (CFC) jaws for interception of the primary beam halo;
- Secondary collimators (TCSG) with robust CFC jaws for interception of the secondary beam halo;

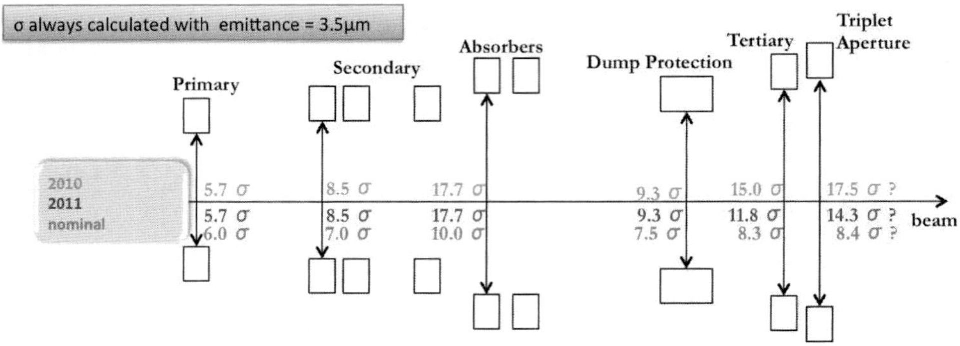


Fig. 22. Relative positions of the LHC collimators. (Courtesy of S. Redaelli)

- Tungsten based absorbers (TCLA) at the end of the cleaning insertions for protecting the SC arcs;
- Tungsten based absorbers (TCT) for protection and cleaning at the triplets in the experimental insertions.

There are also several other collimators which play an important role for local protection against injection and dump faults bringing the total number to more than one hundred installed in the LHC.

The LHC collimation design has taken into account many varied requirements [6]:

- Robustness against the high power normal regular beam losses (500 kW over 10 s) and in case of accidental scenarios (2 MJ shock impact in 200 ns);
- Cleaning inefficiency ($2 \times 10^{-5}$ per m);
- Coupling impedance (with collimation gaps of down to 2.5 mm and a total installed collimator jaw length of about 48 m per beam);
- Mechanical precision (jaw flatness) and actuation (e.g. precise motors);
- Operability, including radiation optimization for fast handling of radioactive collimators.

The performance of the LHC is strongly dependent on the cleaning efficiency and the coupling impedance, both of which may impose an upper limit on the maximum beam intensity. The required local cleaning inefficiency at 7 TeV is about $2 \times 10^{-5}$ m$^{-1}$ for nominal intensity of $3 \times 10^{14}$ p per beam.

### 4.6.3 *Machine protection*

The LHC machine protection system [14] must protect the machine and detector components in case of any failure and under all conditions.

Protection implies dumping the beam and triggering the dissipation of the electro-magnetic stored energy in the magnets in a safe and timely way, as well as prevention of injection under dangerous conditions.

In addition the system must:

- Allow efficient operation. The system should only trigger beam dumps when it is essential, unnecessary triggers of the beam dump must be minimally acceptable.
- Provide post analysis of each failure event to allow continuous improvements in the overall protection and avoidance of similar events in future runs. In case of beam dump or failure in the powering systems, correct diagnostic messages must be made available to the operations team. In case of complex failures involving multiple alarms, the sequence of events must be identifiable.

The machine interlocks system is part of this protection and include two systems, the powering interlock system [15] and the beam interlock system (see Fig. 23).

#### 4.6.3.1 Powering interlock system

Protection during magnet powering prevents damage to the elements in the electrical circuits, including all magnets, normal conducting and superconducting cables, current leads and power converters.

Protection relies on several dedicated sub-systems, such as the quench protection system, the energy extraction system and the powering interlock system. The powering interlock system permits powering of magnet's if and only if several safety-defined conditions are met.

In the case of a failure, the powering interlock system produces a "power abort" (a safe stop of the powering), a discharge of the magnet's stored energy and a trigger for a beam abort.

It is also required that magnets can be powered independently of beam operation; for example during commissioning and for testing equipment.

#### 4.6.3.2 Beam interlock system

Due to the large and varied number of types of beam losses and failures, a strategy was developed for beam dumps for each type of possible beam loss. Since there are about 10,000 magnets powered in more than 1600 electrical circuits with more than 1700 power converters, the most probable cause for beam losses are failures in the magnet powering system. However, there can be many other possible types of failures. Examples are aperture limitations in

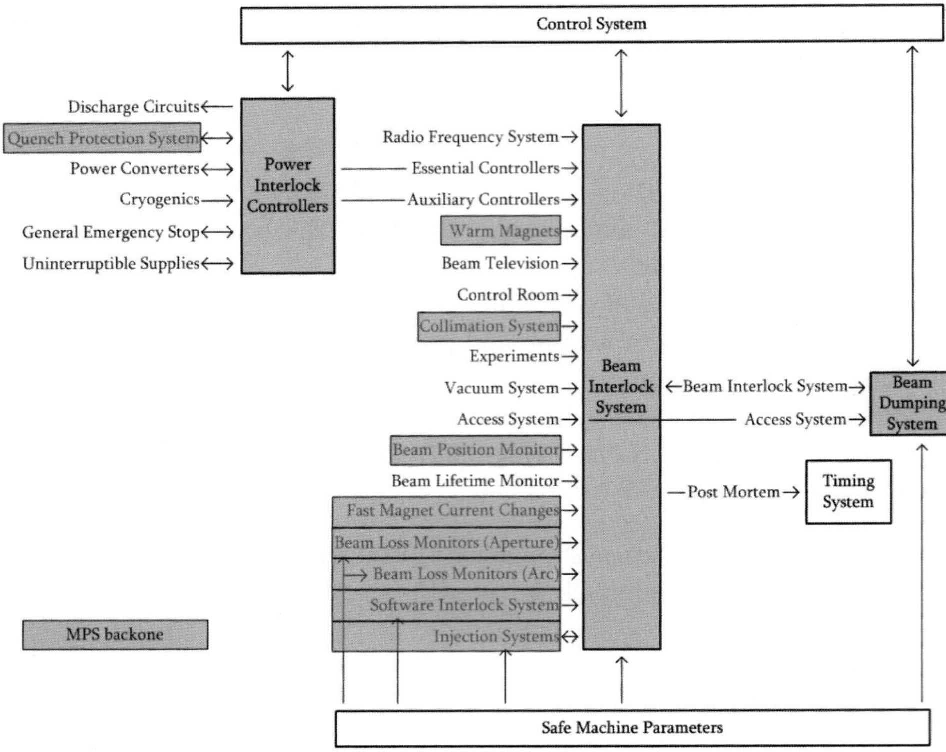


Fig. 23. Machine protection backbone. (Courtesy of M. Zerlauth and B. Todd)

one of the two beam vacuum pipe, helium feed-throughs, interconnects, RF shielding, and moveable devices such as vacuum valves, collimator jaws, and wire scanners can be in a wrong position and obstruct the beam passage. The beam interlock system is also interfaced with the powering interlock system and many other systems such as safety access, and the LHC detectors.

As part of the beam dump strategy, for each type of failure, the time needed (reaction time, following the failure) to produce unacceptably high (dangerous) losses was evaluated. Of course this time depends on the failure mode and several other parameters such as energy, optics and collimator settings. The reaction times were grouped into different categories:

- Ultra-fast failures, (single turn) during injection or beam dump;
- Very fast losses (less than 5 ms);
- Relatively fast losses (more than 5 ms);
- Continuous losses, reducing significantly the lifetime.

For single turn failures during injection and/or extraction, machine protection relies on collimators and dilutors which must be correctly positioned with respect to the beam. The beam interlock system ensures that beam cannot be injected if the collimators and dilutors are not correctly positioned.

Very fast losses: The fastest losses can come from failure of a Dl normal conducting bending magnet in IR1 and IR5. With this failure mode, the beam displacement can reach a collimator within a few turns. Another very fast beam loss can result from a quench of a superconducting dipole magnet. In this case the beam is expected to touch collimator jaws after several tens of turns.

For failures producing very fast beam displacements or emittance blow-up, the beam losses close to aperture limitations (collimators, beam dilutors, low-beta quadrupoles, etc.) are detected by beam loss monitors. The loss monitors trigger a beam dump when the beam loss exceeds a pre-defined threshold. It is also possible to detect rapid beam position changes, and trigger the beam dump when these changes exceed a pre-defined rate.

For fast losses, the beam loss monitors around the machine and interlock signals from equipment (in case of hardware failure) are used to generate beam dump requests; these complement the fast beam loss and beam position monitors signals.

For steady losses, the beam loss monitors and the heat load at collimators are the safety mechanism. In addition a super-reliable beam lifetime detector will be implemented as an additional source of security.

An important analysis tool is provided by the "post mortem" system which records all relevant parameters with a time stamp allowing by post-event analysis, the understanding of the series of events leading to a power or beam abort. This tool uses accurate synchronised time stamping of all equipment attached to the LHC.

### 4.6.4 *Validation and analysis*

The hardware of the machine protection system can only assure complete protection under very well-known and controlled beam conditions. There are several parameters involved in this control, including the beam dump extraction channel and the hierarchical positions of the collimation system.

In order to dump the beam cleanly, the position (closed orbit) of the beam must be well centred at the azimuthal position of the extraction channel of the beam dump system. Tolerances have been evaluated for this positioning and beam position monitors measure the transverse position on a continuous basis. If the measured beam position exceeds the specified tolerances then the beam is dumped.

The collimation system is primordial to the protection of the machine. Efficient beam collimation of large amplitude particles and protection of the machine relies on the precise set-up of the three-stage collimation system, with the primary collimators defining the aperture bottle neck of the LHC. In this way all primary particle losses should be absorbed by the primary collimators. The procedures for setting the collimators and the validation of the settings are complicated and in the early days of operation were very time consuming. The procedure involves empirically positioning each collimator (of which there are more than 100 units) with respect to the beam and subsequent validation of the cleaning efficiency by "beam loss maps". The loss maps are produced following the procedure for positioning of the collimators and involve increasing the transverse beam size (blow-up) which provokes relative beam losses that are measured all around the LHC circumference using the beam loss monitors. A typical loss map is shown in Fig. 24 where the cleaning efficiency was measured to be 99.993%. This validation procedure is repeated at regular intervals during the running periods or when there is a change which may affect the performance of the collimator and machine protection systems.

In summary, the beam interlock system is crucial for safe operation with beam. It permits injection into the LHC when all safety systems are ready. For circulating beam, when abnormal events occur, the interlock system transmits beam dump request from other systems to the beam dumping system.

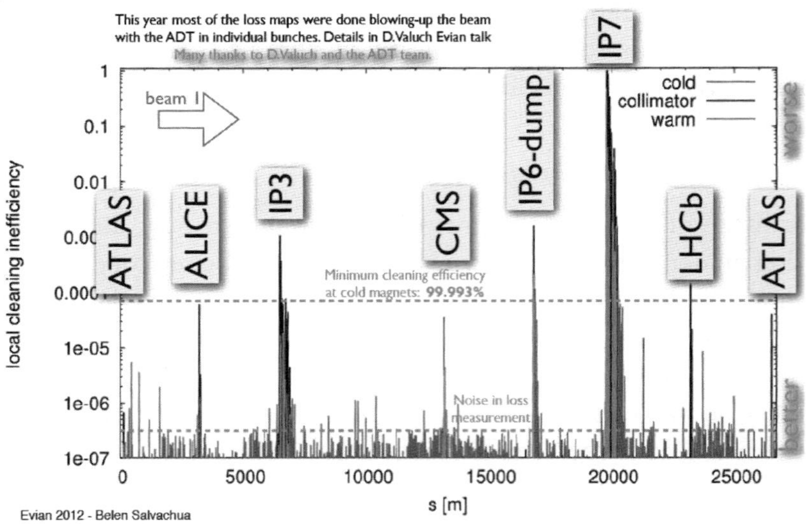


Fig. 24. A typical loss map.

### 4.7 *Cryogenic system*

In order to maximise the field strength of the superconducting magnets with NbTi windings, the LHC is the first ever collider to operate at temperatures below 2 K. The superconducting magnet windings in the arcs, the dispersion suppressors and the inner triplets are immersed in a pressurised bath of superfluid helium at a pressure of about 0.13 MPa (1.3 bar) and a maximum temperature of 1.9 K.

In accelerator applications, the superconductor operates at a fraction of its critical temperature in order to preserve current-carrying capability at high field (magnets) and to limit AC losses (RF cavities). This imposes the use of helium for the cooling of the low temperature superconductors. Other additional important benefits of operating the accelerator beam pipes at low temperature are the achievement of low vacuum pressure through cryogenic pumping, and the reduction of wall resistance which controls image-current losses and transverse impedance. The specific heat of the superconducting alloy and its copper matrix falls rapidly with temperature. Consequently, the full benefit in terms of stability margin of operation at 1.9 K requires making effective use of the transport properties of superfluid helium, which has a maximum thermal conductivity at this temperature. The low bulk viscosity enables the coolant to permeate the core of the magnet windings, while the very large specific heat, combined with the enormous heat conductivity has a powerful stabilising effect against thermal disturbances.

The cryogenic system [16, 17] cools the huge cold mass of the LHC, $37 \times 10^6$ kg in a maximum time of 15 days. The system is capable of coping with the resistive transitions of the superconducting magnets, which may occur in the machine, while minimising the loss of cryogen and system perturbations. The design also limits the propagation of heat to the neighbouring magnets and allows recovery in a time that does not seriously impact on the operational availability of the LHC. The cryogenic system allows a reasonable fast cool-down and warm-up time for limited lengths of cryo-magnet strings, to give a reasonable time for repairing or exchanging a defective unit. The foot-print and schematic layout are shown in Fig. 25.

The present LHC cryogenic system is the largest of its type in the world and the evolution of the CERN installation is shown in Fig. 26. The availability and reliability of this system are crucial to the performance of the LHC, since any major failure entails a long recovery time due to the time required to regain the low temperatures needed for a large number of magnets (1.9 K). The performance of this enormous system has been excellent following the usual teething problems at the start of operation of the LHC collider. The

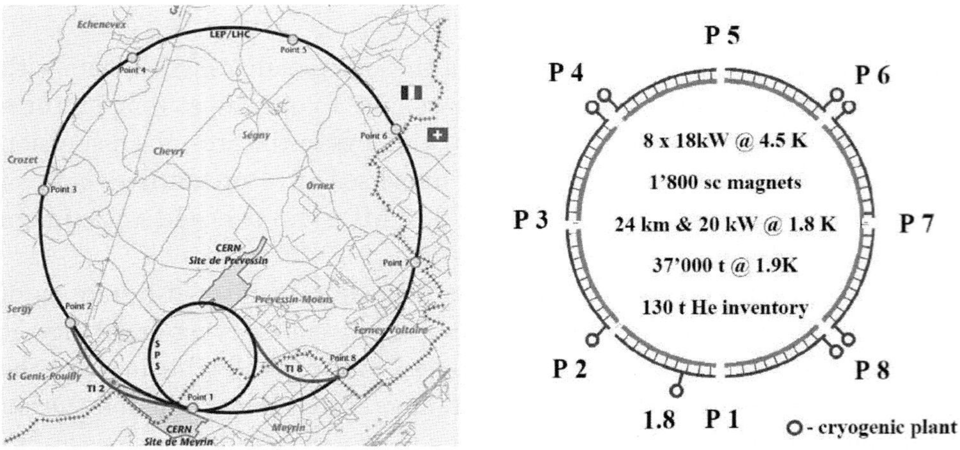


Fig. 25. The geographical distribution of the CERN cryogenic plants (left) and the schematic outlay of the systems for the LHC machine and detectors (right). (Courtesy of P. Lebrun and L. Tavian)

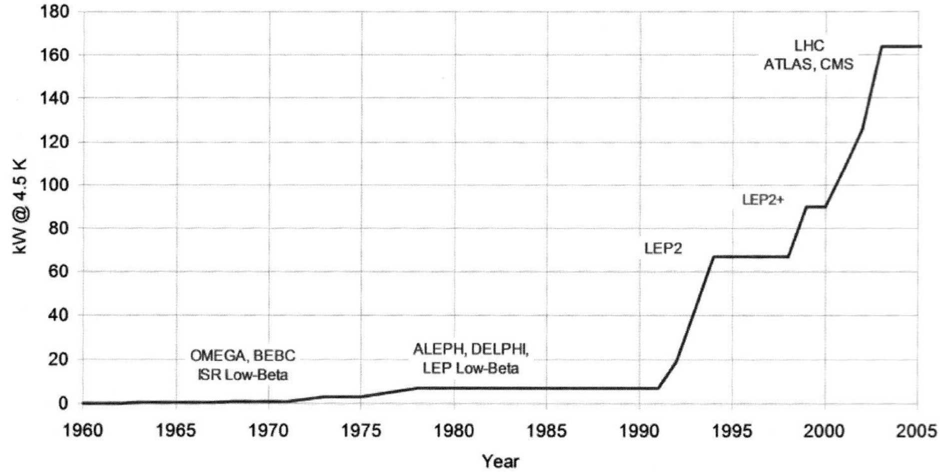


Fig. 26. Evolution of CERN installed cryogenic power. (Courtesy of L. Tavian)

exceptional availability [18] has been achieved in spite of the fact that, during the three years, the LHC has been operating for long periods without the foreseen maintenance stops.

### 4.8 *Transfer, injection and extraction*

Clean, efficient beam transfer and injection [19] to the LHC is of the utmost importance since poor injection can result in poor performance [20] (increased transverse beam size), or even worse, in beam damage to the components of the collider. LHC extraction (dump or abort) is even more critical [21] in that the beams have maximum destructive stored energy and the beam extraction system is the final "safety net" in the machine protection system.

The LHC uses the "single-turn" injection and extraction method because of the low beam losses and low emittance dilution which can be achieved with this type of injection. This technique is rather straight-forward and involves the use of a septum (or series of septa) to deflect the beam into or out of the accelerator aperture, a kicker to deflect the beam onto or away from the closed orbit and a closed orbit bump to reduce the required kick strength.

Fast kickers and magnetic septa are basically dipole magnets and follow the same principles as ordinary bending elements. However they impose many additional and sometimes conflicting design requirements and necessitate many differing technologies including high voltage, vacuum, pulse forming networks, accurate timing, beam impedance, as well as the magnet technology. They are also often used in combination with special protection elements and safety critical systems (such as beam abort) and in these cases redundancy becomes an issue. In addition they are typically purpose-built single elements or produced in small series.

Injection into the LHC takes place in the combined experimental and injection insertions IR2 and IR8 [6]. The transfer line TI2 steers the beam to a point ~150 m left of IP2 for injection into Ring 1 and TI8 steers the beam ~160 m right of IP8 for injection into Ring 2 (this IP is displaced from the injection point).

In both insertions the beam arrives at the LHC on the outside and below the machine plane. The beam is directed by a last series of dipoles, located in the LHC tunnel, towards a series of five "Lambertson" type septum magnets which deflect the beam horizontally under the outer ring. A series of four MKI kicker magnets deflects the beam vertically onto the desired orbit. An injection beam stopper, TDI, is placed 15 m upstream of the superconducting recombination dipole D1 in order to allow proper injection set-up with pilot bunches and to protect the LHC in case of malfunctioning of the injection kickers. (See layout for IR8 in Fig. 27.)

Two collimators, TCLI, near the superconducting quadrupole Q6 on the other side of the insertion provides further protection against injection errors. The geometrical layout and arrangement is almost identical for both injections schemes. However the beam optics vary between IR2 and IR8 due to the displaced interaction point in IR8.

Ideally in a "septum" magnet there is zero field on one side of the septum (the circulating beam side) and high field on the other (the injected or

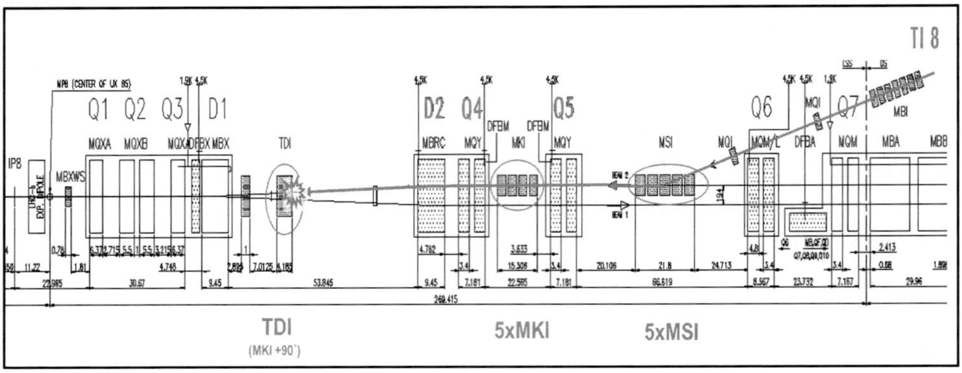


Fig. 27. Layout of injection: right of IP8 in the H plane. (Courtesy of V. Mertens and B. Goddard)

extracted beam side). The septum should be as thin as possible to minimize the particle losses on the septum while the beam is deflected across it. Clearly the stray field in the "zero" field region is more critical as the circulating beam passes through this region.

For LHC injection, five septum magnets of two different types (MSIA with a septum thickness of 6 mm and MSIB with 15.5 mm), each with a septum core length of 4 metres, deflect the injected beam horizontally by 12 mrad under the outer ring. For injection into Ring 1 the septa are located in RA23 between Q6 and Q5, and for Ring 2 in RA87, each system stretching over almost 22 metres.

Injection into each ring of the LHC requires four fast pulsed kicker magnets (MKI), located in RA23 between Q5 and Q4 for Ring 1 and in RA87 for Ring 2. The pulse generators and part of the power and control electronics are located in the adjacent underground galleries, UA23 and UA87. Injection onto the machine orbit requires a deflection of 0.85 mrad which corresponds to an integrated field of 1.2 Tm. In order to maintain the injected beam emittance blow-up to specifications, the flat top ripple of the field pulse must stay below $\pm 0.5\%$.

The LHC accumulated beam consists of 12 batches with 11 gaps of $0.94\,\mu$s for the injection kicker rise-time and one gap of 3 $\mu$s for the beam dump rise-time. The injection kickers have corresponding rise times of 0.9 $\mu$s and fall times of 3 $\mu$s each for fields between 0.5% and 95% of nominal.

### 4.8.1 *The beam dump system*

The beam dump system is designed to extract the LHC beams with minimum extraction losses, transport and treat the extracted beam so that it can be safely disposed of on an external absorber (dump block). Due to the enormous destructive capabilities of the high energy beams, the beam dump systems are the most safety critical systems in the LHC and must therefore meet extremely high reliability criteria [21]. Loss-free extraction imposes a gap (abort gap) in the beam longitudinal distribution of sufficient duration to allow the extraction kicker (MKD) field to rise from zero to its nominal value. In addition the optics parameters (orbit, beta value, emittance, beam energy) must remain within tolerances (around the dump region) if the beam is to be extracted without unacceptable losses. If for example the beam orbit drifts towards the maximum of the permissible range, either due to a real orbit drift or due to detection errors, the beam is automatically dumped.

Both beam dump systems are installed around IR6 and each system is comprised of:

- 15 extraction kicker magnets (MKD);
- 15 steel septum magnets (MSD) of three types;
- 10 modules of two types of dilution kicker magnets between the MSD and Q4;
- The beam dump proper comprising the TDE core assembly and associated steel and concrete shielding, situated in a beam dump cavern ~750 m from the centre of the septum magnets;
- The TCDS and TCDQ diluter elements, immediately upstream of the MSD and Q4 respectively.

The MKD kickers deflect (see Fig. 28) the entire beam horizontally into the high-field gap of the MSD septum which provides a vertical deflection to raise the beam above the LHC machine cryostat before the beginning of the arc sections. The dilution kickers sweep the beam in spiral shaped form and after the appropriate drift distance it is absorbed in the dump block core (TDE) assembly. The TCDS and TCDQ protect the machine elements from a beam abort that is not synchronised with the particle-free beam gap (asynchronous damp).

The beam dump system is used to abort the beam in a planned or unplanned way and forms a vital part of the machine protection system. Due to the criticality of the system, rigorous reliability studies and measurements have been carried out at the design stage.

The beam dump system is designed to be Safety Integration Level 4 (SIL4), which is specified as less than $10^{-8}$ probability of failure per hour of operation. This would result in a maximum of 1 failure in 20,000 years

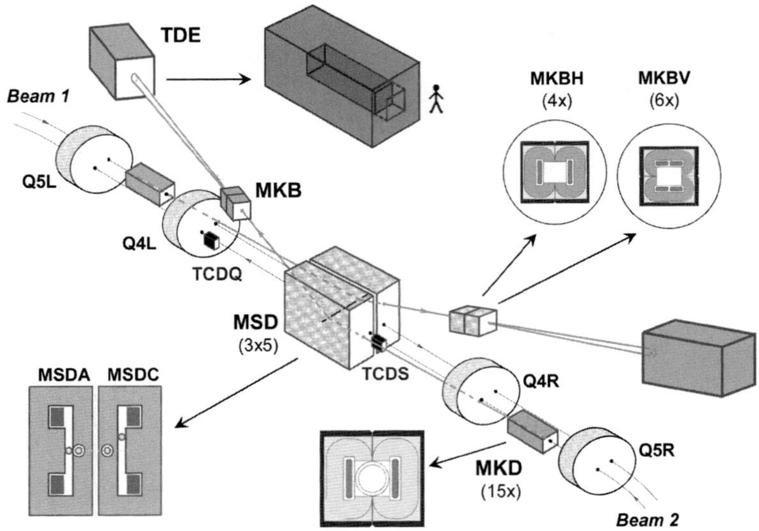


Fig. 28. Layout of beam dump elements. (Courtesy of B. Goddard and V. Mertens)

assuming 5000 h operation per year. The reliability analysis [21] shows that these specifications are met, under the assumptions made.

The overall LHC machine protection system is designed to be SIL3, which is less than $10^{-7}$ failures per hour, corresponding to one failure in around 2000 years. This assumed low failure rate is achieved by the use of high quality components, redundancy in components and in signal paths, continuous surveillance, rigorous follow-up and the imposition of rigorous validation tests before injecting beam in the LHC. However even with these measures implemented, estimates for the extraction kicker system (MKD) imposed full redundancy of one complete MKD kicker and its generator. Consequently the beam can safely be extracted with only 14 out of 15 kicker magnets pulsing. It has also been understood that this assumed failure rate needs to be obtained for the complete system, including the LHC beam interlock system and the LHC beam energy meter. The beam energy meter is one of the most critical elements since underestimating the beam energy could send the full LHC beam into one of the arcs adjacent to the beam dump insertion.

The beam dump system has been operating with excellent reliability for the three years of operation.

### 4.9 *Radio frequency system*

The LHC injected beam is captured, accelerated, and stored using the 400 MHz fixed frequency superconducting cavity system [22] which allows very high accelerating gradients. At superconducting temperatures the RF surface resistance is not exactly zero; however it still allows quality (Q) factors of the order of $10^{10}$. It should be remembered, however, that even with such low power losses in the SC cavities, cooling at 2 K is very inefficient. As a rule of thumb, 1 W lost at 2 K requires refrigeration power of about 1 kW.

The design choices of SC cavities at this frequency resulted from the requirements on the longitudinal emittance at 7 TeV and the maximum bunch length allowed by luminosity, and led to a required maximum voltage of 16 MV per beam. The required longitudinal emittance in collision is defined by intra-beam scattering lifetime in the presence of synchrotron damping, RF lifetime and instability threshold considerations. To provide this emittance, controlled emittance increase is employed (bunches from the SPS) during acceleration by excitation of the RF by band-limited noise.

The two independent RF systems each provide around 16 MV circumferential voltage at high energy, while at injection about 8 MV is needed. The frequency of 400 MHz is close to that of LEP, 352 MHz, which allowed the use of the same proven technology of sputtering niobium onto a copper substrate for the manufacture of the cavities. This technology has the advantage of cost but more importantly cavity quenching is much reduced, since local heat is quickly conducted away by the copper. In addition, niobium sputtered cavities are insensitive to the earth's magnetic field and special magnetic shielding, needed for solid niobium cavities, is not necessary. However one disadvantage is that the maximum accelerating fields achieved with this technique are lower than that obtained with bulk niobium cavities.

An RF cryo-module consists of four cavities, each equipped with their helium tank, tuner, HOM couplers and power coupler. This compact design reduces static thermal losses and results in reduced space requirements. In addition the modular design allows faster replacements in case of problems.

The decision to separate the RF systems for each beam was taken following considerations to curtail the power levels at the RF power couplers. The real power supplied to the beams is relatively small ($\sim$275 kW per beam) whereas the installed power required to control the beams is much greater. The use of separate RF systems per beam reduces these power requirements by nearly a factor of two (due to the transient beam-loading) and allows the possibility of independent control of the beams, for example for beams of different species. However, since the standard distance between beams (194 mm) is insufficient to allow separation of the RF systems, the beam separation was increased in the RF region to 420 mm by using special SC dipoles. With this separation and by "staggering" the cavities longitudinally, the other beam can pass outside the cavity, but must still pass through the cryostat.

Even with separate RF systems, the peak power requirements for the couplers are significantly higher than previously achieved on SC cavities. A large tuning range is required to compensate the average reactive beam component, and each RF system has thus eight cavities, grouped by four in one common cryostat.

Each cavity is driven by an individual RF system with klystron, circulator and load. Feedback loops around each cavity allow precise control of the field in each cavity, to allow for the high-intensity LHC proton beam. The use of one klystron per cavity also avoids coupling between cavities and pondero-motive oscillations that plagued the LEP RF system where one klystron supplied eight cavities. The cavities are tuned to the fixed frequency harmonic of the revolution frequency by a mechanical system producing controlled elastic deformation.

Originally the design of the RF system foresaw an additional 200 MHz system [23] to improve the RF capture efficiency at injection energy. However, following machine studies and an upgrade programme in the SPS which led to a reduction in the coupling impedance, improved control of beam loading, and better beam instability control, it was concluded that the 400 MHz LHC system was sufficient to capture and accelerate the beams with minimal losses. This has proven to be a good decision with the intensities used so far (2013) in the LHC. However, space remains allocated in the machine and certain preparatory work has been done in case such a system becomes necessary in the future with higher intensities.

The different RF systems are concentrated around the centre of Point 4 and extend from the UX45 cavern into the tunnel on either side. The klystron power plant and racks of equipment for the different systems are in both the UX45 and US45 sides of the cavern. Figure 29 shows (left) the installed SC 400 MHz cavity system in the LHC tunnel and (right) the waveguide and power distribution plant in the underground cavern.

### 4.10 *Transverse beam feedback*

The beams in accelerators are inherently unstable. Many mechanisms are used in order to maintain the beam and its quality for many hours of circulation in the vacuum chamber. A crucial feedback system for the LHC is depicted in Fig. 30 (left). The transverse oscillations which appear at

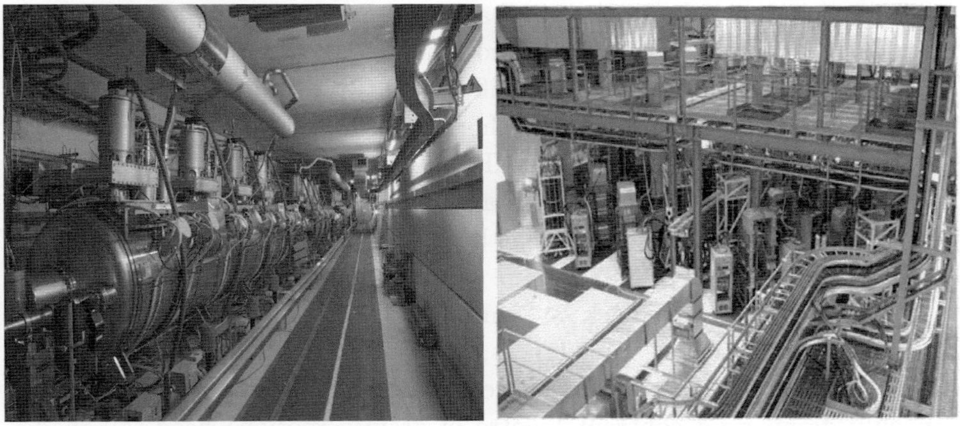

Fig. 29. LHC Point 4: (left) 400 MHz SC cavities and (right) LHC RF power plant in the underground cavern. (Courtesy of O. Brunner)

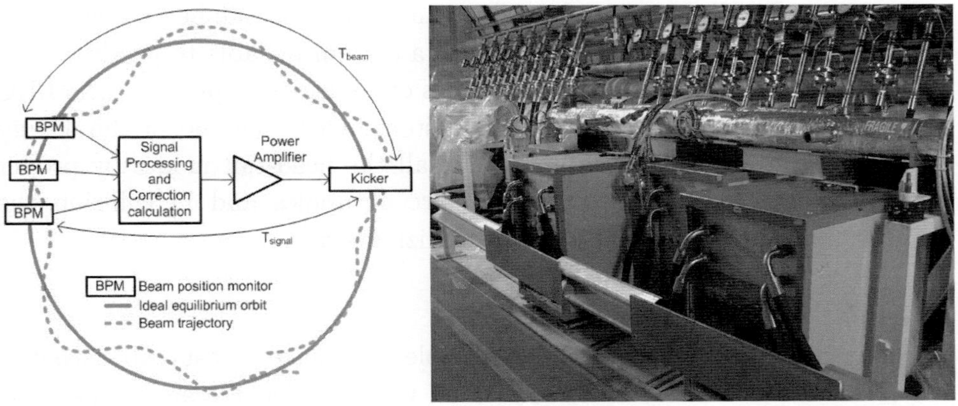


Fig. 30. Schematic of the transverse damping system (left) and photograph of damper kickers in tunnel (right). (Courtesy of W. Hofle)

the onset of instability are measured in one location of the circumference by a suitable beam position monitor (BPM). This signal is processed electronically and a corrective "kick" signal is generated, amplified and applied to the beam by a transverse electro-magnetic field ("kicker"). Figure 30 (right) shows the system as installed in the LHC tunnel.

The LHC transverse feedback system [24] performs three main functions: it damps transverse injection errors, prevents transverse coupled-bunch instabilities (dipole modes) and can be used to excite transverse oscillations in individual bunches for beam measurements such as the validation of the machine protection system by loss maps.

There are four independent feedback systems, one per plane and one per ring. Each system is comprised of two modules, each with two kicker units. Each kicker unit has one power amplifier with two tetrodes installed directly under the kicker vacuum tank. The horizontal kickers and power amplifiers for Ring 1 are installed left of the IP4 and the vertical kickers and power amplifiers are to the right. The installation for Ring 2 is asymmetric with respect to Ring 1. Tunnel space has been allocated for a future upgrade of one extra module per ring and plane to boost the total system.

The transverse feedback system damps coupled-bunch instabilities caused by narrow-band impedances, and in particular the resistive wall instability at injection and during acceleration. Since the feedback system can only damp rigid coherent (centre of gravity motion) head-tail modes, higher order head-tail modes must be stabilized by Landau damping or by chromaticity.

The feedback system is also beneficial in reducing injection oscillations before they can "filament" and produce an increase in the beam emittance.

The inherent noise excitation in such a system is liable to produce slow emittance increase during the many hours of collisions in physics. The LHC strategy is to switch off the feedback system before the end of the energy ramp and to ensure Landau damping of all the head-tail modes by proper control of the tune spread using the arc octupoles and in collision the beam–beam tune spread has a very stabilizing effect.

### 4.11 *Beam instrumentation*

Galileo said "measure what is measureable and make measureable what is not". This is particularly true when faced with collider operation where the performance of the machine is often driven by the precision and parameters which can be measured. The aim of the beam instrumentation physicist or engineer is to provide the diagnostic equipment for the observation of particle beams with the precision required to operate and improve the accelerators and their associated transfer lines. Beam instrumentation and diagnostics combines the disciplines of accelerator physics with mechanical, electronic and software engineering. A non-exhaustive list of the properties and parameters which need to be measured in the LHC [25] is given below:

- Beam intensity (total and per bunch);
- Beam losses all around the machine (machine protection);
- Beam position at regular intervals in the transfer lines and along the circumference of the beam pipe;
- Longitudinal beam profiles;

- Transverse beam profiles;
- Beam optics parameters (tunes, chromaticity, betatron coupling and $\beta$ functions);
- Beam aperture and measurements of non-linearities;
- Luminosity in collision.

A very brief description of some of the measurement systems [6] associated with this list is presented in the subsections below.

### 4.11.1 *Bunch and beam intensity*

The measurement of beam current or bunch intensity [26] is crucial in any accelerator. This is usually done by means of a "beam current transformer" (BCT). In order for the transformer to interact with the magnetic field of the beam, it has to be placed over a ceramic gap in the vacuum chamber. To keep the impedance seen by the beam as low as possible, an RF bypass (either a thin metallic coating or external capacitors on the ceramic) is required for the high frequency wall current components.

Beam current transformers of two different kinds provide intensity measurements for the LHC rings as well as for the transfer lines from the SPS to LHC and from the LHC in the Dumps channel. The transformers are installed in sections where the vacuum chamber is at room temperature and where the beams are separated.

The "DC" current transformers measure the average intensity of the circulating beam and can also be used to evaluate the beam lifetime. Because of their operational importance, two of the devices are installed in each ring. With an intensity of $4.8 \times 10^{14}$ protons and a lifetime of 25 h the decay rate is $5 \times 10^9$ protons/s. With a measurement time of 10 s this decay can be measured with about 1% precision. The fast beam current transformer (FBCT) measures the charge of each LHC bunch, and is intended for low intensity beams where the DCCT accuracy is limited. The measurement precision for the pilot beam of $5 \times 10^9$ protons in a single bunch is around 5%, and the same transformers are used to measure the circulating bunches by averaging the acquired bunch intensities over 20 ms giving an approximate precision of around 1% for pilot beam.

### 4.11.2 *Beam losses (BLM)*

Superconducting magnets are prone to quenches or even damage if subjected to the losses of a small fraction of the total LHC beam. Protection of the magnets is assured by detecting lost beam protons and aborting the remaining beam before the losses reach a level which would cause a quench or damage. Such protection is crucial for the reliable operation of the LHC due to the long recovery time following damage of a magnet (several weeks). In addition to the quench prevention and damage protection, the loss detection allows the observation of local aperture restrictions, orbit distortion, beam oscillations, particle, diffusion, etc.

Ionisation chambers located outside the magnet cryostat measure [27, 28] the energy deposition resulting from secondary showers of particles generated by loss of particles inside the vacuum chamber. Flux detectors are placed at locations of probable beam losses so as to measure a representative fraction of the secondary particle flux. Primary protons are most likely to be lost at aperture bottle-necks; this occurs at each arc quadrupole at injection energy and at the triplets at 7 TeV (in both cases where the $\beta$ functions are maximum). The beam loss monitors are located on either side of the magnets, at positions about 1 m downstream of the most likely loss locations. The ionisation chambers have a volume of about 1 litre and the inner parallel electrodes are separated by 5 mm. The chamber is filled with $N_2$ under normal pressure and a bias voltage of 1500 V.

The settings of the thresholds are based on simulations. The BLMs are grouped in families defined by their locations of use and their time resolution.

### 4.11.3 *Beam position measurement (BPM, sometimes referred to as "pick-ups")*

The orbit and trajectory measurement system of the LHC consists of 516 monitors per LHC ring; each monitor is capable of measuring the beam position in both the horizontal and vertical planes. There are three different types of monitor: 24 mm button electrode monitors, 34 mm button electrode monitors and 120 mm strip-line monitors.

The majority (860 of the 1032) of the monitors are of the arc type which are composed of four 24 mm diameter button electrode feed-throughs mounted orthogonally in a 48 mm inner diameter beam pipe. The electrodes are curved to follow the beam pipe aperture and are retracted by 0.5 mm to protect the buttons from direct synchrotron radiation from the main bending magnets.

The "inner triplet" BPMs in all interaction regions are equipped with 120 mm, directional strip-line couplers, which are capable of distinguishing between counter-rotating beams in the same beam pipe.

The collimation insertions in Points 3 and 7 are equipped with warm, 34 mm diameter button electrodes which are an enlarged version of the arc BPM button. This type of electrode is also used in the matching sections on either side of the four interaction regions as well.

The "closed orbit" of a single beam is defined by the measurement of the beam position in each of the 516 BPMs and the closed orbit software takes this information and calculates changes to the magnet correctors which minimises the excursions of this closed orbit. Operation of the LHC is crucially dependent on the performance of the closed orbit system and in fact LHC would not work without the orbit feedback [29]. In 2012 the channel availability of the orbit system was 97% with an orbit resolution of 10 $\mu$m, and a fill to fill reproducibility of 50 $\mu$m.

The machine is also protected by BPMs which are interlocked to ensure that the orbit remains within $\pm 4$ mm of the ideal transverse position at the location of the extraction septum. If the orbit deviation exceeds this tolerance then the beam is dumped in order to avoid a situation where the beam has a position which is unsafe for extraction.

### 4.11.4 *Transverse beam profiles*

There are several different operational needs for the various transverse beam profile monitors.

- High sensitivity, single-pass monitors with reduced demands on accuracy;
- Intermediate sensitivity "few-pass monitors" (20 turns) for the injected beams and for calibration or matching studies, and producing limited transverse blow-up;
- Circulating beam monitors, working over the whole intensity range, and with little blow-up. These monitors are typically located where the beam profile is large (at high $\beta$ values), and include monitors which measure the synchrotron radiation coming from superconducting undulators as well as gas monitors. Flying wire scanners are also installed and used for calibration of the other transverse monitors as well as for measuring the distribution of particles in the "tails" of the beam;
- Circulating beam "tail" monitors are optimised to scan low beam densities.

### 4.11.5 *Diagnostics of transverse beam motion*

The stability of a particle beam is dependent on many parameters related to the "optics" of the machine. For example non-linear resonances occur when the number of betatron oscillations per turn ("tune") is an integer fraction. Consequently it is of the utmost importance to measure quantities like the betatron tune, chromaticity (tune dependence on particle energy) and transverse coupling, so that corrective action can be taken to correct the optics functions by way of the power converters driving the magnetic fields.

## 5 Initial beam commissioning

The scheduled start-up with beam on September 10, 2008 was an overwhelming success. Both beams made a full machine turn within hours of the scheduled time, and one beam was captured by the RF system.

### 5.1 *The accident*

Following this impressive beginning, a technical problem arose with an electrical transformer which necessitated stopping commissioning for a few days. During this stop it was decided to test the last octant (sector 34) up to 9.3 kA, i.e. 10% above the dipole current required for operation at 5 TeV per beam. At 8.7 kA a resistive zone developed in the dipole bus-bar magnet interconnects. This led to thermal runaway in one of the magnet interconnects, followed by the development of an electrical arc, initially across the interconnect, later puncturing the helium enclosure, and finally puncturing both beam pipes. The release of the helium caused a pressure wave over a region of more than 400 metres resulting in damage to magnets, interconnects and pollution of the ultra-high vacuum system.

#### 5.1.1 *Post-mortem of the accident*

An inquiry by CERN specialists [30] indicated several causes of the substantial damage to the machine:

- There was an absence of solder on the offending magnet interconnect giving a contact resistance of 220 n$\Omega$ (design $\sim$1 n$\Omega$);
- There was poor electrical contact between the superconducting (SC) cable and the copper stabilizing bus-bar;
- The fault detection of the interconnect was not sensitive enough;
- The pressure relief ports were under-dimensioned for an accident of this magnitude;
- The anchorage of the magnets to the tunnel floor was inadequate.

Figure 31 shows the first page of a three-page fault tree describing in detail the evolution of events immediately following the thermal runaway [31].

### 5.2 *The repair*

Following the initial investigation of the resulting damage, a crash programme was set up to repair and consolidate the LHC. The teams included

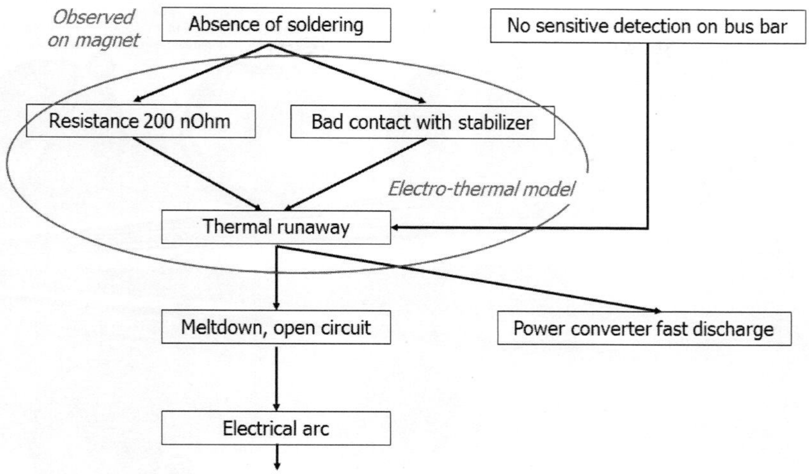


Fig. 31. First page of fault tree of sequence of events immediately following the accident. (Courtesy of P. Lebrun)

many CERN partners, collaborators, detector people as well as the accelerator sector. The task was enormous and required not only the repair of the damaged components but equally importantly re-engineering many elements so that such an accident would never occur in the future.

Figure 32 shows a schematic of the main elements of the repair in the damaged part of the tunnel. A total of 39 dipole magnets (marked 2 in diagram) were replaced. In addition, 14 quadrupole magnets were replaced (marked 1 in diagram) and a total of 54 damaged magnet interconnects needed full repair with around 150 extra needing partial repair (marked 3). About 5 km of ultra-high vacuum beam tube (marked 4) required the removal of small pieces of super-insulation and black soot followed by careful cleaning [32]. A new longitudinal restraining system (marked 5) was designed and installed on 50 quadrupole magnets. All existing flanges on the magnets were equipped with additional pressure relief ports (typically 10 cm diameter) and 20 cm flanges were cut on dipoles and equipped with double size pressure relief ports. In total 900 helium pressure relief ports were added (marked 6).

A major task was the upgrade of the magnet protection system which had been shown to be insensitive as a protection of the interconnect splices. The new design is 3000 times more sensitive than the older system and involves 6500 new detectors (marked 7 in Fig. 32) and the installation of more than 250 km of cable. A major added advantage of the new magnet protection system is that it provides continuous measurement, to sub-n$\Omega$ precision, of the resistance of all inter-magnet splices in the machine.

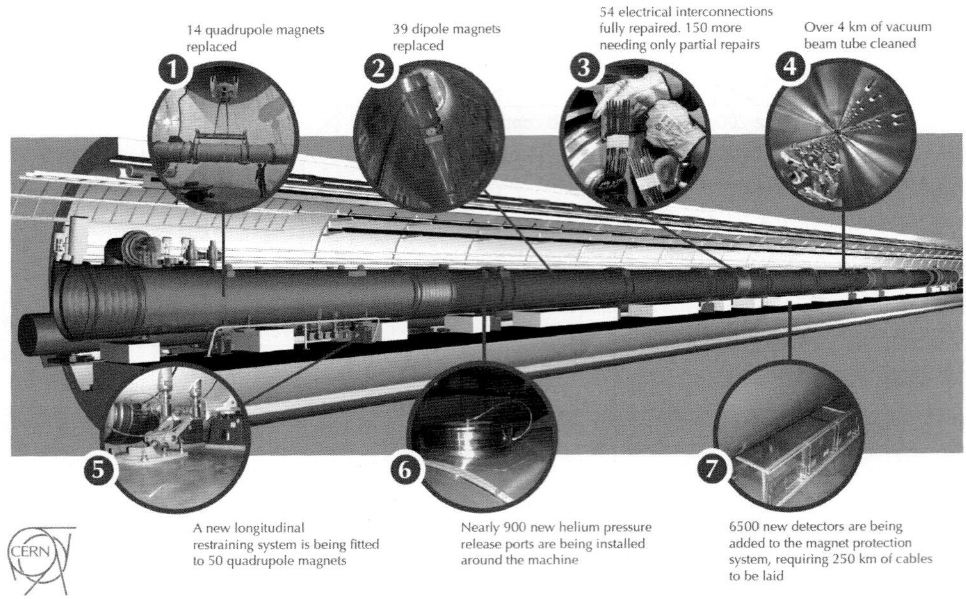


Fig. 32. Schematic of the main elements of the repair.

## 5.3 *The copper stabilizer bus-bars*

The completed inter-magnet bus-bar splice is designed for two separate functions. Firstly, at superconducting (SC) temperatures, to provide perfect electrical contact between the joined SC cable braids; and secondly, at non-SC temperatures (in the event of a quench for example) to ensure electrical continuity across the copper sheath. Hence in the event of a quench when the SC cable has a finite resistance, the copper sheath "shunts" the current away from the SC cable while the stored energy is being extracted from the magnet. In this way the SC cable is protected in case of a quench by the large cross-section of the copper stabilizer.

The components of an inter-magnet bus-bar is depicted in Fig. 33. The solder has two important functions, firstly to ensure good electrical contact between the two SC cable braids and secondly to ensure electrical continuity between the two copper profiles and the left copper sheath (bus-bar), and the right copper sheath. A little thought should indicate that if there is no electrical continuity across the copper stabilizer then the decaying current following a quench will flow through the SC cable (which is no longer superconducting) and possibly cause thermal runaway.

The quality of a splice is therefore determined by the resistance across the splice measured both at superconducting temperatures and at non-SC temperatures. The enhanced quality assurance introduced during the repair

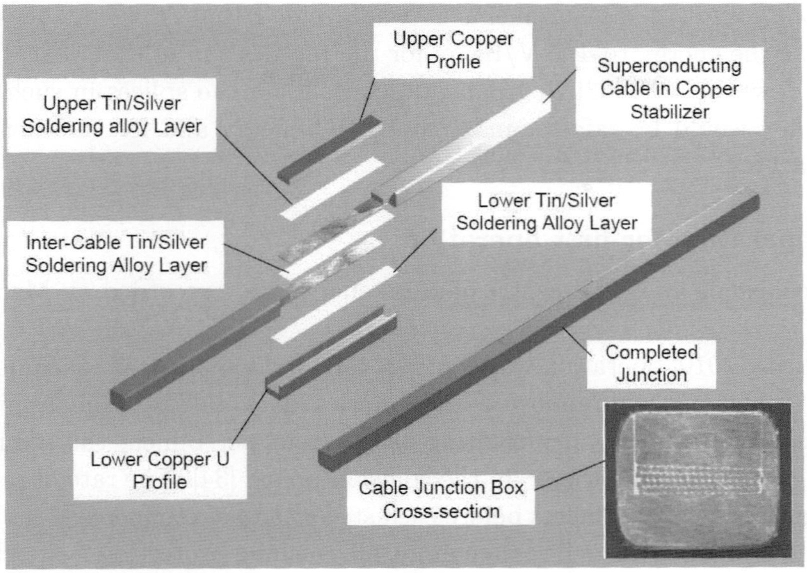


Fig. 33. Exploded diagram of an inter-magnet bus-bar.

of sector 3-4 revealed new concerns about the copper bus-bar in which the superconductor is embedded. Tests demonstrated that the soldering process could cause discontinuities in the copper part of the bus-bars and produce voids which prevent contact between the superconducting cable and the copper stabiliser.

Consequently, in 2009, a campaign was started to measure the splice resistance at room temperature. The "in-situ" measurement technique had a limited precision corresponding to about a factor of three higher than the resistance of a perfect splice. Consequently this technique could only be used to identify "outliers" which had a significantly increased resistance. On identification of outliers a much more precise measurement (which involved the overhead of opening the interconnect) could be performed. In this way all significant outliers were identified and repaired. However the limited precision of the "in-situ" measurement implied that some splices may have resistances significantly above the design value. Calculations, simulations and experimentation revealed that the maximum current that can flow in the interconnects without causing thermal runaway was less than the currents needed for maximum beam energy. Consequently it was decided to operate LHC (for a limited period) at a safe energy compatible with the existing situation of the already installed splices.

For operation in 2010 and 2011, the decision was taken to operate for data taking at 3.5 TeV per beam. Early 2012, with more experience as well as more complete measurements it was decided that it would be safe to increase the energy to 4 TeV/beam for the physics run of that year, then in a shutdown in 2013 [33] to repair and consolidate the splices in such a way that they would be safe for maximum LHC energy and for the lifetime of the machine.

## 6 Preparation for first operation

First collisions at 7 TeV centre-of-mass energy were recorded on March 30, 2010.

During 2010, operation was continuously divided between machine studies to increase the luminosity and physics data taking. The luminosity increase concentrated on increasing the current per bunch, the number of bunches and reducing the $\beta^*$ in the collision points [34]. The rate of progress was impressive, nevertheless before each step was taken to increase the intensity and hence the stored beam energy, all machine protection systems were validated up to the necessary level.

The LHC machine protection system [35] is probably one of the most intricate accelerator protection systems ever operated. It relies on very stringent control of the optics of the machine, both locally and globally. For all protection devices there are closed orbit constraints as well as $\beta$ value constraints. Hence, any effects which are intensity dependent (e.g. closed orbit measurement, $\beta^*$ measurement,...) complicate the procedure for finalizing the protection.

The collimation system [36, 37] is part of the machine protection system and the collimators must intercept almost the totality of any beam losses if they are to protect the rest of the machine. The hierarchy of the collimation system (primary, secondary, and tertiary) must be respected. This implies a stringent control of the $\beta$ functions and the closed orbits at the collimators and at all location with tight aperture margins, at all beam energies and throughout the "squeeze" of the low $\beta$ insertions. The $\beta$ functions are measured and corrected and the orbits are subjected to a feedback control with threshold limits. There is a watch-dog monitoring the orbit deviations at the locations of the collimators and if the limits are violated, the beam is dumped. A similar system is employed at the location of the beam dump extraction kicker. The tolerances set on the $\beta$ "beating" is around 20% and Fig. 34 shows first measurements made at beam energies between 1.5 and 3.5 TeV. In recent months (2012) the $\beta$ "beating" has been corrected to below 7% [38], a factor of three better than the specifications.

In order to measure the cleaning efficiency of the collimation system, "loss maps" (see earlier section) are made by provoking beam losses around

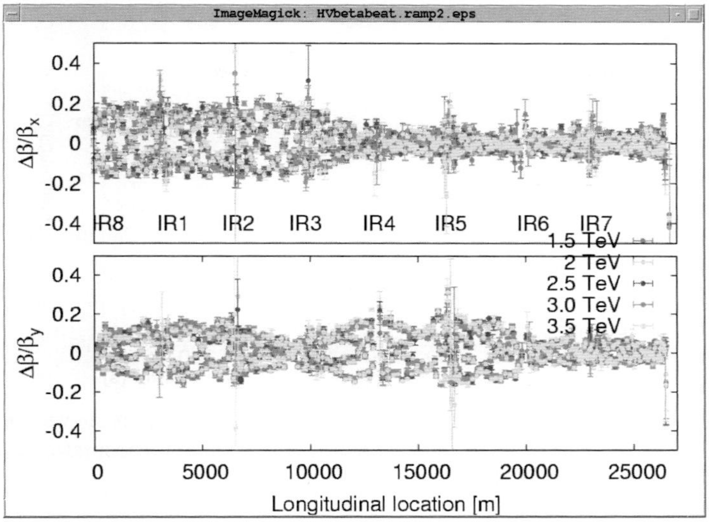


Fig. 34. Measurement of the $\beta$ beating at different energies.

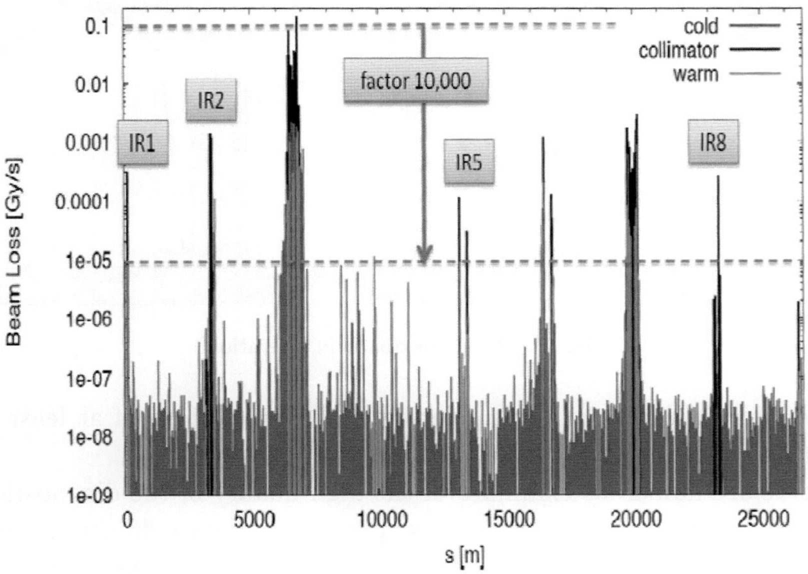


Fig. 35. Loss maps for collimation.

the circumference. When the collimators are well set up and the hierarchy of losses is correct, the vast majority of all losses are localised at the collimator. Figure 35 shows one of the early measurements where it can be seen that the ratio between the losses at the collimators and the worst cold location is about a factor of 10,000. It has been found that the optics and closed

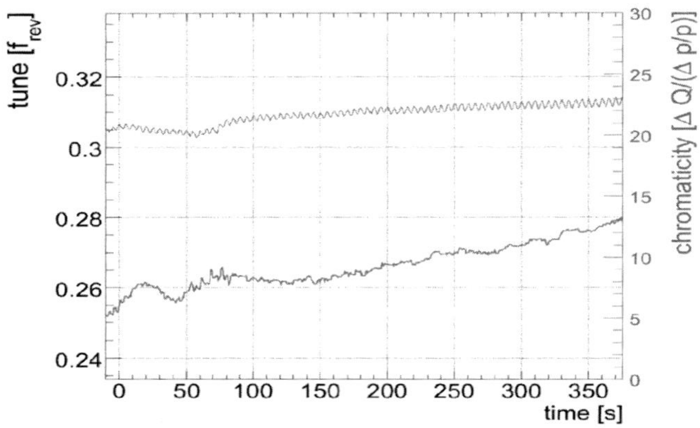


Fig. 36. Tune and chromaticity measurements: the upper blue plot shows the modulation of the tune, and from the peak to peak of the modulation the chromaticity is calculated and shown in the lower red plot.

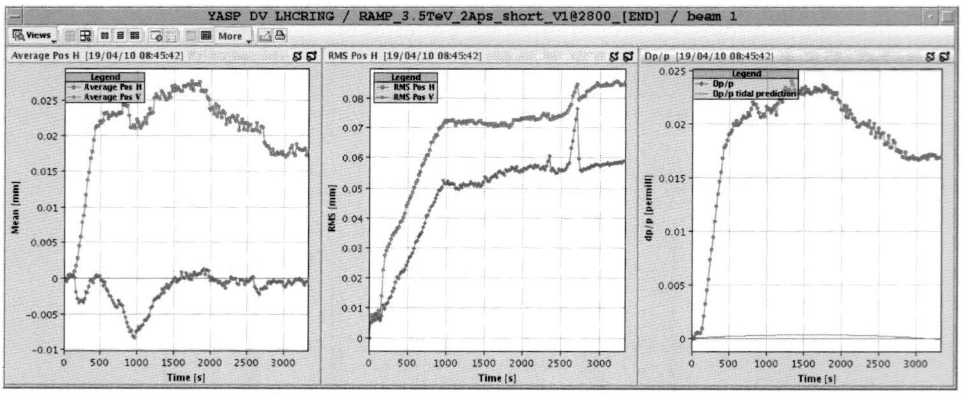


Fig. 37. Orbit feedback in operation.

orbit remain within acceptable tolerances [34] for periods of at least weeks without compromising the overall collimator hierarchy.

There are numerous examples of the high quality of the diagnostics and its implementation into the controls applications. Figure 36 shows one such example. In order to measure the chromaticity during the energy ramp, the beam momentum is modulated by varying the RF frequency and the transverse tunes are measured by a phase lock loop.

Another early example is shown in Fig. 37, which shows the operation of orbit feedback during the energy ramp. The three plots show the mean, RMS orbit distortion and the momentum deviation. The maximum RMS orbit change during the ramp is 0.08 mm compared to the design specifications of 0.5 mm RMS and ±4 mm peak orbit excursions.

## 7 Operational performance in 2010

### 7.1 *Protons*

As previously stated LHC had been operated in shared mode between beam commissioning and physics data taking during 2010. The equation for the luminosity for collisions with round beams (equal beam emittances and $\beta^*$ values for both beams in both planes) is given below:

$$L = \frac{n_b \cdot N_{\text{bunch},1} \cdot N_{\text{bunch},2} \cdot f_{\text{rev}}}{4\pi \cdot \beta^* \cdot \varepsilon_n} \cdot R(\phi, \beta^*, \varepsilon_n, \sigma_s)$$

where:

- $n_b$ is the number of bunches per beam;
- $N$ represents the number of protons in a single bunch;
- $\beta*$ is the insertion region focusing parameter;
- $\varepsilon_n$ is the normalised emittance (related to the cross-section dimension of the beam); and
- $R$ is the interaction region geometric factor.

Table 3 shows the evolution of the beam performance from the first collisions in March until the middle of August (the machine status is in the row event number, and at each event there were several days of operation under these conditions). During the first period (events 1 to 6), the total number of bunches (column marked $n_b$) was increased from the initial value of 2 to 13 and the interaction region focusing parameter $\beta^*$ was reduced to 2 m for most of the events. This progression in parameters produced an increase in the peak luminosity of around a factor of 250 during this first data taking period.

For the second period (events 7 to 12), the number of protons per bunch "$N_{\text{bunch}}$" was increased to near the design value of $1.15 \times 10^{11}$, the $\beta^*$ was set to 3.5 m and the number of bunches was progressively increased from 3 to 48. This progression resulted in peak luminosities reaching close to $10^{31}$ cm$^{-2}$s$^{-1}$. The stored energy (in megajoules, MJ) in each beam reached above 1 MJ in event 11. From experience in previous accelerators it is known that this amount of stored energy is sufficient to puncture and melt fairly large sections of the vacuum chamber. Increase in the stored beam energy from this point onwards was only authorized after a rigorous protocol in the validation of the machine protection system.

Table 3. Evolution of beam performance in 2010 ($N_c$ is the number of bunches colliding).

| Event | TeV | $\beta^*$ | $n_b$ | $N_{\text{bunch}}$ | MJ | $N_c$ | Peak luminosity | Date |
|---|---|---|---|---|---|---|---|---|
| 1 | 3.5 | 10 | 2 | 1.00E+10 | 0.01 | 1 | 8.9E+26 | 30 March 2010 |
| 2 | 3.5 | 10 | 2 | 2.00E+10 | 0.02 | 1 | 3.6E+27 | 02 April 2010 |
| 3 | 3.5 | 2 | 2 | 2.00E+10 | 0.02 | 1 | 1.8E+28 | 10 April 2010 |
| 4 | 3.5 | 2 | 4 | 2.00E+10 | 0.05 | 2 | 3.6E+28 | 19 April 2010 |
| 5 | 3.5 | 2 | 6 | 2.00E+10 | 0.07 | 4 | 7.1E+28 | 15 May 2010 |
| 6 | 3.5 | 2 | 13 | 2.60E+10 | 0.19 | 8 | 2.4E+29 | 22 May 2010 |
| 7 | 3.5 | 3.5 | 3 | 1.10E+11 | 0.19 | 2 | 6.1E+29 | 26 June 2010 |
| 8 | 3.5 | 3.5 | 6 | 1.00E+11 | 0.34 | 4 | 1.0E+30 | 02 July 2010 |
| 9 | 3.5 | 3.5 | 8 | 9.00E+10 | 0.41 | 6 | 1.2E+30 | 12 July 2010 |
| 10 | 3.5 | 3.5 | 13 | 9.00E+10 | 0.66 | 8 | 1.6E+30 | 15 July 2010 |
| 11 | 3.5 | 3.5 | 25 | 1.00E+11 | 1.41 | 16 | 4.1E+30 | 30 July 2010 |
| 12 | 3.5 | 3.5 | 48 | 1.00E+11 | 2.71 | 36 | 9.1E+30 | 19 August 2010 |

Table 4. Performance evolution with bunch trains.

| $n_b$ | $N_{\text{bunch}}$ | MJ | $N_c$ | Peak luminosity (design parameters) | Maximum luminosity (measured) | Pile up (from measured luminosity) | Date |
|---|---|---|---|---|---|---|---|
| 56 | 1.10E+11 | 3.5 | 47 | 1.20E+31 | 2.00E+31 | 1.91 | 23/09/2010 |
| 104 | 1.10E+11 | 6.5 | 93 | 2.38E+31 | 3.50E+31 | 1.80 | 25/09/2010 |
| 152 | 1.10E+11 | 9.4 | 140 | 3.58E+31 | 5.00E+31 | 1.76 | 29/09/2010 |
| 204 | 1.10E+11 | 12.7 | 186 | 4.76E+31 | 7.00E+31 | 1.83 | 04/10/2010 |
| 248 | 1.10E+11 | 15.4 | 233 | 5.97E+31 | 1.03E+32 | 2.22 | 14/10/2010 |
| 312 | 1.10E+11 | 19.4 | 295 | 7.55E+31 | 1.50E+32 | 2.57 | 16/10/2010 |
| 368 | 1.15E+11 | 23.9 | 348 | 9.74E+31 | 2.05E+32 | 2.97 | 25/10/2010 |

In the third period of operation in 2010 "trains" of bunches from the injector were used. This necessitated even more rigorous control and validation of the machine protection system for the injection of bunch trains. As can be seen in Table 4 the number of bunches per beam was increased from 56 to 368 from the end of September to the end of October, and the resulting peak luminosity reached $2 \times 10^{32}$ cm$^{-2}$s$^{-1}$.

Figure 38 shows the peak and integrated luminosity evolution during 2010 for proton operation. The initial goal of a peak luminosity of $10^{32}$ cm$^{-2}$s$^{-1}$ was exceeded by more than a factor of 2, and the integrated luminosity delivered to the experiments was 45 pb$^{-1}$. Following the series of fills with 368 bunches per beam, operation was switched to collisions of lead ions.

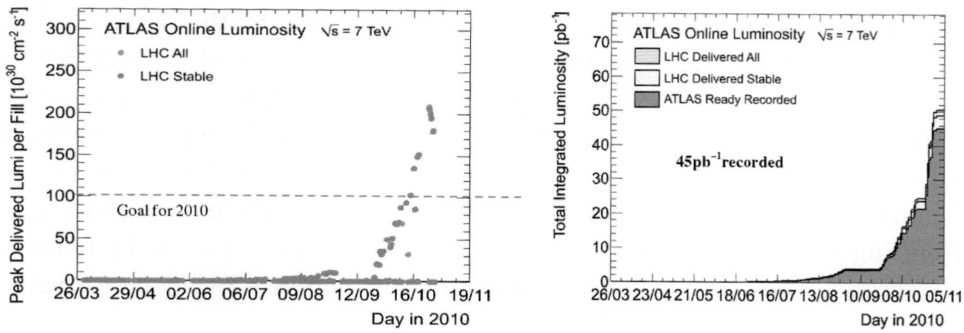


Fig. 38. Peak and integrated luminosity during 2010.

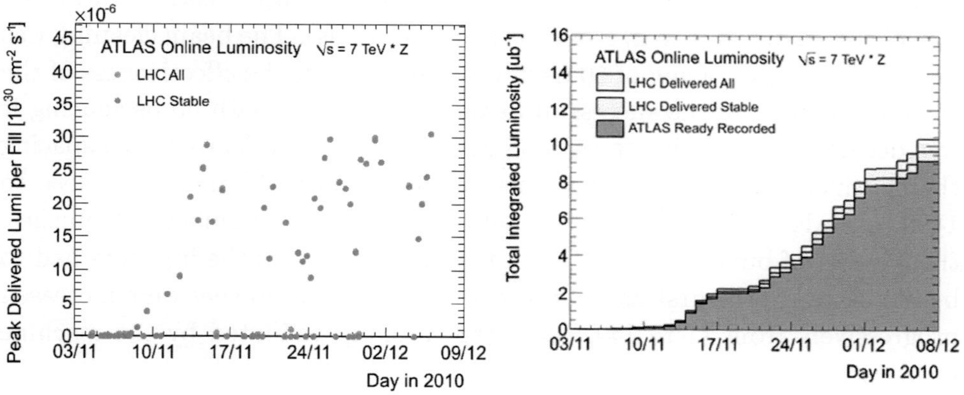


Fig. 39. Ions (2010) peak and integrated luminosities as a function of date.

### 7.2 *Lead ions*

The change-over from protons to lead ions went extremely efficiently, taking only four days to produce colliding lead ions. Operation of the LHC with lead ions was performed for about four weeks and produced luminosities well above expectations. The peak and integrated luminosities are shown as a function of date in Fig. 39.

## 8 Operational performance in 2011

### 8.1 *Protons (2011)*

Contrary to the mode of operation in 2010, the strategy for operation in 2011 was to separate operations for physics data taking from the machine studies. The goal set for 2011 was to produce an integrated luminosity of at least 1 $fb^{-1}$, more than 20 times that achieved in 2010. Towards the end of 2010 while injecting several hundred bunches per beam, clear signs of beam and vacuum instabilities appeared which are produced by the electron cloud effect. This effect had been predicted for the LHC [6] and the mitigation foreseen was to perform "beam cleaning" or "scrubbing". Beam cleaning requires large numbers of high intensity bunches circulating for extended periods so as to generate an electron cloud build-up but still allow stable beam operation below the heat load threshold for the cryogenic system. The resulting surface treatment by the electron bombardment eventually reduces the secondary electron yield of the vacuum tube thereby allowing increasingly higher intensities of circulating protons.

The global strategy for operation in 2011 was firstly to re-establish the good performance of 2010 with up to several hundred bunches, followed by beam cleaning and then to increase the number of bunches towards 900 (maximum achievable with a bunch spacing of 75 ns). The beam cleaning was done with bunch spacing of 50 ns in order to increase the effectiveness of the cleaning. Following the successful beam cleaning runs with 50 ns spacing it was decided that physics operation could continue with 50 ns bunch spacing thereby increasing the potential maximum number of bunches to close to 1400. In order to allow the accumulation of such large numbers of bunches, the number of bunches in the bunch trains coming from the injectors had to be increased. In the first half of the year the trains of bunches were increased progressively from 12 to 144, each step carefully validated by the machine protection panel.

Table 5 shows the evolution of the number of bunches per beam in the LHC during the period from mid-March until mid-August. The maximum number of bunches reached was 1380 (this is the maximum achievable with a bunch spacing of 50 ns) with a total number of protons reaching $1.68 \times 10^{14}$ and a peak luminosity of $2.4 \times 10^{33}$ cm$^{-2}$s$^{-1}$. The energy stored in this high intensity beam is more than 100 MJ.

Figure 40 shows the evolution of the integrated (left) and peak luminosity (right) during 2011. It can be seen that the 2011 goal of 1 fb$^{-1}$ has been achieved very early in the year and exceeded by around a factor of 6 before the end of the 2011 run.

### 8.2 *Lead ions (2011)*

Once again the change-over from protons to lead ions went extremely efficiently, taking less than three days to produce colliding lead ions. Operation of the LHC with lead ions [39] was performed for about four weeks and produced luminosities well above expectations. The peak and integrated

Table 5. Evolution of peak performance in 2011.

| Fill number | Date | Bunch spacing | Number of bunches | Peak luminosity ($10^{33}$ cm$^{-2}$s$^{-1}$) | Total number of protons per beam ($10^{14}$) |
|---|---|---|---|---|---|
| 1635 | 18 March 2011 | 75 | 32 | 0.03 | 0.04 |
| 1637 | 19 March 2011 | 75 | 64 | 0.06 | 0.07 |
| 1644 | 22 March 2011 | 75 | 136 | 0.17 | 0.16 |
| 1645 | 22 March 2011 | 75 | 200 | 0.25 | 0.24 |
| 1712 | 15 April 2011 | 50 | 228 | 0.24 | 0.29 |
| 1716 | 16 April 2011 | 50 | 336 | 0.35 | 0.42 |
| 1739 | 26 April 2011 | 50 | 480 | 0.51 | 0.58 |
| 1749 | 30 April 2011 | 50 | 624 | 0.72 | 0.76 |
| 1755 | 02 May 2011 | 50 | 768 | 0.83 | 0.93 |
| 1809 | 27 May 2011 | 50 | 912 | 1.10 | 1.15 |
| 1815 | 29 May 2011 | 50 | 1092 | 1.27 | 1.33 |
| 1901 | 27 June 2011 | 50 | 1236 | 1.25 | 1.64 |
| 2032 | 18 August 2011 | 50 | 1380 | 2.40 | 1.68 |

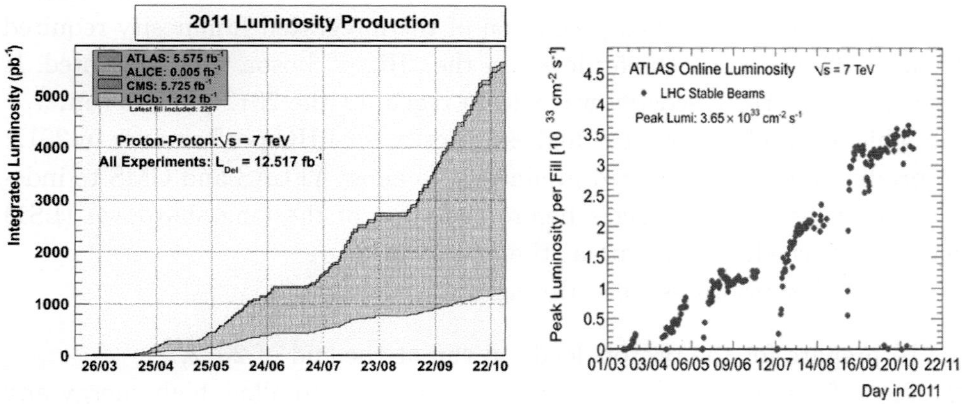


Fig. 40. Integrated luminosity (left) and peak luminosity per fill (right) during 2011.

luminosities are shown as a function of date in Fig. 41 and a comparison with 2010 is given as the dotted line.

## 9 Operational performance in 2012

At the end of 2011, in preparation for the LHC Performance Workshop (Chamonix retreat), an estimate was made of the evolution of the LHC luminosity on a day-to-day basis for the planned running days of 2012. This detailed estimate was based on the performance during the last part of 2011 and extrapolated to the likely beam parameters to be used in 2012 (beam

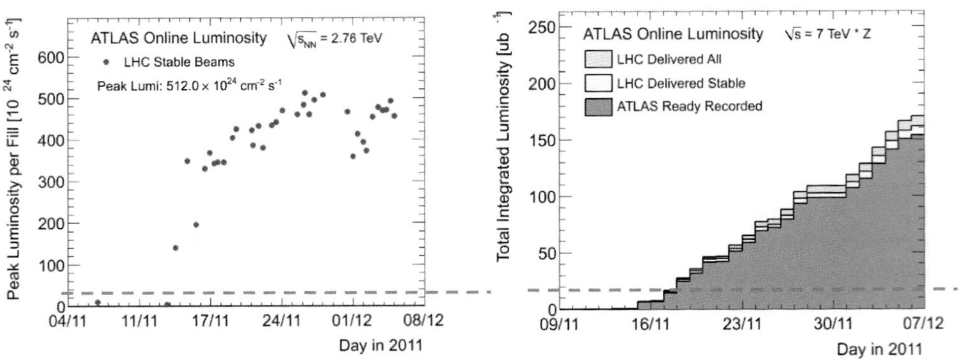


Fig. 41. Ions (2011) peak and integrated luminosities as a function of date.

energy, $\beta^*$, etc.). These estimates were presented to the experimenters as well as the CERN scientific committees in November and December 2011. The predicted integrated luminosity for the 2012 planned running period was above 15 $\text{fb}^{-1}$ provided the beam energy was increased to 4 TeV [40]. Two months later, during the Chamonix retreat, this predicted integrated luminosity [41] and a simplified prediction of the integrated luminosity required to allow discovery (at the $5\sigma$ level) of the "Higgs" boson were presented. It was clear that the discovery was within reach for the 2012 physics run.

This result clearly set the highest priority for LHC performance in 2012: to produce enough integrated luminosity to allow ATLAS and CMS to independently discover the Higgs before the start of the long shutdown (LS1) which was scheduled to start at the end of 2012.

Two other priorities were also agreed upon:

1. To prepare for the proton–lead ion run at the end of 2012;
2. To perform all necessary machine experiments to allow high energy and useful high luminosity running after LS1.

The performance of the LHC got off to a slow start at the beginning of 2012 but then improved dramatically to reach almost exactly the predicted value of around 6.5 $\text{fb}^{-1}$ at the end of June. The "technical stop" at the end of June had been planned because of the important annual summer high energy physic conference, this time to be held in Melbourne, Australia. The experimenters worked day and night to analyze the large amount of stored data coming from the first part of the 2012 run, and on July 4, 2012 announced that both ATLAS and CMS could announce a signal of a "Higgs-like" boson at the discovery level of $5\sigma$. The discovery had been made with significantly less integrated luminosity than estimated due to enormous

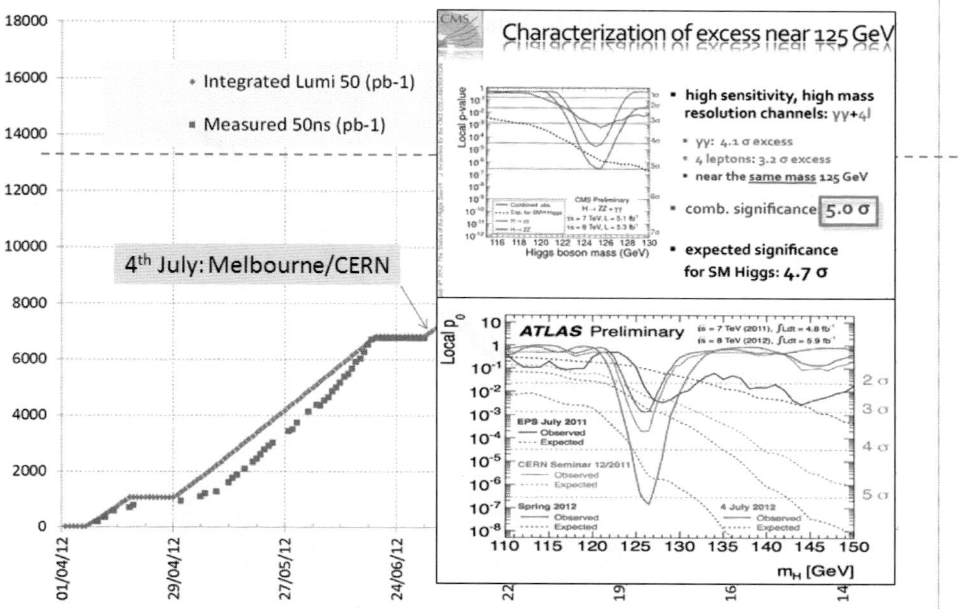


Fig. 42. Comparison of the predicted and actual luminosity until the middle of 2012 with the signals presented to the world on 4th July by both experiments.

improvements in the detector efficiency as well as substantial improvements in the data analysis. Figure 42 shows the comparison of the predicted and actual luminosity in the middle of 2012 as well as the signals presented to the world on 4th July by both experiments.

Following this momentous result it was decided to prolong the data taking with proton collisions to the end of 2012 and delay the run with protons colliding with lead ions until the beginning of 2013. Figure 43 shows the evolution of the measured integrated luminosity superimposed on the predicted.

Figure 44 shows the evolution of the luminosity in ATLAS and CMS during the totality of the running in 2012 with an integrated luminosity of 23.2 $\text{fb}^{-1}$ in each experiment.

## 10 Collisions of protons with lead ions in 2013

Following the discovery of a "Higgs-like" boson in July 2012, it was decided to prolong the 2012 data taking by a further two months in order to collect more data on the new discovery. This meant that the proton–lead collisions were delayed until the beginning of 2013. This highly unusual mode of collider operation was not foreseen in the baseline design of the LHC: the two-in-one magnet design enforces equal rigidity and therefore unequal revolution frequencies, during injection and ramp. Nevertheless, after a successful pilot

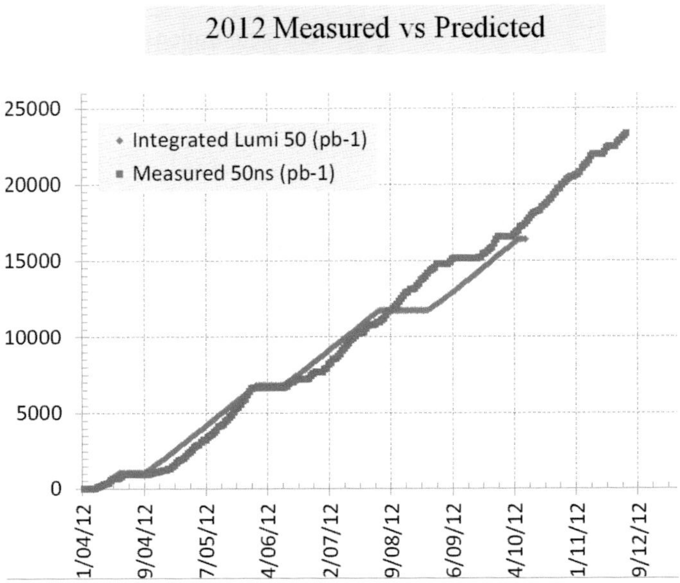


Fig. 43. Predicted and measured performance.

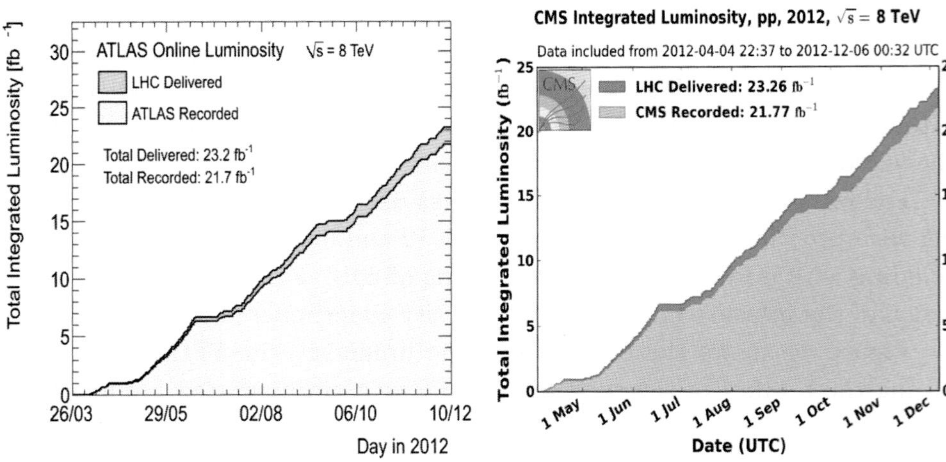


Fig. 44. CMS/ATLAS integrated luminosity.

physics fill in 2012, the LHC provided 31 $nb^{-1}$ of proton–lead (p–Pb) luminosity per experiment, at an energy of 5.02 TeV per colliding nucleon pair, with several variations of the operating conditions in early 2013 [42]. Figure 45 plots the rise of the integrated luminosity with time. Together with a companion p–p run at 2.76 TeV, this was the last physics before the present long shutdown.

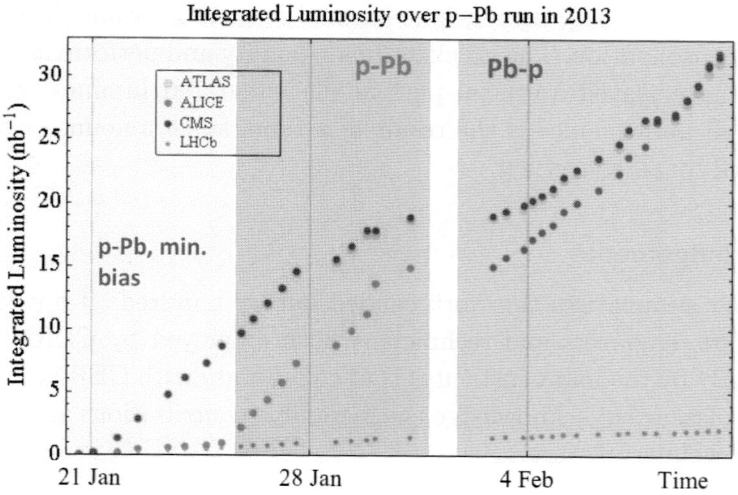


Fig. 45. Proton–lead performance in early 2013.

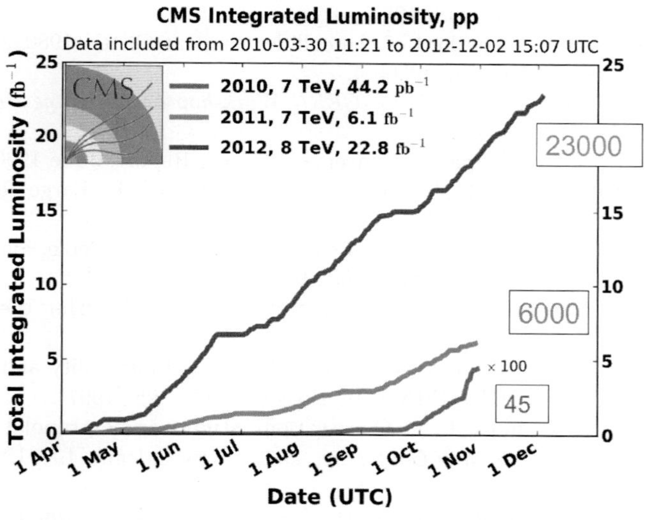


Fig. 46. Integrated luminosity for Run 1 (2010–2012).

## 11 Summary of Run 1 (2010–2012)

Finally, Fig. 46 shows the yearly luminosity for the period 2010 until the end of 2012. The plot for 2010 is amplified by a factor of 100 in order to make it more visible with respect to the zero line.

During Run 1 (2010–2012), the LHC has performed impeccably and produced a huge amount of data for physics as well as valuable experience and understanding. The overall system performance is excellent, with many user-friendly tools, and the machine has had reasonable availability following focussed consolidation. The excellent beam quality and performance from the injectors have played a crucial part in the integrated luminosity achieved. The overall performance is the result of a remarkable amount of effort and team-work by all concerned.

## Acknowledgements

This paper summarises the work carried out by hundreds if not thousands of scientists, engineers and technicians both employed by CERN and very importantly by the many institutes that collaborate with CERN. It is a great personal pleasure to acknowledge the incredible contributions and dedication of such a wonderful team.

## References

[1] S. Myers and W. Schnell; Preliminary Performance Estimates for a LEP Proton Collider, LEP Note 440, April 1983.
[2] H. Schopper; LEP — The lord of the collider rings at CERN 1980–2000, published by Springer, ISBN 978-3-540-89300-4.
[3] CERN 84-10, *Proceedings of ECFA-CERN Workshop on the Large Hadron Collider*, Lausanne, 21–27 March 1984.
[4] SSC-SR-1051; SSC Site Specific Conceptual Design Report, June 1990.
[5] The LHC Study Group; CERN 91-03; Design Study of the Large Hadron Collider (LHC) (pink book), May 1991.
[6] S. Myers, O. Bruning, P. Collier, P. Lebrun, R. Ostojic, J. Poole, and P. Proudlock (eds.); LHC Design Report, CERN-2004-003 (2004).
[7] E. Métral, A. Burov; Review of the instabilities observed during the 2012 run and actions taken, Evian (2012).
[8] F. Zimmermann; A simulation study of electron-cloud instability and beam-induced multipacting in the LHC, CERN LHC Project Report 95 (1997).
[9] T. Baer *et al.*; UFOs in the LHC: Observations, studies and extrapolations, *3rd International Particle Accelerator Conference 2012* , New Orleans, LA, USA, 20–25 May 2012, p. 3936.
[10] D. Missiaen *et al.*; The alignment of the LHC, *Particle Accelerator Conference 2009*, Vancouver, Canada, 04–08 May 2009.
[11] V. Baglin *et al.*; Vacuum performance and lessons for 2012, *Chamonix 2012 Workshop on LHC Performance*, Chamonix, France, 6–10 Feb 2012, pp. 74–80.
[12] R. Assmann *et al.*; Collimation settings and performance in 2011 and 2012, *Chamonix 2012 Workshop on LHC Performance*, Chamonix, France, 6–10 Feb 2012, pp. 183–188.
[13] R. Assmann; Operational experience with LHC collimation, *Particle Accelerator Conference 2009*, Vancouver, Canada, 04–08 May 2009.
[14] R. Schmidt *et al.*; First operational experience with the LHC machine protection system when operating with beam energies beyond the 100 MJ range, *3rd International Particle Accelerator Conference 2012*, New Orleans, LA, USA, 20–25 May 2012, p. 4062.

[15] M. Zerlauth; Powering interlocks, *Chamonix 2009 Workshop on LHC Performance*, Chamonix, France, 2–6 Feb 2009, pp. 248–250.

[16] P. Lebrun, L. Tavian, and S. Claudet; Development of large-capacity refrigeration at 1.8 K for the Large Hadron Collider at CERN, LHC-Project-Report-6; CERN-LHC-Project-Report-6 (1996).

[17] P. Lebrun *et al.*; Cryogenic system for the Large Hadron Collider (LHC), *5th European Particle Accelerator Conference*, Sitges, Barcelona, Spain, 10–14 Jun 1996, pp. 361–363.

[18] S. Claudet *et al.*; Cryogenics system: Strategy to achieve nominal performance and reliable operation, *Chamonix 2012 Workshop on LHC Performance*, Chamonix, France, 6–10 Feb 2012, pp. 236–240.

[19] E. Weisse, V. Mertens, and A. Hilaire; Beam transfer to and injection into LHC, *6th European Particle Accelerator Conference*, Stockholm, Sweden, 22–26 Jun 1998, pp. 2117–2119.

[20] B. Goddard *et al.*; Beamloss at injection into the LHC, *2nd International Particle Accelerator Conference*, San Sebastian, Spain, 4–9 Sep 2011.

[21] B. Goddard *et al.*; Reliability issues on the LHC beam dumping system, *9th European Particle Accelerator Conference*, Lucerne, Switzerland, 5–9 Jul 2004, p. 563; See also R. Filippini, Dependability analysis of a safety critical system: the LHC beam dumping system at CERN, CERN-THESIS-2006-054 - Pisa U., 2006.

[22] D. Boussard *et al.*; The LHC superconducting cavities, *18th Biennial Particle Accelerator Conference*, New York, NY, USA, 29 Mar–2 Apr 1999, pp. 946–8.

[23] D. Boussard *et al.*; Design Considerations for the LHC 200 MHz RF System, CERN-LHC-Project-Report-368 (2000).

[24] W. Hofle *et al.*; LHC transverse feedback system and its hardware commissioning, *11th European Particle Accelerator Conference*, Genoa, Italy, 23–27 Jun 2008.

[25] R. Jones; First results from the LHC beam instrumentation systems, *Chamonix 2009 Workshop on LHC Performance*, Chamonix, France, 2 – 6 Feb 2009, pp. 226–231.

[26] J.-J. Gras *et al.*; Optimization of the LHC beam current transformers for accurate luminosity determination *2nd International Particle Accelerator Conference*, San Sebastian, Spain, 4–9 Sep 2011, p. TUPC155.

[27] B. Dehning *et al.*; LHC beam loss monitor system design, *10th Beam Instrumentation Workshop*, Upton, NY, USA, 6–9 May 2002, *AIP Conf. Proc.* **648**, 229–236 (2002).

[28] B. Dehning *et al.*; Overview of LHC beam loss measurements, *2nd International Particle Accelerator Conference*, San Sebastian, Spain, 4–9 Sep 2011.

[29] R. Steinhagen; Real-time beam control at the LHC, *2011 Particle Accelerator Conference*, New York City, NY, USA, 28 Mar–1 Apr 2011, also CERN-BE-2011-020.

[30] P. Lebrun *et al.*; Report of the task force on the Incident of 19th September 2008 at the LHC, LHC-PROJECT-Report-1168 March 2009.

[31] P. Lebrun; Private communication 5th March, 2009.

[32] J.-M. Jimenez; Recovering about 5 km of LHC beam vacuum system after sector 3-4 incident, *1st International Particle Accelerator Conference*, Kyoto, Japan, 23–28 May 2010.

[33] F. Bordry; The first long shutdown of the LHC, *4th International Particle Accelerator Conference*, Shanghai, China, 12–17 May 2013.

[34] M. Lamont; The first years of LHC for luminosity production, *4th International Particle Accelerator Conference*, Shanghai, China, 12–17 May 2013.

[35] D. Wollmann *et al.*; Machine protection at the LHC-experience of three years running and outlook for operation at nominal energy, *4th International Particle Accelerator Conference*, Shanghai, China, 12–17 May 2013, pp. 3246–8.

[36] E. Quaranta *et al.*; Cleaning inefficiency of the LHC collimation system during the energy ramp: simulations and measurements, *4th International Particle Accelerator Conference*, Shanghai, China, 12–17 May 2013, pp. 975–7.
[37] S. Redaelli and A. Masi; Analysis of failures of the LHC collimators during the 2010-2013 operation, *4th International Particle Accelerator Conference*, Shanghai, China, 12–17 May 2013, pp. 981–3.
[38] R. Tomas *et al.*; Record low $\beta$-beating in the LHC, *Phys. Rev. ST Accel. Beams* **15**, 091001 (2012).
[39] J. Jowett *et al.*; Luminosity and beam parameter evolution for lead ion beams in the LHC, *2nd International Particle Accelerator Conference*, San Sebastian, Spain, 4–9 Sep 2011, also CERN-ATS-2011-144.
[40] S. Myers; LHC 2011, 2012 and 2015" Talks Presented to CMS (21 November 2011) and ATLAS (22 November 2011).
[41] S. Myers and F. Zimmermann; *Summary of the LHC Performance Workshop*, 2012. Published version from CERN.
[42] J. Jowett *et al.*; Proton–nucleus collisions in the LHC, *4th International Particle Accelerator Conference*, Shanghai, China, 12–17 May 2013, pp. 49–51.

# Chapter 23

# The LHC heavy-ion programme: The energy frontier of nuclear collisions

John M. Jowett (CERN)

The greater part of the beam time at CERN's Large Hadron Collider (LHC) is devoted to colliding proton beams for the purposes of elementary particle physics at the highest available energies per colliding nucleon [Myers (2015)]. However a substantial fraction — about one month per operating year — is devoted to colliding the nuclei of heavy atoms with each other or, sometimes, with protons. The much larger total energies in these "ultra-relativistic heavy-ion" collisions produce tiny droplets of strongly-interacting matter, the quark–gluon plasma, at the highest densities and temperatures available in the laboratory. The LHC thus recreates the substance that filled the universe in the first microseconds of its history and exposes it to detailed scrutiny with its array of extraordinarily capable detectors.

In a total of about two months running time in 2010–11, the LHC exceeded its anticipated design luminosity for collisions between two beams of lead nuclei (at half design energy) by a factor 2.

In addition, in 2012–13, the LHC operated in an unanticipated — and almost unprecedented — mode to provide collisions between beams of protons and beams of lead nuclei with equivalent luminosity.

In each of these two modes, the gains in centre-of-mass energy, with respect to previous collisions of similar types, were the largest in the history of particle accelerators. A rich harvest of physics results, including several surprises, has emerged from this first stage of the LHC heavy-ion programme.

## 1 History and physics

The history of ion beams (other than protons) at CERN [Haseroth (2004)] began with the acceleration of deuterons in the 50 MeV proton Linac 1 in 1964. The development of sources, in particular the electron-cyclotron resonance type (ECR), RF quadrupoles, linacs and synchrotrons allowed beams of increasingly heavy species ($O^{8+}$,$S^{16+}$) to be generated and accelerated in the following decades. High points were the world's first colliding ion beams (deuterons) in the ISR [Asbo-Hansen *et al.* (1977)] and the variety of species culminating in lead, delivered by the PS and SPS to their fixed-target experiments.

Driven by theoretical expectations of the QCD phase transition to the quark–gluon plasma at sufficiently high energy density over an extended volume, and the tantalising results emerging from the first heavy-ion collisions at fixed-target energies, the ALICE collaboration formed in the early 1990s and submitted their Letter of Intent [ALICE Collaboration (1993); CERN Courier (1993)] for a specialised experiment at the LHC. ALICE was designed to have specialised capabilities, including the capacity to handle the enormous multiplicities of secondary particles expected in these collisions. The outpouring of results from RHIC [Fischer (2013)] in the early years of the present century further reinforced the interest in the LHC heavy-ion programme.

The "general-purpose" detectors, ATLAS and CMS, have powerful complementary capabilities for the study of heavy-ion collisions and also participate fully in the heavy-ion programme of the LHC. The asymmetric LHCb detector, and the specialised forward detectors TOTEM, ALFA and LHCf joined the programme for proton-nucleus collisions in 2012–13.

### 1.1 *Hot and dense matter*

Physicists say, with perhaps a little understatement, that the Pb–Pb collisions at the LHC produce hot and dense matter, quoting temperature and density values in convenient nuclear units. So, for example, the temperature of the thermal photon spectrum emitted by the quark–gluon plasma, as measured by the ALICE experiment [Wilde (2013); Klasen *et al.* (2013)], is $T_{\rm ALICE} = 304\,{\rm MeV}/k_B = 3.5 \times 10^{12}\,{\rm K}$, where $k_B$ is Boltzmann's constant.[1]

It is worth trying to get a feel for just how hot and dense this matter is. So, for example, we know that the temperature generated by nuclear fusion reactions at the core of the Sun, a hot place by human standards, is about $T_\odot = 1.6 \times 10^7\,{\rm K}$ but this is a factor $T_{\rm ALICE}/T_\odot = 200,000$ times cooler than the temperature measured in the LHC's Pb–Pb collisions, $T_{\rm ALICE}$, the highest temperature ever achieved in the laboratory.

Similarly, the energy density in the Pb–Pb collisions is quoted in nuclear units as $u_{\rm QGP} \simeq 15\,{\rm GeV/fm^3}$, where $1\,{\rm fm} = 10^{-15}\,{\rm m}$ is a convenient unit for

[1]Hydrodynamic models suggest that even higher temperatures occur in the initial phase of the collisions; we quote only the measured values here.

measuring the size of nuclei. To get a feeling for this, we can consider the total amount of electrical energy generated by all the power stations and other generating capacity of all the countries of Europe, running for a whole year, namely $U_{\mathrm{Ey}} = 3.6 \times 10^{12}\,\mathrm{kWh}$. Then we can at least try to imagine pumping all that energy into a sphere of radius $r$ and calculate the value of $r$ that would give a similar energy density to the LHC heavy-ion collisions

$$\frac{U_{\mathrm{Ey}}}{(4/3)\pi r^3} = u_{\mathrm{QGP}} \Rightarrow r = 1.1 \times 10^{-6}\,\mathrm{m}, \tag{1}$$

So the hot sphere would be the size of a speck of very fine dust! It would nevertheless be quite a heavy speck with mass given by $U_{\mathrm{Ey}}/c^2 = 143\,\mathrm{kg}$. In terms of mass density, $u_{\mathrm{QGP}}/c^2$ turns out to be over $10^{15}$ times the density of metallic lead.[2]

Of course, a fill of the LHC only takes a small fraction of the output of a single power station for a few hours and only about $10^{-10}$ of this energy (or about $10^{-23}U_{\mathrm{Ey}}$), goes into a single Pb–Pb collision.[3] The LHC achieves an astounding *concentration* of energy.

In these unimaginably extreme conditions, similar to those that prevailed in the first microseconds of cosmic history, nuclear matter as we know it does not exist. Above a temperature of $T_c \approx 160/k_B\,\mathrm{MeV}$, quarks and gluons inside the nucleons (i.e., protons or neutrons) are deconfined, forming the quark-gluon plasma (QGP). This occupies a volume that is nevertheless large enough to be considered thermalised bulk matter with meaningful thermodynamic and hydrodynamic properties such as temperature, flow and viscosity. The LHC experiments have confirmed the discovery at RHIC [Fischer (2013)] that this new state of matter is the most nearly perfect liquid, with the lowest viscosity, found in Nature. As it expands and cools, it condenses back into a hadron gas. From the distributions of hadrons emerging from this "freeze-out" stage, the experiments can infer many properties of the QGP. Further information is carried by muons and photons. As at RHIC, the "soft" physics of particles produced with low transverse momenta $p_\perp \lesssim 2\,\mathrm{GeV}/c$ continues to be well-modelled by relativistic viscous hydrodynamics.

The step in energy from RHIC to LHC was about a factor 13 for the first run in 2010. It is hardly surprising that there were a number of new phenomena and surprises as a new energy scale and level of experimental

[2]This large ratio partly reflects the smallness of the nuclear radius compared to the interatomic spacing; the energy density in collisions exceeds the mass density of a stationary nucleus by about two orders of magnitude.

[3]Curiously, global annual electricity production seems to amount to about 1 mole of LHC Pb–Pb collisions.

precision became accessible [Muller *et al.* (2012)]. The higher energies created many more rare "hard probes" of the hot medium. These include jets, groups of high-$p_\perp$ particles emitted within narrow cones and bound states of heavy quark–anti-quark pairs such as charmonium ($J/\psi$) or bottomonium ($\Upsilon$). Jets arise from violent scattering events among the quarks and gluons of the incoming nuclei and became spectacularly visible at the LHC. Quenching (dissipation) of the jets as they travel through the QGP medium yields information about its properties. Other new features of heavy-ion collisions at the LHC include massive particles such as $Z$ and $W$ bosons. The QGP is transparent to these new hard probes which do not couple via the strong interactions. Events like these can be very prominent (see Figure 2).

Just as the electromagnetic force is attenuated by the mobile charges in a normal plasma (ionised gas), the familar Debye screening, the colour force between quark and anti-quark pairs is attenuated by the QGP. This is detected in the more rapid decay or "melting" of quarkonium bound states, another signal of deconfinement seen at the LHC.

The LHC also extends CERN's long tradition of creating anti-matter. Like RHIC, it has produced a number of specimens of the heaviest anti-matter nucleus detected so far, $^4\overline{\mathrm{He}}$, as well as hyper-matter such as the hypertriton $^3_\Lambda\mathrm{H}$ and anti-hypertriton $^3_{\overline{\Lambda}}\overline{\mathrm{H}}$ [Martin (2013)]. Searches continue for further exotic nuclei.

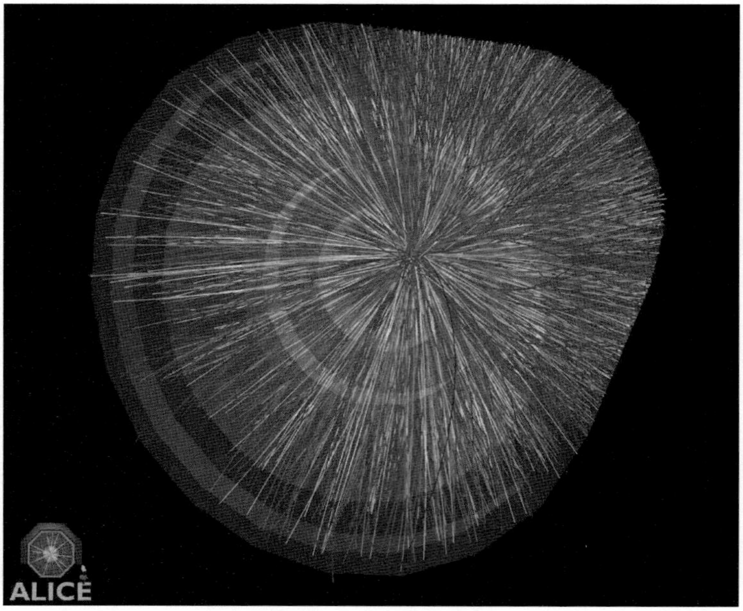


Fig. 1. One of the first high multiplicity events from ALICE in the 2010 Pb–Pb run.

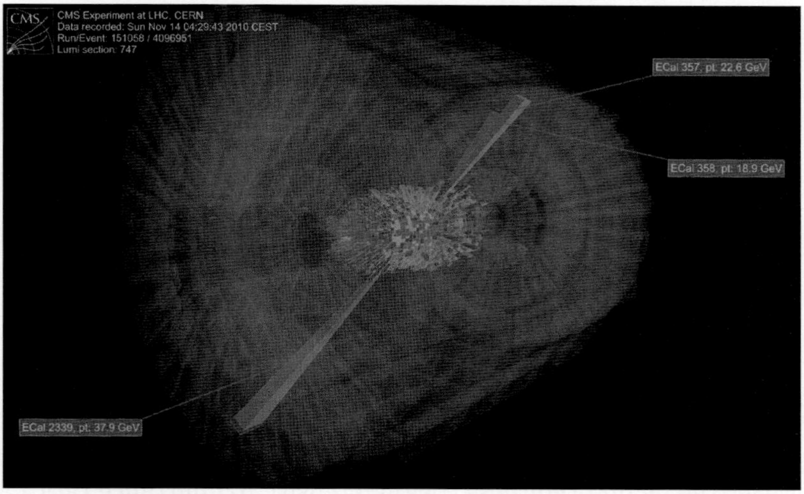


Fig. 2. One of the first $Z$ bosons detected in heavy-ion collisions decays to an $e^+e^-$ pair in the CMS detector.

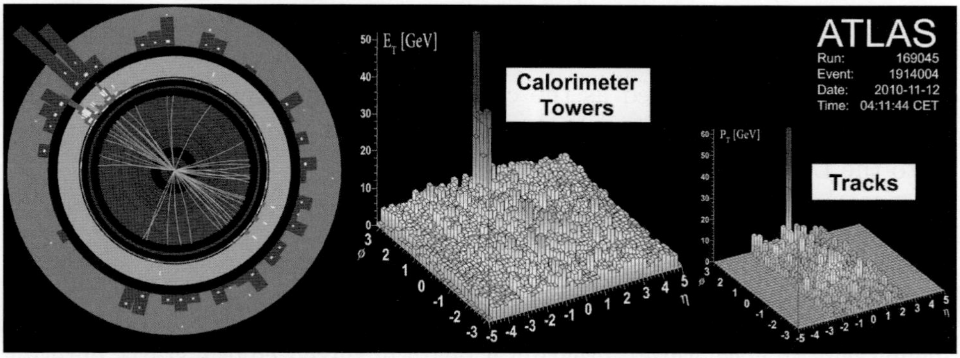


Fig. 3. One of the first highly asymmetric dijet events recorded by ATLAS in LHC Pb–Pb collisions. One very visible jet has a transverse energy above 100 GeV but there is no evident jet recoiling in the opposite direction, showing that it has been absorbed as it propagates a larger distance in the hot, dense medium.

But the most unexpected results emerged from the proton–lead collisions (see Section 4).

## 2 CERN's heavy ion injectors

At the turn of the century, the CERN complex was providing lead ion beams to fixed-target experiments at the SPS. These beams were generated in an electron-cyclotron resonance (ECR) ion source, accelerated via the heavy-ion Linac3 through the PS Booster (PSB), PS and SPS.

Table 1. Charge state, kinetic energies per nucleon and bunch populations of Pb beams at extraction (or top energy) in the chain of heavy ion accelerators at CERN in 2013 [Manglunki *et al.* (2013)].

| Accelerator | Ion charge state $Q$ | Kinetic energy per nucleon | Population of LHC bunch $N_b/10^7$ |
|---|---|---|---|
| ECR ion source | 27+ | 2.5 keV | 860 |
| RFQ/LINAC3 | 54+ | 4.2 MeV | 57 |
| LEIR | 54+ | 72.2 MeV | 55 |
| PS | 82+ | 5.9 GeV | 38 |
| SPS | 82+ | 177 GeV | 22 |
| LHC | 82+ | 1.58 TeV | 20 |

However the beam brightness (beam intensity to emittance ratio) achievable via the PSB, while adequate for fixed-target experiments, fell short of LHC requirements by a factor of about 30. This led to the proposal [Benedikt *et al.* (2004); Chanel (2002); Lefèvre and Möhl (1993)] to modify the mothballed Low Energy Anti-proton Ring (LEAR) into the Low Energy Ion Ring (LEIR) to take the place of the PSB in the heavy-ion injector chain. LEIR construction was launched as the centrepiece of the Ions for LHC (ILHC) Project [Benedikt *et al.* (2004)]. Key parameters of the chain of accelerators up to the LHC are summarised in Table 1.

ILHC also included a new ECR source, intended to double the output from 100 to 200 $\mu$A, The ECR ion source creates the beam by vaporising samples of isotopically pure $^{208}$Pb in an oven. An applied microwave field resonates with the atomic electrons, ionising the lead vapour and the resulting plasma can be accelerated, creating a beam of mixed charge states entering the heavy-ion Linac3. The $^{208}\mathrm{Pb}^{27+}$ state is selected in the acceleration process and after a first stripping foil and RF quadrupole, the $^{208}\mathrm{Pb}^{54+}$ charge state is injected into LEIR. These processes amount to selecting only a small fraction of the original beam. Charge states have to be kept low in the low-energy stages of the acceleration process in order to minimise space-charge and other harmful effects.

Commissioning of LEIR took place in 2005 [Beloshitsky *et al.* (2006)] (Figure 4). The 200 $\mu$s long pulses from Linac3 are equivalent to 70 times the LEIR circumference so that a sophisticated scheme to inject by stacking in the LEIR phase space is required. This process is repeated up to six times with application of a powerful electron cooling system in all three phase planes so as to squeeze these ions into very small emittances to achieve the required phase space density corresponding to a normalised transverse

Fig. 4. The LEIR ring during commissioning in 2005.

emittance of $\varepsilon_n = 0.7\,\mu$m. When this process is complete, the beam is bunched into half the ring and transferred to the PS synchrotron.

From the PS onwards, through the larger SPS synchrotron to the LHC, the ions follow the same path as proton beams. However there are considerable differences in the acceleration scheme and rather complex RF manipulations of the bunches [Manglunki *et al.* (2013)] which have varied from run to run. In the PS, the original design for bunch-splitting was not used in first years of LHC operation.

Full stripping is necessary at the high energy end of the chain in order to reach the highest energy in the available magnetic fields. The beam passes through a second stripping foil in the transfer line from the PS to the SPS, removing the remaining electrons from some of the ions and creating the final $^{208}Pb^{82+}$ nuclear beams that will be collided in the LHC.

Unlike protons, the Pb beams have to cross transition energy in the SPS ramp. Several batches from the PS are accumulated over several seconds to make a train in the SPS. Since the RF accelerating cavities of the SPS are incapable of the frequency variation necessary to accelerate the beams from injection to top energy in the conventional way — in which the RF frequency is a constant integer harmonic of the varying revolution frequency — a special fixed-frequency acceleration technique with effectively non-integer harmonic number was developed [Boussard *et al.* (1989); Boussard *et al.*

(1995)]. This scheme does tend to generate more RF noise on the injection plateau of the SPS which can be as long as 40 s. Together with strong intra-beam scattering effects, this results in emittance blow-up and beam losses, imprinting a pattern on the SPS bunch trains (the bunches injected first suffer these effects for longer) that remains as a major performance limitation of the LHC.

Since accumulation of several such trains in the LHC can take up to about half an hour, a similar pattern from train to train emerges in the LHC. The result is that there is a broad variation of intensities, beam sizes and, therefore, luminosities among the pairs of bunches that eventually collide.

This is an extremely sketchy account of the many complexities of the LHC's heavy-ion injector chain and we refer the reader to [Angoletta *et al.* (2011); Benedikt *et al.* (2004); Manglunki *et al.* (2013)] for further details.

The filling patterns for protons are normally quite different as the bunch spacings are determined in the proton pre-injector chain (Linac2, PS Booster). The hybrid p–Pb collisions require a further level of complexity as both proton and heavy-ion chains have to be operated together in special schemes designed to produce matching filling patterns in the LHC [Jowett and Carli (2006); Manglunki *et al.* (2013)].

## 3 Heavy ions in the LHC main rings

### 3.1 *Similarities with proton beams*

In a given time-independent but spatially varying applied magnetic field $\mathbf{B}(\mathbf{x})$ (e.g., that generated by the magnets of the LHC), particles with different charge and mass can follow the same trajectories provided their initial momenta are related by $\mathbf{p_1}/Q_1 = \mathbf{p_2}/Q_2$.[4] From this it would follow that the injection trajectories, closed orbits and magnetic optics of protons and heavy ions should be identical provided their magnetic rigidities $|\mathbf{p}/Q|$ are the same. If, however, electric fields $\mathbf{E}(\mathbf{x}, t)$ (e.g., the time-dependent ones of the RF cavities) are present, this equivalence no longer holds unless further conditions are met. In practice, these amount to the reduction of the main RF frequency for synchronism with the slightly slower Pb nuclei, so ensuring that the appropriate 4-dimensional projection of the closed orbit in the full 6-dimensional phase space is the same as for protons.

This means that the momentum of a $^{208}Pb^{82+}$ nucleus at the full design magnetic field is related to the momentum of a proton in the same field by

[4] This fact is often glibly cited as an obvious consequence of the Lorentz force equation. Nevertheless the exercise of deriving it is instructive and makes a good physics examination question.

$p_{\mathrm{Pb}} = Zp_p = 82 \times 7 = 574\,\mathrm{TeV}/c$ where $Z = 82$ is the atomic number and $A = 208$ the nucleon number. This is the primary quantity appearing in all physical calculations, e.g., solution of the Lorentz force equation. However it is conventional to quote the corresponding momentum per nucleon $p_{\mathrm{Pb}}/A = (82/208) \times 7 = 2.76\,\mathrm{TeV}/c$ although care has to be exercised in using this in calculations. For the extremely relativistic beams of the LHC, one can blur the distinction between energy and momentum ($E \simeq pc$). However in lower energy machines, where the ions may not be fully stripped so $p_{\mathrm{Pb}} = Qp_p < Zp_p$, the *kinetic energy per nucleon* is usually quoted (as in Table 1). This is given by

$$E_{Kn} = \frac{\sqrt{p_{\mathrm{Pb}}^2 c^2 + m_{\mathrm{Pb}}^2 c^4} - m_{\mathrm{Pb}} c^2}{A} \tag{2}$$

where $m_{\mathrm{Pb}}$ is the mass of the lead ion, and is much less directly related to the primary kinematic quantities. Care has to be taken in using this value as input to any physical calculation. At the LHC, we have adopted the convention of writing

$$E_{\mathrm{Pb}} \simeq c p_{\mathrm{Pb}} = 7Z\,\mathrm{TeV} = 2.76A\,\mathrm{TeV} \tag{3}$$

which is accurate, formally correct and very convenient when discussing a machine that accelerates both species at different times. The "Page 1" display of the LHC, for example, can convert the electrical current in the bending magnets to beam energy, without knowledge of the particle species, provided the unevaluated symbol $Z$ is inserted. Operators and other viewers immediately know where the machine is in its magnetic cycle or ramp. In the proton–lead runs, this form of energy display was valid for both beams.

The two-in-one magnet design of the LHC means that the main bending magnetic fields in the two rings are equal. For physics purposes, collisions between heavy ions $(Z_1, A_1)$ and $(Z_2, A_2)$ are usually characterised in terms of the Lorentz-invariant centre-of-mass energy per colliding nucleon pair

$$\sqrt{s_{\mathrm{NN}}} \simeq 2\sqrt{\frac{E_1 E_2}{A_1 A_2}} = 2E_p\sqrt{\frac{Z_1 Z_2}{A_1 A_2}} \tag{4}$$

where $E_p$ is the energy that a proton would have circulating in either ring of the LHC. Note that in the case of unequal species, the centre-of-mass of colliding nucleon pairs will have a rapidity shift in the *opposite direction* to that of the centre-of-mass of the entire colliding system.

In practice, the equal rigidity principle was a key element of the strategy [Jowett (2012)] for setting up the orbits and optical configurations for the heavy-ion runs. Starting with a Pb beam of the same magnetic rigidity and initial injection conditions as the protons would minimise the changes to the established proton magnetic configuration, reduce the time taken for the routine initial commissioning steps (achieving circulating beam, ramp, squeeze) and allow us to move quickly on to dealing with the more substantial differences between heavy ions and protons.

### 3.2 *Differences from proton beams*

The total charge per heavy-ion bunch in the LHC is generally less than 10% of that of the proton bunches. Accordingly, a number of the intensity-dependent effects that limit proton–proton performance (electron cloud, collective instabilities, conventional beam–beam effects, ...) are of little consequence for the heavy-ion performance. On the other hand, the low bunch charges mean that the beam intrumentation of the LHC often has to work near its lower limits of sensitivity.

Most of the differences between protons and heavy ions in the LHC result from effects that depend on higher powers of the much larger charge $Z = 82$ of the lead nuclei. The larger mass (206.4 times the proton mass) also plays a role.

Multiple small-angle Coulomb scattering within the bunches, the so-called *intra-beam scattering* is strongly enhanced by large $Z$ and is much more significant for Pb beams than proton beams. It causes a slow blow-up of the beam emittances and some particle losses, particularly at injection energy. On the other hand, synchrotron radiation damping also depends on $Z$ and is about twice as fast for Pb beams as for protons [Jowett (2004)].

Beam lifetime in other heavy-ion machines is often strongly affected by scattering on residual gas. Thanks to the cryo-pumping in the LHC this is a very small effect [Jowett (2004)].

In p–p operation, the LHC relies on an elaborate two-stage collimation system, based on the diffractive scattering of protons in the primary collimator material, to clean the beam of halo particles. Hadronic fragmentation and electromagnetic dissociation of nuclei impinging on the primary collimators mean that the system works like a single-stage system and collimation efficiency is poorer, with a whole spectrum of nuclides being lost into superconducting magnets [Braun *et al.* (2014, 2004); Pshenichnov *et al.* (2001)]. This constitutes a limit on total intensity rather than luminosity.

Perhaps the most striking difference between heavy-ion collisions and proton collisions is related to the beam lifetime from collisions. The nuclear cross section (some 8 b for Pb–Pb) is only a small fraction of the total of around 500 b for removing particles from beam. This large additional cross section [Baltz *et al.* (1996); Braun *et al.* (2014); Jowett *et al.* (2004); Meier *et al.* (2001)] is due to electromagnetic interactions that are strongly enhanced because of the large nuclear charge. As a result the beam populations are rapidly eaten away by the collisions. This so-called burn-off results in a rather short luminosity lifetime in heavy-ion fills of the LHC [Bruce *et al.* (2010); Jowett *et al.* (2004)]. Further, the more experiments take collisions, the shorter it becomes.

In the Fermi–Weizsacker–Williams picture [Jackson (1999)], the electromagnetic fields surrounding the nuclei can be described as a beam of quasi-real photons which interact with the nuclei of the opposing beam or with each other (these processes also have important physics interest [Baltz *et al.* (2008)]). In particular, the photon–photon interactions create large numbers of electron–positron pairs. The largest contribution to the cross section for removal of particles from the beam comes from the so-called bound-free pair production process in which the electron is created in a bound-state of one of the colliding nuclei. It was pointed out [Klein (2001)] that the beam of one-electron ions thus generated could possibly quench superconducting magnets in the LHC, leading to a limit on luminosity never encountered in any previous collider. This effect was analysed in some detail with increasingly realistic models of the LHC [Bruce *et al.* (2009); Jowett *et al.* (2004); Jowett (2004); Jowett *et al.* (2003)] and was seen to be a major concern. However the luminosity limit depends crucially on the level of energy deposition which will cause the LHC magnets to quench and recent measurements [Redaelli *et al.* (2012)] and other estimates suggest that it will only occur for luminosity values above design. Countermeasures are being implemented [Jowett *et al.* (2014)] for the future upgrade of the LHC.

### 3.3 *First run in 2010*

The first Pb–Pb run of the LHC was allotted a single month at the end of 2010. Although a rapid-commissioning strategy had long been foreseen [Jowett (2005)] to maximise time available for physics and minimise risk, it is fair to say that there was considerable apprehension about the prospects of commissioning a new collider mode and delivering a decent amount of luminosity (a few $\mu b^{-1}$ had been promised) within that time frame.

As in the preceding proton run in 2010, the beam energy was $3.5Z$ TeV, half the design. The first Pb beams were injected into the LHC around 20:00 on Thursday, 4 November 2010. With some care taken to establish a proton orbit using similar charge per bunch (hence dynamic range of beam position

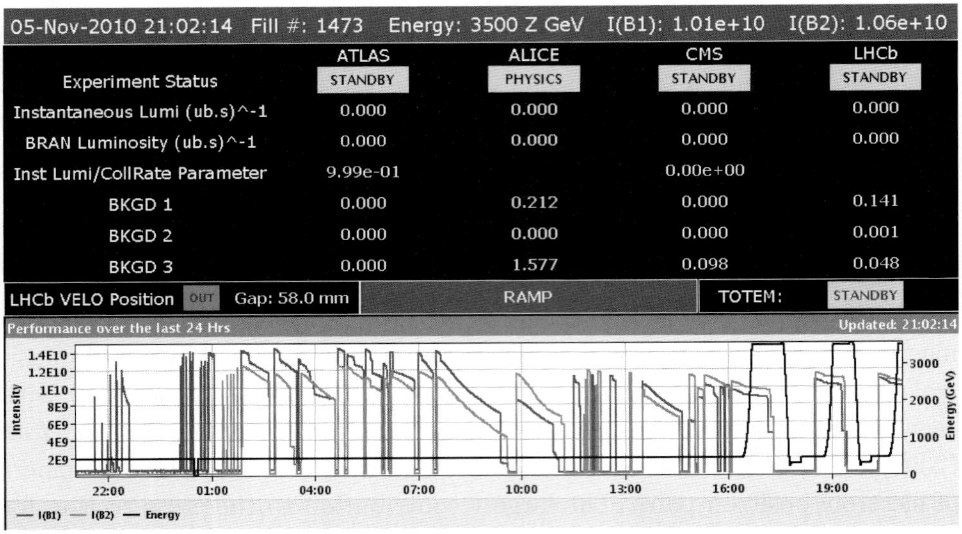


Fig. 5. The LHC status display showing the charge of a single bunch in each beam (Beam 1 in blue, Beam 2 in red) and the beam energy (black). During these first 24 hours of Pb beam commissioning in 2010, one can see the first injections, RF capture of circulating beams in both rings, the success of measures taken to improve the beam lifetime and the first three ramps of both beams together, to $3.5Z$ TeV $= 1.38A$ TeV, smashing the previous energy record ($0.1A$ TeV).

monitors), the very first circulating beam was achieved within an hour with no orbit steering. The predicted change to the RF frequency then captured the beams (the first 24 hours are shown in Figure 5). The energy ramp to $3.5Z$ TeV and squeeze of the collision optics to $\beta^* = 3.5$ m in the three experiments then followed rapidly. Despite eight hours of down time on the Saturday, first collisions were obtained in the small hours of the following Sunday morning, just 53.5 hours after first injection [Jowett *et al.* (2011)]. After a few shifts to set up the collimation system, "Stable Beams" were declared for physics data-taking at 11:20 on the Monday morning.

In the following days, the number of bunches per beam, $k_b$ changed on every single fill, through $k_b = 2, 5, 17, 69, 121$, injecting single bunches or batches of four from the SPS in variants of the "Early" filling scheme [Benedikt *et al.* (2004)]. In the last few days of the run, injection of batches of eight bunches allowed $k_b = 137$. The rapid rise in peak luminosity is shown in the left panel of Figure 7.

Commissioning was fast because the principle of equal magnetic rigidity was exploited to take over the same injection optics, ramp and "squeeze" to the low-$\beta$ optics in ATLAS and CMS. However a special reduced crossing angle, resulting from the combination of their muon spectrometer orbit bump and externally applied bumps, had to be set up for the ALICE experiment. The purpose of this was to ensure that their Zero-Degree Calorimeter, a detector set between the two LHC beam pipes just after they separate, could intercept the "spectator neutrons" resulting from the fragmentation of the non-overlapping parts of the colliding nuclei. This is a crucial measurement allowing the *centrality*, i.e., the extent to which the colliding nuclei overlap, to be determined. The ZDC also allows cross sections of some of the electromagnetic processes mentioned in Section 3.2 to be measured [Oppedisano (2011)].

This was the one and only time that it was possible to directly take over the magnetic optics used in the preceding proton run; from 2011 onwards significant changes to the optics had to be made for every run.

### 3.4 *The leap in performance in 2011*

A number of improvements were implemented to increase the luminosity in 2011. The injector performance [Manglunki *et al.* (2012)] was enhanced in various ways and trains of 24 lead bunches were injected for the first time, giving a total of 358 bunches in the LHC. However this introduced a new feature: the very strong variations of the intensity and emittance along the bunch trains (Figure 6) [Schaumann (2014)], imprinted by the

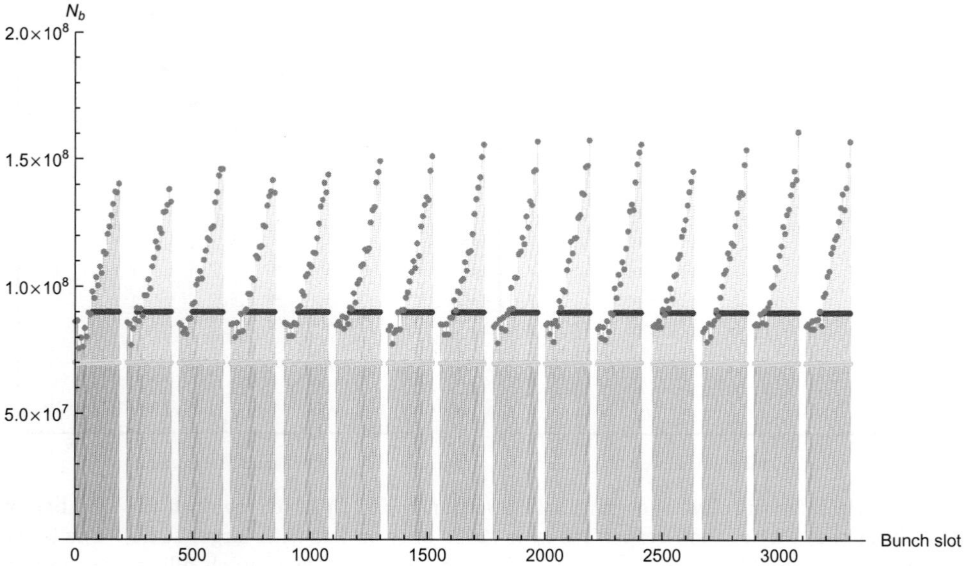


Fig. 6. Bunch intensities in the LHC in Fill 2351. The green points show the design single bunch intensity, the blue ones an intensity that could be regarded as constant along the bunch trains. The actual intensities are shown in red [Jowett *et al.* (2012)].

effects of intra-beam scattering, RF noise and possibly space-charge along the injection plateau of the SPS. Nevertheless the mean-square value of the single-bunch intensity was 2.4 times the design value [Brüning *et al.* (2004); Jowett (2004)], translating directly into a similar factor in the luminosity.

In the LHC itself, the $\beta^*$ values at the interaction points were reduced considerably, contributing another factor 3.5 to the peak luminosity [Jowett *et al.* (2012)].

As in 2010, the commissioning strategy ensured a rapid rise in peak luminosity as shown in the right panel of Figure 7. A gain of an order of magnitude over the 2010 luminosity was achieved within three days and the overall gain in integrated luminosity in the run (which itself was 15% shorter) was a factor 15. Table 3 summarises the principal parameters for comparison with Table 2.

The interruptions of luminosity production visible in Figure 7 are due to down time of major systems, ion source refills and changes of the operating conditions, mainly reversals of the ALICE spectrometer polarity.

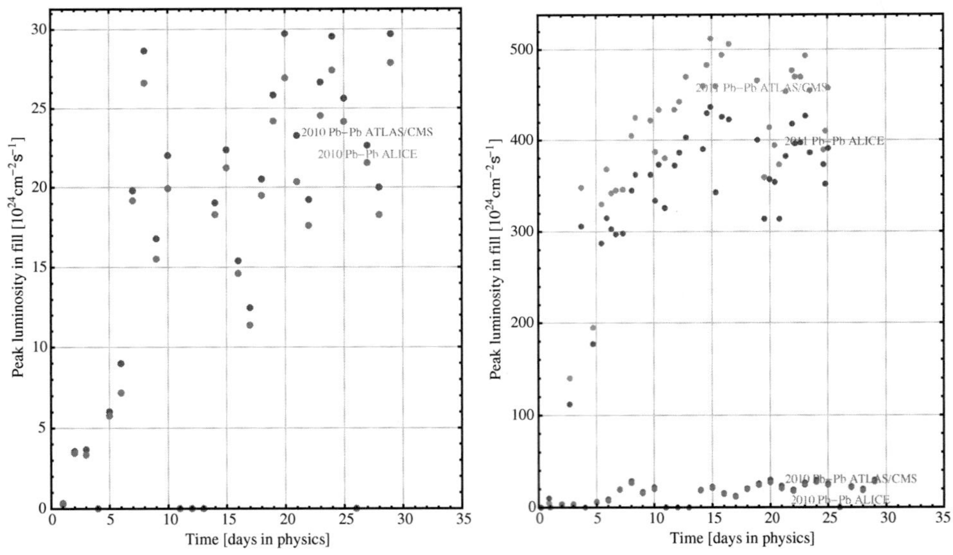


Fig. 7. Peak luminosity in every fill in 2010 and 2011, plotted against the number of days from the start of their respective runs (2010 only on left with larger scale). The lower luminosity for ALICE is a consequence of the symmetry of the bunch filling scheme and abort gap, which leads to somewhat fewer colliding bunch pairs than in ATLAS and CMS.

Table 2. Effective parameters (averaging over bunch-to-bunch and horizontal-vertical variations) at peak luminosity in Fill 1541 in 2010 (values in parentheses are design values at full energy).

| | | |
|---|---|---|
| Beam energy | $E_b$ | $1.38(2.76)A$ TeV $= 3.5(7)Z$ TeV |
| Collision energy | $\sqrt{s_{\rm NN}}$ | 2.76 (5.52) TeV |
| Ions/bunch | $N_b$ | $11.2(7) \times 10^7$ |
| No. of bunches per beam | $k_b$ | 137 (592) |
| Normalised emittance | $\varepsilon_n$ | $2(1.5)\ \mu$m |
| Optical function at IP | $\beta^*$ | 3.5 (0.5/0.55) m |
| Luminosity | $L$ | $3(100) \times 10^{25}\ {\rm cm}^{-2}{\rm s}^{-1}$ |

Table 3. Average parameters (averaging over the variations along the bunch trains at peak luminosity in Fill 2351 in 2011 (values in parentheses are design) [Jowett *et al.* (2012)].

| | | |
|---|---|---|
| Beam energy | $E_b$ | $1.38(2.76)A$ TeV $= 3.5(7)Z$ TeV |
| Collision energy | $\sqrt{s_{\rm NN}}$ | 2.76 (5.52) TeV |
| Ions/bunch | $N_b$ | $10.7(7) \times 10^7$ |
| No. of bunches per beam | $k_b$ | 358 (592) |
| Normalised emittance | $\varepsilon_n$ | $1.5(1.5)\ \mu$m |
| Optical function at IP | $\beta^*$ | 1.0 (0.5/0.55) m |
| Luminosity | $L$ | $50(100) \times 10^{25}\ {\rm cm}^{-2}{\rm s}^{-1}$ |

## 4 Proton–nucleus collisions

### 4.1 *History*

The hybrid p–Pb collisions were not included in the baseline design of the LHC [Brüning *et al.* (2004)]. Nevertheless they are of crucial importance to the physics programme [Salgado *et al.* (2012)], both as a reference for interpretation of the nucleus–nucleus collisions and for their potential to illuminate the partonic structure of matter at low Bjorken $x$. They cannot be achieved in the same way that d–Au collisions were at RHIC [Fischer (2013); Satogata *et al.* (2003)]. There the revolution frequencies of the two beams could be kept equal by varying the magnetic fields of the two rings independently. This is precluded by the two-in-one magnet design of the LHC which imposes equal magnetic fields and unequal frequencies on the two beams at injection and during the ramp, until the energy is high enough that the differences can be absorbed by small, opposite displacements of the central orbits [Jowett *et al.* (2013); Jowett and Carli (2006)]. At injection energy this manoeuvre would place the beams outside the vacuum chamber.

At injection energy therefore the protons made eight more turns of the LHC ring per minutes than the lead ions. With unequal revolution frequencies, the modulation of the long-range beam–beam forces while bunch encounter points moved slowly along the interaction regions. Experience at RHIC and other machines suggested that this could cause beam instabilities.

Nevertheless, first discussions in a workshop at CERN in 2005, in [Jowett and Carli (2006)] and a demonstration of p–Pb injection and acceleration on 31 October 2011 showed that p–Pb operation should be feasible in LHC and plans for a full operational run [Jowett *et al.* (2012)] were agreed in early 2012 and the necessary changes to the LHC hardware and controls software were implemented.

The operational cycle would involve injection and ramp with unequal RF frequencies in the two beams and a so-called "cogging" process to equalise frequencies at top energy and rotate the encounter points between bunches to their proper positions at the centres of the detectors. For collisions, the beams would be displaced radially inwards (Pb) and outwards (p) to "off-momentum" orbits to ensure equal revolution times.

### 4.2 *Pilot run in 2012*

When Pb beams were next available, a pilot run with p–Pb collisions was designed to test the new operational procedures and provide the experiments with an opportunity to set up their triggers sufficiently in advance of the main production physics run a few months later. Operating conditions were carefully chosen to satisfy machine protection constraints yet still allow a fast-track to physics conditions. These were achieved in a single fill, within 9 h of injecting the first Pb bunch of 2012 on 13 September. Injection of 15 single bunches provided eight colliding pairs in each of the four experiments (LHCb participated in the heavy-ion programme for the first time) with three bunches sacrificed for an essential off-momentum collimation setup of the unsqueezed optics at flat-top. Table 4 gives key parameters in collision conditions.

This first run, although it lasted just a few hours, opened up a new physics programme for the LHC and immediately led to new and quite unexpected physics results [CERN Courier (2012, 2013)]. Figure 8 shows a typical event display.

### 4.3 *Proton–lead run in 2013*

The first full physics run with p–Pb collisions was finally scheduled in January 2013 with the goal of increasing the luminosity by three orders of magnitude over the pilot run. In the meantime there had been no opportunity for

Table 4. Indicative parameters of peak performance in the p–Pb pilot and Pb–p production runs. Some numbers are averages because of the wide distribution of individual bunch parameters. Sets of four values correspond to the interaction points IP1(ATLAS), IP2(ALICE), IP5(CMS), IP8 (LHCb).

| | 2012 pilot | 2013 production |
|---|---|---|
| Beam energy | $4Z$ TeV | $4Z$ TeV |
| Collision energy $\sqrt{s_{\rm NN}}$ | 5.02 TeV | 5.02 TeV |
| Collisions | $(8, 8, 8, 8)$ | $(296, 288, 296, 39)$ |
| $\beta^*$/m | $(11, 10, 11, 10)$ | $(0.8, 0.8, 0.8, 2.0)$ |
| Protons/bunch | $1.2 \times 10^{10}$ | $1.6 \times 10^{10}$ |
| Pb ions/bunch | $7 \times 10^7$ | $12 \times 10^7$ |
| Normalised emittance (p) | $1.7\,\mu$m | $2\,\mu$m |
| Normalised emittance (Pb) | $1.2\,\mu$m | $1.5\,\mu$m |
| Luminosity | $1 \times 10^{26}\,\rm cm^{-2}s^{-1}$ | $(1.12, 1.01, 1.16, 0.05) \times 10^{29}\,\rm cm^{-2}s^{-1}$ |

any further development of the p–Pb mode. The complexity of the run was unprecedented with requests from the experiments for a number of changes of operation mode such as initial operation of ALICE at very low luminosity for minimum-bias data-taking, reversal of the beams half-way through the run and polarity switches of the ALICE and LHCb solenoids.

The need to provide matching bunch patterns of the two beams in the LHC made new demands on the CERN injector complex [Manglunki *et al.* (2013)]. A complex filling scheme with alternating 200 and 225 ns bunch

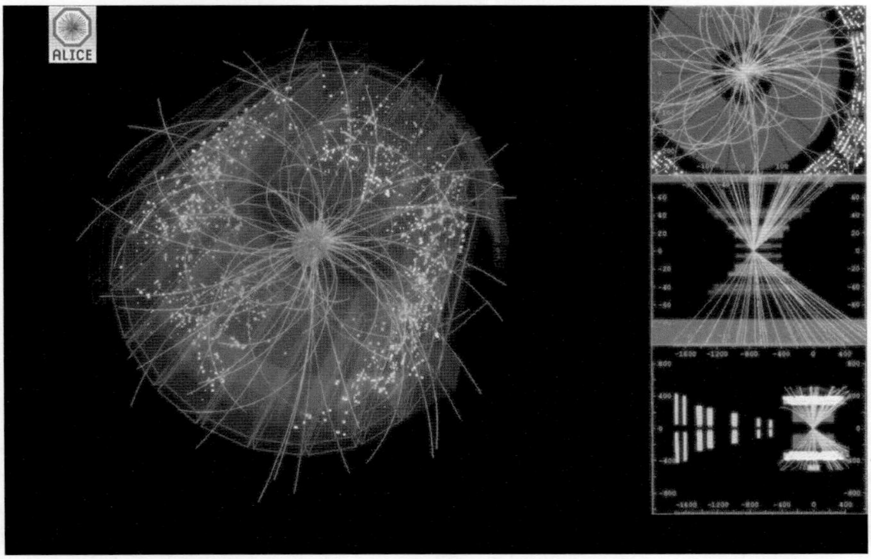


Fig. 8. A p–Pb event in the ALICE detector.

spacings was devised to fill the LHC optimally from the two independent pre-injector chains and was modified to allow collisions in the LHCb experiment which joined the heavy-ion programme for the first time. Nevertheless the performance of the heavy-ion injectors set new records [Manglunki *et al.* (2013)].

Thanks to meticulous planning and the quality of the hardware and controls of the LHC and its injectors, the commissioning of this new mode of operation was achieved within 10 days of the restart of the LHC after the

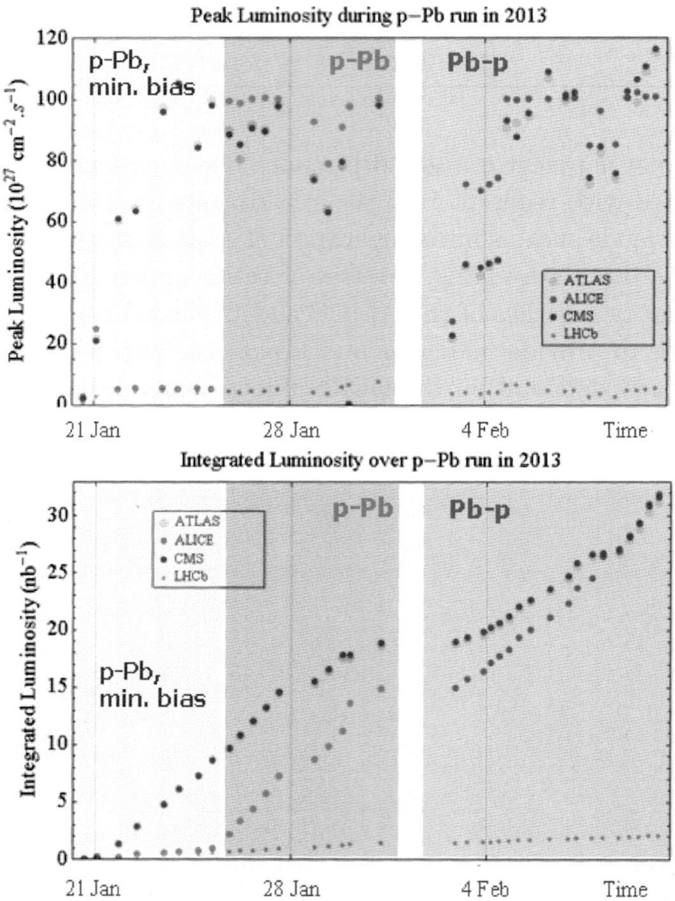


Fig. 9. Full instantaneous luminosity $1 \times 10^{29}$ cm$^{-2}$s$^{-1}$ already reached with the first fill with full filling scheme. Levelling in ALICE at $1 \times 10^{29}$ cm$^{-2}$s$^{-1}$ in almost all standard fills. Two catch-up fills done with IP1 and 5 separated, allowing ALICE to catch up after initial minimum-bias. Van der Meer scans done in both configurations. Final integrated luminosity above experiment request of 30 nb$^{-1}$. The run ended with record peak luminosity of $1 \times 10^{29}$ cm$^{-2}$s$^{-1}$, record turn around of 2.37 h.

technical stop at the end of 2012. The unequal frequency acceleration and cogging processes were fully automated and worked quickly and smoothly.

For this run, the optics was substantially new with a squeeze sequence to low $\beta^*$ values in all four experiments for the first time [Jowett *et al.* (2013)]. The squeeze sequence also incorporated a special correction of the chromatic effects due to opposite displacements of the two beams [Versteegen *et al.* (2013)]. Table 4 gives a comparative summary of the main beam parameters and compares them to the pilot run in 2012.

Figure 9 shows the development of peak and integrated luminosity during the run, indicating some of the major changes in operating conditions and other events.

LHC Run 1 ended with a few days devoted to collecting p–p data at the centre-of-mass energy corresponding to the p–Pb run.

Finally, Figure 10 summarises the integrated luminosity accumulation for the heavy-ion programme in the whole of LHC Run 1.

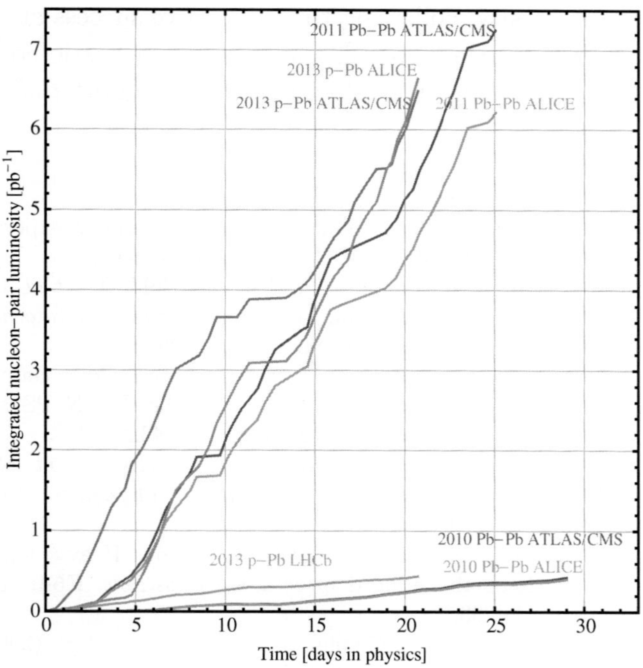


Fig. 10. Integrated nucleon–nucleon luminosity ($= A_1 A_2 L$) for the Pb–Pb runs in 2010 and 2011 and the p–Pb run in 2013, shown for each experiment as a function of days since the start of the respective run. There is a good match between the final Pb–Pb and p–Pb integrated luminosities for ALICE, ATLAS and CMS. Note the lower but still substantial p–Pb luminosity for LHCb in 2013. The single p–Pb fill in 2012 is not shown.

## 5 Future prospects at full energy

At the time of writing, it seems clear [Jowett *et al.* (2014)] that the Pb beam produced by the injectors for the 2013 p–Pb run, plus the boost in energy and somewhat reduced $\beta^*$, will allow the initial goal of $1\,\mathrm{nb}^{-1}$ of integrated luminosity to be attained during LHC Run 2, the period from 2015 until the second long shutdown expected in 2018.

During that shutdown, some upgrades of the injectors and LHC will allow more bunches to be stored with a basic bunch spacing reduced to 50 ns. The effects of bound-free pair production will be mitigated by various strategies including the first application of new superconducting magnet technology to install special collimators in the dispersion suppressor sections around the experiments. With these measures the peak luminosity is expected to exceed the original LHC design value by a factor 4–5. However the rapid luminosity decay will shorten the fills putting a premium on rapid turn-around time between dumping one fill and colliding the next. Beyond injecting still more beam current, the only way to improve this would be to shrink the beam emittance during the fill, allowing a larger fraction of the injected beam to be converted into luminosity. Following the impressive success at RHIC [Blaskiewicz *et al.* (2010); Fischer (2013)], the idea of stochastic cooling is being explored for LHC also [Jowett *et al.* (2014); Schaumann *et al.* (2013)].

## References

ALICE Collaboration (1993). Letter of intent for a large ion collider experiment [ALICE], Tech. Rep. CERN-LHCC-93-016. LHCC-I-4, CERN, Geneva.

Angoletta, M., Baudrenghien, P., Bellodi, G., Blas, A., Bohl, T. *et al.* (2011). Ions for LHC: Performance of the injector chain, *Proceedings of the 2011 International Particle Accelerator Conference*, San Sebastian, Spain, pp. 2529–2531.

Asbo-Hansen, P., Barbalat, O., Boussard, D., Boutheon, M., Gareyte, J. *et al.* (1977). Acceleration and stacking of deuterons in the CERN PS and ISR, *IEEE Trans.Nucl.Sci.* **24**, 1557–1560, doi:10.1109/TNS.1977.4329009.

Baltz, A., Baur, G., d'Enterria, D., Frankfurt, L., Gelis, F. *et al.* (2008). The physics of ultraperipheral collisions at the LHC, *Phys.Rept.* **458**, 1–171, doi: 10.1016/j.physrep.2007.12.001.

Baltz, A., Rhoades-Brown, M. and Weneser, J. (1996). Heavy ion partial beam lifetimes due to Coulomb induced processes, *Phys.Rev.* **E54**, 4233–4239, doi: 10.1103/PhysRevE.54.4233.

Beloshitsky, P., Bojtar, L., Carli, C., Chanel, M., Cornelis, K. *et al.* (2006). LEIR commissioning, *Conf.Proc.* **C060626**, 1876–1878.

Benedikt, M., Collier, P., Mertens, V., Poole, J. and Schindl, K. (2004). LHC Design Report. 3. The LHC injector chain.

Blaskiewicz, M., Brennan, J. and Mernick, K. (2010). Three-dimensional stochastic cooling in the Relativistic Heavy Ion Collider, *Phys.Rev.Lett.* **105**, 094801, doi: 10.1103/PhysRevLett.105.094801.

Boussard, D., Brennan, J. M. and Linnecar, T. P. R. (1989). Fixed frequency acceleration in the SPS, Tech. Rep. CERN-SPS-89-49-ARF, CERN, Geneva.

Braun, H. H., Fassò, A., Ferrari, A., Jowett, M. J., Sala, R. P. and Smirnov, I. G. (2014). Hadronic and electromagnetic fragmentation of ultrarelativistic heavy ions at LHC, *Phys. Rev. ST Accel. Beams* **17**, 021006, doi:10.1103/PhysRevSTAB.17.021006.

Braun, H.-H., Assmann, R. W., Ferrari, A., Jeanneret, J.-B., Jowett, J. M. and Pshenichnov, I. A. (2004). Collimation of heavy ion beams in LHC, *Proceedings of the 2004 European Particle Accelerator Conference*, Lucerne, Switzerland, p. 551.

Bruce, R., Gilardoni, S., Jowett, J. and Bocian, D. (2009). Beam losses from ultra-peripheral nuclear collisions between $^{208}Pb^{82+}$ ions in the Large Hadron Collider and their alleviation, *Phys.Rev.ST Accel.Beams* **12**, 071002, doi: 10.1103/PhysRevSTAB.12.071002.

Bruce, R., Jowett, J., Blaskiewicz, M. and Fischer, W. (2010). Time evolution of the luminosity of colliding heavy-ion beams in BNL Relativistic Heavy Ion Collider and CERN Large Hadron Collider, *Phys.Rev.ST Accel.Beams* **13**, 091001, doi: 10.1103/PhysRevSTAB.13.091001.

Brüning, O. S., Collier, P., Lebrun, P., Myers, S., Ostojic, R. *et al.* (2004). LHC Design Report, Vol. 1. The LHC Main Ring.

CERN Courier (1993). ALICE in the looking-glass, *CERN Courier* **July/August 1993**, 4–5.

CERN Courier (2012). Proton-ion collisions at the LHC, *CERN Courier* **Vol. 52, No. 9, November 2012**, 1, 6, 8.

CERN Courier (2013). Mysterious long-range correlations seen in pPb collisions, *CERN Courier* **Vol. 53, No. 1, January 2013**, 9.

Chanel, M. (2002). LEIR: The low energy ion ring at CERN, *Nucl. Instrum. Methods Phys. Res., A* **532**, 137–143. CERN-PS-2002-015-AE. CERN-AB-2003-086-ABP.

Boussard, D. *et al.* (1995). Non integer harmonic number acceleration of lead ions in the CERN SPS, *Proceedings of the 1995 Particle Accelerator Conference*, Dallas, Texas, p. 1506.

Fischer, W. (2015). The Relativisitc Heavy Ion Collider, *Challenges and Goals for Accelerators in the XXI Century* (World Scientific, Singapore), p. 345.

Haseroth, H. (2004). Heavy ions in the CERN accelerators, *Phys.Rept.* **403–404**, 27–32, doi:10.1016/j.physrep.2004.08.008.

Jackson, J. D. (1999). *Classical electrodynamics; 3rd ed.* (Wiley, New York, NY).

Jowett, J. (2005). LHC operation for heavy ions, *Proceedings of the LHC Performance Workshop, Chamonix XIV*, CERN-AB-2005-014, p. 88.

Jowett, J., Alemany-Fernandez, R., Baudrenghien, P., Jacquet, D., Lamont, M., Manglunki, D., Redaelli, S., Sapinski, M., Schaumann, M., Solfaroli Camillocci, M., Tomàs, R., Uythoven, J., Valuch, D., Versteegen, R. and Wenninger, J. (2013). Proton–nucleus collisions in the LHC, *Proceedings of the 2013 International Particle Accelerator Conference*, Shanghai, China, CERN-ACC-2013-0057, pp. 49–51.

Jowett, J., Arduini, G., Assmann, R., Baudrenghien, P., Carli, C. *et al.* (2011). First run of the LHC as a heavy-ion collider, *Proceedings of the 2011 International Particle Accelerator Conference*, San Sebastian, Spain, pp. 1837–1839.

Jowett, J., Braun, H.-H., Gresham, M., Mahner, E., Nicholson, A. *et al.* (2004). Limits to the performance of the LHC with ion beams, *Proceedings of the 2004 European Particle Accelerator Conference*, Lucerne, Switzerland, pp. 578–580.

Jowett, J. M. (2004). Ions in the LHC ring, *Proceedings of the LHC Performance Workshop, Chamonix XIII*, CERN-AB-2004-014-ADM.

Jowett, J. M. (2012). Ions in the LHC, *Proceedings of the Chamonix 2012 Workshop on LHC Performance*, CERN-ATS-2009-001.

Jowett, J. M., Alemany-Fernandez, R., Assmann, R., Baudrenghien, P., Bellodi, G., Hancock, S., Lamont, M., Manglunki, D., Redaelli, S., Sapinski, M., Schaumann, M., Solfaroli, M., Versteegen, R., Wenninger, J. and Wollmann, D. (2012). Heavy ions in 2012 and the programme up to 2022, *Proceedings of the Chamonix 2012 Workshop on LHC Performance*, CERN-2012-006.

Jowett, J. M. and Carli, C. (2006). The LHC as a proton–nucleus collider, *Proceedings of the 2006 European Particle Accelerator Conference*, Edinburgh, UK, pp. 550–552.

Jowett, J. M., Jeanneret, J. B. and Schindl, K. (2003). Heavy ion beams in the LHC, *Proceedings of the 2003 Particle Accelerator Conference*, Portland, Oregon, TPPB029, p. 1682.

Jowett, J. M., Schaumann, M. and Versteegen, R. (2014). Heavy ion operation from Run 2 to HL-LHC, *Proceedings of the Review of LHC and Injector Upgrade Plans*, Archamps, France.

Klasen, M., Klein-Büsing, C., König, F. and Wessels, J. (2013). How robust is a thermal photon interpretation of the ALICE low-$p_T$ data? *JHEP* **1310**, 119, doi: 10.1007/JHEP10(2013)119.

Klein, S. R. (2001). Localized beam pipe heating due to $e^-$ capture and nuclear excitation in heavy ion colliders, *Nucl.Instrum.Meth.* **A459**, 51–57, doi:10.1016/S0168-9002(00)00995-5.

Lefèvre, P. and Möhl, D. (1993). A low energy accumulation ring of ions for LHC (a feasibility study), Tech. Rep. CERN-PS-93-62. CERN-PS-93-62-DI. LHC-NOTE-259. CERN-LHC-Note-259, CERN, Geneva.

Manglunki, D., Angoletta, M., Bartosik, H., Bellodi, G., Blas, A., Bodendorfer, M., Bohl, T., Carli, C., Carlier, E., Cettour-Cave, S., Cornelis, K., Damerau, H., Findlay, A., Gilardoni, S., Hancock, S., Jowett, J., Küchler, D., O'Neil, M., Papaphilippou, Y., Pasinelli, S., Scrivens, R., Tranquille, G., Vandorpe, B., Wehrle, U. and Wenninger, J. (2013). The first LHC p–Pb run: Performance of the heavy ion production complex, *Proceedings of the 2013 International Particle Accelerator Conference*, Shanghai, China, CERN-ACC-2013-0035, pp. 2648–2650.

Manglunki, D., Bartosik, H., Papaphilippou, Y., Angoletta, M., Bellodi, G., Blas, A. *et al.* (2012). Performance of the CERN heavy ion production complex, *Proceedings of the 2012 International Particle Accelerator Conference*, New Orleans, USA, **C1205201**, pp. 3752–3754.

Martin, N. (2013). (Anti-)matter and hyper-matter production at the LHC with ALICE, *J.Phys.Conf.Ser.* **455**, 012007, doi:10.1088/1742-6596/455/1/012007.

Meier, H., Halabuka, Z., Hencken, K., Trautmann, D. and Baur, G. (2001). Bound free electron positron pair production in relativistic heavy ion collisions, *Phys.Rev.* **A63**, 032713, doi:10.1103/PhysRevA.63.032713.

Muller, B., Schukraft, J. and Wyslouch, B. (2012). First results from Pb+Pb collisions at the LHC, *Ann.Rev.Nucl.Part.Sci.* **62**, 361–386, doi:10.1146/annurev-nucl-102711-094910.

Myers, S. (2015). The Large Hadron Collider, *Challenges and Goals for Accelerators in the XXI Century* (World Scientific, Singapore), p. 365.

Oppedisano, C. (2011). Measurement of the electromagnetic dissociation cross section of Pb nuclei at $\sqrt{s_{\rm NN}} = 2.76$ TeV, *J.Phys.* **G38**, 124174, doi:10.1088/0954-3899/38/12/124174.

Redaelli, S., Assmann, R., Bellodi, G., Brodzinski, K., Bruce, R., Burkart, F., Cauchi, M., Deboy, D., Dehning, B., Holzer, E., Jowett, J., Lari, L., Nebot del Busto, E., Pojer, M.,

Priebe, A., Rossi, A., Schmidt, R., Sapinski, M., Schaumann, M., Solfaroli Camollocci, M., Valentino, G., Versteegen, R., Wenninger, J., Wollmann, D. and Zerlauth, Z. (2012). Quench tests at the Large Hadron Collider with collimation losses at 3.5 Z TeV, *Proceedings of the HB2012 Workshop*, Beijing, China, MOP245, p. 157.

Salgado, C., Alvarez-Muniz, J., Arleo, F., Armesto, N., Botje, M. *et al.* (2012). Proton–nucleus collisions at the LHC: Scientific opportunities and requirements, *J.Phys.* **G39**, 015010, doi:10.1088/0954-3899/39/1/015010.

Satogata, T. *et al.* (2003). Commissioning of RHIC deuteron-gold collisions, *Proceedings of the 2003 Particle Accelerator Conference*, Portland, Oregon, pp. 1706–1708.

Schaumann, M. (2014). Semi-empirical model for optimising future heavy-ion luminosity of the LHC, *Proceedings of the 2014 International Particle Accelerator Conference*, Dresden, Germany, CERN-ACC-2014-0084, pp. 1033–1035.

Schaumann, M., Jowett, J. and Blaskiewicz, M. (2013). Potential of stochastic cooling of heavy ions in the LHC, *Proceedings of the COOL2013 Workshop*, Murren, Switzerland, CERN-ATS-2013-043, pp. 76–78.

Pshenichnov, I. P. *et al.* (2001). Mutual heavy ion dissociation in peripheral collisions at ultrarelativistic energies, *Phys. Rev. C* **64**, p. 024903.

Versteegen, R., Bruce, R., Jowett, J., McAteer, M., McLean, E., Langner, A., Levinsen, Y., Persson, T., Redaelli, S., Salvachua, B., Skowronski, P., Solfaroli Camillocci, M., Tomás, R., Valentino, G., Wenninger, J. and White, S. (2013). Operating the LHC off-momentum for p–Pb collisions, *Proceedings of the 2013 International Particle Accelerator Conference*, Shanghai, China, CERN-ACC-2013-0180, pp. 1439–1441.

Wilde, M. (2013). Measurement of direct photons in pp and Pb–Pb Collisions with ALICE, *Nucl.Phys.A904-905* **2013**, 573c–576c, doi:10.1016/j.nuclphysa.2013.02.079.

## Chapter 24

# The LHC upgrade plan and technology challenges

Lucio Rossi and Oliver Brüning (CERN)

## 1 Objectives

### 1.1 *Context*

After three years of physics running, and two years of Long Shutdown (LS1), LHC has restarted collision in spring 2015 at 13 TeV (centre-of-mass). It is expected that the LHC will reach the **design luminosity**[1] **of $10^{34}$ $cm^{-2}s^{-1}$** during 2015 or beginning of 2016. This peak value should give a total integrated luminosity over one full year of about 40 $fb^{-1}$. In the period 2015–2020 LHC will hopefully increase the peak luminosity: indeed margins have been taken in the design to allow, in principle, reaching about 2 times the nominal design performance. The baseline programme for the next ten years is depicted in Fig. 1, while Fig. 2 shows the graphs of the possible evolution of peak and integrated luminosity.

A few years after 2020 some critical components of the accelerator will reach the radiation damage limit and others will experience reliability lessening because of vulnerability to radiation, wear and high intensity beam operation. Therefore, important consolidation actions are required before 2020, just to keep the LHC running with a good availability. Further, the statistical gain in running the accelerator without an additional considerable luminosity increase beyond its design value will become marginal. The running time necessary to half the statistical error in the measurements will be more than ten years in 2022, before LS3. Therefore, around that time, to maintain scientific progress and to explore its full capacity, the LHC will need to have a decisive increase of its luminosity.

[1]**Luminosity** is the number of collisions per unit of cross section (in square centimetre) per second, $cm^{-2}$ $s^{-1}$.

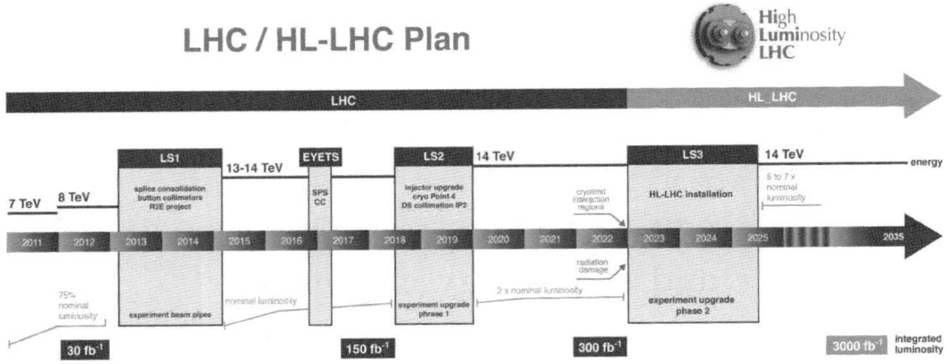


Fig. 1. LHC baseline plan for the next ten years, in terms of energy of the collisions (upper line) and of luminosity (lower lines). The first long shutdown during 2013–14 is to allow design parameters of beam energy and luminosity. The second one, 2018–2019, is for secure luminosity and reliability as well as to upgrade the LHC injectors. Integrated luminosity foreseen after each run is indicated, too.

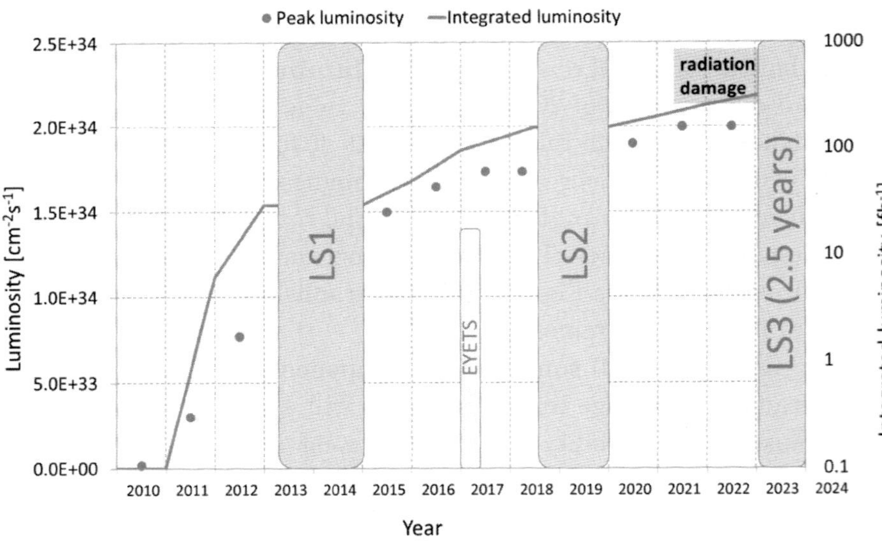


Fig. 2. Possible peak luminosity evolution (till the expected limit) with best forecast consequent integrated luminosity (courtesy of M. Lamont, CERN). Superimposed the three long shutdowns, with the extended year-end technical stop decided for 2016–17 and the area where we expect that radiation damage will call for changing of the low-$\beta$ quadrupole triplet (also called inner triplet).

This new phase in the LHC life, referred to as High Luminosity LHC (HL-LHC) has the scope of preparing the machine to attain the astonishing threshold of 3000 $fb^{-1}$ during 10–12 years of operation.

All hadron colliders in the world, before LHC, have produced a total integrated luminosity of about 10 $fb^{-1}$. The LHC has delivered in the first three years of operation 2010–12 about 30 $fb^{-1}$ and has been designed to deliver about 300 $fb^{-1}$ in its first 10 years of life. The design goals of 300 $fb^{-1}$ for the present LHC configuration and 3000 $fb^{-1}$ for the HL-LHC machine, are matched with the two general purpose experimental collaborations, ATLAS and CMS and their respective upgrade programs. The High Luminosity LHC represents a major and extremely challenging upgrade project. For its successful realization a number of **key novel technologies** have to be developed, validated and integrated [1, 2]. The project has been established by the CERN director of Accelerators at the end of 2010 and the international collaboration initiated with the FP7 Design Study **HiLumi LHC** which was approved by the European Commission (EC) in 2011 and officially started in November 2011 [3].

The project is now the first priority of CERN, as stated by the Strategy update for HEP group approved by CERN Council in a special session held on 30 May 2013 in Brussels [4].

## 1.2 *Objectives of the luminosity upgrade and relation to the LHC baseline programme*

### 1.2.1 *HL-LHC goals*

The LHC is designed for a 14 TeV collision energy [5,6], i.e. 7 TeV per proton beam, and a peak luminosity for each of the two general purpose experiments (ATLAS and CMS) of $1 \times 10^{34}$ $cm^{-2}s^{-1}$ (this value being called also the nominal design luminosity). This luminosity associated with a bunch spacing of 25 ns (2808 bunches per beam) gives an average value of 27 events/crossing,[2] or pile up. This is one of the first results of the LHC: a revision of the total hadronic cross section at 8 TeV and 14 TeV CM.

**The main objective of High Luminosity LHC** is to determine a set of beam parameters and hardware configuration that will enable the LHC to reach the following targets:

(1) A peak luminosity of $\mathbf{5 \times 10^{34}\ cm^{-2}s^{-1}}$ **with levelling**, allowing:
(2) An integrated luminosity of $\mathbf{250\ fb^{-1}}$ per year, enabling the goal of $\mathbf{3000\ fb^{-1}}$ within twelve years after the upgrade. This luminosity is about ten times the luminosity reach of the first twelve years of the LHC lifetime.

The above goal has been set with the companion LHC detector upgrade program based on the hypothesis of a peak average pile up of 140 events per crossing, and parameter sets have devised for both 25 ns and 50 ns

[2]Early design parameter list was quoting 19 events/crossing for the nominal luminosity.

bunch spacing. The experiments will design the new upgraded detectors capable to "digest" a maximum of 200 events/crossing both for keeping a reasonable margin against shortfall (including a possible run of machine at 50 ns, should 25 ns become too difficult) and for taking into account the inevitable fluctuations around the average value.

### 1.2.2 *LHC luminosity and experience gained from present operation*

Luminosity is, together with the collision energy, the most important parameter of a collider, because it is proportional to the number of useful events. For physics purposes, luminosity integrated over time is the relevant parameter: however integrated luminosity does not depend only on the collider performance, but also on many operational parameters (machine availability and quality of technical services, long stops or short breaks required by maintenances, etc.). The expectations in terms of integrated luminosity in Fig. 1 are based on LHC capability and general run parameters (peak luminosity, burn-off rate of protons, duration of a run, etc.) and on the running experience at CERN, which includes the operational factors previously mentioned. As a rule of thumb, it is estimated that the LHC running with a peak luminosity of $10^{34}\,\mathrm{cm}^{-2}\mathrm{s}^{-1}$ will produce about $40\,\mathrm{fb}^{-1}$ of integrated luminosity per year for each of the two general purpose experiments, ATLAS and CMS.

The LHC has provided collisions at 8 TeV CM, with proton beam energies of 4 TeV, compared to its nominal design energy of 7 TeV, and with total beam currents of about 0.4 A (70% of the nominal design value but with only half the nominal number of bunches).Once the magnet interconnections have been consolidated and the beam energy limits removed, as well as some radiation-to-electronic (R2E) intensity limits mitigated during LS1 in 2013–14, the design luminosity will hopefully be attained and eventually overcome in 2015. Once outstanding beam intensity limits in the injector chain and the LHC have been removed [the bottle necks of the LHC injector chains are addressed in a dedicated project, the LHC Injector complex Upgrade project (LIU)] during an additional long shutdown following the LS1, the LHC will hopefully head toward the so-called "ultimate" design luminosity, which is about twice the nominal luminosity, i.e., $2.15 \times 10^{34}\,\mathrm{cm}^{-2}\mathrm{s}^{-1}$. This ultimate luminosity performance was planned to be reached by increasing the bunch population from 1.15 to $1.7 \times 10^{11}$ protons, with a bunch spacing of 25 ns (beam current increases from 0.58 A to 0.86 A). Transforming this ultimate peak performance into a doubling of the annual integrated luminosity will be however very difficult: it is more likely that it will be around 60–70 $\mathrm{fb}^{-1}$/y.

Actually the performance of 2012 at 4 TeV beam energy is already potentially beyond the nominal design luminosity, when it is scaled to operation at 7 TeV beam energy. We run stable collisions at surprisingly high beam–beam tune shift values: values of $\Delta Q_{b-b} > 0.03$ have routinely been reached, which is three times the conservative value of 0.01 that was taken for the nominal design LHC tune shift, and even a factor two beyond the value foreseen for the "ultimate" design performance of the machine. This better than anticipated machine performance allowed the use of intense bunches with $1.5-1.6 \times 10^{11}$ protons/bunch spaced by 50 ns, instead of nominal $1.15 \times 10^{11}$ protons/bunch spaced by 25 ns, with less than nominal beam emittances, an optimization that has come naturally because the luminosity depends quadratically on the bunch population and only linearly on the number of bunches.

The (instantaneous) luminosity $L$ can be expressed as (neglecting the hourglass effect at this stage as it is only relevant for $\beta^*$ much smaller than the bunch length):

$$L = \gamma \frac{n_b N^2 f_{\text{rev}}}{4\pi \beta^* \varepsilon_n} R, \quad R = 1/\sqrt{1 + \frac{\theta_c \sigma_z}{2\sigma}}$$

where

- $\gamma$ is the proton beam energy in unit of rest mass;
- $n_b$ is the number of bunches in the machine: 1380 for 50 ns spacing and 2808 for 25 ns;
- $N$ is the bunch population. $N_{\text{nominal 25 ns}}$: $1.15 \times 10^{11}$ p ($\Rightarrow$0.58 A of beam current at 2808 bunches);
- $f_{\text{rev}}$ is the revolution frequency (11.2 kHz);
- $\beta^*$ is the beam beta function (focal length) at the collision point (nominal design: 0.55 m);
- $\varepsilon_n$ is the transverse normalized emittance (nominal design: 3.75 $\mu$m);
- $R$ is a luminosity geometrical reduction factor (0.85 at 0.55 m of $\beta^*$, down to 0.5 at 0.25 m);
- $\theta_c$ is the full crossing angle between colliding beam (285 $\mu$rad as nominal design);
- $\sigma$, $\sigma_z$ are the transverse and longitudinal r.m.s. size, respectively (16.7 $\mu$m and 7.55 cm).

The use of 50 ns spaced bunches avoids the regime of e-clouds, with a negligible time required for scrubbing the beam pipe wall, while the use of the 25 ns would have required at least 2–3 weeks of beam scrubbing and conditioning, time that would have been missed for the physics operation. In addition to this, the better than expected emittance delivered by the LHC injectors at 50 ns (a factor two lower with respect to the nominal emittance values, partly foreseen as margin) could be well maintained (ca. 30% emittance growth from injection to collision) and used in LHC: this has further enhanced the luminosity reach, making the 50 ns spacing a key ingredient for the success of LHC during its first running period, despite the increased event pile up in the experiments. However, also in this domain the positive news from LHC operation is that luminosity levelling is possible and actually easier than foreseen: levelling by a simple transverse beam separation has been used already in 2011–12 for LHCb collision point (at relatively low luminosity). The peak luminosity in Run I was $7.6 \times 10^{33}$ and levelling was done for LHCb to $1.4 \times 10^{33}$ with $\beta^* = 3$ m (instead of 0.6 m in IP1 and IP5). In addition to the luminosity levelling via transverse beam offset, the levelling by $\beta^*$ variation during the physics run has also been successfully tested in 2012 operation. Despite of all this bias in favour of the 50 ns, the "nominal design" configuration with 25 ns bunch separation remains the baseline for next LHC run and for the HL-LHC, mainly for mitigating the pile up problem in the experiments. Indeed 50 ns with higher bunch population enhances the number of events per crossing, with a loss of quality of the data taking by the experiments which can be detrimental to the physics reach of the LHC.

### 1.2.3 *Present luminosity limitations*

There are various expected limitations to a continuous increase in luminosity, either in beam characteristics (injector chain, beam impedance, e-cloud and beam–beam interactions in the LHC) or in technical systems (see Section 1.4). Mitigation of potential performance limitations arising from the LHC injector complex are addressed by the companion CERN project: LHC Injector Upgrade Project (LIU), which should be completed in 2018 (LS2). Any potential limitations coming from the LHC injector complex put aside, it is expected that the LHC will reach a performance limitation from the beam current, from cleaning efficiency at 350 MJ beam stored energy, from e-cloud effects and the beam–beam interactions and from the acceptable pile-up level in the experiments. A performance increase of the LHC beyond the ultimate luminosity (the ultimate bunch intensity of $1.7 \times 10^{11}$ protons with nominal emittance and nominal optics at the interaction points corresponds to a performance level of about $2 \times 10^{34}$ cm$^{-2}$s$^{-1}$) will require significant hardware and beam parameter modifications with respect to the designed LHC configurations.

### 1.3 *Technical reasons for an upgrade through a performance improving consolidation*

As mentioned above, the LHC performs extremely well in terms of luminosity. However, its astonishingly good performance should not hide the fact that the LHC remains a very complex and somehow "fragile" machine,vulnerable to breakdown of various systems and wear out of many components. Many systems were not designed for maximum flexibility in operation — with the experience gained in the LHC construction and operation we do know now, to a certain extent, what is best. Interventions and equipment replacement will not be limited simply to anticipate breakdown or wear out of equipment: the new equipment will be designed to improve peak performance and machine availability, which both concur to integrated luminosity, and to increase flexibility and ease of repair and maintenance, an important objective for a machine operating at 200 times the beam stored energy and 50 to 100 times the luminosity of previous hadron colliders.

#### 1.3.1 *Magnets*

Radiation damage for equipment exposed to the radiation near the experiments is directly proportional to the integrated luminosity and is expected to occur around an integrated luminosity of 300 $\mathrm{fb}^{-1}$. As shown in Fig. 2, this value is expected to be reached within a few of years of the post LS1 operation of the LHC. Collision debrises are mainly intercepted by the TAS and other absorbers. However a good fraction of the debris escapes and is absorbed inside the quadrupole cold mass of the low-beta triplets, and the corrector magnets lodged in the same cold mass, with two main effects: (1) heat deposition that may limit with various mechanism the performance of the superconducting magnets by increasing the conductor temperature; (2) radiation damage, especially to insulation but also to metallic components.

The first effect, heat deposition, puts a limit on the peak luminosity at about $1.7 \times 10^{34}\,\mathrm{cm}^{-2}\mathrm{s}^{-1}$; the uncertainty is estimated around 25%, probably in the conservative sense, so the ultimate peak luminosity is probably within reach in the present configuration but not at all certain, based on this hardware limitation.

The second effect, which just scales with the dose, and therefore with the integrated luminosity, calls for a replacement of the inner triplet once too much radiation has been accumulated. The quadrupole magnet itself may withstand 400–500 $\mathrm{fb}^{-1}$and even more but their reliability may seriously decrease after 300 $\mathrm{fb}^{-1}$, at which the peak does in the insulation is calculated to be about 25 MGy. The corrector magnets placed in the quadrupole cold masses receive more or less the same dose. In particular a nested type corrector package controlling the final steering of the beam into collision, despite its dose be about half the one of the other magnets, is likely to wear out already at 300 $fb^{-1}$ because of the particular mechanical torque to which it is submitted during operation. In any case damage must be anticipated because the most likely way of failing is through sudden electric breakdown, entailing serious and long repairs. Replacement of whole or part of the low-beta triplet system must be envisaged before damage, because it is a long intervention, requiring more than one year shutdown and must be coupled with a planned major detector upgrade. Furthermore the replacement must be coupled with an improvement of the quadrupole aperture, which is today the bottleneck of the LHC luminosity, to give room to an increase of the luminosity via lower $\beta^*$. Furthermore, the whole Interaction Region (IR) needs to be redesigned with larger D1/D2 (the pair of recombination/separation dipole magnets), with a new DFBX, the cryo-distribution electrical feedbox of the low-beta triplet (considered today the most fragile of the critical equipment) and with much better access to various equipment for maintenance.

### 1.3.2 *Cryogenics*

An important consolidation of the LHC cryo-plant is adding a new helium refrigerator in Point 4, to separate the cooling of the four LHC superconducting RF modules from the magnet cooling circuit. The present coupling has two adverse effects: (i) it greatly reduces the flexibility of intervention: a stop of cryogenics of the magnets halts the cryogenics of the RF and vice versa, which is detrimental to the machine availability and then to the integrated luminosity; (ii) the triplet at left of IP5 (CMS) is cooled by the refrigerator of P4, which has to cool not only an arc and a long straight section, like in the other LHC sectors, but here it has to cool also the RF cryomodules (SCRF), reducing the cooling power which is available for the triplet at Point 5 Left. When the machine is to operate near nominal condition, the coupling may limit the luminosity of CMS and also the operation with 25 ns bunch spacing (which requires more cooling power in the arcs because of the e-cloud heat load). It is foreseen to implement the new cryo-plant and the full separation between SCRF and magnets cooling already in 2018 during LS2.

A further consolidation, which is deemed necessary in the long term, is the separation between the cooling of the inner triplets and of all or part of the matching section magnets from the arc, which are individually powered. This coupling means that an intervention in the triplet region requires warm up of the entire arc (an operation of 3 months and not without risk). When running at high luminosity and full energy this coupling will be detrimental to the LHC availability. A full separation is the first necessary step, followed by a final consolidation with new cryo-plants, dedicated to the triplet region, to fully decouple the IR zones from the arcs. Probably the new cryogenic system of the interaction regions will extend to include all the Long Straight Sections (LSS), including all individually powered magnets of the matching section (MS) and eventually the crab cavities (see next sections).

### 1.3.3 *Collimation*

The collimation system is operating very well, according to design. However the severe wear out by aging, imposed by beam impacts, will become tougher and tougher and will eventually require a renovation of parts of the system. However, this renovation must be done with new materials and new concepts to cope with higher than foreseen energy density of the LHC and with the necessity to reduce collimator impedance. As previously mentioned, one of the reasons of the very high performance of the LHC is a transverse emittance half the nominal value, which has an impact on the operation and wear of the collimation jaws, among the most delicate equipment of the whole LHC since they have to withstand the primary beam. All new collimators will be equipped with Beam Position Monitors ("button" collimators), in order to improve accuracy and time of setting up the jaws, and will be designed for a low impedance to make possible to increase the beam current and then luminosity. This consolidation will concern the momentum and betatron cleaning in P3 and P7, as well as the tertiary collimators protecting the triplets. Any small gain in triplet aperture and performance must be accompanied by an adequate consolidation or modification of the collimation system.

A second area that will require a special attention to the collimation system is the Dispersion Suppressor (DS): here a leakage of off-momentum particles into the first and second main superconducting dipole has already been identified as a possible LHC performance limitation. Various concepts of collimation for the cold area (the DS is part of the continuous cryostat) have been studied. An international review called in 2011 has advised to postpone installation of this new collimation system to 2018, while supporting all necessary studies to well evaluate the problem for future LHC beam conditions and the associate R&D to identify the best solution. The most promising concept is to substitute an LHC main dipole with a dipole of equal bending strength (121 Tm) obtained by a higher field (11 T) and shorter length (11 m) than those of the LHC dipoles (8.3 T and 14.2 m). The room gained is sufficient for placing special collimators. This new 11 T dipole, which is developed in a collaboration between CERN and Fermilab, might become the first magnet breaking the 10 T barrier in an accelerator. For the collimation system two options are under investigation: cryo-collimators operating at 20–60 K (with some features similar to the ones of the SIS100 FAIR project) and more classical room temperature collimators with a special, very compact, cold-warm-cold bypass. The system is likely necessary for ions, in IP2 DS, and most probably for IP1 and IP5 with luminosity well above nominal. The actual need for P3 and P7, where we have the standard collimators that also generate off-momentum particles, is under assessment.

Coming back to the collimation system itself, one should notice that safe handling of a beam current beyond 0.5 A, with beta function at collision beyond the design value will constitute a progression into new territory. For beam collimation, 75 collimators need to be precisely aligned in a dynamic mode with a precision of $\sim$10 $\mu$m, in order to assure the protection of the triplet. The protection of the triplet must be accomplished during the large change of the collision beam parameter ($\beta^*$ passing from 10 m to 10–15 cm, see the next section on $\beta^*$ reach), which will be one of the most critical phases of HL-LHC operation; just the beam halo itself could be well beyond the damage limit. Since the collimation system must be renovated anyway, more precise and more powerful material collimators must be implemented.

### 1.3.4 *R2E and SC links for remote cold powering*

Many electronic components of the LHC are vulnerable to single event upset, which is one of the most frequent causes of LHC unavailability, already at present energy and luminosity. Among the most difficult equipment to protect against R2E are the magnet power converters. They are bulky equipment and further shielding is almost impossible. While a considerable effort is under way to study how to replace the radiation sensitive electronic boards with rad-hard cards, another more radical solution is also pursued: relocation the power supplies and associate DFBs (electrical feed-boxes, delicate equipment today in line with the continuous cryostat) from the tunnel to the ground surface. This will solve radically the R2E power converter problem and will make much easier all interventions on the electrical circuits: most of the interventions in the tunnel require electrical consignation of the magnets and disconnection of the delicate HTS current leads; both operations will be much easier on surface. Not only the LHC availability will be greatly improved (because of less stop and much faster intervention, without tunnel access) but radiation dose to people will be lessened. Relocation of all power supplies to the surface would be very expensive and almost impossible. So we have identified the regions where R2E levels are the highest. In all other places we will replace the most sensitive electronic boards with rad-hard ones.

Relocation to ground surface is only possible thanks to a novel technology: superconducting links (SCLs) made out of high temperature superconductors (HTS) (YBCO or Bi-2223) or $MgB_2$ superconductors. The high operating temperature of these materials offers a large margin of stability, 100 to 1000 higher than classical NbTi, and a wide temperature excursion, from 4 K to 20 K in case of the $MgB_2$ (favoured because of the much lower cost) and 4 to 40 K for HTS. SCLs will allow a 40 kA, 700 m horizontal link in P7 and 200 kA, 300 m link in IP1 and IP5. The vertical jump is of course the 100 m of the LHC tunnel depth. Water cooled resistive cable are ruled out because of the dissipated power of 30 MW (40% of the total LHC power consumption) and because they would require new power converters of much higher power than the present ones.

### 1.3.5 *QPS, machine protection and remote manipulation*

Other systems will need vigorous consolidation to assure the LHC running for long time in conditions between nominal and ultimate luminosity, i.e., collecting between 40 and 80 $\mathrm{fb}^{-1}$ per year. Even without the HL-LHC luminosity increase to 250 $\mathrm{fb}^{-1}$/y, just running at 80 $\mathrm{fb}^{-1}$/y will not be possible in the present LHC without a vigorous consolidation.

The first system requiring consolidation is the *Quench Protection System* (QPS) of the superconducting magnets. Based on a design of almost twenty years ago, it lacks flexibility to face operating conditions somewhat different than foreseen. Despite a large consolidation and upgrade program implemented in LS1, to face the harsh condition of HL-LHC, the QPS will need a complete overhaul around 2020. This includes among other measures change of the quench loop from a simple loop to a triple loop with voting mechanism (so-called 2oo3 loop), redesign and relocation of all sensitive electronics, possibly to the surface.

*Machine protection*: improving vulnerability to kicker sparks and asynchronous beam dumps. The kicker system is, together with the collimation system, the main safety net against severe beam induced damage. The system today is at its limit and will need modifications and improvements to guarantee a probability of accident with equipment damage as low as $10^{-5}$ to $10^{-4}$. Continuous renovation is simply not sufficient and we saw already during the first running period of the LHC signs of necessity for improvements (to reduce heating), especially in the injection system. Not only the kicker system, but also the interlock system needs to be fully renovated at around 2020.

*Remote handling*: The LHC has not been designed specifically for remote handling. However the level of activation from 2020 onwards, and potentially even earlier, requires a careful study and development of special equipment to allow replacing collimators, magnets, vacuum components etc., according to the ALARA (As Low As Reasonably Achievable) principle. The first challenge will be the substitution of collimators; another big challenge will be the replacement of the inner triplet magnets and associated cryogenics and vacuum equipment. The higher the luminosity, the higher the necessity of interventions and the less time operators can stay in contact with this equipment. While full robotics is difficult to implement, given the actual conditions in the LHC tunnel, remote manipulation and supervision is a key for minimizing the radiation dose to operators.

## 2 Upgrading the performance to the high luminosity LHC goals

### 2.1 *Luminosity levelling and virtual peak luminosity*

Both the consideration of energy deposition by collision debris in the interaction region magnets, and the necessity to limit the peak pile up in the experimental detector, impose *a priori* a limitation of the peak luminosity. The consequence is that the HL-LHC operation will have to rely on luminosity levelling. As shown in Fig. 3 (left), the luminosity profile without levelling quickly decreases from the initial peak value, due to "proton burning" (protons consumed in the collisions). By designing the collider to operate with a constant luminosity, i.e. "levelling" it and suppressing its decay for a good part of the fill, the average luminosity is almost the same as the one for a run without levelling, see Fig. 3 (right), however with the advantage that the maximum peak luminosity is only a fraction of that without levelling.

Indeed pile up and degraded performance by intense radiation are serious limitations in the high luminosity regime. Coping with a peak luminosity substantially higher than $5\times10^{34}\,\mathrm{cm^{-2}s^{-1}}$ may become impossible and therefore levelling is a key ingredient of the HL-LHC baseline.

The concept of luminosity levelling introduces a new parameter: the virtual peak luminosity, i.e. the luminosity that could be "virtually" reached at the beginning of the run without levelling. Levelling means acting on one or more of the parameters controlling the (instantaneous) luminosity: by

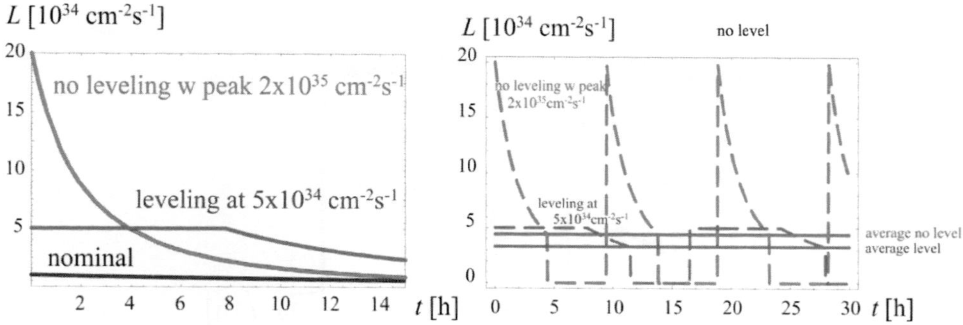


Fig. 3. Left: Luminosity profile for a single long run starting at nominal peak luminosity (black line), with upgrade no levelling (red line) and with levelling (dotted line). Right: Luminosity profile with optimized run time, without and with levelling (blue and red dashed lines), and average luminosity in both cases (solid lines).

detuning the chosen parameter(s) the luminosity is kept fixed at the chosen levelled value. Then the same parameter(s) is(are) slowly retuned to its ideal value in a way that the gain just compensates the proton burn-off due to collisions (and potential other phenomena decreasing the luminosity, like beam losses, emittance increase etc.). It is clear from this reasoning and from Fig. 3 that the higher the virtual luminosity, the longer the collider can be operated at constant levelled luminosity. The ratio $k =$ *Virtual Peak Luminosity/Levelled Luminosity* is therefore one of the parameters to be maximized in the HL-LHC configuration.

Once all the parameters are fully retuned to compensate the proton burning (and other possible effects) the luminosity decreases again with time since levelling is no longer possible, as shown in Fig. 3, until the run is terminated to restart another luminosity cycle.

## 2.2 *Integrated luminosity and availability*

The upgrade has one main objective: an integrated luminosity, proportional to the discovery potential, to be reached in a "reasonable" time: 3000 fb$^{-1}$ in 10–12 years. This fixes the goal of the annual integrated luminosity to 250 fb$^{-1}$ (300 fb$^{-1}$ will be pursued, if possible). The fact that the maximum levelled luminosity is limited, means that to maximize the integrated value one needs to maximize the run length, which can be obtained by filling the maximum number of proton, i.e. by maximizing the beam intensity: $I_{\text{beam}} = n_b \times N$. Other key factors for maximizing the integrated luminosity are a short average machine turnaround time (we assume 5 hours in the following), the optimization of the luminosity decay time in a run (see Fig. 4) and the overall machine efficiency. With HL-LHC parameters, the cycle depicted

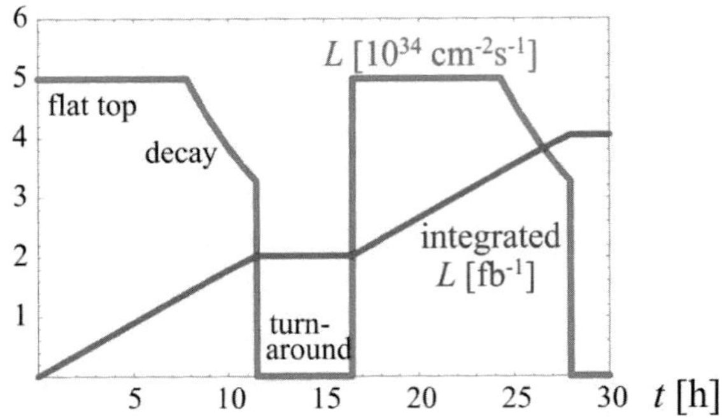


Fig. 4. Luminosity cycle for HL-LHC with levelling and a short decay (optimized for integrated luminosity). The set of parameters generating cycle are the 25 ns column of Table 1, stretched.

in Fig. 4 can yield more than $3\,\text{fb}^{-1}$/day! However, the HL-LHC will not always perform as in Fig 4. Some runs will be prematurely aborted, with beam dump required by BLM spikes and equipment failures, either true or spurious. And any equipment failure during the machine turnaround might entail a longer than anticipated turnaround time (time from end of physics to physics again, all included). In certain cases extra time is required to solve problems, for tunnel access or for the cryogenic system recovery. All this can be summarized in the overall machine "efficiency" defined as the ratio between actual time spent in physics production and the physics time of the ideal cycle. In practice in a luminosity levelled operation this is the same as the ratio between actual integrated luminosity and integrated luminosity obtained with a continuous ideal cycle.

HL-LHC with 150 days of physics needs an efficiency of ca. 50% for a scenario with 25 ns bunch spacing. During the 2011 run the efficiency varied without luminosity levelling between 20% and 40%. Clearly for the integrated luminosity the efficiency counts almost as much as the virtual peak performance. Achieving an efficiency higher than the one of the present LHC, with a (levelled) luminosity five times the nominal one, and with beam currents larger than ultimate (see the next section),will be a real challenge. The issue of efficiency calls for a further challenge: we need to not only increase the peak performance, but also to decrease the downtime of the LHC by reducing the number of faults and by mitigating their impact on the machine availability. For this reason the project must foresee a vigorous consolidation effort for the high intensity and high luminosity regime and must increase the reliability of all systems: in one word, the High Luminosity LHC must be, also, a High Availability LHC.

## 2.3 *Upgrade parameters*

### 2.3.1 *Bunch spacing*

Although the 25 ns bunch spacing remains the baseline scenario for the HL-LHC, 50 ns is kept as a viable alternative, in case the e-cloud or other unforeseen effects undermine the 25 ns performance. However, the companion LHC detector upgrade project is designing for an average pile up around 140: for 25 ns spacing this means about $5 \times 10^{34}$ of luminosity, while at 50 ns this means limiting the levelling luminosity to half this value, if the pile up is to be the same. This translates into a longer run time for the 50 ns scenario and inevitably also to an even higher machine efficiency, unless a higher pile up can be accepted for 50 ns bunch spacing. New concepts like pile-up density per bunch length are being explored to identify the actual limitations for the different bunch beam configurations. However these studies can also result in even stricter limitation for the HL-LHC as compared to the nominal LHC, since the luminous region in the HL-LHC tends to be shorter than in the nominal LHC. Experience with the LHC shows that the best set of parameters for actual operation is difficult to predict at this stage (before we obtained operational experience with the LHC at 7 TeV). The upgrade studies should therefore provide the required HL-LHC performance over a wide range of parameters, and the machine and experiments will find eventually the best set of parameters in actual operations once the LHC probes the operation at maximum energy and with above nominal beam intensities.

### 2.3.2 *Beam current*

The total beam current may be a hard limit in the LHC since many systems are affected by it in a direct way: RF power systems and RF cavities, collimation, cryogenics, kicker magnets, vacuum system, beam diagnostics, etc., and other systems in an indirect way, mainly through an increase of the R2E events, like quench detection system of the SC magnets, and virtually all controllers.

Radiation effects put aside, all systems have been designed in principle for $I_{\mathrm{beam}} = 0.86$ A, the so called "ultimate" beam current. However this is still to be experimentally validated and for the performance goal of the HL-LHC we need to go beyond the ultimate value by 30% with 25 ns bunch spacing. In principle we should be able to do so, profiting from margins and better than designed performance of some systems (for example, the cryostat insulation losses). The LHC operation at full beam energy in 2015–17 will push the beam current above the nominal value of 0.58 A. The operational experience in Run II will therefore tell us the actual margin we do have on $I_{\text{beam}}$. As a general consideration one can state that 50 ns bunch spacing is slightly less demanding since it requires for the same peak luminosity a factor $\sqrt{2}$ less beam current than the 25 ns bunch spacing. However this translates inevitably to $1/\sqrt{2}$ less protons circulating in the machine and therefore inevitably to a shorter levelling time, which requires in turn a higher machine efficiency to attain the same target value of integrated luminosity.

### 2.3.3 *Emittance and bunch population*

Transverse emittance has already been better than the design value during the first running period of the LHC, thanks to a better than anticipated performance of the LHC injectors, their transfer lines and good emittance preservation in the LHC and along the whole accelerator chain. Together with the absence of a measured beam–beam limit during the first operation phase of the LHC, the smaller emittance reach for the 50 ns bunch spacing as compared to the 25 ns bunch spacing (thanks to the different beam preparations for the two bunch spacings in the LHC injector complex) offers an very attractive performance reach. Actually the beam emittance is better than expected for both 25 and 50 ns operation modes, however the gain is accentuated for 50 ns. What actually matters for the HL-LHC performance is to increase the beam brightness, the ratio of the bunch intensity over the transverse emittance, a property that must be maximized at beginning of the beam generation and then preserved throughout the entire injector chain and LHC itself. In other words: it is a global property. The LIU project has as primary objective to increase the brightness at the extraction from the SPS, basically increasing the number of protons per bunch by a factor two above nominal while keeping the emittance close to the present low value. It is worth noticing that for the injectors the 50 ns HL-LHC parameters (Table 1) is the more difficult scenario to comply with (assuming that the e-cloud effects in the SPS, which are more accentuated for the 25 ns, can be mitigated).

### 2.3.4 $\beta^*$ *and cancelling the reduction factor R*

The classical route to a luminosity upgrade is to reduce $\beta^*$ by means of larger aperture triplet magnets for a given gradient or longer, larger aperture low-$\beta$ triplet quadrupoles with a reduced gradient. With respect to reducing the emittance, this requires only a local intervention, as compared to modifications of the whole machine and injector chain. The conventional approach to a triplet upgrade is to keep the overall triplet length constant which leads to an increase of the peak field in the quadrupole magnets when increasing the aperture at constant gradient. However, a new design approach has been chosen for the HL-LHC upgrade: increasing the magnet aperture up to a maximum allowed by physics limits (mainly the peak field, determined by the type of superconducting technology) and by space constraints (i.e. the triplet length). However a reduction in $\beta^*$ implies an increase of beam sizes over the whole matching section where the optics of the collision points is matched to the regular optics of the LHC arc cells and a wider crossing angle over the region where both beams share the same vacuum system in order to minimize the long range beam–beam (LRbb) effects. So a reduction in $\beta^*$ implies not only larger triplet magnets but also larger separation/recombination dipoles and larger and/or modified matching section quadrupoles. Stronger chromatic aberrations coming from the larger $\beta$-functions inside the triplet magnets may exceed the values that can be corrected with the existing correction circuits in the LHC. The maximum acceptable beta function is also limited by the possibility to match the optics to the regular beta functions of the arcs. A previous study has shown that a practical limit in the LHC arises for $\beta^* = 30-40$ cm (depending on the actual length and strength of the upgraded low beta quadrupole magnets) compared to the 55 cm foreseen in nominal operation. However a novel scheme has been recently proposed to overcome these practical limitations of the LHC matching section. The scheme called Achromatic Telescopic Squeeze (ATS) uses the adjacent arcs as enhanced matching sections and the increase of the beta functions in those arcs to boost at constant strength the efficiency of the lattice sextupole magnets which are used for the chromatic correction inside the triplet magnets with high beta functions. In this way a $\beta^*$ value of 15 cm and even 10 cm can be envisaged and a flat optics with a $\beta^*$ as low as 5 cm in the plane perpendicular to the crossing plane is enabled. Actually, $\beta^* = 10$ cm has been recently attained in a machine development run dedicated to test the ATS principle. In order to be compatible with such small $\beta^*$ values, the quadrupole magnets need to double their aperture, with a peak field 50% above the present LHC triplet magnets, requiring a new superconducting magnet technology based on $Nb_3Sn$ (see Section 3.1).

The drawback of very small $\beta^*$ values is that it requires a larger crossing angle for a constant normalized beam separation, which entails in turn a reduction of the luminosity via the geometrical factor $R$, see luminosity expression in Section 1.2.2. Figure 5 shows the geometrical reduction factor

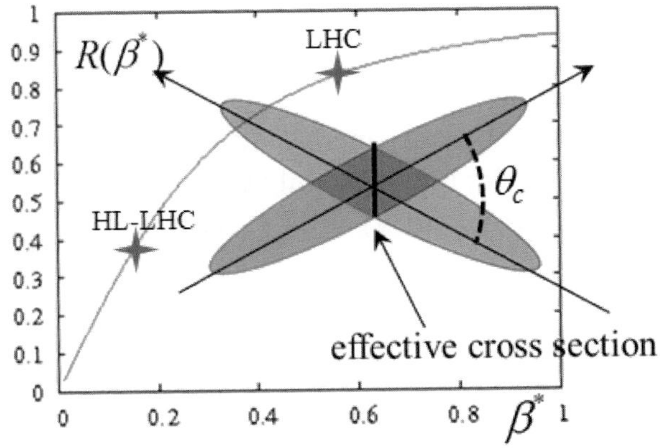


Fig. 5. Behaviour of geometrical reduction factor of luminosity vs. $\beta^*$ with indicated two operating points: nominal LHC and HL-LHC. The sketch of bunch crossing illustrates the reduction in luminous regions which provides the mechanism for the loss in luminosity for an operation with crossing angle.

vs. $\beta^*$ for a constant normalized LRbb separation. To restore the full gain of operating with low $\beta^*$, we have three options:

1. Reduce the beam separation at the parasitic encounters: for the HL-LHC studies we assume a safe $11\sigma$ beam separation, however a preliminary test in 2012 — not yet conclusive — has shown that 7–$8\sigma$ separation might be acceptable and even beneficial by slowly cleaning away the halo particles);
2. Compensate the LRbb by electric wire which is under study but not yet proven;
3. Compensate for the crossing angle at the beam collision points.

Probably a mix of the first two options can be a viable solution, however the most efficient and elegant solution for compensating the geometrical reduction factor is option 3 via the use of crab cavities.

Special RF "crab" cavities are capable of generating a strong transverse electric field used to give a torque to the beam. In this way the beams do not suffer from overlap reduction due to the crossing angle ($\theta_c$), as shown in Fig. 5: a crab cavity just rotates each bunch such that it collides head on with bunches of the other beam, overlapping perfectly at the collision point as illustrated in Fig. 6. In this way the crossing angle is maintained over the long drift space in the common vacuum beam pipe avoiding the LRbb interactions, but the geometrical reduction is totally suppressed. Of course the same opposite kick must be given to the beam at the opposite side of the collision point. Crab cavities have been successfully tested for the first time in the $e^+e^-$ KEKB accelerator in Japan. However, their feasibility for hadron

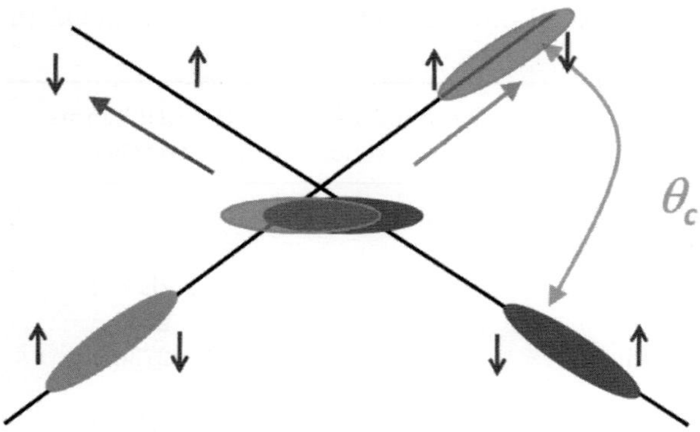


Fig. 6. Effect of the crab cavity on the beam (small arrows indicate the torque on the beam by transverse varying RF field).

beam operation still has to be demonstrated. Crab cavities make accessible the full performance reach of the operation with small $\beta^*$ that the ATS scheme and the large inner triplet quadrupoles can generate: their primary function is boosting the virtual peak luminosity for attaining the full HL-LHC performance. Initially they were also thought to be a tool for luminosity levelling, since changing the cavity voltage provides a straightforward local control of the beam rotation and therefore the instantaneous luminosity. However the use of crab cavity for luminosity levelling inevitably reduces the length of luminous region such that the event pile-up density per unit length of the luminous region remains approximately constant. Since the performance limitations of the experiments seem to be linked rather to the pile-up density than to the total number of events per bunch crossing, the preferred levelling mechanism is at the moment a levelling via a dynamic $\beta^*$ variation during operation (the method has been tested in special machine studies but not yet during nominal physics operation).

### 2.3.5 *Table of HL-LHC parameters*

Table 1 lists the main parameters for the HL-LHC upgrade for different bunch configurations. For convenience, the first parameter column lists the nominal LHC parameters as a reference. The other columns show the upgrade parameters both for 25 ns and for 50 ns bunch spacings. As mentioned above: 25 ns is our operation target and 50 ns is being maintained as a fall-back solution. For the 25 ns bunch spacing we list what the LIU project considers as stretched beam parameters, being the most ambitious objective that might enable $300\,\mathrm{fb}^{-1}$ per year and are approximately 10% above the intensity of the nominal LIU target bunch intensities. The efficiency of 50% is needed to reach the goal of $250\,\mathrm{fb}^{-1}/\mathrm{y}$. The efficiency of the LHC in 2012

Table 1. Parameters for HL-LHC compared with LHC nominal (in bold the most critical ones).

| Parameter | Nominal LHC (design report) | HL-LHC 25 ns (stretched) | HL-LHC 50 ns |
|---|---|---|---|
| Beam energy in collision [TeV] | 7 | 7 | 7 |
| $N_b$ | 1.15E+11 | 2.2E+11 | 3.5E+11 |
| $n_b$ | 2808 | 2748 | 1374 |
| Number of collisions in IP1 and IP5 (Ref. 1) | 2808 | 2736 | 1368 |
| $N_{\text{tot}}$ | 3.2E+14 | 6.0E+14 | 4.9E+14 |
| Beam current [A] | 0.58 | **1.09** | 0.89 |
| Crossing angle [$\mu$ rad] | 285 | 590 | 590 |
| Beam separation [$\sigma$] | 9.4 | 12.5 | 11.4 |
| $\beta^*$ [m] | 0.55 | **0.15** | 0.15 |
| $\varepsilon_n$ [$\mu$m] | 3.75 | **2.50** | 3 |
| $\varepsilon_L$ [eVs] | 2.50 | 2.50 | 2.50 |
| r.m.s. energy spread | 1.13E-04 | 1.13E-04 | 1.13E-04 |
| r.m.s. bunch length [m] | 7.55E-02 | 7.55E-02 | 7.55E-02 |
| IBS horizontal [h] | 80 → 106 | 18.5 | 17.2 |
| IBS longitudinal [h] | 61 → 60 | 20.4 | 16.1 |
| Piwinski parameter | 0.65 | 3.14 | 2.87 |
| Total loss factor R0 without crab cavity | 0.836 | 0.305 | 0.331 |
| Total loss factor R1 with crab cavity | (0.981) | 0.829 | 0.838 |
| Beam–beam/IP without crab cavity | 3.1E-03 | 3.3E-03 | 4.7E-03 |
| Beam–beam/IP with crab cavity | 3.8E-03 | 1.1E-02 | 1.4E-02 |
| Peak luminosity without crab cavity [$\text{cm}^{-2}\text{s}^{-1}$] | 1.00E+34 | 7.18E+34 | 8.44E+34 |
| Virtual luminosity with crab cavity: $L_{\text{peak}} \times$ R1/R0 [$\text{cm}^{-2}\text{s}^{-1}$] | (1.18E+34) | **19.54E+34** | 21.38E+34 |
| Events/crossing without levelling and without crab cavity | 27 | 198 | 454 |
| Levelled luminosity [$\text{cm}^{-2}\text{s}^{-1}$] | — | 5.00E+34 (Ref. 5) | 2.50E+34 |
| Events/crossing (with leveling and crab cavities for HL-LHC) (Ref. 8) | 27 | 138 | 135 |
| Peak line density of pile up event [event/mm] (max over stable beams) | 0.21 | 1.25 | 1.20 |

(*Continued*)

Table 1. (*Continued*)

| Parameter | Nominal LHC (design report) | HL-LHC 25 ns (stretched) | HL-LHC 50 ns |
|---|---|---|---|
| Leveling time [h] (assuming no emittance growth) (Ref. 8) | — | **8.3** | **18.0** |
| Number of collisions in IP2/IP8 | 2808 | 2452/2524 (Ref. 7) | 0/1262 (Ref. 4) |
| $N_b$ at LHC injection (Ref. 2) | 1.20E+11 | 2.30E+11 | 3.68E+11 |
| $n_b$ at injection | 288 | 288 | 144 |
| $N_{\text{tot}}$ at injection | 3.46E+13 | 6.62E+13 | 5.30E+13 |
| $\varepsilon_n$ at SPS extraction [$\mu$m] (Ref. 3) | 3.40 | 2.00 | 2.30 |

after the initial period was around 25–40%, so the efficiency proposed for the 25 ns is reasonable, nevertheless a big leap forward is required on increasing availability (as previously mentioned) and turnaround time (time from end of physics to next start of physics). The 50 ns option requires an even higher efficiency level. A measure that can help to reach the HL-LHC performance goal at 50 ns, and help for operation with 25 ns bunch spacing, is increasing the number of days for proton physics: in Table 1 we assume 150 days, maybe this number can be increased to 180–200, at the expense of ion runs and of allocated machine development time. A margin that is not considered in the table is the possibility to work at $\beta^*$ of 10 cm, which thanks to the ATS and larger $Nb_3Sn$ quadrupoles could be within reach.

## 3 Hardware modifications and technology challenges of HL-LHC

While the LHC is the summit of 30 years of hadron collider evolution, its high luminosity upgrade will open the gate for new technologies and new concepts that will likely mark the next generation of colliders, either for hadrons or for leptons. New high field magnets, beyond 10 T will be a real breakthrough for collider technology as well as crab cavity for protons.

In this section we review the hardware that needs to be modified, rebuilt, or completely changed for the HL-LHC, which will require works in many LHC points, see Fig. 7. In all cases, with the notable exception of the crab cavities, this new hardware is very much entangled with the improving consolidation plan that needs to go on to maintain LHC in an efficient status.

### 3.1 *High field magnets for the interaction regions*

The present LHC constitutes the summit of 30 years of developments in the domain of superconducting magnet technologies. NbTi based magnets are pushed to their limits: very compact two-in-one magnets provide 8.3 T operating field by using superfluid helium cooling (magnets are designed, and many have been tested, up to 9 T). The plot in Fig. 8 illustrates the progress over the years from the resistive magnet era to the jump in performance required by HL-LHC.

#### 3.1.1 *Superconductor*

The upgrade heavily relies on the success of the advanced $Nb_3Sn$ technology, since NbTi superconductor cannot go beyond 9 T. $Nb_3Sn$ has been under development for more than ten years and has now reached a maturity that allows designs of real equipment based on it. $Nb_3Sn$ has been used in solenoids for NMR spectroscopy for more than 20 years. ITER is now using $Nb_3Sn$ on a very large scale of 400 tonnes (similar to the HL-LHC

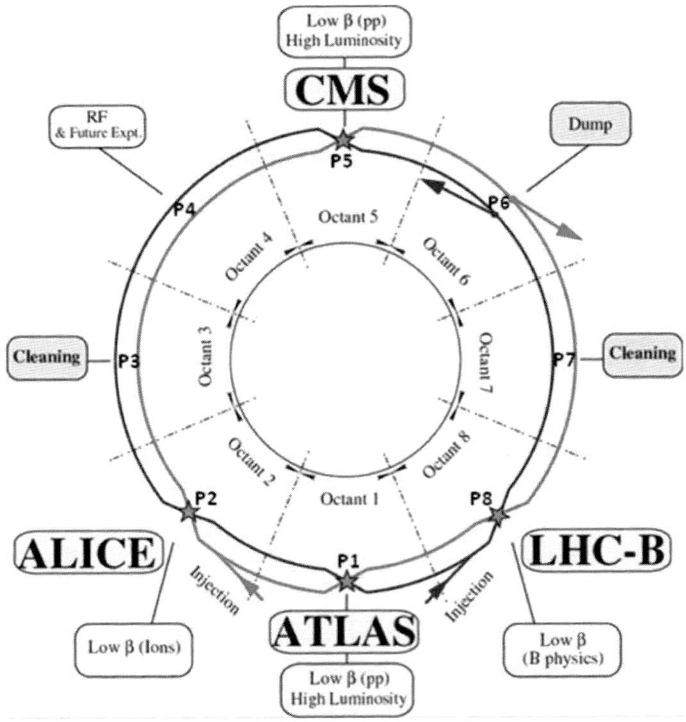


Fig. 7. Schematic layout of the LHC, indicating the points of beam collision or beam services (P1 to P8). HL-LHC will require deep modifications of at least 1.2 km of the accelerator section in P1, P2, P4, P5, and potentially additional areas in P3 and P7.

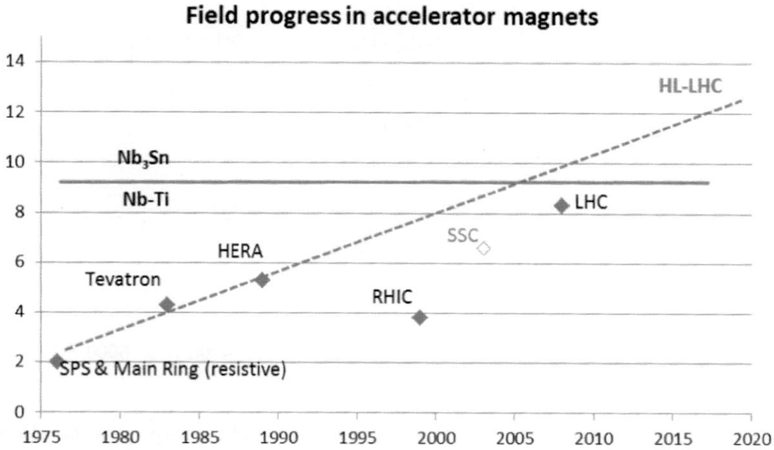


Fig. 8. Progress of accelerator magnets for hadron colliders: from 2 to 9 tesla is the realm of NbTi; beyond 9 tesla, $Nb_3Sn$ is needed.

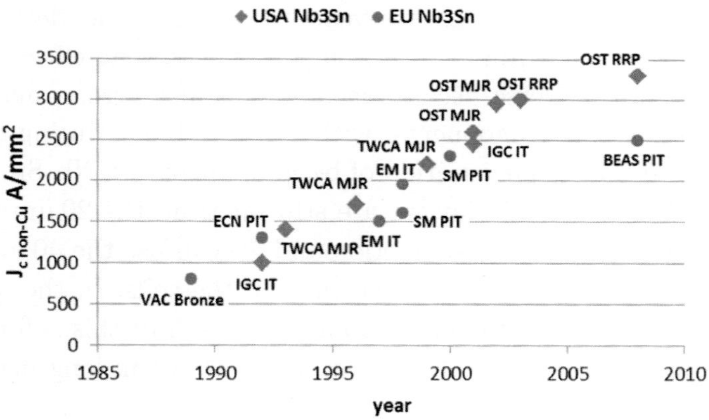


Fig. 9. Progress in critical current density in non-copper $Nb_3Sn$ strands. The best values are quoted, usually real availability in technical conductor for magnet follows 1–3 year later. The DOE conductor program in USA has double the $J_c$ in ten years.

scale), making a decisive step in industrializing the process. However, for accelerators we need a current density between 2.5 and 3 times that used in the ITER coils: from value of $J_c \cong 800-1000$ A/mm$^2$ requested for ITER, to $J_c \cong 2500-3000$ A/mm$^2$ for the HL-LHC magnets. As shown in Fig. 9 the progress has been steady in the last fifteen years, especially thanks to the DOE supported Conductor Development Program in the USA. Europe has followed with some delay, because of the heavy engagement for LHC construction that has taken almost all budget and human resources from 1995 to 2010.

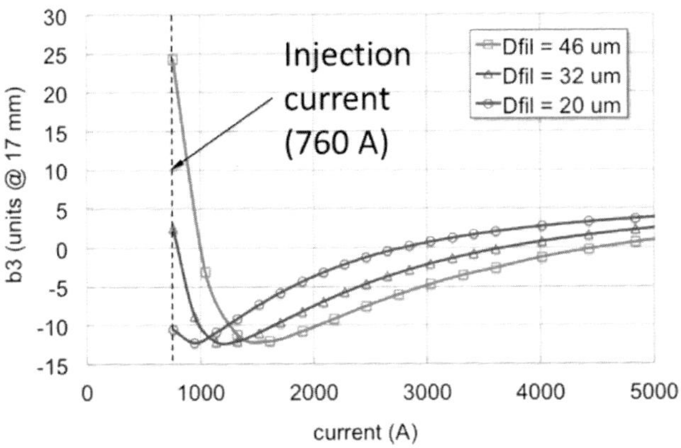


Fig. 10. Variation of the sextupole components during LHC ramp cycle for different size of the $Nb_3Sn$ filaments in the 11 T dipole under design for the LHC dispersion suppressor upgrade (courtesy of B. Auchmann and L. Bottura, CERN).

As shown in Fig. 9, $J_c$ is in the desired range since almost ten years ago, however it took more than five years to decrease the effective filament diameter from over 100 $\mu$m down to 50 $\mu$m or less. The effect of large filament diameters is adverse because it generates filed errors, mainly — but not only — sextupole components, with big variability during the current ramp. Figure 10 shows the variation of $b_3$ component of a $Nb_3Sn$ 11 T dipole for a typical LHC cycle. Today we are still far from the 20 $\mu$m that would constitute a good base for $b_3$ correction and we can use the 40 $\mu$m filaments only by making use of iron corrective shim in the coils. In the main dipole magnets one unit $b_3$ corresponds to 45 units of $Q$', but this is for 1200 15 m long magnets and here we are talking of a couple of 5 m long magnets.

### 3.1.2 *High field magnets*

Once the conductor becomes available, the magnet field progresses steadily. Figure 11 shows the progress of the maximum field in short dipole magnets (what we call a model, typically 1 m long or less). The values in Fig. 11 are actually the record fields, obtained after many quenches and in conditions that are far from being operative in an accelerator. That is why the values for NbTi magnets of Fig. 11 are much higher than the ones in Fig. 8 where operating fields obtained in mass production are reported: usually effective operating fields are 10% to 20% less than the record fields, and follow typically 3–5 years later.

Two high field magnet R&D programs are running in parallel for the High Luminosity LHC project: the 11 T dipole project, aiming at removing the collimation limitation described in Section 1.1.3, and the new inner

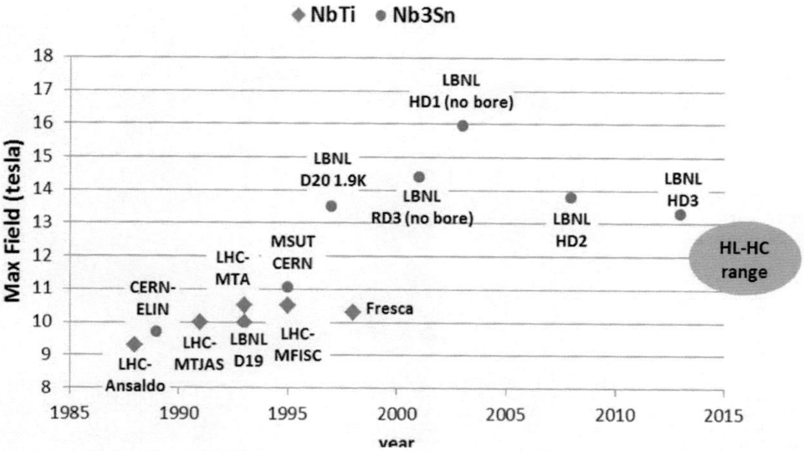


Fig. 11. Evolution of record fields in short magnet models for both NbTi and $Nb_3Sn$ technologies together with an indication of the field region required for the HL-LHC project. NbTi dipoles all have 50 mm bore, except Fresca which features an 88 mm aperture. Early $Nb_3Sn$ dipoles have 50 mm bore while the HD series features an aperture of 40 mm.

triplet quadrupole project as the main tool for reducing $\beta^*$ and providing aperture for shielding so that the magnets can survive a factor ten increase in radiation. We will discuss only the R&D for the quadrupoles, as this is a key ingredient for the high luminosity machine. The R&D has been so far carried out by the US LARP (LHC Accelerator Research Program). Launched in 2004, this program, besides developing the superconductor, has passed through a number of successful steps to go beyond the 70 mm diameter and 205 T/m gradient ($G$) of the present LHC triplet in NbTi technology:

1. Development of short magnet model of 90 mm aperture ($G > 200$ T/m)
2. Development of long magnets of 90 mm aperture ($G > 200$ T/m)
3. Development of short magnet model of 120 mm aperture ($G > 170$ T/m)
4. Development of short magnets of final aperture $\phi = 150$ mm ($G >$ 140 T/m)
5. Development of a first prototype unit: 4.5 m long, 150 mm, 140 T/m.

Step 1 has been accomplished very successfully by 2009 with the magnet series Technological Quadrupoles (TQS), based on a new magnet structure called shell plus bladders and keys, first developed for the LBNL high field magnet program. Not only the requirement on the field was well exceeded, but also in one important test it was shown that the RRP $Nb_3Sn$ can withstand up to 200 MPa of compressive stress with minor reduction of the performance. Based on this result the HiLumi magnets aim at a compressive stress below 170 MPa in operation. There has been also a series of TQCs,

Fig. 12. The 3.6 m long quadrupole LQ03 ready for test. Target and measured quench values are shown on the picture (courtesy of G. Ambrosio, FNAL).

with classical collar technology, not selected for the quadrupoles but is the base for the 11 T dipole program.

Step 2 has been accomplished in 2009–2012. Three 3.6 m long quadrupoles with shell, bladders and keys have been manufactured and tested (the first "long" accelerator magnet in $Nb_3Sn$, ever). The first magnet reached up to 220 T/m, well beyond the goal, with some training. The last, LQ03 shown in Fig. 12, has reached 210 T/m but with little training, going immediately to 206 T/m at first quench and at 4.2 K.

Step 3, whose quadrupole models are named HQ, is near completion: HQ01 has been tested in five assemblies. The first results have been moderate, however each version has brought improvement and the last one went well beyond the required gradient, see Fig. 13. However various problems in coil technology, made more critical by the large coil size, have required a HQ02 to assure that all problems have been well understood; HQ02 is at present under test (status of May 2013). A HQ03 will also be manufactured to verify repeatability.

Step 4 will start in the summer of 2013 after the decision is made for the 150 mm aperture for the HL-LHC quadrupole inner triplet magnets, a decision that maximize the $\beta^*$ reach. The cross section has been designed; see sketch of Fig. 14 and tooling is being purchased.

Step 5 is the final in order to produce a demonstrator of final length, about 4.5 m. The longest magnets of 9 m will be actually composed of two 4.5 m quadrupole cold mass in the same cryostat. However the middle

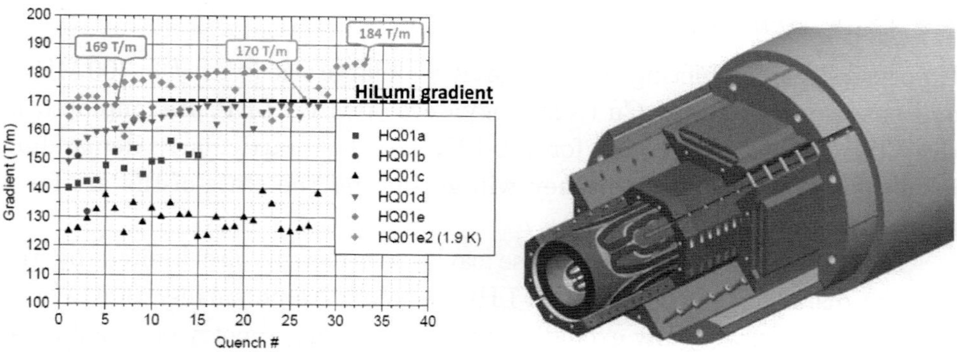


Fig. 13. Left: Quench results of the HQ01 quadrupole models. Each differs either because a few coils have been replaced, with minor improvement in insulation and shimming, or because different pre-stress has been applied (data: LARP). Right: Sketch showing the various component of the HQ structure the inner tube is the coil package, not detailed here (courtesy of G.L. Sabbi, LBNL).

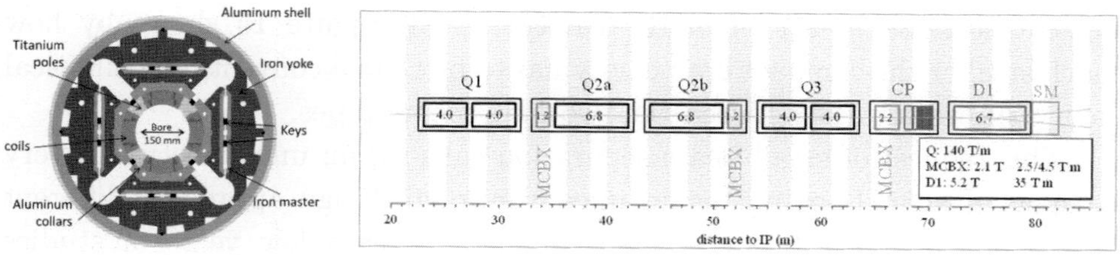


Fig. 14. Left: Cross section of the 150 mm aperture quadrupole of the inner triplet. Right: Layout of the HL-LHC IR inner triplet, correctors and D1 (courtesy of P. Ferracin and E. Todesco, CERN).

quadrupole, Q2a and Q2b will be composed of two full length cold mass of about 8 m each, see Fig. 14 where the baseline layout recently approved for the interaction region of HL-LHC is shown.

Special tungsten shielding will be placed inside the inner bore to limit the radiation deposition to the same level of the nominal LHC, about 30 MGy, despite the ten times higher integrated luminosity. $Nb_3Sn$ has also a higher temperature margin than NbTi, making it easier to deal with heat deposition issues, which, again thanks to the shielding, should not be higher than in the present LHC. The field quality is still a factor two to five worse than that of the NbTi magnets of the nominal LHC but is steadily improving. Fortunately the field quality of these quadrupoles is required only at collision and not at injection energy, making the problem of magnetization less relevant than for the main dipoles.

### 3.2 *Crab cavities (CC)*

Elliptical superconducting (SC) RF cavities of large size (with $f = 400\,\text{MHz}$), and based on Nb coated Cu technology developed for LEP, are employed in the LHC. The crab cavities for HL-LHC are not particularly demanding in terms of voltage, however they will go beyond the state-of-the art for a number of reasons.

The first is that the transverse cavity dimensions are limited by the 194 mm distance between the two LHC beams, a value smaller that $\lambda/4$ of 400 MHz wave, practically excluding the well-known geometry of an elliptical cavity, see Fig. 15. For the accelerating cavities, a special region around Point 4 was created in the LHC, in which the beam separation is increased, by the use of magnetic doglegs, to 400 mm in order to allow the installation of the elliptical 400 MHz accelerating cavities. Compact crab cavities have to be installed on both sides of each high luminosity Points 1 and 5, without additional doglegs. Their design goes therefore definitely well beyond the present state-of-the-art of SCRF cavity design. Figure 15 shows by how much smaller the compact crab cavity has to be than a conventional elliptical cavity, in order to fit between the two LHC beam pipes.

This request of small beam separation calls for an unconventional, very compact design. Looking for an unconventional design approach, different design options were proposed and pursued. After a few years of studies

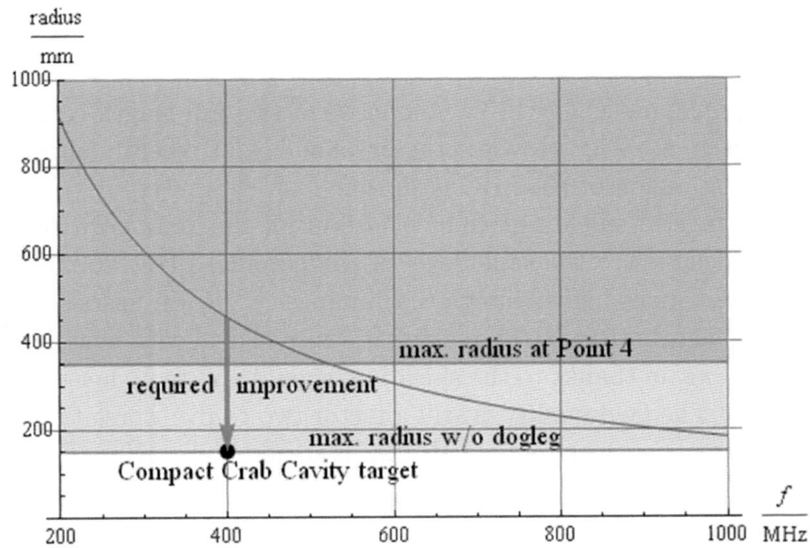


Fig. 15. Size of compact crab cavities (dot): they are more than a factor 2 smaller in transverse dimension than classical elliptical cavities (see their radius vs. frequency curve).

a down-selection process has reduced the options to three realistic design options, see Fig. 16.

Very recently the first full test on such a cavity has been completed for the RF-dipole type: it went well above the target of 3.4 MV transverse voltage $V_T$, see Fig. 17, quenching at 7 MV. From the graph one can project

Fig. 16. The three types of compact crab cavities that are being pursued at the moment for the HL-LHC.

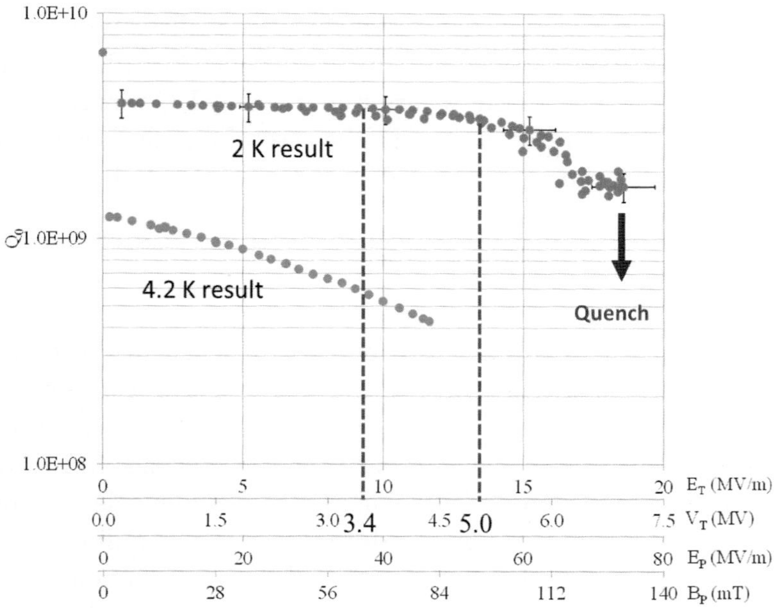


Fig. 17. Results of the first full test of a crab cavity: the RF dipole (ODU, SLAC, LARP). Vertical lines indicate the target voltage (3.4 MV) and actual usable voltage (5 MV). Beside transverse gradient and voltage, on the horizontal axis, the peak gradient and field are reported too. The test has been carried out at JLab. (Courtesy of J. Delayen, ODU)

that operation at 5 MV seems to be within reach: with such a voltage we can relax the optics and probably remove from the HL-LHC layout an additional quadrupole (Q7b) in the matching section. This first result is very encouraging, though many other questions remain open, like the lower than expected $Q$ value (a factor two lower than expected, see the red dot in Fig. 17), the integration in a very compact cryostat, etc.

One further critical point about the use of CC is the actual effect on the proton beam in terms of noise induced beam emittance growth, something that cannot be scaled from experience on the e-ring at KEK. To test this point, and to test the actual accuracy in phase synchronization between CC on the opposite sides of the collision point (another critical feature to preserve beam quality) and to test the whole concept, it is foreseen to perform an operational test with beam in the SPS around 2018.

One big concern is understanding the various failure modes of the crab cavity system, which must be studied in detail in order to allow safe operation of the machine. A CC failure will be the fastest failure mode in the LHC, with the notable exception of the dumping kickers. This point is under study and ideas on how to cope with it are under scrutiny.

Should the crab cavity turn out not to be usable for the HL-LHC, we have a mitigation scheme to avoid losing a big factor in luminosity. This scheme is based on colliding flat beams at the smallest possible crossing angle, by pushing to the extreme the compensation for the LRbb interactions by electric wire. For the moment we are studying only DC wire compensation, since we need a "moderate" action. In the case of a scheme without CC, a vigorous study is on pulsed wire which in principle should be more effective in the compensation action.

## Conclusions

The High Luminosity LHC plans to prepare the machine to reach about $3000\,\text{fb}^{-1}$ by mid 2030s. It relies on novel accelerator concepts, like the ATS optics and crab cavities, and on a breakthrough in accelerator technology, like high field superconducting magnets (beyond 10 T) and compact superconducting crab cavities. Many other technologies are involved, like advanced collimators, SC links, advanced remote handling etc. A more detailed description of the HL-LHC configuration, its operation and its technology advance can be found in the preliminary version of the Preliminary Design Report [8] and in the High Luminosity LHC book [9]. The high luminosity machine is a medium size project, evaluated in total at a cost of about 1 BCHF and implying deep changes in about 1.2 km of the present

LHC ring. Besides its physics goal, this upgrade can pave the way to a larger project like a higher energy LHC, which is based on further enhancement of the same technologies.


## Acknowledgment

The project is partially supported by the EC as *FP7 HiLumi LHC Design Study* under grant no. 284404 and by Director, Office of Science, High Energy Physics, U.S. D.O.E. under Contract DE-AC02-05CH11231 (LARP program).


## References


[1] L. Rossi and O. Brüning, High Luminosity Large Hadron Collider A description for the European Strategy Preparatory Group, CERN-ATS-2012-236 (2012).

[2] L. Rossi, LHC upgrade plans: Options and strategy, CERN-ATS-2011-257(CERN, Geneva, 2011); also in *2nd International Particle Accelerator Conference*, San Sebastian, Spain, 4–9 Sep 2011, TUYA02.

[3] European Commission FP7 — HiLumi LHC Design Study, under Grant Agreement no. 284404.

[4] The European Strategy for Particle Physics Update 2013, CERN-Council-S/106, adopted at a special session in the Brussels on 30 May 2013, http://cern.ch/council/en/EuropeanStrategy/ESParticlePhysics.html.

[5] O. Brüning *et al.* (eds.), *LHC Design Report*, Vol. I, The LHC Main Ring (CERN, Geneva, 2004).

[6] S. Myers, The Large Hadron Collider, in *Challenges and Goals for Accelerators in the XXI Century*, Chapter 22 (World Scientific, 2015).

[7] R. Garoby, The LHC injector complex upgrade: Existing challenges and upgrade plans, in *Challenges and Goals for Accelerators in the XXI Century*, Chapter 25 (World Scientific, 2015)

[8] The HiLumi LHC Collaboration, HL-LHC Preliminary Design Report (Deliverable Report D1.5), CERN-ACC-2014-0300, https://cds.cern.ch/record/1972604/files/CERN-ACC-2014-0300.pdf.

[9] O. Brüning and L. Rossi (eds.), *The High Luminosity Large Hadron Collider*, Advanced Series on Directions in High Energy Physics (World Scientific, 2015).

# Chapter 25

# The LHC injector complex upgrade: Existing challenges and upgrade plans

Roland Garoby (CERN)

## 1 Introduction

Luminosity in the LHC ($L_{\text{LHC}}$) depends upon beam characteristics and machine parameters according to the following formula:

$$L_{\text{LHC}} = \left(\frac{\gamma}{4\pi}\frac{1}{\beta^*}f_{\text{rev}}F\right)\cdot\left(n_bN_b\cdot\frac{N_b}{\varepsilon_n}\right)$$

where $\gamma$ is the usual relativistic factor, $\beta^*$ the betatron function at the Interaction Point, $f_{\text{rev}}$ the beam revolution frequency, $F$ a form factor depending upon the geometry of the bunch crossing, $n_b$ the number of bunches per ring, $N_b$ the number of protons per bunch and $\varepsilon_n$ the normalised transverse emittance of the beam (assumed round).

While the first term in this formula exclusively depends on the collider itself, the second term results from the injectors' characteristics and can only degrade in the LHC. Indeed, beam intensity $n_bN_b$ as well as beam brightness $N_b/\varepsilon_n$ can only decrease after injection.

As a typical illustration, the excellent performance of the injector complex ($1.65\cdot 10^{11}$ p/b with 50 ns bunch spacing within emittances of 1.6 mm.mrad at ejection from the SPS) has been an essential ingredient to the results obtained until 2012 in the LHC. To reach the baseline goal of the High Luminosity Upgrade of the LHC (HL-LHC [1]) ($2.3\cdot 10^{11}$ p/b with 25 ns bunch spacing within emittances of 2.1 mm.mrad at ejection from the SPS, assuming 5% beam loss and 20% blow-up in the LHC), the injected beam must be improved by a factor three in terms of $n_bN_b(N_b/\varepsilon_n)$ within 10 years.

# 2 Present LHC proton injector complex

## 2.1 *Description*

The CERN set of accelerators is sketched in Fig. 1. The LHC injector complex is composed of 6 accelerators [Linac2 (50 MeV), PSB (1.4 GeV), PS (25 GeV) and SPS (450 GeV) for protons, plus Linac3 and LEIR for other ions] which were initially commissioned with beam many years ago (in 1959 for the PS). Before the LHC start-up in 2008, these machines were subjects of a specific set of upgrades and consolidations addressing their limitations [2].

In the transverse phase planes, space charge is the main concern. The induced tune spread which is directly proportional to $1/\beta\gamma^2 \cdot (N_b/\varepsilon_n)$ represents a basic limitation for beam brightness both in the PSB and in the PS. It is brought to an acceptable level in the PSB by dividing the intensity per pulse $N_b$ by a factor two, filling the PS with 2 batches instead of a single one. This was made possible by operating the PSB on harmonic 1 with a single bunch per ring. To reduce the effect of space charge in the PS, where the first batch of bunches stays at injection energy during 1.2 s, the transfer energy from the PSB has been brought up to 1.4 GeV (1.5 times the $\beta\gamma^2$ at 1 GeV).

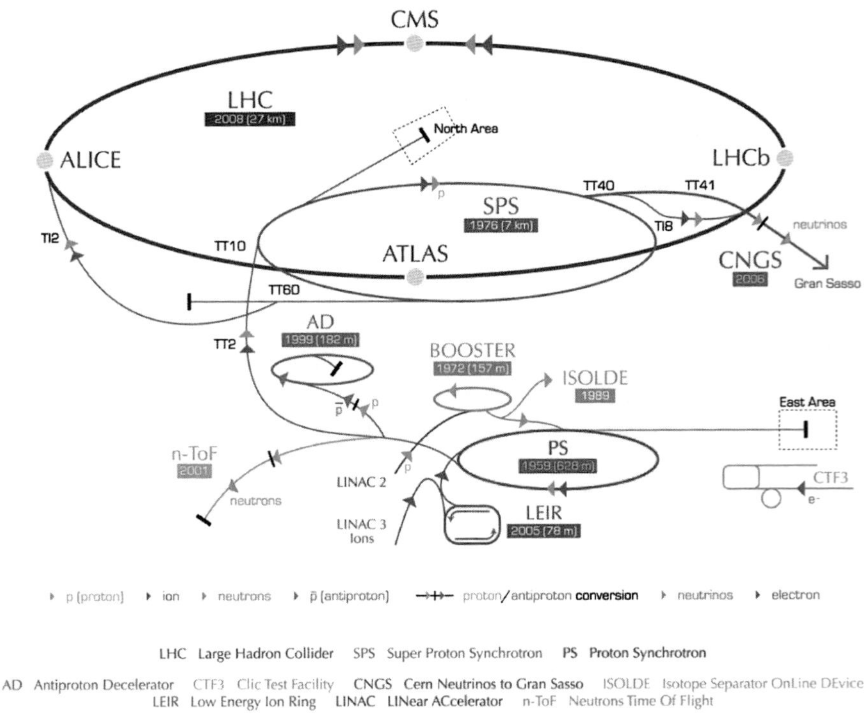


Fig. 1. CERN accelerator complex in 2012.

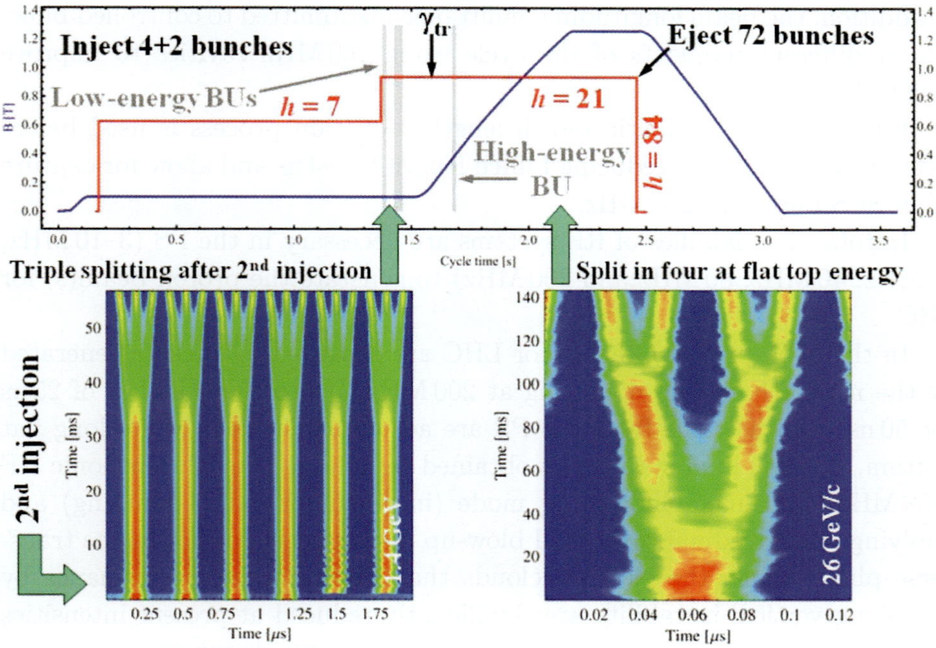


Fig. 2. Longitudinal bunch splitting to generate a 25 ns bunch train in the PS. (Top) B field (blue) and beam current (red) during a PS cycle; (Bottom) 3D displays of longitudinal density at 1.4 GeV (left) and at 25 GeV (right).

In the longitudinal phase plane, the long and intense bunches delivered by the PSB are transformed into trains of bunches spaced by 25 ns (or 50 ns) at ejection from the PS [3, 4], as sketched in Fig. 2. This is obtained with quasi-adiabatic bunch splitting gymnastics which keep the beam bunched and under control of the RF. As a result, the gap without beam corresponding to the empty bucket at injection (6 PSB bunches being sent to the PS on h = 7) is preserved and used for the rise-time of the ejection kicker, avoiding beam loss at ejection. Moreover, shorter bunch trains can be obtained simply with less bunches from the PSB.

Multiple splitting steps are used:

— Splitting in three is done at injection energy (1.4 GeV) combining the simultaneous use of three RF systems on harmonics 7, 14 and 21. At the end of the process, the beam is held on h = 21 on which it is accelerated up to top energy.
— Splitting in four takes place at 25 GeV, in two successive steps, using RF systems on h = 21 and 42 for the first step, and on h = 42 and 84 for the second one. Without this last step, bunch spacing is 50 ns.

In addition, the beam longitudinal emittance is submitted to controlled blow-ups at different moments of the cycle using 200 MHz cavities to improve longitudinal stability.

Finally, a non-adiabatic bunch length reduction process is used before ejection to the SPS for reducing bunch length to ~4 ns and allow for capture in an SPS bucket at 200 MHz.

In total, five families of RF systems are necessary in the PS (3–10 MHz, 20 MHz, 40 MHz, 80 MHz and 200 MHz) to generate the proton beam(s) for LHC.

In the SPS, proton bunches for LHC are captured in buckets generated by the main RF system operating at 200 MHz. Up to four batches of 25 ns (or 50 ns) bunch trains from the PS are accumulated on a 10.8 s long flat bottom. Longitudinal stability is obtained by adding the 4th harmonic RF (800 MHz) in Bunch Shortening mode (increasing Landau damping) and applying a longitudinal controlled blow-up during acceleration. In the transverse phase plane, the electron clouds that were limiting performance by provoking vertical instability are significantly reduced at present intensities thanks to the scrubbing of the surface of the vacuum chamber.

## 2.2 *Present performance and future needs*

The beam characteristics delivered at injection in the LHC before its first long shutdown in 2013 are summarized in the first column of Table 1. The corresponding brightness at injection in LHC is 20% higher than the "nominal" value considered in the LHC Design Report [4] for a bunch spacing of 25 ns. During this first phase of operation, 50 ns spacing has however been preferred, with approximately twice the ultimate brightness and the ultimate intensity per bunch ($\sim 1.7 \cdot 10^{11}$ p/b) at 450 GeV. In spite of transverse blow-up in the LHC (central column in Table 1), it allowed to regularly reach 75% of the nominal peak luminosity ($7.5 \cdot 10^{33}$ instead of $10^{34}$ cm$^{-2}$s$^{-1}$), mostly

Table 1. Beam characteristics in 2012.

| | 50 ns bunch trains at LHC injection | 50 ns bunch trains at start of collisions | 25 ns bunch trains at LHC injection |
|---|---|---|---|
| Number of bunches ($n_b$) | 1374 | 1374 | 2808 |
| Protons/bunch ($N_b$) | $1.65 \cdot 10^{11}$ | $1.6 \cdot 10^{11}$ | $1.1 \cdot 10^{11}$ |
| Norm. trans. emittance ($\varepsilon_n$) [mm.mrad] | 1.6 | 2.4 | 2.8 |

Table 2. Beam characteristics for the High Luminosity LHC project.

| | 25 ns bunch trains at of collisions | 25 ns bunch trains at injection (estimate) | 50 ns* bunch trains at start of collisions | 50 ns* bunch trains at injection (estimate) |
|---|---|---|---|---|
| Number of bunches ($n_b$) | 2808 | 2808 | 1404 | 1404 |
| Protons/bunch ($N_b$) | $2.2 \cdot 10^{11}$ | $2.3 \cdot 10^{11}$ | $3.5 \cdot 10^{11}$ | $3.7 \cdot 10^{11}$ |
| Norm. trans. emittance ($\varepsilon_n$) [mm.mrad] | 2.5 | 2.1 | 3 | 2.5 |

*The 50 ns scenario is a back-up, in case fundamental limitations in LHC (e.g. due to electron clouds or total intensity) are encountered with the 25 ns baseline parameters.

compensating the effect of the larger physical emittance due to the reduced beam energy (4 instead of 7 TeV).

With these beam characteristics, the injector complex is performing as foreseen but without any margin. For the High Luminosity LHC (HL-LHC) project, which aims at accumulating ~250 fb$^{-1}$/year, beam characteristics in collision have to progress to the level described in Table 2. Assuming 20% emittance blow-up and 5% beam loss between injection and collision in LHC [1], the beam intensity required from the injectors (Table 2) has to double in the baseline case (25 ns) and the brightness shall almost triple. The presently identified limitations in the injectors are illustrated in Fig. 3 together with the achieved and expected beam performances.

In the coordinate system emittance versus intensity, a constant space-charge induced tune spread is represented by a straight line passing through

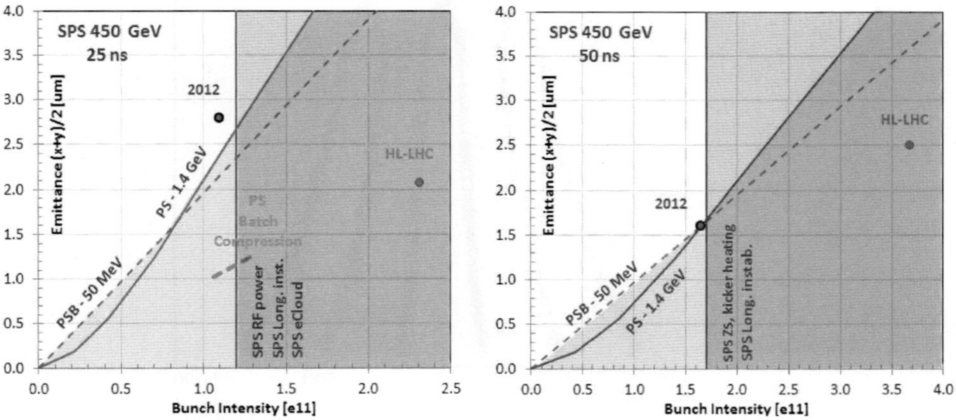


Fig. 3. Performance and limitations at SPS ejection, in 2012, of the LHC proton injector complex for 25 ns (left) and 50 ns (right) bunch spacing.

the origin. Below that line the space-charge tune spread is excessive. The curve corresponding to the PS is not a straight line because it takes into account the energy spread assuming a constant longitudinal emittance. The other limitations restrict the maximum intensity per bunch, which corresponds to a vertical line parallel to the y-axis. For 25 ns bunch spacing, $1.2 \cdot 10^{11}$ p/b is the maximum intensity in the SPS because of the available RF power and because of longitudinal coupled bunch instabilities. The limit due to electron clouds is nowadays beyond this intensity. For 50 ns, the main limitations result from heat dissipated in the equipment because of the beam image current and from longitudinal instabilities ($N_b < 1.7 \cdot 10^{11}$ p/b).

Before the implementation of the upgrades described in the following part of this document, new sophisticated beam gymnastics have been proposed for generating in the PS 25 ns batches with a brightness similar to 50 ns [5]. The principle is to split the PSB beam in less bunches while keeping spacing at 25 ns. For that purpose, the batch of PSB bunches that fills most of the PS circumference at injection is first accelerated to an intermediate energy (typically 2.5 GeV) where space-charge is sufficiently reduced and then compressed into a smaller fraction of the circumference. A typical scenario is illustrated in Fig. 4: (i) beam is injected and captured in the PS as shown in Fig. 2 (bottom left), (ii) after acceleration up to an intermediate energy it is compressed in a 50% smaller fraction of the circumference by adiabatically increasing the harmonic number holding the beam from h = 7 to h = 14,

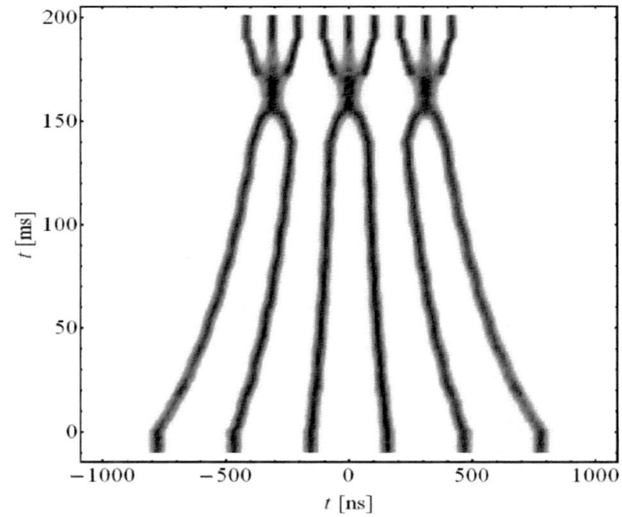


Fig. 4. 3D displays (simulated) of longitudinal density in the PS during batch compression.

(iii) bunches are merged two by two which results in three bunches on h = 7, and finally (iv) triple splitting is applied, generating nine bunches on h = 21. Beam is then accelerated up to high energy where splitting in four is taking place like nowadays (bottom right of Fig. 2). Compared to the present process, the PS will then provide only 36 bunches with 25 ns spacing, instead of 72, increasing the filling time of the LHC and decreasing slightly the filling factor in the collider because of the gaps required for kickers' rise time in the SPS and LHC.

The corresponding beam characteristics at LHC injection, shown as a dashed green line in Fig. 3 (left), can potentially increase luminosity by a factor ~1.4 with respect to the 50 ns scheme while reducing the number of events per crossing and hence easing operation of the detectors in the experiments. The interest of the scheme will however depend upon the LHC capability to preserve the small transverse emittances.

## 3 Upgrade plans for the LHC proton injector complex

### 3.1 *Transverse phase planes*

The primary limitation due to the space-charge induced tune spread in the PSB and in the PS will be addressed by increasing the minimum energy. In the case of the PSB, a new linac is in construction (Linac4 [6]) which will bring injection energy up to 160 MeV kinetic, doubling $\beta\gamma^2$ with respect to the present 50 MeV Linac2. The main parameters of Linac4 are summarized in Table 3, and its layout and connection to the PSB [7] are shown in Fig. 5. For the same $|\Delta Q_y|$ as today (0.44), the higher injection energy is expected to allow for a brightness of $1.8 \cdot 10^{12}$ p/mm.mrad in the PSB, twice the present level, corresponding to a potential brightness of $10^{11}$ p/mm.mrad for 25 ns bunch spacing at ejection from the SPS (respectively $2 \cdot 10^{11}$ p/mm.mrad for

Table 3. Linac4 beam parameters.

| | |
|---|---|
| Ion species | $H^-$ |
| Output energy | 160 MeV |
| Bunch frequency | 352.2 MHz |
| Maximum repetition rate | 2 Hz |
| Beam pulse length | 400 $\mu$s |
| Chopping scheme | 222/133 transmitted bunches/empty buckets |
| Mean pulse current | 40 mA |
| Maximum number of particles per pulse | $1.0 \cdot 10^{14}$ |
| Number of particles per bunch | $1.14 \cdot 10^{9}$ |
| Transverse emittance | $0.4\pi$ mm.mrad (rms) |

Fig. 5. Layout of Linac4 and connection to the PSB.

50 ns). Charge exchange injection will replace multi-turn betatron stacking, increasing the efficiency up to ~98% and providing the means to tailor the transverse distribution of protons circulating in the PSB. Painting is also foreseen in the longitudinal phase plane, to maximize capture efficiency and to optimize the longitudinal particle distribution.

In the case of the PS, the beam transfer energy will be increased from 1.4 to 2 GeV kinetic, increasing $\beta\gamma^2$ and decreasing the space-charge tune spread by a factor ~1.6. This energy is attainable in the PSB [7], provided that a number of equipments are upgraded or redesigned, like the power supply for the main dipoles. Likewise, in the PS, important modifications and new equipment must be added for beam injection at 2 GeV [8] and the existing transverse damper will be renovated to avoid transverse instabilities, providing more flexibility in the choice of the tunes at low energy and hopefully stabilizing the beam on the high energy flat top.

Beyond these major changes, an extensive campaign is in progress for optimizing the transverse tunes and improving the compensation of resonances [9]. As a result, operation with larger vertical tune spreads than today is foreseen to be manageable in all synchrotrons, and especially in the PS.

In the SPS, the tunes have recently been changed from an integer part of 26 ("Q26") to an integer part of 20 ("Q20"), reducing the transition energy and enhancing the slip factor $|\eta| = |1/\gamma_t^2 - 1/\gamma^2|$ to increase the thresholds of longitudinal and Transverse Mode Coupling Instabilities (TMCI) [10, 11]. With this optics, operation with a space charge tune shift in excess of 0.15 is expected to be manageable, corresponding to a brightness of $\sim 10^{11}$ p/mm.mrad at SPS ejection, matched to the capability of the upgraded PSB for 25 ns bunch spacing.

### 3.2 *Longitudinal phase plane*

The PSB is not expected to suffer from limitations in the longitudinal phase plane when providing the high brightness beams for LHC. A major renovation of the main RF systems is however required to guarantee a reliable operation during the full lifetime of the LHC and to let other users [e.g. ISOLDE] benefit from the higher intensity beams allowed with Linac4.

In the PS, the measures presently used to stabilize the beam in the longitudinal phase plane (controlled longitudinal blow-up and coupled bunch instability damper) cannot handle a bunch intensity larger than $\sim 1.7 \cdot 10^{11}$ p/bunch, both for 25 and 50 ns bunch spacing. This limitation will be addressed by a new longitudinal damper using a dedicated "broad band" cavity, aimed at bringing the instability threshold beyond $3 \cdot 10^{11}$ p/bunch. Moreover, transient beam loading in the five families of RF systems will increase with beam intensity, degrading the quality of the multiple beam gymnastics. Fast RF feedback systems will therefore be upgraded on all high power RF systems, and one-turn delay feedbacks will be renovated on the 3–10 MHz ferrite cavities and implemented on the other cavities. More RF voltage at 40 MHz will be installed to improve longitudinal capture efficiency in the SPS [12]. The combined effect of all these actions is expected to allow for the operational availability of $3 \cdot 10^{11}$ p/bunch at PS ejection.

In the SPS, two new 1.6 MW RF power plants will be installed, doubling the available power at 200 MHz, and the cavities will be reorganized into six assemblies (four today), reducing the beam impedance. This will allow the acceleration of a beam current of 3 A, and the availability of 10 MV on the high energy flat top, before ejection. Up to $2.3 \cdot 10^{11}$ p/bunch with 25 ns bunch spacing could then be transferred to the LHC. Longitudinal stability of the beam in the SPS is presently obtained through the combined effects of controlled longitudinal blow-up up to 0.6 eVs and 800 MHz RF voltage used in bunch shortening mode. The instability threshold will increase with the new Q20 optics thanks to the increased slip factor $|\eta|$, although this will be balanced by the smaller longitudinal emittance imposed by the reduced acceptance of the buckets. The lower impedance of the 200 MHz RF system after its reorganization will also be beneficial, as well as the planned renovation of the low and high power equipment of the 800 MHz system. The present estimate is that $2.3 \cdot 10^{11}$ p/bunch with 25 ns bunch spacing and $3.5 \cdot 10^{11}$ p/bunch with 50 ns should be attainable. Such intensities might require transferring longer bunches (1.6–1.8 ns) to the LHC where mitigation measures have to be studied [13].

### 3.3 *Electron cloud*

Electron cloud formation is observed in the PS on the 25 ns beam a few milliseconds before ejection and a transverse instability has repeatedly been diagnosed at the same time. Although not presently affecting performance, it is a subject of theoretical and experimental investigation to determine the risk with the future beam characteristics and to prepare cures or mitigation measures.

In the SPS electron clouds have been a major concern as soon as an LHC-like beam has been injected [14]. They trigger vacuum pressure rises, instabilities, beam losses and transverse emittance blow-up. As a result of the major effort invested in modelling/simulation and in experimental tests, cures and mitigation measures have been developed. Scrubbing is showing an interesting potential, as demonstrated by the continuous improvement of the SPS since the beginning of operation for LHC. It suffers however from degradation whenever the vacuum chambers are exposed to atmosphere and the minimum obtainable Secondary Electron Yield (SEY) is limited, depending upon the nature and cleanliness of the vacuum chamber. Coating of the vacuum chamber with a low SEY material would be a perfect cure, completely avoiding the appearance of electron clouds. Amorphous carbon is especially efficient in that respect and adequate coating processes for the SPS vacuum chambers have been developed and experimentally demonstrated. The use of clearing electrodes has also been considered, but no satisfying engineering solution has been found which would not reduce the available aperture. In any case, getting rid of the electron cloud limitations in the SPS is considered as feasible, either with scrubbing or with amorphous carbon coating [15].

### 3.4 *Other upgrades*

The equipment in all accelerators must match the increased level of performance and be capable to operate reliably:

— New beam instrumentation has to be developed for measuring with adequate accuracy beams of reduced size and high brightness and intensity. The capability to detect and quantify the intensity in "spurious" bunches in the PS and in the SPS is an important and challenging need.
— New beam intercepting and protection devices have to be built to withstand impact from the higher brightness/higher intensity beam. That concerns beam dumps in all machines, as well as the SPS scraper system

for halo shaping and the devices in the SPS to LHC transfer lines protecting the LHC.

— A number of power supplies need to be replaced because of aging and/or because of more demanding specifications.

— Civil engineering and building construction are also necessary for radiation shielding (PS injection and ejection sectors) and to host new large size equipment (PSB new main power supply and new SPS high power RF amplifiers).

Very expensive items like the main dipoles are not planned to be changed, but their ageing will be carefully monitored and spares have to be available.

## 4 Estimated performance of the upgraded LHC proton injector complex

The performance reach of the LHC proton injector complex after the improvements described in the previous section are graphically represented in Fig. 6. Compared e.g. to the present situation with 25 ns (Fig. 3), the intensity per bunch is 70% higher and brightness is more then doubled.

The baseline option preferred by the LHC experiments is 25 ns bunch spacing. It is also preferable for the injectors because the beam characteristics expected by the HL-LHC project (yellow dot) are approximately compatible with all identified limitations, except with the SPS one at $2 \cdot 10^{11}$ p/bunch due to beam loading and longitudinal instabilities.

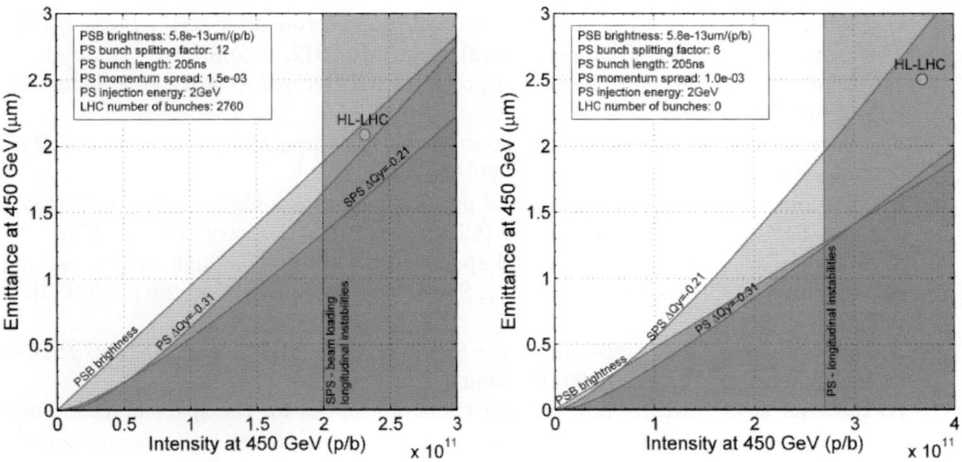


Fig. 6. Performance and limitations at SPS ejection of the upgraded LHC proton injector complex for 25 ns (left) and 50 ns (right) bunch spacing.

As a spare solution, in case the 25 ns beam cannot be used in the LHC (e.g. because of electron cloud or beam intensity), 50 ns bunch spacing could be considered. More limitations would then have to be faced in the injectors:

— in the PS, mainly because of longitudinal instability, with an estimated limit of $2.7 \cdot 10^{11}$ p/bunch, while the HL-LHC specification is at $3.5 \cdot 10^{11}$ p/bunch.
— in the SPS, because of longitudinal instability and because of space charge (the tune spread will reach 0.22 on the injection flat porch).

The performances shown in Fig. 6 are however only estimates which are likely to be regularly revised during the 10 years duration of the injectors' upgrade programme. As past experience with the CERN accelerators has shown, it is not unreasonable to hope that, as a result of the intense effort invested both in theory and in beam experiments [16], beam characteristics will finally exceed the present expectation and meet the present HL-LHC requirements. The possibility cannot be discarded either that the HL-LHC beam specifications will evolve as experience with the collider progresses.

## References

[1] O.S. Bruning (2012), "HL-LHC parameter space and scenarios", *Proceedings of Chamonix 2012 workshop on LHC Performance*, CERN-2012-006, p. 315.
[2] F. Blas *et al.* (1997), "Conversion of the PS complex as LHC proton pre-injector", CERN-PS-DI-97-048-DI, http://cdsweb.cern.ch/record/328735/files/ps-97-048.pdf.
[3] R. Garoby (1999), "Status of the nominal proton beam for LHC in the PS", CERN-PS-99-013-RF, http://cdsweb.cern.ch/record/382971/files/ps-99-013.pdf.
[4] LHC Design Report (2004), v.3: The LHC Injection Chain, pp. 5–8 and 45–52, CERN-2004-003-V-3.
[5] H. Damerau (2012), "Performance potential of the injectors after LS1", *Proc. of LHC Performance Workshop* (Chamonix 2012), February 2012, p. 268.
[6] M. Vretenar *et al.* (2011), "The Linac4 project at CERN", *IPAC2011*, San Sebastian, Sept. 2011, TUOAA03, p. 900.
[7] K. Hanke *et al.* (2010), "PS booster energy upgrade, feasibility study", https://edms.cern.ch/document/1082646/3.
[8] S. Gilardoni *et al.* (2011), "PS potential performance with a higher injection energy", *Proc. of LHC Performance Workshop* (Chamonix 2011), January 2011, p. 349.
[9] S. Gilardoni *et al.* (2011), "Tune and space charge studies for high brightness and high intensity at CERN PS", *IPAC2011*, San Sebastian, Spain, Sept. 2011, MOPS014, p. 625.
[10] E. Shaposhnikova (2011), "Lessons from SPS studies in 2010", *Proc. of LHC Performance Workshop* (Chamonix 2011), January 2011, p. 359.
[11] H. Bartosik *et al.* (2012), "Low gamma transition optics for the sps: simulation and experimental results for high brightness beams", *Proc. of HB2012 Workshop*, Beijing, China, Sept. 2012, WEO1B01.
[12] H. Timko *et al.* (2012), "Longitudinal beam loss studies of the CERN PS to SPS transfer", *Proc. of HB2012 workshop*, Beijing, China, Sept. 2012, WEO1C03.

[13] L. Drosdal *et al.* (2011), CERN-ATS-Note-2011-063 MD.
[14] G. Rumolo *et al.* (2007), *Proc. of CARE-HH-APD Beam'07 workshop*, CERN 2007.
[15] J.M. Jimenez (2012), "SPS: Scrubbing or coating?", *Proc. of LHC Performance Workshop* (Chamonix 2012), February 2012, p. 339.
[16] G. Rumolo *et al.* (2012), "Summary of the LIU beam studies review", CERN-ATS-Note-2012-083 PERF.

# Chapter 26

# HE-LHC and FCC: The quest for increasing the energy beyond the LHC

Frank Zimmermann (CERN)

In the 21st century circular proton–proton (pp) or proton–antiproton (p$\bar{\text{p}}$) colliders are the main, and possibly the only, experimental tools available for exploring particle physics in the energy range of tens of TeV.

The bending radius $\rho$ of a relativistic particle of charge $e$ and momentum $p$ in a magnetic field of strength $B$ is given by $p = eB\rho$. Accordingly there are only two approaches for raising the energy of pp (or p$\bar{\text{p}}$) collisions beyond those at the LHC, namely increase the magnetic field of the bending magnets above the LHC's 8.33 T or increase the ring circumference and, thereby, the radius $\rho$.

The magnets of the present LHC are made from Nb-Ti superconductor, which supports a maximum field of about 10 T. Figure 1 illustrates the existence of several other superconductors (SCs) with large critical currents up to significantly higher magnetic fields. $Nb_3Sn$ superconductor can reach a magnetic field of 16 T or higher. The production of $Nb_3Sn$ cables is well advanced, and the installation of a few $Nb_3Sn$ magnets is planned for the HL-LHC, which will represent an important milestone towards a higher-energy machine. *High temperature superconductor* (*HTS*) materials like the bismuth copper oxide BSCCO, in the form of Bi-2212, or yttrium copper oxide YBCO, in the form of Y-123, may withstand even much higher fields of up to 45 T; other materials of interest for constructing future affordable high-field magnets are the conventional SC $MgB_2$, discovered in 2001, and iron-based SCs, discovered in 2006. R&D on *HTS* wires is ongoing, but there is still a long road ahead for *HTS* high-field magnets; challenges include mechanical weakness, magnetization, AC losses, quench detection and protection techniques, unit length and cost. R&D on high-field SC magnets was pushed forward by earlier studies for a Very Large Hadron Collider (VLHC) [1] in the United States and by the international fusion project ITER. In Japan,

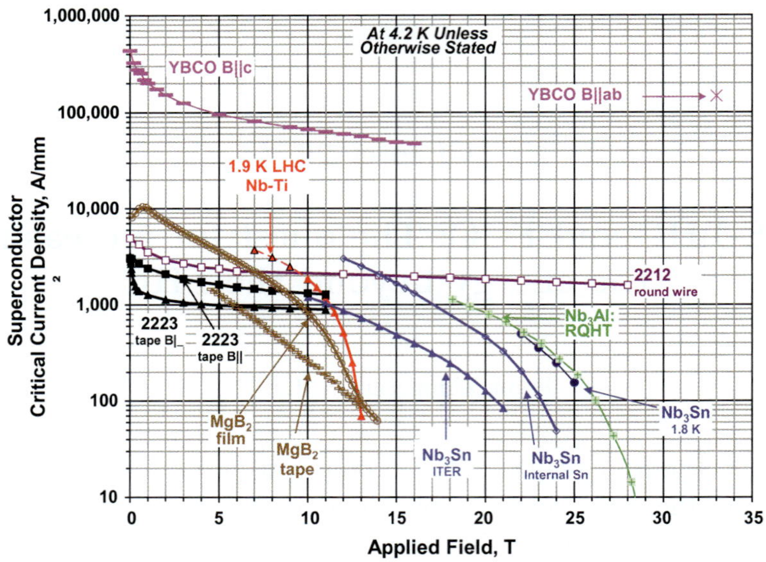


Fig. 1. Critical current density for SC cables as a function of applied field [3].

efforts are underway to develop a 13-T demonstration magnet based on another superconductor, $Nb_3Al$, which could be an alternative to $Nb_3Sn$. Most ambitiously, an ongoing EC-funded effort is directed at building and testing an *HTS* dipole insert coil for a 13-T dipole background magnet aiming at a field increase of about 6 T (i.e. targeting a total field of up to 19 T) [2].

The High-Energy LHC (HE-LHC) is a proposed new accelerator in the existing 27-km LHC tunnel [4, 5]. Utilizing new dipole magnets with a nominal field up to 20 T (2.5 times the LHC field), it can provide pp collisions with a CM energy of 33.5 TeV, assuming the same main-bend filling factor as in the present LHC (about 66% of the circumference). The HE-LHC magnets are based on an optimized hybrid-coil design, comprising blocks made from Nb-Ti, two types of $Nb_3Sn$, and *HTS*, respectively, which minimizes the cost while maximizing the performance. Figure 2 shows the block layout of such a dipole. Without the *HTS* the maximum field would be only about 15 T. The ratio between peak field and central field is 1.03. The coil full aperture is 40 mm, and the beam half aperture, including margins for vacuum tube and beam screen, equal to 13 mm. Concerning a 3D magnet design, the rectangular block layout would favor a solution based on flat racetrack coils with flared ends.

In the LHC, the coil stress due to electromagnetic forces is of the order of 70 MPa. Going to 20 T with the same current density brings stresses to

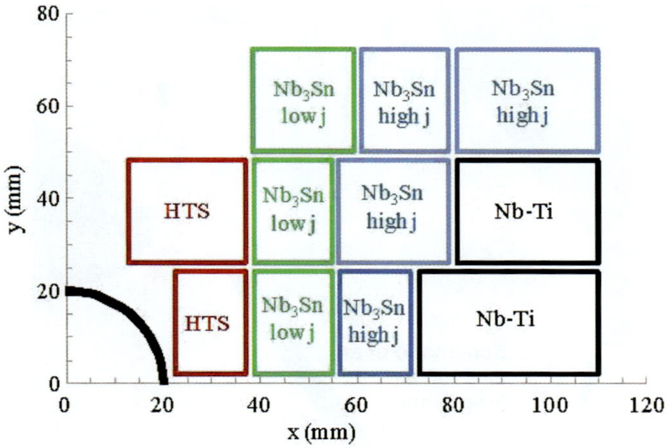


Fig. 2. Block layout of the 20-T hybrid Nb-Ti/$Nb_3Sn$/*HTS* dipole-magnet coil design for the HE-LHC [6]. Only one quarter of one aperture is shown.

the 150–200 MPa level, above which considerable degradation of the $Nb_3Sn$ material may start to occur.

In the HE-LHC magnets a large amount of electromagnetic energy is stored, about 25 GJ in total. For safe energy extraction the HE-LHC might need to be sectorized into 32 parts with separate powering, in order to keep the energy per sector below 1 GJ, to be compared with 8 sectors for the present LHC.

The Future Circular Collider (FCC, or, more precisely FCC-hh — formerly VHE-LHC) [7, 8] is an alternative proposal based on a larger ring, which would allow exploring much higher beam energies still, up to 100 TeV in the centre-of-mass frame. This FCC would be installed in a new tunnel of about 100-km circumference, which could also accommodate a high-luminosity circular $e^+e^-$ collider (FCC-ee), which would serve as Higgs, Z, W and top factory. Specifically, the larger tunnel would enable pp collisions of about 50 TeV CM with the present 8.3-T LHC magnets, of 100 TeV CM with 16-T magnets, and of 130 TeV with 20-T HE-LHC type magnets. Preliminary geological and environmental studies for identifying the best locations in the Geneva area for such a large infrastructure have been performed [8,9]. Figure 3 presents one of the possible locations.

In Table 1, key parameters for HE-LHC and FCC-hh are compiled and compared with those of the LHC and its high-luminosity upgrade (HL-LHC).

For the HE-LHC and FCC-hh, tentatively an initial luminosity value of $5 \times 10^{34}$ cm$^{-2}$s$^{-1}$ has been chosen, i.e., the same value as for the HL-LHC. At the higher energy this luminosity is easily attained if the interaction

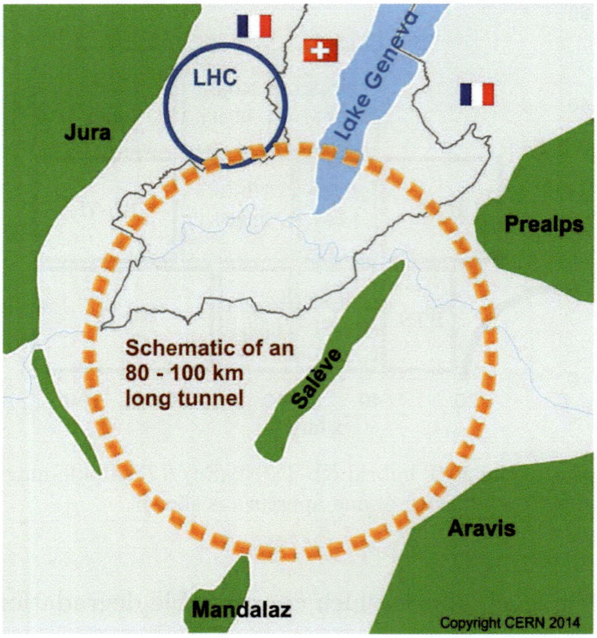


Fig. 3. Schematic of an 80–100 km tunnel infrastructure in the Geneva basin together with the existing LHC tunnel [7, 8].

regions are adequately shielded against the more energetic luminosity debris. In Table 1, the bunch spacing has been kept at the nominal LHC design value of 25 ns, taking into account event pile up, electron cloud and machine protection. However, for HE-LHC and FCC-hh much shorter bunch spacings, e.g. 5 ns or 2.5 ns, can also be considered. Thanks to the strong radiation damping, at these bunch spacings even higher luminosities could be delivered than for 25 ns, with a reduced event pile up in the experimental detectors. For bunch spacings below 5 ns the electron-cloud build up becomes more benign. Additional electron-cloud mitigation measures (coatings and clearing electrodes) are also being considered.

Conservatively, for FCC-hh and HE-LHC a total beam–beam tune shift of 0.01 has been adopted, just as for the LHC design. However, more than three times higher values have already been obtained at the actual LHC without any harmful consequences for beam lifetime, luminosity lifetime, or beam emittance. In order to maintain Landau damping and to keep the bunch length comparable to the one of the LHC, for HE-LHC and FCC-hh the longitudinal emittance ($4\pi\sigma_t\sigma_E$) is increased from 2.5 eVs to about 3.8 eVs and 9 eVs, respectively. The crossing angle is chosen to provide a separation of $12\sigma$ at the parasitic encounters closest to the IP, which is higher

Table 1. Parameters of LHC, HL-LHC, HE-LHC, and FCC-hh.

| Parameter | LHC | HL-LHC | HE-LHC | FCC-hh |
|---|---|---|---|---|
| CM energy [TeV] | 14 | 14 | 33 | 100 |
| circumference $C$ [km] | 26.7 | 26.7 | 26.7 | 100 |
| dipole field [T] | 8.33 | 8.33 | 20 | 16 |
| dipole coil aperture [mm] | 56 | 56 | 40 | 40 |
| beam half aperture [cm] | $\sim 2$ | $\sim 2$ | 1.3 | 1.3 |
| injection energy [TeV] | 0.45 | 0.45 | >1.0 | >3.0 |
| no. of bunches $n_b$ | 2808 | 2808 | 2808 | 10600 |
| bunch population $N_b$ [$10^{11}$] | 1.15 | 2.2 | 0.94 | 1.0 |
| init. transv. norm. emit. [$\mu$m] | 3.75 | 2.5 | 1.38 | 2.2 |
| initial longitudinal emit. [eVs] | 2.5 | 2.5 | 3.8 | 9 |
| no. IPs contributing to tune shift | 3 | 2 | 2 | 2 |
| max. total beam–beam tune shift | 0.01 | 0.015 | 0.01 | 0.01 |
| beam circulating current [A] | 0.584 | 1.12 | 0.478 | 0.5 |
| rms bunch length [cm] | 7.55 | 7.55 | 7.55 | 8 |
| IP beta function [m] | 0.55 | 0.15 (min.) | 0.35 | 1.1 |
| rms IP spot size [$\mu$m] | 16.7 | 7.1 (min.) | 5.2 | 6.8 |
| full crossing angle [$\mu$rad] | 285 | 590 | 185 | 74 |
| stored beam energy [MJ] | 362 | 694 | 701 | 8400 |
| SR power per ring [kW] | 3.6 | 7.3 | 96.2 | 2400 |
| arc SR heat load [W/m/aperture] | 0.17 | 0.33 | 4.35 | 28.4 |
| energy loss per turn [keV] | 6.7 | 6.7 | 201 | 4600 |
| critical photon energy [eV] | 44 | 44 | 575 | 4300 |
| photon flux [$10^{17}$/m/s] | 1.0 | 2.0 | 1.9 | 1.8 |
| longit. SR emit. damping time [h] | 12.9 | 12.9 | 1.0 | 0.54 |
| horiz. SR emit. damping time [h] | 25.8 | 25.8 | 2.0 | 1.08 |
| init. longit. IBS emit. rise time [h] | 57 | 23.3 | 40 | 290 |
| init. horiz. IBS emit. rise time [h] | 103 | 10.4 | 20 | 270 |
| peak events per crossing | 27 | 135 (lev.) | 147 | 171 |
| total/inelastic cross section [mb] | 111/85 | | 129/93 | 153/108 |
| peak luminosity [$10^{34}$ cm$^{-2}$s$^{-1}$] | 1.0 | 5.0 | 5.0 | 5.0 |
| beam lifetime due to burn off [h] | 45 | 15.4 | 5.7 | 19.1 |
| optimum run time [h] | 15.2 | 10.2 | 5.8 | 12.1 |
| opt. av. int. luminosity/day [fb$^{-1}$] | 0.47 | 2.8 | 1.4 | 2.2 |

than the $9.5\sigma$ separation of the nominal LHC, and ensures that long-range beam–beam effects are negligible.

The remaining parameters have been determined so as to comply with all the aforementioned conditions. Both flat and round beam options are considered. A quasi-continuous controlled longitudinal blow up during the store is assumed, maintaining a constant bunch length in the presence of strong radiation damping. For both HE-LHC and VHE-LHC the intrabeam scattering rise times are much longer than the radiation damping times. The fast shrinkage of the transverse emittances under the influence of the radiation damping would lead to excessive beam–beam tune shifts. Therefore, noise excitation must also be applied in the transverse plane, so that the emittances shrink more gradually, following the reduction in intensity due to consumption in collision, so as to keep the horizontal and vertical tune shifts the same and at their initial moderate value. For the optimum duration of a physics store, no equilibrium between synchrotron radiation and intrabeam scattering will ever be reached, and, therefore, noise injection is required throughout the store.

The time evolution of luminosity and beam parameters in collision is completely determined by the above assumptions [10,11]. Namely, considering round Gaussian beams ($\epsilon_x = \epsilon_y \equiv \epsilon_N/\gamma$, $\beta_x^* = \beta_y^* \equiv \beta^*$, $\sigma_x^* = \sigma_y^* \equiv \sigma^*$), colliding in two interaction points (IPs) at a crossing angle $\theta_c$ with alternating planes of crossing (horizontally in IP and vertically in the other), the luminosity is [12]

$$L \approx \frac{f_{\rm rev} n_b N_b^2 \gamma}{4\pi\beta^* \epsilon_N} \frac{1}{\sqrt{1+\left(\frac{\theta_c \sigma_z}{2\sigma^*}\right)^2}} = \frac{f_{\rm rev}}{2r_p} \frac{n_b N_b}{\beta^*} \Delta Q_{\rm tot} \gamma \,, \tag{1}$$

where $\Delta Q_{\rm tot}$ denotes the total beam–beam tune shift (sum over the two IPs), and we have assumed that the "hourglass effect" can be neglected, i.e. $\beta^* \gg \sigma_z$. At constant $\Delta Q_{\rm tot}$ the luminosity increases linearly with the total beam intensity ($n_b N_b$) and with the beam energy ($\gamma$), and it scales with the inverse of $\beta^*$. Since the emittances are continually adjusted — by appropriate noise excitation — so as to keep $\Delta Q_{\rm tot}$ constant, the intensity decreases exponentially and so does the luminosity. The latter varies as $L = L_0 \exp(-t/\tau_L)$ with an exponential luminosity decay time

$$\frac{1}{\tau_L} = \sigma_{\rm tot} n_{\rm IP} \frac{f_{\rm rev}}{r_p \beta^*} \Delta Q_{\rm tot} \gamma \,, \tag{2}$$

where $\sigma_{\rm tot}$ denotes the total pp collision cross section (about 100–150 mbarn) and $n_{\rm IP}$ the number of interaction points (assumed to be 2). Introducing the average time interval between successive collisions runs $t_{\rm ta}$, also called "turnaround time", and the mean run time $T_{\rm run}$, the average luminosity becomes

$$< L >_{\rm ave} = L_0 \tau_L \frac{1 - e^{-T_{\rm run}/\tau_L}}{T_{\rm run} + t_{\rm ta}} \,. \tag{3}$$

The maximum average luminosity is reached at the optimum run time $T_{\rm run,opt}$, satisfying the implicit equation

$$T_{\rm run,opt} = \tau_L \ln\left(1 + \frac{T_{\rm run,opt} + t_{\rm ta}}{\tau_L}\right). \tag{4}$$

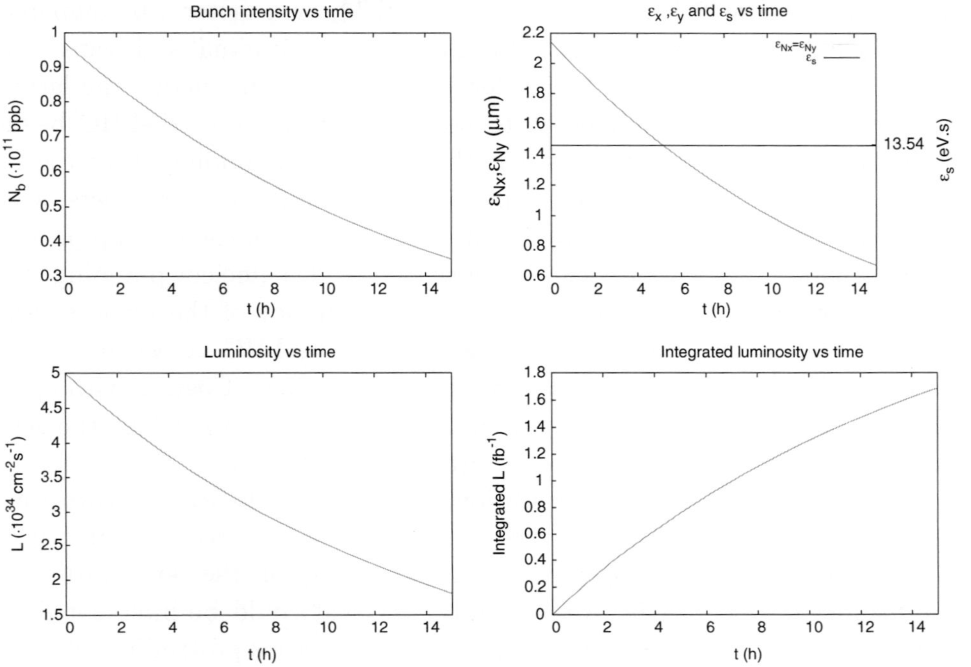


Fig. 4. Expected time evolution during an FCC-hh physics store at 100 TeV CM: bunch intensity (top left), transverse and longitudinal emittances (top right), instantaneous luminosity (bottom left), and integrated luminosity (bottom right) as a function of time spent in collision. In this example round beams are considered, and controlled noise excitation is assumed in both the transverse and longitudinal planes such as to maintain a constant total beam–beam tune shift of 0.01 (sum of two interaction points) and a constant bunch length of 7.7 cm, respectively. For computing the beam lifetime, a total pp cross section of 153 mbarn is assumed.

Figure 4 shows an example time evolution of various parameters during round-beam collisions for the FCC-hh (VHE-LHC). In this example, for a turnaround of 5 hours, the optimum run time is about 11 hours.

One important novel feature of highest-energy proton storage rings is the high synchrotron radiation power of close to 100 and 2400 kW for HE-LHC and FCC-hh, respectively, to be contrasted with a few kW at the LHC or HL-LHC. This power translates into a heat load per meter and aperture of 4.4 W/m for HE-LHC or 28 W/m for FCC-hh.

The HE-LHC heat load could be absorbed on a beam screen (BS) inside the cold magnets, as for the present LHC, but at a higher BS temperature of 40–60 K instead of 4.6–20 K [13]. Under this assumption, the existing LHC cryoplants would provide approximately the required cooling capacity. Compared with the LHC, the resistivity $\rho$ of the Cu-coated beam screen is enhanced due to its higher temperature (about a factor 4–8 in resistivity for RRR values of 100–200) and due to the larger magnetoresistance (at

20-T field a factor 2.5 compared with zero field, or a factor 1.6 compared with the 8.33-T field of the LHC) [14]. Since the longitudinal and transverse impedances scale as $\sqrt{\rho}$, and considering the high beam energy, the larger copper resistivity is not expected to be a problem for the HE-LHC beam stability. Vacuum considerations, e.g. the proper cryopumping of hydrogen, require an operating temperature below 2–3 K for the cold magnets, surrounding the beam screen. With the smaller aperture of the beam pipe and a larger photon desorption yield, in order to provide sufficient pumping and to avoid pressure instabilities, either the transparency of the beam screen must be significantly increased compared with the LHC or warm photon absorbers be introduced in the magnet interconnections. Coating with open metallic foams has been suggested as a possible means to reduce the secondary emission yield and to improve the vacum stability [15].

Also for the FCC-hh the absorption of the much enhanced synchrotron-radiation power could be performed at a higher beam-screen temperature, e.g., at 50 or 100 K, the optimum value depending on the temperature of the magnet cold bore [16]. Alternatively, the heat could be intercepted at room temperature, e.g. by using dedicated photon stops protruding into the beam tube at the end of each (short) dipole magnet. Such photon stops are routinely used in storage-ring light sources. They were also being considered in the design of the VLHC [17], for which a photon-stop cryo-experiment demonstrated the concept. The photon stops would allow removing the beam screen from inside the magnets, gaining precious aperture. In addition, a 100-V bias voltage applied to the photon stops would suppress electron emission and electron-cloud formation. The total longitudinal and transverse impedance is expected to remain acceptable despite the large number of photon stops [17]. A small fraction of photons reemitted by X-ray fluorescence and escaping from the stops can be minimized by appropriate surface coating [17].

Collimation is a critical subsystem of the present LHC. Some collimators are the elements closest to the beam and, therefore, are hit first by parts of the beam in case of a failure, e.g. an accidentally triggered asynchronous beam dump. To avoid destruction, the primary and secondary LHC collimators are fabricated from extremely robust fibre-reinforced carbon (CFC). Since for HE-LHC and FCC-hh the energy density of the beam is much higher than for the LHC, even more robust composite materials may need to be considered. A key function of the LHC collimation system is to minimize beam loss in the cold regions of the machine and thereby to support a poor beam lifetime without quenching any superconducting magnet. The performance of the system is characterized by its "cleaning efficiency," which can be both simulated and measured. Nuclear processes during the scattering of primary beam particles inside a collimator jaw depend on the beam energy. The cross section of single-diffractive scattering increases and, as a result the collimation cleaning efficiency decreases with energy. Higher energy also implies smaller beam sizes, and full collimator gaps of 1 mm or less, at HE-LHC and FCC-hh, which calls for higher precision in collimator control, setup and reproducibility. Lastly, the warm magnets installed in the present LHC collimator insertions are close to the technological limit. A higher beam energy may require stronger elements, perhaps shielded superconducting magnets.

To relax the aperture requirements at injection and to reduce the energy swing to a value similar to the present LHC, the injection energy of the HE-LHC should be above 1 TeV, and for the FCC-hh above 3 TeV. Several options are being considered. For the HE-LHC the 6.9-km SPS could be replaced by a new machine with fast ramping SC magnets reaching a peak field of 4.5 T, similar to magnets developed for the SIS300 synchrotron of the FAIR project at GSI. The FCC-hh could conceivably use a reconfigured, faster cycling LHC as its injector. Another solution would be a new injector ring based on superferric transmission-line magnets with a maximum field of 2 T [18] installed in the same 100-km tunnel as the FCC-hh itself.

In conclusion, thanks to the smaller emittances, higher energy, and strong synchrotron radiation damping, both HE-LHC and FCC-hh are easy machines from an operational and beam-dynamics point of view. Controlled noise injection will be needed in both the transverse and longitudinal planes, providing a simple form of luminosity leveling and beam–beam tune-shift control. For either machine daily integrated luminosities of about 2/fb are expected and higher values are possible, limited primarily by the maximum acceptable pile up and data-taking rate of the experiments. For HE-LHC the cryogenics cooling capacity required is near the limit of the capacity of the existing LHC refrigerators, assuming an optimized beam-screen temperature around 50 K. The FCC-hh will also require a higher beam-screen temperature, of either 50 or 100 K, or the use of discrete warm photon stops. At the higher beam energies collimation becomes more challenging. The key ingredient of the HE-LHC, however, is the 20-T magnet with its underlying combination of superconductors, while the baseline design for FCC-hh requires dipole magnets with a field of 16 T. The performance and availability of $Nb_3Sn$ magnet coils is essential for both machines. FCC-hh requires a new 100-km tunnel, which in an initial phase could accommodate an ultra-high luminosity circular $e^+e^-$ collider operating at CM energies up to 350 or 500 GeV, and could also support highest-energy highest-luminosity electron–proton and electron–ion collisions (FCC-he).

## References


[1] http://www.vlhc.org
[2] http://eucard.web.cern.ch/eucard/activities/research/WP7
[3] Lee, P. (FSU), compilation of $j_c$ data; www.magnet.fsu.edu/magnettechnology/research/asc/plots.html; P. McIntyre *et al.* (2011), in Ref. [5]
[4] Assmann, R., *et al.* (2010). First Thoughts on a Higher-Energy LHC, CERN ATS Report, CERN-ATS-2010-177 (2010)
[5] *Proceedings EuCARD AccNet-EuroLumi Workshop: The High-Energy Large Hadron Collider, 'HE-LHC-2010,'* 14–16 October 2010, Malta (eds. E. Todesco and F. Zimmermann), CERN Yellow Report, CERN-2011-003
[6] Rossi, L. and Todesco, E. (2010). Conceptual Design of 20-T Dipoles for High-Energy LHC, in Ref. [5]
[7] Benedikt, M., *et al.* (2014). Challenges for Highest Energy Circular Colliders, *Proc. IPAC2014*, Dresden
[8] http://cern.ch/fcc
[9] Osborne, J.A. and Waaijer, C.S. (2012). Pre-Feasability Assessment for an 80 km Tunnel Project at CERN, submission no. 165, *Open Symposium — European Strategy Preparatory Group*, 10–12 September 2012, Cracow, Poland
[10] Dominguez, O. and Zimmermann, F. (2011). Luminosity Optimization for a Higher-Energy LHC, *Proc. IPAC2011*, San Sebastian
[11] Dominguez, O. (2014). Electron Cloud Studies for the LHC and Future Proton Colliders, EuCARD Monograph, Vol. 23 (eds. Vretenar, M. and Romaniuk, R.S.)
[12] Ruggiero, F. and Zimmermann, F. (2002). Luminosity Optimization Near the Beam-Beam Limit by Increasing Bunch Length or Crossing Angle, *PRST-AB* **5**, 061001
[13] Delikaris, D. and Tavian, L. (2011). Heat Loads and Cryogenics for HE-LHC, in Ref. [5]
[14] Metral, E. (2011). Beam Screen Issues, in Ref. [5].
[15] Croce, R.P., Petracca, S., and Stabile, A. (2013). Open Cell Metal Foams for Beam Liners?, e-Print arXiv:1307.4779 [physics.acc-ph]
[16] Lebrun, P., and Tavian, L. (2014). Beyond the Large Hadron Collider: A First Look at Cryogenics for CERN Future Circular Colliders, *Proc. 25th Int. Cryogenic Conference & Int. Cryogenic Materials Conference 2014*, Enschede, Netherlands, 7–11 July 2014
[17] Bauer, P., *et al.* (2001). Synchrotron Radiation Issues in the VLHC, *Proc. PAC2001*, Chicago; (2002). Synchrotron Radiation Absorbers for Hadron Colliders, *Proc. EPAC 2002*, Paris; (2003). Report on the First Photon Stop Cryogenic Design Experiment, FERMILAB TD-03-021
[18] Foster, G.W., *et al.* (1999). Conductor Design for the VLHC Transmission Line Magnet, *Proc. PAC1999*, New York

# Chapter 27

# Electron–ion collider eRHIC

Vladimir N. Litvinenko (Stony Brook University)

In this article, we describe our planned future electron–ion collider (EIC), based on the existing Relativistic Heavy Ion Collider (RHIC) hadron facility, with two intersecting superconducting rings, each 3.8 km in circumference [1]. We plan to add a polarized electron beam with energy tunable within the 5–30-GeV range to collide with variety of species in the existing RHIC accelerator complex, from polarized protons with a maximum energy of 250 GeV, to heavy, fully striped ions with energies up to 100 GeV/u.

## 1 Introduction

Brookhaven's innovative design (Fig. 1) is based on one of the RHIC's hadron rings and a multi-pass energy-recovery linac (ERL). Using the ERL as the electron accelerator assures our success in ultimately reaching high luminosity up to $10^{34}-10^{35}\,\mathrm{cm^{-2}sec^{-1}}$ and a center-of-mass (CM) energy range from 30 GeV to 200 GeV.

The eRHIC will support the collision of highly polarized electrons with polarized protons or $He^3$ ions, or with un-polarized heavy-ion beams up to uranium. The eRHIC will offer up to three interaction regions for electron–hadron collisions. If needed, a dedicated ring would deliver polarized positrons for the ERL to collide with the ions. The luminosity of these collisions will be modest.

Cost is the major factor in assuring the realization of the EIC facility, and hence, our design for the eRHIC is highly cost effective. It fully utilizes the existing RHIC facility whose replacement cost is about two billion US dollars.

Furthermore, the eRHIC's ERL is located inside the existing RHIC tunnel, which significantly reduces cost of its civil contraction. The extremely small size of the electron beam in the ERL allows us to install cost-effective small (few mm) gap magnets and a vacuum chamber for its recirculating loops [2] or novel permanent-magnet FFAG arcs [3] accommodating multiple turns.

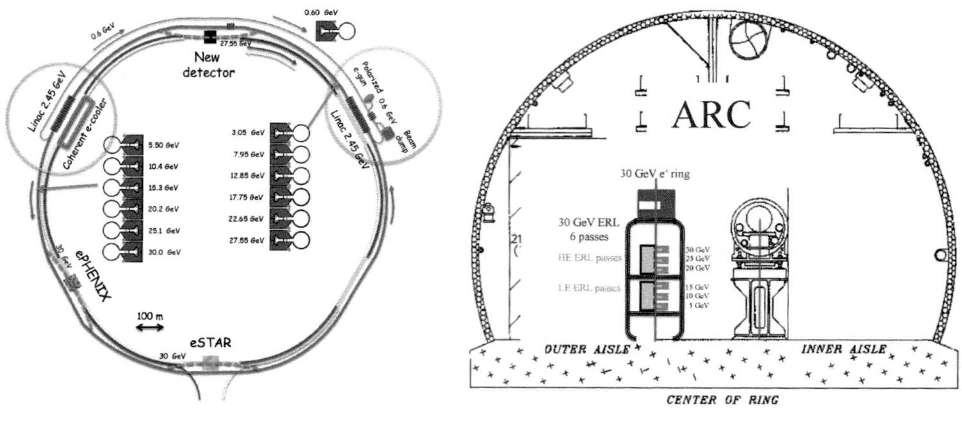


Fig. 1. (a) Layout of the ERL-based, all-in-RHIC-tunnel, 30 GeV × 325 GeV high-energy high-luminosity eRHIC. (b) Location of eRHIC's six recirculation arcs in the RHIC tunnel.

The eRHIC's ERL has natural staging scenario of increasing, in stages, the electron beam's maximum energy from an initial value of 5–10 GeV to its final energy of 30 GeV by adding additional cavities to its SRF linacs and increasing settings in the magnet's power supplies. The eRHIC design includes several scenarios for staging its luminosity and its detectors. For example, the two existing RHIC detectors are considering upgrades to serve as initial detector(s) for eRHIC.

The first phase of eRHIC aiming at keeping the cost below US$ 0.5B, will provide luminosity in the range of $10^{33}$–$10^{34}\,\mathrm{cm}^{-2}\mathrm{sec}^{-1}$ with a CM energy range from 30 to 100 GeV. We detail the eRHIC's performance in Section 5.

## 2 Choice of the scheme for the EIC

Since the first article on eRHIC published in 2001 [4], its design underwent several iterations. Initially, the main option (the so-called ring–ring (RR) design) was based on an electron ring, with the linac–ring (LR) option as its backup. In 2004, we published the detailed "eRHIC 0th-Order Design Report" [5]. After comprehensive explorations, we found that an LR eRHIC has about 10-fold higher luminosity than the RR; hence, since 2007, the LR, with its natural staging strategy and full transparency for polarized electrons, became the main choice for eRHIC. In 2009, we completed technical studies of the design and dynamics for MeRHIC with a 3-pass 4-GeV ERL. We learned much from this evaluation, completed a bottom-up cost estimate for this $350M machine, but then shelved the design.

eRHIC is based on a ring–linac collider design, which, within a given set of conditions, is superior to an optimized ring–ring design. In eRHIC, the electron beam is provided by a six-pass superconducting RF (SRF) energy-recovery linac (ERL) that accelerates polarized electrons to the maximum energy, collides them with the hadron beam, and then decelerates them by removing their energy; finally, they are dumped at very low energy of a few MeV.

The single use of the electron beam, in contrast with its repetitive use in a ring–ring collider, allowed us to increase by two orders of magnitude the beam–beam effects on the electrons. Detailed studies of eRHIC design had revealed that a linac–ring configuration assures a 10- to 50-fold higher luminosity than does the optimized ring–ring design [6]. These advantages especially are evident for high electron-beam energies. Here, we briefly review the main arguments and considerations that resulted in our switching from the early ring–ring design [5] for the eRHIC to the linac–ring one.

The most generic formula for collider luminosity is the well-known one,

$$L = f_c \frac{N_1 N_2}{4\pi\sigma_x\sigma_y} \cdot h, \tag{1}$$

where $f_c$ is the frequency of the bunch collision, $N_{1,2}$ represent the number of particles per bunch in the corresponding beams, $\sigma_{x,y} = \sqrt{\beta^*_{1,x,y}\varepsilon_{1,x,y}} = \sqrt{\beta^*_{2,x,y}\varepsilon_{2,x,y}}$ are transverse beam sizes,[1] and $h \sim 1$ is a luminosity-suppression factor. The latter can be kept near unity with the proper design of the IR and choice of the bunches' length. From simply observing formula (1), it is apparent that luminosity can be enhanced by increasing the frequency of collisions, the number of particles in bunches, or by reducing transverse beam sizes at the collision point. Since the limitations in luminosity are similar for flat and round beams, for simplicity, we focus here on round beams with equal emittances $\varepsilon$ and $\beta^*$: $\sigma_x = \sigma_y = \sqrt{\beta^*\varepsilon}$. Hence, we have

$$L = f_c \frac{N_1 N_2}{4\pi\beta^*\varepsilon} \cdot h. \tag{2}$$

In practice all of these increases have confines that often are specific to a type of collider. For example, restrictions for lepton colliders differ from those for

[1] It was proven experimentally that both beams should have same transverse size at the point of collision.

hadron colliders. The EIC, being a lepton–hadron collider, has limitations from both sides.

First, in contrast with lepton colliders wherein the collision frequency can be as high as 500 MHz [7], that in hadron colliders is restrained by the need to have a detector trigger to avoid an otherwise intolerable background. At present, LHC detectors have the fastest electronics, supporting a collision frequency up to 40 MHz. Furthermore, existing RHIC detectors limit this frequency to 10 MHz. The EIC detector would have the same or nearly the same limitation in the collision frequency as do hadron detectors. The eRHIC design takes these boundaries into consideration: The first eRHIC phase will support a collision rate of 9 MHz, while its ultimate performance (discussed later) could be extended to 56 MHz, so increasing six-fold the attainable luminosity.

Second, the intensities and densities of the colliding beams are limited by non-linear effects occurring during the beam–beam collisions. In storage rings, these beam–beam effects are characterized by the tune shift

$$\xi_1 = \frac{N_2}{\gamma_1}\frac{r_1}{4\pi\varepsilon}, \quad \xi_2 = \frac{N_1}{\gamma_2}\frac{r_2}{4\pi\varepsilon}; \tag{3}$$

$\gamma_{1,2} = E_{1,2}/m_{1,2}c^2$ are the beam's relativistic factors, and $r_e = e^2/m_e c^2$, $r_p = e^2/m_p c^2$ are the corresponding classical radii of colliding particles that we assume to be electrons and protons.[2] In a ring–ring collider both of the beam–beam tune shifts are limited:

$$\frac{N_e}{4\pi\varepsilon} \le \frac{\gamma_p \cdot \xi_{p,\max}}{r_p}, \quad \frac{N_p}{4\pi\varepsilon} \le \frac{\gamma_e \cdot \xi_{e,\max}}{r_e}; \tag{4}$$

with $\xi_{e,\max} \le 0.1$, $\xi_{p,\max} \le 0.03$. This limitation means that in a ring–ring EIC both the increase of the number of colliding particles and reduction of the beam emittance are limited by the above, and the maximum attainable luminosity can be written as

$$L_{\max,R-R} \le f_{c,\max} \cdot \min\left\{\gamma_e \frac{N_e}{\beta^* r_e}\xi_{e,\max}, \gamma_h \frac{N_p}{\beta^* r_p}\xi_{p,\max}\right\} \sim \frac{\varepsilon}{\beta^*}. \tag{5}$$

Hence, the remaining optimization would require either increasing the beam's emittance or reducing $\beta^*$. Since the $\varepsilon/\beta^*$ represents nothing else but the opening solid angle of the beams at the interaction point, increasing

[2] While the generalization for the case of ions with charge $Ze$ and atomic number $A$ is straightforward, to assure the clarity of the concept we use single-charge particles.

emittance and reducing $\beta^*$ would create problems with the acceptance of the final focusing elements and the detector's hermeticity.

In contrast, the single use of the electron beam in the linac–ring EIC collider removes the limitation on the beam effect imposed on the electron beam (i.e., in eRHIC $\xi_e \sim 10$) and luminosity is only limited by available parameters of the hadron beam:

$$L_{\max,L-R} \leq \gamma_e \frac{N_p}{\beta^* r_p} \xi_{p,\max} \cdot h. \tag{6}$$

In hadron storage rings operating at hundreds of GeV, the final focusing quadrupoles impose serious restrictions on the attainable $\beta^*$. Opening their apertures assures reducing focusing strength and, therefore, opposes the reduction of $\beta^*$. In contrast with lepton colliders where $\beta^*$ of few mm was achieved, the smallest $\beta^* = 25\,\text{cm}$ was attained in TEVATRON. New superconducting quadrupoles developed for LHC upgrade should allow about a 5-fold reduction of $\beta^*$ at eRHIC energies, while keeping chromatic effects under control [15]. Hence, we assume $\beta^* = 5\,\text{cm}$ for eRHIC operations.

Third, in contrast with 100-GeV-scale hadron beams, electrons at 10–30 GeV lose much energy via synchrotron radiation, which can seriously limit the attainable luminosity in ring–ring EIC:

$$P_{SR} = \frac{4\pi}{3} f_c N_e \cdot \frac{e^2}{\rho} \gamma_e^4, \tag{7}$$

where $\rho$ is the bending radius, and $I_e = e f_c N_e$ is the electron beam current. Imposing a reasonable limit on the power of synchrotron radiation (which must be compensated for by the RF system with about 50% plug efficiency) in the EIC, we can derive the limitation on the maximum attainable luminosity in ring–ring EIC:

$$L_{\max,R-R} \leq \gamma_e^{-3} \frac{3}{4\pi} \frac{P_{SR} \cdot \rho}{m_e c^2} \frac{\xi_{e,\max}}{\beta^* r_e^2}. \tag{8}$$

Thus, restricting synchrotron losses for 20-GeV electron beam to 10 MW in a reasonably sized storage collider with a 200-m bending radius (i.e. ~2 km in circumference) limits the maximum attainable luminosity to $2.45 \cdot 10^{33}\,\text{cm}^{-2}\text{sec}^{-1}$, subject to the overall limitation in Eq. (5). We assumed $\beta^* = 5\,\text{cm}$ for this estimation.

The luminosity of the linac–ring would continue be limited by Eq. (6) and can exceed the ring–ring limit by one to two orders of magnitude. Hence, we conclude that within a given set of parameters, the linac–ring EIC always would have higher luminosity than the ring–ring version.

An asymmetric IR design as well as issues with beam stability may apply further limitations on the EIC's luminosity. The eRHIC team undertook an in-depth comparison of highly optimized ring–ring and linac–ring scenarios for eRHIC and found that the luminosity in the latter would exceed that in the former by a factor from ten to fifty, depending on the e-beam's energy.

## 3 Choice of the scheme for the eRHIC

### 3.1 *Injector*

As shown in Fig. 1, an electron gun will provide fresh electron beams. In Phase I, we will employ a 50-mA polarized electron gun, based either on single large-sized GaAs cathode [9] (Fig. 2 (a)), or on a Gatling gun [10, 11], an approach combining beams from a large array of GaAs cathodes (Fig. 2 (b)). Illuminated by a circular polarized IR laser light, a strained or a super-lattice GaAs cathode will produce longitudinally polarized electrons with polarization as high as 85–90%. The direction of electron's spin can be flipped on a bunch-to-bunch basis by changing the helicity of the laser photons.

If needed, we will utilize a dedicated un-polarized SRF electron gun, similar to that designed for BNL's R&D ERL [12] to generate a significantly higher beam current (up to 250 mA CW).

Thereafter, the electrons will be accelerated in a pre-injector linac and then will pass six times around RHIC tunnel, gaining energy from two superconducting RF (SRF) linacs located in two of RHIC's straight sections (see Fig. 1(a), wherein the linacs are located in the 2- and 10-o'clock straight sections). They can accommodate SRF 703-MHz linacs up to maximum length

(a) (b)

Fig. 2. Two candidates for eRHIC polarized electron gun: (a) With a large-sized GaAs cathode gun; (b) Gatling gun, combining beams from an array of 24 GaAs cathodes.

of 201 m that suffice for a 2.45-GeV linac operating with a real-estate gradient of 12.45 MeV per meter, corresponding to 20.4-MeV gain per 5-cell 703-MHz cavity.

### 3.2 *The main ERL*

While we will install the eRHIC magnets from the start of operations, the top energy of electron beam will be raised in stages by increasing the length (and the energy gains) of each linac in the ERL chain. At the final stage with six passes, the two main linacs each will have energy gain of 2.45 GeV, while the injection SRF linac will provide 0.6 GeV of energy. At all intermediate stages, the energy gains of all linacs will be proportionally lower, i.e., for the 10-GeV stage, the e-beam will be injected at 0.2 GeV into the main ERL, and each main linac will provide a gain of 0.817 GeV.

We plan to build the eRHIC's linacs from modules comprising six 5-cell 703-MHz SRF cavities. Figure 3 is a 3D rendering of such modules with the HOM-dumped 5-cell cavities.

At their peak energy, the electrons collide with hadrons and then the same linacs recover their energy. The latter process is assured by the additional 180-degree delay of the electrons at the peak energy; such a delay switches acceleration to deceleration.

Dedicated combiners and splitters assure that beams at all energies pass through the same linacs while propagating in their individual beam-lines around the arcs. Figure 4 depicts the arrangement in the 10-o'clock straight section; there is a similar system in the 2-o'clock section.

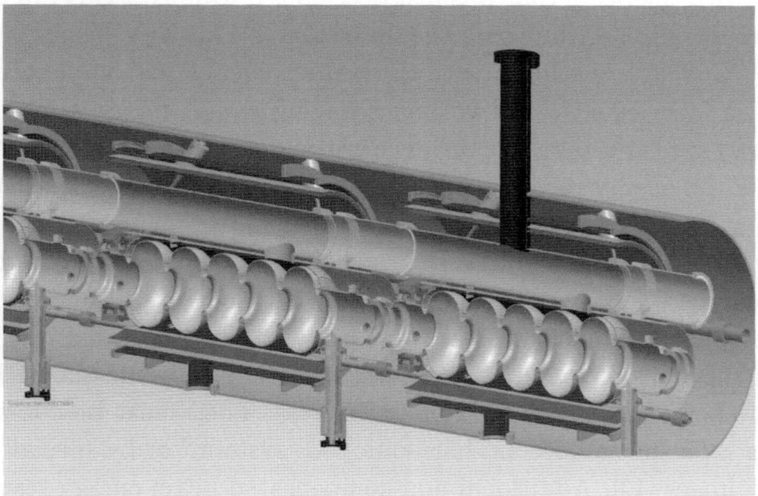

Fig. 3. A cross-section of eRHIC SRF cryomodule showing two 5-cell SRF cavities.

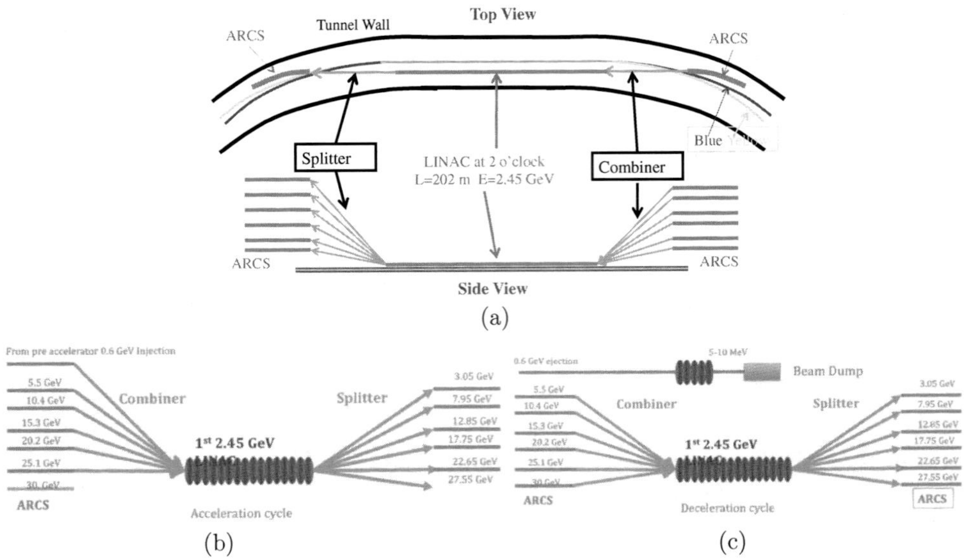


Fig. 4. Scheme for the combiners and the splitters providing for 6-pass acceleration and 6-pass deceleration of the electron beam in eRHIC's ERL. The beams are separated vertically. (a) Overall layout with top and side views of the 10-o'clock RHIC straight section with the eRHIC linac; (b) action of the combiner and the splitter for accelerating beams; and, (c) their action for decelerating beams.

Except at their peak energy, the accelerating and decelerating beams share the arcs, though separated in time. For example, electron beams at 15.3 GeV traverse the same arc between IP2 and IP10, wherein the energy of accelerating beam increases to 17.75 GeV. It enters the 17.75-GeV arc together with the beam that was just decelerated from 20.2 GeV. In contrast, after passing through the linac, the decelerating 15.3-GeV beam passes into the 12.85-GeV arc, sharing it with the beam that was just accelerated in the same linac from 10.4 GeV. Two linacs having equal energy gains maintain this important ratio between the accelerating and decelerating beams. The process of the energy recovery in SRF linacs is extremely efficient, such that only about one kilowatt of RF power per 2.45-GeV linac is absorbed by the SRF surfaces. Main part of the RF transmitter power is reactive and is used to combat the micro-phonic effects in the SRF cavities.

The main beam-energy losses come from synchrotron radiation, resistive losses in the walls of vacuum chambers, and HOM losses in the SRF linacs. Figure 4 shows the values for this power loss. They must be compensated for either by a special (second-harmonic) RF system,[3] or by specially tuning the main linacs [13]. Additional non-compensated beam energy results from

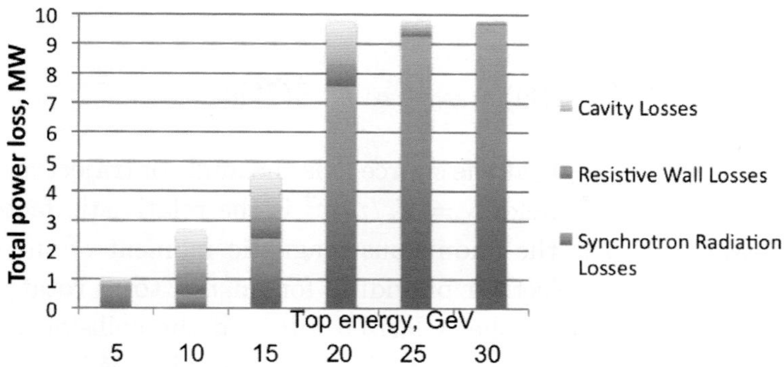


Fig. 5. Electron beam's power loss for various top energies of eRHIC operating with polarized electrons. Note, the losses for synchrotron radiation are kept at a fixed level for e-beam energies above 20 GeV by proportionally reducing the electron beam's current to about the fourth power of the energy.

dumping the beam at about 10 MeV; this energy is generated by the pre-injector.

The size of the electron beam in ERL is so small that the sizes of the vertical gaps in the arcs can be about a few mm; hence, this warrants our using small-gap magnets. They are an important cost-saving factor for eRHIC; we discuss the prototyping of such magnets in Section 6. The vacuum pipe will be made from extruded aluminum with a typical keyhole antechamber design characteristic of modern light sources. In practice, the minimal vertical gap of the vacuum chamber (and, therefore, that of the magnets) is likely to be influenced by the tolerable wakefield effects from resistive walls and roughness; their exact value will be determined by complete theoretical and experimental studies.

### 3.3 *Preserving polarization*

We will preserve in the ERL the high degree of the electrons' polarization originating from the polarized electron gun [14], and provide the desirable direction, i.e., longitudinal, of the electron's polarization at the interaction point (IP). The easiest (and most economical) way of doing so is to keep the spin in the horizontal plane. In this condition, the angle between the direction of electron's velocity and its spin grows according a very simple

[3]Since accelerating and decelerating beams at the same energy experience identical losses but separated by 180 degrees of the main ERL RF, using second (or any even) harmonic of the main RF would provide for identical energy loss compensation for the accelerating and decelerating beams. This restriction is not applicable to the beam at the peak energy that propagates alone in the arc.

equation:

$$\varphi(\Theta) = \varphi_0 + \alpha \int_0^{\Theta} \gamma(\theta) d\theta, \tag{9}$$

where $\varphi_0$ is the initial angle at the source, $\theta$ is the angle of trajectory rotation in the bending magnetic field, $\gamma = E_e/m_e c^2$ is the relativistic factor of the electron beam, and $\alpha$ is the anomalous magnetic moment of the electron. Selecting the energy of electron providing for an $m\pi$ total rotation angle, where $m$ is integer between the polarized gun and the collision point, will ensure the longitudinal polarization of electrons at the IP.[4] With six passes in the ERL and layout shown in Fig. 6, the required condition will be satisfied at IP6 for collisions at electron energies of $E_e = N \cdot 0.07216$ GeV, where $N$ is an integer. This signifies that tuning the energy for 0.24% of a peak energy of 30 GeV will assure such a condition.

RHIC is the only high-energy polarized proton collider. It had demonstrated polarization of protons at the collision energy of 250 GeV at the 60% level. There are plans, and means (including but not limited to use of additional Siberian snakes), to bring the polarization to the 70% level, which we plan to use in eRHIC. Proton polarization at the IP is controlled by spin rotators and can be directed either longitudinally or transversely. The direction of proton beam in RHIC is controlled, and can be switched on

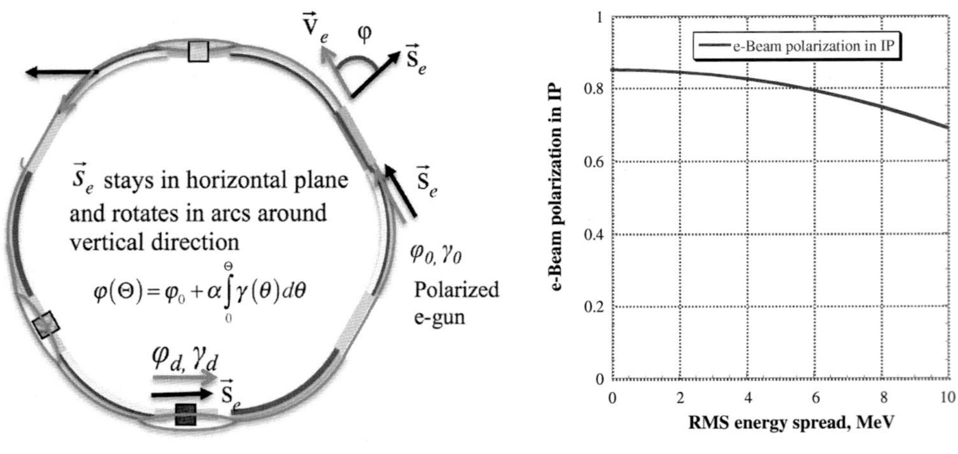


Fig. 6. (a) Electron spin dynamics in eRHIC; (b) Degree of longitudinal polarization as a function of RMS energy-spread averaged along the six paths.

[4]There is no need for the transverse polarization of electrons in exploring the physics processes of interest.

a bunch-by-bunch basis. As discussed above, the direction of electron spin will be changed by reversing the helicity of the laser photons in the gun on the bunch-by-bunch basis to provide any desirable spin-bunch pattern. This flexibility affords an important opportunity to lower systematic errors in data analysis.

This option is impossible to achieve in a ring–ring scenario wherein the polarization and depolarization of electron beams depend upon their spin direction.

### 3.4 *Arcs lattice*

The eRHIC's arc lattice has two components, viz., that of the Blue hadron ring and of the ERL lattice. The lattice of RHIC's Blue ring would be modified significantly in the IR straight sections. We discuss this in the next section. The lattice of six passes for eRHIC's ERL is based on a low-emittance near-isochronous lattice module. The concept of such a lattice originated from the early work of Dejan Trbojevic [15]. In addition to having an excellent filling factor, this lattice supports the fine-tuning of the $R_{56}$ elements in the transport matrix, so supporting the perfect isochronism of the complete paths. Figure 7 illustrates the main building block of the arc lattice. Similar blocks at both sides of the arc lattice make it perfectly achromatic. The lattice of the regular arcs is identical for all passes, independent of their energy. The differences arise only from the splitters and combiners in the SRF linac straights, as well as from the by-pass sections in the other straights.

As evident from Fig. 4, the ERL linacs will be located inside the RHIC rings, while ERL arcs are located outside. This transition, as well as other

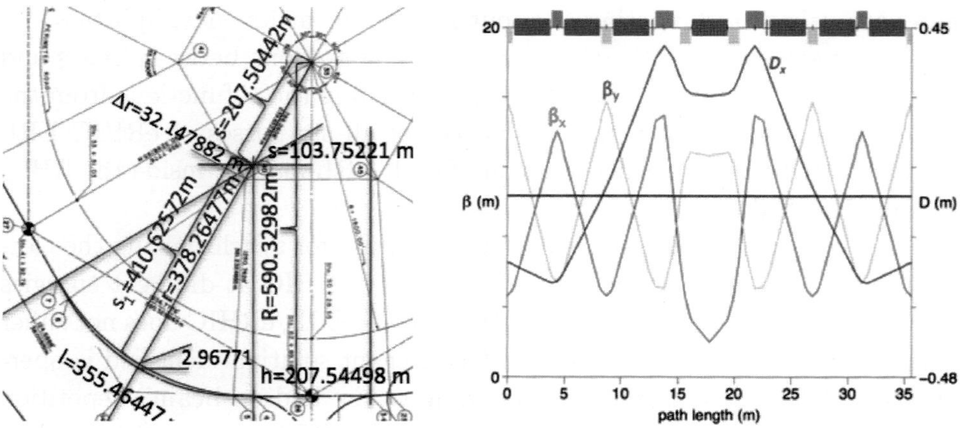


(a) (b)

Fig. 7. Geometry (a) and lattice functions (b) for the standard building block.

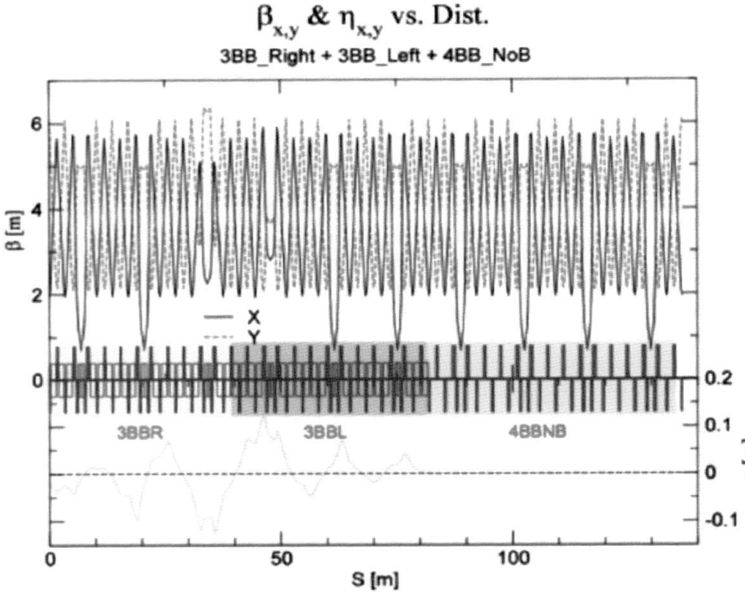


Fig. 8. Half of the bilaterally symmetric lattice of the by-pass around eRHIC detector at 6 o'clock.

peculiarities of the RHIC tunnel's geometry, are accommodated by using two types of the same basic section (Fig. 7) with slightly different radii of curvature. Similar basic blocks are used for the straight passes and for by-passes around the detectors. Figure 8 shows such a design for the by-pass around the eSTAR detector.

Presently, we are considering using a linac lattice without quadrupoles and with values of $\beta$-function of about 200 meters at its ends. Splitters and combiners serve an additional role as matching sections between linacs and arcs. Figure 9 shows the 30-GeV splitter matching the $\beta$-functions from the linac to the arcs. At present, the lattices of all six passes of eRHIC ERL are completed, and the exact location of each ERL magnet inside the RHIC tunnel identified.

One very important issue is finding a solution for synchronizing the electron beam with the hadron beam circulating in RHIC at different energies from 50 to 325 GeV/u. Because it is based on the ERL, eRHIC does not suffer from standard ring–ring limitations. One elegant solution identified is operating the RHIC at energies corresponding to the hadron beam's repetition frequency, i.e., various sub-harmonics of the ERL RF frequency (Fig. 10(b)). The remaining tunability of $\pm 8$ cm can be realized by changing the length of a hadron or an electron beam by-pass.

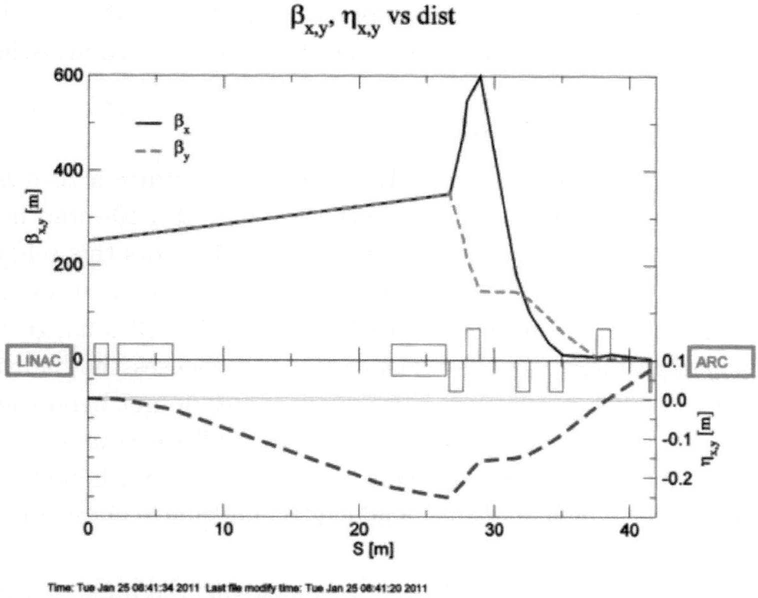


Fig. 9. Lattice of the 30-GeV splitter matching the optical functions of the SRF linac and the arc.

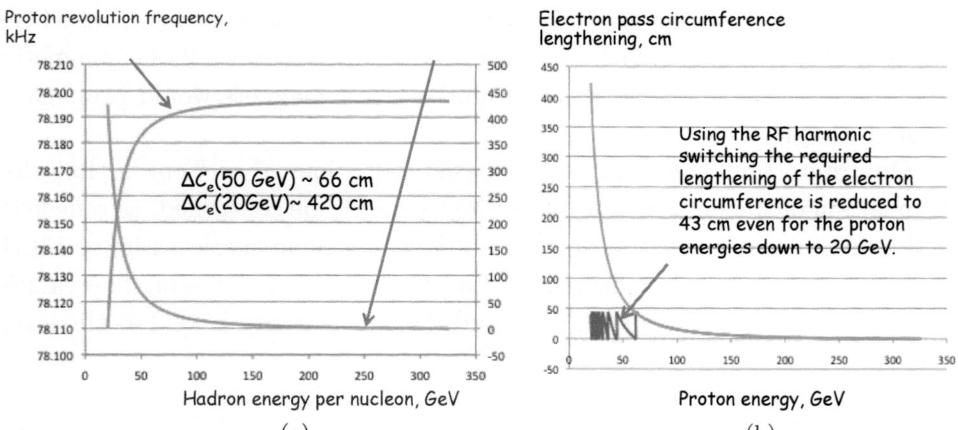


(a) (b)

Fig. 10. (a) Change in the revolution frequency of hadron beams in the RHIC as function of their energy; (b) Red line — the required change for the e-beam circumference without harmonic switching (i.e. ring–ring case); Blue — the same curve with switching the harmonic number.

We explored many issues in beam dynamics for the eRHIC ERL, identifying no major deterrents [16]. We detailed the effects of synchrotron radiation (both its energy spread and growth of emittance wakefields from SRF linacs, resistive walls, and the transverse beam's (TBBU's) stability. One remaining question is about the effect on energy spread of the wakefields from the wall's roughness. These issues and their remedies are under active investigation.

## 4 eRHIC interaction region

The current high-luminosity eRHIC IR design incorporates a 10-mrad crab-crossing scheme; thus, hadrons traverse the detector at a 10-mrad horizontal angle, while electrons go straight through. Figure 11 plots this scheme. The hadron beam is focused to $\beta^* = 5\,\text{cm}$ by a special triplet, wherein first magnet is a combined function magnet (1.95 m long with 2.23-T magnetic fields and a 88-T/m gradient). It has two functions; it focuses the hadron beam while bending it by 4 mrad. Two other quadrupoles do not bend the hadron beam but serve only for focusing it. Importantly, all three magnets provide zero magnetic fields along the electron beam's trajectory. Quadrupoles for this IR require very high gradients, and can be built only via modern superconducting technology [17, 18].

This configuration guarantees the absence of harmful high-energy X-rays from synchrotron radiation. Further, the electron beam is brought into the collision via a 130-meter long merging system. The radiation from regular bending magnets would be absorbed. The last 60 meters of the merging system use only soft bends: The downwards magnets have strength of 84 Gs (for 30-GeV beam), and the final part of the bend uses only a 24-Gs magnetic field. Only 1.9 W of soft radiation from the latter magnets would propagate through the detector.

One important factor in the IR design with low $\beta^* = 5\,\text{cm}$ is that the chromatism of the hadron's optics in the IR should be controlled, as reflected in the maximum $\beta$-function of the final focusing quadrupoles. Figure 13(a) shows the designed $\beta$ and dispersion functions for hadron beam. The values of $\beta$-function are kept under 2.5 km, and the chromaticity is held at the level

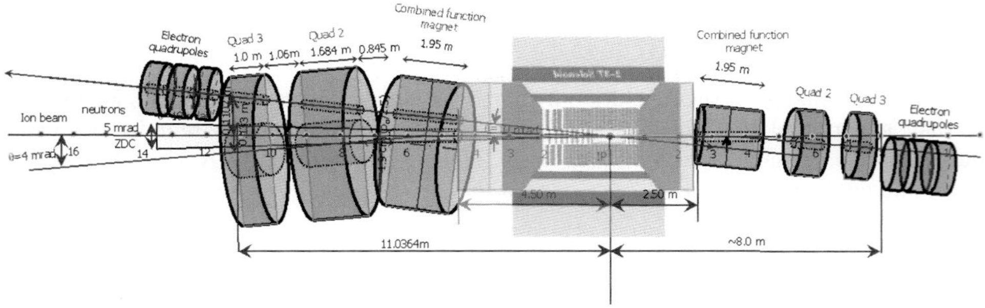


Fig. 11. Layout of the eRHIC IR — vertical scale is expanded for visibility. Electron beam propagates with 10-mrad angle through the field-free regions inside the hadron magnets.

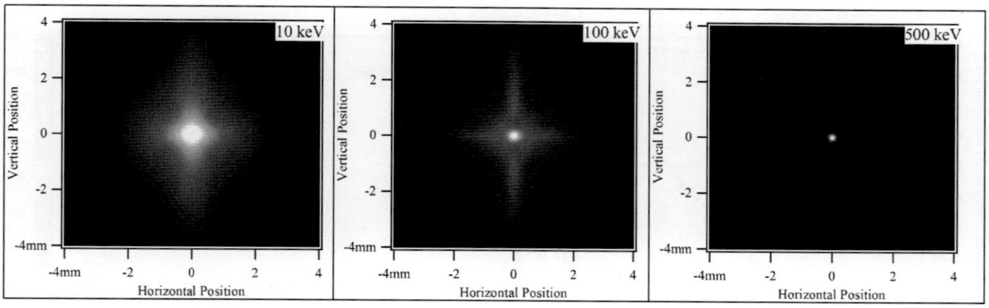


Fig. 12. Distribution of synchrotron radiation from the final focusing triplet in the center of the IP.

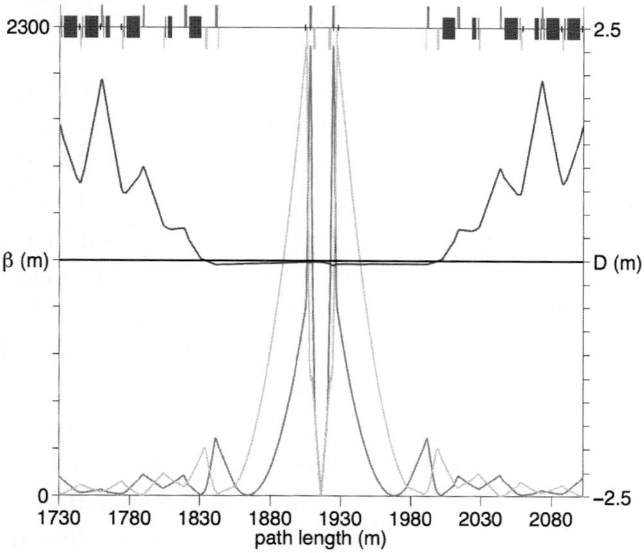


Fig. 13. A symmetric hadron beam's optics at the eRHIC IR. The 5-cm $\beta^*$ is matched into the RHIC's arc lattice with 90-degree phase advance.

typical for RHIC operations with $\beta^* \sim 1\,\text{m}$. We are starting a full-fledged program of tracking of hadron beams in the RHIC, including characterizing beam–beam effects and all known nonlinearities of RHIC magnets: we do not anticipate any serious chromatic effects originating from our IR design. Furthermore, we introduced the bending field in the first quadrupole for the hadrons, thereby to separate them from the neutrons. Physicists considering processes of interest for electron–ion collider (EIC) science requested our installing this configuration.

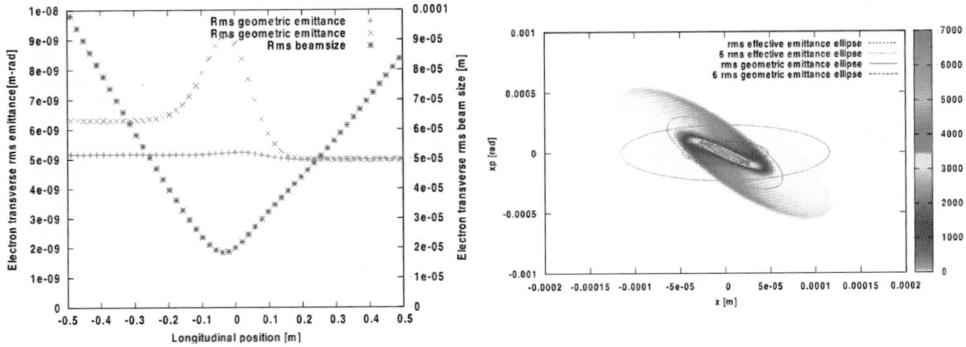


(a) (b)

Fig. 14. (a) The optimized e-beam envelope during collision with the hadron beam in eRHIC; (b) distribution of electrons with initial Gaussian distribution after colliding with the hadron beam in eRHIC.

Since the electrons are used only once, the optics for them is much less constrained, and hence, does not present any technical or scientific challenges; therefore, even though it is designed, we do not describe it here.

Finally, beam–beam effects play important role in the eRHIC's performance. While we will control these effects on the hadron beam, i.e., we will limit the total tune shift for hadrons to about 0.015, the electron beam is used only once, and will be strongly disrupted during its single collision with the hadron beam. Consequently, the electrons are strongly focused by the hadron beam (pinch effects), and the e-beam's emittance grows by about a factor of two (disruption) during the collision. These effects, illustrated in Fig. 14, do not represent a serious problem, but will be studied carefully and taken into account in designing the optics and the aperture.

One of the important effects arising in a linac–ring collision is a potential for the so-called kink instability. Our detailed studies showed that using broad-band feedback suppresses this potentially dangerous instability within all ranges of the eRHIC parameters [19].

More details on the lattice and IR design appear in Reference [20].

## 5 eRHIC luminosity

As we discussed above, eRHIC energy as well as its luminosity will be staged. Hence, here we initially describe the Phase I eRHIC performance. After detailed studies, and considering the expense of its construction and running costs, we set the following limits for the eRHIC:

1. Beam-intensity limits:
   For protons: $4 \times 10^{11}$

For Au ions: $3 \times 10^9$
Electron-beam current: 50 mA

2. Minimum $\beta^* = 5$ cm for all species
3. Space-charge tune spread is compensated by an electron beam [24]
4. Maximum proton (ion) beam–beam parameter: 0.015
5. Coherent e-cooling will cool and maintain the hadron beam at
   a. Hadron beam 95% normalized emittance, mm mrad: 1.2
   b. RMS bunch length 4.9 cm
6. Synchrotron radiation's intensity limit is defined at 10 MW
7. Collision repetition rate: 9 MHz

With 50 mA of beam current, the Phase I eRHIC luminosity does not depend on the e-beam's energy but is linearly proportional to that of the hadrons. Table 1 lists the typical eRHIC Phase I's luminosity.

Table 1. Phase I eRHIC luminosity.

| | e | p | $^2He^3$ | $^{79}Au^{197}$ | $^{92}U^{238}$ |
|---|---|---|---|---|---|
| Energy, GeV | 10 | 250 | 167 | 100 | 100 |
| CM energy, GeV | | 100 | 82 | 63 | 63 |
| Number of bunches/distance between bunches, nsec | 107 | 111 | 111 | 111 | 111 |
| Bunch intensity (nucleons) | $0.24 \times 10^{11}$ | $4 \times 10^{11}$ | $6 \times 10^{11}$ | $6 \times 10^{11}$ | $6.3 \times 10^{11}$ |
| Bunch charge, nC | 5.8 | 64 | 60 | 39 | 40 |
| Beam current, A | 0.05 | 0.556 | 0.556 | 0.335 | 0.338 |
| Normalized emittance of hadrons 95%, mm mrad | | 1.2 | 1.2 | 1.2 | 1.2 |
| Normalized emittance of electrons, RMS, mm mrad | | 16 | 24 | 40 | 40 |
| Polarization, % | 80 | 70 | 70 | none | none |
| RMS bunch length, cm | 0.2 | 5 | 5 | 5 | 5 |
| $\beta^*$, cm | 5 | 5 | 5 | 5 | 5 |
| Luminosity per nucleon, $cm^{-2}sec^{-1}$ | | $\mathbf{2.7 \times 10^{34}}$ | $\mathbf{2.7 \times 10^{34}}$ | $\mathbf{1.6 \times 10^{34}}$ | $\mathbf{1.7 \times 10^{34}}$ |

We included in the luminosity numbers the hourglass effect of 0.851 and the e-beam's pinch effect. This effect raises luminosity by 20 to 30%, depending on the ratio between the energies of the electron and hadron beams. For simplicity we use the lowest value of 1.2 in the table.

An increase in the e-beam's energy up to 20 GeV would not affect luminosity; above it, the SR would exceed the 10 MW level so that the e-beam current and luminosity must be reduced inversely in proportion to the fourth power of its energy. Figure 15 illustrates the eRHIC Phase I luminosity in polarized e–p collision as functions of the particles' energy $E_e$ and $E_p$, as well as the CM energy and $E_e/E_p$.

Ultimately, we could raise the eRHIC's luminosity and the CM energy reach by increasing the collision frequency to 56 MHz and taking advantage of up to a 25% enhancement of RHIC's energy. Table 2 shows the ultimate reach of the eRHIC luminosity that necessitates an enhancement in the beams' currents.

We included the hourglass effect of 0.851 and 1.2 for the pinch effect into the luminosity numbers. With these beam parameters, the 10 MW limit for synchrotron radiation power will be attained at 15 GeV e-beam's energy, above which the luminosity would fall with the decline in the power of the e-beam's energy. The plots in Fig. 16 depict the dependence of luminosity on electron energy for the peak energy of the hadron beams, and on the hadron energy for an electron energy of 15 GeV or less.

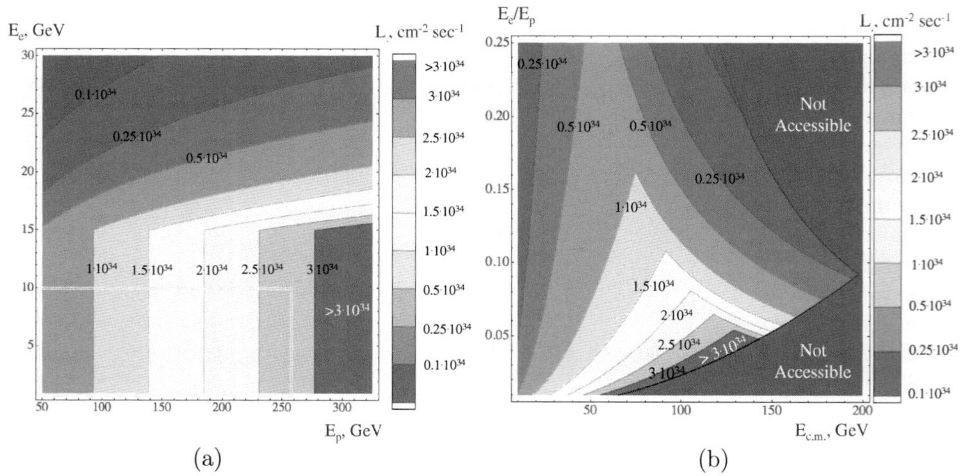


Fig. 15. The contour plots of eRHIC luminosity with Phase I beam parameters a function of the electron and proton energies (a), and the CM energy and the ratio of the e-beam's and the proton beam's energies (b). The box in (a) shows the reach of energy in Phase I .

Table 2. Ultimate eRHIC luminosity.

| | e | p | $^{2}He^{3}$ | $^{79}Au^{197}$ | $^{92}U^{238}$ |
|---|---|---|---|---|---|
| Energy, GeV | ≤15 | 325 | 215 | 130 | 130 |
| CM energy, GeV | | 80–161 | 131 | 102 | 102 |
| Number of bunches/distance between bunches, nsec | 18 | 666 | 666 | 666 | 666 |
| Bunch intensity (nucleons) | $0.24 \times 10^{11}$ | $4 \times 10^{11}$ | $6 \times 10^{11}$ | $6 \times 10^{11}$ | $6.3 \times 10^{11}$ |
| Bunch charge, nC | 3.8 | 64 | 60 | 39 | 40 |
| Beam current, A | 0.22 | 3.33 | 3.33 | 2.00 | 2.03 |
| Normalized emittance of hadrons 95%, mm mrad | | 1.2 | 1.2 | 1.2 | 1.2 |
| Normalized emittance of electrons, RMS, mm mrad | Matches hadron beam | 5.8–23 | 7–35 | 12–57 | 12–57 |
| Polarization, % | 80 | 70 | 70 | none | none |
| RMS bunch length, cm | 0.2 | 4.9 | 4.9 | 4.9 | 4.9 |
| $\beta^*$, cm | 5 | 5 | 5 | 5 | 5 |
| Luminosity per nucleon, $cm^{-2}sec^{-1}$ | | $\mathbf{1.4 \times 10^{35}}$ | $\mathbf{1.4 \times 10^{35}}$ | $\mathbf{0.84 \times 10^{35}}$ | $\mathbf{0.88 \times 10^{35}}$ |

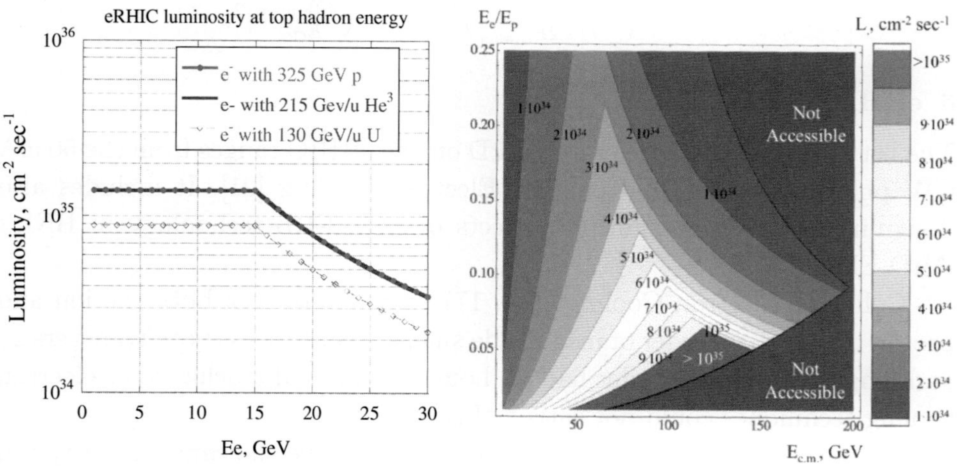


(a) (b)

Fig. 16. The dependence of the ultimate eRHIC luminosity on the e-beam's energy (a). The luminosity contour is plotted as function of the CM energy and the ratio of the energies of the e-beam and the proton beam (b).

Finally, eRHIC's maximum luminosity also can be calculated using the following formulae:

$$L_{e,p} = 1.40 \cdot 10^{35} \cdot \frac{E_p}{325\,\text{GeV}} \cdot \min\left(1, \left(\frac{15\,\text{GeV}}{E_e}\right)^4\right),$$

$$L_{\text{e,He3}} = 1.39 \cdot 10^{35} \cdot \frac{E_{\text{He3}}/u}{215\,\text{GeV}} \cdot \min\left(1, \left(\frac{15\,\text{GeV}}{E_e}\right)^4\right),$$

$$L_{e,U} = 0.884 \cdot 10^{35} \cdot \frac{E_U/u}{130\,\text{GeV}} \cdot \min\left(1, \left(\frac{15\,\text{GeV}}{E_e}\right)^4\right),$$

as function of the beam's energies. Using these formulae

$$L_{e,p}(E_{\text{CM}}, \alpha) = 1.40 \cdot 10^{35} \cdot \frac{E_{\text{CM}}}{\sqrt{\alpha} \cdot 650\,\text{GeV}} \cdot \min\left(1, \left(\frac{30\,\text{GeV}}{\sqrt{\alpha} \cdot E_{CM}}\right)^4\right),$$

$$L_{\text{e,He3}}\,(E_{\text{CM}}, \alpha) = 1.39 \cdot 10^{35} \cdot \frac{E_{\text{CM}}}{\sqrt{\alpha} \cdot 430\,\text{GeV}} \cdot \min\left(1, \left(\frac{30\,\text{GeV}}{\sqrt{\alpha} \cdot E_{\text{CM}}}\right)^4\right),$$

$$L_{e,U}\,(E_{\text{CM}}, \alpha) = 0.884 \cdot 10^{35} \cdot \frac{E_{\text{CM}}}{\sqrt{\alpha} \cdot 260\,\text{GeV}} \cdot \min\left(1, \left(\frac{30\,\text{GeV}}{\sqrt{\alpha} \cdot E_{\text{CM}}}\right)^4\right),$$

the luminosity is given as function of the CM energy $E_{\text{CM}} \cong 2\sqrt{E_e E_h}$ and the beam energy ratio $\alpha = E_e/E_h$ with natural kinematic limits of $E_{\text{CM}} \leq 2\min(\sqrt{\alpha}E_{h,\max}, E_{e,\max}/\sqrt{\alpha})$, $(\frac{E_{\text{CM}}}{2E_{h,\max}})^2 \leq \alpha \leq (\frac{2E_{e,\max}}{E_{\text{CM}}})^2$.

## 6 eRHIC R&D

The list of the needed accelerator R&D on the eRHIC ranges from the 50 mA CW polarized source to Coherent Electron Cooling [21]. It includes also designing and testing multiple aspects of SRF ERL technology in BNL's R&D ERL [22].

Coherent Electron Cooling (Fig. 17) promises to cool both the ion and proton beams to an order of magnitude smaller emittances (both transversely and longitudinally) in under half an hour. Traditional stochastic or electron cooling techniques could not satisfy this demand.

In the CeC, the electron and hadron beams have the same velocity, and co-propagate, in a vacuum, along a straight line in the modulator and the kicker; this is achieved by selecting the energy of electrons such that the relativistic factors of the two beams are identical.

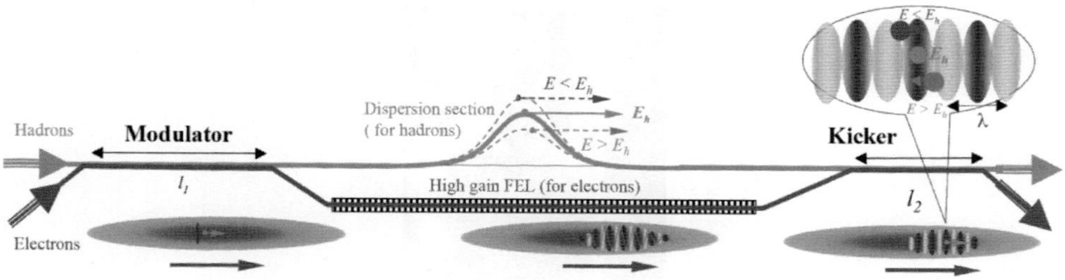


Fig. 17. A general schematic of the Coherent Electron Cooler comprising three sections: A modulator, an FEL plus a dispersion section, and a kicker. For clarity, the size of the FEL wavelength, $\lambda$, is exaggerated grossly.

The CeC works as follows: In the modulator, each hadron (with charge, $Ze$, and atomic number, $A$) induces density modulation in electron beam that is amplified in the high-gain FEL; in the kicker, the hadrons interact with the beam's self-induced electric field and experience energy kicks toward their central energy. The process reduces the hadrons' energy spread, i.e., it cools the hadron beam.

Being a novel unverified technique, the CeC will be tested in a proof-of-principle experiment (CeC PoP) at RHIC in collaboration with scientists from Daresbury Lab, BINP and Tech-X [23]. Table 3 lists main parameters of the test and eRHIC CeC systems.

Table 3. CeC estimates.

| Parameter | CeC PoP | eRHIC |
|---|---|---|
| Species | Au | p |
| Energy, GeV | 40 | 250 |
| Particles per bunch | $10^9$ | $2 \times 10^{11}$ |
| $\varepsilon_{\mathrm{n}}$, mm mrad | 2 | 0.2 |
| Energy spread | $3.7 \times 10^{-4}$ | $10^{-4}$ |
| RMS bunch length, nsec | 3.5 | 0.27 |
| e-beam energy, MeV | 21.8 | 136.2 |
| Peak current, A | 75 | 50 |
| $\varepsilon_n$ (electron), mm mrad | 5 | 1 |
| RMS bunch length (electron), nsec | 0.05 | 0.27 |
| Modulator, m | 3 | 10 |
| Kicker, m | 3 | 10 |
| FEL length, m | 7.5 | 9 |
| $\lambda_{\mathrm{w}}$, cm | 4 | 3 |
| $\lambda_{\mathrm{o}}, \mu$m | 13.8 | 0.422 |
| $\mathrm{a_w}$ | 0.5 | 1 |
| FEL gain | 100 | 3 |
| CeC bandwidth, Hz | $6.2 \times 10^{11}$ | $1.1 \times 10^{13}$ |
| Cooling time, hours, estimate | 0.1 | 0.12 |

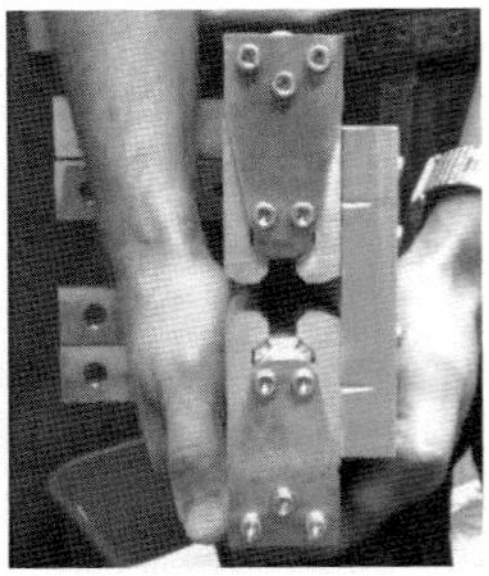

(a)

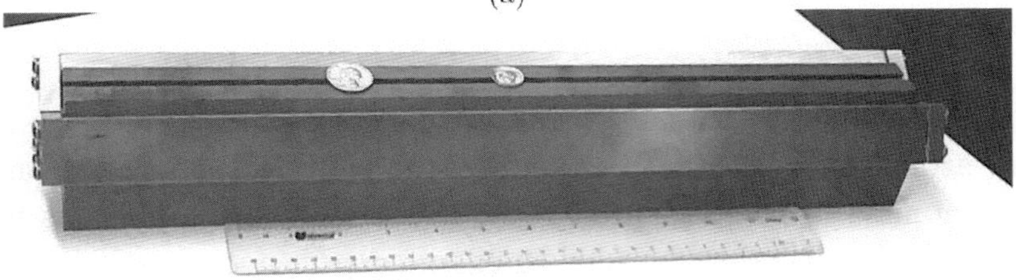

(b)

Fig. 18. (a) A prototype of eRHIC quadrupole with 1-cm gap; (b) Assembled prototype of eRHIC dipole magnet with 5-mm gap.

Other important R&D effort focuses on designing and prototyping small-gap magnets and a vacuum chamber for cost-effective eRHIC arcs [2]. In addition to their energy efficiency and cheapness, small-gap magnets assure a very high gradient as room-temperature quadrupole magnets. Figure 18 shows two such prototypes; they were carefully tested and their fields were mapped using high-precision magnetic measurements. While the quality of their dipole field is close to satisfying our requirements, the quadrupole prototype was not manufactured to our specifications. We will continue this study, making new prototypes employing various manufacturers and techniques. At the moment, Budker Institute of Nuclear Physics (Novosibirsk, Russia) is manufacturing second generation of prototype quadrupoles and dipoles for eRHIC.

Cooling the hadron beams in eRHIC significantly increases the space-charge tune shift to a level that dedicated compensation by an electron beam is required [24]. The detailed studies and simulations of this scheme are also part of eRHIC R&D.

The FFAG arc, including those with a permanent magnet lattice, may provide an inexpensive option for eRHIC. We are intensively investigating this option for Phase I of eRHIC.

Another part of our R&D encompasses testing the RHIC in the various modes that will be required for the eRHIC's operation.

## Conclusions and Acknowledgements

BNL is making steady progress in designing the high-energy, high-luminosity cost-effective electron–ion collider eRHIC and plan to continue our R&D projects and studies of various effects and processes (see for example [25,26]). So far, we have not encountered a problem in our proposed design that we cannot resolve. Being an ERL-based collider, eRHIC offers a natural staging of the electron beam's energy from 10 to 30 GeV.

This paper describes eRHIC design circa-2013. At the time of publishing, the eRHIC linac–ring design has significantly evolved towards more cost effective option [27].

The described eRHIC design is the result of intense work by a large group of scientists and engineers for many years. I would like to acknowledge contributions of all people involved, and whose ideas and research results are described in this paper. This work was supported by Brookhaven Science Associates, LLC under Contract No. DE-AC02-98CH10886 with the U.S. Department of Energy.

## References

[1] RHIC, http://www.bnl.gov/rhic/

[2] Y. Hao, P. He, A.K. Jain, V.N. Litvinenko, G. Mahler, W. Meng, J. Tuozzolo, *Proceedings of First International Particle Accelerator Conference,* IPAC'10, Kyoto, Japan, May 23–28, 2010, p. 1614

[3] D. Trbojevic, S.J. Berg, I. Ben-Zvi, M. Blaskiewicz, V. Litvinenko, W. MacKay, V. Ptitsyn, T. Roser, and A.G. Ruggiero, *Proceedings of Particle Accelerator Conference (PAC 07),* Albuquerque, New Mexico, 2007, p. 3205

[4] I. Ben-Zvi, J. Kewisch, J. Murphy and S. Peggs, *Nuclear Instruments and Methods in Physics Research A* **463**, 94 (2001)

[5] M. Farkhondeh and V. Ptitsyn, (eds.), eRHIC — 0th Order Design Report, BNL, 2004, http://www.bnl.gov/cad/eRhic/eRHIC_ZDR.asp

[6] V. Ptitsyn, J. Beebe-Wang, I. Ben-Zvi, A. Fedotov, W. Fischer, V.N. Litvinenko, W.W. MacKay, C. Montag, S. Ozaki, T. Roser, S. Tepikian, D. Trbojevic, W. Franklin, W. Graves, R. Milner, B. Surrow, C. Tschalaer, E. Tsentalovich, J. van der Laan, A.V. Otboev, Yu.M. Shatunov, D.P. Barber, *Proceedings of EPAC 2006,* Edinburgh, Scotland, June 26–30, 2006, p. 676, http://accelconf.web.cern.ch/AccelConf/e06/PAPERS/ MOPLS058.PDF

[7] SuperB Conceptual Design Report, http://superb.infn.it/cdr

[8] D. Trbojevic, J. Beebe-Wang, Y. Hao, D. Kayran, V.N. Litvinenko, Y. Luo, V. Ptitsyn, and N. Tsoupas, eRHIC Interaction Region Design, *Proc. of Second International Particle Accelerator Conference, IPAC'11* San Sebastian, Spain, September 4–9, 2011, p. 3729, http://accelconf.web.cern.ch/AccelConf/IPAC2011/papers/thpz020.pdf

[9] E. Tsentalovich, High Intensity Polarized Electron Gun Studies at MIT-Bates, *18th International Spin Physics Symposium Charlottesville*, Virginia, 6–11 October 2008, p. 997

[10] V.N. Litvinenko, Gatling Gun: High Average Polarized Current Injector for eRHIC, C-A/AP/417 note, January 2011, http://www.cadops.bnl.gov/AP/ap_notes/ap_note_417.pdf

[11] X. Chang, I. Ben-Zvi, J. Kewisch, V. Litvinenko, A.I. Pikin, V. Ptitsyn, T. Rao, B. Sheehy, J. Skaritka, Q. Wu, E. Wang, T. Xin, A Multiple Cathode Gun Design for the eRHIC Polarized Electron Source, *Proceedings of 2011 Particle Accelerator Conference,* New York, USA, March 25–April 1, 2011, p. 1969, http://accelconf.web.cern.ch/AccelConf/PAC2011/papers/wep263.pdf

[12] R. Calaga, I. Ben-Zvi, M. Blaskiewicz, X. Chang, D. Kayran, V. Litvinenko, *Physica C* **441**, 159 (2006)

[13] Y. Hao, D. Kayran, V.N. Litvinenko, V. Ptitsyn, D. Trbojevic, N. Tsoupas, ERL option for *LHeC*, CERN-LHeC-Note-2010-010 ACC (2010) http://cdsweb.cern.ch/record/1323297/files/ERL%20for%20LHeC%20-%20BNL%20paper%20-%20January%202011.pdf

[14] V.N. Litvinenko, L. Ahrens, M. Bai, J. Beebe-Wang, I. Ben-Zvi, M. Blaskiewicz, J.M. Brennan, R. Calaga, X. Chang, A.V. Fedotov, W. Fischer, D. Kayran, J. Kewisch, W.W. MacKay, C. Montag, B. Parker, S. Peggs, V. Ptitsyn, T. Roser, A. Ruggiero, T. Satogata, B. Surrow, S. Tepikian, D. Trbojevic, V. Yakimenko, S.Y. Zhang, M. Farkhondeh, A. Deshpande, *Proc. of 2005 Particle Accelerator Conference,* May 2005, Knoxville, TN, 2768 http://cern.ch/AccelConf/p05/PAPERS/TPPP043.PDF

[15] D. Trbojevic, E.D. Courant, A. Garren, AIP Conference Proceedings, V. 530, (2000) p. 333

[16] G. Wang, M. Blaskiewicz, A.V. Fedotov, Y. Hao, J. Kewisch, V.N. Litvinenko, E. Pozdeyev, V. Ptitsyn, *Proceedings of First International Particle Accelerator Conference, IPAC'10,* Kyoto, Japan, May 23–28, 2010, p. 1889, http://accelconf.web.cern.ch/ AccelConf/IPAC10/papers/tupec075.pdf

[17] H. Felice, S. Caspi, D. Cheng, D. Dietderich, P. Ferracin, R. Hafalia, J. Joseph, J. Lizarazo, G.L. Sabbi, X. Wang, M. Anerella, A.K. Ghosh, J. Schmalzle, P. Wanderer, G. Ambrosio, R. Bossert, A.V. Zlobin, *Proceedings of First International Particle Accelerator Conference, IPAC'10,* Kyoto, Japan, May 23–28, 2010, p. 403, http://accelconf.web. cern.ch/AccelConf/IPAC10/papers/mopeb059.pdf

[18] P. Wanderer, *IEEE Appl. Superconductivity,* **19**, (3) 1208 (2009)

[19] Y. Hao, V.N. Litvinenko and V. Ptitsyn, Kink Instability Suppression with Stochastic Cooling Pickup and Kicker, *Proceedings of IPAC12,* New Orleans, Louisiana, USA, 2012

[20] D. Trbojevic, J. Beebe-Wang, Y. Hao, D. Kayran, V.N. Litvinenko, V. Ptitsyn, N. Tsoupas, and Y. Luo, *Proc. of Second International Particle Accelerator Conference,* IPAC'11, San Sebastian, Spain, September 4–9, 2011, p. 3729, http://accelconf.web.cern.ch/AccelConf/IPAC2011/papers/thpz020.pdf

[21] V.N. Litvinenko, Y.S. Derbenev, *Physical Review Letters* **102**, 114801 (2009)

[22] I. Ben-Zvi, Z. Altinbas, D. Beavis, S. Belomestnykh, J. Dai, L. DeSanto, D. Gassner, L. Hammons, Ha. Hahn, A. Hershcovitch, A. Jain, P. Jain, J. Jamilkowski, D. Kayran, N. Laloudakis, R. Lambiase, D. Lederle, E. Lessard, X. Liang, V. Litvinenko, C. Liu, G. Mahler, M. Mapes, G. McIntyre, R. Michnoff, W. Meng, T. Miller, D. Pate, P. Pile, D. Phillips, S. Pontieri, J. Reich, T. Roser, M. Ruiz Osés, C. Schultheiss, T. Seda, B. Sheehy, J. Smedley, T. Rao, K. Smith, R. Than, R. Todd, J. Tuozzolo, E. Wang, D. Weiss, M. Wilinski, W. Xu, A. Zaltsman, The Status of the BNL R&D ERL, *ICFA Beam Dynamics Newsletter* No. 58 (2012)

[23] I. Pinayev, S.A. Belomestnykh, I. Ben-Zvi, K.A. Brown, J.C. Brutus, L. DeSanto, A. Elizarov, C. Folz, D.M. Gassner, Y. Hao, R.L. Hulsart, Y.C. Jing, D. Kayran, R.F. Lambiase, V. Litvinenko, G.J. Mahler, M. Mapes, W. Meng, R.J. Michnoff,

T.A. Miller, M.G. Minty, P. Orfin, A. Pendzick, F. Randazzo, T. Rao, T. Roser, J. Sandberg, B. Sheehy, J. Skaritka, K.S. Smith, L. Snydstrup, R. Than, R.J. Todd, J.E. Tuozzolo, G. Wang, D. Weiss, M. Wilinski, W. Xu, A. Zaltsman, G.I. Bell, J.R. Cary, K. Paul, B.T. Schwartz, S.D. Webb, C.H. Boulware, T.L. Grimm, R. Jecks, N. Miller, M.A. Kholopov, P. Vobly, M. Poelker, *Proceedings of IPAC2013,* Shanghai, China, 12–17 May, 2013, p. 1535, http://accelconf.web.cern.ch/AccelConf/IPAC2013/papers/tupfi081.pdf

[24] V.N. Litvinenko and G. Wang, Compensating Tune Spread Induced by Space Charge in Bunched Beams, *Phys. Rev. ST Accel. Beams* **17**, 114401 (2014)

[25] Y. Hao and V. Ptitsyn, *Physics Review Special Topics: Accelerators and Beams* **13**, 071003 (2010)

[26] Y. Hao, V.N. Litvinenko, V. Ptitsyn, *Proceedings of 2009 Particle Accelerator Conference,* Vancouver, Canada, May 4–8, 2009, p. 2868, http://accelconf.web.cern.ch/AccelConf/PAC2009/papers/we6pfp057.pdf

[27] E.C. Aschenauer *et al.*, eRHIC Design Studies: An Electron–Ion Collider at BNL, arXiv:1409.1633

## Chapter 28

# LHeC: A TeV energy scale lepton–hadron collider using the LHC infrastructure

Oliver Brüning (CERN) and Max Klein (Liverpool University)

This chapter describes the conceptual design proposal for a future lepton–hadron collider of unprecedented luminosity and energy, which is based on the existing LHC infrastructure.

## 1 Introduction

The Large Hadron electron Collider (LHeC) project provides the unique possibility of exploring lepton–proton collisions in the TeV range (center-of-mass frame, CM), for the study of new phenomena in the partonic structure of protons and nuclei, precision Higgs physics and the search for physics beyond the Standard Model of particle physics [1,2]. The LHeC may become the first electron–ion collider ever built. The LHeC is designed to use one of the hadron beams of the LHC in a synchronous operation mode. It therefore represents an important opportunity for a further exploitation of the existing LHC infrastructure and its massive investment already taken and to come. Achieving CM collision energies in the TeV range with a 7 TeV energy proton beam demands lepton beam energies significantly larger than the electron beam energy of HERA (27.6 GeV) [3], the first ep collider built. The presented conceptual design of the LHeC [1] is based on a lepton beam energy of 60 GeV, but also considered an option of much higher lepton energy (140 GeV) for exploring the high energy CM regime.

Following a first study of the LHeC [4], a more comprehensive design study was initiated by the CERN Scientific Program Committee (SPC) in autumn 2007, followed by the mandate from the CERN directorate and the European Committee for Future Accelerators (ECFA) to provide a conceptual design study on the physics, the accelerator and a detector for the LHeC. This mandate was soon after also supported by the Nuclear Physics European Collaboration Committee (NuPECC). An international LHeC study group was formed, today comprising about 200 physicists from 75 institutes, which developed the "Conceptual Design Report" (CDR) [1] conducted under the auspices of CERN, ECFA and NuPECC with four workshops, see [5], in the time between September 2008 (held at Divonne, Switzerland) and June 2012 (Chavannes-de-Bogis, Switzerland). Prior to its publication in July 2012, the CDR was peer reviewed by more than twenty experts on the various LHeC topics who had been invited by the CERN directorate for scrutinizing the design.

The LHeC has been designed for exploitation in parallel with the HL-LHC operation over a time scale of approximately a decade. Synchronous pp and ep operation provides the possibility for collecting a total integrated luminosity of the order of $100\,\text{fb}^{-1}$ with peak ep luminosities of the order of $L = 10^{33}\,\text{cm}^{-2}\,\text{s}^{-1}$. The luminosity prospects are thus exceeding the HERA achievements by about two orders of magnitude. The total electrical power consumption of the electron facility was limited to 100 MW. The CDR describes in some detail two options for the LHeC: a so-called Ring–Ring (R–R) and a Linac–Ring (L–R) option, which are conceptually different in the realization of the electron beam, and are both sketched here. The electron beam energy was set to 60 GeV, for both R–R and L–R, a value between the beam energies of LEP-I and LEP-II and not too demanding in either of the two configurations. This value may be altered in a further design.

## 2 Ring–Ring option

The Ring–Ring configuration features the installation of a new electron synchrotron storage ring inside the LHC tunnel, on top of the existing LHC ring. The R–R option is technically rather straightforward (in between the technologies of the LEP-I and LEP-II projects). It yet is challenging, as it requires integration work inside a tunnel with an already operational accelerator. New bypasses have to be build around the existing experiments for the HL-LHC accelerator, which are large enough to house the RF infrastructure for the electron beam. Figure 1 illustrates the schematic layout of the LHC with the additional lepton beam bypasses around the ATLAS and CMS experiments [1]. Each bypass has a total length of ca. 1.2 km and provides a maximum beam separation of about 20 m with respect to the proton beam Interaction Points (IP). Figure 2 shows the schematic bypass layout around the CMS experiment in the LHC Interaction Region 5. The ATLAS bypass concept exploits the existing survey gallery in IP1 in order to minimize the size of the bypass extension around the large ATLAS cavern. It thus consists of two separate parts left and right to the IR1 of about 500 m each, see Fig. 3.

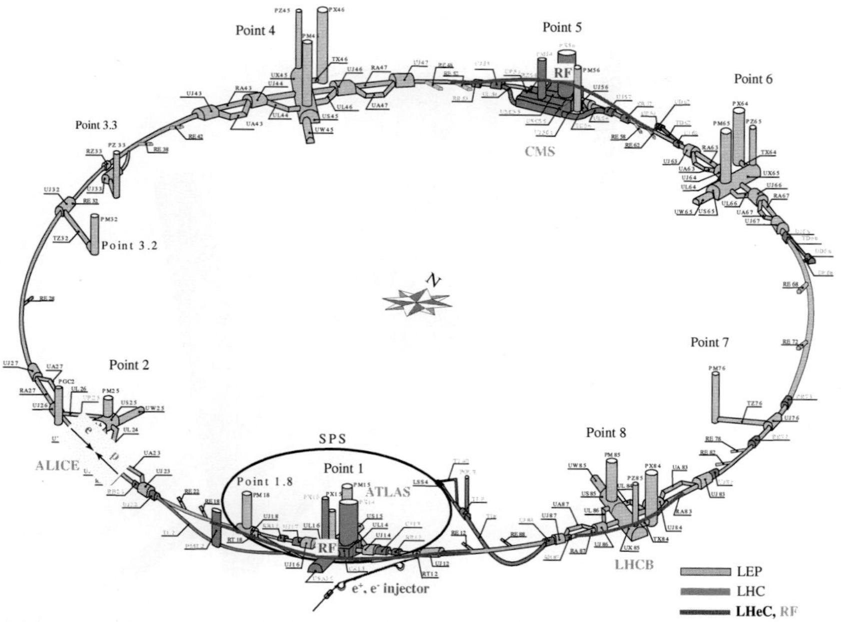


Fig. 1. Schematic layout of the LHC with the additional lepton beam bypasses (marked in blue) around the ATLAS and CMS experiments for the Ring–Ring option of the LHeC and an ep interaction at Point 2.

Fig. 2. Schematic layout of the CMS bypass in IP5: Top view. The bypass has about 2 times 600 m length. The RF is installed in a parallel tunnel of 120 m length, at a distance of about 8 m to the bypass tunnel.

Figure 4 illustrates a typical cross section of the LHC tunnel with the LHC equipment and the space required for the additional components of the lepton storage ring. The position of the lepton synchrotron magnets is shifted inwards with respect to the LHC magnets in order to compensate for the path length increase due to the bypasses and to generate a circumference for the LHeC lepton beam equal to that for the proton beams of the LHC machine.

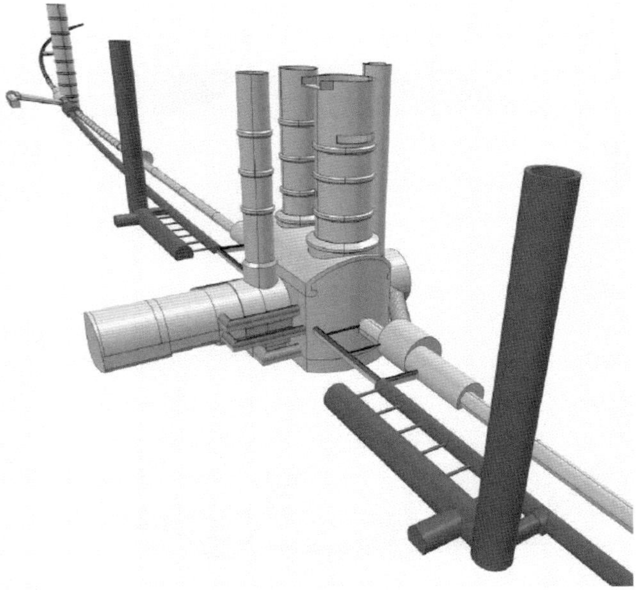

Fig. 3. Schematic layout of the LHeC bypass (blue) of ATLAS (grey) in IR1. The bypass has two parts and the two 60 m long parallel tunnels are designed to house part of the RF for the electron beam. Each side is accessed with a separate shaft.

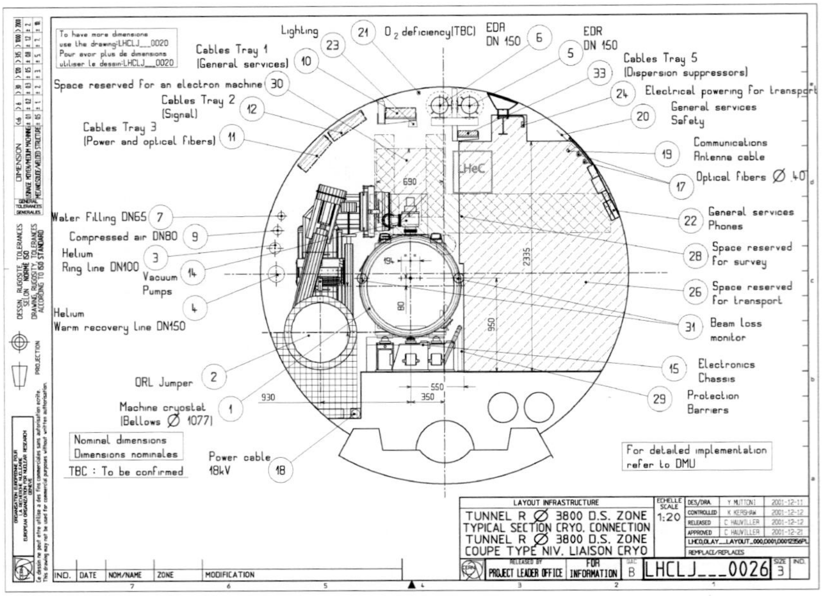


Fig. 4. Typical cross section of the LHC tunnel with the LHC equipment and the space required for the additional components for the electron synchrotron (red box), adjusted to ensure equal circumferences of the e and the p beams.

Fig. 5. View on the LHC installation in Interaction Region 4, illustrating the tight space restrictions for the installation of an additional electron storage ring at certain places inside the LHC tunnel.

As an example of space limitations for the installation of a new lepton storage ring, Fig. 5 shows the existing LHC installations in IR4. However, lattice and beam optics could be designed in spite of such space constraints exploiting asymmetric FODO cells [1]. The interference with the existing LHC installations is large but as discussed in [1] not insurmountable. The main challenge for the R–R solution would be the time and scheduling and the risk for its installation.

The interaction region for the LHeC, running synchronously with the LHC, has the novel feature of accommodating three beams: the colliding proton and lepton beams and the non-colliding second proton beam of the LHC. A schematic view of the interaction region layout for the R–R solution is shown in Fig. 6. The low-beta electron beam quadrupoles are placed close to the detector and are shifted in the horizontal plane to generate an additional dipole deflection field for separating the lepton beam from the proton beam left and right from the IP, making them effectively act as combined function magnets. They are followed by additional dipole separation magnets for the electron beam (light blue colored elements in Fig. 6) and then by low-beta quadrupole magnets for the proton beam (in red).

Figure 7 shows the schematic cross section of the first quadrupole magnet next to the ep IP that is acting on the proton beam. It features a low field

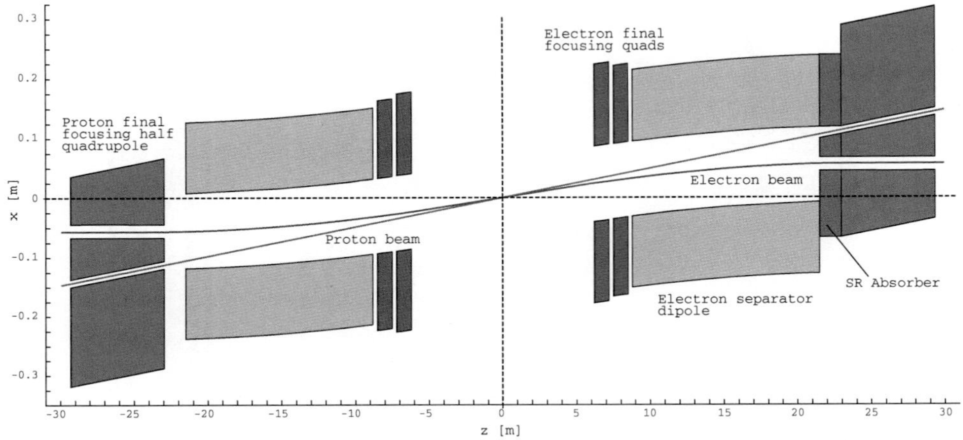


Fig. 6. Schematic layout of the LHeC interaction region in the R–R machine configuration with maximum coverage of the polar angle acceptance by the detector placed between $z = \pm 6\,\mathrm{m}$.

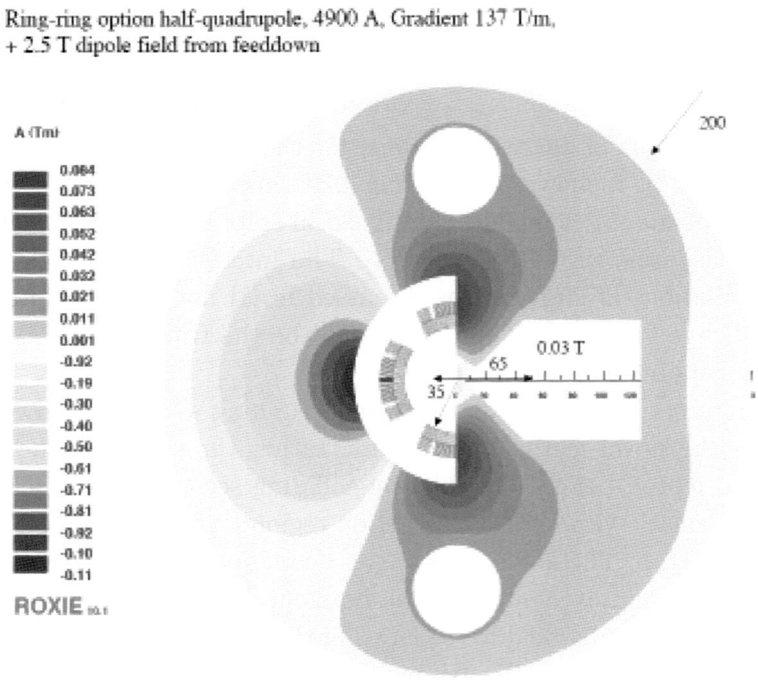


Fig. 7. Cross section of the first low-beta superconducting half quadrupole in the proton lattice. The lepton beam will pass on the right-hand side of the mirror plate in a low field region.

aperture (right-hand side) for the lower energy electron beam, a high field region (left-hand side) for the colliding high-energy proton beam and two field free apertures for the non-colliding proton beam (top and bottom of the magnet cross section in Fig. 7). Table 1 summarizes the optics parameters for the electron beam while Table 2 lists the main beam parameters and performance projections for the LHeC R–R configuration indicating that a peak luminosity of about $L = 10^{33}\,\mathrm{cm}^{-2}\,\mathrm{s}^{-1}$ is reachable. Note that this projection is based on the so-called "ultimate" proton beam parameters of the LHC. Higher performance levels should be attainable when using the nominal HL-LHC proton beam parameters.

The Ring–Ring option of the LHeC requires also the construction of a new electron beam injector complex. The LHeC CDR [1] considers a compact and efficient 10 GeV injector based on the principle of a recirculating linac, taking advantage of the studies for ELFE at CERN [6]. Such a low injection energy requires dipole magnets with a well reproducible, low magnetic field, for which prototypes had been successfully built both by BINP Novosibirsk and CERN [1].

Table 1. Optics parameters of the LHeC e-ring lattice.

| Electron beam energy | 60 GeV |
|---|---|
| Phase advance per half cell, H/V | 180°/120° |
| Cell length | 106.881 m |
| Dipole fill factor | 0.75 |
| Damping partition $J_x/J_y/J_e$ | 1.5/1/1.5 |
| Coupling constant $\kappa$ | 0.5 |
| Horizontal emittance (no coupling) | 3.96 nm |
| Horizontal emittance ($\kappa = 0.5$) | 2.97 nm |
| Vertical emittance ($\kappa = 0.5$) | 1.49 nm |

Table 2. LHeC R–R performance estimate and main parameters.

| Parameter | Value |
|---|---|
| Electron beam energy | 60 GeV |
| Proton beam energy | 7 TeV |
| $e^+$, $e^-$ intensity per bunch | $2 \times 10^{10}$ |
| p intensity per bunch | $1.7 \times 10^{11}$ |
| #bunches | 2808 |
| ep luminosity (HL-layout) | $0.8 \times 10^{33}\,\mathrm{cm}^{-2}\,\mathrm{s}^{-1}$ |
| Total wall plug power | 100 MW |
| Transverse lepton emittances $\varepsilon_{x,y}$(conservative estimate) | 5.0 nm, 2.5 nm |

## 3 Linac–Ring option

The Linac–Ring option requires a new linear accelerator for the electron beam that intersects in one location with the existing LHC machine. Several options have been considered for the linear accelerator (pulsed, recirculating and Energy Recovery Linac configurations). These provide a range of energy and luminosity combinations. The baseline option for the LHeC CDR is a recirculating 60 GeV Energy Recovery Linac (ERL) which allows for high luminosity operation. It is based on a modest civil engineering effort and directly compares with the LHeC Ring–Ring option. A pulsed linac option provides still an interesting option for maximizing the energy reach of the LHeC (at the cost of a reduced peak luminosity performance) as could be demanded by findings at the LHC. Table 3 summarizes key parameters for both options. The 60 GeV ERL version is capable of reaching a luminosity as high as the R–R option. First considerations have been made as to possibly reach a luminosity level of $10^{34}\,\mathrm{cm}^{-2}\,\mathrm{s}^{-1}$, which would enhance the potential of the LHeC for precision Higgs measurements [2]. It is notable, as was recently pointed out [7], that essentially the same machine in a 4-pass regime and going to 80 GeV has an interesting application as a cost effective photon–photon collider for the study of the newly observed particle at 125 GeV.

Figure 8 shows the schematic layout of the 60 GeV ERL option and Fig. 9 shows a 3D civil engineering illustration for the underground installation. It features two 1 km long superconducting 10 GeV linacs and a total of six return arcs with a radius of curvature of ca. 1 km that are installed in two

Table 3. Key parameters for two options for the L–R version of the LHeC.

| | Operation mode | |
|---|---|---|
| Parameters | CW | Pulsed |
| Beam energy [GeV] | 60 | 140 |
| Peak luminosity [$\mathrm{cm}^{-2}\mathrm{s}^{-1}$] | $10^{33}$ | $4 \times 10^{31}$ |
| Cavity gradient [MV/m] | 20 | 32 |
| RF power loss [W/cavity] | 13–37 | 11 |
| W per W (1.8 K to RT) | 700 | 700 |
| Cavity $Q_0$ | $2.5 \times 10^{10}$ | $2.5 \times 10^{10}$ |
| Power loss/GeV at RT | 0.51–1.44 | 0.24 |
| RF length [km] | 2 | 7.9 |
| Total length (including return arcs) [km] | 9 | 7.9 |
| Beam current [mA] | 6.4 | 0.27 |
| Repetition rate [Hz] | — | 10 |
| Pulse length [ms] | — | 5 |

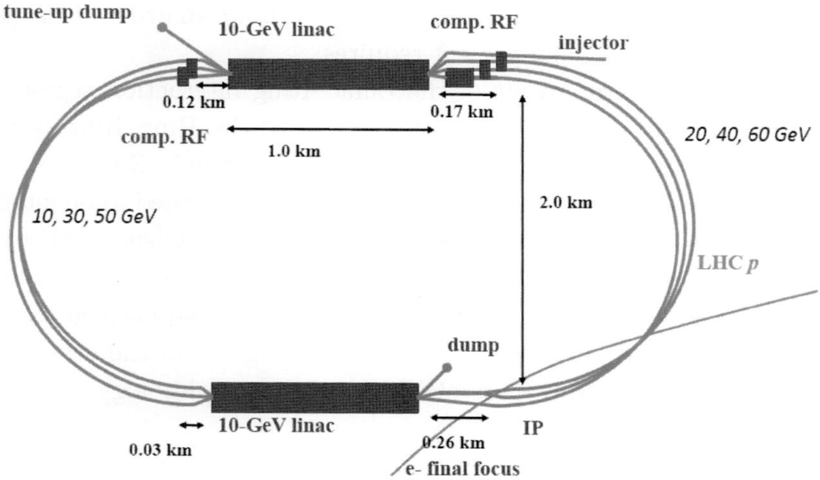


Fig. 8. Schematic layout of the 60 GeV ERL option.

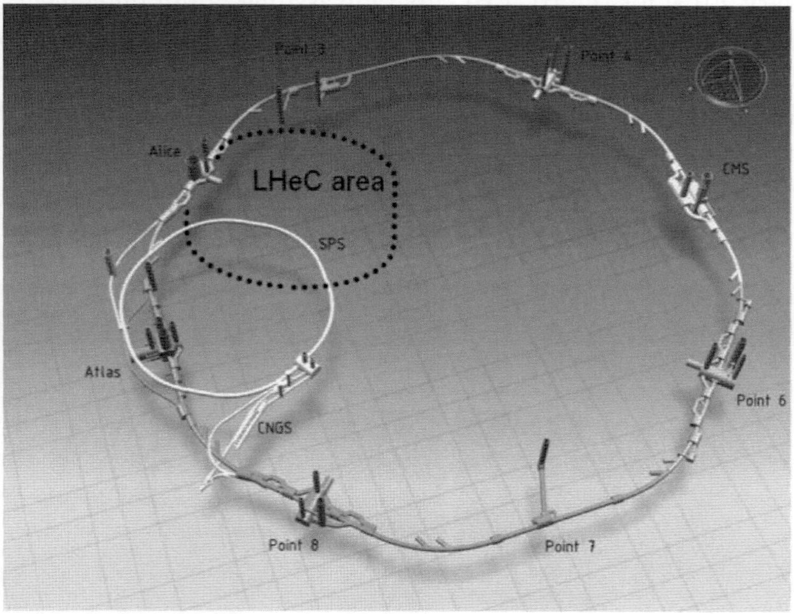


Fig. 9. Civil engineering illustration for the underground installation of the ERL Linac–Ring option with the existing LHC and SPS (white ring near the ATLAS covern) installations.

half circular tunnel sections. The ERL infrastructure requires a total of ca. 9 km underground tunnel length (a size comparable to the SPS installation at CERN or HERA at DESY) and features a total bending arc length of ca. 20 km when accounting for the six separate return arcs. The total number of accelerator components is about 6000, which is about an order of magnitude less than what the CLIC500 project requires.

While the schematic layout of the Linac–Ring interaction region of the LHeC looks on first glance similar to the one of the Ring–Ring option, it requires slightly different IR features. Most notably, the L–R option requires additional dipole magnets in the interaction region (designed to be integrated in the central detector space together with the detector solenoid [1]) to bring the electron beam to collision with the proton beam after the last accelerating passage in the ERL. Moreover, higher gradient requirements arise for the superconducting half-quadrupole low-beta magnets of the proton beam, which most likely require the use of $Nb_3Sn$ magnet technology.

## 4 LHeC planning and timeline

The LHeC study assumes that the LHC hardware will reach the end of its lifetime by the end of the HL-LHC upgrade phase. The current planning (as of January 2013) for the HL-LHC upgrade assumes a full implementation of the HL-LHC upgrade by the end of 2023. Assuming a total exploitation time of ca. 10 years for both the HL-LHC and the LHeC project, this implies that the LHeC project should start operation approximately parallel to the HL-LHC exploitation phase [8]. Figure 10 shows the resulting planning schedule for the LHeC project, which lead to launching dedicated R&D work on key technical systems by 2012. The goal for the LHeC proposal had been set by

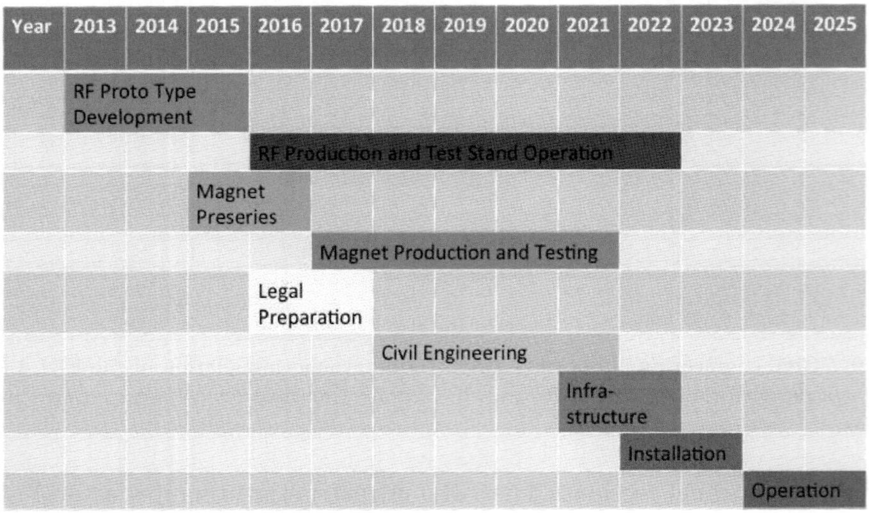


Fig. 10. LHeC project planning schedule [1] as considered for the current (1/2013) LHC shutdown schedule and with the goal of maximizing the synchronous ep/pp operation time during the HL-LHC phase.

CERN at the 2012 LHeC Workshop to develop and drive R&D activities for key technical components and concepts such that a final decision on the LHeC project implementation can be taken by the time the LHC reaches its full energy and large statistics at 13 TeV CM energy will be analyzed. In parallel the detector concept is being further developed.

Both the R–R and the L–R LHeC options have been demonstrated to be technically feasible [1]. The Linac–Ring option has the key advantage that its civil engineering and installation work can be performed mostly decoupled from the LHC operation. This strategic advantage of the Linac–Ring option has led to the decision to concentrate the technical R&D work for the LHeC during the years 2013 to 2015 on the Linac–Ring option. As this involves the most demanding parts (cavity and IR magnet in particular) it remains possible to consider all the ep collision options eventually, by 2016 in the light of then expected conclusive first high energy LHC results. The total number of and the main specifications for the normal conducting magnets in the arcs and SC RF cavities are comparable for both LHeC options.

## 5 Key technical systems

The R&D effort for key technical systems will focus on the following systems for the coming two to three years:

- Superconducting RF technology for the development of cavities with high $Q_0$ for CW operation. The LHeC CDR is based on the assumption of 720 MHz superconducting RF technology with a gradient of approximately 20 MV/m in CW mode. However, other frequency choices are still being investigated (e.g. 800 MHz and 1.3 GHz) and might eventually turn out to be more suitable for the LHeC project.
- Superconducting magnet technology for the development of $Nb_3Sn$ magnet technology for quadrupole designs with mirror cross sections with apertures for high as well as low magnetic field configurations.
- Optimization for the design of normal conducting magnets suited for the return arcs of the Energy Recovery options with multiple magnet systems per arc.
- Design of an LHeC Energy Recovery Linac test facility for studying and testing the various technical components and building up operational experience for ERL facilities in the future.
- Full optics and layout integration of the LHeC into the HL-LHC project.
- Civil engineering studies for the Linac–Ring option.

- Design of the vacuum system for the experimental insertion of the LHeC.
- Detector development.

## 6 Summary

The high energy, intense hadron beams of the LHC provide a unique opportunity for complementing the pp and AA physics program of the LHC, and possibly that of a pure lepton collider, with deep inelastic ep and eA scattering experiments at the energy frontier. The LHeC design [1], briefly characterized here, has been made for synchronous operation with the LHC, which is considered to have an operation lifetime until the mid thirties with an interesting maximum luminosity program. Unless future LHC results lead to new dramatic insight, the chosen energy of 60 GeV and the high luminosity, likely in excess of a hundred times that of the first ep collider HERA, promise to be a most reliable basis for an innovative physics program. The most likely realization of the LHeC is then with a 9 km circumference Energy Recovery Linac configuration using multiple passes through two 1 km superconducting linear accelerators, for which a test facility is under design. An important design and prototype phase is planned until 2016 in order to facilitate a final decision on the LHeC when the high energy data of the LHC will eventually help clarifying the priorities for future High Energy Particle Physics projects and of CERN. The LHeC reaches its high performance owing to the uniqueness of the LHC hadron beams for which most substantial investments have been taken over many years. Not to use these for the LHeC "at some point during the LHC lifetime appears to be a waste" [9].

## References

[1] LHeC Study Group, A Large Hadron Electron Collider at CERN, *J. Phys. G: Nucl. Part. Phys.* **39** (2012) 075001 [arXiv:1206.2913(2012)].
[2] LHeC Study Group, On the Relation of the LHeC and LHC [arXiv:1211.5102(2012)].
[3] F. Willeke, The HERA lepton–proton collider, in *this book*, Chapter 16.
[4] J. Dainton, M. Klein, P. Newman, E. Perez and F. Willeke, Deep Inelastic Scattering at the LHC, *JINST* **1** (2006) P10001 [hep-ex/0603016].
[5] LHeC website, http://cern.ch/lhec.
[6] K. Aulenbacher, B. Aune, J. Aysto, J. Baldy, H. Burkardt *et al.*, ELFE at CDR, Conceptual Design Report, CERN, Geneva (1999).
[7] S.A. Bogacz *et al.*, SAPPHiRE: A Small Gamma–Gamma Higgs Factory [arXiv: 1208.2827].
[8] S. Myers, Talk at EPS2011, Grenoble, France.
[9] G. Altarelli, LHeC Workshop, Divonne, 2008, see [5].

# Chapter 29

# Linear colliders: ILC and CLIC

Jean-Pierre Delahaye (CERN)

## 1 Motivation for linear colliders

CERN's latest and foremost accelerator, the LHC, will probe a new "terascale" energy region and provide a rich program of physics at a new high-energy frontier over the coming years. In this new energy domain, it will study the validity of the Standard Model and explore the possibilities for physics beyond the Standard Model, such as super-symmetry, extra dimensions and new gauge bosons. The discovery potential is huge and will set the direction for future high-energy colliders. Particle physicists worldwide supported by ICFA [1] have reached a consensus that the results of the LHC will need to be complemented by experiments at a precision facility, preferably a lepton collider in the tera-electron-volt (TeV) energy range. The required beam collision energy range will be better defined following Physics requirements based on LHC results when substantial integrated luminosity will have been accumulated, tentatively by 2013–15.

The highest energy of lepton collisions so far, 209 GeV, was reached with electron–positron colliding in LEP at CERN. In spite of its 27 km diameter, beam energy was limited by synchrotron radiation losses just compensated by the most powerful superconducting RF system built so far and providing up to 3640 MV per revolution. Since synchrotron radiation is inversely proportional to the bending radius and to the fourth power of the particle mass, two alternatives are being explored to overcome this limitation and build a terascale lepton collider:

- use muons with a mass 207 times larger than electrons. The feasibility of muon colliders is being studied [2] addressing critical challenges especially the limited muon lifetime (2 $\mu$s), their production in large emittances requiring developments of novel cooling methods, and the large background induced along the whole process by muon disintegration.

- mitigate particle trajectory bends in $e^{+/-}$ linear colliders where two opposing linear accelerators accelerate the particles to their final energy in one pass before focusing and collision in a central interaction point inside a detector.

## 2 Challenges of high energy, high luminosity linear colliders

In order to reach high luminosity and high energy, linear collider parameters are very different from those of circular colliders because of intrinsic differences. Indeed, in a linear collider:

- electron and positron colliding beams are accelerated by a dedicated linac in one single pass.
- after acceleration, the two beams collide once only.

### 2.1 *Energy*

The colliding beam energy of a linear collider is given by:

$$E = 2E_b = FE_aL$$

where $L$ is the total length of each linac, $E_b$ is the beam energy at collision, $F$ the filling factor of accelerating structures along the linac and $E_a$ the average (beam loaded) accelerating field of the accelerating structures. The major challenge lies in developing accelerating structures at a reasonable cost per MeV energy gain, able to accelerate several MW of beam power with high field and high efficiency and installed in the linacs with a high filling factor.

### 2.2 *Luminosity*

In a linear collider where the particles collide only once, the luminosity is given by the standard formula:

$$L = H_D \frac{N^2}{4\pi\sigma_x\sigma_y} n_b f_r = \frac{H_D}{4\pi}\frac{N}{\sigma_x}\frac{1}{\sigma_y}P_b$$

It only depends on a small number of parameters namely the number of particles per bunch $N$, the number of bunches per beam pulse $n_b$, the repetition rate of beam pulses $f_r$, the beam power $P_b$, the transverse beam sizes at collisions $\sigma_{x,y}$ and the luminosity enhancement factor $H_D$, due to the pinch effect during collisions of high density bunches. In order to limit the beam power, the beam sizes $\sigma_{x,y}$ have to be extremely small in the nm range so as to compensate for the low repetition rate as compared to a circular collider which benefits from high collision rate due to the multiple collisions. As a consequence, each beam exerts a strong electro–magnetic force on the other beam, which is focusing in case of electron–positron collisions. This disruption can shrink the beam size significantly during collision, the so-called pinch effect with an increase of the luminosity by the factor $H_{D,}$ typically between 1 and 2. But the bending of the particles' trajectories stimulates them to radiate so-called beamstrahlung photons, a process similar to synchrotron radiation. Consequently, not all collisions take place at the nominal centre-of-mass energy. Hence, the beam parameters are chosen in order to limit the beamstrahlung and to achieve an acceptable luminosity spectrum for the experiment.

The factor $N/\sigma_x$ in the equation of the luminosity above is a measure of the beamstrahlung, which can conveniently be described with the beamstrahlung parameter $Y$, the ratio of the average critical energy $h\omega_c$ to the beam energy $E$. In the classical limit $Y \ll 1$, which is applicable in the terascale energy range, the number of beamstrahlung photons emitted per beam particle $n_\gamma$ and their average energy $E_\gamma$ can be approximated as $n_\gamma \propto N\gamma/(\sigma_x+\sigma_y)$. Hence, flat beams with $\sigma_x \gg \sigma_y$ are favored to maximize luminosity ($\propto N/(\sigma_x\sigma_y)$) while limiting the beamstrahlung ($\propto N/(\sigma_x+\sigma_y) \approx N/\sigma_x$). Typically one aims for $n_\gamma \leq 1-2$ to maximize luminosity while maintaining the degradation of the luminosity spectrum due to beamstrahlung comparable to the degradation by initial state radiation. Hence, the machine is designed such that the optimum value of $N/\sigma_x$ can be reached. The vertical beam size depends on the vertical beta-function and emittance at the interaction point $\sigma_y = \sqrt{\beta_y\varepsilon_y/\gamma}$. The beta-function cannot be made much smaller than the bunch length in order to avoid luminosity losses due to the hourglass effect. Thus the vertical emittance is reduced as much as possible, with limits arising from the lattice designs and dynamic and static imperfections in the beam transport system. A small value of $\sigma_y$ also has a strong impact on the beam–beam collision dynamics and tolerances.

Expressing the beam power in terms of the wall plug power consumption with $n_{ac}$, the wall plug to beam transfer efficiency, and assuming the optimization above, the luminosity becomes:

$$L \propto n_\gamma \frac{n_{ac}P_{ac}}{\sqrt{\gamma\varepsilon_y\sigma_z}}$$

The figure of merit of linear colliders independent of the adopted technology is defined [3] as the luminosity normalized to the wall plug power consumption and the background at collision (beamstrahlung photons):

$$\text{FoM} = \frac{L}{n_\gamma P_{ac}} \propto \frac{n_{ac}}{\sqrt{\varepsilon_y\sigma_z}}$$

which favors high wall plug to beam efficiency, small beam vertical emittance and short bunch length.

### 2.3 *Beam and technological challenges*

A performing high energy, high luminosity linear collider therefore requests beams and accelerator systems with parameters well above the present state of the art, namely:

- Accelerate large beam powers (several MW) with accelerating fields and wall plug to beam transfer efficiency as high as possible
- Generate beam transverse emittances as low as possible (in the nm range)
- Preserve the beam emittances during acceleration and focusing in the final focus which imposes tight alignment (in the $\mu$m range) and stabilisation (in the nm range) of the accelerator components
- Compress the bunch to small length in the tens of microns range
- Focus the beam to extremely small dimensions especially in the vertical plane (in the nm range).

## 3 CLIC & ILC, two complementary technologies

### 3.1 *Introduction*

Following the successful development and operation of the 100 GeV SLAC Linear Collider (SLC) [4], the only linear collider built so far and described in Chapter 5, about 25 years of R&D exploring various alternatives have greatly advanced understanding of the projected performance of linear colliders. Collaborations have been formed to develop two complementary technologies with different potential energy reach and performances:

- Following the ICFA recommendation [5] for a linear collider in the TeV energy range, the design of the International Linear Collider (ILC) [6] has been launched by a global collaboration coordinated by the Global Design Effort (GDE). The ILC is based on beam acceleration by RF superconducting cavities with initial colliding beam energy of 500 GeV, upgradable to 1 TeV. The ILC Reference Design Report (RDR) [7], authored by 325 institutes, was published in 2007 and a Technical Design Report (TDR) has been published in 2013.
- The Compact Linear Collider (CLIC) [8] study is exploring the possibility of a linear collider with a multi-TeV energy range through the development of Two Beam Acceleration, a novel technology. The feasibility of this technology over energies ranging up to 3 TeV is under

Table 1. Basic parameters for ILC and CLIC.

| | ILC | CLIC |
|---|---|---|
| Centre-of-mass energy | 500 GeV (upgradable to 1 TeV) | 3 TeV (first stage at 500 GeV) |
| Total luminosity ($cm^{-2}s^{-1}$) | $2.0(1.5) \cdot 10^{34}$ | $5.9(2.0) \cdot 10^{34}$ |
| Total site length (km) | 31 | 48.3 |
| Loaded accel. gradient (MV/m) | 31.5 (33) | 100 (80) |
| Main linac techno. & RF frequency | superconducting @ 1.3 GHz | normal conducting @ 12 GHz |
| Beam power/beam (MW) | 10 | 14 |
| Bunch charge ($10^9\, e^{+/-}$) | 20 | 3.72 |
| Bunch separation (ns) | 369 | 0.5 |
| Beam pulse duration ($\mu$s) | 960 | 0.156 |
| Repetition rate (Hz) | 5 (4) | 50 |
| Hor./vert. norm. emitt ($10^{-6}/10^{-9}$) | 10/40 | 0.66/20 |
| Hor./vert. IP beam size (nm) | 639/5.3 | 40/1 |
| Beamstrahlung photon/electron | 1.3 | 2.2 |
| Wall plug to beam transfer eff. (%) | 8.7 | 4.8 |
| Total power consumption (MW) | 230 | 582 |

study by a global collaboration [9] consisting of 44 institutes. The CLIC Conceptual Design Report (CDR) has been published in 2012 [10].

These two studies, with basic parameters shown in Table 1, aim to devise the most appropriate facility to complement the LHC. A collaboration between CLIC and ILC that takes advantage of the overlapping portions of the two schemes has been launched and has proven to be extremely fruitful.

### 3.2 *ILC design*

The design of the International Linear Collider (ILC) is based on 1.3 GHz superconducting RF technology (SCRF), thus taking advantage of the high RF to beam transfer efficiency of the well proven superconducting technology. The configuration of the linac power and utility infrastructure is conventional and is based on klystron sources and waveguide distribution with hybrid matching circuits and RF loads. The configuration has been demonstrated and is well established in each of three regions: Europe (at CERN-LEP and DESY), Asia (at KEK — TRISTAN and KEKB), and America (at Jefferson Lab and SNS). The European FEL project [11], being implemented at DESY corresponding to about a tenth of the ILC linac with a very similar technology, constitutes an ideal validation and industrialization of the ILC technology.

A train of 2625 bunches is accelerated during a 1 ms macro-pulse at a repetition rate of 5 Hz. At the average gradient of 31.5 MV/m, each 1 meter long cavity delivers 300 KW to the beam close to 100% efficiency. Additional RF power (23%) is available as needed for higher gradient operation, stabilization feedback system overhead and to compensate for waveguide distribution system losses. The long time scale of the 960 $\mu$s macro-pulse, with 369 ns between bunches, provides the time needed for effective intra-train trajectory, energy and interaction region collision feedback resulting in very relaxed mechanical vibration tolerances. Accelerating cavity positioning tolerances are also relaxed due to the very large 76 mm diameter clear aperture of the accelerating cavities. The overall layout of a 500 GeV ILC is shown on Figure 1 with two centrally-positioned detectors in a push-pull working mode located within the electron and positron damping rings. It also shows the mid-linac undulator-based positron source.

### 3.3 *CLIC design*

The CLIC design is based on a novel Two Beam Acceleration (TBA) scheme where a high intensity drive beam running all along the linacs accelerating the main beams. The CLIC main linacs are made of normal-conducting structures resonating at RF frequency of 12 GHz with average accelerating fields of 100 MV/m resulting from a trade-off between the linac extension, therefore the civil engineering investment cost and the required RF power, therefore its exploitation cost. The X band frequency of 12 GHz also results

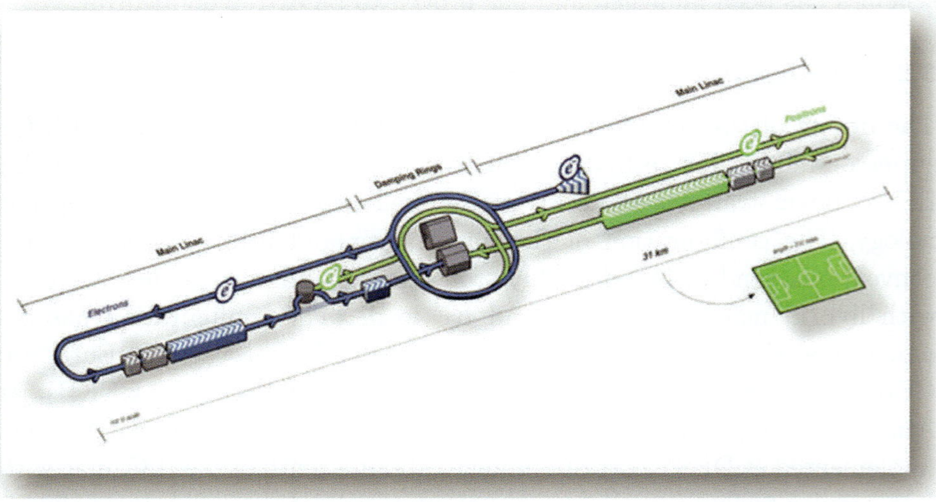


Fig. 1. ILC overall layout.

from a trade-off between the required RF power (scaling with inverse square of RF frequency) and the corresponding wake-fields which limit the charge per bunch, therefore the luminosity.

The RF power of 275 MW/m which is necessary to feed the main linac accelerating structures with high field is efficiently generated by the TBA scheme where the energy of a high intensity drive beam is converted into RF power by specially designed Power Extraction and Transfer Structures (PETS). The 95 A drive beam is generated from a 139 $\mu$s long train of bunches accelerated by a 2.4 GeV normal conducting linac at low intensity and low frequency working in fully loaded mode with 94% RF to beam efficiency. The train is then compressed by funnelling in a delay loop and two combiner rings, multiplying the beam intensity and frequency by a factor 24 thus providing series of trains with the required 95 A current and 12 GHz bunch repetition frequency. Each train is then used to power one 870 m long sector of the main linac providing a main beam acceleration of 62.5 GeV. Upgrade in energy by adding sectors powered by additional drive beam generated by the same drive beam generation complex is particularly cost effective.

The overall layout with a 3 TeV CLIC facility equipped with two detectors in push-pull mode is shown in the right part of Figure 2 whereas the principle of the two-beam scheme is displayed on the left part.

## 3.4 *R&D*

### 3.4.1 *ILC specific*

Superconducting RF technology was chosen by the International Technology Review Panel in 2004 in part because it has a well established industrial infrastructure. Excellent quality ILC-style cavities are available from fabrication companies in each of three regions: Americas, Asia and Europe.

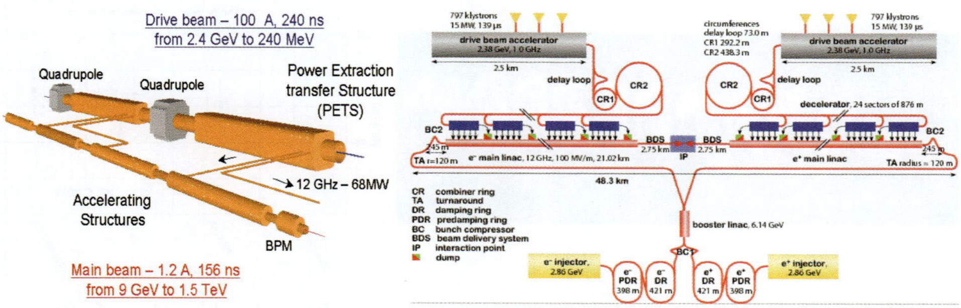


Fig. 2. CLIC two-beam schema and overall layout.

Linacs based on this technology are now under construction in each region, the largest being the 1 km EU-XFEL, scheduled to start operation in 2014. Design development is focused on the application of the technology, on cost reduction and preparation for mass production. The ILC linac R&D effort is aimed at developing and sharing the optimum fabrication and surface treatment process information to vendors to allow each to perfect their production infrastructure. Superconducting RF linac technology is used for many purposes and was highlighted in each section of the recently published US Department of Energy Report on "Accelerators for America's Future", [12]. Ongoing global development investment is about 100 M$ per year. The cost-effectiveness of SCRF technology is expected to steadily improve from this strong investment.

The global cavity test team has devised a standard design and standard process criteria to use for assessing performance so that performance can be compared between collaboration members. Of the cavities passing these criteria, more than half exceed the specified average gradient and cryogenic loss for low-power test, 35 MV/m, indicating the standards are well understood. Cavity gradient test results have been cross-calibrated in the beam test facility at DESY. R&D is underway to support operation in the TeV range aimed at: 1) reducing cavity fabrication cost through sheet-metal forming technology; 2) surface coating techniques, as pioneered at LEP; and 3) increasing the gradient performance with alternate cavity shapes. For the latter, a gradient performance improvement of 45% (52 MV/m) has been observed [13]. A picture of one standard superconducting structure and of the progress in accelerating field and yield after RF conditioning is shown in Figure 3.

R&D is also underway on various non-linac related subsystem technologies, such as the electron/positron sources, damping rings, and beam

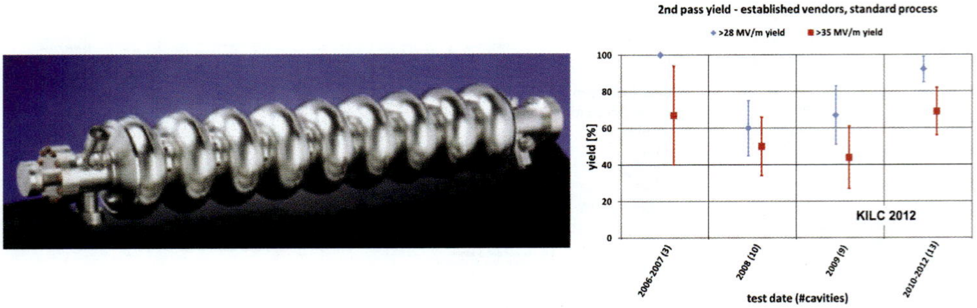


Fig. 3. An ILC superconducting 1.3 GHz RF cavity (left) and achieved performances (right).

delivery. Many of these R&D topics are common with CLIC and are performed in close collaboration.

### 3.4.2 *CLIC specifics*

The feasibility of the novel two-beam scheme is being addressed in an ambitious CLIC Test Facility (CTF3) [14] which consists of a complex of accelerators for drive beam generation and a CLIC Experimental Area (CLEX) (left part of Figure 4). The drive beam is used to test the Two Beam Acceleration scheme of a probe beam with standard modules integrating all necessary components including RF structures, quadrupoles, instrumentation, vacuum beam alignment and stabilisation (right part of Figure 4).

An accelerating gradient of 100 MV/m with the specified breakdown rate of $3 \times 10^{-7}$/m has been demonstrated in a few accelerating structures equipped without high order modes damping features as shown on Figure 5. Prototype of PETS structures already achieved the required performance of 136 MW during 140 ns with low breakdown rate. It is equipped with ON/OFF mechanism in order to allow RF conditioning and adjust the RF power delivered during operation to the acceleration structures which demonstrated excellent performances.

Novel methods of alignment in the micron range and stabilisation in the nano-meter range are being tested on specific test benches prior to integration in prototypes of Two Beam Modules. A few fully equipped Two Beam Modules, including focusing quadrupoles, beam instrumentation, alignment and stabilization systems are being built for tests first in the laboratory and then with beam in the experimental hall CLEX of CTF3.

### 3.4.3 *Common issues*

Apart from the linacs based on different technologies, ILC and CLIC have similar supporting sub-systems. This is especially so for the beam delivery

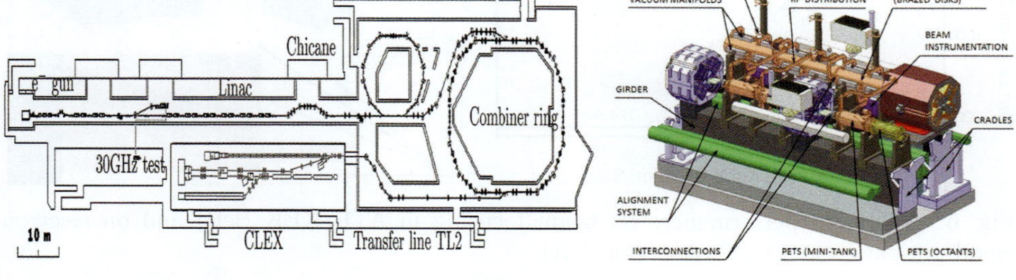


Fig. 4. CLIC Test Facility (CTF3) and a two-beam module.

system, the machine detector interface, the two detector push-pull arrangement and the civil engineering and conventional facilities. To take advantage of the overlapping aspects of the two studies, common working groups have been set-up and actively addressing common issues for both studies including beam dynamics, low beam emittance generation, positron generation, beam delivery system, cost and schedule, and detectors. Issues of low emittance beam generation, electron cloud collective instabilities, and high demagnification beam optics in the interaction region are being tested in test facilities supported by linear collider groups, notably the Accelerator Test Facility (ATF and ATF2 [15]) at KEK and the CESR-Test Accelerator at Cornell [16] as shown on Figure 6.

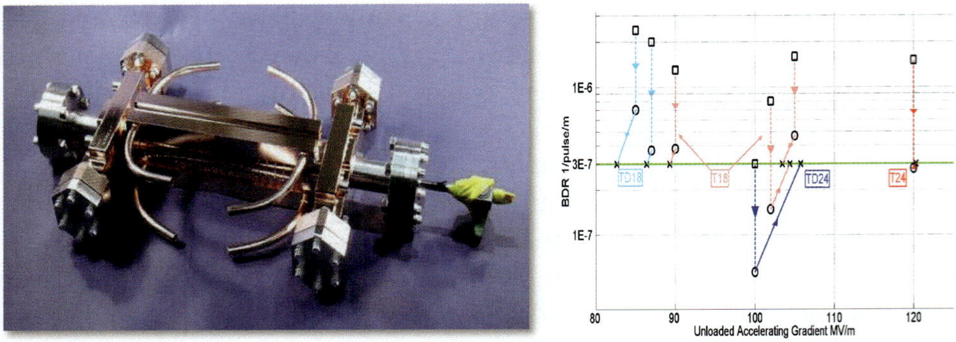


Fig. 5. A CLIC normal-conducting 12 GHz RF cavity (left) and achieved performances (right).

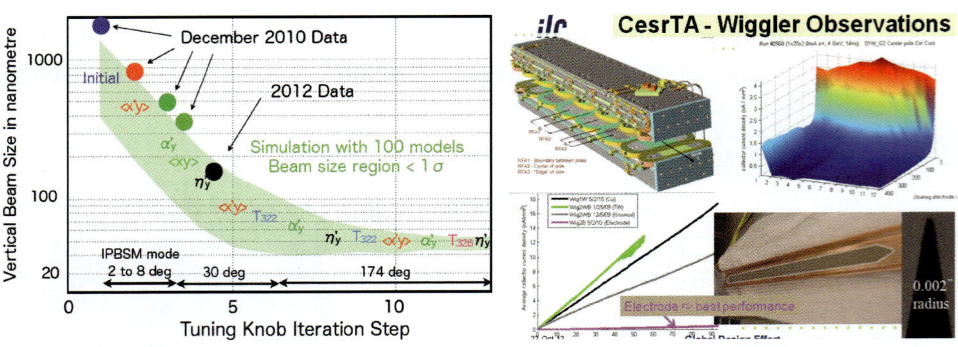


Fig. 6. Achieved performances on beam focusing in ATF/KEK (left) and on electron cloud mitigation at CesrTA/Cornell (left).

## 4 Status and prospective

The ILC Technical Design Report has been published in 2013 with an application requesting project approval following soon after since the technology is demonstrated to be mature enough in the TeV energy range. A CLIC Conceptual Design has been made available mid-2012 assessing the feasibility of this novel technology and extending the energy reach of linear colliders into the multi-TeV energy range. However, R&D towards a technical design is still necessary before aiming at a CLIC Project Implementation Plan by 2016. The optimum energy range for the two designs is thus complementary. The required beam collision energy range will be defined following Physics requirements emerging from LHC results when substantial integrated luminosity will have been accumulated, tentatively by 2013–15. The most appropriate technology for the future facility complementing the LHC will then be defined taking into account risk, technical maturity, performance, and cost.

A new linear collider organization has been recently set-up by the International Committee for Future Accelerators (ICFA) including a Linear Collider Board to oversee the CLIC & ILC accelerators and the Linear Collider Physics and Detectors studies, thus fostering further the CLIC–ILC collaboration towards a single linear collider community.

## References

[1] "ICFA Statement on Linear Colliders" (1999): http://www.fnal.gov/directorate/icfa/icfa_r&d.html (1993) and http://www.fnal.gov/directorate/icfa/icfa_LCstatement.html (1999)
[2] The Muon Accelerator Program (MAP): http://map.fnal.gov/
[3] J.P. Delahaye, G. Guignard, T. Raubenheimer, I. Wilson (1999): Scaling laws for $e^+/e^-$ linear colliders. *Nucl. Instrum. Meth.* **A421**, 369, doi:10.1016/S0168-9002(98)01132-2
[4] SLC: http://www-sldnt.slac.stanford.edu/alr/slc.htm
[5] "ICFA Statement on Linear Colliders" (2004): http://www.fnal.gov/directorate/icfa/icfa_LCstatement0204.html
[6] ILC: http://www.linearcollider.org/
[7] CLIC: http://clic-study.web.cern.ch/
[8] ILC RDR (2007): http://www.linearcollider.org/about/Publications/Reference-Design-Report
[9] http://clic-study.web.cern.ch/content/list-members
[10] CLIC CDR: http://clic-study.web.cern.ch/content/conceptual-design-report
[11] XFEL: http://xfel.desy.de/
[12] http://www.acceleratorsamerica.org/report/index.html
[13] F. Furuta *et al.* (2006): in *EPAC'06*, Edinburgh, June 2006, MOPLS084
[14] CLIC Test Facility, CTF3: http://clic-study.web.cern.ch/content/ctf3-0

[15] Y. Honda *et al.* (2004): *Phys. Rev. Lett.* **92**, 054802; P. Bambade *et al.* (2010): *Phys. Rev. ST Accel. Beams* **13**, 042801

[16] M. Palmer *et al.* (2009): in *Proceedings of the 23rd Particle Accelerator Conference*, Vancouver, Canada, 2009, http://trshare.triumf.ca/~pac09proc/Proceedings/papers/fr1rai02.pdf

# Chapter 30

# Circular lepton colliders as an option for a Higgs factory: The highest energy circular lepton collider

Frank Zimmermann (CERN)

With a maximum centre-of-mass energy of 209 GeV, LEP2, in operation at CERN until 2001, has been the highest energy $e^+e^-$ collider so far. The discovery, in 2012 by two LHC experiments, of a Higgs-like boson at an energy reachable by a collider slightly more energetic than LEP2, together with the excellent performance achieved in the two B factories PEP-II and KEKB during the first decade of the 21st century, have led to new proposals for a next-generation circular $e^+e^-$ collider at the energy frontier [1–5]. In order to serve as a Higgs factory such a collider needs to be able to operate at least at a centre-of-mass energy of 240 GeV (for efficient $e^+e^- \to ZH$ production), i.e. 15% above the LEP2 peak energy. Reaching even higher energies, e.g. 350 GeV centre-of-mass (CM), for $t\bar{t}$ production or 500 GeV for $ZHH$ and $Zt\bar{t}$ studies would be possible for a new ring of larger circumference.

In 2011 and 2012, several concrete proposals of high-energy circular $e^+e^-$ colliders have emerged. For example, as the least expensive option, "LEP3" was proposed to be installed in the LEP-LHC tunnel, the existence of which, together with the associated infrastructure and the LHC detectors, could be an attractive and ecomonical starting point. Another option is a storage-ring collider with three (or four) times the LEP or LHC circumference, originally called "triple LEP," or TLEP. The design of such a machine is now being pursued in the frame of the Future Circular Collider (FCC) study under the name FCC-ee [7, 8]. Similar studies of circular $e^+e^-$ Higgs factories are being pursued in China (CepC), Japan (Super-TRISTAN), Russia, and in the US [5].

The FCC-ee could deliver extremely high luminosities for CM energies below 240 GeV and it could also operate above the $t\bar{t}$ threshold at a CM energy of 350 GeV with a luminosity of still a few $10^{33}$ cm$^{-2}$s$^{-1}$ at each of

Table 1. Parameters of circular $e^+e^-$ colliders, comparing actual values for LEP2 with design values for LEP3 (26.7 km) and FCC-ee (100 km) at various energies of interest. The value for the rms bunch length refers to non-colliding bunches. In collision the bunch length increases as a result of beamstrahlung, an effect which is taken into account for the FCC-ee luminosity numbers. The beam lifetime quoted is due to the unavoidable radiative Bhabha scattering. The FCC-ee option for 90 GeV marked with the superindex * refers to a low-emittance crab-waist scheme [9].

| | LEP2 | LEP3 | | FCC-ee | | | |
|---|---|---|---|---|---|---|---|
| CM energy [GeV] | 209 | 90 | 240 | 90 | 90* | 240 | 350 |
| number of IPs | 4 | 4 | | 4 | | | |
| no. bunches | 4 | 400 | 4 | 16700 | 29791 | 1360 | 98 |
| $e^\pm$/bunch [$10^{11}$] | 4.2 | 5 | 10 | 1.8 | 1.0 | 0.46 | 1.4 |
| $\epsilon_x$ [nm] | 48 | 25 | 25 | 29 | 0.14 | 1 | 2 |
| $\epsilon_x/\epsilon_y$ [1000] | 0.2 | 0.25 | 0.25 | 0.5 | 0.1 | 0.5 | 1 |
| $\beta_y^*$ [cm] | 5 | 0.1 | | 0.1 | | | |
| $\sigma_{z,\mathrm{rms}}^{\mathrm{SR}}$ [mm] | 16.1 | 2.3 | 3.1 | 1.6 | 2.7 | 0.8 | 1.2 |
| $L$ [$10^{33}$ cm$^{-2}$s$^{-1}$] | 0.1 | 250 | 10 | 280 | 2120 | 60 | 18 |
| beam lifetime [min] | 360 | 14 | 8 | 298 | 39 | 29 | 21 |
| circumference [km] | 26.7 | 26.7 | | 100 | | | |
| $\Delta E$/turn [GeV] | 3.5 | 0.1 | 7 | 0.03 | 0.03 | 1.7 | 7.6 |
| RF voltage [GV] | 3.6 | 1 | 12 | 2.5 | 0.54 | 5.5 | 11 |
| RF freq. [MHz] | 352 | 700 | | 800 | 300 | 800 (or 400) | |
| SR power [MW] | 23 | 100 | | 100 | | | |

up to four collision points. The same two machines could deliver an unprecedentedly high luminosity, of up to several $10^{36}$ cm$^{-2}$s$^{-1}$ for FCC-ee at the $Z$ pole (91 GeV CM), and also operate at the $WW$ threshold (160 GeV CM).

Table 1 compares parameters for LEP3 and FCC-ee with those of LEP2. Figure 1 illustrates the dependence of FCC-ee luminosity on beam energy. The table and figure also include some proposed alternative parameter sets based on lower transverse emittance and crab-waist collisions [9], which at the $Z$ pole may provide almost ten times higher luminosity than the baseline configuration.

For both FCC-ee and LEP3, on the $Z$ resonance significant beam polarisation, up to 80% [6], appears possible, but this requires further investigation, especially when operating with very high luminosity. In FCC-ee some transverse polarization, allowing for precise energy calibration, may be obtained also at the $WW$ threshold.

Figure 2 compares the expected measurement precision for various Higgs-boson couplings from the LHC, LEP3, FCC-ee and linear $e^+e^-$ colliders, as

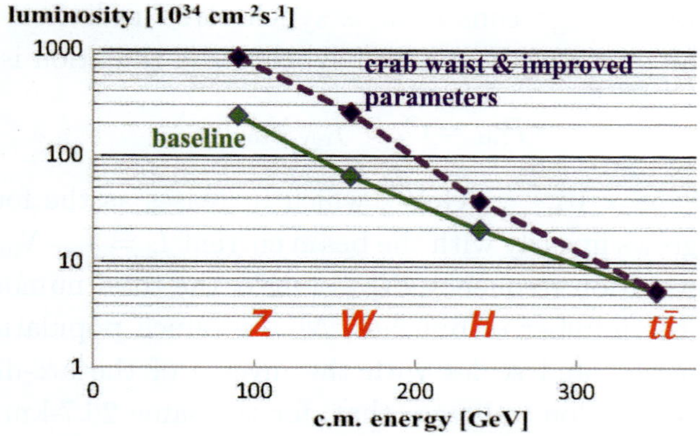


Fig. 1. Total luminosity of FCC-ee with 4 collisions points versus beam energy [7]. The improved parameters are discussed in Ref. [9].

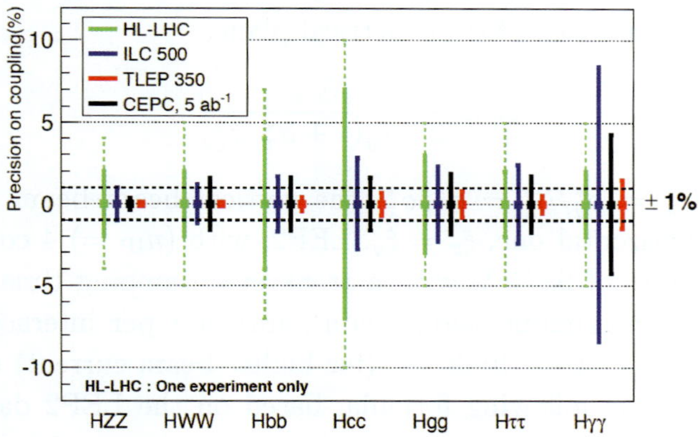


Fig. 2. Expected precision of Higgs coupling measurements at the LHC and various $e^+e^-$ colliders [5, 10–12].

compiled in [5, 10–12]. According to this comparison, FCC-ee (TLEP) offers the most precise measurements.

Further arguments in favor of a circular Higgs factory are the comparatively moderate extrapolation from previous colliders (LEP2, KEKB, PEP-II, SuperKEKB), similarities with modern synchrotron light sources, reliance on well-known and established technologies, and large potential synergy with a future high-energy proton collider such as the FCC-hh, as well as with a highest-energy electron–hadron collider (FCC-he). Indeed, the FCC-hh could share the tunnel, cryogenics, parts of the experimental detectors and, perhaps, even its injector arc magnets with an $e^+e^-$ Higgs factory like FCC-ee.

An important design constraint is synchrotron radiation. In a circular storage ring the power of the emitted synchrotron radiation is given by

$$P_{\rm SR} = C_\gamma E^4 f_{\rm rev} N_{\rm tot}/\rho_b \;, \tag{1}$$

where $C_\gamma = 8.858 \times 10^{-5}$ m/GeV$^3$, which increases as the fourth power of beam energy, grows linearly with the beam current $I_e = f_{\rm rev} N_{\rm tot} e$ (where $f_{\rm rev}$ denotes the revolution frequency, $N_{\rm tot} = n_b N_b$ the total number of particles per beam, $n_b$ the number of bunches, $N_b$ the bunch population, and $e$ the elementary charge), and scales with the inverse of the arc-dipole bending radius $\rho_b$. This equation indicates that, for the same 26.7-km ring size and the synchrotron radiation (SR) power the beam current at 120 GeV beam energy would be about 60% of the LEP2 current (at 104.5 GeV).

Traditionally the beam–beam limit of storage-ring colliders can, to first approximation, be characterized by a maximum value for the beam–beam parameter defined as, e.g. for the vertical plane,

$$\xi_y \equiv \frac{N_b r_e}{2\pi\gamma\epsilon_y(1+\sigma_x^*/\sigma_y^*)} \;. \tag{2}$$

Choosing $\epsilon_y/\epsilon_x = \beta_y^*/\beta_x^* = \sigma_y^*/\sigma_x^*$, the vertical beam–beam parameter is equal to the horizontal one, $\xi_x = \xi_y$. LEP2, with ($n_{\rm IP} =$) 4 collision points at a beam energy of 98 GeV with a transverse damping time of 73 turns, demonstrated a maximum beam–beam parameter per interaction point of 0.083, and an asymptotic limit $\xi_y^\infty$ (for higher beam current) of 0.115 [13]. From Ref. [13], the following formula, based on the LEP2 data, describes the variation of the asymptotic beam–beam parameter with beam energy, bending radius, and number of collision points:

$$\xi_y^\infty \approx 0.115 \left(\frac{150\, C_\gamma E^3}{n_{\rm IP}\rho_b}\right)^{0.4} \;. \tag{3}$$

Concerning the parameter choice, for a given electric power and RF-system efficiency, Eq. (1) determines the maximum beam current. For the FCC-ee at a beam energy of 175 GeV the latter amounts to a few mA as for LEP2. Maximum luminosity at high energy is then reached with a small number of bunches, where the maximum bunch intensity, close to $10^{12}$ particles, may be limited by collective effects. The FCC-ee design aims at providing a high-energy horizontal emittance similar to, or smaller than, the one of LEP2, which is needed to approach the beam–beam limit of Eqs. (2) and (3). The luminosity is further maximized by squeezing the vertical IP beta function, $\beta_y^*$, down towards 1 mm.

For lower beam energy, assuming a constant electric power, the beam current can be increased. At the $Z$ pole the beam current then reaches values of order 1 A, which allows for extremely high luminosities. With a fixed optics, the geometric emittance varies like the square of the energy. Therefore, to enable operation at these lower beam energies the arc optics may need to be changed (i.e. the cell-length could be increased by switching off arc quadrupoles) so that the emittance does not shrink too much and that the beam–beam tune shift does not exceed its expected limiting value. The latter could also be accomplished by the crab-waist scheme, in this case keeping the same arc optics at all energies.

The lifetime of colliding beams at LEP2 was determined by radiative Bhabha scattering, with a cross section of about 0.2 barn for a momentum acceptance of about 1% [14, 15]. Radiative Bhabha scattering will also be the major and unavoidable limitation of the beam lifetime for any higher-energy higher-luminosity ring collider. At several orders of magnitude higher luminosity than LEP2, in 4 collision points, the beam lifetime for LEP3 or FCC-ee will be between 5 and 30 minutes, i.e. a circular Higgs factory will efficiently burn the beams to produce physics. The cross section has a logarithmic dependence on the beam energy and on the momentum acceptance. A factor 3 increase in momentum acceptance would improve the beam lifetime due to radiative Bhabha scattering by about 35%.

In view of this short a beam lifetime, to achieve a high average luminosity the circular Higgs factory should be a double ring, as sketched in Fig. 3. Namely the collider ring operating at constant energy is complemented by a second ring installed in the same tunnel (or alternatively by a recirculating linear accelerator) to 'top-up' the collider continuously, as is routinely done

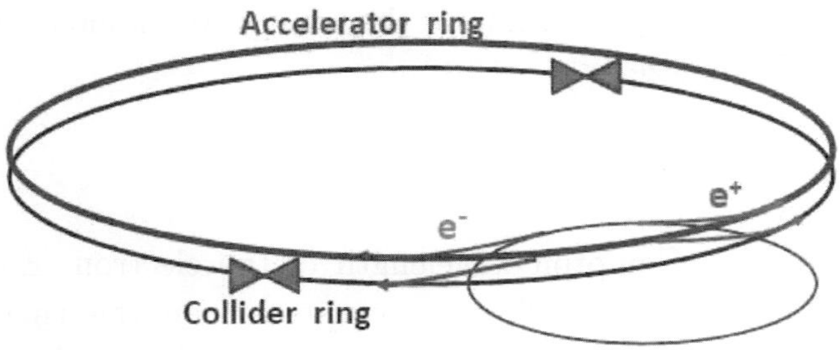


Fig. 3. Sketch of circular Higgs factory based on a double ring [1]: a first ring accelerates electrons and positrons up to the operating energy, e.g. 120 GeV, and injects them, at a few seconds interval, into the low-emittance collider ring, which includes up to four high-luminosity interaction points.

in modern synchrotron light sources and also was the case in the two high-luminosity B factories PEP-II and KEKB. If the top-up interval is short compared with the beam lifetime the average luminosity can be close to the peak luminosity. Assuming the same energy ramp speed as for the CERN SPS the injector ring could have a cycle time well below 10 s. The LEP injector complex delivered positrons at a rate of order $10^{11}$ per second [16]. Such a rate would be adequate for operation at 240 or 350 GeV CM. However, high-luminosity running at the $Z$ pole would require a much higher rate of positron production.

Aside from Bhabha scattering another effect may restrict the lifetime, namely beamstrahlung [17] — synchrotron radiation in the field of the opposing beam emitted during the collisions. The far tail of the beamstrahlung energy spectrum corresponds to beam particles which lose a significant fraction of their energy. Considering Gaussian bunches, the minimum effective bending radius at the center of the collision is of order $\rho_{b,\text{coll}} \approx \gamma\sigma_x\sigma_z/(r_e N_b)$, which allows estimating the critical energy of the beamstrahlung as

$$\frac{E_{c,\text{bs}}}{E} \approx \frac{\gamma r_E^2 N_b}{\alpha\sigma_x\sigma_z}\,, \tag{4}$$

where $\alpha$ denotes the fine structure constant. This critical energy depends on the horizontal beam size and bunch length, but not on the (much smaller) vertical beam dimension. For a reasonable beam lifetime from beamstrahlung the critical energy should be sufficiently small, namely [17]

$$E_{c,\text{bs}} \leq 0.1\eta E \tag{5}$$

where $\eta$ denotes the relative momentum acceptance, and the 0.1 is a numerical factor related to the required value of the exponent in the expression for the synchrotron radiation spectrum [17]. Combining the previous two equations yields the following condition for the product of momentum acceptance and horizontal beam size

$$\eta\sigma_x \geq 10\; r_e ƛ_c \frac{N\gamma}{\sigma_z} \tag{6}$$

with $ƛ_c$ the reduced Compton wavelength of the electron. Eq. (6) reveals how the beamstrahlung introduces a constraint on the ring momentum acceptance and on the horizontal IP beam size, and how these two can be traded against each other. In particular, an acceptable beam lifetime from beamstrahlung can be achieved by either of the following two approaches or a combination thereof: (1) providing enough longitudinal acceptance (RF voltage) and a sufficiently large off-momentum dynamic aperture over a radiation damping time (typically 10–100 turns for the machines considered); (2) increasing the horizontal beam size (via enlarged $\beta_x^*$) while decreasing the vertical emittance by the same factor, which efficiently suppresses the beamstrahlung, but holds the luminosity and the beam–beam tune shift constant.

Vertical emittance tuning and associated tolerances will be an important aspect of a circular Higgs factory. LEP2 operated with an emittance ratio of 200. It should be possible to increase this value with more and better beam-position monitors, using improved steering algorithms. The SuperKEKB design aims at a ratio of 400. Modern light sources achieve values of $10^4$, albeit without any collisions.

Most importantly, the luminosity of the circular Higgs factory is given by

$$L = f_{\mathrm{rev}} \frac{n_b N_b^2}{4\pi\sigma_x^*\sigma_y^*} R_L, \tag{7}$$

where $R_L = \sqrt{2/\pi}\, a e^{a^2} \mathrm{K}_0(a^2)$, with $a \equiv \beta_y^*/(\sqrt{2}\sigma_z)$ denoting the geometric luminosity-reduction factor due the hourglass effect. Using Eqs. (1) and (2), the luminosity (7) can be rewritten as

$$L = \frac{1}{2C_\gamma r_e m_e c^2} P_{\mathrm{SR}} \frac{\rho_b}{E^3} \frac{\xi_y}{\beta_y^*} \left( \frac{\sigma_y^*}{\sigma_x^*} + 1 \right) R \tag{8}$$

In view of the asymptotic limit (3), for constant synchrotron-radiation power $P_{\mathrm{SR}}$, the luminosity scales with energy and bending radius as $L \propto \rho_b^{0.6}/E^{1.8}$. A possible optimization strategy for circular Higgs factories is discussed in Ref. [18].

The synchrotron-radiation heat load per unit length of about 800 W/m per beam for FCC-ee is 10 times smaller than the heat load accommodated in routine operation at PEP-II and at SPEAR3. Therefore, vacuum stability and cooling are not expected to be major issues. However, the critical photon energy at FCC-ee is much higher than for the aforementioned two machines, namely of the order of 1 MeV, so that neutron production becomes a concern [19]. For operation at the $t\bar{t}$ threshold the FCC-ee photon spectrum will, in fact, be quite similar to the one for LEP2 at its maximum energy. Adequate shielding will be important, using materials with high neutron threshold. This shielding may be facilitated by introducing separate arcs for the electron and positron beams, since in a given vacuum chamber the synchrotron radiation will then only come from one direction. The shielding, as well as the cooling, could be concentrated at discrete photon absorbers protruding into the vacuum chamber, e.g. located behind every arc bending magnet. Having separate arcs for the electron and positron beams will offer other, possibly more important advantages. So will this separation avoid parasitic collisions at low beam energy, when operating with a large number of bunches, as well as optics problems associated with a different "energy sawtooth" of the two beams at high energy [20].

The most critical hardware component of FCC-ee is the radio-frequency (RF) system. At the $t\bar{t}$ threshold an RF voltage of 11–12 GV needs to be provided. The 700–800 MHz SC cavity technology being developed for SPL and ESS seems to be a good choice, as would be a frequency around 400 MHz, already used in the LHC and planned for eRHIC. Pertinent R&D is ongoing at BNL, CERN, and ESS for cavities and components in this frequency range. A frequency of 802 MHz would be synergetic with the harmonic RF systems for SPS and LHC as well as with the contemplated main RF system of the LHeC.

With about 20 MV/m cavity gradient, for FCC-ee at 175 GeV beam energy, the total cryomodule length is estimated to be about 1000 m, which is similar to the past 812 m of cryomodule length at LEP2. With a cavity $Q_0$ value of $2 \times 10^{10}$ the total dynamic heat load at 1.9 K would be 25 kW, comparable to the existing LHC cryoplant capacity.

The large beam energy loss from synchrotron radiation implies that a power close to 200 kW needs to be provided for each $\sim$1-m long cavity. This appears well within reach of present technology [21]: Water and air-cooled fundamental power couplers developed for ESS and SPL have been tested up to 1.2 MW with 10% duty cycle. LHC windows are routinely tested above 500 kW in cw.

The average higher-order-mode (HOM) power per cavity would be of order 5 kW for FCC-ee at 240 GeV CM. A design and prototype exists for eRHIC which can handle up to 7.5 kW, and another one for KEKB with an expected 15 kW of HOM power.

A posssible RF power source would be klystrons, which are available on the market in the 400–800 MHz frequency range from three different companies, and show 62–65% efficiency. The high voltage power converters for the RF system could be thyristor-based with 95% efficiency, but non-optimum AC power quality, or switched-mode converters with better quality, albeit only 90% efficiency. The klystron efficiency of about 65% is reached only if the klystron is run at saturation without headroom for RF feedback, as in LEP1 and LEP2. Also taking into account the RF distribution losses of 5 to 7% in waveguides and circulators, the overall RF efficiency (wall to beam) is estimated to be between 54% and 58%. The total electric power for the RF system and the cryo system combined is then extrapolated at about 210–220 MW for FCC-ee at the $t\bar{t}$ threshold, and potentially much less at lower energies [21]. At energies of 240 GeV (CM) and below, where the maximum voltage per beam does not exceed 6 GV, the RF system could be decoupled between the two beams, allowing for a complete separation of the electron and positron beams, except for the primary interaction points, with no parasitic collisions anywhere in the machine.

At the maximum beam energy, 175 GeV, the collider-ring dipole magnets need to produce a field of around 500 G. At 46 GeV the field is only 140 G. At the possible injection energy of 20 GeV the dipole field of the collider ring would be 60 G; at 10 GeV it would be only 30 G. The challenge for the designer will be to provide energy-efficient, inexpensive low-field magnets. The strength and length of the quadrupole magnets depend on the detailed optics design. Parameter optimization will aim at minimizing the electric power required for all the magnets.

SuperKEKB, the beam commissioning of which will start in 2015, is set to explore and demonstrate many of the key concepts of a circular Higgs factory: Its $\beta_y^*$ is only 0.3 mm, or 3–4 times smaller than the design value for FCC-ee or LEP3. Its beam lifetime — limited by Touschek scattering — will be about 5 minutes, even shorter than for FCC-ee, and also maintained by top-up injection. To even achieve so short a lifetime, SuperKEKB will need to establish a sufficiently large off-momentum dynamic aperture of $\eta \approx \pm 1.5\%$. The SuperKEKB design emittance ratio $\epsilon_x/\epsilon_y$ of about 400 is a factor of 2 better than obtained at LEP2 and similar to the values assumed for FCC-ee. The design $e^+$ production rate for SuperKEKB of $2.5 \times 10^{12}$/s is 5–500 times higher than required by FCC-ee (depending on the beam energy).

In conclusion, a circular Higgs factory could be the best complementary machine to the LHC, allowing highest-precision measurements of Higgs properties and stringent tests of SM closure. FCC-ee is based on a well known technology, supported by much progress in $e^+e^-$ circular factories and light sources in the two decades following the start of LEP, and with fairly reliable performance predictions. A circular Higgs factory would be a first step in a long-term vision for particle physics, offering many synergies, and sharing much of the infrastructure, with a subsequent hadron collider (FCC-hh) delivering *pp* collisions at 100 TeV CM.

## References

[1] Blondel, A. and Zimmermann, F. (2011). A high luminosity $e^+e^-$ collider in the LHC tunnel to study the Higgs boson, CERN-OPEN-2011-047, arXiv:1112.2518v1

[2] Oide, K. (2012). SuperTRISTAN: A possibility of ring collider for Higgs factory, KEK Seminar, 13 February
[3] Zimmermann, F., Koratzinos, M., Blondel, A. and Zanetti, M. (2012). A high luminosity $e^+e^-$ collider in the LHC tunnel to study the Higgs boson, *Proc. IPAC'12*, New Orleans, p. 2005
[4] Blondel, A. *et al.* (2012). LEP3: A high luminosity $e^+e^-$ collider to study the Higgs boson, arXiv:1208.0504, submitted to PRST-AB.
[5] ICFA Beam Dynamics Workshop on Accelerators for a Higgs Factory: Linear vs. Circular "HF2012," 14–16 Nov 2012, Fermilab, Batavia, IL, USA; website *indico.fnal.gov/conferenceDisplay.py?confId=5775*
[6] Abelleira Fernandez, J.L. *et al.* (2012). *J. Phys. G: Nucl. Part. Phys.* **39**, 075001.
[7] Benedikt, M. *et al.* (2014). Challenges for highest energy circular colliders, *Proc. IPAC2014*, Dresden
[8] `http://cern.ch/fcc`
[9] Bogomyagkov, A. *et al.* (2014). Beam–beam effects investigation and parameters optimization for a circular $e^+e^-$ collider at very high energies, *PRST-AB* **17**, 041004.
[10] Ellis, J. (2013). Summary and conclusions, *Nobel Symposium on LHC Results*, Uppsala, 17 May 2013
[11] Bicer, M. *et al.* (2014). First look at the physics case of TLEP, *J. High Energy Physics*, 164.
[12] Rui, M. (2014). Higgs measurements at $\mathrm{e^+e^-}$ colliders, *Proc. ICHEP'14*, Valencia, arXiv:1411.5606
[13] Assmann, R. and Cornelis, K. (2000). The beam–beam interaction in the presence of strong radiation damping, *Proc. EPAC2000*, Vienna, p. 1187
[14] Burkhardt, H. and Kleiss, R. (1994). Beam lifetimes in LEP, *Proc. EPAC1994*, London, p. 1353; Burkhardt, H. (1999). Beam lifetime and beam tails in LEP, CERN-SL-99-061-AP
[15] Assmann, R.W. (2001). LEP operation and performance with electron–positron collisions at 209 GeV, presented at *11th Workshop of the LHC*, Chamonix, France, 15–19 January 2001
[16] Bossart, R. *et al.* (1990). The LEP injector linac, CERN-PS-90-56-LP
[17] Telnov, V. (2012). Restriction on the energy and luminosity of $e^+e^-$ storage rings due to beamstrahlung, arXiv:1203.6563
[18] Talman, R. (2013). Circular $e^+e^-$ Higgs factory, *Higgs Factory Muon Collider Workshop*, UCLA, 21–23 March 2013
[19] Fasso, A. (2013). TLEP3 — How radiation issues could be studied through FLUKA simulations. Possible approaches/mitigation schemes, *3rd EuCARD TLEP Workshop*, CERN, 10 January 2013
[20] Oide, K. (2012). SuperTRISTAN, at *ICFA Beam Dynamics Workshop "HF2012,"* 14–16 Nov 2012, Fermilab, Batavia, IL, USA; website *indico.fnal.gov/conference Display.py?confId=5775*
[21] Butterworth, A. and Jensen, E. (2013). RF system for TLEP, *4th EuCARD-AccNet TLEP Workshop*, CERN, 4–5 April 2013

# Part 3

# Particle Accelerators for the XXI Century

# Chapter 31

# HIE-ISOLDE: The future of radioactive beam physics at CERN

Yacine Kadi*, Yorick Blumenfeld*, Richard Catherall*, Walter Delsolaro Venturini*, Maria Jose Garcia Borge*,†, Mark Huyse‡ and Piet Van Duppen‡
*CERN
†CSIC, Spain
‡Katholic University, Belgium

## 1 Introduction

The High Intensity and Energy (HIE)-ISOLDE project aims at several important upgrades of the present ISOLDE radioactive beam facility at CERN [1]. The main focus lies in the energy upgrade of the post-accelerated radionuclide beams from 3 MeV/u up to 10 MeV/u through the addition of a superconducting linac. This will open the possibility of many new types of experiments including transfer reactions throughout the nuclear chart. Phase 1 of this upgrade involves the design, construction, installation and commissioning of two high-$\beta$ cryomodules downstream of REX-ISOLDE, the existing post-accelerator previously discussed in Chapter 18. Each cryomodule houses five high-$\beta$ superconducting cavities and one superconducting solenoid. Prototypes of the Nb-sputtered Quarter Wave Resonators (QWRs) cavities for the new superconducting linear accelerator have been manufactured and are undergoing RF cold tests, whilst cryomodule assembly is underway with installation and first commissioning expected in 2015.

The project also aims at improving the target and front-end part of ISOLDE [2] to fully benefit from potential upgrades of the existing CERN proton injectors e.g. LINAC4 and upgrade in energy of the PS Booster.

## 2 Physics reach of HIE-ISOLDE

### 2.1 *Nuclear structure physics*

In spite of 100 years of investigation of the atomic nucleus, nuclear models still face essential questions: What is the nature of the nucleonic matter?

What are the effects of varying the degrees of freedom of the nucleus, such as the internal excitation energy, the angular momentum, the proton to neutron ratio or the temperature? Nuclei far from stability, which are the object of study at ISOLDE, tend to present new phenomena, such as new types of nucleonic aggregations or key interactions not present in stable nuclei. Their experimental study is of paramount importance for obtaining a unified description of the atomic nucleus. Measurements of the ground-state properties are done using traps and laser spectroscopy techniques, while excited states can be approached through radioactive decay and nuclear reactions.

The research program with nuclear reactions will be based on the new superconducting linac and will concentrate on regions of the nuclear chart which are ideal testing grounds for a number of important issues: e.g. fundamental interaction studies using $N \sim Z$ nuclei, the evolution of shell structure in the neutron-rich regions around $Z = 20$, 28 and also $N = 82$, 126 and the relation between collectivity and single-particle behavior in the neutron-deficient region around $Z = 82$. A large part of this research program will be covered by MINIBALL, an array of high purity Germanium detectors for gamma-particle detection in Coulomb excitation reactions and an upgraded Si-strip array (T-REX) used for transfer reactions. Other instrumentation is planned. A Helical Orbit Spectrometer (HELIOS) will represent a new approach to measurements of light ions from nuclear reactions and overcome the limitation in energy and angular resolution of the conventional approaches. A new generation active target detector (ACTAR) will allow exploiting direct, transfer and resonant nuclear reactions even at beam intensities as low as a few pps, complementing the existing setups.

## 2.2 *Nuclear astrophysics*

Nuclear astrophysics searches to explain the chemical element abundance in the Universe. Although the element production up to the iron region has been subject to many studies, the questions of energy production in stars and of nuclear reactions producing elements beyond iron have not yet been answered. Nuclear processes in stars and stellar explosions are held responsible for the production of these heavy elements but it has turned out to be a tremendous challenge to match specific events to the observed element abundance patterns. The pathways of such reactions clearly involve exotic nuclei very far from stability, but nuclear theory is still far from being able to predict their properties. Within the scientific opportunities envisaged for HIE-ISOLDE, it will become possible in the near future to study capture reactions originating from novae explosions or X-ray bursts. In both cases, not only the proton-capture process and the rapid proton-capture process (r-process and rp-process, respectively) need to be investigated but also beta decay and reaction rates of other processes, like for example (p,$\alpha$) reactions will be studied.

In this light, the recent proposal of installing a low-energy heavy-ion storage ring at an ISOL facility for the first time is of great importance. The post-accelerated high-intensity beams available from HIE-ISOLDE will be cooled and stored using the TSR ring. This will not only allow for an extremely rich scientific program in nuclear and atomic physics, but especially in nuclear astrophysics: due to high temperatures and densities, the nuclides involved in stellar nucleosynthesis in stars are — as a rule — not in a neutral atomic charge state but highly ionized. Since the TSR will be able to provide ions in various atomic charge states up to bare nuclei, experimental investigations on these radioactive decays will become possible.

The r-process is thought to be the major production mechanism of most elements heavier than iron and runs through the extremely neutron-rich regions of the nuclear chart. Although sophisticated astrophysical models for the r-process have evolved in recent years, the astrophysical conditions for a successful r-process have not been identified yet, and r-process model predictions still suffer from large uncertainties. An alternative theory is the process of neutronization occurring in the high-density crust of neutron stars. The stable nuclei known on earth cannot form in the gaseous environment of neutrons and electrons but nuclei lying far out on the neutron-rich side of the nuclear chart become so-called equilibrium nuclei. As stable nuclides within the neutron-star crust they can contribute to the elemental abundance. In both cases, precise mass values are important input parameters to constrain models of stellar element composition and to test their predictive power in comparison with observations. Since ISOLDE will profit from up to 2 GeV proton energy enhancing spallation as well as proton currents of up to 5 $\mu$A enhancing fission products, there will be an ideal opportunity to measure the masses of very neutron-rich nuclei. These will also help to improve mass models which are used to predict masses of nuclei participating in the astrophysical creation process whenever experimental values are not (yet) available. In addition, the s-process (slow neutron-capture process) relies on neutron-capture cross sections as the key ingredient for stellar model calculations, where HIE-ISOLDE could supply high-purity, radioactive samples for experimental studies.

### 2.3 *Low energy particle physics and solid state physics*

At the WITCH (Weak Interaction Trap for CHarged Particles) setup the Penning trap technology is combined with a retardation spectrometer with the aim of determining the beta neutrino angular correlation coefficient $a$ achieving a relative precision of the order of 0.5%. At the moment $a$ is known with a precision of about 1% for just a few transitions and agrees with the Standard Model value. A deviation would probe an interaction which is mediated by a charged scalar boson, as e.g. a charged Higgs particle. In the future it is planned to increase the resolution of the WITCH spectrometer by implementing cryogenic traps, which would reduce the ion energy distribution in the decay trap and thereby one of the major systematic effects of the setup. An energy calibration by trapping nuclei with a strong (monoenergetic) electron capture branch such as the rare earth isotopes $^{144}$Eu and $^{140}$Pm will further contribute to the understanding of the systematic effects at WITCH. Therefore, intensity and purity of these beams need to be significantly improved. Having these beams available, a search for heavy neutrinos becomes feasible by looking for small kinks in the recoil energy distribution. With WITCH one should be able to identify admixtures in the electron capture decay of below 0.5% from heavy neutrinos within a few days of measurement time.

The observation of a non-zero electric-dipole moment (EDM) indicates T-violation beyond the Standard Model. Atoms with nuclei having degenerate parity doublets and large collective octupole deformation will provide the most stringent limits on EDMs. An experimental programme is planned to measure EDMs in odd Ra octupole nuclei at ISOLDE, and will be complemented by the necessary nuclear structure measurements carried out using HIE-ISOLDE.

Solid state physics at ISOLDE is an umbrella term for an experimental program that embraces the study of materials, biophysics and biochemistry. Over the course of the last 15 years, dedicated laboratories and equipment have been developed and installed at and near the ISOLDE complex and the labs — in addition to many other techniques and preparation areas — which host probably the world's largest collection of perturbed angular correlation apparatuses. However, progress in solid state and materials science is rapid: e.g. graphene was only isolated in 2004, and the labs as currently installed reflect the trends and scientific priorities at the time of their installation 10 years ago *viz.* bulk materials such as semiconductors and superconductors.

Bulk materials still constitute a significant proportion of the research carried out at ISOLDE. However the existing laboratories are no longer up-to-date to support first class research in the exciting new areas of nanomaterials and biophysics. The increasing demands from a chemical viewpoint would necessitate the establishment of a dedicated chemical laboratory for the study of biophysics, biochemistry and also medical physics. For nanomaterials, clean-room facilities with additional space for handling and characterization would be desirable. The cause of materials and biological research at ISOLDE would be greatly enhanced by the modernization and expansion of the current laboratories.

## 3 The intensity upgrade: HIE-ISOLDE design study

At the forefront of radioactive ion beam production with its diversity of available beams, the ISOLDE facility continues to strive for higher beam intensities and improved beam quality. While beam intensities can be increased by improving the efficiency of target and ion source systems, one obvious approach is to increase the driver beam intensity. The future LINAC4 at CERN [3, 4] will be able to deliver to ISOLDE up to $1 \times 10^{14}$ compared to the present $3.3 \times 10^{13}$ protons per pulse. In parallel, plans to increase the PS Booster energy to 2 GeV are under way [5] and a proposal to extend this energy upgrade to ISOLDE has been submitted to the CERN Research Board. Finally the Design Study has to consider that a 900 ms super cycle, compared to the existing 1.2 s, may be available in the future. Table 1 summarizes the possible scenarios.

Clearly an increase in primary-beam intensity and energy will have an impact on the operation and maintenance of the existing facility. Issues to be addressed include the resistance of target materials, target lifetime, incidental target heating, shielding, air activation, vacuum and interventions, all of which are covered within the HIE-ISOLDE Design Study. Figure 1 shows the layout of one of the current frontend.

The frontends have been subject to HT breakdowns owing to the incidental pollution of the extraction electrodes during operation, and the only option has been to change the highly contaminated extraction electrode tip. The proposal implies mounting the extraction electrode tip inside the target unit. Acceleration would then be provided by a two-stage extraction process

Table 1. Projected beam parameters considered within the HIE-ISOLDE Design Study. Based on ISOLDE receiving 50% of available proton pulses from the PS Booster.

| Protons/pulse | Intensity ($\mu$A) | Energy (GeV) | Cycle (s) | Power (kW) |
|---|---|---|---|---|
| $3.3 \times 10^{13}$ | 2.2 | 1.4 | 1.2 | 3.1 |
| $1 \times 10^{14}$ | 6.7 | 1.4 | 1.2 | 9.3 |
| $1 \times 10^{14}$ | 6.7 | 2.0 | 1.2 | 13.3 |
| $1 \times 10^{14}$ | 8.9 | 1.4 | 0.9 | 12.4 |
| $1 \times 10^{14}$ | 8.9 | 2.0 | 0.9 | 17.8 |

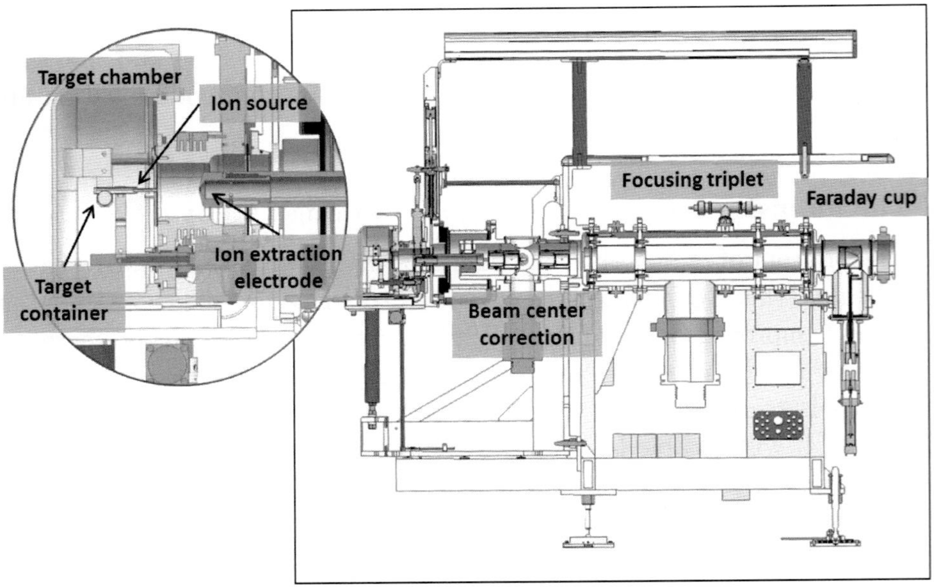


Fig. 1. Frontend 6 (as currently installed) [7].

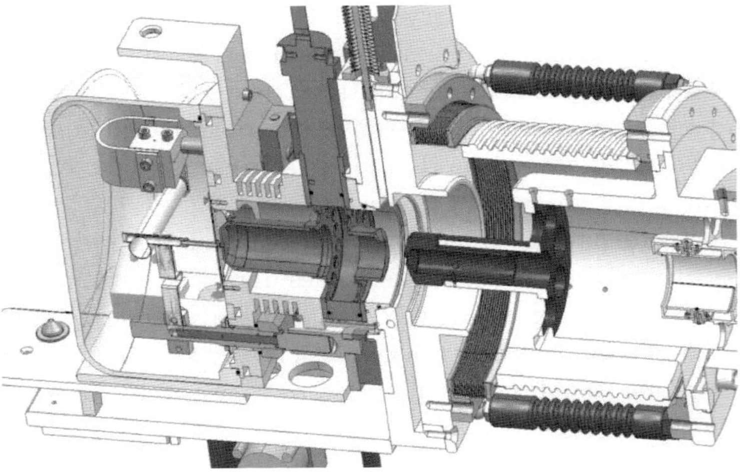

Fig. 2. A proposed pre-extraction prototype with the ion source (green) at 60 kV, the intermediate electrode (red) at 57 kV and the fixed frontend electrode (blue) at earth potential. Courtesy of J. Montano [7].

with one potential on the tip and a second on the fixed extraction electrode inside the frontend, see Fig. 2. Besides having the advantage of renewing the extraction electrode tip after each target change, this approach will also eliminate the existing mechanics of the electrode movement; a potential source of failure.

With an increase in proton-beam intensity, the load on the HT modulator will exceed the 5 mA limit of the present design. Different types of high-voltage switches have been tested and preference has been given to commercial semiconductor switches. A test bench simulating a dynamic electrical load similar to the conditions generated by a p-beam impinging on a high-$Z$ target has been built and tests continue.

The target team is also working on the thermodynamics of the target heating process. The goal is to reduce the required power for target heating and to produce a uniform heat distribution across the target system to minimize radio-isotope production losses through temperature variations. Materials are also being addressed. Potential future target material samples of SiC and $Al_2O_3$, with a tailor-made microstructure using the ice-templating technique (Fig. 3), have already been subjected to proton-beam bombardment at ISOLDE and CERN's HiRadMat facility [6]. These samples have

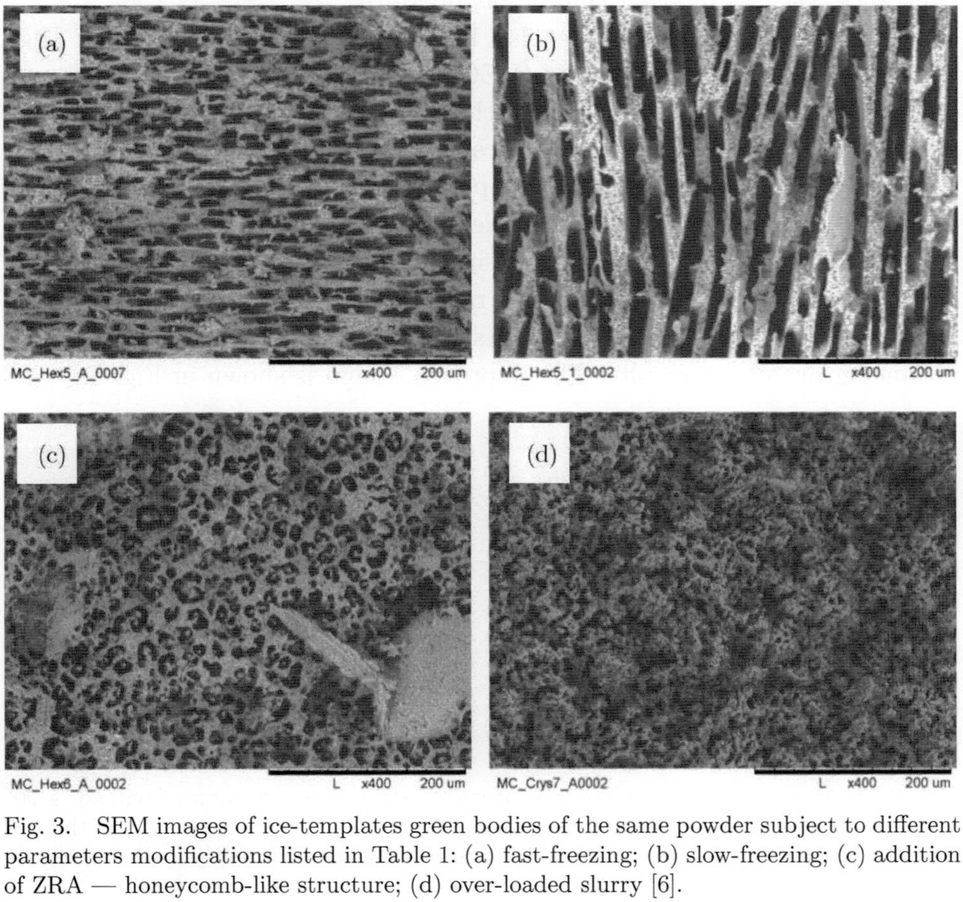


Fig. 3. SEM images of ice-templates green bodies of the same powder subject to different parameters modifications listed in Table 1: (a) fast-freezing; (b) slow-freezing; (c) addition of ZRA — honeycomb-like structure; (d) over-loaded slurry [6].

undergone several tests to characterize and evaluate their structure after irradiation.

The beam quality section looks at the required upgrades and techniques for improving beam purity. This is addressed partly by improving the neutron converter design to further increase the ratio of specific neutron-rich fission isotopes to neighbouring isobars.

The second part of the Design Study considers the improvement in radioactive ion beam (RIB) parameters; namely the beam emmitance and the mass resolution. This involves the re-design of the High Resolution Separator (HRS) layout to include changes in magnet design and the integration of a newly designed RFQ Cooler before the separator magnets. An off-line separator is currently built within the Design Study and will be used to implement and validate different concepts as well as testing a prototype RFQ Cooler, see Fig. 4. Mechanical modifications of the RFQCB are being simulated to optimize critical parameters relevant to beam transmission and emittance, such as internal helium gas pressures.

The beam quality upgrade also includes the development of a High Energy Current and Compression ($HEC^2$) EBIS charge breeder with the goal of improving the repetition rate of the pulsed beam ejected from the ion source [14]. To exploit the full physics program at ISOLDE and match the demanding injection requirement of a possible future TSR@ISOLDE [15], the new EBIS design should provide an electron energy of up to 150 keV and an electron-beam density of $10^4$ $Acm^{-2}$ in order to produce sufficiently highly charged ions within a given repetition rate. In collaboration with Brookhaven National Laboratory, preliminary tests of a new design are on-going and are producing promising results. The improved design is shown in Fig. 5.

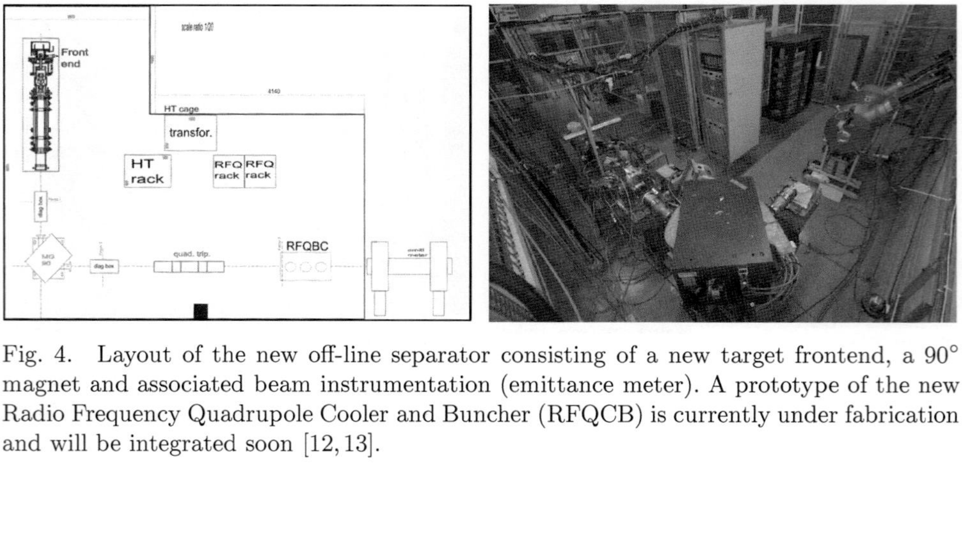


Fig. 4. Layout of the new off-line separator consisting of a new target frontend, a 90° magnet and associated beam instrumentation (emittance meter). A prototype of the new Radio Frequency Quadrupole Cooler and Buncher (RFQCB) is currently under fabrication and will be integrated soon [12, 13].

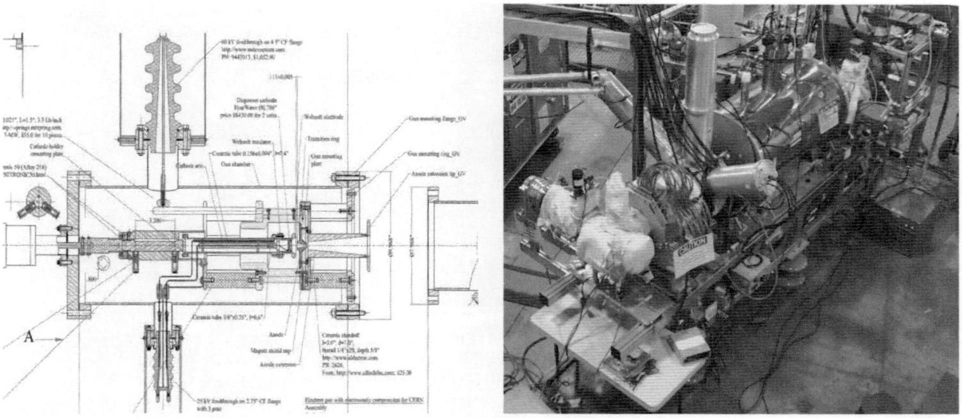

Fig. 5. Improved design of the High Energy Current and Compression (HEC$^2$) electron beam for charge breeding (left). On 7 Aug 2013 HEC$^2$ was moved to BNL, assembled and installed on the Test EBIS stand.

## 4 The energy upgrade: HIE-ISOLDE superconducting linac

The REX post-accelerator, in operation since 2001, is being upgraded (HIE-ISOLDE) to provide energies up to 10 MeV/u for $A/q = 4.5$. The post-accelerated radioactive beams will allow multi-step Coulomb excitation studies for a wide range of nuclei. The major components of this upgrade include a new superconducting (SC) linear accelerator (linac) based on Quarter Wave Resonators (QWRs) for the post-acceleration and the necessary 4.5 K cryogenic station for LHe.

The linac upgrade will be staged [16] in order to deliver higher beam energies to the experiments as soon as possible, with future upgrade stages ensuring a wide range of energy variability and providing an optional ~100 ns bunch spacing. The first stage of the upgrade involves the design, construction, installation and commissioning of two cryomodules downstream of REX, the existing post-accelerator. These cryomodules will each house five high-$\beta$ ($\beta_g = 10 : 3\%$) SC cavities and one SC solenoid. Extra cryomodules will be added to the beam line in a modular fashion until all six cryomodules, including two cryomodules housing six low-$\beta$ ($\beta_g = 6 : 3\%$) SC cavities and two SC solenoids, are online. The upgrade will be completed with a final stage that will see the linac extended in order to make room to pre-bunch the beam into the existing RFQ accelerator at a sub-harmonic frequency below 101.28 MHz, allowing the bunch spacing to be increased without significant loss in transmission; time-of-flight particle detection will then be viable at the experiments. Also foreseen is a beam chopper to reject the background

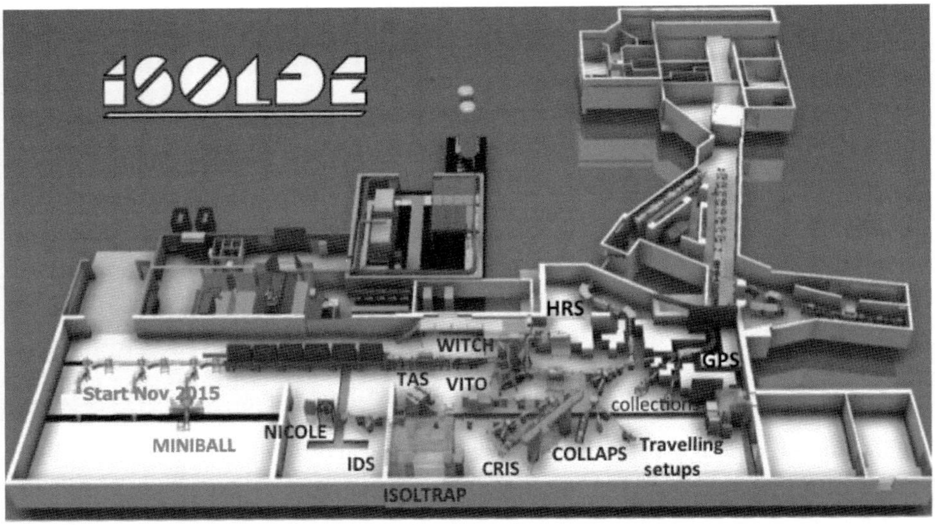


Fig. 6. Layout of the ISOLDE facility. It displays the existing experimental lines for low energy physics and the HIE-ISOLDE linac with its first two experimental beam lines, one holding MINIBALL and the other the travelling experiments, and the two extra buildings hosting the cryogenic plant.

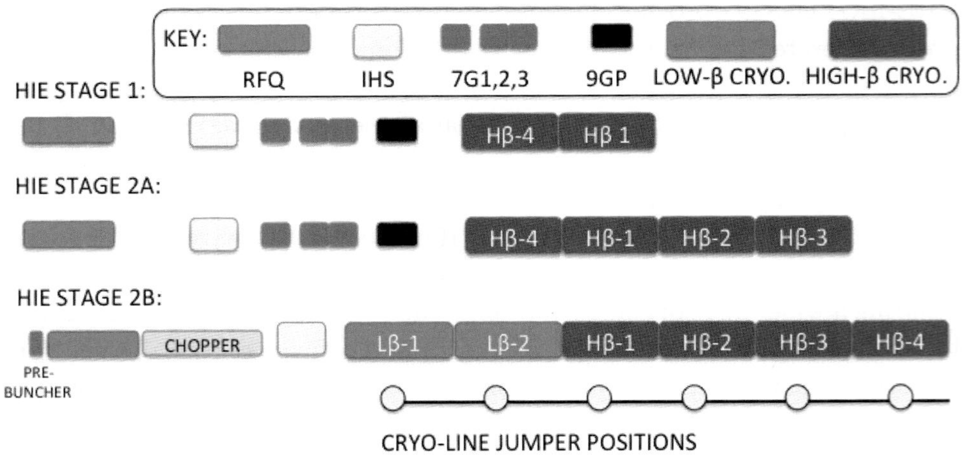


Fig. 7. A schematic of the staged installation of the HIE-ISOLDE SC linac (Existing REX structures: RFQ, IHS: 20-gap IH-structure, 7GX: 7-gap split-ring cavities, 9GP: 9-gap IH-structure).

of populated satellite bunches either side of the main sub-harmonic beam pulses. The staged installation of the linac is shown schematically in Fig. 7.

The beam dynamics studies [17,18] arrived at a compact lattice, based on cavities operating at a gradient of 6 MV/m, designed both due to the limited space in the experimental hall and to maximise the dynamical acceptance

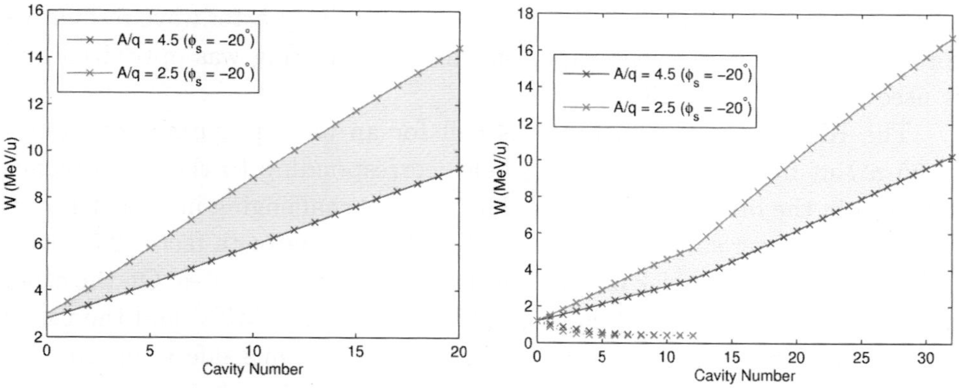


(a) Stages 1 and 2a

(b) Stage 2b

Fig. 8. Energy reach of post-accelerated beams for each stage of the upgrade as a function of the $A/q$ acceptance.

and optical performance of the machine. These considerations influenced the specification of the SC solenoids, alignment system, beam diagnostics system, steering magnets and cryomodules. The superconducting linac is designed to provide a total accelerating voltage of 39.6 MV with an average synchronous phase $\Phi_s$ of $-20$ deg, i.e. the minimum voltage required in order to achieve a final energy of at least 10 MeV/u with the heaviest beams ($A/q = 4.5$). The energy range accessible with the new SC linac is shown for the each stage in Fig. 8, including the decelerated beam that opens a range of energy not previously accessible with REX because of the fixed velocity profile of the 20-gap IH-structure (IHS).

A research and development programme looking at all the different aspects of the SC linac started in 2008 and continued throughout 2011. In particular, the R&D effort has focused on the development of the high-$\beta$ cavity, for which it has been decided to adopt technology based on copper cavities sputter-coated with niobium. The required infrastructure has been determined and the integration of the SC linac and High Energy Beam Transfer (HEBT) lines inside the existing experimental hall has been finalised.

### 4.1 *Superconducting RF cavity R&D*

The technology of Nb sputtering on copper was selected for the Quarter Wave Resonators (QWR) needed for HIE-ISOLDE project [19] at CERN, where it had been invented [20], and originally developed for the LEP cavities [21]. More specifically this choice made reference to the sputtered QWR developed at INFN-LNL for the ALPI linac [22]. The state of the art SRF surfaces made at CERN were magnetron sputtered Nb films on elliptical cavities (for LEP and for future linac colliders), while in the case of the quarter wave resonators for heavy ions, different in shape, bias diode sputtering was optimized and is used at INFN-LNL [23].

The HIE-ISOLDE specifications call for an average surface resistance of 65 n$\Omega$ at 6 MV/m accelerating field, corresponding to the performance achieved in the best ALPI cavities. This is quite challenging in consideration of the much larger surface of the HIE-ISOLDE resonator (a factor 2.5), with the consequent higher risk of spurious defects on the substrates. On the other hand, the lower operating frequency (101 MHz vs. 160 MHz) and the use of clean rooms for cavity preparations should provide some safety margin.

Work at CERN to establish a complete production chain for Nb sputtered QWR started in 2008 [24]. By the end of 2009 the chemistry, coating and RF testing infrastructures were operational, the first prototype cavity had been produced, and it was ready to be coated. A first design relying on rolled sheets and extensive electron beam welding was initially adopted (with 4 prototypes built) for the early phases of the program. In the second time a new version, based on machining from bulk with minimal number of EB welds, was produced and 3 prototypes were manufactured. The shape of the helium reservoir was also modified. As a result, the sensitivity to the helium pressure fluctuations was decreased by two orders of magnitude down to 0.02 Hz/mbar. A detailed report on the copper substrates is in these proceedings [25, 26].

In recent years much effort has been put into the optimisation of the coating process. The latest cavities went up to 6 MV/m at 7 W, 30% above

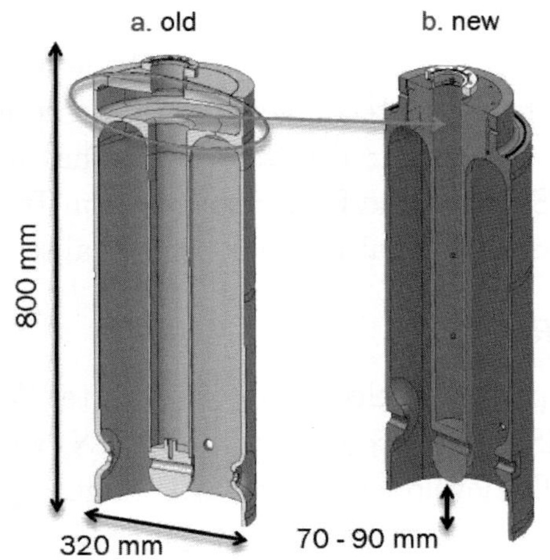


| Design Parameter | Value |
|---|---|
| $f_0$ (MHz) | 101.28 |
| $\beta_g$ (%) | 10.3 |
| $E_{acc}$ (MV/m) = $V_0/L_a$ | 6 |
| $L_a$ (m) | 0.3 |
| $R_{shunt}/Q_0$ ($\Omega$) | 550 |
| $E_{peak}/E_{acc}$ | 5.6 |
| $H_{peak}/E_{acc}$ (G/(MV/m)) | 100 |
| $Q_0$ (at 6 MV/m, P = 10 W) | 5 X $10^8$ |
| $\Gamma = R_s Q_0$ ($\Omega$) | 31 |
| $U/E^2_{acc}$ (mJ/(MV/m))$^2$ | 210 |
| T (K) | 4.5 |

Fig. 9. HIE-ISOLDE high-beta copper substrates: (left) rolled and welded prototype; (right) bulk machined prototype (inner conductor shown in magenta).

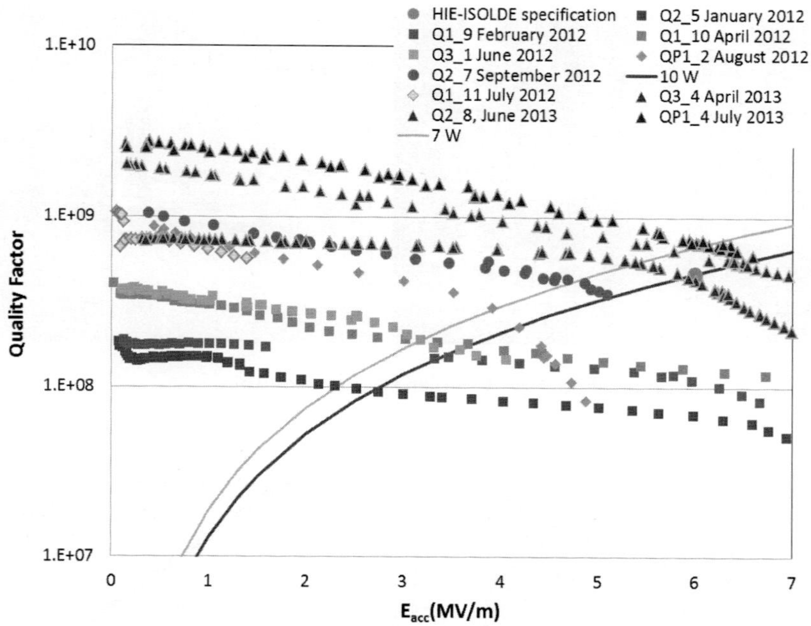


Fig. 10. $Q$ vs. $E_{\rm acc}$ curves of bias diode coated cavities [30].

the design goal. More details on the diode coating procedure at CERN are given in [27–29]. The evolution of the RF performance in vertical cryostat at 4.5 K is displayed in Fig. 10.

The level of performance reached so far is sufficient to start the series production, but work will continue in order to secure more margins and cover possible performance losses from the vertical tests to the linac operation.

Fundamental power coupler and tuning systems were designed and prototypes are available [31, 32]. However, a continuing effort is being put in reviewing and improving the existing designs, both on economic and on technical grounds. The first model of fundamental power coupler showed some mechanical problems after several thermal cycles between warm and operational temperatures. At the end of 2011 a redesign of this element was launched and the first prototype was successfully tested in March 2012. The new design is based on a stainless steel external body to minimize the thermal load on the cavity and a displacement system and guidance rails made of stainless steel and brass to ensure positioning of the antenna in the cryogenic environment (Fig. 11).

Two prototypes of the simplified tuning plates have been designed, manufactured and coated with niobium at CERN. They were subsequently installed on the cavity and tested at 4.5 K. Both plates were able to tune the

Fig. 11. Coupler system evolution.

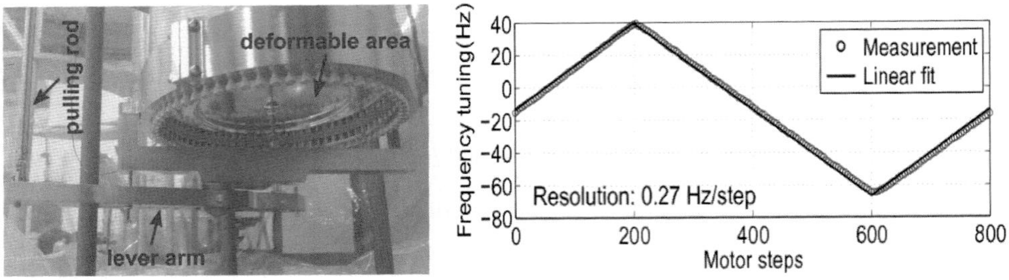


Fig. 12. Tuning system: Measurement results of the frequency tuning for the Cu OFE plate.

cavity frequency linearly with the stepper motor control. The field and power requirements for HIE-ISOLDE were reached for both plates. The resolutions of both plates were measured to be much better than 0.5 Hz/step required by LLRF. The measurement results fit simulations perfectly for both coarse range and sensitivity. Given the large margin left for the resolution, we plan to enlarge the deformable area of the plate to increase the coarse range and lower the forces applied on the plate.

### 4.2 *The RF system*

As mentioned, the HIE-ISOLDE accelerating system will be based on independently phased quarter wave resonators, working at 101.28 MHz. A sketch of the RF system is depicted in Fig. 13.

A negligible beam loading and the insensitivity of the rigid copper cavity substrates to vibrations allow running the resonators at a relatively high QLOADED, typically $1 \times 10^7$ to $5 \times 10^7$ (10 down to 2 Hz bandwidth). Therefore the cavities shall be powered by standard FM transmitter amplifiers,

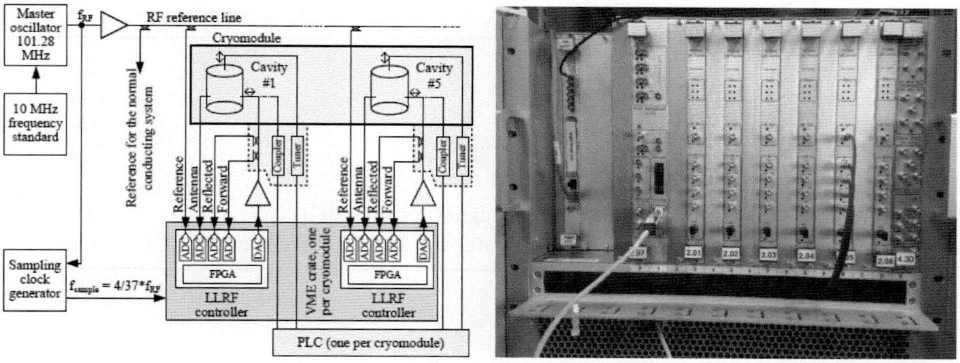


Fig. 13. Sketch of the HIE-ISOLDE RF system (left) — LLRF system for a complete cryomodule (6 cavities).

commercially available, with an available power of 700 W at 1 dB saturation point.

The low-level RF (LLRF) system excites and keeps the cavity field stable at the requested operating point. The system is of the generator-driven type, and has a self-excited loop mode to find and track the resonance frequency of the cavity at start-up. The values of the accelerating voltage and phase can be easily changed to adapt the linac for a new ion species in a matter of seconds. Another function of the LLRF system is to provide an automatic cavity conditioning tool during the cavity cool down process and after extended idle periods.

The LLRF system is based on VME form factor modules. Advances in the ADC and DAC technology allowed using a direct RF sampling approach, together with a direct digital quadrature demodulation. After digital processing, the cavity excitation signal is directly generated in the RF domain by a fast DAC. More information on the system can be found in [33]. Due to space and cost constraints, the LLRF system will be housed in 14 shielded racks nearby the linac, instead of a traditional "Faraday" cage. The LLRF and RF power racks have been installed in spring 2014 and the cabling campaign has started in September 2014. A complete RF system for the first cryomodule will be installed in spring 2015.

### 4.3 *The superconducting solenoid*

Transverse beam focusing at the HIE-ISOLDE linac will be provided by superconducting solenoids [34], integrated in the common vacuum cryostats. The magnets are based on Nb-Ti superconducting graded coils surrounded by an iron yoke, and enclosed in a stainless steel vessel filled with liquid helium

Table 2. HIE-ISOLDE SC solenoid main parameters.

| | |
|---|---|
| Bore diameter [mm] | 31 |
| Magnet length [mm] | 312 |
| Nominal operating current $I_{\mathrm{nom}}$ [A] | 115.5 |
| Peak field at solenoid magnetic center [T] | 7.86 |
| Integrated square field at $I_{\mathrm{nom}}$ [$T^2$m] | 13.5 |
| Stray field at 230 mm from solenoid magnetic centre [G] | 12 |
| Remanent magnetization at 230 mm from solenoid magnetic centre [G] | 0.26 |

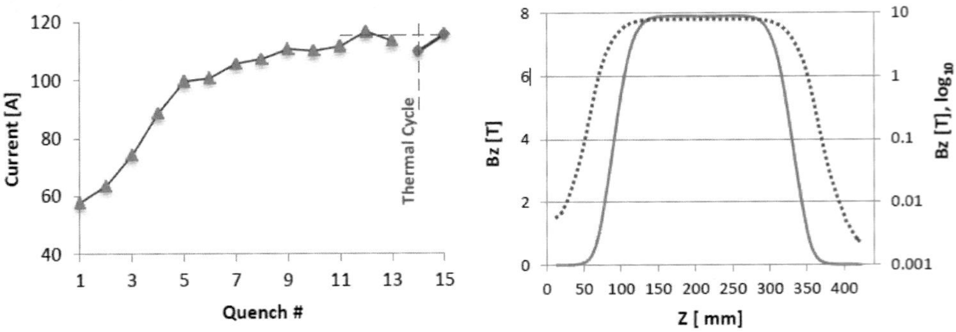


Fig. 14. Training performance of the first solenoid (left) — Magnetic field along the beam axis at Inom (dotted line: logarithmic scale).

at 4.5 K. The operational point is at 81% of the load line, the coils are passively protected with bypass diodes and parallel resistors, and the powering Nb-Ti leads are cryostable. The presence of the neighbouring superconducting cavities imposes tight specifications on the remanent magnetization (to avoid flux trapping) and on the stray field at nominal current (to avoid overcoming the lower critical field of the Nb layers on the RF surfaces). Table 2 lists the main design parameters of the solenoid.

The series production started in industry in 2014. Four coils have been produced. Figure 14 shows the training performance of the first magnet achieving the nominal field. A small detraining was observed after the second thermal cycle, but the magnet reached again nominal field after 2 quenches. The magnetic field was measured by means of axial scans with a Hall Probe. The measured field profile is shown in Fig. 14. The maximum stray field at the cavity position is well within the specifications.

### 4.4 *High-β cryo-module assembly*

The design of the cryomodule is detailed in [35]. A stainless steel vacuum vessel contains the cavities and the solenoid, surrounded by a thermal screen

Fig. 15. HIE-ISOLDE Superconducting Solenoid#1 (Courtesy of DANFYSIK).

which is actively cooled with helium gas at 50 K and 12 bar at the inlet. The active elements are supported close to the beam axis on precise reference surfaces which in turn are mechanically connected to the top plate. Liquid helium at 4.5 K is supplied to a common reservoir and distributed to cavities, solenoid and support frame. In order to minimize the drift length between cavities and the overall length of the machine a common vacuum was chosen for the beam and cryogenic insulation. Insulation will be guaranteed by a heat screen at 75 K. A vacuum of $10^{-8}$ mbar after cryopumping is necessary for optimal operation. The solenoids need to be aligned with stringent precision ($\pm 0.15$ mm, $\pm 1\sigma$ tolerance) and a system of independent adjustment under vacuum for the cavity and solenoid assembly is foreseen. A position monitoring system based on CCD cameras has been developed and validated.

A 3D model of the HIE-ISOLDE cryomodule is shown in Fig. 16.

The elements of the cryomodule, together with specially developed assembly tooling, are being procured through several contracts with industry which are coming to the delivery phase in a staggered fashion as from August 2014. Several key elements are produced in the CERN main workshop. The assembly of the cryomodule components will take place in a class 100 clean room, which was set up and commissioned at CERN for this purpose. Detailed procedures are being elaborated for the assembly work, which will be interleaved with a thorough set of quality checks covering cleanliness, leak tightness, alignment, and electrical quality assurance. This work is starting in August 2014 with the requalification of the clean room after the installation therein of a class 100 compatible precision assembly tower.

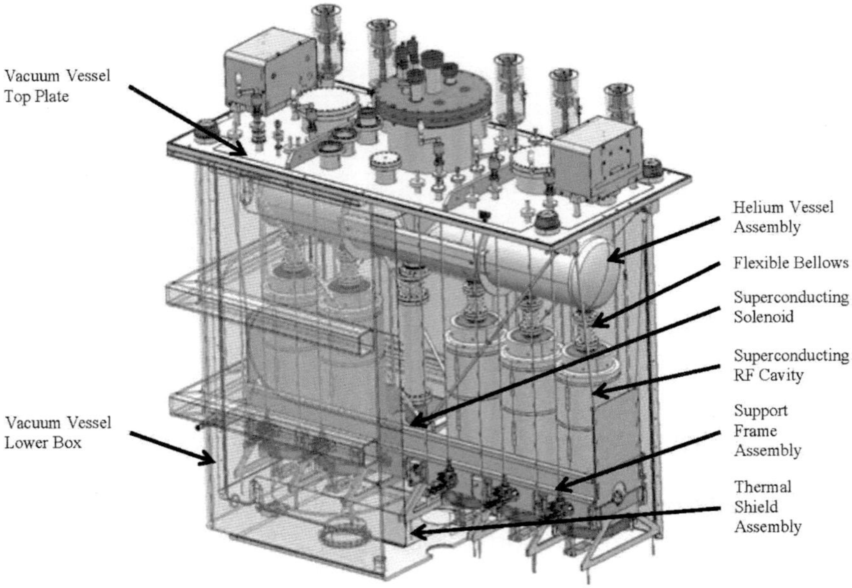


Fig. 16. General assembly view of the complete high-beta cryomodule.

Fig. 17. Elements of the high-beta cryomodule procured in industry (left & centre) and produced at CERN central workshop (right).

## 4.5 *Cryomodule testing*

After assembly and the final qualification tests at warm, the finished cryomodule will be transported with a dedicated tool to a nearby Horizontal Test Facility for a complete set of qualification tests in operational conditions, prior to installation in the linac tunnel. The test campaign will cover the vacuum and cryogenics performance, monitoring and adjustment of the alignment of the active elements, conditioning and RF measurement of the superconducting cavities, power tests of the superconducting solenoid, and

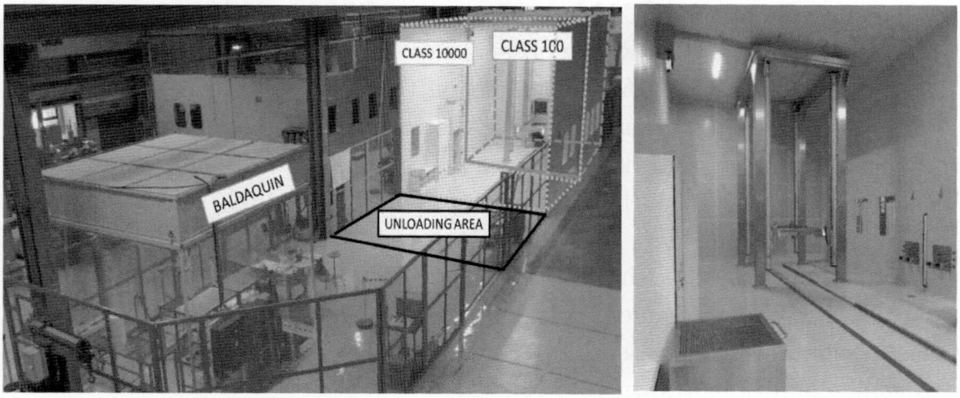


Fig. 18. Cryomodule assembly area equipped with specific toolings (CERN-SM18).

commissioning of the LLRF and of the tuning systems. Cold testing of the first cryomodule is scheduled in February 2015.

### 4.6 *Survey and monitoring of the SC linac*

To run the linac in optimum conditions, the active components, cavities and solenoid, must be aligned and monitored on the REX Nominal Beam Line (NBL) within a precision of 0.3 and 0.15 mm respectively at one sigma level along directions perpendicular to the beam [36].

As sketched in Fig. 19, the Monitoring and Alignment Tracking for HIE-ISOLDE (MATHILDE) system [37] uses a set of newly developed double-sided HIE-ISOLDE Brandeis CCD Angle Monitor (HBCAM) [38]. These sensors are similar to calibrated cameras equipped with laser diodes that enable reciprocal sensor observations as well as light spots or targets measurement. HBCAMs are fixed to metrological tables in order to create a close geometrical network link to the Nominal Beam Line by reference pillars. Two external lines of sight, one on each side of the cryomodule, are created and

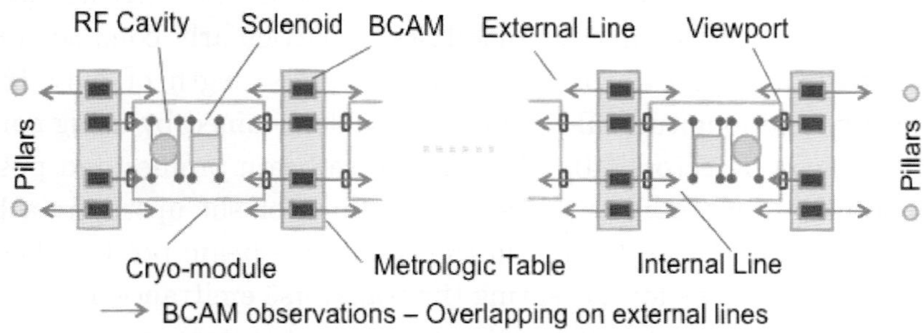


Fig. 19. Sketch of the alignment system — Top view.

act like a frame. The HBCAMs belonging to the internal lines are placed in front of viewports and allow the observation of targets attached to the active elements and of the HBCAMs situated on the previous and next table.

Targets must be compatible with vacuum and cryogenic conditions, be measurable from several directions and have a narrow shape (6 mm max). The design is based on the properties of prototype high refraction index (~2) 4 mm diameter glass ball lens, which were manufactured and tested at CERN. The precision of the reconstructed target displacement measured by a HBCAM is about 10 microns in object space at one sigma level. HBCAM measurements were successfully performed on targets placed in vacuum and cryogenic conditions (down to 5 K).

The effects of viewport crossing have been modeled, verified by tests and corrections were implemented in the software routines [39]. Tests with an early computation shell of the Monitoring and Alignment Tracking for HIE-ISOLDE Software (MATHIS), still under development, have shown that the one sigma precision for the reconstruction of a full set of seven metrological tables is within 20 microns.

The full set of HBCAM sensors and associated electronic equipment is already procured. The target support is designed and a price inquiry is on-going. The metrological tables are in the final design phase.

### 4.7 *Beam instrumentation*

New beam diagnostics devices have been developed for the HIE-ISOLDE project, in particular for the measurements of the intensity, energy, transverse and longitudinal profiles, and transverse emittance of the stable pilot beams. The instruments will be integrated in octagonal-shaped vacuum chambers with five radially distributed ports available for the installation of instruments or collimating devices plus a port for vacuum pumping. The beam intensities, in the range between 1 pA and 1 nA, will be measured by means of Faraday cups (FC). Due to constraints on the longitudinal space available in between the linac cryomodules, a particularly compact FC was designed with an aperture of 30 mm and overall length of only 16 mm. Beam energy and longitudinal profiles will be determined by using commercial PIPS silicon detectors [40]. The transverse beam profiles and positions will be obtained by scanning a V-shaped collimator slit upstream a FC. A resolution of the order of 0.1 mm in the transverse beam position has been estimated. Two options for measuring the transverse emittance are available: one using the existing REX-ISOLDE slit and grid system, the other by the combined use of two scanning slits and a FC. Experimental tests with beam

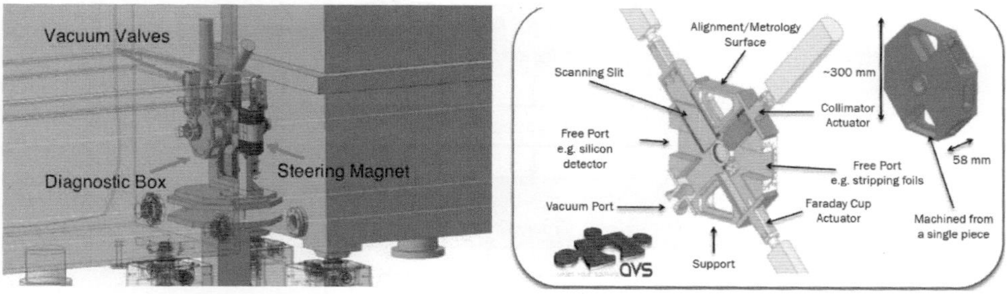


Fig. 20. 3D model of the region between cryomodules and of the short diagnostic box (courtesy of AVS technology).

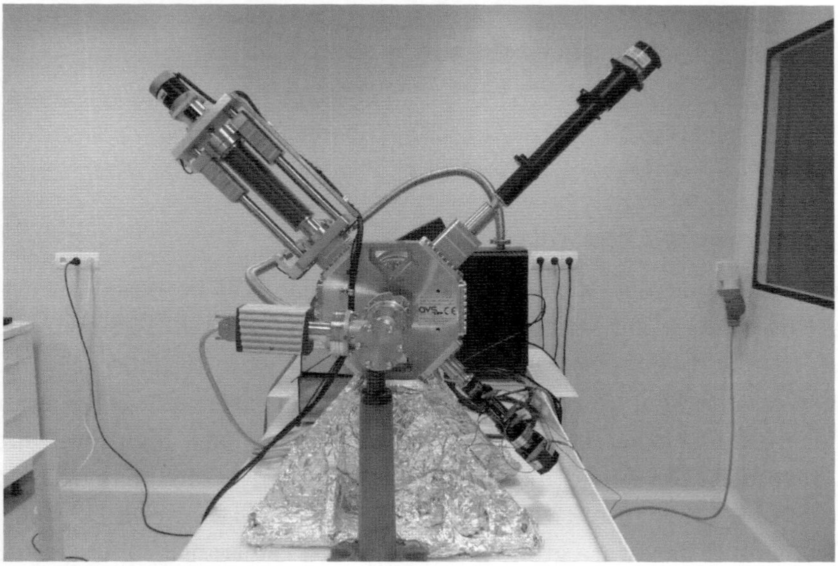

Fig. 21. Prototype short diagnostic box being tested (courtesy of AVS technology).

of all the mentioned devices have been carried out successfully between 2011 and 2013; more details can be found in [41]. A new VME card was designed for controlling the actuators and the FC, and also a new frontend preamplifier for the beam intensity measurements. As the tight longitudinal constraint is not present in the HEBT section, longer boxes will be installed there, allowing for the use of a standard FC. Contracts have been signed with industry for the production of the fully assembled diagnostic boxes (six short and nine long), electronic modules are also under final production. The installation of the equipment is scheduled to start at the beginning of 2015.

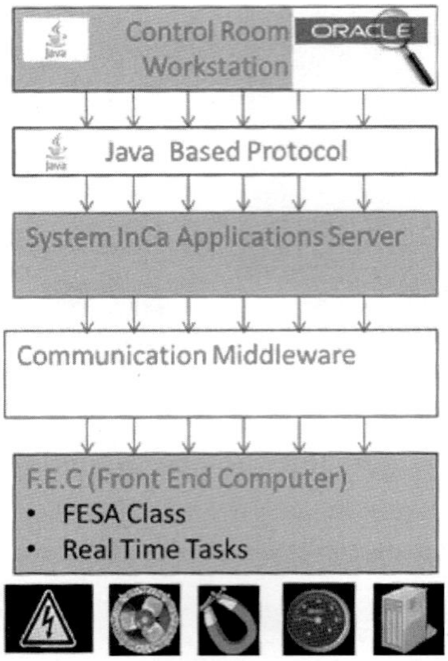


Fig. 22. Three-tier CERN control architecture.

### 4.8 *Software and controls*

Software and controls at the HIE-ISOLDE linac will be based on a three-tier architecture widely used in CERN accelerators [42]. The machine will rely on the controls infrastructure already deployed in the injector complex. Cryogenics and vacuum controls will be as well based on CERN standard solutions used in other machines. For beam operations, high level applications have been developed including a settings generator, beam diagnostics tools, and tools for automatic cavity phasing as $A/q$ is varied [43, 44].

Dry runs of hardware groups, starting with the transfer lines, will begin early 2015 using the operational software.

## 5 Installation and future upgrade of experimental facility

The new HIE-ISOLDE superconducting linear accelerator requires a major increase of equipment to the existing facility's infrastructure. Two new surface buildings have been constructed in order to house the helium compressor station and the helium refrigerator cold box (Fig. 23). Ground breaking started at the end of summer 2011 with the preparation of the site and the construction of a new lock to allow access of personnel and material to the experimental hall. The construction works were completed by the third quarter of 2012 after which the installation of the electrical systems and main services took over.

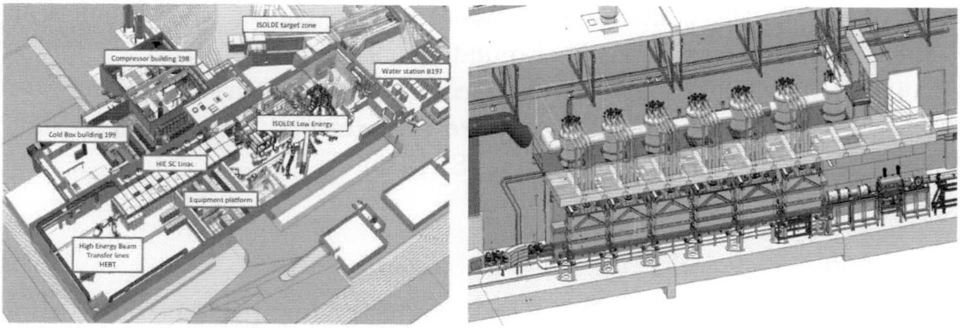


Fig. 23. New HIE-ISOLDE buildings (left), Cryo Cold Line and Jumper Boxes (right).

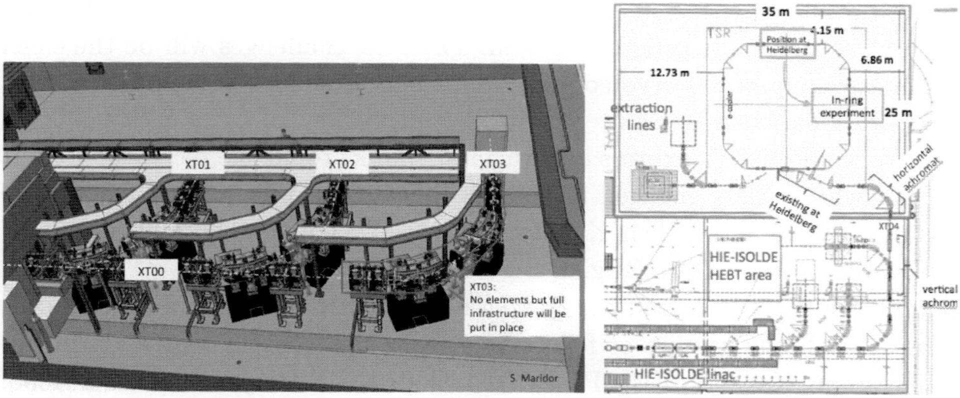


Fig. 24. Layout of the High Energy Beam Transfer lines in the ISOLDE experimental hall.

The cryogenic station installation started in the first quarter of 2014. The LHe liquefier will be installed in a separate light construction building as close as possible to the linac in order to minimize the length of the LHe distribution line. This will enable an easier and more stable operation of the cryogenic system.

The cryogenic system to be supplied includes one cryogenic transfer line, which will link the cold box and the different interconnecting ("jumper") boxes feeding from the top, the six cryomodules of the new SC linac.

A new high-energy beam transfer line (HEBT) will bring the beam into the existing extension of the ISOLDE experimental hall.

The MINIBALL segmented Ge array and T-REX experimental setups already operational at REX will be used intensively when the first beams are delivered in 2015. An open beam line will also be available for traveling experiments. In this context, an active-target experiment for resonant scattering and transfer reactions is also under consideration.

The HEBT will be extended in a second stage of the installation foreseen during 2017/18 to accommodate a third experimental station and to allow for the connection to a storage ring (TSR) [15].

## 6 Outlook

Work is in full swing at CERN and in industry towards the construction and commissioning of the HIE-ISOLDE linac, foreseen in summer 2015. The main technical issues regarding the performance of the superconducting elements have been solved. All the main components of the machine are being produced, either at CERN or in industry, and the work on the general infrastructure is progressing well. The next challenges will be the clean room assembly of the cryomodules, and their subsequent qualification tests in operational conditions.

We would like to express our warmest thanks to all the technical staff at CERN who is daily engaged in making the work progress. We acknowledge funding from the Swedish Knut and Alice Wallenberg Foundation (KAW 2005-0121) and from the Belgian Big Science program of the FWO (Research Foundation Flanders) and the Research Council K.U. Leuven. We acknowledge as well the receipt of fellowships from the CATHI Marie Curie Initial Training Network: EU-FP7-PEOPLE-2010-ITN Project number 264330. Support from the Spanish Programme Industry for Science from CDTI is also acknowledged.

## References

[1] M. Lindroos and T. Nilsson, CERN Yellow Report, CERN-2006-013.
[2] A. Herlert, The ISOLDE facility, *Nuclear Physics News*, Vol. 20, No. 4, 2010.
[3] Linac4 Technical Design Report, CERN-AB-2006-084 ABP/RF.
[4] M. Vretenar *et al.*, Progress in the construction of Linac4 at CERN, *26$^{th}$ International Linear Accelerator Conference, LINAC12*, Tel Aviv, 2012.
[5] R. Garoby *et al.*, Plans for the upgrade of the LHC injectors, *IPAC2012 Conf. Proc.* C1205201 (2012) TUXA02; http://www.JACoW.org.
[6] M. Czapski *et al.*, *Nucl. Instrum. Meth. B* **317** (2013) 385–388.
[7] J. Montano *et al.*, Design upgrade of the ISOLDE target unit for HIE-ISOLDE, *Proceedings of EMIS 2012 Conference.*
[8] R. Catherall *et al.*, Radioactive ions produced by neutron-induced fission at ISOLDE, *Nucl. Instrum. Meth. B* **204** (2003) 235–239.
[9] M. Hermann *et al.*, Simulations of the HIE-ISOLDE radio frequency quadrupole cooler and buncher using the Monte Carlo test particle code Molflow+, *Proceedings of EMIS 2012 Conference.*

[10] B. Jonson *et al.*, ISOLDE PS Booster facility at CERN; Experiments with slow radioactive beams, *Nuclear Physics News*, Vol 3, No. 2, 1993.
[11] A. Sarrió Martínez *et al.*, Upgrade strategies for the Proton Synchrotron Booster dump at CERN, *IPAC2012 Conf. Proc.* C1205201 (2012) THPPP016; http://www.JACoW.org.
[12] C. Babcock *et al.*, *Nucl. Instrum. Meth. B* **317** (2013) 484–487.
[13] M. Augustin *et al.*, Design upgrade of the mass separator magnets for HIE-ISOLDE, *Proceedings of EMIS 2012 Conference.*
[14] A. Shornikov *et al.*, *Nucl. Instrum. Meth. B* **317** (2013) 395–398.
[15] M. Grieser *et al.*, *Eur. Phys. J.*, Special Topics, **207** (2012) 1–117.
[16] M. Lindroos, P.A. Butler, M. Huyse and K. Riisager, HIE-ISOLDE, *Nucl. Instrum. Meth. B* **266** (2008) 4687–4691.
[17] M.A. Fraser, R.M. Jones, M. Pasini, Beam dynamics design studies of a superconducting radioactive ion beam postaccelerator, *Phys. Rev. ST Accel. Beams* **14** (2011) 020102.
[18] M.A. Fraser, Beam dynamics studies of the ISOLDE post-accelerator for the high intensity and energy upgrade, PhD Thesis (University of Manchester), Vol. 13 in EuCARD Series (2012).
[19] M. Pasini *et al.*, A SC upgrade for the REX-ISOLDE accelerator at CERN, *Proceedings of LINAC08*, Victoria, BC, Canada, 2008.
[20] C. Benvenuti, N. Circelli and M. Hauer, Niobium films for superconducting accelerating cavities, *Appl. Phys. Lett.* **45** (1984) 583.
[21] C. Benvenuti *et al.*, Superconducting niobium sputter-coated copper cavity modules for the LEP energy upgrade, *Proceedings of the Particle Accelerators Conference, PAC1991*, Vol. 2, pp. 1023–1025.
[22] V. Palmieri, V.L. Ruzinov, S. Stark *et al.*, New results on niobium sputtered copper quarter wave resonators, *Proceedings of the Sixth Workshop on RF Superconductivity*, Newport News, Virginia, 1993.
[23] A.M. Porcellato *et al.*, Performance of ALPI new medium beta resonators, *Proceedings of HIAT 2012*, Chicago, IL, USA.
[24] G. Lanza *et al.*, The HIE ISOLDE superconducting cavities: Surface treatment and niobium thin film coating, *Proceedings of SRF09*, p. 801.
[25] A. D'Elia, R.M. Jones, and M. Pasini, HIE-ISOLDE high beta cavity study and measurement, *Proceedings of SRF09*, p. 609.
[26] L. Alberty *et al.*, The copper substrate developments for the HIE-ISOLDE high-beta QWRs, *Proceedings of SRF13.*
[27] N. Jecklin *et al.*, Niobium coatings for the HIE ISOLDE QWR superconducting accelerating cavities, *Proceedings of SRF13.*
[28] A. Sublet *et al.*, Preliminary results of Nb thin film coating for HIE-ISOLDE SRF cavities obtained by magnetron sputtering, *Proceedings of SRF13.*
[29] A.M. Sublet *et al.*, Thin film coating optimization for HIE-ISOLDE SRF cavities: Coating parameters study and film characterization, *Proceedings of SRF13.*
[30] W. Venturini Delsolaro *et al.*, Status of the superconducting RF activities for the HIE-ISOLDE Project, *Proceedings of LINAC12*, Tel Aviv, Israel, September 2012.
[31] A. D'Elia, Design and characterisation of the power coupler line for the HIE-ISOLDE high-beta cavity, HIE-ISOLDE-Project-Note-011, 16 August 2011.
[32] P. Zhang *et al.*, The tuning system for the HIE ISOLDE high beta Quarter Wave Resonator, *Proceedings of SRF13.*
[33] M. Elias, Digital measurement system for the HIE-ISOLDE superconducting accelerating cavities, CERN-THESIS-2013-165.

[34] G. Nielsen, N. Hauge, E. Krauthammer, K. Thaarup, L.O. Baandrup, and A. Baurichter, Superconducting solenoids for RF accelerators and electron guns, *IEEE Transactions on Applied Superconductivity* **24** (2014) 4100205.

[35] L. Williams *et al.*, Design of the high beta cryomodule for the HIE-ISOLDE upgrade at CERN, *IPAC2011*, San Sebastian, Spain.

[36] M.A. Fraser, R.M. Jones and M. Pasini, Beam dynamics design studies of a superconducting radioactive ion beam post accelerator, *Phys. Rev. ST Accel. Beams* **14** (2011) 020102.

[37] J.-C. Gayde, G. Kautzmann *et al.*, HIE ISOLDE alignment and monitoring system technical design and project status, *IWAA2012*, Chicago, 2012.

[38] Brandeis CCD Angle Monitor (BCAM), http://alignment.hep.brandeis.edu.

[39] J.-C. Gayde, G. Kautzmann *et al.*, The HIE-ISOLDE alignment and monitoring system software and test mock-up, *IWAA2012*, Chicago, 2012.

[40] F. Zocca *et al.*, Development of a silicon detector monitor for the HIE-ISOLDE superconducting upgrade of the REX-ISOLDE heavy-ion linac, *Nucl. Instrum. Meth. A* **672** (2012) 21.

[41] E.D. Cantero, W. Andreazza, E. Bravin, and A. Sosa, Status of beam diagnostics for the HIE-ISOLDE linac, to be published in *IBIC2014*, California, 2014.

[42] LHC design report, p. 381.

[43] D. Lanaia, M.A. Fraser, D. Voulot, Overview of the new high level software applications developed for the HIE ISOLDE linac, *Proceedings of LINAC14*.

[44] M. Fraser, S. Haastrup, J.C. Broere, D. Lanaia, D. Valuch, and D. Voulot, Understanding the error tolerances required to automatically phase the HIE ISOLDE linac, *Proceedings of LINAC14*.

# Chapter 32

# FAIR project at GSI

Oliver Kester[*,†], Peter Spiller[*] and Horst Stoecker[*,†,‡]
[*]GSI Helmholtzzentrum fuer Schwerionenforschung GmbH, Darmstadt
[†]Helmholtz International Center for FAIR and
Goethe Universitaet Frankfurt am Main
[‡]Frankfurt Institute for Advanced Studies FIAS

FAIR — the Facility for Antiproton and Ion Research in Europe — constructed at GSI Helmholtzzentrum für Schwerionenforschung GmbH in Darmstadt comprises an international centre of heavy ion accelerators that will drive heavy ion and antimatter research (FBTR, 2006). FAIR will provide worldwide unique accelerator and experimental facilities allowing a large variety of fore-front research in physics and applied science. FAIR will deliver antiproton and ion beams of unprecedented intensities and qualities. The main part of the FAIR facility is a sophisticated and cost efficient accelerator system, which delivers parallel beams to different experiments of the FAIR experimental collaborations — APPA, NuSTAR, CBM and PANDA. The accelerated primary beams will then be employed to create new, often highly exotic particles in a series of parallel experimental programs. Experiments with exotic particles will explore fundamental processes which are expected to have taken place in the early phases and still happen in the on-going evolution of the Universe. These processes produced the basic constituents of matter and overall structure we observe today.

The scientific program of FAIR can be summarized as "cosmic matter in the laboratory" (FAIR-EXP, 2005). It is based on four scientific pillars. The "compressed baryonic matter — CBM" experiment (CBM, 2005) will investigate hadrons and nuclei under extreme pressure in the lab. Nuclear densities similar to those in the interior of neutron stars can be created, which allows for spectroscopy of multi-messenger observables, e.g. photons, hadrons, dileptons, multi-strange and charmed particles and antiparticles as well as (anti-)clusters thereof. NuSTAR encompasses the research on nuclei far from stability. FAIR will deliver insight into nucleosynthesis and hence the understanding of the creation of elements in the Universe. Up to 7000 types of nuclei are possible although most of them are rather unstable, especially those with high proportions of either protons or neutrons. Nevertheless, their complex and sometimes unusual structures can show fascinating behaviour, which throws light on how the forces that hold nuclei together work. Although existing for only a very short moment, some of these rare isotopes play a critical role in the nuclear reactions that build up the elements in stars. While the lighter elements are formed during a star's lifetime, the heavier elements beyond iron in the periodic table are thought to be created mostly in the death-throes of supermassive stars when they explode as supernovae. This is explored in the framework of NuSTAR — Nuclear Structure, Astrophysics and Reactions (NuSTAR, 2005). APPA is devoted to exploration and discoveries in atomic physics, plasma physics, material science and biophysics. The collaboration will investigate matter under extreme conditions concerning density, temperature and pressure. The results will foster the understanding of astrophysical plasmas and the understanding of radiation damages and modification of materials and biological samples. Highly-charged heavy ions that have been stripped of practically all their electrons offer the possibility of testing the theory describing the interaction at very high electromagnetic fields (quantum electrodynamics, QED). Ions will be slowed down and caught in special 'traps' for precision studies. PANDA (Antiproton Annihilation in Darmstadt) (PANDA, 2005) will explore the complex structure of and the dynamic processes in hadrons. The collaboration will search for unusual hadrons containing both strange and charm quarks, and even particles made of just gluons (glueballs), which will act as sensitive probes of the theory of the strong force (quantum chromodynamics, QCD).

Many questions cannot be addressed with the present accelerator facilities; therefore, more powerful and sophisticated accelerators are required to address cutting edge physics. The FAIR facility consists of a carefully designed configuration of accelerators for generation of high-quality primary beams, and for creating new particles by colliding or bombarding the beams on specially tailored targets. The FAIR accelerators will be supplied with ion beams by the GSI accelerator facilities, which presently undergo an upgrade program that addresses all major technical systems from the ions sources to the SIS18 synchrotron extraction.

## 1 The FAIR accelerator facility

An overview of the final FAIR/GSI accelerator complex is shown in Fig. 1. The full FAIR facility will consist of eight circular accelerators with up to

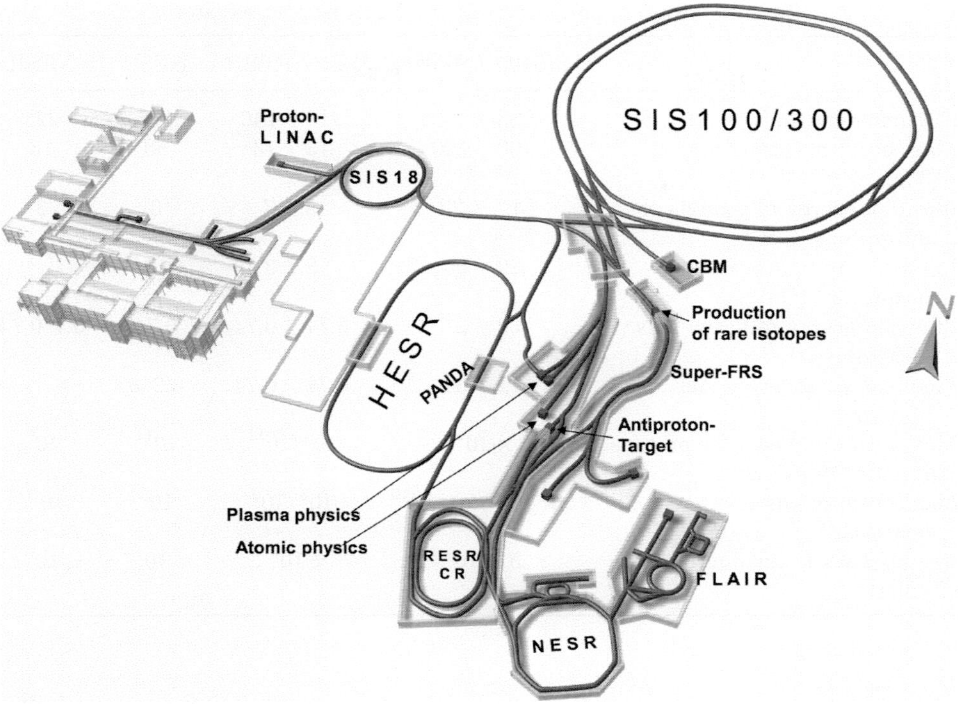


Fig. 1. GSI Helmholtzzentrum für Schwerionenforschung GmbH in Darmstadt together with the FAIR accelerator facility (sketched in red). The existing injectors (UNILAC and SIS18) are sketched in blue. The working horse of the facility will be the SIS100 that generates the primary beams for experiments of secondary particle production.

1,100 meters circumference, two linear accelerators and about 3.5 kilometers of beam lines. The existing GSI accelerators provide heavy ions with energies up to 2 AGeV, i.e. up to 95% of the speed of light. For FAIR, the Research Centre Jülich (ZFJ) will build the HESR — High Energy Storage Ring — for the research with high energy antiprotons using the PANDA detector. Budker Institut of Nuclear Physics (BINP) in Novosibirsk is in charge of the Collector Ring (CR). GSI is in charge of the design of all other machines and of the common systems (controls, cryogenics etc.) which are constructed in consortia with international partners. The main parameters of the FAIR ring machines can be found in Table 1.

The driver accelerator of FAIR is the fast ramping, superconducting heavy ion synchrotron — SIS100 — that allows the acceleration of the most intense beams of stable elements from protons (29 GeV) to uranium (11 AGeV). SIS100 is installed in a 20 m deep tunnel, which is designed for the installation of the SIS300 synchrotron in a later stage of the project. In the full version of FAIR this superconducting stretcher ring allows a further boost of the energy of primary beams to 90 GeV (p) and 34 AGeV (U) and

Table 1. System parameters of the FAIR ring accelerators.

| | SIS100 / SIS300 | CR / RESR | HESR | NESR |
|---|---|---|---|---|
| Circumference [m] | 1083 / 1083 | 215 / 240 | 575 | 223 |
| Max. beam magnetic rigidity [Tm] | 100 / 300 | 13 / 13 | 50 | 13 |
| Injection energy of protons or antiprotons [GeV] | 4.7 / 29 | 3 / 3 | 3 | 3 |
| Final energy of protons or antiprotons [GeV] | 29 / 90 | 3 / 3 | 14 | |
| Injection energy of heavy ions [GeV/u] | 0.2 / 2.7 | 0.74 / 0.74 | 0.74 | 0.1–0.74 |
| Final energy of heavy ions [GeV/u] | 11 / 34 | 0.74 / 0.74 | 0.2–5 | |
| Max. beam intensity for protons or antiprotons /cycle | $2.5 \times 10^{13}/-$ | $10^8/10^{11}$ | $10^{11}$ | $10^9$ |
| Max. beam intensity of ions/cycle | $5 \times 10^{11}/2 \times 10^{10}$ | $10^8/10^9$ | $10^8$ | $10^{10}$ |
| Required static vacuum pressure [mbar] | $< 5 \times 10^{-12}$ | $< 10^{-9}$ | $< 10^{-9}$ | $< 10^{-11}$ |

will provide quasi-continuous beams by slow extraction over 100 seconds. The CBM, Plasma, and Biomat experiments are directly supplied with primary beams from the SIS100. Two target stations for the generation of secondary beams ($\bar{p}$ and RIBs) allow the conversion of primary ions into secondary particles. The intensities of secondary beams will be increased by a factor of 1,000–10,000 as compared to presently available beams. The secondary beams may be stopped in gas cells or accumulated in the FAIR storage rings for precision experiments. Thus, isotopic clean beams of all elements of the periodic table, exotic nuclides and antiprotons are available, which will make FAIR a worldwide unique facility.

The production mechanisms of secondary particle beams lead to large momentum spreads and phase space distributions and thus to poor beam quality. In the full extension of FAIR, the storage rings CR and RESR will accumulate the secondary beams and improve their quality by stochastic and electron cooling. High beam quality is essential for the performance of precision experiments, hence active phase space cooling in transverse and longitudinal direction of the beams in the storage rings (CR, RESR, HESR and NESR) will be performed. The storage rings HESR and NESR host a large fraction of the experiment platforms for a variety of different experiments. The product of both high beam intensity and excellent beam quality of the FAIR accelerators is unique.

The challenges of the FAIR accelerators address different areas of accelerator physics and technology and determine presently the technical limits. The planned high beam intensities require superconducting magnets for both SIS100 and SIS300 that can be ramped to maximum field in a few seconds. Such fast field ramps in superconducting magnets have not yet been realised. Powerful and sophisticated, inductively loaded resonator structures and power amplifier systems must accelerate the ions synchronously with the ramps of the magnetic field and have to generate short intense single ion pulses for the production of secondary particles. Beam loss at high beam intensities is an issue for the lifetime of accelerator components, thus new radiation-resistant materials and components will be applied. To keep the beam losses on acceptable low levels, a careful modelling of the circular accelerators is required, taking into account realistic magnetic field distributions, space-charge and impedance effects. Extremely low residual gas pressures have to be reached in the kilometre-long vacuum pipes to minimize beam losses by collisions of primary and secondary particles with residual gas atoms.

## 2 The heavy ion synchrotron SIS100

The synchrotron SIS100, as the main new driver accelerator of the FAIR project, accelerates heavy ion beams with much higher intensities to higher energies than it is presently possible with SIS18. Higher beam energies are specifically required for the production of antiprotons by bombarding nickel-targets with proton beams. To provide the necessary injector currents for SIS18, a new 70 MeV, 35 mA proton linear accelerator will be built. The final energy of the ions in SIS100 is chosen such that both radioactive ions and antiprotons are produced with highest efficiency in the targets. A short acceleration cycle of just two seconds ensures sufficiently high average intensities. Due to the fast ramping with 4 Tesla per second, the magnets and their vacuum chambers are subject to heating by eddy currents. The eddy current heating was a real challenge for the development of appropriate superconducting magnets with cryogenic vacuum chambers with wall thicknesses of only 0.3 mm. The cryogenic vacuum chambers are used as a huge system of cryogenic pumps and contribute significantly to the required extremely low vacuum pressure. Beam losses by stripping via collisions of the ions with atoms of residual gases are sufficiently low at an expected residual gas pressure below $10^{-11}$ mbar.

With the aim for higher intensities and consequently increasing space-charge effects, the charge state of heavy ions must be reduced. Charge exchange processes in a machine cycle depend on their cross sections as a function of the beam kinetic energy. For instance the charge exchange of

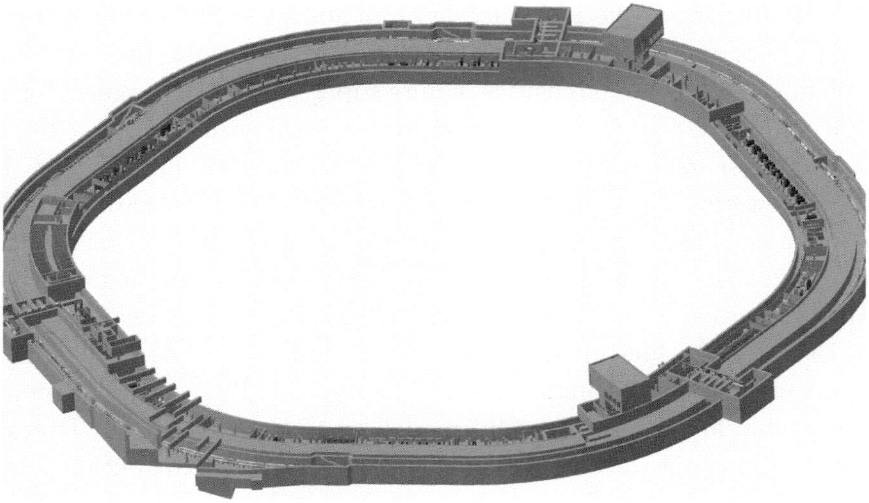

Fig. 2. The FAIR superconducting synchrotrons SIS100/300 in a 20 m underground tunnel with the (inner) parallel supply tunnel.

incompletely ionized ions, such as $U^{28+}$, due to impact ionization by collisions with residual gas atoms causes beam losses in the synchrotrons at comparably low beam energies. Losses of ions on the walls of the accelerator vacuum tube generate the so-called "dynamic vacuum effect" which may lead to tremendous ionization yields and consequently a drastic reduction of the ion intensity in the ring (Reich-Sprenger *et al.*, 2004). Therefore, injection into SIS18 and SIS100 is critical as the ions circulate at low beam energies, where the cross sections for ionization are high (Spiller, 2010). In case of a change in the charge state, the ions hit the wall of the beam pipe and at the point of impact, gas molecules are desorbed from the surface with a yield of about $10^4$/ion. The pressure is locally increased and a kind of avalanche effect is initiated. The strength of the charge exchange processes is strongly linked to the real machine cycle. Since the cross sections for ionization decrease significantly with energy, long term operation at low energies (e.g. injection plateaus) should be avoided, while fast acceleration with high ramp rates is mandatory. This rule is the reason for the fast ramping of the FAIR synchrotrons.

For an optimum operation of individual experiments and for a multiple-beam operation, two scenarios are foreseen concerning the extraction of ions from SIS18 and the injection into SIS100: The transfer of four pulses from the SIS18 to SIS100 within one second with subsequent acceleration, and the transfer of a single pulse to SIS100, which is immediately accelerated. An essential prerequisite for all scenarios is the rapid acceleration of the ions that requires high acceleration RF-voltages. In addition, a dedicated RF-system is needed to compress the beam after acceleration in a two-step process into a short pulse. Due to the low resonance frequency of ring resonators and the strict space limitation in all rings, the compact design of these cavities is mandatory and requires modern magnetic alloy (MA)-ring-core cavities (Hülsmann *et al.*, 2004). Short beam pulses are required for the production of secondary beams with maximum efficiency. The beam in SIS100 can be extracted either as a short pulse of less than 100 ns for injection into the storage rings or for plasma physics experiments, or in a long spill with duration of several seconds for fixed target experiments like CBM.

## 3 The FAIR targets and separators

The multi-stage superconducting fragment separator (Super-FRS) is the work horse of the NuSTAR experiment program (see Fig. 3). The Super-FRS will be the most powerful in-flight separator for exotic nuclei up to relativistic energies (Geissel *et al.*, 2003). Rare isotopes of all elements up to uranium can be produced, spatially separated and within some hundred nanoseconds analysed in the CR running in an isochronous mode. Thus very short-lived nuclei can be studied. The produced fragment beams consist of a very wide variety of different isotopes from the entire area of the nuclear chart. The

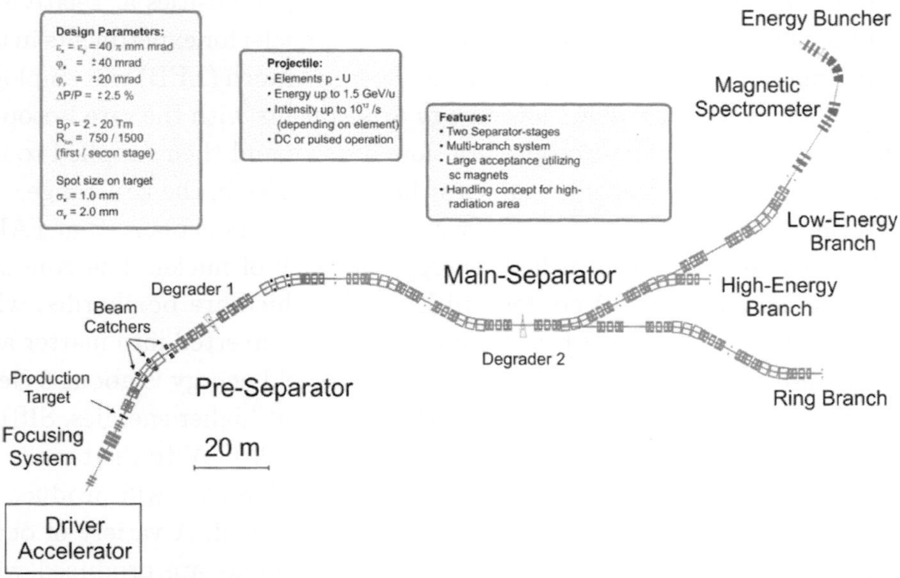


Fig. 3. Layout of the proposed superconducting fragment separator (Super-FRS) for the production, separation, and investigation of exotic nuclei. Spatially separated rare-isotope beams are delivered to the experimental areas via three different branches.

Super-FRS is a large-acceptance superconducting fragment separator with three branches serving different experimental areas including the storage-ring complex of FAIR. The layout of the Super-FRS consists of magnets designed for a magnetic rigidity of $B\rho_{max}$ of 20 Tm. Approximately 10% of the primary beams provided by SIS100 will be converted in a special target into exotic isotopes. In the production process of such exotic beams the kinetic energy is approximately preserved. The remaining 90% of the primary beam is selectively dumped in special beam catchers made of graphite and iron. The in-flight production provides secondary beams with high kinetic energies. Nevertheless, their large phase space volume requires huge magnets with enormous apertures and high field gradients. Ultimately, this demands the use of massive, superferric magnets. Due to the strong exposure to radiation in the vicinity of the production target, radiation-resistant magnets with ceramic insulation must be used. The beam power in the pulse, which is supplied by SIS100, is 200 GW, which can cause the destruction of the production targets. Therefore, the Super-FRS target consists of a rotating wheel which distributes the beam power over a large target area. The operating principle of the Super-FRS is based on a combination of magnetic field analysis and element-dependent energy loss in specially shaped degraders. Super-FRS has three branches which are the links to the different experimental stations and the collector ring CR. At the high energy branch (HEB) kinematically complete reaction studies are carried out using highest beam intensities at relativistic energies. Super-FRS is also used to provide exotic nuclei for experiments in the storage rings of the FAIR project. The low energy branch (LEB) will employ a gas stopping cell which allows low energy experiments with the rare isotopes. In the gas cell, the exotic isotopes are slowed down and then supplied to the experiments for spectroscopic studies at beam energies in the keV range.

With beams of antiprotons, a variety of experiments is planned at FAIR. Antiprotons are produced in high energy collisions of nuclei. The common technique uses a set of 10 cm long nickel rods, which are bombarded with proton beams. A fraction of the collision energy is converted into matter and thereby into a proton–antiproton pair. The threshold energy is about 6 GeV, however it is more efficient to generate antiprotons at higher energies. SIS100 will deliver highest intensities of proton beams with 29 GeV to the target. At 29 GeV beam energy, one out of hundred-thousand protons will produce an antiproton. About $10^8$ antiprotons per spill are expected. A variety of other secondary particles such as neutrons, protons, or pions are produced with intensities that exceed the number of antiprotons by orders of magnitude. Therefore, the antiprotons must be separated with high efficiency from the background particles as well as from the remains of the primary beam. The concept for the production of $\bar{p}$ beams at FAIR is basically determined by the luminosity requirement of $L = 2 \times 10^{32}\,\mathrm{cm}^{-2}\mathrm{s}^{-1}$ for experiments with cooled antiproton beams in the HESR. In the PANDA experiment, collision of the $\bar{p}$ beam with the internal hydrogen gas-target will be possible in the kinetic energy range from 0.8 GeV to 14 GeV (Franzke *et al.*, 2002). The maximum consumption rate is expected to be $10^7$ per second which is to be compensated by a corresponding net rate of the $\bar{p}$ production and accumulation.

## 4 The FAIR storage rings

One of the most important rings for the preparation of secondary particle beams is the Collector Ring (CR). The CR is a high acceptance ring with full aperture injection and extraction kickers, RF cavities for bunch rotation, adiabatic de-bunching and re-bunching, and a dedicated stochastic cooling system. The CR will collect and cool secondary particle beams that emerge from the production targets and have a large spread in beam energy and a huge spatial extension. The injection kicker design guarantees that the full ring acceptance is available for the incoming hot secondary beams. The beams coming from the production targets are not suitable for precision experiments. The use of beam cooling improves the quality of the secondary beams by several orders of magnitude and the well prepared beams can be then transported into subsequent storage rings for use in experiments. The corresponding accumulation method in the HESR is sketched in Fig. 4. The distribution of particles over the ring and the transverse momentum distribution are shown. While already injected particles are cooled, a coasting beam is formed. Using barrier buckets, a gap is opened that additional particles can be injected. The subsequent cooling reduces the transverse momentum of the injected particle and generates space for further injections. This accumulation scenario has been demonstrated for the first time in the ESR of GSI (Steck *et al.*, 2010).

Also in the HESR, antiprotons can be collected using the method described above and beams with increased intensity are available for the PANDA experiment. The stochastic cooling system of the CR can be used for antiprotons and radioactive beams as well. In addition, the CR allows mass measurements of short-lived radioactive ions using the isochronous mode. In this special ion-optical mode, the revolution time of the particles is independent of their momentum and with a special detector mounted in the storage ring, the mass of the ions can be determined directly from their orbital period in the ring. The CR has to perform stochastic precooling of

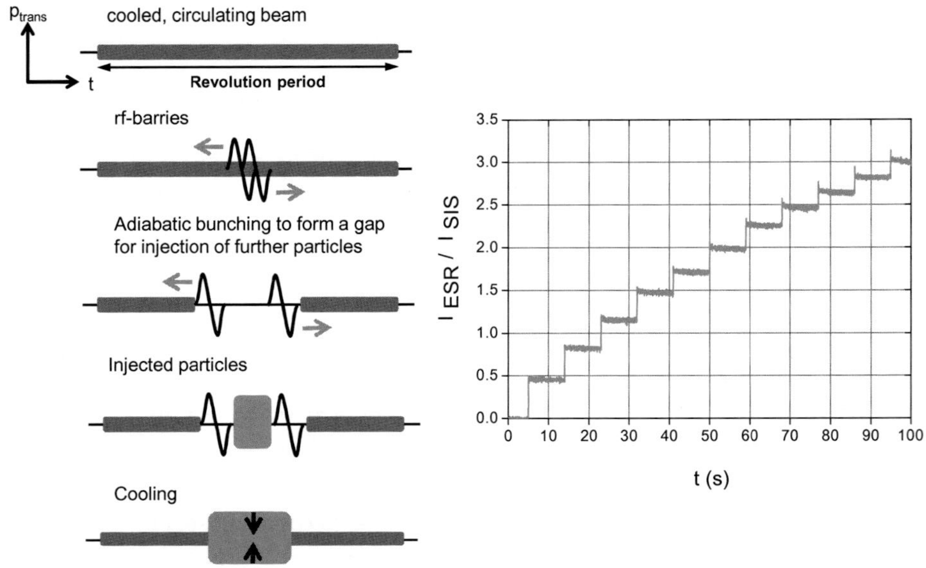


Fig. 4. Collection and cooling scenario for the accumulation of particles in the HESR and NESR (left) and measurements to test the method in the ESR of GSI (right). The particles are spread over the entire circumference of the ring and cooled. RF-barriers establish a gap in the particle distribution in which more particles can be injected.

secondary beams at a fixed kinetic energy of 740 MeV/u for radioactive isotopes and 3 GeV for antiprotons which may be transfered in a later stage to the RESR storage ring system. A special procedure is applied in case of the longitudinal phase space. Since bunches emerging from the production target are very short (<50 ns) but show a significant large momentum spread, phase space rotation with subsequent adiabatic de-bunching will be applied. For the fast de-bunching, a total maximum voltage of 200 kV for the bunch rotation cavities is available. For a later upgrade up to a maximum voltage of 400 kV, installation place is reserved in one of the straight sections for the five additional cavities. The CR stochastic cooling system is a technical challenge. It consists of three pick-up tanks in straight sections with zero dispersion, three kicker tanks in straight sections with zero dispersion, and a pick-up tank at high dispersion. All systems will work in the frequency band of 1–2 GHz. The distance between the electrodes and the shrinking beam will be kept constant by means of a fast mechanical adjustment of the pick-up electrodes. The microwave noise floor is minimized by using electrodes and preamplifiers operating at a temperature of about 20 K.

In the sequence of the storage rings for secondary beams the RESR follows the CR storage ring and is accommodated in the same building. The main purpose of the RESR is the accumulation of antiprotons from the CR. The operation of the RESR is simplified by the fact that the secondary beams are precooled by the stochastic cooling system of the CR before they are transferred to the RESR. The second task of the RESR will be the fast deceleration of RIBs from the injection energy of 740 MeV/u to energies between 100 MeV/u and 500 MeV/u within 1 s. Therefore, the huge dipole magnets have to be ramped at a rate of 1 T/s. In this way the RESR is able to provide short-lived rare isotope beams at low energies for the electron–ion collision experiments in the NESR. The NESR (Dimopoulou *et al.*, 2007), which shall be realized in a later stage of the FAIR project, will provide low energy ion and antiproton beams. The NESR is conceived as a versatile storage ring employed in the experimental program with stored ion beams, both stable and radioactive, and in the preparation of low energy antiprotons. The NESR can decelerate antiprotons down to 30 MeV and heavy ions down to 4 MeV/u and provide those to FLAIR, the facility for low energy antiproton and ion research (FLAIR, 2005). Therefore, the NESR will be equipped with stochastic cooling as well as electron cooling to keep the beam emittance small while slowing down the ions. This machine is designed to house high resolution spectrometers, internal targets, laser experiments and an electron ring for scattering experiments. The accumulation procedure described above will also be used in this ring.

Beside the option to generate high intensity beams of arbitrary stable elements, the FAIR complex provides a unique combination of primary and secondary beam facilities, with fixed target experiments, production targets and internal targets in storage rings, separators and storage rings. The individual accelerator sections are perfectly matched to each other by appropriate beam manipulations and thereby provide the experimentalist a huge range of parameters and properties of high quality beams.

## References

CBM, http://www.fair-center.eu/for-users/experiments/cbm.html

Dimopoulou, C. *et al.* (2007), *Phys. Rev. ST-AB* **10**, 020101

FAIR-EXP, http://www.fair-center.eu/public/experiment-program.html

FBTR, FAIR Baseline Technical Report, March 2006, http://www.fair-center.eu/fileadmin/fair/publications_FAIR/FAIR_BTR_6.pdf

FGP, FAIR Green Paper — The Modularized Start Version, October 2009.

FLAIR, http://www.flairatfair.eu/

Franzke, B. *et al.*, Conceptual design for an internal target experiments with antiprotons up to 15 GeV/c, *Proc. of the 8th European Particle Accelerator Conference 2002*, Paris, THPLE079.

Geissel, H. *et al.* (2003), *Nucl. Instr. and Meth. B* **204**, 71.

Hülsmann, P., Hutter, G. and Vinzenz, W., The bunch compressor system for SIS18 at GSI, *Proceedings of EPAC 2004*, Luzern, Schweiz, p. 1165.
NuSTAR, http://www.fair-center.eu/for-users/experiments/nustar.html
PANDA, http://www-panda.gsi.de/
Reich-Sprenger, H., Kollmus, H., Krämer, A., Bender, M. , Kurdal, J., Bellachiamo, M.C., Spiller, P., R&D vacuum issues of the future GSI accelerator facilities, *Proceedings of EPAC 2004*, Luzern, Schweiz, p. 1657.
Spiller, P., Towards the high intensity limit in the FAIR project — Present status and future challenges, *Proceedings of HB2010*, Morschach, Switzerland, MOIC01.
Steck, M. *et al.*, Demonstration of longitudinal stacking in the ESR with barrier buckets and stochastic cooling, *Proc. of the COOL'11*, Alushta, Ukraine, TUPS20.

# Chapter 33

# ELENA antiproton facility

Wolfgang Bartmann, Pavel Belochitskii, Horst Breuker, François Butin, Christian Carli, Tommy Eriksson, Stephen Maury, Walter Oelert, Sergio Pasinelli and Gerard Tranquille (CERN)

## 1 Motivation to build ELENA

Extra Low Energy Antiproton Ring (ELENA) is to be an extra ring used for post decelerating antiprotons received from AD. The kinetic energy of beam extracted from AD is 5.3 MeV. Most of AD experiments need antiprotons at few orders of magnitude lower energy. Now an antiproton energy reduction is made by passing through the sequence of degraders. After passing each degrader, the antiprotons slow down while their energy spread is essentially increased. The last degrader usually is placed as close as possible to a trap. The typical trap voltage is about 5 kV. Due to very large energy spread after passing through degraders, only a small portion of all extracted antiprotons (about 0.3%) is captured inside the trap.

With building of ELENA the number of antiprotons delivered to experiments will be increased by one to two orders of magnitude. Instead of deceleration in foils, beam will be decelerated in a small ring provided with an electron cooler. The electron cooler serves to reduce beam phase space, which is blown up during deceleration and counteracting beam blow up caused by scattering on residual gas and by intra-beam scattering (IBS) caused by repulsive Coulomb forces of antiprotons.

## 2 From initial ideas to machine project

Initially ELENA was proposed for post deceleration of antiprotons from Low Energy Ion Ring (LEAR). The very small ring of circumference 7.85 m, equipped with an electron cooler, was planned for deceleration from 5 MeV to 200 keV beam kinetic energy, followed by beam slow extraction [1].

This idea was resurrected [2] with the growing interest for antihydrogen study program, which is being developed successfully by AD experiments [3].

More detailed studies have been performed, followed by cost estimation for building ELENA [4,5]. A careful investigation of possible options for ELENA layout has been done, and the cheapest solution inside the AD hall has been chosen. Finally, in June 2011 ELENA was approved as a CERN Project.

## 3 Choice of ELENA extraction energy

The ideal case would be to deliver antiprotons to the experiments with energies of about 5 keV, which perfectly fits trapping requirements. Unfortunately, this is not possible for many reasons. One of which is the beam space charge effect which disturbs machine optics due to extra defocusing and set limit on beam intensity

$$N \sim \varepsilon\beta^2\gamma^3/l_b$$

where $\varepsilon$ is the beam emittance, $\gamma$ is the relativistic factor close to 1, $\beta = v/c$ and $l_b$ is the bunch length. In ELENA this limit is essential at the end of bunch compression right before beam extraction, because a short bunch length about 1.3 m is required for efficient trapping. The maximal beam emittance is limited by trapping efficiency as well, typically experiments need beam size $\sigma_{x,y} \approx 1$ mm at the beam line focal point.

Another important intensity limitation is intra-beam scattering (IBS), which blows the transverse and the longitudinal emittances with growth rate $1/\tau \sim \exp(N \cdot \beta^{-3})$. The beam emittance blow up due to multiple scattering of antiprotons on the molecules of residual gas scales as $\Delta\varepsilon \sim 1/\gamma^{-2}\beta^3$ which impose strong requirements to high vacuum at low energies. Another challenge at low energy is building of electron cooler with ultra cold electron beam.

The careful considerations show that the kinetic energy of ejected antiprotons at 100 keV is an optimal choice. In addition to fitting requirements mentioned above, it allows to build electrostatic lines for beam transport to users, which is a much cheaper option compared with magnetic lines. Another big advantage of using electrostatic lines is a fast switching between beam destinations, which allows extracting each bunch to different user during one turn of beam in the ring.

## 4 ELENA layout and optics

Placing ELENA ring inside of AD ring (see Fig. 1) allows using existing experimental areas and making beam lines from AD to ELENA and

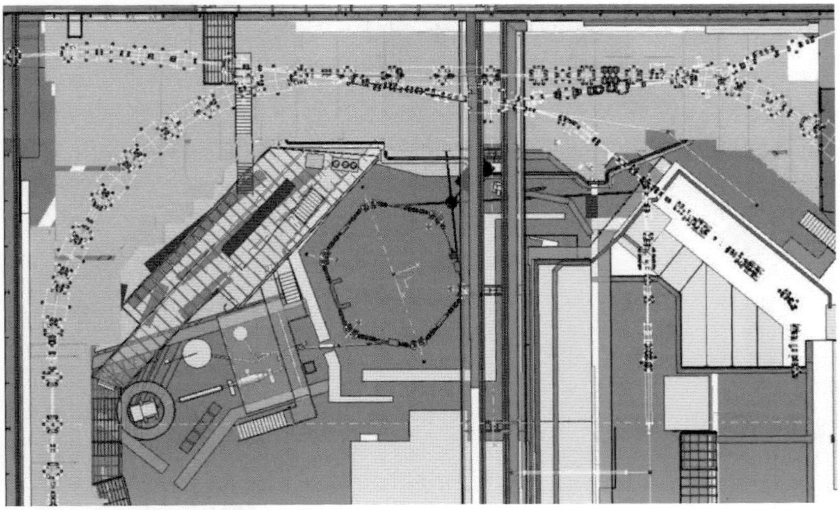

Fig. 1. ELENA in AD hall.

from ELENA to experiments reasonably short. On the other side, a lot of constraints have to be taken into account like existing AD equipment and barracks of AD users, availability of crane for installation, and others. Special attention has to be paid to the crossing of ELENA injection and extraction lines. The placement of $H^-$ source (which is planned to be used for machine commissioning and, possibly, for start-ups) in this overcrowded area makes layout design difficult. Along with optimization of position of the centre of ELENA, the ring has to be rotated to ease beam transport in electrostatic lines.

The ring circumference is 30.4 m, which is 1/6 of AD ring. The ELENA optics has been prepared in view of several important constraints. One of them is to fit in existing experimental area and to provide solutions for beam injection from AD and beam extraction into two experimental areas. Another task is to provide enough space for equipment of all the systems. Two long straight sections are used for electron cooling and for beam injection. Six folder optics was found optimal for beam extraction in two experimental areas (Fig. 2).

Tunes $Q_x = 2.3$, $Q_y = 1.3$ have been chosen to provide the biggest area free from the low order resonances to relax intensity limit imposed by space charge at extraction energy. Beta function values in cooling section have to be suitable for fast cooling of antiproton beam, antiproton beam alignment

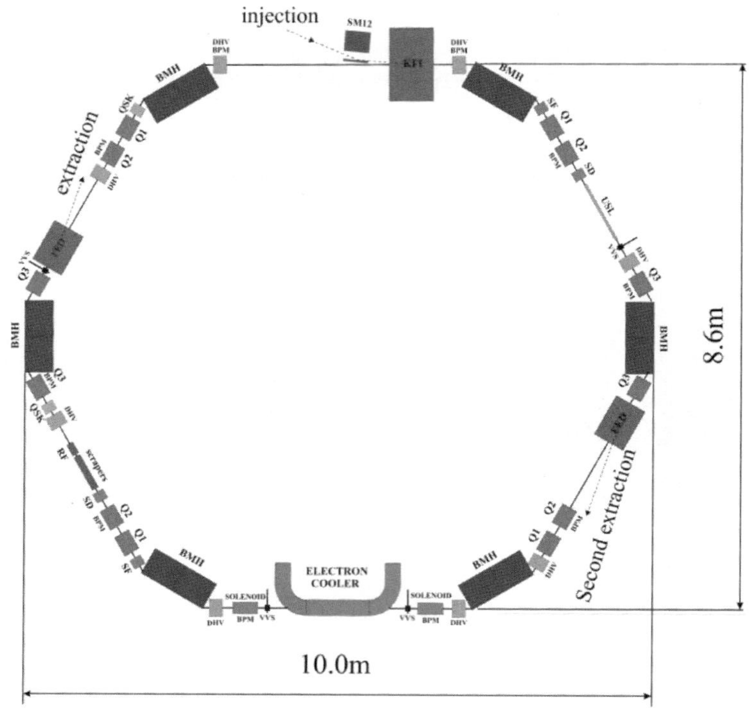


Fig. 2. Schematic layout of ELENA ring.

with respect to electron beam is foreseen. Dispersion in electron cooler should not be large, otherwise beam losses could occur.

Two beam extractions are performed by electrostatic deflectors, placed in two short straight sections. Due to low momentum of extracting beam extraction septum is not needed.

## 5 ELENA cycle

ELENA cycle is shown in Fig. 3. The bunched beam is injected on plateau, and then decelerated down to momentum of about 35 MeV/c. The cooling at injection energy can be skipped, beam is already cooled in AD right before extraction. The cooling at intermediate momentum is required to counteract beam emittance blow up and avoid losses due to limited acceptance. Then beam is decelerated again down to extraction momentum of 13.7 MeV/c. Here the final cooling is applied to prepare antiprotons with parameters required by experiments.

The cooler will be similar in size to the device that was built for the S-LSR ring [6] in Japan and will incorporate adiabatic expansion to reduce the electron beam temperature, as well as electrostatic bending plates for

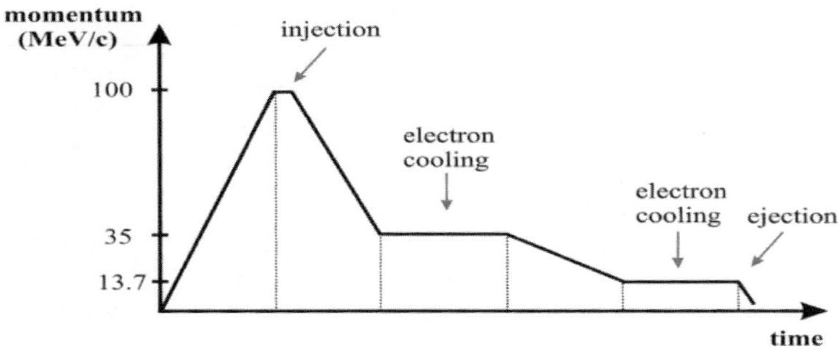


Fig. 3. Schematic view of ELENA cycle.

Table 1. The main parameters of the cooler.

| | | |
|---|---|---|
| Momentum, MeV/c | 35 | 13.7 |
| Electron beam energy, eV | 355 | 55 |
| Electron beam current, mA | 5 | 2 |
| Electron beam density, $m^{-3}$ | $1.38 \times 10^{12}$ | $1.41 \times 10^{12}$ |
| Magnetic field in gun, G | 1000 | |
| Magnetic field in drift, G | 100 | |
| Expansion factor | 10 | |
| Cathode radius, mm | 8 | |
| Electron beam radius, mm | 25 | |
| $\beta_x/\beta_y/D_x$, m | 2.2/2.1/1.5 | |

efficient collection of the electron beam. The optimization of the gun geometry will ensure a minimum transverse temperature of the electron beam and cooling simulations using Betacool [7] have been made to fine-tune the final parameters of the cooler.

The cooling will be extended to bunch compression process to counteract beam blow up caused by IBS and to reduce beam momentum spread, which is increased at the same time as bunch length is reduced. The small momentum spread is vital to build compact deflectors in electrostatic transfer lines, where dispersion could reach big values at some points, thus requiring big apertures and high voltage solution.

## 6 Beam extraction and main machine parameters

The basic scenario is to extract 4 bunches to 4 experiments during one turn. This will increase significantly time available for physics for each of the users. At the same time main intensity limitations like space charge and IBS are relaxed as well. Yet extraction of smaller number of bunches at $h = 1$, 2 or 3 in case of smaller number of experiments is foreseen as well. Fast electrostatic

Table 2. The main ELENA parameters.

| | |
|---|---|
| Momentum range, MeV/c | 100–13.7 |
| Energy range, MeV | 5.3–0.1 |
| Circumference, m | 30.4 |
| Intensity of injected beam | $3 \times 10^7$ |
| Intensity of ejected beam | $1.8 \times 10^7$ |
| Number of extracted bunches | 1 to 4 |
| Emittances (h/v) at 100 keV, $h = 4$, $\pi$·mm·mrad [95%] | 4/4 |
| $\Delta p/p$ of extracted beam, $h = 4$ [95%] | $2 \times 10^{-3}$ |
| Bunch length at 100 keV, m/ns | 1.3/300 |
| Required (dynamic) vacuum, Torr | $3 \times 10^{-12}$ |

deflectors in ELENA transfer lines provide true destination for each bunch. The extraction of 1 or 2 bunches might be performed with reduced number of particles in bunch, or with bigger emittances due to space charge limit.

## 7 Beam instrumentation

To measure the closed orbit during the deceleration cycle, 10 electrostatic beam position monitors will be installed inside the ring quadrupoles, compensation solenoids and in two orbit correctors in injection section. Their expected resolution is 0.1 mm with an accuracy of 0.3 to 0.5 mm. The tune measurement system based on the BBQ systems used on other rings [8] will use one pick-up to provide the tune evolution throughout the cycle.

As in the AD, it is planned to use a high sensitivity longitudinal Schottky pick-up to measure the beam intensity. Nevertheless, investigations are being made to check the feasibility of using a superconducting DCCT to make this low intensity measurement. This Schottky pick-up will also monitor the longitudinal beam cooling.

Scrapers coupled to scintillators will be used to destructively measure the transverse profile (and hence emittance) of the circulating beam. The use of ionisation profile monitors has also been considered to measure the emittance evolution throughout the deceleration cycle. However they have the detrimental effect of inducing a strong transverse kick on the beam and the gas injection system required to increase the ionisation rate will cause a large bump in pressure around a significant proportion of the circumference. Their use would be limited to the machine commissioning and specific machine development sessions.

## 8 ELENA transfer lines

The ELENA transfer lines will transport a 5.3 MeV beam from AD to ELENA and after deceleration and cooling in the ring a 100 keV beam to the experiments. Following a strong demand for antiproton experiments two extractions from the ring are foreseen to optimize the available area for existing and new experimental areas.

For the transfer lines from ELENA to the experiments the low beam rigidity of 45 mT·m (magnetic) or 200 kV (electrical) allows for choosing between magnetic and electrostatic elements for the beam transport. Economic construction of electrodes and power converters, low power consumption, no hysteresis and hence easy operation, no cooling needs and in particular good possibilities of shielding elements against magnetic stray fields favoured electrostatic elements. The same reasons lead to the choice of an electrostatic extraction kicker with a variable pulse length from 400 ns up to 7 $\mu$s to facilitate extracting either single bunches or the full beam. The nominal operational scenario foresees extraction of all four bunches at once from the machine in order to avoid emittance growth due to intra-beam scattering at the extraction flat bottom. Thus, the transfer lines need to be equipped with fast switches to split the bunch train into the respective experiments' channels. These switches are standardized with the extraction kicker elements and will be complemented with DC bending elements to reach the full required deflection.

The geometry of the lines is mainly constrained by the existing shielding structure and the equipment of the experiments in the hall. Special care is taken with the transfer line geometry close to the strong solenoids of the experiments because of stray fields. In particular since the experiments will receive beam from ELENA at the same time and not sequentially as presently from AD, any experimental magnet ramping has to be treated as dynamic error source for the other beam lines. Magnetic measurements were performed at the stray field hot spots and fed back into beam trajectory calculations. In order to mitigate these trajectory variations $\mu$-metal layers will be mounted around the beam pipe to shield the beam against the stray fields.

The focusing structure consists of a repeated FODO cell with 1.4 m drift length between quadrupoles and matching sections at the beginning of the extraction transfer line and close to the focal point at the experiments. The tight constraints on the line geometry demand compact and strong bending elements with a bending radius of 200 mm for 90° deflection angle. The voltage distribution between inner and outer electrode is optimized with

Fig. 4. Electrostatic triplet installed in the ASACUSA experiment, courtesy D. Barna.

respect to the element's focusing properties and its effect on the dispersion. The antiprotons lose or gain energy depending on the shape of the fringe field and thus experience different focal strengths in the bending elements. The strong dispersion introduced by these elements limits the maximum acceptable momentum spread by the transfer lines.

For commissioning purposes it is foreseen to install an external source delivering $H^-$ ions and protons at 100 keV. The protons circulating in reverse sense should allow for commissioning the magnetic elements and the optics of the ring while the $H^-$ ions circulating in the same direction as the antiprotons could allow for commissioning the electron cooler and the ELENA-to-experiments transfer lines. The injection of 100 keV $H^-$ ions into the 5.3 MeV antiproton AD-to-ELENA transfer line and the injection of 100 keV protons into the 100 keV antiproton extraction channel in reverse direction presents an interesting challenge for the beam transfer.

The design of the electrostatic quadrupoles is based on a prototype used in the ASACUSA experiment to focus 130 keV antiprotons, see Fig. 4.

## 9 ELENA experimental areas

The ELENA machine will deliver beams to the existing AD experiments, but also to a number of new experiments.

### 9.1 *Existing experimental areas*

The existing experimental areas (hosting namely ALPHA, ASACUSA, ATRAP and AEGIS experiments) are planned to be kept for the foreseeable future. These experiments or their upgraded versions will use the beams delivered from ELENA instead of AD when the existing transfer lines are dismounted to be replaced by new ones.

It is planned to upgrade a number of infrastructure features of the existing experimental areas, in particular:

- Relocate the control rooms into an adjacent building (No. 93): AEGIS experiment control room is already there and ATRAP control room removal is planned for 2013. ASACUSA and ALPHA control room removals remain to be confirmed.
- Re-allocate the space currently used by the control rooms to either preparation areas or rack space. The displacement of ATRAP cleaning room to the space currently used by the control room is planned for 2013.
- Implement Oxygen Deficiency Hazard (ODH) detectors. This is already the case for AEGIS, and will be studied in detail for the other areas in 2013.

The ASACUSA experimental area may accommodate a second beam line, parallel to the existing ones when the ELENA transfer lines are installed.

A visitor platform has been installed that allows viewing the existing experiments from short distance, whatever the beam conditions are. The visit circuit allows even to see the AD machine, but only during shutdowns due to radiation environment.

### 9.2 *New experimental areas*

It is planned at this stage to accommodate space for at least one new experiment exploiting a beam extracted from sector 3 of ELENA. The corresponding experiment GBar was approved in May 2012, and is planned to be installed in parallel with ELENA machine, to become operational in 2017.

The corresponding experimental area will partly use the space currently occupied by the workshop and partly the location of existing racks. The project of this new experimental area was made possible by the approval of an extension of the AD hall, where the AD kicker generators will take place as of 2015.

The geometry of the GBar experimental area is not finely defined yet, as it strongly depends on the experiment to be built there, which is currently being designed. The latest proposal that minimizes floor space and beam bending in the transfer line is shown in Fig. 5. The new GBar experimental area will be surrounded by a layer of concrete shielding, 0.4 m thick, with a separate access from the ELENA machine perimeter.

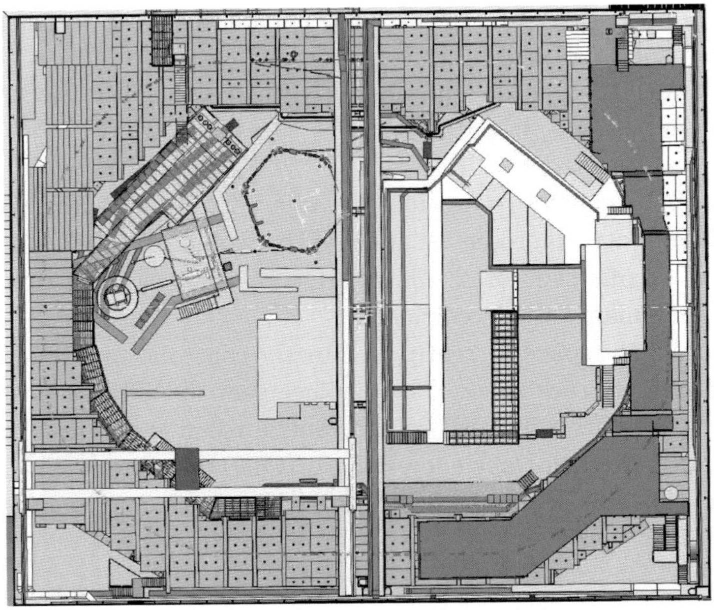

Fig. 5. Existing and future AD/ELENA experimental areas in AD hall.

Studies have been initiated to accommodate space for possible extra experiments.

In particular, the space currently occupied by the ATRAP cleaning room (soon to be moved to the upper floor) could profitably be used for a small dimension experiment. It is important that the experiment installed there remains compact and relatively light as no crane access is possible in this region. A proposal has been submitted by a collaboration from RIKEN to measure the magnetic moment of antiprotons. This experiment could potentially make use of this location (see location 1 in Fig. 5).

Other possibilities exist for more experiments, but it appears that making space for sizeable experimental areas would require in depth modification of the organization of the AD hall, in particular in the area currently partly occupied by the ASACUSA laser hut (see location 2 in Fig. 5). These should be considered in the longer term.

It is anticipated to allow visitors to have close views both onto ELENA machine and onto the new experiments. At this end, viewing platforms are being designed at mid height, with appreciable views from above the shielding of all new equipment. The low energy of the beams is an advantage, permitting to keep radiation doses compatible with general public visits, whatever the beam conditions.

## 10 Conclusion

With the new ELENA facility, a further deceleration down to 100 keV of the antiproton beam injected from the AD at 5 MeV is a very challenging and promising project. The beam intensity provided for the experiments will be increased up to two orders of magnitude and several experiments will operate simultaneously. A new experimental area is created which opens the possibility to install new experiments. This investment guarantees that the antiproton physics will continue at least for the next two decades. However, to ensure this long period all upstream machines should be consolidated.

In view of the lack of resources such as finances and manpower the project will have to be supported partly by external laboratories.

## References

[1] Herr, H. (1984), A small deceleration ring for extra low energy antiprotons (ELENA), CERN-PSCC-82-3; PSCC-P-52-Add-1, CERN, Geneva.

[2] Belochitskii, P. (2004), Extra Low Energy Antiproton Ring (ELENA) for antiproton deceleration after the AD, *SPCS Villars Meetings on a Future Fixed Target Programme at CERN*, Villars-sur-Ollon, September 22–28, 2004.

[3] Oelert, W., Gabrielse, G., Hayano, R., Holzscheiter, M., Hangst, J., Lichard, P. (2009), ELENA: An upgrade to the Antiproton Decelerator, CERN-SPSC-2009-026; SPSC-P-338.

[4] Angoletta, M.-E., Barnes, M., Beuret, A., Belochitskii, P., Borburgh, J., Bourquin, P., Buzio, M., Cornuet, D., Eriksson, T., Fowler, T. *et al.* (2007), ELENA, a preliminary cost and feasibility study, CERN-AB-2007-079.

[5] Angoletta, M.-E., Belochitskii, P., Borburgh, J., Breuker, H., Buzio, M., Capatina, O., Chiggiato, P., Ducimetière, L., Eriksson, T., Fowler, T. *et al.* (2010), ELENA — An updated cost and feasibility study, CERN-BE-2010-029.

[6] Shirai, T. *et al.* (2006), S-LSR, cooler ring development at Kyoyo University, *Proc. of Cool2005 Workshop*, AIP Conf. Proc. **821**, 103.

[7] Sidorin, A. *et al.* (2006), Betacool program for simulation of beam, dynamics in storage rings, *NIM A* **558**, 325–328.

[8] Gasior, M. (2012), Tune measurement throughout the AD cycle, ELENA beam instr. review, CERN, April 12, 2012.

# Part 3

# Particle Accelerators for the XXI Century

## Chapter 34

# CNGS: A conventional neutrino long-baseline $\nu_\tau$ appearance experiment

Ilias Efthymiopoulos and Edda Gschwendtner (CERN)

### 1 Introduction

Neutrinos, although they are the second most abundant particle in the Universe (after photons), are extremely hard to detect. They are weakly interacting and able to pass through matter without interaction, even at the size of the Earth. In order to detect them, large volume (mass) detectors and high flux of neutrinos are required.

In the Standard Model they are included as massless spin 1/2 particles, and for several decades the search of their properties have been the subject of many experiments worldwide. The discovery at the end of the last century that neutrinos that come from the Sun or generated in the Earth atmosphere change flavour as they propagate in space, known as neutrino oscillations, implies that neutrinos contrary to the SM prediction have mass. The observed neutrino oscillations can be described by a flavour-mixing mechanism characterised by a complex $3 \times 3$ matrix, similar to the one used for the quarks. The early experiments that followed this discovery measured the neutrino mixing angles $\theta_{12}$, and $\theta_{23}$, by measuring excess or disappearance of muon ($\nu_\mu$) or electron ($\nu_e$) neutrinos. The neutrino oscillation probability $P$ in the standard two-flavour mixing scheme may be expressed in a simplified form by the equation:

$$P = \sin^2 2\theta \sin^2(1.27 \times (L/E) \times \Delta m^2),$$

where $L$ is the mean distance between the detector and the neutrino source in km, $E$ is the average neutrino energy in GeV, $\theta$ is the mixing angle for the two flavours concerned, and $\Delta m^2$ the neutrino mass difference squared in $eV^2$. For a given distance $L$ from the source, the energy of the neutrino beam, or the energy of the focused mesons, can be varied to address a $\Delta m^2$

range and vice versa, within the allowed values for the detector capabilities and particle flux.

The CNGS beam and experiments were proposed to give an indisputable answer to the neutrino oscillations by measuring the transitions $\nu_\mu \rightarrow \nu_\tau$. The CERN Neutrinos to Gran Sasso (CNGS) project was proposed at the end of the 20th century to study the neutrino oscillations on a long-baseline of 730 km [Elsener *et al.* (1998)]. In the initial proposal the project consisted of the neutrino beam at CERN, a near detector located at about 1.5 km from the target and the far detectors at 730 km distance in the underground Grand Sasso Laboratory in Italy. This proposal included in addition a broad physics reach like the possibility of using the near detector to seed light into the LSND signal [Athanassopoulos (1997)] and sterile neutrino search. However, mainly due to funding limitations, the near detector cavern was later removed from the project, leaving the $\nu_\tau$ appearance signal the main physics goal for the experiments.

## 2 Neutrino production with accelerators

Proton accelerators are used to create controlled and intense neutrino $\nu_\mu$ beams in the so-called "conventional" method as tertiary particles with a tunable mean energy. The principle of the neutrino beam production is illustrated in Fig. 1 for the case of the CNGS beam. An intense proton beam is directed towards a target where secondary charged mesons ($\pi$ and $K$) are produced. A set of toroidal magnets capture and focus these secondary particles to produce a parallel beam that enters a long volume where they decay to neutrinos:

$$\pi^{+(-)} \rightarrow \mu^{+(-)} + \nu_\mu(\overline{\nu}_\mu),$$
$$K^{+(-)} \rightarrow \mu^{+(-)} + \nu_\mu(\overline{\nu}_\mu).$$

A massive absorber at the end of the decay volume stops all remaining particles (mainly non-interacting protons from the primary beam) leaving

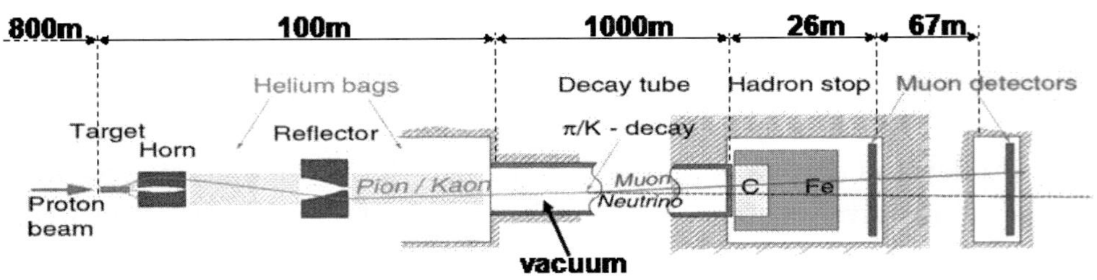


Fig. 1. Schematic representation of the production of an accelerator-made neutrino beam.

only muons and the neutrino beam through. The muon beam is measured in muon detectors before it is stopped in the Earth. The produced neutrino beam consists mainly of muon neutrinos $\nu_\mu$ with a small ($\sim 2\%$) contamination of electron neutrinos ($\nu_e$) from the Kaon and muon decay. The whole setup of the primary and the secondary beam elements is directed towards the far large-volume detector located underground few hundreds of kilometres away to detect the $\nu_\mu \to \nu_x$ oscillations, where $x$ is $e$ or $\tau$. CNGS is a wide-band high-energy neutrino beam. In Fig. 2 the flux of the four neutrino beam components is shown. The average energy of the $\nu_\mu$ that are sent to Gran Sasso is about 17 GeV. Accelerator neutrino configurations, like CNGS,

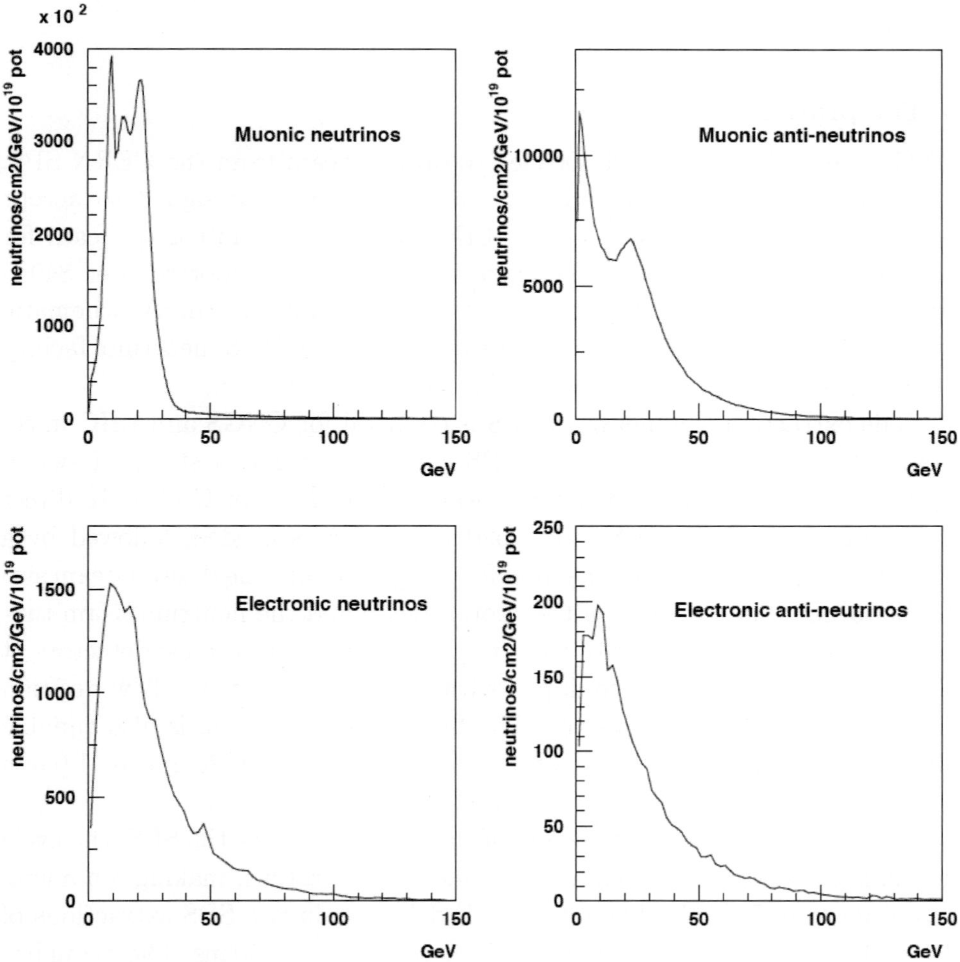


Fig. 2. Expected $\nu_\mu$ fluence spectrum at Gran Sasso, compared to the product of oscillation probability times $\nu_\tau$ cross-section.

where $L \sim 1000$ km, are defined as long-baseline experiments. For CNGS the distance between the source at CERN and the detectors at Grand Sasso Lab is measured [Jones *et al.* (2013)] to be 732 km. The neutrinos from the beam will spread out to a couple of orders of magnitude larger in radius than the far detector. Only neutrinos emitted at zero angle from the parent well-focused pions will reach the far on-axis detector having an energy of approximately 43% of the pion energy. The small angular acceptance, combined with flux reduction at the far detector with the square of the distance ($\sim 1/L^2$), the small interaction cross-section of the neutrinos in the detector mass, and that the neutrino beam is in fact a tertiary beam, result in requiring a high-power proton beam to produce adequate flux for the experiment. CNGS profits from the high-intensity and high-energy beam from SPS at 400 GeV/c to produce this required neutrino flux.

## 3 The proton beam

CNGS uses a fast extracted 400 GeV/c proton beam from the CERN SPS. The extraction channel installed at the SPS LSS4 is designed to accept both the high-brightness beam for LHC and the high-intensity beam for CNGS [Kain *et al.* (2007)]. The extracted beam is transported over 840 m onto a graphite target located in an underground cavern. A schematic overview of the extracted proton beam and the CNGS neutrino facility underground structures is shown in Fig. 3.

The extraction channel from SPS is common for CNGS and LHC injection. After about 100 m from the SPS extraction point, a string of switch magnets is used to direct the beam towards either LHC or CNGS. To direct the beam towards Grand Sasso, a 620 m long arc is needed, followed by a straight section of 120 m for the final focusing to obtain the desired beam size at the target. The last part of the proton beam and the neutrino beam tunnels have a downward slope of 5.6% required to point to the far detectors at Grand Sasso. The beam is equipped with steering elements to allow position and angle steering at the target. The final focusing system is also tunable allowing a variable beam spot at the target from $\sigma = 0.25$ mm to 1.0 mm with the nominal value of $\sigma = 0.50$ mm.

The nominal design parameters of the CNGS beam in the SPS, the cycle length is 6 s and the required intensity $4.8 \times 10^{13}$ protons, making a nominal beam power of 510 kW. The intensity is delivered in two SPS extractions of $2.4 \times 10^{13}$ protons each lasting for 10.5 $\mu$s, separated by 50 ms. The beam has a micro-structure of bunches with 5 ns bunch spacing originating from the SPS RF system and fill the entire circumference of SPS (23 $\mu$s) except for

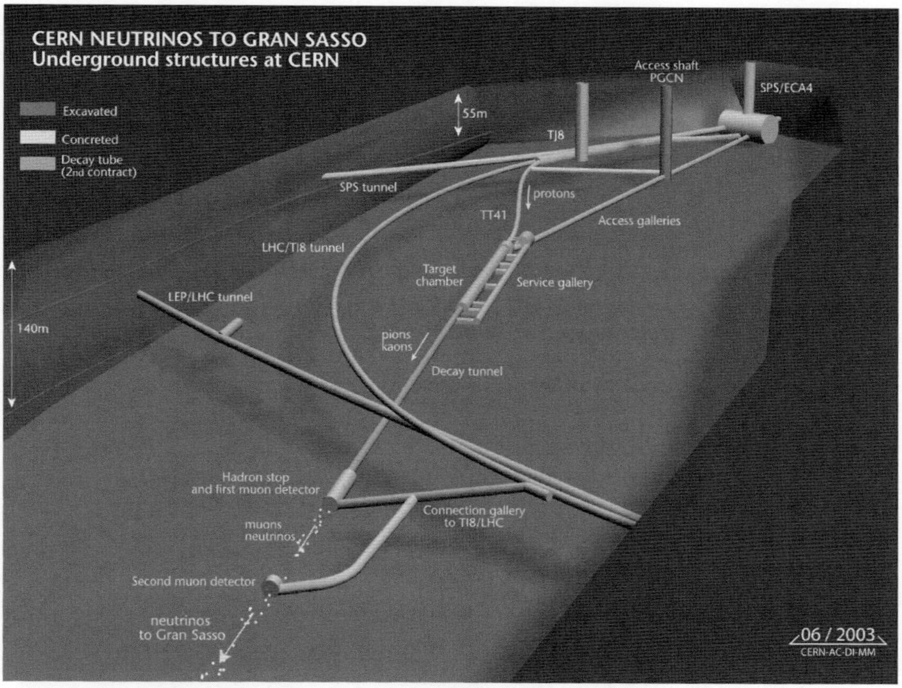


Fig. 3. Layout of the CNGS underground structures.

two $\sim 1\mu$s gaps required to accommodate the rise and fall time of the fast extraction system to LSS4. In Table 1 the main parameters of the proton beam are listed. For a typical SPS year, with 200 days of operation and 60% beam sharing with other SPS physics users, $4.5 \times 10^{19}$ protons on target are nominally expected for CNGS.

The magnetic system of the proton beam line consists of 73 dipole magnets (nominal field 1.7 T at 400 GeV/c) [Schirm *et al.* (2006)], 20 quadrupole magnets (nominal gradient of 40 T/m) and 12 corrector magnets (maximum deflection angle of 80 $\mu$rad). The 5.6% slope of the proton beam, which is needed to direct the beam towards Gran Sasso, is provided by 32 horizontal bending magnets tilted by 12.8° (see Fig. 4). Although the nominal beam energy for CNGS is 400 GeV, the beam line could operate at a lower energy down to 350 GeV, limited by the vertical aperture in the main bending magnets. The instrumentation provided in the proton beam line allows to accurately track the position of the high-intenstiy beam and monitor its profile and intensity at various locations [Jensen (2006)]. The beam position is measured using 23 Beam Position Monitors (BPM); all but the last one equipped with button electrodes for the two planes. The last one is an innovative strip-line coupler pick-up monitor in air, mechanically coupled to the

Table 1. Nominal design parameters of the CNGS proton beam at 400 GeV/c.

| Parameter | Nominal Value | Unit |
|---|---|---|
| Normalized emittance | H = 12, V = 7 | $\mu$m |
| Physical emittance | H = 28, V = 16 | nm |
| Momentum spread | $0.07 \pm 20$ | % |
| Extraction number per cycle | 2 | — |
| $\Delta T$ between extractions | 50 | ms |
| Extraction batch length | 10.5 | $\mu$s |
| Number of bunches per extraction | 2100 | — |
| Intensity per extraction | $2.4 \times 10^{13}$ | protons |
| Bunch length ($4\sigma$) | 2 | ns |
| Bunch spacing | 5 | ns |
| Beam beta at focus | H = 10, V = 20 | m |
| Beam sizes at focus (400 GeV, $1\sigma$) | 0.5 | mm |

Fig. 4. Photo of the 800 m proton beam transfer line for CNGS with the tilted magnets.

target structure. It provides information on the beam position as seen by the first target rod with a $\pm 0.35$ mm precision. For the proton beam profile, 8 optical transition radiation (OTR) monitors equipped with either a 7 $\mu$m carbon screen for high-intensity operation, or a 12 $\mu$m titanium screen for a low-intensity beam are used. The OTR radiation at the beam passage is captured by CCD cameras to provide the beam image. To measure the beam intensity, two Beam Current Transformers (BCT) are used at the beginning and end of the proton line. Beam Loss Monitors (BLM), ionisation chambers filled with $N_2$ are used to monitor the beam losses along the line.

## 4 The CNGS facility

The tunnel structures for CNGS are located deep underground at about 50 m deeper than SPS and about 100 m from the surface due to the angle required directing the 400 GeV/c proton beam to Gran Sasso (Fig. 5). A single tunnel about 1 km long gives access to the neutrino beam areas. The design, and in particular the operation of this high-intensity beam facility is optimised for being deep underground mainly for what concerns the handling of the equipment, containment of radiation and environmental impact. A large cavern, see Fig. 3, contains the target station and the two magnetic focusing elements, called horn and reflector. It extends to 100 m from the target where the 994 m long decay pipe begins. A parallel service tunnel allows direct access to the equipment in the target hall for services and interventions.

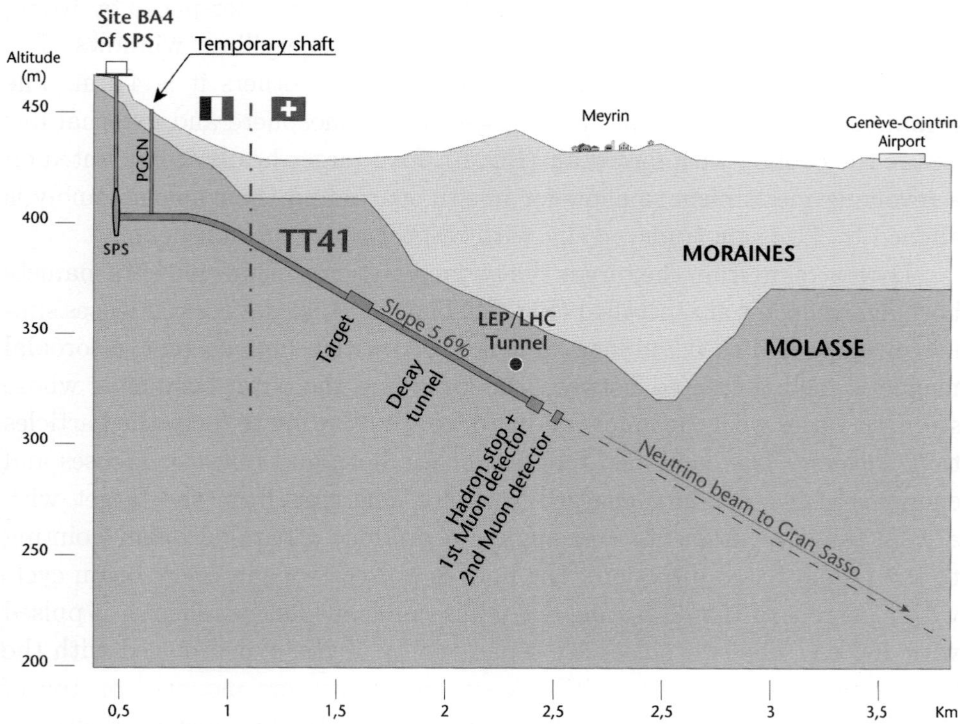


Fig. 5. Sketch showing the vertical arrangement of the CNGS neutrino beam tunnels.

(a) (b)

Fig. 6. The CNGS target. (a) Detail of a target head with the graphite rods, the support structure and the external aluminium cylinder, (b) photo during the installation of the target unit with the five heads in its place in the beam. The side iron and marble shielding and the forced-air duct at the bottom are visible. In the final installation a top shielding with remotely movable blocks is installed.

The target used in CNGS is based on design using graphite as prime material due to its thermomechanical properties [Peraire *et al.* (1996)]. A target head consists of 13 graphite rods each 10 cm long interspaced by 10 cm, placed in an aluminium cylinder closed off with beryllium windows. The diameter of the first two rods is 5 mm and of the others it is 4 mm. The cylinders are maintained in an inert gas (He) atmosphere and external fins assure the cooling with forced air (Fig. 6). Five target heads are mounted on a revolving mechanism to allow for in-situ spares and the whole assembly is aligned to $\pm 0.1$ mm transversally with respect to the proton beam.

Downstream from the target the two magnetic focusing elements, namely horn and reflector are installed (Fig. 7). They are toroidal coaxial lenses similar in shape and each 7 m long. When pulsed with a high current, a toroidal magnetic field is created between the inner and the outer conductor whose strength varies with the inner conductor shape allowing to focus the particles that traverse. The combined function of the horn and reflector focuses and energy-selects secondary charged particles emerging from the target with a wide range of momenta and angles to produce a parallel beam pointing to the Grand Sasso direction. The horn is pulsed twice in every beam cycle with a current of 150 kA lasting few milliseconds, while the reflector is pulsed with 180 kA. The two pulses are separated by 50 ms, synchronised with the two beam extractions from SPS. A system of sprayers mounted on top of the outer conductor continuously injects water onto the inner conductor to remove the induced heat from the electrical current and from the radiation.

Fig. 7. Photo of the horn during its installation in the tunnel.

As for the target, the horn and reflector design was optimised for remote handling and maintenance operations in the radiation harsh environment of CNGS. By changing the sign of the electric field in the horn and reflector positive or negative hadrons can be focused, producing respectively a neutrino or anti-neutrino beam. Remotely controlled mechanisms allow the re-alignment of the target head and the horn in the transverse plane and relative angle.

The space between the horn and reflector is determined to optimise the wide-band high-energy $\nu_\mu$ beam. In the areas between the horn and reflector a tube filled with helium (31 m long, 1.2 m diameter, at 20 mbar overpressure) is installed to minimise the particle interactions and radioisotope production in the cavern. A second helium tube 41 m long is installed downstream from the reflector covering the distance to the end of the target cavern where the decay tube starts. The CNGS decay tube is made of steel, 18 mm thick and 994 m long, with a diameter of 2.45 m, surrounded by 50 cm concrete in its full length. The decay tube construction was excellent, allowing to easily maintain it under vacuum at less than 1 mbar. The entrance window of the decay pipe is made of a 3 mm thick titanium foil, while the downstream window is made of steel and has a thickness of 50 mm and is water-cooled. At the end of the decay pipe is the hadron stop. It is a massive assembly of

Fig. 8. The hadron stop during its construction.

blocks 18 m long, the first 3 m of graphite and the rest of cast iron, water-cooled designed to absorb up to 100 kW of power from the non-interacting part of the proton beam and all the secondary hadrons (see Fig. 8). A set of two detection stations separated by 67 m downstream from the hadron stop measure the muons in the beam line before they are absorbed further downstream in about 500 m of rock. The muon detectors can provide the measurement of the beam profile in both the horizontal and vertical plane, thus providing useful information on its direction and quality.

The secondary beam elements are designed to withstand the nominal beam intensity from SPS of $2 \times 2.4 \times 10^{13}$ protons on target (pot) every 6 s corresponding to a yearly integrated total of $4.5 \times 10^{19}$ pot resulting in a harsh radiation environment. In the design phase of the facility it was considered important to provide some margin for possible improvements or upgrades in the accelerator chain and therefore build components — as far as technically and financially affordable — for possible higher intensities than the nominal. In Table 2 the expected yearly delivery of protons on target for the different operation modes are described. The "ultimate" beam intensity of $2 \times 3.5 \times 10^{13}$ pot per 6 s cycle, corresponding to a beam power of 750 kW in dedicated operation of SPS for CNGS, is basically to be considered as

Table 2. Estimates of annual proton beam delivery for different scenario of SPS operation used for the design of the CNGS facility, as explained in the text. The estimates are for injection of two PS batches per SPS cycle of 6 s.

| | | | | 200 days (pot per year) | | |
|---|---|---|---|---|---|---|
| Beam type | Protons per PS batch | Protons per SPS cycle | Efficiency Sharing | 100% no | 55% no | 55% 60% |
| Nominal | $2.4 \times 10^{13}$ | $4.8 \times 10^{13}$ | | $1.38 \times 10^{20}$ | $7.6 \times 10^{19}$ | $4.56 \times 10^{19}$ |
| Ultimate | $3.5 \times 10^{13}$ | $7.0 \times 10^{13}$ | | $2.02 \times 10^{20}$ | $1.11 \times 10^{20}$ | $6.65 \times 10^{20}$ |

the limit for the instantaneous beam intensity rather than the continuous operation and integrated intensity. For the design of the target head the ultimate pulse intensity of $7.0 \times 10^{13}$ protons, corresponding to a maximum of 750 kW of beam power, was used as guideline considering the associated thermal stresses, deposited energy (up to 1.4 kW) and radiation environment. The concept of the target magazine with the five in-situ spares was also designed for the same intensity, offering the additional possibility to continue without interruption even after a single failure.

Since the early stage in the design, optimisation processes for radiation protection issues have been embedded, and followed all along the operation of the facility. Extensive simulations of the facility using the FLUKA code [Bohlen *et al.* (2014)] for dose estimates, physics performance and particle fluences have been made, using the maximum nominal integrated intensity of $7.6 \times 10^{19}$ pot per year. Examples of this optimisation process can be found in the choice of materials, like the use of marble shielding; of technical solutions, like the installation of a fully remotely controlled overhead crane with automatic plug-in system for the removal of the shielding blocks and the elements of the neutrino line; and of procedures, as the extensive study of the horn and reflector replacement in case of failure validated with real-life practice sessions before startup. However the operation of such high-intensity beam facilities remains a very challenging task as proven in CNGS and also other neutrino facilities in the past [Efthymiopoulos *et al.* (2009)].

## 5 Operational Performance

The construction of the facility started in 2000 and was completed as scheduled in 2006. The beam was successfully commissioned in July 2006 [Meddahi *et al.* (2007)], followed by the first physics operations for the experiments. After solving the initial technical problems in 2006 and 2007, the facility had a smooth operation for five years until the end of 2012 delivering $1.824 \times 10^{20}$

Fig. 9. Photo of the CNGS target cavern with the target station in the forefront. The slope of the tunnel required to point to Grand Sasso is visible.

protons on target, corresponding to more than 81% of the approved program [Gschwendtner *et al.* (2013)]. In Fig. 10 the development of the integrated yearly and total statistics for the CNGS beam is shown. The record year for CNGS was 2011, where although running in parallel to LHC and the SPS fixed target program, $4.84 \times 10^{19}$ protons on target were delivered, slightly above the expected $4.5 \times 10^{19}$ from the project proposal. The nominal beam power of SPS for the CNGS beam of $4.5 \times 10^{13}$ 400 GeV protons every 6 s corresponds to 510 kW in dedicated mode of operation. The actual average beam power during the CNGS operation is lower, approximately 300 kW, primarily due to intensity limitations in order to maintain the beam losses in the Proton Synchrotron (PS) and the SPS within the authorized limits, and secondly due to beam sharing in the SPS with the LHC, the fixed target physics program and machine developments studies. However in certain periods as shown in Fig. 11 the SPS was used in dedicated mode for CNGS with an intensity reaching about 85% of the nominal, corresponding to a beam power of 405 kW that is a world record for conventional neutrino beams. The sustained maximum during a day was 330 kW. The efficiency of the accelerator complex increased over the years; the integrated efficiency in

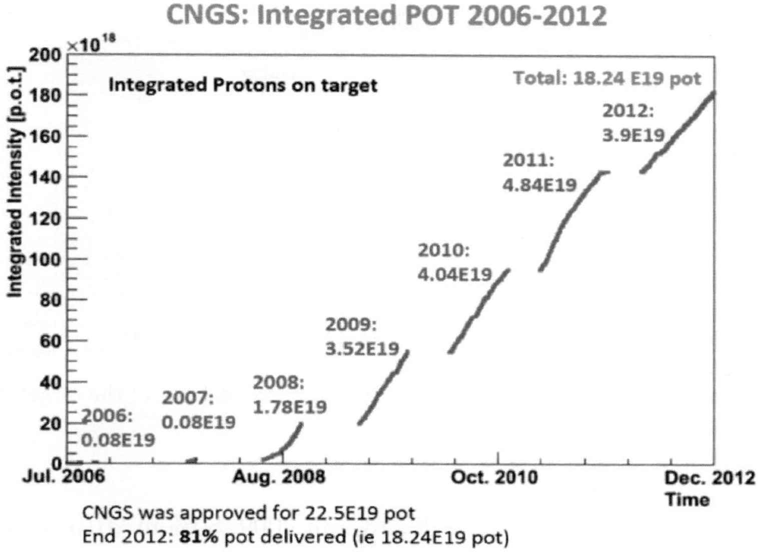


Fig. 10. Integrated number of protons delivered from SPS to the CNGS target.

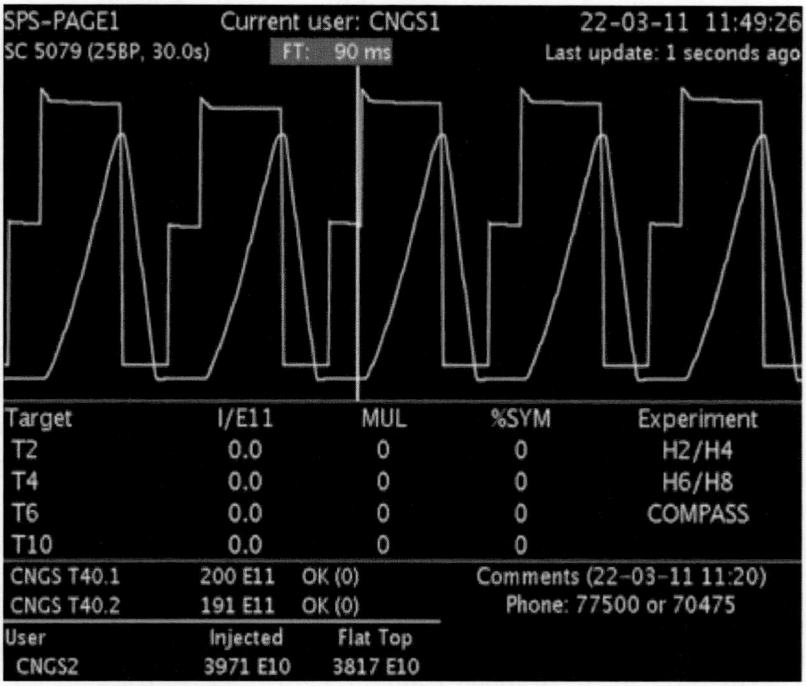


Fig. 11. SPS super-cycle configuration with five 6 s CNGS cycles, or 100% duty cycle of the neutrino beam. The intensity per cycle is $3.8 \times 10^{13}$ protons, corresponding to a beam power of **405 kW**.

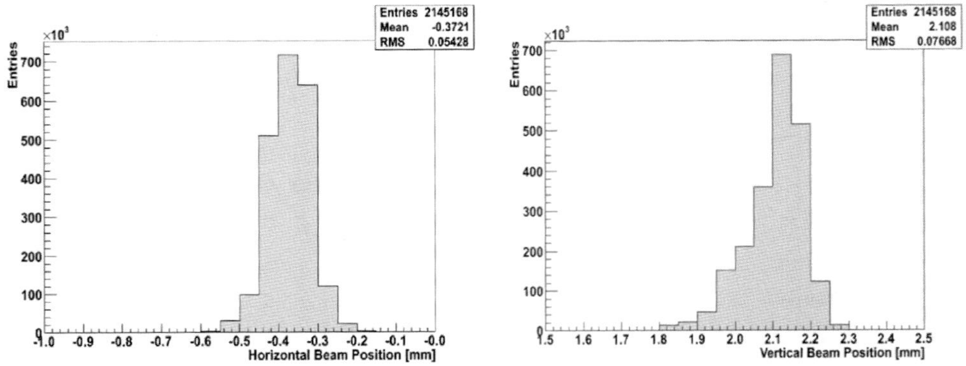


Fig. 12. Horizontal and vertical beam position measured in front of the target during the CNGS operation.

the SPS for CNGS was 61% in 2008, 73% in 2009, 80% in 2010, 79% in 2011 and 82% in 2012.

The overall beam performance and the stability of the CNGS primary proton beam line on the target position was excellent throughout the five years of operation. The beam position stability over the full running period of CNGS was very good (Figure 12) with an rms of 54 (77) $\mu$m in the horizontal (vertical) direction for more than 2 million extractions over the five years of operation.

Following the initial beam setup at commissioning the beam stayed well tuned and stable throughout the CNGS operation. The maximum beam excursion measured to be well within the $\pm 4$ mm allowed by the magnet aperture constraints and the losses along the line were minimal. Only small beam steering corrections per week were necessary.

The two muon detection stations placed in the forward direction, the first right after the hadron stop and the second at 67 m distance, proven to be the most sensitive and useful instruments in the quality monitoring of the neutrino beam, responding to any misalignment between the proton beam, the target and the horn, as well as any deterioration of the target and the horn. In each station, 42 muon detectors are installed to measure the muons created in association with the $\nu_\mu$ in the decay of the secondary hadrons. The muon detectors are assembled in a cross-shaped array to provide the vertical and horizontal muon profiles and the total muon intensity (see Fig. 13) [Marsili *et al.* (2008)]. The total height/width covered by the detectors is 2.7 m. In addition, an identical monitor is installed on a motorized support, downstream from the fixed ones to allow cross-calibration of the fixed monitors and to probe the muon profile where there is no fixed monitor.

Fig. 13. Right: An LHC ionization chamber (BLM monitor) used for the CNGS muon station. Left: The muon monitoring station equipped with 42 ionization chambers.

The beam profiles from the muon stations provide on-line feedback for the quality control of the neutrino beam. The muon detectors must measure up to $10^8$ muons/cm$^2$ during one extraction, i.e. during $10.5\,\mu$s. To cope with such a high instantaneous rate, nitrogen-filled sealed ionization chambers are used. These ionization chambers are also used as beam loss monitors (BLMs) at the Large Hadron Collider (see Fig. 13). Since the chambers are separated by 67 m of rock, the muon energy spectrum is different in the two chambers. Only muons with an energy above 20 GeV (50 GeV) reach the first (second) muon detector chamber. The first muon detector station is very sensitive to misalignment between the target and the horn, while the second station gives information on proton–target alignment. Hence scanning the proton beam across the target and the horn and comparing the muon monitor profiles provides optimal secondary particle production efficiency and a precise alignment of the beam w.r.t. the target and the horn. The muon monitor

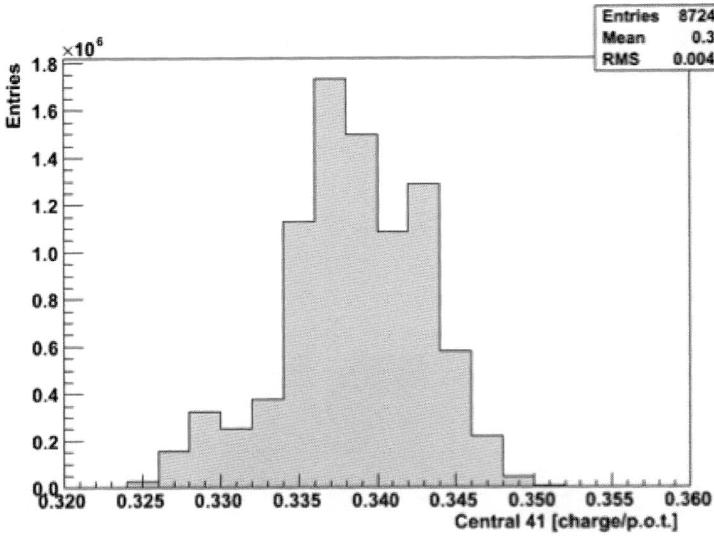


Fig. 14. Central muon detector signal during the five years of physics for CNGS.

reproducibility for all the detectors was measured to be better than 1% rms. The response of the different muon monitors is identical within less than 1%, as cross-checked with the motorized monitor.

Fig. 14 shows the muon detector signal in the centre of the detector station in the first muon pit during the five years of CNGS operation. The mean value is 0.338 collected charges/protons on target, with an rms of 0.005, demonstrating the remarkable stability of the secondary beam line elements. The stability of the signal amplitudes in the muon detectors also shows the target yield remained constant and no deterioration of the target has been observed. The effect of deteriorated horn capacitors was seen in the muon yield, though only at the 1% level.

## 6 Future conventional neutrino beams

CNGS, like the present generation of accelerator-based neutrino oscillation experiments like NuMI [Childress and Strait (2013)] in the United States and T2K [Scholberg (2013)] in Japan, was designed to confirm the neutrino oscillations using man-made neutrinos, and measure the mixing angles, $\theta_{13}$ in particular. With that goal achieved [Abe (2011); An (2012); Ahn (2012)], upgrades of the existing installations to high-power (thus more $\nu$-flux) coupled to a new generation and larger mass of neutrino detectors are under study. In a first step, operation with an increased proton beam power to around 700–750 kW is planned, to be followed on a second stage with a further increase up to 1.3–2 MW of beam power. The new high-power and high-precision neutrino sources (better neutrino beam quality and less systematic uncertainties in $\nu$-flavor content) would allow these new experiments to provide answers on the neutrino mass hierarchy (MH), the presence (or not) and study of leptonic charge-parity (CP) violation, and further precision studies of the neutrino mixing parameters.

The T2K facility at J-PARC provides a neutrino beam to the Super-Kamiokande detector at 295 km baseline. Stable operation to 300 kW was demonstrated in 2012 providing sufficient statistics for the measurement of the $\theta_{13}$ neutrino mixing angle. The upgrade plan for the accelerator chain to reach 750 kW of beam power includes increasing beam current and energy at the LINAC, and increased RF power and repetition rate at the main ring. The target cavern and neutrino beam infrastructure designed and built for MW operation is an asset, however R&D for the target and horn(s) may be required. In parallel to the beam, upgrade plans for the detectors include the new Hyper-K water Cherenkov detector, with 590 kt fiducial mass at the same baseline of 295 km but 2.5 deg off-axis, and the 100 kt Liquid-Argon detector at the Okinoshima island at 658 km baseline and 0.78 deg off-axis.

The NUMI facility at the Fermi National Laboratory in USA provides a neutrino beam to the far detector at the Sudan mine at 735 km baseline. The facility concluded in 2012 seven years of operation with $1.57 \times 10^{21}$ protons on target, including periods of running with anti-neutrinos. In the 2012–2013 shutdown the neutrino beam and accelerator complex was upgraded to allow operation at 700 kW for the NOvA experiment, at a slightly longer baseline of 810 km at 14 mrad [NOvA Collaboration (2007)]. The upgraded NUMI line has a new target and capture system to focus lower energy neutrinos, capable to operate with the increased intensity and repetition rate from the accelerators ($4.9 \times 10^{12}$ protons per 1.333 s cycle). In continuation to the neutrino physics program, FNAL plans the construction of a new Long-Baseline Neutrino Experiment (LBNE) [Adams (2013)] that provides a beam to far detectors at 1300 km distance. The new beam line will initially use the same beam as the upgraded NUMI line at 700 kW and in a later stage the beam power would be increased to 2.3 MW with beam from the upgraded injector chain at FNAL with Project-X [Altmannshofer *et al.* (2012)].

At CERN, building on the foreseen LHC Injectors Upgrade (LIU) [Coupard *et al.* (2014)], the conventional neutrino beam technology of CNGS could remain competitive for long baseline facilities, once optimized to produce the softer, few GeV, neutrino beams favoured by the recent measurement of $\theta_{13}$. Such a beam, successor to CNGS, has been proposed under the

LAGUNA–LBNO [Stahl (2012)] collaboration to a long-baseline (2300 km) detector site located in the Pyhasalmi mine in Finland. The potential of the injector chain to provide the "ultimate" CNGS beam intensity of a nominal power 750 kW or $1(1.4) \times 10^{20}$ protons per year in dedicated (shared) mode is addressed, which in $\sim$ 10 years of operation combined with the foreseen large-volume 20 kt Liquid Argon detector could provide conclusive determination of the neutrino mass hierarchy. In the LBNO design study the option of a new high-power proton synchroton (HP-PS) is included, to be used in a second phase as proton injector for the neutrino beam. This new HP-PS accelerator would use the low-power SPL extension of CERN's LINAC-4 as injector, have a repetition rate of 1 Hz and pulse intensity of $2.5(1.7) \times 10^{14}$ protons for a top energy of 50 (75) GeV to provide a 2 MW proton beam [Papaphilippou *et al.* (2013)]. The use of super ferric fast ramping magnets is considered along with special optics design and a challenging collimation and RF system.

It is interesting that in all facilities the upgrade plans follow the same pattern and would have similar challenges to face, equally shared between the proton drivers, the neutrino beam, and the experiments. The proton drivers in either facility would require major upgrades in the existing accelerators and in some cases new technologies and magnets to be developed. For the neutrino beams in all cases solid targets are considered, however substantial R&D would be required to design robust solutions that can afford the thermal shock and energy deposition of the 2 MW beams. Finally the large detectors of either water Cherenkov or Liquid Argon technology have to be build underground, including equally performant challenging near detectors working at high particle rates. As such large-scale programs require large investments to be realised, coordination and collaboration at a global scale would be required, as outlined in the recent update of the European Strategy for the High-Energy Physics, where close collaboration between the European and US future neutrino program is encouraged.

## Acknowledgements

The design and operation of high-intensity beam facilities like CNGS is a very challenging task. The successful operation of the CNGS neutrino beam was only possible thanks to the motivation and commitment of many colleagues and technicians at CERN and collaborators world-wide to whom the authors are particularly indebted. Particular thanks to Konrad Elsener, Malika Meddahi and the members of the CNGS project team for the successful completion of the design and construction project, embedding all the past experience of CERN in building neutrino beams, and to Karel Cornelis, Jorg Wenninger and Heinz Vincke for the continuous attention and optimisation during the beam operation.

## References

Abe, Y. *et al.* (2011). Indication for the disappearance of reactor electron antineutrinos in the Double Chooz experiment, Tech. Rep., arXiv:1112.6353.

Adams, C. *et al.* (2013). Scientific opportunities with the Long-Baseline Neutrino Experiment (LBNE), arXiv: hep-ph/1307.7335.

Ahn, J. *et al.* (2012). Observation of reactor electron antineutrino disappearance in the RENO experiment, Tech. Rep., arXiv:1204.0626.

Altmannshofer, W., Carena, M., Fox, P., Henderson, S., Holmes, S., Kim, Y., Kopp, J., Kronfeld, A., Lykken, J., Quigg, C. and Tschirhar, R. (2012). Physics opportunities with Stage 1 of Project X, http://www.fnal.gov.

An, F. P. *et al.* (2012). Observation of electron-antineutrino disappearance at Daya Bay, Tech. Rep., arXiv:1203.1669.

Athanassopoulos, C. *et al.* (1997). Evidence for $\nu_\mu \to \nu_e$ neutrino oscillations from LSND, *Phys. Rev. Lett.* **81**, 1774.

Bohlen, T., Cerutti, F., Chin, M., Fasso, A., Ferrari, A., Ortega, P., Mairani, A., Sala, P., Smirnov, G., and Vlachoudis, V. (2014). The FLUKA code: Developments and challenges for high energy and medical applications *Nuclear Data Sheets* **120**, 211–214.

Childress, S. and Strait, J. (2013). Long baseline neutrino beams at Fermilab, in *Proc. XIII International Workshop on Neutrino Factories, Super beams and Beta Beams, (NUFACT11)*, JoP, Conference Series, Vol. 408 (IOPscience).

Coupard, J., Damerau, H., Funken, A., Garoby, R., Gilardoni, S., Goddard, B., Hanke, K., Lombardi, A., Manglunki, D., Meddahi, M., Mikulec, B., Rumolo, G., Shaposhnikova, E. and Vretenar, M. (2014). LHC Injectors Upgrade Technical Design Report, Tech. Rep. CERN-ACC-2014-0337, CERN, Geneva.

Efthymiopoulos, I., Cornelis, K., Ferrari, A., Gschwendtner, E., Kadi, Y., Masi, A., Pardons, A., Vincke, H., Wenninger, J., Autiero, D., Sala, P. and Guglielmi, A. (2009). First year physics at CNGS, presented at PAC09, Vancouver, Canada, 4-8 May 2009, Tech. Rep. EuCARD-CON-2009-013.

Elsener, K., Acquistapace, G., Baldy, J. L., Ball, A. E., Bonnal, P., Buhler-Broglin, M., Carminati, F., Cennini, E., Ereditato, A., Falaleev, V. P., Faugeras, P. E., Ferrari, A., Fortuna, G., Fo, L., Genand, R., Grant, A. L., Henny, L., Hilaire, A., Hbner, K., Inigo-Golfin, J., Kissler, K. H., Lopez-Hernandez, L. A., Maugain, J.-M., Mayoud, M., Migliozzi, P., Missiaen, D., Palladino, V., Papadopoulos, I. M., Pietropaolo, F., Praire, S., Rangod, S., Revol, J. P. C., Roche, J., Sala, P. R., Sanelli, C., Stevenson, G. R., Tomat, B., Tsesmelis, E., Valbuena, R., Vincke, H. H., Weisse, E. and Wilhelmsson, M. (1998). The CERN neutrino beam to Gran Sasso (NGS): conceptual technical design CERN, Geneva.

Gschwendtner, E., Cornelis, K., Efthymiopoulos, I., Pardons, A., Vincke, A. and Wenninger, J. (2013). CNGS, CERN Neutrinos to Gran Sasso, five years of running a 500 kilowatt neutrino beam facility at CERN, Tech. Rep., CERN, *iPAC13 Conference Proceedings.*

Jensen, L. (2006). Beam instrumentation for the CNGS facility, Tech. Rep. CERN-AB-Note-2006-022, CERN, Geneva.

Jones, M., Missiaen, D. and Efthymiopoulos, I. (2013). The distance from CERN to LNGS, Tech. Rep. CERN-ACC-2013-0118, CERN, Geneva.

Kain, V., Carlier, E., Gaxiola, E., Goddard, B., Gschwendtner, E., Gourber-Pace, M., Meddahi, M., Vincke, H., Vincke, H. H. and Wenninger, J. (2007). High intensity commissioning of the SPS LSS4 extraction for CNGS, CERN-AB-2007-048.

Marsili, A., Dehning, B., Ferioli, G., Gschwendtner, E., Holzer, E. B. and Kramer, D. (2008). CNGS muon monitors, Tech. Rep. CERN-AB-Note-2008-044-BI, CERN, Geneva.

Meddahi, M., Cornelis, K., Elsener, K., Gschwendtner, E., Herr, W., Kain, V., Lamont, M. and Wenninger, J. (2007). CERN Neutrinos to Gran Sasso (CNGS): Results from commissioning, CERN-AB-2007-039.

NOvA Collaboration, Ayres, D. *et al.* (2007). NOvA proposal to build a 30 kiloton off-axis detector to study neutrino oscillations in the Fermilab NuMI beamline, Fermilab-Proposal-0929.

Papaphilippou, Y., Alabau-Gonzalvo, J., Alekou, A., Antoniou, F., Benedikt, M., Efthymiopoulos, I., Garoby, R., Gerigk, F., Goddard, B., Lazaridis, C., Parfenova, A., Shaposhnikova, E. and Steerenberg, R. (2013). Design options of a high-power synchrotron for LAGUNA-LBNO, CERN-ACC-2013-0125.

Peraire, S., Ross, M. and Zazula, J. M. (1996). The SPS target station for chorus and nomad neutrino experiments, CERN-SL-96-044-BT.

Schirm, K. M., Kalbreier, W., Anashin, V., Kiselev, O., Maraev, V., Ogurtsov, A., Pupkov, Y., Ruvinsky, E., Zhilyaev, K., Konstantinov, Y. S., Kosjakin, M. and Peregud, V. (2006). The bending magnets for the proton transfer line of CNGS, Tech. Rep. CERN-AT-2006-003, CERN, Geneva, work supported in part by the European Community-Research infrastructure activity under the FP6 Program.

Scholberg, K. *et al.* (2013). The T2K experiment, in *Proc. XIII International Workshop on Neutrino Factories, Super Beams and Beta Beams, (NUFACT11)*, JoP, Conference Series, Vol. 309 (IOPscience).

Stahl, A. *et al.* (2012). Expression of interest for a very long baseline neutrino experiment (LBNO), CERN-SPSC-2012-021, SPSC-EOI-007.

# Chapter 35

# European Spallation Source

Mohammad Eshraqi, David McGinnis and Mats Lindroos (ESS)

## 1 Neutron usage and historical background

The neutron beams are being used extensively in science and industry to investigate the properties of matter. The beams of neutrons are used to monitor the structure of the matter in atomic levels using scattering methods. These methods give a high precision information about matter and the high penetration property of neutron makes it the only probe for these measurements.

Most existing neutron sources in Europe are based on nuclear reactors. This approach has been taken to its limits set by both technical issues such as cooling but also by licensing which is non-trivial for any facility using fissile material. Compared to existing spallation sources ESS will be 30 times brighter and it will be the first spallation source with a time averaged flux of neutrons as high as the best research reactors. Europe has today over 5000 researchers who use neutrons and this community is asking for a new intense source of neutrons.

The need for the European Spallation Source (ESS) [Perlado *et al.* (1989)] was articulated 20 years ago but a decision to build it in Lund in Sweden was only taken in May 2009. A series of meetings organized in 1991 and 1992 by Forchungszentrum Jülich (FZ-J) and Rutherford Appleton Laboratory (RAL) explored the basis for an advanced accelerator driven pulsed spallation source, which later formed the basis for the specification for ESS [Bohn *et al.* (2002)]. The decision to site it in Lund was the final stage of a process initiated by the European Commission and steered by the European Strategy Forum on Research Infrastructures (ESFRI). ESFRI was created in 2002 and is a strategic instrument of EC to develop the scientific integration of Europe and to strengthen its international outreach [ESFRI (2011)].

## 2 Spallation

Spallation is a nuclear process in which neutrons at different energies are emitted in several stages following the bombardment of heavy nuclei with highly energetic particles. For a more detailed discussion of the spallation process itself, the target, the moderators and the physics at such facilities see (for example) [Willis and Carlile (2009)]. However, it is worth noting that there are other ways to produce neutrons with accelerators, for example photofission induced by an intense electron beam. The spallation process is the most practical and feasible way of producing neutrons for a reasonable effort (or simply cost) of the neutron source cooling system. Research reactors also require fissile material handling, potentially a major constraint for both handling and licensing [Schopper and Myers (2013)].

## 3 History of spallation sources

The first ever spallation source was built and operated at the Argonne National Laboratory in the USA, ZING-P. It was based on a synchrotron and provided a short pulse well suited for Time-Of-Flight (TOF) techniques at the instruments. It was at Argonne followed by the Intense Pulsed Neutron Source (IPNS), in Japan by the KEK Neutron Science Center (KENS) and the ISIS neutron facility at Rutherford in UK. All of them were or are based on synchrotrons, as is the high power, most recent spallation facility of J-PARC at Tokai in Japan.

The first spallation source based on a linear accelerator combined with a storage ring to provide very high intensity short pulses of neutrons for the experiments was the Los Alamos Meson Physics Facility (LAMPF) in the USA, it later became the Los Alamos Neutron Science Center (LANCSE). Today SNS has superseded LANCSE in intensity using a superconducting proton linear accelerator combined with a storage for short pulse generation.

The European Spallation Source (ESS) will provide even higher intensities but is focusing on instrument development to be able to use the longer linac pulses directly for spallation and to by-pass the use of a costly and possibly limiting storage ring [Mezei (1993, 1997)]. Any circular accelerator will have an intensity limit set by the detuning caused by space charge. The detuning will eventually cause beam instabilities, which will cause beam losses and at some ultimate limit make it impossible to further increase the intensity [Schopper and Myers (2013)].

## 4 The ESS facility

The spallation cross section for protons on heavy nuclei increases as a function of proton energy up to several tens of GeV [van der Meer *et al.* (2004)]. Nonetheless it is generally agreed that a kinetic proton energy between 1–3 GeV is optimal for practical target and moderator designs, and in order to keep the shielding requirements reasonable.

The ESS accelerator design has gone through several evolutions. In 2002 a fully normal conducting design was proposed [Gardner *et al.* (2002)] in parallel to a pre-dominantly superconducting design derived from the CONCERT project [Aune *et al.* (2001)]. In 2003 a new design was presented [Bohn *et al.* (2002)] with two front-ends consisting of ion source, radio frequency quadrupole (RFQ) and a drift-tube linac (DTL) up to 20 MeV. The beams of the two front-ends were merged at 20 MeV and further accelerated in normal conducting structures up to 400 MeV and superconducting structures up to 1334 MeV. The facility had both a long pulse target station and a storage and accumulation ring for a short pulse target. To enable low loss space charge injection into the accumulator ring $H^-$ ions were proposed to be accelerated at a total beam power of 10 MW. In 2009, the competing ESS-Bilbao [Bermejo *et al.* (2009)] team and the ESS-Scandinavia [Lindroos *et al.* (2009); Peggs *et al.* (2009)] team both proposed a simplified design with only one front-end to avoid the complex funnelling stage and higher beam energy to lower the required beam current. The requirement for a short pulse target station was also dropped to reduce cost. This makes it possible to accelerate $H^+$ ions, which should reduce losses due to intra-beam stripping of $H^-$ ions and simplifies the ion source design.

The ESS has the ambitious goal of becoming a sustainable research facility with zero release of carbon dioxide. This will be achieved through a combination of actions, but with the linac being the most energy hungry part of ESS, the energy efficient design of the RF power sources, the cryogenics systems and high-Q cavities are important issues.

### 4.1 *Main parameters*

The design of an accelerator is affected by several factors, the requirements, the performance and the cost. These three criteria had influenced the design of the ESS linac. However, a linac could not be costed properly unless it is designed, therefore a post-design analysis of the linac design choices is required which in turn leads to the next iteration of the linac design. As it was mentioned earlier the ESS linac has gone through several iterations converging to a linac which respects the goals and the budget. The next paragraph summarizes the choices and trade-offs in the design.

The achievable gradient of the state-of-the-art accelerating structures, both normal conducting and superconducting, is limited due to several factors, among them, the surface break-down, field emission, and Q-value drop to name a few. As a result the length of the linac is a linear function of its final energy. In the case of ESS, the final beam power is fixed to 5 MW, and since the beam power is the product of beam current and beam energy, the beam current is inversely proportional to the beam energy. This favors higher beam currents in favor of cheaper linacs, but higher beam current leads to stronger space-charge forces which increases the chances of halo production and eventually losses. Moreover accelerating a beam of higher current puts a higher demand on power couplers that deliver the RF power to the cavity. These couplers are limited on the amount of power they can deliver safely to the beam without breakdown. One of the aspects of optimizing the design is fine tuning the current not to cause excessive loss and activation, and putting high demands on RF sources and power transmission on one hand and minimizing the costs on the other hand.

The next parameter to be chosen in the design of the linac is the choice of structures, specially normal conductor versus superconductor or a mix of them and in this case optimizing the transition energy from normal conducting to superconducting. The duty factor of the linac, *i.e.* the percentage of time during which there is a beam in the linac, and the pulse length are two of the main parameters which set this value, the former through energy consumption and the latter by setting the surface break-down in normal conductors. In normal conducting structures a significant amount of RF power is lost in the cavity walls through ohmic losses, this value can easily exceed 50% of the total power, this power is needed only when there is a beam in the linac. These linacs are usually cheaper to build compared to superconducting linacs, but their operation costs are higher, and in a world moving towards energy sustainability are not the most energy efficient. On a more applied point of view, the normal conducting accelerating structures are having a smaller bore radius (max beam aperture $< 2$ cm) compared to superconducting cavities (min bore radius $> 3$ cm). Also the peak surface field in normal conducting cavities is inversely proportional to pulse length [Pritzkau (2001)] and for long pulse lengths the safe accelerating fields result in long linacs, *e.g.* $> 700$ m in case of 2 GeV linac of ESS. In superconducting structures on the other hand RF ohmic losses are negligible, however, since the available superconductors are only effective in cryogenic temperatures, *e.g.* very close to absolute zero (2–4.5 K), the required energy to cool down the structures to these temperatures and keeping them cooled during operation should be considered. The structure should be kept at the cryogenics constantly to compensate the static heat load (a function of temperature, constant) and the dynamics heat load (a function of temperature and cavity field), which demands a continuous operation of cryogenics plant. Since the peak surface field is independent of pulse length, for long pulse structures, *e.g.* ESS, the linac could be designed to be much shorter, *e.g.* $\sim$400 m for the 2 GeV linac of ESS. The pulsed energy consumption of RF sources versus the quasi-constant energy consumption of the cryogenics sets the transition energy from the normal conducting to superconducting to be inversely proportional to duty factor. On the extremes, in continuous wave (CW) linacs the transition energy is chosen to be as low as possible, while for low duty factor linacs there is a high chance that the linac is completely normal conducting.

The RF frequency is yet another parameter to be chosen. The ohmic losses in normal conducting cavities go down with frequency and the peak surface field increases with frequency, both favoring a high frequency for the normal conducting structures. But space-charge forces increase with frequency due to smaller longitudinal beam size, the machining tolerances gets tighter at higher frequencies and since the length of accelerator components are proportional to the product of particle velocity in wavelength, higher frequencies demands smaller components, *e.g.* quadrupoles, which are technologically impossible to built at very small sizes.

These factors define a range of frequencies for ion linacs which lies between 100−400 MHz, and the designer should choose a frequency which satisfies the physics and engineering needs and finding a frequency where there is already RF sources and equipment available is a plus. For the same reason almost all the linacs which are designed in the twenty first century in Europe have a frequency which is a multiple of 352.21 MHz, a frequency earlier used by the Large Electron Positron collider at CERN and led to extensive R&D in the RF domain in that frequency. At higher energies where the beam has a higher velocity the space-charge forces weaken and one can choose a multiple of bunch frequency. A higher frequency reduces the transverse size of the cavities and needs smaller RF equipment, therefore in most of the high power linacs there is at least one frequency jump to benefit from these facts. As a result of a series of optimisation and iterations the ESS linac beam current is set to 62.5 mA and the final energy is fixed to

2.0 GeV to provide 125 MW of peak power to the target. More than 95% of acceleration is done in superconducting structures. The linac accelerates a 2.86 ms long pulse with a repetition rate of 14 Hz, resulting in a 4% duty cycle and giving an average beam power of 5 MW with a high demand on availability of the facility (>95%). The former two parameters are set by the experiments, instrument design and scientific performance of ESS. Two RF frequencies, 352.21 MHz up to around 200 MeV and twice of that afterwards (704.42 MHz) are used for acceleration.

## 5 The ESS linac

The configuration of the current baseline linac is shown schematically in Fig. 1, and selected linac parameters are listed in Table 1 [Eshraqi *et al.* (2013)]. The warm linac design has been done in a collaboration consisting of ESS, INFN Catania, CEA Saclay, ESS-Bilbao and INFN Legnaro; the superconducting cavities and their cryomodules are developed at IPN Orsay and CEA Saclay; and the HEBT is designed by ISA in Aarhus.

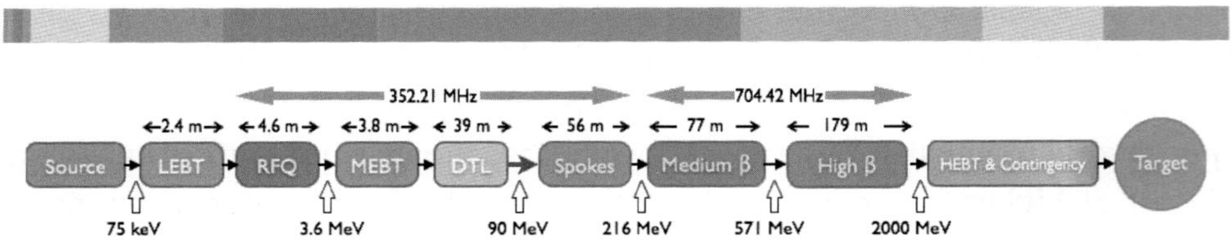


Fig. 1. Top: Scaled layout of ESS. The normal conducting structures are indicated by warm colors and the blue themes sectors indicate the superconducting structures. From left: Ion source, LEBT, RFQ, MEBT, DTL, spoke, medium $\beta$, high $\beta$, and HEBT (consisting of contingency and beam transport, dogleg, and expander). Bottom: Blob layout of the ESS linac.

Table 1. ESS linac parameters.

| | Energy (MeV) | $\mathrm{No.}_{\mathrm{modules}}$ | $\mathrm{No.}_{\mathrm{cavities}}$ | $\beta_G$ | $T$(K) | $L_{\mathrm{Period}}$ (m) |
|---|---|---|---|---|---|---|
| Source | 0.075 | 1 | 0 | — | ∼300 | — |
| LEBT | 0.075 | — | 0 | — | ∼300 | — |
| RFQ | 3.6 | 1 | 1 | — | ∼300 | Varying |
| MEBT | 3.6 | — | 3 | — | ∼300 | — |
| DTL | 90 | 5 | 5 | — | ∼300 | Varying |
| Spoke | 216 | 13 | 26 | 0.50* | ∼2 | 4.28 |
| Medium $\beta$ | 570 | 9 | 36 | 0.67 | ∼2 | 8.56 |
| High $\beta$ | 2000 | 21 | 84 | 0.86 | ∼2 | 8.56 |
| HEBT | 2000 | — | 0 | — | ∼300 | — |

$^*\beta_{opt}$.

A proton beam of less than 80 mA is produced in a pulsed microwave-discharge source on a platform at 75 kV. A low-energy beam transport, LEBT, with two solenoid magnets as focusing elements brings the beam to the entrance of the RFQ. The LEBT has a chopper that cuts away the beam while the proton pulses from the ion source stabilize, preventing a beam with off-nominal parameters from being accelerated in the RFQ and lost at high energy. The 4-vane RFQ accelerates the beam to 3.6 MeV with small losses and a minimal emittance growth. It is designed specifically for ESS but it is based on the IPHI RFQ at Saclay. The RF frequency of the RFQ and the warm linac is 352.21 MHz. After the RFQ there is a medium-energy beam transport, MEBT, with three buncher cavities and 10 quadrupole magnets. The MEBT has several different functions: it has optics to match and steer the beam from the RFQ into the drift-tube linac, it has a comprehensive set of beam-instrumentation devices, it has a chopper which acts faster than the LEBT chopper since space-charge neutralization is not an issue in the MEBT, and it allows collimation of the transverse particle distribution. A drift-tube linac, DTL, with five tanks which each are $\sim$7.5 m long takes the beam from 3.6 MeV to 90 MeV. It has a FODO structure with permanent-magnet quadrupoles. Every second drift tube is empty or used for steering magnets and beam-position monitors.

The superconducting linac has three types of cavities: double-spoke resonators, six-cell medium-beta elliptical cavities and five-cell high-beta elliptical cavities. The linac has 13 spoke cryomodules with two double-spoke resonators in each, and between the cryomodules there are warm quadrupole doublets. The spoke resonators operate at 352.21 MHz like the warm linac, but then there is a frequency doubling to the 704.42 MHz at the elliptical cavities. There are 9 medium-beta cryomodules with four cavities in each and quadrupole doublets in between, and there are 21 high-beta cryomodules with four cavities in each and quadrupole doublets between every cryomodule. The period length in the medium-beta and high-beta section is exactly the same to allow swapping the cryomodules in case of low performing medium-beta cavities.

All accelerating structures will be powered by klystrons or IOTs, except the spoke resonators where tetrodes might be used. The possibility of using Solid State Amplifiers is yet under study. With one klystron per elliptical cavity plus a few for the warm linac, there will be more than 130 large tubes and more than 65 modulators depending on how many tube each modulator is feeding. A tube could be a klystron, an IOT or a tetrode. The density of components in the klystron building would become too high if these were

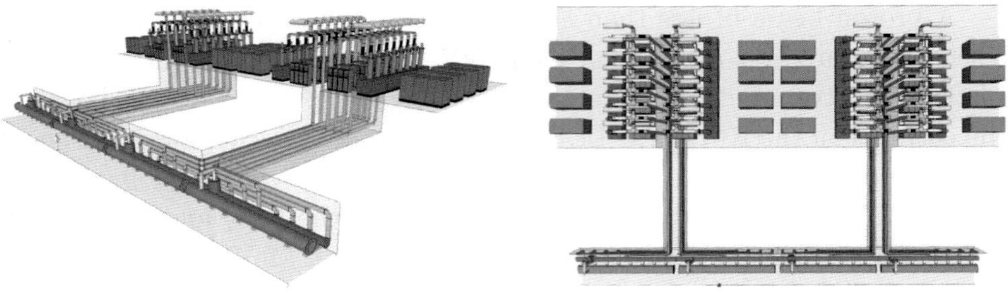

Fig. 2. The ESS RF sources in and the RF waveguide stub lay-out. The 3D view on left and the view from top on right.

to be positioned linearly. Instead they will be located in groups of eight klystrons and four modulators across the klystron building (Fig. 2).

After the last cryomodule there is the contingency and upgrade section where additional cryomodules can be installed to compensate for a shortfall in linac performance and/or a power upgrade of the facility. The additional tunnel would also make it possible to bend the beam out for a second target station. Then the beam is brought from the tunnel to the spallation target at the surface through two vertical bends and an expansion section. Quadrupole and octupole magnets or a raster scanning system are used to blow the beam up onto the desired profile of the proton-beam window and the target window.

## 6 Beam physics

The European Spallation Source requires a high current proton linac to accelerate protons to be used for the spallation process from which high flux of pulsed neutrons are generated. The 5 MW accelerator delivers beams of 2.0 GeV to the target in long pulses of 2.86 ms, with a repetition rate of 14 Hz [Bohn *et al.* (2002); Peggs (2011)]. Beam current is 62.5 mA, which at 352.21 MHz is equivalent to $\sim 1 \times 10^9$ protons per bunch.

Such a high current and power sets a strict limit on beam losses. Both hands-on maintenance and machine protection set a strict limit on beam losses and have been a concern in every high power linac [Jameson (1993); Wangler *et al.* (1998)]. Therefore it is crucial, especially for high power accelerators, to design a linac which does not excite particles to form beam halo and also keeps the emittance growth to a minimum to avoid losing the particles that otherwise get too close to either the transverse acceptance or the longitudinal separatrix and eventually escape the bucket and get lost. The ESS linac is designed carefully to minimize such effects all along the linac and transfer lines. A recent study relaxed the losses in the low energy part of the linac, mainly in the RFQ and MEBT [Tchelidze and Stovall (2013)], from the conventional 1 W/m.

### 6.1 *Design choices*

The first stage for the design of the linac is choosing the right parameters for a given final energy and power on top of the main accelerator parameters which were discussed in Section 4.1. Among these parameters there are the accelerating gradient of the cavities and the power per cavity. Choosing the right accelerating gradient is of great importance, since an over specified value which will not be reached, will result in a linac which will be shorter on paper, but will not be able to bring the beam to its final energy. On the other hand an under specified value may result in a linac which will be unnecessarily long and expensive. One way to have a best initial estimate of the optimum gradient is to look at the available state-of-the-art cavities worldwide. Plotting the accelerating gradients of the cavities normalized by their peak surface field versus the geometric beta of the cavities gives meaningful results, the longer the cavity the higher gradient one can expect (Fig. 3). The geometric beta dependence of the ratio of the peak surface electric field to the accelerating electric field could be formulated, not uniquely, by Eq. (1).

$$E_{\text{acc}} = \frac{E_{\text{peak}}}{1.95/\beta_G + 1.15 \cdot \beta_G - 1}, \tag{1}$$

where $E_{\text{peak}}$ is the peak electric surface field on the cavity surface.

The second parameter, the power per cavity, is limited by the power source and the power coupler capabilities. In the case of the ESS linac, the output of the RF power sources is much higher than the power on which

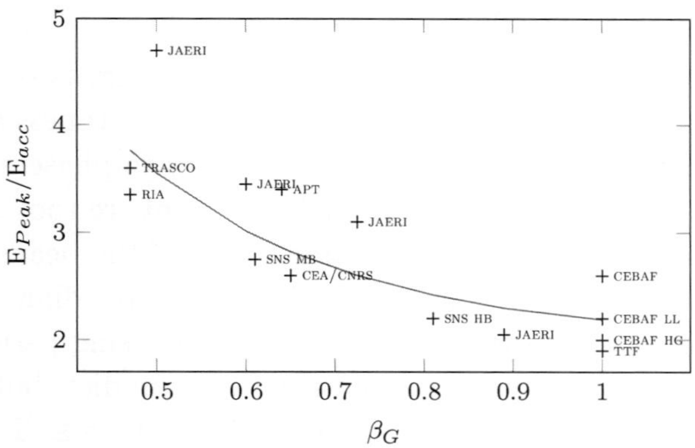


Fig. 3. Ratio of peak surface electric field to accelerating electric field versus geometric beta of the cavities.

the RF power couplers have been tested. Therefore the power coupler defines the maximum power per cavity, and is 1.1 MW per coupler. For superconducting cavities, this parameter is related to the geometric beta of the cavity, $\beta_G$; number of cells of the cavity, $N_{\text{cells}}$; maximum accelerating gradient, $E_0.TTF$; and beam current, $I$, through:

$$P = 0.5\beta_G \cdot \lambda \cdot N_{\text{cells}} \cdot E_0.TTF \cdot I. \tag{2}$$

One can see that for a given available power, $P$, current, and accelerating gradient, the product of number of cells and geometric beta is limited.

In a superconducting linac, usually few accelerating cavity types are used for a wide range of energies. Since these cavities are having a fixed geometry for a range of energies, the acceleration efficiency is not the same in the whole energy range due to the variation of transit time factor with energy. Most of the superconducting cavities work in the $\pi$-mode, meaning that the adjacent cells of the cavity are 180° out of phase with respect to each other. This is called the main accelerating mode. However, the other modes, which are $k\pi/N_{\text{cell}}$ $\{k = 1, \ldots, N_{\text{cell}} - 1\}$ apart could be excited too. These modes, usually known as Same Order Modes, could be excited as strongly as the main mode at transitions if the transitions are not chosen carefully and can decelerate the beam. Therefore it is important to choose the right cavities, number of cells, and geometric betas, as well as the right transition energies between these structures. This is important not only to increase the efficiency of the whole accelerator as a system, but also to improve the beam quality by avoiding the parasitic modes which otherwise could be excited at the extremities of each section.

### 6.2 *Beam physics criteria*

The design of the linac is focused on improving the integrity of the accelerator as a single structure. One of the guidelines is the smoothness and monotonic variation of the average phase advance. The average phase advance, or the phase advance per meter, is proportional to the square root of the external forces on the beam. Internal space-charge forces of the beam tend to redistribute the beam particles in specific distributions depending on the strength of external forces. Within less than a couple of focusing periods in a linear focusing channel a beam reaches a quasi-equilibrium distribution which will not change significantly during the accelerating process. The main cause of this redistribution is that particles at the edge of the beam redistribute themselves to screen the effect of external forces. The higher the force the deeper is this shielding layer, approximately equal to the Debye length [Wangler (2008)]. Therefore the beam takes a quasi-Gaussian distribution under strong focusing forces and a more uniform distribution in weaker focusing channels. There is an energy difference between the uniform, lowest energy state of the distribution and any other distribution. In any transition where the forces are not varied smoothly this energy is released and converted to beam emittance, usually accompanied by halo development.

At the same time the tune depression, the ratio of the phase advance with current to the phase advance without current, is being kept above 0.4. This choice limits the number of mismatch resonances to only two, from which one is always present irrespective of the tune depression and to avoid the second resonance one should keep the tune depression above $\sim 0.6$ [Lagniel (1996)].

Theoretically for beams with negligible current the phase advance per period must be limited to 180°. However, when the beam current is not negligible anymore and consequently the tune depression is less than 1, both phase advance with and without current should be either greater or lower than 90°, otherwise beam will become unstable and significant halo and therefore losses may occur. Due to this reason the zero-current transverse phase advance per focusing period is limited to 87° in the ESS linac to reduce the percentage of the beam that due to their phase and energy spread would otherwise have had a phase advance exceeding 90° per period. Since the tune depression is very close to 0.4, unless the phase advance ratios in the three planes are chosen carefully the tune depression will be less than 0.4 at least in one plane [Eshraqi and Lagniel (2013)]. This limits the zero-current longitudinal phase advance per period to 85°.

Though every section of the accelerator is designed separately, having included in the design to match the phase advance of neighboring structures, a further phase advance smoothness matching is performed during the integration to assure a premium beam quality.

Finally after performing the single particle simulations including the criteria a set of start-to-end simulations are performed with 100,000 macroparticles using the 3D PICNIC space-charge routine. The space-charge kick is applied 15 times per $\beta \cdot \lambda$ and the space-charge forces within the beam are calculated using a $10 \times 10 \times 10$ mesh. Increasing the frequency of kicks, or the number of mesh points more than this limit slows down the calculation and does not enhance the quality of the results. The code TraceWin [Duperrier *et al.* (2002)] is used for the single and multi-particle simulations, and another in-house code has been used for validating the single particle simulations, this code "ESS Linac Simulator" [Laface *et al.* (2013)] is being expanded to be able to run in multi-particle mode. To have a distribution which is not a pure analytic distribution a Gaussian beam truncated at $5\sigma$ is generated at the RFQ input and transported through the linac and HEBT to the target wheel surface.

## 7 The front-end and the normal conducting linac

The front-end of the ESS linac is composed of an ion source, a couple of normal-conducting accelerating structures and transport lines. The accelerating structures are the radio frequency quadrupole, RFQ, and the drift tube linac, DTL, and the transport lines are low energy beam transport and medium energy beam transport, upstream and downstream of the RFQ respectively.

Though the normal-conducting linac covers less than 5% of the energy range covered by the linac, its performance is of great importance since it defines the beam quality throughout the whole linac. The front-end is able to define properly the required time structure of the pulse, and keep the emittance growth and halo generation minimized. Achieving these qualities assures minimized beam losses in the rest of the linac and increases the safety and reliability of the accelerator.

### 7.1 *Ion source and LEBT*

The initial stage of any accelerator is creating the charged particles. In the ESS linac a microwave discharge ion source, MDIS, will be employed for the production of the high intensity proton beam in macro-pulses of up to 3 ms. The proton energy will be 75 keV and a current up to 80 mA is expected with a normalised rms emittance of 0.2 $\pi$.mm.mrad at the RFQ entrance. The absence of hot filaments in the microwave discharge ion sources significantly increases the mean time between failures with reliability close to 100% and high current stability as already experienced by other ion sources [Gobin *et al.* (2004)]. The low energy beam transport (LEBT) is composed of two magnetic solenoids together with all the necessary equipment to prepare and match the proton beam out of the source to the radio frequency quadrupole entrance while keeping the emittance growth due to non-linearities in solenoid field and the space-charge effects to a minimum. The LEBT will also house a chopper with fast rise and fall times in order to remove the heads and tails of the beam macro-pulses maintaining rise and fall times in the order of hundreds of nanoseconds.

### 7.2 *RFQ*

The ESS radio-frequency quadrupole, RFQ, will accelerate and bunch proton beams from 75 keV at the LEBT exit to 3.6 MeV before the MEBT entrance. The beam current is 62.5 mA for 4% duty cycle. The RFQ is a 4-vane-type structure running at 352.21 MHz and roughly 4.5 m in length divided into 5 segments of less than 1 m each. The peak electric field on the vane surface has been limited to a Kilpatrick value [Kilpatrick (1957)] of 1.8. A transmission of more than 99% is foreseen for the nominal current and high quality beams will be delivered to the downstream structures of the accelerator.

### 7.3 *MEBT*

The major challenges of this part of the accelerator is to maintain the beam quality (low emittance and minimized halo), limit the beam losses downstream the linac and maximize the ESS reliability. The considered versatile MEBT is designed to achieve four main goals:

(1) To contain a chopper and its corresponding beam dump that could serve in the commissioning as well as in the ramp up phases,
(2) To collimate by means of two collimator sections,
(3) To measure the beam phase and profile between the RFQ and the DTL, along with other beam properties,
(4) To match the RFQ output beam to the DTL in all the three planes.

For this purpose a set of ten quadrupoles is used to match the beam characteristics transversely, combined with three 352.21 MHz *buncher* cavities, which are used to adjust the beam in order to fulfill the required longitudinal parameters.

### 7.4 *DTL*

The drift-tube linac increases energy of the 62.5 mA beam to 90 MeV in five tanks. The transverse focusing is achieved by permanent magnet quadrupoles arranged in a FODO lattice leaving half the drift tubes empty for diagnostics and steerers with a maximum field of 1 mT·m to correct the beam trajectory and reduce the losses. The Kilpatrick value is limited to 1.4 for reliable operation of the accelerator. This safe value is chosen because at the first cells of the DTL the magnetic field of the quadrupoles has a strong component perpendicular to the drift tube face. When the magnetic and electric fields are parallel a lower Kilpatrick limit should be chosen to avoid sparking [Moretti *et al.* (2008)].

## 8 Superconducting linac

The DTL is followed by a differential pumping section. The purpose of this section is to provide a vacuum enhancement from that of the last DTL tank, a normal-conducting structure, to the first spoke cavity, a superconducting structure, in order to limit the gas flow from the former structure to the latter. This is done using three vacuum pumps (ion pumps) which are located in the short area between the two structures.

The optimum transition energy to superconducting structures is in the range of 70–100 MeV in the case of ESS linac from the energy consumption as well as beam dynamics performance point of view. Superconducting cavities can give almost three time the average gradient of a normal-conducting structure for the ESS pulse length. The bore radius of the cavities is 28 mm in spoke cavities and is more than 50 mm in the elliptical cavities, these are twice and four times bigger than the aperture of the normal-conducting linac respectively and therefore allow halo particles to go through instead of intercepting them.

The superconducting part of the linac consists of three families of SC cavities. The beam dynamics has been discussed in the previous section. We will here give some insight to the specific issues relating the superconducting technology for a high power linac like ESS.

### 8.1 *Cavities*

The three families of SC cavities intended for use in the ESS linac are:

- Double spoke cavities operating at 352.21 MHz,
- Medium-$\beta$ elliptical cavities and
- High-$\beta$ elliptical cavities both types operating at 704.42 MHz.

#### 8.1.1 *Double spoke cavities*

The proton beam will be accelerated from 90 MeV to 216 MeV by 26 spoke cavities. These will be installed in 13 cryomodules (*i.e.* two cavities per cryomodule), operating at 2 K. It should be noted that ESS will be one of the world's first accelerators to make use of spoke cavities.

Figure 4 shows a cross-sectional view of a spoke cavity designed for use in ESS. These cavities are designed to have an optimal beam velocity, $\beta_{\rm opt} = 0.50$, and a peak accelerating gradient, $E_{\rm acc} = 9$ MV/m. The design of these cavities is driven by some quite conservative limits on the maximum surface fields — $B_{\rm peak} \leq 70$ mT, $E_{\rm peak} \leq 45$ MV/m — in order to minimize the risk associated with this new technology. The expected quality factor, $Q_0 \geq 2 \times 10^9$, at the nominal gradient.

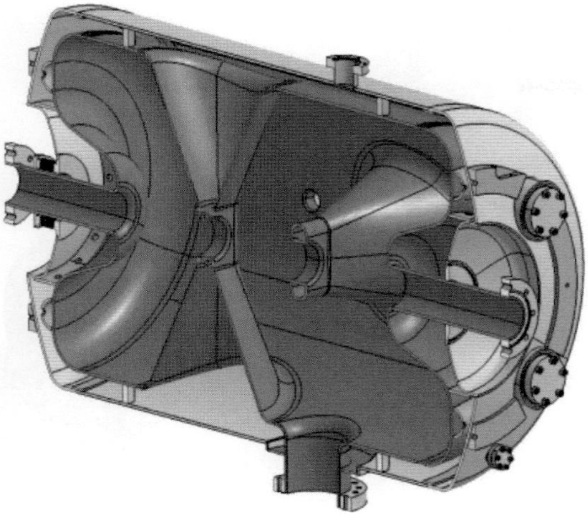

Fig. 4. A cross-section through a spoke cavity installed within its cryo-vessel.

### 8.1.2 *Medium-β cavities*

Following the spokes, the proton beam is then accelerated from 216 MeV to 570 MeV by 36 SC elliptical cavities. This is the first section to use the higher frequency of 704.42 MHz. These cavities contain six resonant cells operating in the $\pi$-mode, and they are designed with a geometric beta, $\beta_G = 0.67$. At the nominal gradient, $E_{\rm acc} = 16.7$ MV/m, the cavity surface will be treated such that a quality factor, $Q_0 \geq 5 \times 10^9$, is achieved.

### 8.1.3 *High-β cavities*

The final section of the ESS linac consists of 84 high-$\beta$ cavities that are used to accelerate the protons from the 570 MeV with which they exit the medium-$\beta$ section, to the final energy of 2.0 GeV.

The high-beta cavities are $\pi$-mode cavities with five resonant cells with a geometric beta, $\beta_G = 0.86$ in the current baseline. At the nominal gradient, $E_{\rm acc} = 20$ MV/m, the cavity surface will be treated such that a quality factor, $Q_0 \geq 6 \times 10^9$, is achieved.

## 8.2 *Cryomodules*

The cryomodules are installed in a ~310 m long ESS cold linac section. There are three families of cryomodules: the spoke resonator cryomodules, the medium and the high-beta elliptical cavity cryomodules. The cryomodules are composed of the cavity packages, the supporting system, the alignment system, the thermal and magnetic shieldings, the internal cryogenic piping and the diagnostic instrumentation. Each elliptical cryomodule contains four

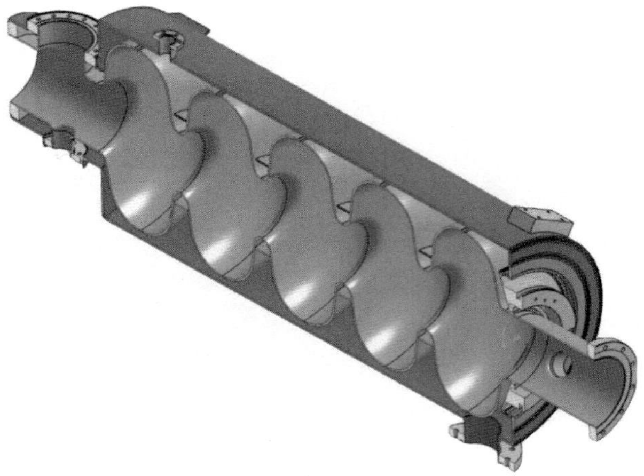

Fig. 5. A cross-section through a high-$\beta$ elliptical cavity installed within its cryo-vessel.

superconducting cavity packages, which include cavities (or resonator), fundamental power coupler, cold tuning system, its helium tank and the RF pick-up coil. Each spoke cryomodule contains two spoke resonator packages.

The cryomodules must be designed to allow for an optimum transfer of energy from the RF system to the beam, assure good alignment of the cavity package to the beam axis and assure that the cavities are kept at the required temperature (approximately 2 K for ESS) for a minimal static and dynamic thermal energy load. A space-frame is used to position and align the cavities in the cryomodule. This type of cryomodule has been previously used in the Spallation Neutron Source, SNS. The elliptical cryomodules have identical length for both medium and high-beta sections. A single type of cryomodule will simplify installation and lower the costs. Adding a sixth cell in the medium-beta cavity package compensates for the difference in length between medium and high-beta cavities.

Warm quadrupole doublets and beam instrumentation systems are installed in between the cryomodules for transverse focusing, orbit correction and to monitor the beam parameters. Figure 6 shows the conceptual design of the spoke resonator cryomodule and Fig. 7 shows the conceptual design of the elliptical cavity cryomodules.

## 9 RF sources

The vast majority of structures in the ESS linac are superconducting cavities, therefore the dominant feature of the ESS RF system is the number of RF stations. The enormous gradients in the superconducting cavities is strong enough to deform the 4 mm thick niobium cavities by exerting a significant

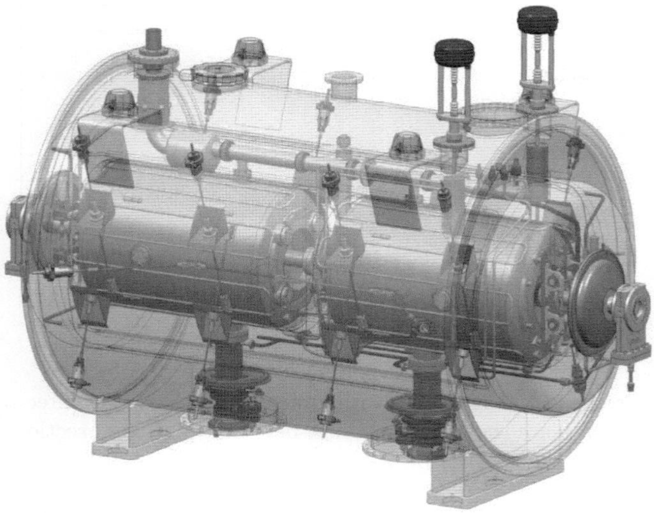

Fig. 6. Spoke cavity cryomodule for ESS.

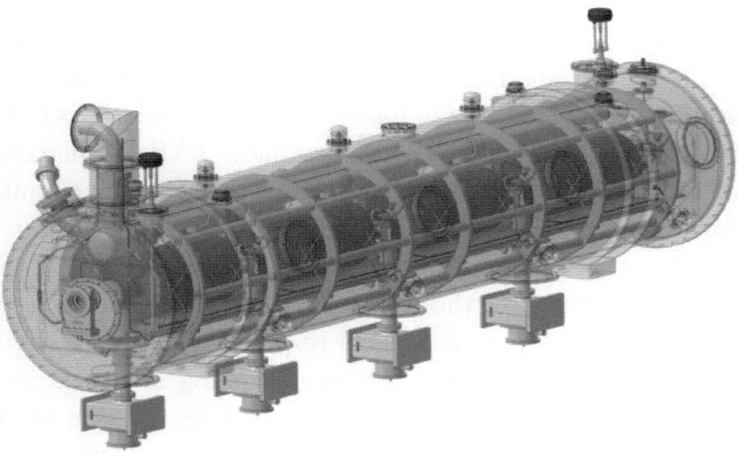

Fig. 7. Elliptical cavity cryomodule for ESS.

radiation pressure. This deformation is called Lorentz detuning and must be compensated, otherwise the deformed cavity will resonate at a different frequency which will destroy beam quality and induce losses. To handle variations in cavity coupling and Lorentz detuning there is one RF power amplifier per superconducting resonator; totaling around 150 individual RF stations for the superconducting cavities. The core of each RF station is an RF power amplifier which is typically a klystron. For the high-beta section

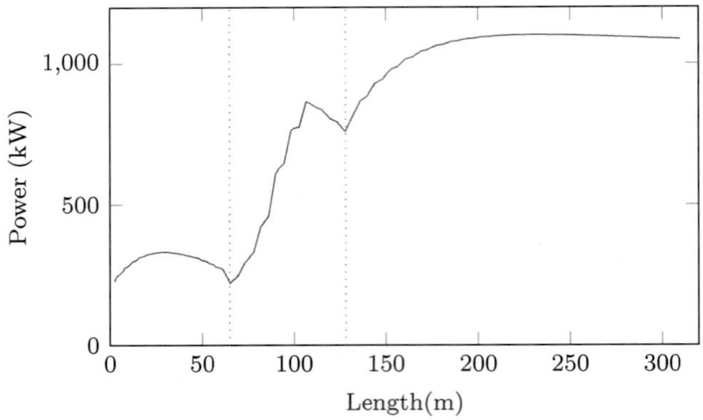


Fig. 8. Power profile in the superconducting linac of the ESS, the three sections divided by vertical dotted lines are the spoke, medium-beta and high-beta section.

of the linac consisting of 84 stations, the peak power required for the cavities is 1100 kW (Fig. 8).

Even though the loaded Q of the cavities is quite high ($7 \times 10^5$) to compensate for heavy beam loading, the Lorentz detuning for ESS can be quite severe. Without any Lorentz-detuning compensation, the resonant frequency of the cavities could shift up to 400 Hz. The bandwidth of the cavity in the absence of the beam, the unloaded cavity bandwidth, is 0.07 Hz, and the bandwidth in the presence of beam is almost 1 kHz. Comparing to the number of oscillation of the RF wave during the 2.86 ms of pulse length one can calculate that even this small difference in frequency can shift the RF wave with respect to bunches by more than one whole RF period. In addition, the beam pulse for the ESS linac is very long and is about a factor of three longer than the mechanical response time of the cavity to the Lorentz detuning. Other methods like pre-detuning of the cavity as done in the SNS linac is not sufficient in the case of ESS. To save on RF power overhead and maintain the beam quality, all superconducting cavities will be equipped with fast piezo-electric tuners to negate the Lorentz detuning of the cavities. The klystrons are operated 30% below the maximum saturated power so that variations in the RF system such as modulator droop, cavity coupling, residual Lorentz detuning, and power loss in the waveguide distribution system can be compensated. Also part of the overhead is used for stable regulation of the feedback loop. The target efficiency of the klystrons at maximum saturated power is 60% which will give an operational efficiency of about 45%. The required maximum saturated power for the klystrons is 1.5 MW. For stability, reliability and economic reasons, the klystrons will be powered in the pulsed cathode configuration.

The klystrons are energized using modulators. The modulators must supply a 3.5 ms long pulse at a rate of 14 Hz. The cathode voltage for each klystron is about 100 kV with a peak current of about 20 A. For a modulator efficiency of 90% and a power factor of 0.9, 120 kVA will be required for each klystron. For economical and space reasons, it has been decided to power two klystrons per modulator which limits the required power per modulator to 240 kVA. To handle the long pulse length of 3.5 ms, the modulators will implement solid-state switches at relatively moderate voltages and the output voltage pulse will be stepped up by a pulse transformer or equivalent technology.

The regulation of each RF system will be done independently with the low level RF system. Using modern digital technology, most of the regulation will be done with adaptive feed-forward algorithms. Feedback regulation will play a secondary role to the adaptive feed-forward so the required power overhead can be minimized.

With the large number of RF systems, it will be un-economical for the high power waveguide of each RF system to have its own penetration to the tunnel. ESS will use a "stub" concept for distributing 16 RF systems into the tunnel as shown in Fig. 2. The stubs will provide access to additional conventional services such as water and power as well. In addition, the radiation shielding issues are best handled with the stub concept.

ESS is also considering the use of Inductive Output Tubes (IOTs) in place of klystrons for the high-beta section of the linac. IOT shows the promise of high efficiency and much lower capital cost. IOTs can be thought of as a cross between klystrons and tetrodes. As in tetrodes, IOTs employ a grid to control electron flow from cathode to collector. These grids are very robust with the advent of pyrolithic carbon technology. Since IOTs are gridded tubes, no pulsed modulation system is needed to energize the cathode. This eliminates costly high voltage switches in the power supply for the IOT. Also a gridded tube can be run in deep class C that can produce very high efficiencies (less than 70%). Also IOTs do not exhibit the severe saturation effect of klystrons. This permits the operation of IOTs at the maximum rated power in feedback loops which increases the efficiency of the RF system significantly. It is estimated that 3–4 MW of power consumption can be saved if klystrons are replaced by IOTs. Like klystrons, IOTs use a resonant output cavity to efficiently couple power from the electron beam at high frequencies. However the gain of an IOT is much lower than a klystron because an IOT uses a single output cavity while a klystron may employ 4–5 stages of cavities. With the advent of low cost solid stage RF amplifiers as pre-drivers, the high gain provided by multi-cavity klystrons is not required. Because of this low gain, IOTs require cathode voltages in the order of 35 kV as compared to 100 kV for high gain klystrons. The lower cathode voltage also reduces the cost of the power supply that energizes the IOT. Current estimates for a power supply for an IOT is 60% the cost of a klystron modulator of comparable power. Currently there are no commercially available IOTs for the ESS power range. However there has been very good progress made in higher order mode multi-beam IOTs in which power levels of 1 MW have been achieved [Wright and Bohlen (2006)]. For a more detailed discussion on the ESS RF system and the potential of IOTs we refer to [Jensen *et al.* (2013); McGinnis (2013)].

## Summary

The European Spallation Source to be built in Lund will be the most powerful linac worldwide. Design and construction of ESS is done through an international collaboration and benefits from the experiences of the similar facilities. The construction of the ESS will start in 2014 and the accelerator will be operational in 2019. The beam power will be increased after the start up and will reach its nominal value of 5 MW by 2025 and scientists will be welcomed to perform experiments at the facility.

## References

Aune, B., Curtoni, A., Desmons, M., Devanz, G., Laclare, J.-L., Lagniel, J.-M., Luong, M., Mosnier, A., Pichoff, N., Safa, H., Travier, C., Uriot, D., Bongardt, K., Maier, R. and Martin, S. (2001). SC proton linac for the concert multi-users facility, in *Proc. Particle Accelerator Conference* (Chicago, USA).

Bermejo, F. J., Lucas, J. and Bustinduy, I. (2009). Baseline design for the ESS-Bilbao superconducting proton linac, in *Proc. Particle Accelerator Conference* (Vancouver, Canada).

Bohn, F. H., Bongardt, K., Carsughi, F., Claver, A., Klausen, K., Desailly, C., Fabi, P., Gardner, I., Hake, C., Laclare, J.-L., Palanque, S., Richter, D. and Tietze-Jaensch, H. (eds.) (2002). *The ESS Project, Vol III, Technical Report* (ESS Council).

Duperrier, R., Pichoff, N. and Uriot, D. (2002). CEA Saclay codes review for high intensities linacs computations, in *Proc. International Conference on Computational Science* (Amsterdam, The Netherlands).

ESFRI (2011). Strategy report on research infrastructures — roadmap 2010, `http://ec.europa.eu/research/esfri`.

Eshraqi, M., Danared, H. and McGinnis, D. (2013). Design options of the ESS linac, in *Proc. International Particle Accelerator Conference* (Shanghai, China).

Eshraqi, M. and Lagniel, J.-M. (2013). On the choice of linac parameters for minimal beam losses, in *Proc. International Particle Accelerator Conference* (Shanghai, China).

Gardner, I. S. K. *et al.* (2002). The ESS 280/560 MHz acceleration system, Tech. rep., ESS report.

Gobin, R., Beauvais, P.-Y., Bogard, D., Charruau, G., Delferriere, O., Menezes, D. D., France, A., Ferdinand, R., Gauthier, Y., Harrault, F., Mattei, P., Benmeziane, K., Leherissier, P., Paquet, J.-Y., Ausset, P., Bousson, S., Gardes, D., Olivier, A., Celona, L. and Sherman, J. (2004). Status of the light ion source developments at CEA/Saclay, *Review of Scientific Instruments* **75**(5), 1414–1416, doi:10.1063/1.1690474.

Jameson, R. A. (1993). Design for low beam loss in accelerators for intense neutron source applications, in *Proc. Particle Accelerator Conference* (Washington D. C., USA).

Jensen, M. R. F., Garoby, R., Gerigk, F., Lindroos, M., McGinnis, D., Montesinos, E. and Sunesson, A. (2013). Applications of high power induction output tubes in high intensity superconducting proton linacs, in *Proc. International Vacuum Electronics Conference* (Paris, France).

Kilpatrick, W. D. (1957). Criterion for vacuum sparking designed to include both RF and DC, *Rev. Sci. Instrum.* **28**, 824.

Laface, E., Eshraqi, M. and Miyamoto, R. (2013). The ESS linac simulator: A first benchmark with TraceWin, in *Proc. International Particle Accelerator Conference* (Shanghai, China).

Lagniel, J.-M. (1996). Halos and chaos in space-charge dominated beams, in *Proc. European Particle Accelerator Conference* (Barcelona, Spain).

Lindroos, M., Peggs, S. and Oyon, C. (2009). The ESS superconducting linear accelerator, in *Proc. Conference on RF Superconductivity* (Berlin, Germany).

McGinnis, D. (2013). The European Spallation Source, in *Proc. International Vacuum Electronics Conference* (Paris, France).

Mezei, F. (1993). Comparison of neutron efficiency of reactor and pulsed source instruments, in *Proc. International Collaboration on Advanced Neutron Sources* (Abingdon, UK.).

Mezei, F. (1997). The raison d'être of long pulse spallation sources, *Journal of Neutron Research* **6**(1–3), 3–32.

Moretti, A., Qian, Z., Norem, J., Torun, Y., Li, D. and Zisman, M. (2008). Effects of high solenoidal magnetic fields on RF accelerating cavities, *Phys. Rev. ST Accel. Beams* **8**, 072001.

Peggs, S. (ed.) (2012). *The Conceptual Design Report*, Tech. report, ESS-2012-001, `http://esss-se/documents/CDR_final_120206.pdf`.

Peggs, S., Calaga, R., Duperrier, R., Eshraqi, M., Jansson, A., Lindroos, M., Papotti, G., Plewinski, F. and Stovall, J. (2009). Conceptual design of the ESS-Scandinavia accelerator and target, in *Proc. Particle Accelerator Conference* (Vancouver, Canada).

Perlado, J. M., Piera, M. and Sanz, J. (1989). Option for spallation neutron sources, *Journal of Fusion Energy* **8**(3–4), 181–192, doi:10.1007/BF01051648.

Pritzkau, D. P. (2001). RF pulsed heating, `http://www.slac.stanford.edu/cgi-wrap/getdoc/slac-r-577.pdf`.

Schopper, H. and Myers, S. (eds.) (2013). *Elementary Particles — Accelerators and Colliders* (Springer), ISBN 978-3-642-23052-3.

Tchelidze, L. and Stovall, J. (2013). Beam loss limits in high power proton linear accelerators, in *Proc. International Particle Accelerator Conference* (Shanghai, China).

van der Meer, K., Goldberg, M. B., Lehmann, E. H., Abderrahim, H. A., Bar, D., Berkovits, D., Daum, M., Dekelver, S., Foucher, Y., Gerber, J., van Gestel, F., Hajdas, W., Linder, H.-P., Malambu, E., Mardor, I., Oeyen, J., Saphier, D., Shor, A., Willekens, M. and Yariv, Y. (2004). Spallation yields of neutrons produced in thick lead/bismuth targets by protons at incident energies of 420 and 590 MeV, *Nuclear Instruments and Methods in Physics Research Section B* **217**(2), 202–220.

Wangler, T. P. (2008). *RF Linear Accelerators*, 2nd edn. (Wiley-VCH).

Wangler, T. P., Gray, E. R., Krawczyk, F. L., Kurennoy, S. S., Lawrence, G. P., Ryne, R. D. and Crandall, K. R. (1998). Basis for low beam loss in the high-current APT linac, in *Proc. LINAC* (Chicago, USA).

Willis, B. T. M. and Carlile, C. J. (2009). *Experimental Neutron Scattering* (Oxford University Press, UK), ISBN 978-0-19-851970-6.

Wright, E. L. and Bohlen, H. (2006). Progress in CPI Microwave Tube Development, *AIP Conference Proceedings* **807**, 145.

# Chapter 36

# Neutrino factory proton driver and target design

Roland Garoby*, Chris Densham†, John Thomason†, Tristan Davenne†, Ottone Caretta†, John J. Back‡

*CERN

†Rutherford Appletan Laboratory

‡University of Warwick

## 1 Introduction

Neutrinos are very elusive particles belonging to the lepton family. They exist in different types corresponding to the different charged leptons, namely electrons, muons and taus. Contrary to electrons, neutrinos hardly interact with matter which makes them very difficult to detect and study. To the best of today's knowledge, neutrinos have hardly any mass and they can change from one type to another (so-called "neutrino oscillation"). Most physicists think that this oscillation occurs because neutrinos have mass.

A Neutrino Factory [1] is a special facility producing a large amount of neutrinos every year (typically $10^{21}$ neutrinos/year). Its main purpose is to study the change of type of neutrinos between the place where they are generated and a remote location. In a Neutrino Factory, neutrinos result from the decay of muons, unstable particles with a mean lifetime of $2.2\,\mu s$ in their rest frame. Sharp beams of high energy neutrinos are obtained at the end of the long straight sections of a multi-GeV muon storage ring. These muons are themselves generated by the decay of pions obtained after interaction of protons with a target. To produce neutrinos at the necessary rate ($\sim 10^{21}$ neutrinos/year), a large number of multi-GeV protons must be sent on target, and hence an accelerator delivering a high power proton beam (a "proton driver") must be used. The nominal characteristics of the proton driver for a Neutrino Factory [2] are shown in Table 1.

Beyond the need for a very high beam power, the peculiarity of the Neutrino Factory is to require the proton beam to be packed in few (3)

Table 1. Proton beam characteristics for a Neutrino Factory.

| Parameter | Value |
|---|---|
| Beam kinetic energy | 5–15 GeV |
| Average beam power | 4 MW |
| Repetition rate | 50 Hz |
| Number of bunches per pulse | 3 |
| Time interval between bunches | $\sim$120 $\mu$s |
| Bunch length (rms) | 1–3 ns |

short bunches of small length ($\sim$ 2 ns rms) every 20 ms (50 Hz rate). This results from the fact that bunch rotation (exchange of energy spread and time length) is used to quickly reduce the very large energy spread of the pions and muons beam in the beam line after the target.

## 2 Proton driver

### 2.1 *Design options*

Among the possible solutions for the design of a proton driver, those that can be realistically expected to meet the requirements of a Neutrino Factory after a limited amount of technological development will be outlined in the following sections. They capitalise on the experience gained during the past decades building and operating proton accelerators in many laboratories worldwide (e.g. BNL and FNAL in the USA, J-PARC in Japan, ISIS in the UK, CERN in Switzerland, etc.). They belong to two categories: (i) the ring-based solutions where most the energy gain is obtained in circular machines and (ii) the linac-based solutions, where the energy gain is obtained in a linear accelerator and the beam time structure is achieved in fixed energy ring(s).

Typical designs are described which correspond to the upgrade of existing facilities, showing that proton drivers can largely benefit from existing assets (staff competence and laboratory infrastructure) and that they can also serve multiple purposes, beyond the needs of a Neutrino Factory.

### 2.2 *Synchrotron-based proton driver*

The ring-based solution is well illustrated by the plans at the Rutherford Appleton Laboratory (RAL) in the UK. RAL is home of ISIS, the world's first large size accelerator driven spallation neutron source. ISIS has two neutron producing target stations (TS-1 and TS-2), driven at 40 Hz and 10 Hz respectively by a 50 Hz, 800 MeV proton beam from a Rapid Cycling

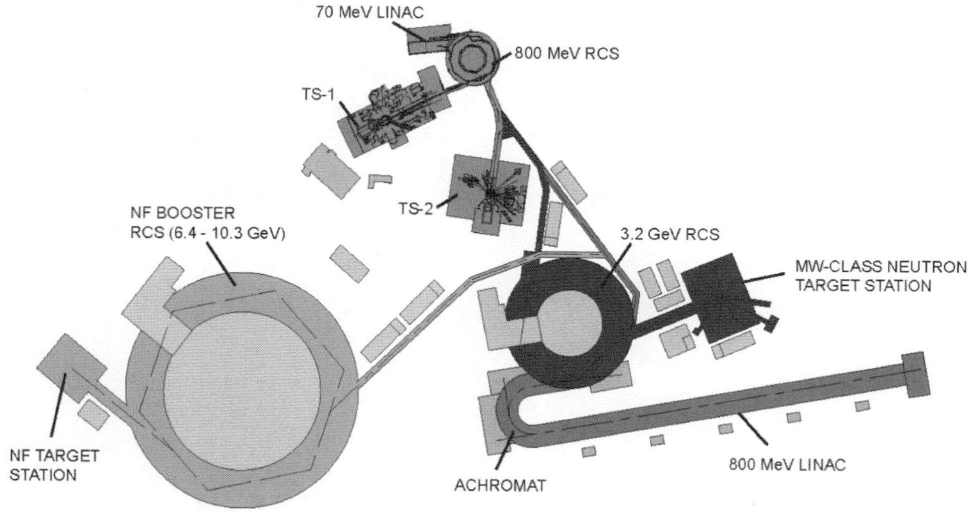


Fig. 1. Conceptual layout of a Neutrino Factory proton driver at RAL, with ISIS (green), 3.2 GeV RCS (blue), 800 MeV linac (red) and dedicated Neutrino Factory booster (orange).

Synchrotron (RCS), which is fed by a 70 MeV $H^-$ drift tube linac (DTL) [3]. Phased upgrades are envisaged which could ultimately provide beam powers of 2–5 MW in the few GeV energy range for a proton driver shared between a short pulse spallation neutron source and the Neutrino Factory. To meet the time structure requirements of a Neutrino Factory, an additional RCS or FFAG booster bridging the gap in proton energy and performing appropriate bunch compression is proposed. The conceptual layout of such a facility at RAL is shown in Figure 1, with different colors representing the components added at the successive upgrade stages.

The first stage of the upgrade path is to replace parts or all of the ISIS 70 MeV $H^-$ injector in order to address obsolescence issues with the present linac, and ensure reliable operation for the foreseeable future. A higher energy linac (up to ~180 MeV) and a new optimised injection system into the present ring should therefore be built, which could increase beam power up to ~0.5 MW [4].

The second stage is a new ~3.2 GeV, 50 Hz RCS that will increase the energy and provide ~1 MW of beam power. This new RCS will require a new building, along with a new MW-class target station. The preferred lattices [5] use either a doublet–triplet design with five superperiods (5SP) or a triplet design with four superperiods (4SP), both of which include features required for fast injection directly from the existing ISIS RCS or for multi-turn charge exchange injection from a future 800 MeV linac. The circumference of this new ring is 9/4 of the size of the ISIS 800 MeV synchrotron in the case

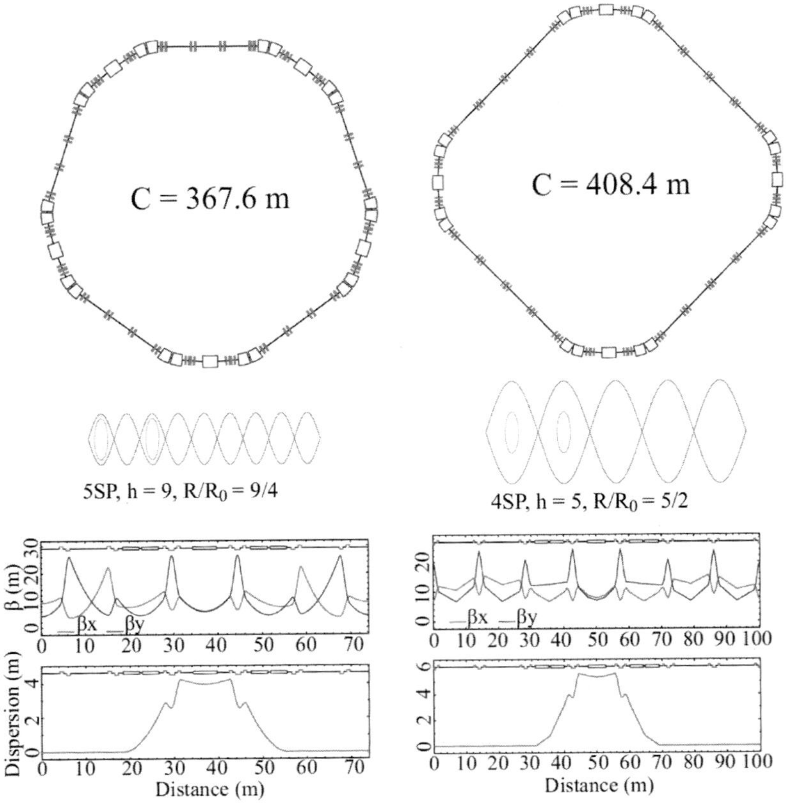


Fig. 2. Schematic of 5SP (top left) and 4SP (top right) designs, bunch to bucket transfer scheme from the ISIS synchrotron (middle) and beta and dispersion functions (bottom).

of the 5SP design, and 5/2 of the existing synchrotron for the 4SP design. Schematics are shown in Figure 2, together with an outline of the longitudinal capture scheme and their respective beta and dispersion functions.

The third upgrade stage is to replace the present ISIS synchrotron with a new 800 MeV $H^-$ linac [6] injecting directly in the 3.2 GeV RCS, which will bring the beam power up to 2–5 MW [5]. Its main characteristics are shown in Table 2.

Up to 75 MeV, normal conducting accelerating structures are used, operating at 324 MHz. Beyond 75 MeV, the beam is accelerated by superconducting elliptical multicell cavities of three different types operating at 628 MHz. More details are shown in Figure 3. A beam line of 87 m with achromatic bending sections (HEBT) is used to transport the 800 MeV beam to the 3.2 GeV RCS.

In the final upgrade stage a dedicated RCS or FFAG booster will be added to accelerate a fraction of the 3.2 GeV bunches and let them

Table 2. Main characteristics of the RAL 800 MeV $H^-$ linac.

| Parameter | Value |
|---|---|
| Beam kinetic energy | 800 MeV |
| Average beam power | 0.5 MW |
| Repetition rate | 30 (upgradable to 50) Hz |
| Bunch frequency | 324 MHz |
| Peak beam current | 43 mA |
| Beam pulse duration | 0.75 ms |
| Linac length | 243 m |

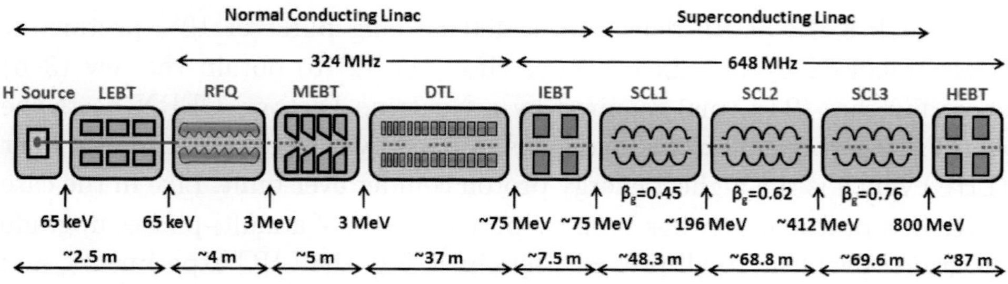


Fig. 3. Schematic layout of the future 800 MeV linac at RAL.

Table 3. Main characteristics of the RAL high energy RCS.

| Parameter | Value |
|---|---|
| Beam kinetic energy | 9.6 GeV |
| Average beam power | 4 MW |
| Repetition rate | 50 Hz |
| RF frequency | 7.149–7.311 MHz |
| RF voltage | 3.7 MV |
| Harmonic number (number of bunches) | 17 (3) |
| Number of protons per bunch | $1.76 \times 10^{13}$ |
| Transition gamma | 13.37 |
| Circumference | 694.352 m |

adiabatically reach the bunch length required before ejection. The characteristics of the beam sent on target by this last ring should hopefully meet the needs of a Neutrino Factory. The main characteristics of this additional ring are given in Table 3.

This synchrotron would only use a fraction of the total number of bunches provided by the 3.2 GeV RCS every 20 ms. In the case of the 5SP design, for example, three bunches would be transferred to the higher energy RCS, and the six other bunches corresponding to a beam power of 2.66 MW could be sent to a neutron production target. Although the feasibility of such an accelerator has been shown by a preliminary study [7], more work is necessary to (i) analyze collective effects for such bunches ($1.76 \times 10^{13}$ p/b within ~0.3 eVs), (ii) estimate beam loss/design mitigation means and (iii) find means to tolerate the energy difference between the bunches ejected 10's of turns apart.

### 2.3 *Linac-based proton driver*

The CERN-proposed proton driver for a Neutrino Factory is a representative linac-based solution. It uses a 5 GeV High Power Superconducting Proton Linac (HP-SPL) [9, 10] delivering a 400 $\mu$s long burst of $10^{14}$ protons at 50 Hz, followed by two fixed energy rings [11, 12] to obtain the few (3–6) short bunches (2 ns rms) required by a Neutrino Factory. CERN hosts the world's largest complex of accelerators, topped by the Large Hadron Collider (LHC) which is the highest energy proton collider ever built. Like in the case of RAL, a proton driver has been proposed as part of a multi-phases upgrade program benefiting to all present users including the LHC experiments, and providing the means to satisfy the needs of future generations of experimental facilities [13].

The first phase of this upgrade is the construction of a new 160 MeV $H^-$ linac ("Linac4") [14] to solve obsolescence issues in the present 50 MeV proton linac ("Linac2") and double the brightness and intensity delivered by the following synchrotron, the PS Booster ("PSB"). The construction of Linac4 is well advanced [15] and beam commissioning is planned in 2014–2015. This new linac is using normal conducting accelerating structures at 352 MHz (Figure 4). It has been designed to be easily upgradable for becoming the front end of a future SPL.

The proposed next phase was to add a low power version of the SPL ("LP-SPL") immediately after Linac4. Combined with a new 50 GeV synchrotron ("PS2") this new pair of accelerators would have replaced three existing ones (Linac2, PSB and PS) and boosted the performance of the

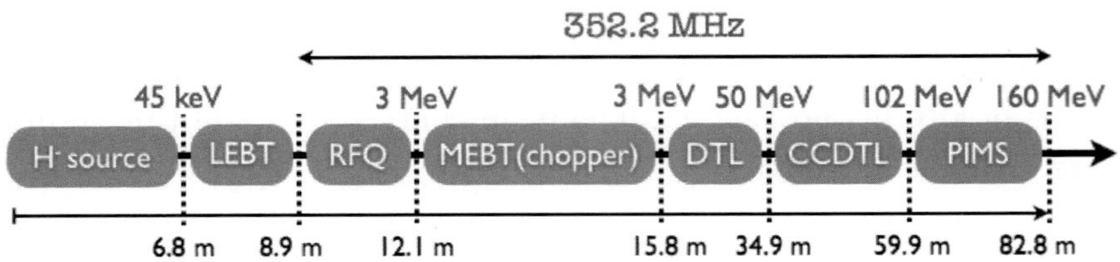


Fig. 4. Schematic layout of Linac4 at CERN.

SPS by injecting beam at twice the present energy. The LP-SPL could have later evolved into a high power multi-MW accelerator, capable to serve users of high power beam at 4–5 GeV, in addition to its role in the LHC injector complex [13].

This intermediate phase is not favored anymore, but the construction of a high power SPL remains a possible option for fulfilling simultaneously the needs of a couple of next generation experimental facilities including a Neutrino Factory. The schematic layout of this superconducting linac using Linac4 as injector is shown in Figure 5. Operating at 704 MHz, twice the frequency of Linac4, it would use only two types of 5-cell elliptical cavities with a geometrical beta of 0.65 and 1 to cover the energy range from 160 MeV to 5 GeV [9, 10].

The parameters of the SPL in a proton driver for a Neutrino Factory are summarised in Table 4. It is followed by two fixed-energy rings which transform the long linac pulse into the required bunches [8, 11, 12].

In the first ring (the accumulator), the chopped linac beam is accumulated in a few long bunches, using charge-exchange injection. The

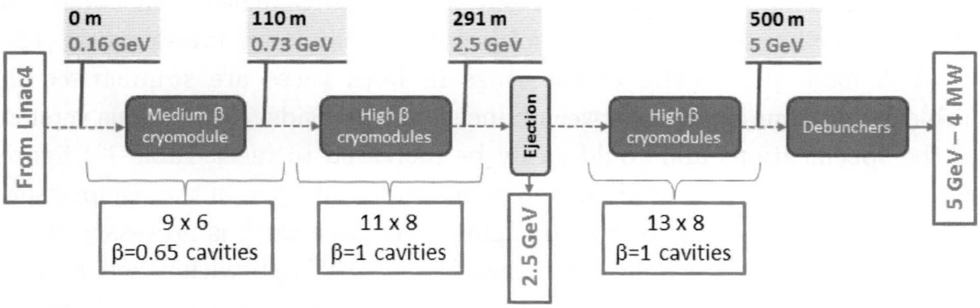


Fig. 5. Schematic layout of the SPL at CERN.

Table 4. Main characteristics of the high power SPL at CERN.

| Parameter | Value |
|---|---|
| Beam kinetic energy | 5 GeV |
| Average beam power | 4 MW |
| Repetition rate | 50 Hz |
| Bunch frequency | 352 MHz |
| Peak (average) beam current | 60 (40) mA |
| Beam pulse duration | 400 $\mu$s |
| Linac length (+ Linac4) | 500 m (+ 90 m) |

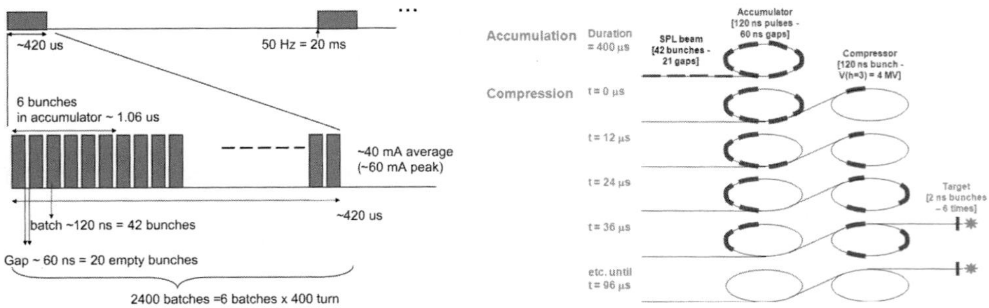


Fig. 6. Principle of bunch generation in a linac-based proton driver for a Neutrino Factory.

accumulator is isochronous to preserve the time structure of the linac beam, and it has no RF system to minimise impedance. Once accumulation is finished, after 400 $\mu$s, bunches are transferred one by one to the compressor ring where bunch rotation takes place with the energy stored in RF cavities. The ratio between the two rings is such as to guarantee the arrival of the successive bunches at the correct location in the compressor. This principle is described in Figure 6 in the case of six bunches. Lattices have been designed for both rings, and beam dynamics studies including collective effects have shown feasibility [8, 11]. The generation of three bunches with twice the intensity per bunch has been the subject of a preliminary investigation [12].

The main parameters of the rings in both cases are summarised in Table 5. The time interval between bunches corresponds to a previous version of the specifications and could easily be increased to meet Table 1's figure with a higher transition gamma and a smaller RF voltage in the compressor.

This design remains very challenging, and more work is necessary to (i) refine the scenario with three bunches ($3.33 \times 10^{13}$ p/b within 2 ns rms at ejection), including collective effects, (ii) develop the technology for stripping and charge-exchange injection at 5 GeV with 4 MW of beam power, (iii) design an adequate collimation system and (iv) design/prototype equipment for fast ejection/injection and RF.

## 3 Target

### 3.1 *Outline of target station*

The target station of a Neutrino Factory is required to convert the incoming proton beam into mesons and to capture and transport both signs as they decay into muons. The mesons (pions) are generated by interaction of the proton beam with a target material with parameters designed to maximise production and minimise reabsorption. Both signs of charged pions can be captured by installing the target within the bore of a ~20 T solenoid [16].

Table 5. Main parameters of the accumulator and compressor rings.

| Ring | Parameter | 6 bunches case | 3 bunches case |
|---|---|---|---|
| Accumulator | Circumference | 318.5 m | 185.8 m |
| | Number of turns of accumulation | 400 | 740 |
| | Tunes (H/V) | 7.77/7.67 | 7.37/5.77 |
| | Transition gamma | Quasi-isochronous | |
| | Magnet technology | Normal conducting | Superconducting |
| Compressor | Circumference | 314.2 m | 200 m |
| | Nb of turns for bunch compression | 36 | 86 |
| | RF voltage | 4 MV | 1.7 MV |
| | RF harmonic number | 3 | 2 |
| | Transition gamma | 2.3 | 2.83 |
| | Tunes (H/V) | 10.79/5.77 | 4.21/2.74 |
| | Magnet technology | Superconducting (dipoles) | Normal conducting |
| | Time interval between bunches | 12 $\mu$s | 30 $\mu$s |

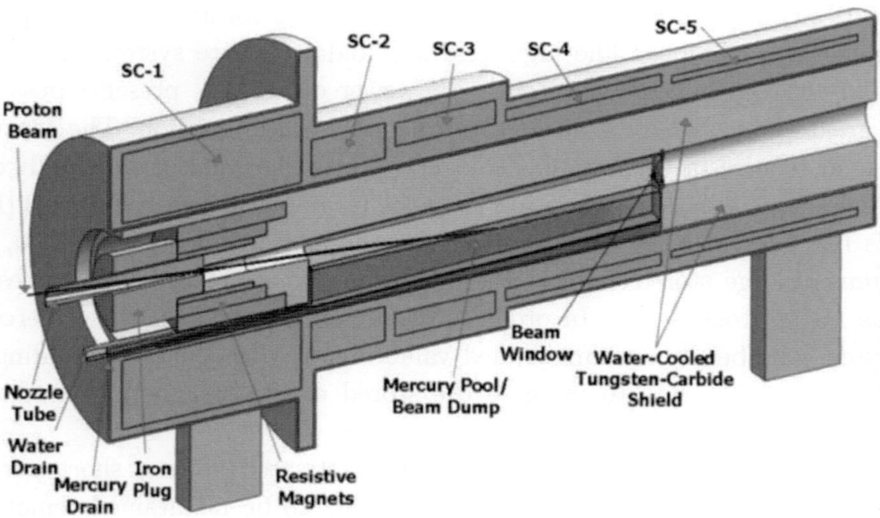


Fig. 7. Sectional view of target station showing superconducting coils (SC-1–5) and resistive inserts in capture solenoid [26].

This comprises a normal conducting 6 T inner coil and a superconducting 14 T outer coil as shown in Figure 7. This is followed by a channel of solenoids tapering down to 1.5 T [17] to adiabatically exchange transverse for longitudinal momentum and transport the beam to the muon front end [1, 2].

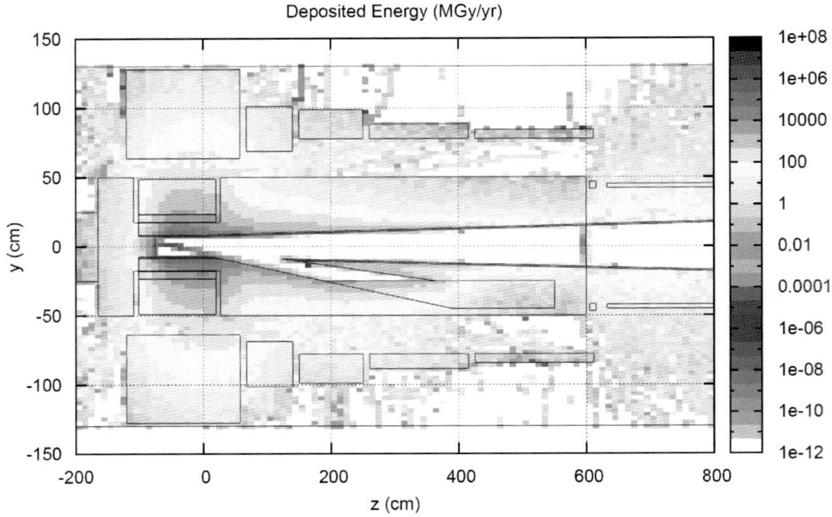


Fig. 8. Energy density deposited in target station elements [18].

This arrangement compares with existing, so-called conventional Neutrino Facilities which utilise magnetic horns to generate a toroidal field to capture a single sign of pions. The high field solenoidal capture system in combination with an incoming proton beam power of ~4 MW presents possibly the greatest technological hurdle for a neutrino target station. The capture solenoid requires a thick high-Z shield to protect the coils from the flux of particles and radiation emanating from the target as shown in Figure 8 [18]. This results in a large bore for the superconducting solenoid of 1.3 m, an extremely large stored energy of 600 MJ and enormous magnetic forces of some 10,000 tonnes [19]. In order to keep the heat load in the superconducting coils below a recommended value of 1 mW/cc, further shielding is required which serves to increase the stored energy, magnetic forces and technological challenge.

The target station will become highly activated after a short period of operation, requiring all the critical elements to be maintained remotely, which has to be taken carefully into account with the design and construction of the facility and will be a significant cost.

The target system itself is also a significant engineering challenge and several technologies have been considered. Key factors in the design and construction of such a facility are the reliability and lifetime of systems and components and these considerations need to be included in the technology selection and design optimisation.

### 3.2 *Target baseline*

The target must be able to dissipate the heat load generated by interaction with the 4 MW proton beam but needs to be compact to maximise the production, capture and transport of low energy pions. The optimum dimensions are around three times the rms beam radius and approximately two nuclear interaction lengths. It must be surrounded by a vacuum or helium environment to provide a free path for the pions to perform spiral trajectories as they traverse the solenoidal magnetic field.

Graphite targets have been used successfully in conventional Neutrino Facilities [20–23] due to favourable thermal and mechanical properties and reasonable resilience to radiation damage. Also, typically only a few % of the beam power from multi-GeV proton beams interacting with the low-Z material is deposited as heat. Nevertheless, a number of failures have occurred in static graphite targets and a Neutrino Factory is proposed to have a beam power an order of magnitude greater than has hitherto been available. It is widely expected that issues of radiation damage [24], shock wave damage, thermal transport and the need to avoid unacceptable compromises to physics performance would require any solid targets that may be found to be feasible to be replaced with a high frequency. Potential solid target technologies are described below. Contained liquid mercury targets have been adopted in the most recent generation of Neutron Facilities [25] with the mercury continuously recirculated to be cooled externally and reformed. Mercury has the added benefit of being a more copious source of mesons compared with graphite. However, depending on the proton energy, around 10–15% of the beam power is deposited in the mercury as heat. The considerably greater pulsed power density in a Neutrino Factory compared with a neutron source means that any container wall would fail due to the intense pressure wave generated by the pulsed beam. Consequently an open liquid mercury jet injected directly into the bore of the solenoid is the baseline target technology for a Neutrino Factory [26]. This was originally proposed for a Muon Collider [16], where a point-like source was paramount and hence a high-Z material with a short nuclear interaction length was required. The mercury jet needs to be injected at a high enough velocity for the proton beam to interact with two nuclear interaction lengths (around 30 cm), and both the beam and the jet need to be at a small angle to the axis of the solenoid to minimise the reabsorption of pions as they spiral out from the surface of the target.

The layout in Figure 7 shows a layout of the mercury jet target together with the incoming proton beam and the solenoid capture coils [27]. The jet

Table 6. Baseline target station parameters [1,28].

| Parameter | Value |
|---|---|
| Target type | Free mercury jet |
| Jet diameter | 8 mm |
| Jet velocity | 20 m/s |
| Jet/solenoid axis angle | 96 mrad |
| Proton beam/solenoid axis angle | 96 mrad |
| Proton beam/jet angle | 27 mrad |
| Capture solenoid field strength | 20 T |
| Front end transport channel field strength | 1.5 T |
| Length of transport channel | 15 m |

is foreseen to enter a pool which also acts as a beam dump and is continuously drained and recirculated. The baseline target system parameters in the International Design Study [1,28] are listed in Table 6.

The MERIT experiment [29,30] carried out at CERN in 2007 tested the response to a pulsed proton beam of a prototype liquid mercury jet injected into the bore of a high field solenoid. This demonstrated that a solenoidal field of 20 T greatly mitigated the beam induced disruption of the mercury jet, and the jet recovered quickly enough that a beam repetition rate of 50 Hz would also be feasible.

The MERIT experiment demonstrated the proof-of-principle of a mercury jet as a neutrino factory target, nevertheless there remain a number of significant technical and site-specific licensing hurdles to be overcome before such a system could be implemented. Technical issues include dealing with the kinetic energy of the high velocity jet of mercury as it enters the beam dump, and the response of the beam dump to non-disrupted proton beam pulses. ANSYS CFX simulations of these effects are shown in Figure 9. Regarding safety and licensing issues, the European Spallation Source project [31] considered the possibility of a mercury target and concluded that the costs just for applying for the license to generate highly radioactive mercury would be prohibitive, with no guarantee of a successful outcome.

### 3.3 *Target alternatives*

#### 3.3.1 *Fluidised tungsten powder*

Fluidised tungsten powder has been proposed [32] as an alternative high-Z target technology to the open mercury jet, with the potential to accommodate extremely high pulsed power densities. A powdered solid can be pumped like a liquid, has excellent heat transfer characteristics, is highly resistant to

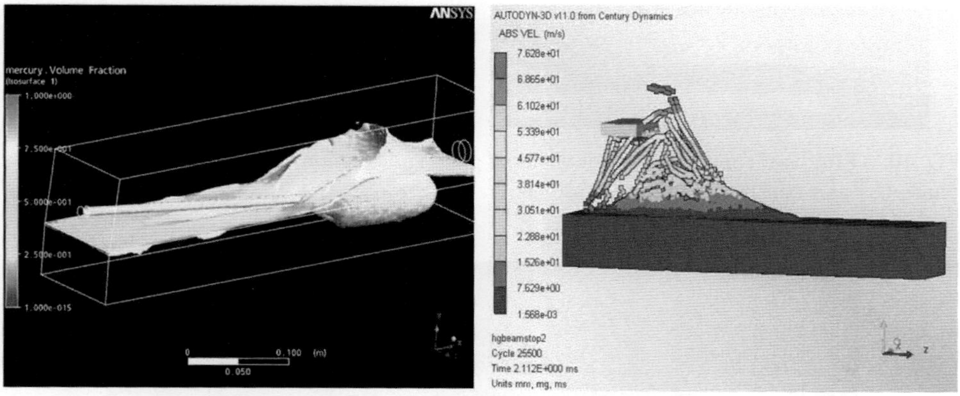


Fig. 9. ANSYS CFX simulation of mercury jet entering static pool (L) and ANSYS AUTODYN simulation of non-disrupted proton beam entering static pool (R).

shock wave induced damage and cannot suffer from cavitation which is a purely liquid phenomenon. A research programme is underway [33] to investigate whether such a technology can combine some of the advantages of a solid target with those of a liquid while avoiding some of the disadvantages of either. The concept is that of tungsten powder fluidised and transported within a pipe using helium as a carrier gas [34], in a configuration that is compatible with the solenoid meson capture system. The target material would pass through the beam and be recirculated and cooled externally.

Fluidised beds and powder jets are a mature technology developed for the conveyance of powders in industry. Such experience has been the starting point in the design of various components in the development of a fluidised powder target test facility. Many of the questions relating to implementation can be investigated experimentally offline. Also, unlike mercury, powdered tungsten is non-toxic. These factors mean that a productive experimental programme has been possible at a relatively modest cost. The bespoke test facility has demonstrated dense-phase fluidisation and lean-phase recirculation of tungsten powder using both air and helium as the carrier gas. Both open jets and contained flows of dense phase powder have been generated in a horizontal configuration suitable for a particle accelerator target such as a Neutrino Factory, as shown in Figure 10. A study carried out to compare the particle production performance of fluidised tungsten assuming a material fraction of 50% compared with the mercury jet was performed, and found a small reduction in yield [18] that could be completely recovered simply by increasing the target length accordingly [35].

Flow contained within a pipe is preferred as this offers the ideal well defined cylindrical geometry for interaction with the proton beam. The

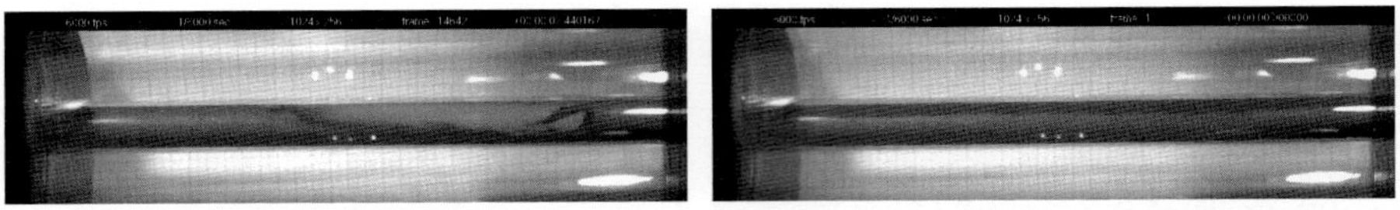

Fig. 10. Fluidised tungsten powder within a glass tube demonstrating discontinuous dense phase (L) and continuous dense phase (R).

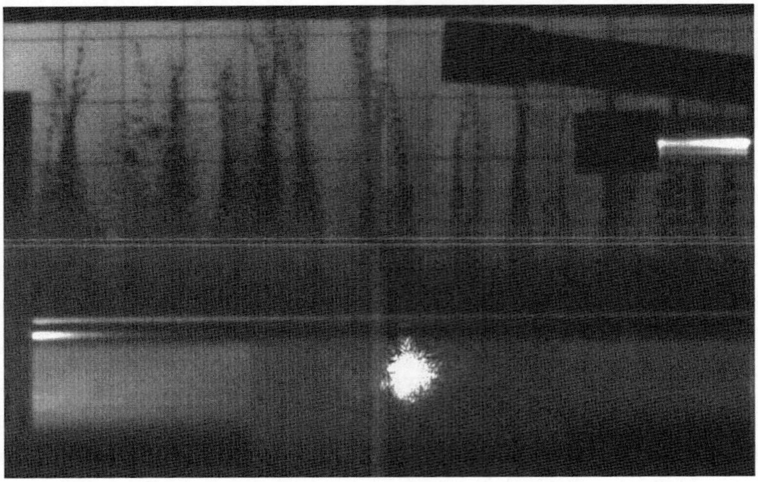

Fig. 11. Response of a trough of sub-150 $\mu$m tungsten powder 30 ms after interaction by a pulse of $2.9 \times 10^{11}$ protons at 440 GeV [36].

material can be recirculated at a lower velocity so will present less risk of erosion, and the containment provided by the pipework significantly controls the radiological hazard presented by the powdered material. However the pipe wall would need to be cooled and would also suffer from radiation damage. The separate grains of powder are inherently resilient to beam induced shock wave damage, and relatively low pressure rises in the helium carrier gas are expected. This hypothesis has been tested in an online experiment [36] using the 440 GeV proton beam available at the HiRadMat facility, CERN [37]. This experiment demonstrated that a static trough of tungsten powder in a helium environment is perturbed by an intense pulsed proton beam, but with a significantly higher threshold and with a response almost two orders of magnitude less than that resulting from a 'mercury thimble' experiment [38] at a similar pulsed power density (Figure 11).

This offline and online experimental programme has demonstrated the feasibility of fluidised tungsten as a high power target technology suitable for a Neutrino Factory. Nevertheless, a number of issues remain to be demonstrated before implementation could be recommended in an actual facility, for example erosion and radiation damage in pipe walls and the handling and disposal of radioactive powder. The technology may share with the mercury baseline some of the concerns regarding complexity, reliability, safety, licensing, handling and cost.

### 3.3.2 *Recirculating tungsten rods*

For a high-Z target material, the high deposited power and power density for the Neutrino Factory proton beam parameters mean that the target material has to be circulated and cooled externally. Recirculating radiatively cooled solid tungsten bars have been suggested [39] as another alternative to the liquid mercury jet and a test programme using pulsed currents to generate stress waves of the magnitude of those that would be generated by the pulsed proton beam in a Neutrino Factory has demonstrated the required fatigue lifetime in the material [40]. However, the materials and engineering difficulties of recirculating solid tungsten bars through the bore of a high field solenoid are daunting [41].

### 3.3.3 *Packed beds*

A study using an acceptance probability histogram at the end of the first 6 m of the solenoid field map [42] has shown that some lower-Z target materials can be as productive as high-Z materials (Figure 12). This, in combination with the engineering challenges of any recirculating high-Z target technology

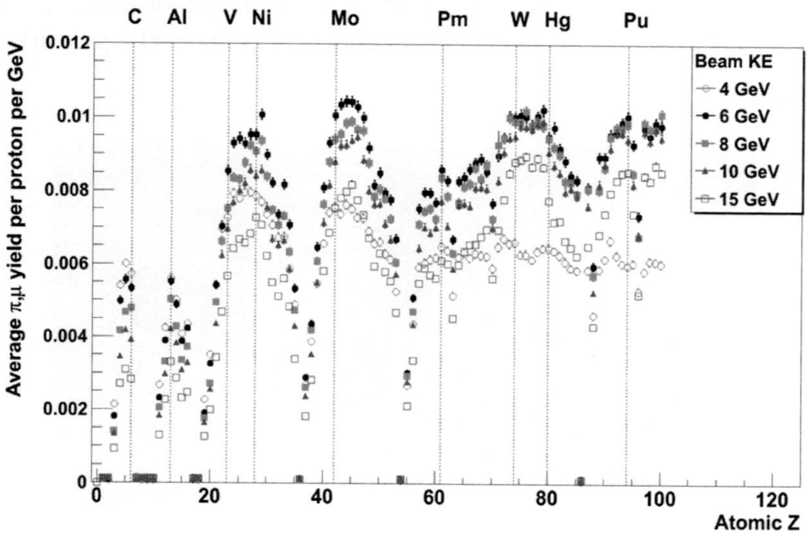


Fig. 12. $\pi$, $\mu$ yield per proton per GeV for different target materials, tracked through solenoid field and within acceptance 6 m downstream of target [42].

and the substantially reduced deposited heat load in a low-Z material, has renewed interest in the idea of exploring the limits of static, low-Z targets for a Neutrino Factory.

At the lower power densities experienced by the targets in existing, so-called conventional neutrino facilities such as T2K [22], a peripherally cooled solid target is most efficient [23] for particle production. For a solid target to dissipate the heat load deposited from a 4 MW proton beam it is necessary to segment the material to increase the heat transfer surface area and also reduce the thermal stresses [20, 21]. A packed bed of target spheres with coolant flowing between them is the optimum segmented stationary target design in terms of heat transfer surface area and power dissipation capability. To maintain the particle yield from a packed bed, it is necessary to compensate for the reduction in material fraction by increasing the target and beam dimensions.

Helium gas is favoured as a coolant to avoid shock waves that result from the pulsed beam interaction with an incompressible liquid coolant, and also because it offers negligible interactions/pion absorption, low activation and good heat transfer properties. Sievers and Pugnat [43] first proposed a static granular target for a Neutrino Factory, cooled by high pressure helium. This idea has been developed significantly [44] with a cross-flow path packed bed design shown in Figure 13, for the application of a neutrino SuperBeam [42] utilizing the High Power Superconducting Proton Linac [9, 10] described above. This design was studied and optimized for four targets operating in parallel within four magnetic horns, with each target of packed titanium spheres subject to a 1 MW beam. This concept may have the potential to be

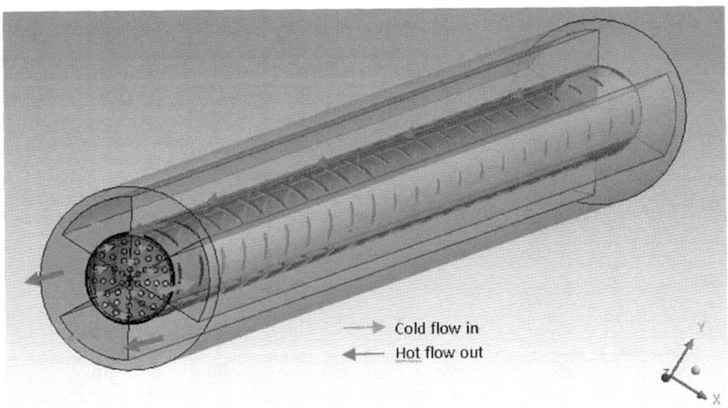


Fig. 13. Concept of a cross-flow helium cooled packed bed target.

extended e.g. to a graphite target although the limits and hazards of high pressure helium on the required beam windows has yet to be fully studied.

## References

[1] International Design Study for a Neutrino Factory (IDS-NF), https://www.ids-nf.org/wiki/FrontPage.

[2] IDS-NF Collaboration, International Design Study for a Neutrino Factory — Interim Design Report, IDS-NF-020.

[3] D.J.S. Findlay, ISIS — Pulsed Neutron and Muon Source, *Proc. PAC'07*, Albuquerque, USA, June 2007, p. 695.

[4] C.M. Warsop *et al.*, Status of Injection Upgrade Studies for the ISIS Synchrotron, *Proc. IPAC'11*, San Sebastian, Spain, Sept. 2011, p. 2760.

[5] J.W.G. Thomason *et al.*, Megawatt Upgrades for the ISIS Facility, *Proc. HB2008 Workshop*, Nashville, USA, August 2008, p. 434.

[6] C. Plostinar, C. Prior, G. Rees, Conceptual Design of a New 800 MeV $H^-$ Linac for ISIS Megawatt Developments, *Proc. IPAC'11*, San Sebastian, Spain, Sept. 2011, p. 2100.

[7] J. Pasternak, L.J. Jenner, Design of an RCS for the Final Stage of Acceleration in a Common Proton Driver for a Neutrino Factory and a Spallation Neutron Source Based on Megawatt Upgrades of ISIS, *Proc. IPAC'11*, San Sebastian, Spain, Sept. 2011, p. 2751.

[8] R. Garoby, E. Benedetto, M. Aiba, M. Meddahi, Linac-Based Proton Driver for a Neutrino Factory, CERN-NUFACT-Note-157 (2009).

[9] M. Baylac, F. Gerigk (eds.), Conceptual Design of the SPL II : A High-Power Superconducting H- Linac at CERN, CERN 2006-006 (2006).

[10] F. Gerigk *et al.*, Layout and Machine Optimization of the SPL at CERN, Proc. LINAC'10 (Tsukuba, Japan, September 2010), CERN-ATS-2010-208 (2010).

[11] M. Aiba, Feasibility Study of Accumulator and Compressor for the 6-Bunches SPL-Based Proton Driver, CERN-AB-2008-060 BI (2008).

[12] M. Aiba, A first Analysis of 3-Bunches and 1-Bunch Scenario for the SPL-Based Proton Driver, CERN-AB-Note-2008-048-BI (2008).

[13] R. Garoby, Synergies between the Needs of LHC, Neutrinos and Radioactive Ion Beams, sLHC Project Report 0034 (2009).

[14] F. Gerigk, M. Vretenar (eds.), Linac4 Technical Design Report, CERN-AB-2006-084 ABP/RF (2006).

[15] M. Vretenar *et al.*, The Linac4 Project at CERN, CERN-ATS-2011-041 (2011).

[16] R.B. Palmer *et al.*, Muon Colliders, *The 9th Advanced ICFA Beam Dynamics Workshop: Beam Dynamics and Technology Issues for Muon Colliders* [477].

[17] H.G. Kirk, N. Souchlas, R.C. Fernow, X.P. Ding, V.B. Graves, R.J. Weggel, K.T. McDonald, C.J. Densham, P. Loveridge, T. Guo *et al.*, A Solenoid Capture System for a Muon Collider, *Conf. Proceedings of PAC2011*, C110328, pp. 1316–1318 (2011).

[18] J. Back, C. Densham, R. Edgecock, G. Prior, Particle Production and Energy Deposition Studies for the Neutrino Factory Target Station, *Phys. Rev. ST Accel. Beams* **16**, 021001 (2013).

[19] P. Loveridge, Technical Challenges of the Neutrino-Factory/Muon-Collider Capture System, presentation at the *Solenoid Capture Workshop*, Brookhaven National Laboratory, 29–30 November 2010.

[20] FNAL NuMI-MINOS — NuMI Technical Design Handbook, http://www-numi.fnal.gov/numwork/tdh/TDH_V2_4.2.3-Target.pdf

[21] P. Hurh *et al.*, High-Power Targets: Experience and Potential for 2 MW, *Conf. Proc. PAC 2011*, C110328, pp. 1496–1500 (2011); PAC-2011-WEODS2.
[22] K. Abe *et al.* (T2K Collaboration), The T2K experiment, *Nucl. Instrum. Meth. A* **659** (1), 106-135 (2011).
[23] C.J. Densham, M. Baldwin, M.D. Fitton *et al.*, Design and Development of the T2K Pion Production Target, *Proceedings of the 23rd Particle Accelerator Conference*, Vancouver, Canada, 2009 (IEEE, Piscataway, NJ, 2009).
[24] N. Simos *et al.*, Irradiation Damage Studies of High Power Accelerator Materials, *J. Nucl. Mater.* **377**, 41–51 (2008).
[25] K. Haga *et al.*, Mercury Target and its Peripheral Devices for 1 MW Spallation Neutron Source, *Proc. ICONE-12*, Washington DC, USA, 2004, Paper 49518.
[26] H.G. Kirk, Targetry for a $\mu^+\mu^-$ Collider, *Proceedings of PAC99*, New York, 1999, edited by C.A. Luccio, W. MacKay, p. 3029.
[27] S. Ozaki, R.B. Palmer, M.S. Zisman, J.C. Gallardo, Feasibility Study-II of a Muon-Based Neutrino Source, Tech. Rep. BNL-52623, Brookhaven National Laboratory, Upton, NY, 2001.
[28] X. Ding, D. Cline, H. Kirk, and J.S. Berg, Optimized Parameters for a Mercury Jet Target, *Proceedings of the 23rd Particle Accelerator Conference* [474], p. 2748.
[29] K T. McDonald *et al.*, The MERIT High-Power Target Experiment at the CERN PS, *Proceedings of IPAC'10*, Kyoto, Japan [472], p. 3527.
[30] I. Efthymiopoulos *et al.*, The MERIT High Power Target Experiment at the CERN PS, *Proceedings of the 23rd Particle Accelerator Conference*, Vancouver, Canada, 2009 (IEEE, Piscataway, NJ, 2009) TU6PFP085.
[31] S. Peggs (ed.) ESS Technical Design Report, Lund, April 2013, ESS-doc-274, https://europeanspallationsource.se/documentation/tdr.pdf.
[32] C.J. Densham, O. Caretta, P. Loveridge, T.W. Davies, R. Woods , The Potential of Fluidised Powder Target Technology in High Power Accelerator Facilities, *Proceedings of the 23rd Particle Accelerator Conference*, Vancouver, Canada, 2009.
[33] O. Caretta *et al.*, Preliminary Experiments on a Fluidised Powder Target, *Proceedings of EPAC08*, Genova, Italy, 2008, WEPP161.
[34] T.W. Davies, O. Caretta, C.J. Densham, R. Woods, The Production and Anatomy of a Tungsten Powder Jet, *Powder Technology* **201** (3), 296–300 (2010).
[35] J.J. Back, pers. comm.
[36] O. Caretta, T. Davenne, C. Densham, M. Fitton, P. Loveridge, J. O'Dell, N. Charitonidis, I. Efthymiopoulos, A. Fabich, L. Rivkin, Response of a Tungsten Powder Target to an Incident High Energy Proton Beam, *Phys. Rev. ST Accel. Beams* **17**, 101005 (2014).
[37] I. Efthymiopoulos *et al.*, HiRadMat: A New Irradiation Facility for Material Testing at CERN, *Proceedings of the 2nd International Particle Accelerator Conference, IPAC-2011*, San Sebastián, Spain (EPS_AG, Spain, 2011), TUPS058.
[38] A. Fabich, S. Gilardoni, M. Benedikt, M. Farhat, E. Robert, J. Lettry, Thermal Shocks and Magnetohydrodynamics in High Power Mercury Jet Targets, *Physics G: Nuclear and Particle Physics* **29**, 1621–1627 (2003).
[39] J.R.J. Bennett *et al.*, Studies of a Target System for a 4-MW, 24-GeV Proton Beam, Tech. Rep. CERN-INTC-2004-016, CERN, Geneva, Switzerland, April, 2004.
[40] T.R. Edgecock, J.R.J. Bennett, S. Brooks, G.P. Skoro, C.N. Booth, Overview of Solid Target Studies for a Neutrino Factory, *Proceedings of IPAC'10*, Kyoto, Japan THPEC089.

[41] O. Caretta, T. Davenne, P. Loveridge, C.J. Densham, Engineering Considerations on Targets for a Neutrino Factory and Muon Collider, *Journal of Nuclear Materials* **433** (1–3), 538–542 (2013).

[42] T.R. Edgecock *et al.*, High Intensity Neutrino Oscillation Facilities in Europe, *Phys. Rev. ST Accel. Beams* **16**, 021002 (2013).

[43] P. Pugnat and P. Sievers, A He-Gas Cooled, Stationary Granular Target, *J. Phys. G: Nucl. Part. Phys.* **29**, 1797–1800 (2003).

[44] Baussen *et al.*, Neutrino Super Beam Based on a Superconducting Proton Linac, *Phys. Rev. ST Accel. Beams* **17**, 031001 (2014).

# Chapter 37

# Neutrino factories

Elena Wildner (CERN)

## 1 Introduction

Neutrinos are produced by many processes in our universe. These elusive particles reach the earth having a certain energy permitting them to react with nuclei in detectors that are specifically designed to probe their properties.

However, to get higher intensities and higher energy neutrinos for better statistics and better physics reach, the use of accelerators is necessary to advance in the field of neutrino research. To produce neutrinos with an accelerator, one needs to send a high power beam onto a target to get particles or isotopes that produce neutrinos with the required properties, by decay. The parent particles have to be collected and prepared for injection into an accelerating structure. Accelerator-based experiments can tune the energy of the produced neutrinos by boosting and controlling the energy of the parent particle. The produced neutrinos will travel the distance between the source and the detector, generally through earth; the distance the neutrino travels through earth, the energy of the neutrino as well as the flavor of the neutrino give important information on their interaction with matter. The position of the physics detector is coupled to the energy of the neutrino, since the neutrino oscillation length varies inversely with the energy. The position of the detector is chosen depending on what kind of physics is being explored. "Short Baseline" experiments (a few km between the target and the detector) need beam powers up to a few hundred kW and longer baseline experiments, having detectors at from a few hundred up to thousands of km, need beam power on target reaching the MW range. "Next Generation" facilities will go up to many MW. Longer baseline experiments address physics related to oscillations of active neutrinos, while short baseline facilities do research related to the search for neutrinos not yet observed, so called sterile neutrinos.

Getting the sensitivities needed for physics today translates to beam-power on target reaching up to 5 MW for some of the proposed facilities. The potential of the neutrino beam can be enhanced by accelerating and storing the parent particles in a decay ring. In the case where muons are used for neutrino production, the pions produced in the target are collected and focused into a decay pipe behind the target, where they decay into muons. Challenging cooling and phase space manipulations of the muons make injection of the muon beam into the accelerator chain possible. Then follows acceleration up to the energy of a storage ring, and the neutrino useful for physics is produced in long straight sections directed to the physics detector. Alternatively, beta active isotopes produced in a target can be collected in an ion source, accelerated and stored in a race track decay ring where beta decay gives neutrinos aimed at the experiment.

Target technology and development is crucial for accelerator-based neutrino facilities. Few target stations are capable of handling more than 1 MW today, upgrade plans for several facilities to have more energy and intensity of the primary beam imply R&D and studies of the accelerators involved.

The availability of the big and expensive detectors, and caverns that could house them, has advantages for accelerator-based neutrino facilities (synergy with other physics fields). However, the interest to take advantage of such opportunities is evaluated case by case.

Some of the possible developments on present neutrino facilities to achieve higher performance and the present baselines of some future neutrino facilities will be described.

## 2 Accelerator developments for high power conventional beams for neutrinos

The Japan Proton Accelerator Research Complex (J-PARC) $\nu$-beam is in operation since 2009 for the T2K experiment, "Tokai to Kamioka", using the Super-Kamiokande detector. The J-PARC accelerator complex is shown in Figure 1. The experiment recently made the first $3\sigma$ measurements of one of the missing parameters for understanding the neutrino oscillations, the mixing angle $\theta_{13}$ [Tanaka (2012)]. T2K uses the conventional technique of interacting a 30 GeV proton beam of about 200 kW, with a repetition period of 2.48 s, with a graphite target. A magnetic horn system collects pions of one charge and focuses them into a decay volume where the neutrino beam is produced. The primary proton beam line consists of three sections: A preparation section to tune beam extracted from the J-PARC Main Ring, an arc section to bend beam towards Kamioka, and a final focusing section

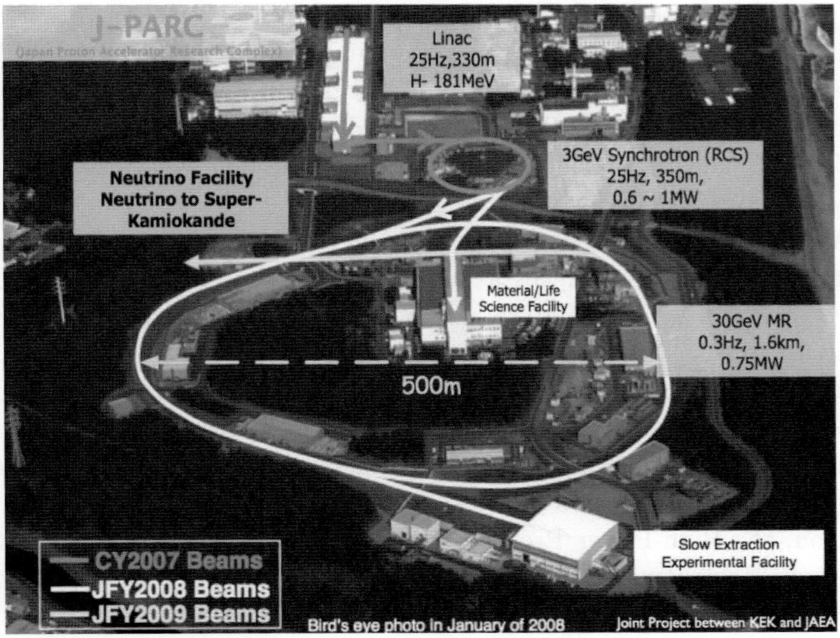


Fig. 1. The J-PARC facility. The proton beam is accelerated to 30 GeV in the Main Ring, then extracted and guided onto a target, producing a neutrino beam that is directed to a detector located in Kamioka, Gifu about 250 km away.

to focus beam onto the target. Normal-conducting magnets are used for the preparation section and final focusing section. In the arc section with radius of 104 m and length of about 150 m, the beam is bent about 80 degrees by 28 Superconducting Combined Function Magnets (SCFM). This is the first attempt to develop a combined function magnet as a superconducting device. It is 3.3 m long, with dipole and quadrupole components being 2.6 and 18.6 T/m, respectively. Two SCFM magnets are assembled inside one cryostat (doublet). A large coil aperture of 173.4 mm is chosen to lower the possibility of a magnetic quench due to beam loss.

To increase the power on target for future experiments addressing CP violation in the leptonic sector and to benefit from the Hyper Kamiokande detector, an upgrade path for the J-PARC Main Ring has been proposed to combine high repetition rate of the cycle and high beam intensity [JPARC (2012)]. The injection kicker magnets need to be replaced (beam coupling impedance) and collimation will be improved to avoid halo beam loss from the 3 GeV RCS. Ongoing upgrades of the linac from 181 MeV to 400 MeV by including a new source and RFQ will increase the current from 30 mA to 50 mA. Higher intensities will be achieved by a second harmonic RF system and profit from R&D on high gradient RF cores. A research program to increase the repetition rate of the Main Ring power supply aims at 1 Hz or more. Eventually the upgrade plans would permit the J-PARC Main Ring beam power to reach 1.7 MW, which would give a considerable potential to the J-PARC neutrino facility. The neutrino beam line would tolerate up to 2 MW without any major upgrade.

At CERN, a 400 GeV proton beam at a maximum repetition rate of one every 6 seconds produces a beam power of 0.5 MW. The LAGUNA-LBNO study [LAGUNA (2012)] addresses a possible upgrade of the beam power up to 0.75 MW, mainly through an increase of the beam intensity, and going to the ultimate repetition rate of one Super Proton Synchrotron (SPS) pulse every 3.6 s. As part of the LAGUNA-LBNO project, conducted within the European EU FP7 framework, CERN is in addition presently studying the feasibility of a 30–50 GeV high-power high-energy proton synchrotron, the High-Power Proton Synchrotron (HP-PS), with the LP-SPL, a low power version of SPL, as injector [Gerigk (2007)], aiming at a beam power of 2 MW. LP-SPL provides $H^-$ ions at an energy of 4 GeV and a repetition rate of 2 Hz with about $1.1 \times 10^{14}$ protons per pulse and a duration of 900 $\mu$s. Designing the HP-PS will need to address transverse and longitudinal beam dynamics, as well as requirements in terms of RF systems and the feasibility of magnets with a 1.7 T peak field and a 4 T/s ramp rate [LAGUNA (2012)].

Fermi National Laboratory (FNAL) has a program of upgrades of the existing 120 GeV Main Injector for the next decade of neutrino physics to achieve 700 MW on target ($4.9 \times 10^{13}$ ppp every 1.33 s). Protons will be injected into the Recycler (for the Tevatron operation the Recycler was used for anti-protons) by slip stacking and accumulated while the Main Injector is ramping. A new project aiming at 2.3 MW on target and a detector at 1300 km, the Long Baseline Neutrino Experiment (LBNE), needs a new proton beam line up to the production target. LBNE would start using the high intensity, 120 GeV proton beam from the Main Injector to produce 700 MW. To reach full power LBNE would need a new proton driver.

## 3 Project X

Project X is a high intensity proton facility project proposal that would support "Intensity Frontier" physics over the next several decades at Fermilab, see Figure 2 Project X would deliver, simultaneously, up to 6 MW of site-wide beam power to multiple experiments, at multiple energies, and with flexible beam formats. The Reference Design [Holmes *et al.* (2012)] is

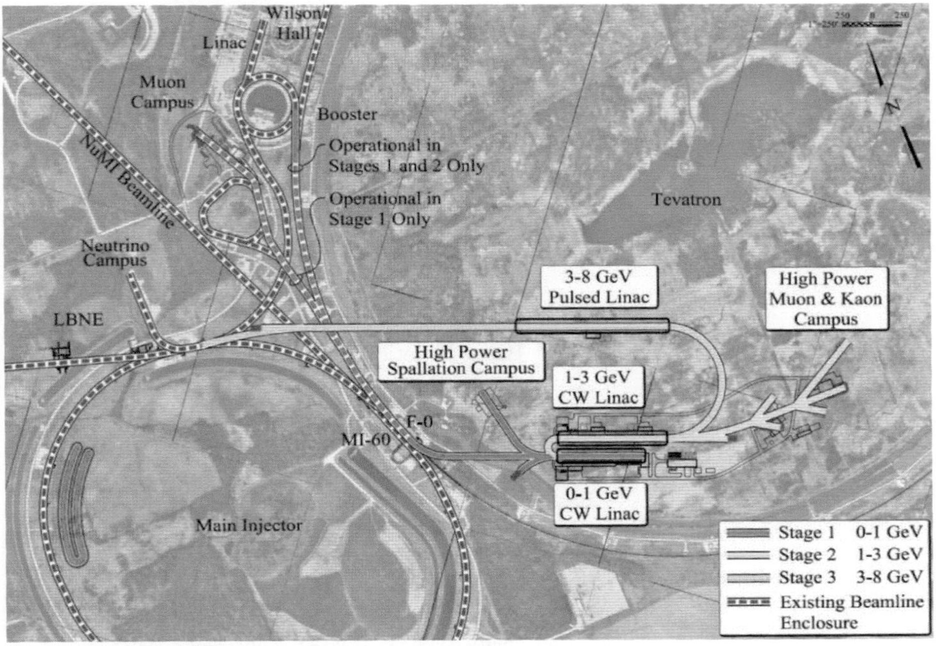


Fig. 2. Project X, a high-power proton facility, will support world-leading programs in long baseline neutrino physics and the physics of rare processes.

based on a continuous wave (CW) superconducting 3 GeV linac providing 1 MW and 3 MW of beam power at 1 and 3 GeV respectively. Apart from about 4% of the beam for the neutrino program, there would be provision for variable beam structures delivered to multiple users simultaneously, like the muon, kaon, nucleons, and nuclei programs.

Roughly 4% of the beam from the CW linac is delivered to the pulsed linac, accelerating a peak current of 1 mA. The pulsed linac accelerates the $H^-$ beam to the 8 GeV injection energy of the existing Recycler and further in the Main Injector. In addition to the support of the long baseline neutrino program, the pulsed linac opens a path toward a future Neutrino Factory or Muon Collider. Upgrades of the Recycler and Main Injector, both in the same tunnel, support 2 MW of beam power at 60 to 120 GeV. $H^-$ beams from the pulsed linac are accumulated in the Recycler, and then transferred and accelerated in the Main Injector. The capacity of the major facility components of Project X are shown in Table 1. To achieve 4 MW on target for the Neutrino Factory, the 3–8 GeV pulsed linac would need to be upgraded.

The primary technical risk element is the front end. The present baseline is to have a CW filament-discharge $H^-$ ion source, giving a current of up to

Table 1. The major components of the Project X facility.

| | | | |
|---|---|---|---|
| CW Linac | | | |
| | Particle Type | $H^-$ | |
| | Beam Kinetic Energy | 3 | GeV |
| | Average Beam Current | 1 | mA |
| | Linac Pulse Rate | CW | |
| | Beam Power to 1 GeV Program | 1 | MW |
| | Beam Power to 3 GeV Program | 2.87 | MW |
| Pulsed Linac | | | |
| | Particle Type | $H^-$ | |
| | Beam Kinetic Energy | 8 | GeV |
| | Average Beam Current | 1 | mA |
| | Linac Pulse Rate | 10 | Hz |
| | Linac Pulse Width | 4.4 | ms |
| | Cycles to Recycler/Main Injector | 6 | |
| | Total Beam Power at 8 GeV | 0.35 | MW |
| | Beam Power to 8 GeV Program[a] | 0.17 | MW |
| Main Injector | | | |
| | Particle Type | p | |
| | Beam Kinetic Energy | 60–120 | GeV |
| | Cycle Time | 0.6–1.2 | s |
| | Beam Power | 2.4 | MW |

[a]Beam power available for an 8 GeV experimental program with simultaneous operations of the Main Injector at 120 GeV. For Main Injector operations at 60 GeV the beam power available for an 8 GeV program is reduced to 0.

10 mA, and an RFQ system followed by a wide-band chopper with $1 \times 10^{-4}$ extinction rate. The chopper would prepare the beam to be accelerated by the CW linac to support the needs of several programs at 1 and 3 GeV. The individual bunch intensity is $14 \times 10^7$ with a 50 ps (FW) bunch length. This system is capable of being cascaded; in particular a second separator system at 3 GeV allows the support of at least three experimental programs at 3 GeV simultaneously with operations of the 1 GeV programs. About 4% of the beam is further accelerated to 8 GeV and 5 ms long pulses will be injected into the Recycler at 10 Hz. The linac performance demands, reaching 1 mA, are significantly smaller than what is accelerated today within the SNS linac [Zhang (2010)]. Nonetheless, appropriate attention needs to be paid to emittance dilution, halo formation, and beam losses. R&D effort is also invested in cavities for different relativistic beta (half wave resonators and single spoke cavities), cryomodules, RF sources for getting the long pulses with CW linac operation and $H^-$ injection into the Recycler. The main R&D elements are coupled to the "Project X Injector Experiment" (PXIE),

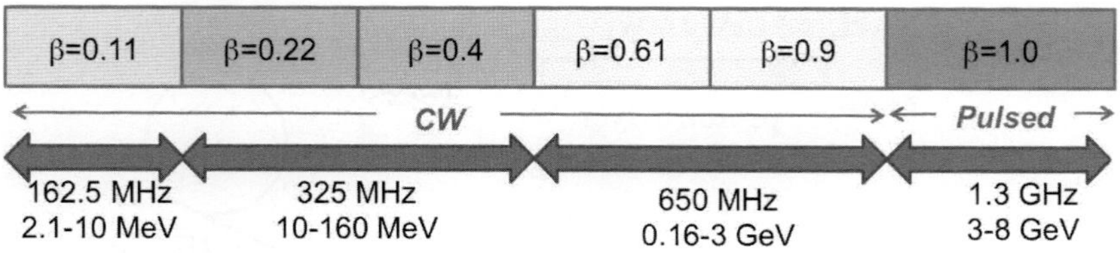


Fig. 3. Technology map for the Project X linacs.

integrating systems test for Project X front end components to validate the concept and technologies.

## 4 EUROnu

The European FP7 funded 4-year project EUROnu [Edgecock *et al.* (2013)] studied three different neutrino facilities for physics after 2020. The Super Beam (SB) and the Beta Beam (BB) are designed to send neutrino beams to a future Water Cherenkov detector located in the Fréjus tunnel, 130 km from CERN. Due to the short baseline (distance to the detector) these two facilities have good sensitivity to the CP violation process. The SB can also, with great benefit for the physics reach, deliver a beam to a longer baseline (use of the second oscillation maximum), where the CP reach is good but also some access to the mass hierarchy is possible. The third facility studied in the EUROnu project, the Neutrino Factory (NF), would use a Magnetized Iron Neutrino Detector (MIND) located at more than 2000 km from the NF. This facility would have the best physics reach, able to determine both the neutrino mass hierarchy (the spectral order of neutrinos) and to study CP violation. In addition it would have the best potential to detect evidence of new physics. The European contribution to the NF was done within EUROnu, as part of the International Design Study for a Neutrino Factory [Abrams (2012)].

The EUROnu SB creates neutrinos by impinging a high power proton beam onto a target and focusing the pions produced towards a far detector using a magnetic horn. The neutrinos come from the decay of pions in a decay tunnel following the target, thus producing a beam in the direction of the tunnel (see Figure 4). The SB uses the High Power Superconducting Proton Linac (HP-SPL) [Brunner *et al.* (2009)] as the proton driver, producing a 4 MW beam at 5 GeV at 50 Hz. The detector (MEMPHYS) would be built in two new caverns in the Fréjus tunnel. The design study [Baussan *et al.* (2012a)] foresees a proton energy of 4.5 GeV and a beam power of 4 MW at 50 Hz repetition frequency with a pulse duration of about 400 $\mu$s. A successive accumulator ring divides the protons into pulses of $\simeq 1$ $\mu$s with a

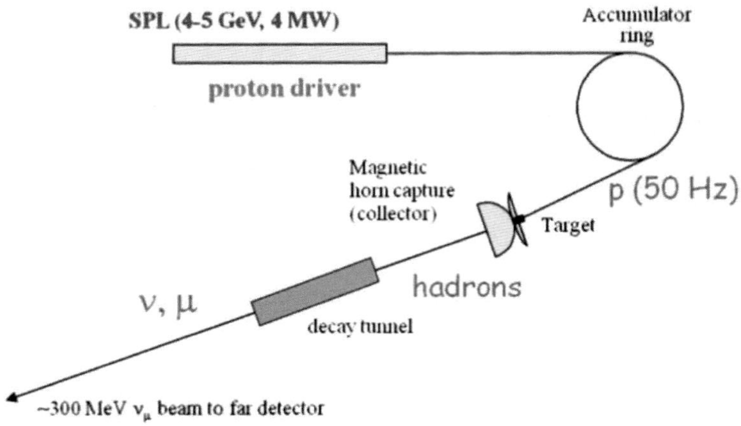


Fig. 4. The layout of the Super Beam.

sub-structure of either 6, 3 or 1 bunches per pulse in order to comply with the pulse duration of the horn current. A charge exchange injection accumulator for the neutrino factory, described in [Garoby *et al.* (2012)], can be used for the SB.

Given the difficulty in producing a single target and horn capable to work in a 4 MW beam, the option taken for the EUROnu SB is to use four of each instead. The beam from the accumulator will then be steered on to each target in turn, so that they all run at 12.5 Hz rather than 50 Hz and receive 1 MW each. For the targets and the horns, this results in a smaller extrapolation from technology already in use. To achieve this, a system of two kicker and four bending magnets has been designed to steer the beam on to each target in turn. The four targets and horns can be seen in Figure 5.

To minimize the production of thermal neutrons and, hence, reduce the heat load and radiation damage to the surrounding horns, the baseline design for the target is a pebble bed, consisting of 3 mm diameter spheres of titanium in a canister 200 mm long. These are cooled by flowing helium gas through vents in the canister at a pressure of 10 bar. This should be sufficient to cool targets up to a few MW. Tests of the cooling system and of the effect of the beam impact will show the feasibility of the target systems. The focusing horn design has been optimized for the CERN to Fréjus neutrino beam. It will employ a single horn around the target, and will not have a reflector. As for the targets, four horns will be used, pulsed at 350 kA. The heating from the powering and the beam loss will result in a maximum of 12 kW on the surface around the target. Water cooling of the outer surface

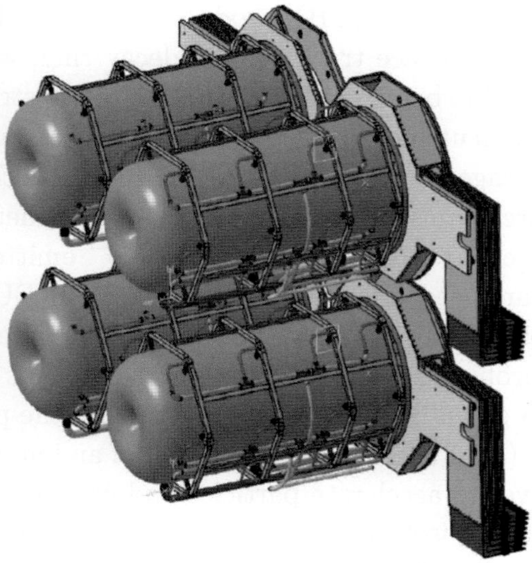

Fig. 5. Conceptual engineering design of the four target and horn system for the Super Beam.

of the horn will remove the heat. The thermal stresses in the horn material resulting from the heating amount to a maximum of 18 MPa. The lifetime of a horn will depend on fatigue in the material and from radiation damage and will need to be verified by prototyping and testing.

The targets and horns would be mounted in a target station which allows the change, storage, and maintenance of targets and horns, in case of failure. To enable this, the target station will have a number of separate sections and the shielding requirements for each is determined by activation studies. The final design also takes into account experience gained with the T2K (see Section 2) target station. It incorporates remote handling facilities, a hot cell for maintenance, and a storage area for old targets and horns, called the morgue. It will allow access to the critical components of the system, for example the power supplies for the horns, and will allow the safe removal of activated components for disposal. The section of the target station that contains the targets and horns is shown in Figure 5.

An opportunity that is investigated is to adapt the SB to work with an $H^-$ beam in parallel (interleaved) with the $H^+$ beam in the linac that is planned for neutron spallation at the European Spallation Source in Lund, Sweden, see [ESS (2012)]. The energy of the beam for spallation is 2.5 GeV, so the challenge is the higher intensities needed. In particular the accumulator will have to deal with space charge and stability issues to reach 4 MW on target. This study, EUROSB, is described in [Baussan *et al.* (2012b)].

The Beta Beam facility (BB) produces neutrino beams by storing radioactive isotopes in a race track shaped decay ring, where the decaying isotopes form a neutrino beam exiting at the end of the straight sections.The production of (anti)neutrinos from the beta decay of radioactive isotopes circulating in a storage ring was proposed in 2002 [Zucchelli (2002)]. Beta Beams produce pure beams of electron neutrinos or antineutrinos, depending on whether the accelerated isotope is a $\beta^+$ or a $\beta^-$ emitter. Both neutrinos and antineutrinos are needed for the physics goals. The EUROnu Beta Beam facility is based on CERN's infrastructure and would reuse some existing accelerators and profit from the injector upgrades. This would reduce the cost compared to a green field site, though constrain the performance of the Beta Beam facility (see Figure 6). It will consist of an ion production system, using a proton driver to accelerate particles and create the required isotope species in a target. The isotopes will then be collected in a 60 GHz electron cyclotron resonance (ECR) source to create a 50 $\mu$s long beam pulse. Ions will be accelerated using a dedicated ion linac to 100 MeV, a rapid cycling synchrotron, the existing Proton Synchrotron, and the Super Proton Synchrotron (SPS), before injection into a race track shaped decay ring [Wildner *et al.* (2012)].

One of the most important issues for Beta Beam is the production, acceleration, and storage of a sufficient flux of isotopes to meet the physics goals. An isotope pair producing antineutrinos or neutrinos with similar reaction

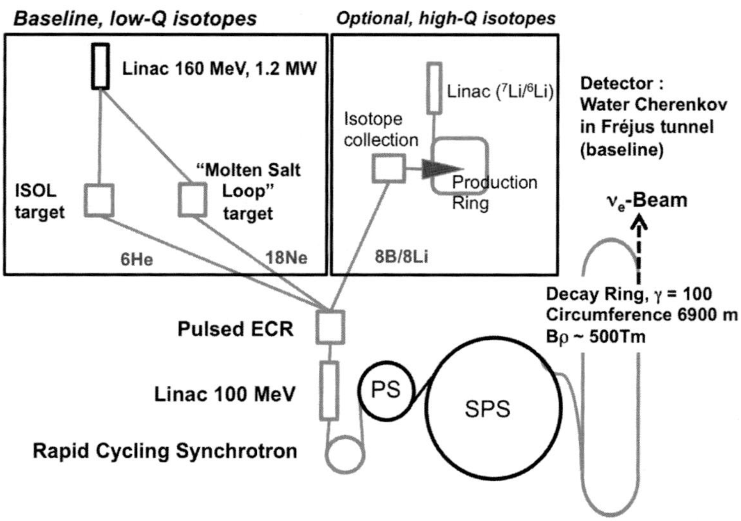


Fig. 6. The Beta Beam layout at CERN. The two options, $^6$He or $^{18}$Ne (left box) and $^8$Li and $^8$B (right box) are shown.

Q-values and lifetimes are needed. The isotope pair that was first studied for neutrino production, in the EURISOL FP6 Design Study [Blumenfeld *et al.* (2007)], is $^6$He and $^{18}$Ne, accelerated to $\gamma = 100$ in the SPS and stored in a decay ring [Benedikt (2011)]. Physics studies have indicated that the required fluxes of these ions are $6 \times 10^{13}$ and $1 \times 10^{13}$ ions/s, respectively. At the end of EURISOL, it looked possible to produce the required flux of $^6$He, but that of $^{18}$Ne looked a factor of 20 too small. This has subsequently been addressed in two ways. The first was to consider a production ring (12 m circumference) with an internal gas jet target [Rubbia *et al.* (2012)] to make an alternative ion pair, $^8$Li and $^8$B. As the neutrinos from the decay of these ions have about five times larger energy than those for $6 \times 10^{13}$ and $1 \times 10^{13}$, the required baseline has to be five times longer and the flux of ions required for the same physics is $1 \times 10^{14}$ ions/s. In this production ring, a 25 MeV beam of $^7$Li and $^6$Li is injected over a gas jet target of d or $^3$He, respectively. To determine the production rate, the double differential cross-sections for both processes, $^7$Li(d,p)$^8$Li and $^6$Li($^3$He,n)$^8$B, have been measured [Vardaci *et al.* (2012)]. Based on these measurements, the production ring has been designed and a prototype device for collection of the ions has been built and tested. However, studies have shown that the thickness of the gas jet target needed to produce the required flux of ions, $1 \times 10^{19}$ atoms/cm$^2$, is four orders of magnitude bigger than any in current use and will create significant problems for the ring vacuum. Alternative production possibilities have been looked at, for example liquid lithium films, but it remains difficult to meet the isotope production goals using $^8$Li and $^8$B.

These results triggered research on a novel $^{18}$Ne production method, using a molten salt loop (NaF) by the reaction $^{19}$F(p,2n)$^{18}$Ne (see Figure 7). Modeling suggests that this could achieve the required production rate with a 160 MeV proton linear accelerator at a current of 7 mA. This would be achievable at CERN with an upgrade of Linac 4 [Vretenar (2012)]. An experiment at ISOLDE at CERN in June 2012 demonstrated that the required flux could be achieved using the molten salt loop technology [Stora *et al.* (2012)]. As a result of the work done so far, the $^6$He and $^{18}$Ne ion pair is the recommended baseline for the Beta Beam.

To accept the intense continuous flux of $^6$He or $^{18}$Ne produced, ionize the gas, and bunch the ions with high efficiency, it is planned to use a 60 GHz pulsed electron cyclotron resonance (ECR) ion source. A prototype device called SEISM (sixty gigahertz ECR ion source using mega-watt magnet structures) has been designed and the magnetic confinement structure successfully built and tested, see Figure 8 [Latrasse (2010)].

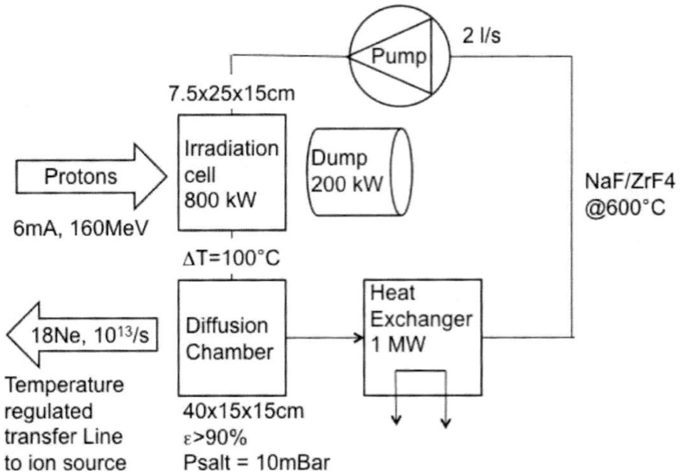


Fig. 7. The molten salt loop $^{18}$Ne isotope production at ISOLDE.

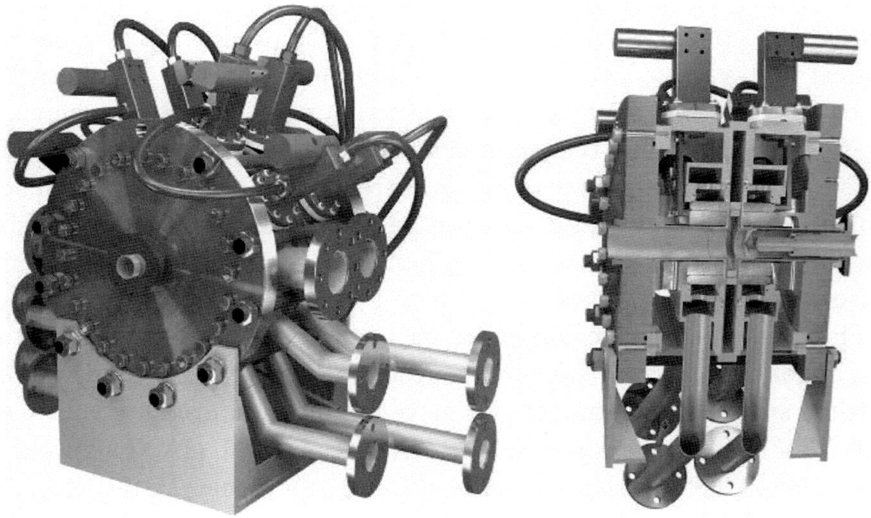

Fig. 8. SEISM ECR source prototype.

As shown in Figure 6, after bunching, the ions will be accelerated to 100 MeV/u using a purpose-built linear accelerator about 110 m long. This will be followed by a rapid cycling synchrotron, 251 m in circumference, that will accelerate the ions to a maximum magnetic rigidity of 8.2 Tm, corresponding to 2.0 GeV protons, 0.37 GeV/u for $^{6}He^{2+}$ and 0.85 GeV/u for $^{18}Ne^{10+}$. Final acceleration of the ion beams will take place in the existing PS and SPS. Simulations of these show that, although not optimal, they can deliver the required performance. Preliminary activation studies have also been done and these show that the effect of the Beta Beams compared to high intensity proton running varies with the component or material being activated, but the rate is never significantly higher and this should not prevent operation. Collective effects studies are only preliminary so far, and show that there are major difficulties; overcoming these would need reconsideration of cycling of the machines, bunch structures and impedances, including the constrains imposed by LHC operation.

The Beta Beam decay ring would have a race track shape, with a total circumference the same size as the SPS, 6.9 km, and a production straight section which is 37% of this size to maximize the neutrino flux by optimising arc magnets [Chancé (2012)]. The Decay Ring is topped up with 20 bunches from the SPS at each cycle. The preferred method of doing this is to use a dual frequency RF and inject new beam at a slightly different energy from that already in the ring. The voltage and phase of the two cavity families will then be varied to perform the merging of the 20 new bunches with the 20 already in the ring. This technique has been simulated and in part successfully tested [Heinrich *et al.* (2011)]. The RF system for the merging in the Decay Ring is challenging; a technical solution is proposed, see in [Wildner *et al.* (2013)]. As the ring will use superconducting magnets, one way to avoid energy deposition in the coil mid-planes would be to use coil-free mid-plane magnets [Wildner (2007)]. Calculations of vacuum conditions need to be performed. Collective effects would ultimately limit the intensity in the ring. Although the intensity limit for transverse Mode Coupling Instabilities is above the required intensity for $^6$He, this is not true for $^{18}$Ne, where it is only about 20% of the required intensity and needs further considerations. The Decay Ring design would still have some freedom to adapt, since this is a new machine. The collimation system is particularly demanding during merging, and has not been designed yet.

As a far detector, the baseline isotopes, $^6$He and $^{18}$Ne could use the MEMPHYS Water Cherenkov detector [Agostino *et al.* (2012)] in the Fréjus tunnel, at a distance of 130 km. Because of the higher energy of the neutrinos, the $^8$Li and $^8$B option would need a detector at some 700 km and may profit from a different detector technology, such as liquid argon [Orme (2010)].

In a Neutrino Factory (NF) the neutrinos are produced from the decay of muons in a storage ring with long straight sections [Abrams (2012)]. The muons are produced by impinging a 4 MW proton beam onto a target and focusing the pions produced into a decay channel using a 20 T superconducting solenoid. To be able to distinguish signal from background, the far detector must be able to separate $\mu^+$ from $\mu^-$ with high efficiency. Therefore,

Table 2. Beta Beam Decay Ring beam parameters for the different stored ions.

| Parameter | Unit | $^{6}He^{2+}$ | $^{18}Ne^{10+}$ | $^{8}Li^{3+}$ | $^{8}He^{5+}$ |
|---|---|---|---|---|---|
| Atomic mass $A_{eff}$ | u | 6.019 | 18.006 | 8.022 | 8.025 |
| $E_{rest}$/ion | GeV | 5.606 | 16.772 | 7.471 | 7.473 |
| $\gamma$ | — | 100 | 100 | 100 | 100 |
| half-life at rest | s | 0.807 | 0.167 | 0.840 | 0.770 |
| $B\rho$ | Tm | 934.87 | 559.27 | 830.64 | 498.50 |
| Nominal annual $\nu_e$ flux | $10^{18}$ | 2.9 | 1.1 | 14.5 | 5.5 |
| Stored beam: | | | | | |
| Nb of stored ions | $10^{13}$ | 9.346 | 7.178 | 48.18 | 16.70 |
| Full energy of beam | MJ | 8.39 | 19.28 | 57.67 | 19.98 |
| Average beam current | A | 1.30 | 4.99 | 10.04 | 5.80 |
| Peak beam current | A | 227.9 | 875.0 | 1762 | 1017 |
| Longitudinal emittance (full) | eVs | 14.4 | 43.3 | 19.3 | 19.3 |
| Bunch length | m | 1.97 | 1.97 | 1.97 | 1.97 |
| Momentum spread (full) | $10^{-3}$ | 2.5 | 2.5 | 2.5 | 2.5 |
| Rel. energy difference injected beam | $10^{-3}$ | 5 | 5 | 5 | 5 |

the baseline detector is a magnetized iron neutrino detector (MIND). The baseline of the NF has changed after the measurement of the $\theta_{13}$ oscillation parameter. The muon energy has been reduced from 25 GeV to 10 GeV and only one decay ring will be used, there were two before; this is a significant simplification. The envisaged neutrino baseline is around 2000 km. The Neutrino Factory layout is shown in Figure 9.

There are several options for a Neutrino Factory proton driver, depending on where the facility would be built. One is a superconducting linear accelerator like the CERN HP-SPL [Brunner *et al.* (2009)], followed by an accumulator and a compressor ring. This is described in Chapter 13. The other option employs a rapidly cycling synchrotron, working at 50 Hz, to accelerate the beam to 10 GeV. This would use a normally conducting linear injector to accelerate the beam to 180 MeV. Project X also includes options for high power proton drivers, as described in Section 3.

The baseline pion production target is a continuous liquid mercury jet. This would be fired across the proton beam at a small angle so that the beam and target overlap for two interaction lengths. The pions produced would be focused by a combined normal and superconducting magnet of 20 T around the target, see Figure 10. Both the beam and target would also be at a small angle to the axis of the solenoidal field, so that the mercury collects in a pool. The magnetic field would be tapered down adiabatically to 1.5 T at the entrance of the pion decay channel, using a succession of

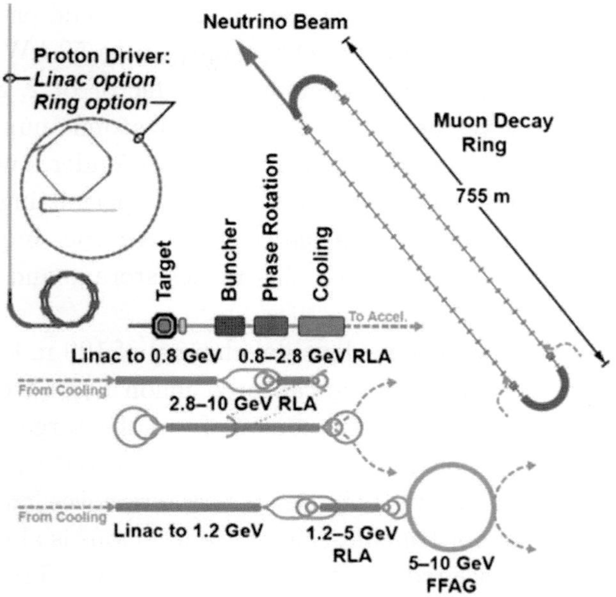


Fig. 9. Conceptual engineering design of the four target and horn system for the Super Beam.

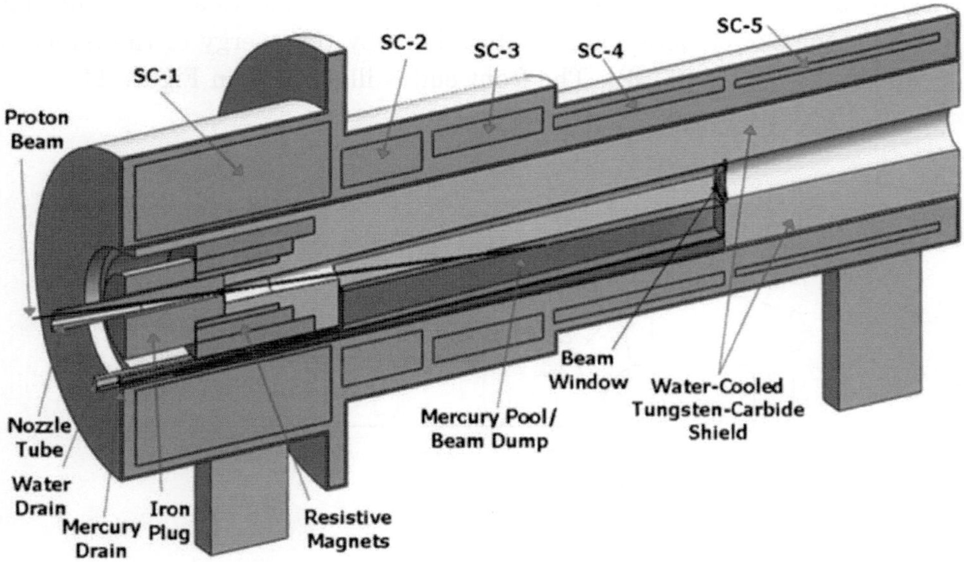


Fig. 10. The Neutrino Factory target.

superconducting coils. However, secondary neutrons would produce too high energy deposition in the superconducting coils, up to 50 kW, and efficient shielding would increase the coil aperture by a factor of 2. Sufficient production rates can however be achieved with lower atomic number elements, giving significantly fewer neutrons. Such targets are under study. Gallium is an interesting candidate with a low melting point, permitting it to be used in liquid form in a similar way to a mercury jet. Another important fact is that it is solid at room temperature; this makes storage and disposal after activation significantly easier.

The pion decay channel is a solenoidal channel of 100 m length, employing 1.5 T magnets to maximize the captured muon flux. A chicane is separating muons from other particles coming from the target, which would increase activation levels of the transport channel by 100 times the acceptable levels for hands-on maintenance. An absorber removes lower momentum particles. The efficiency for transmission of useful muons is about 90%, while the unwanted particles are reduced to a manageable level. The chicane is followed by a section, 33 m long, which bunch the beam using RF structures. This in turn is followed by a phase rotation section 42 m long, which utilizes the correlation between position in the bunch train and energy that has built up by this stage. It uses RF cavities changing phase along the bunch train passage to slow down the faster going muon bunches at the front and speed up the slower ones coming later. Thereby the energy of the bunches is equalized [Neuffer (2013)]. The front end is illustrated in Figure 11.

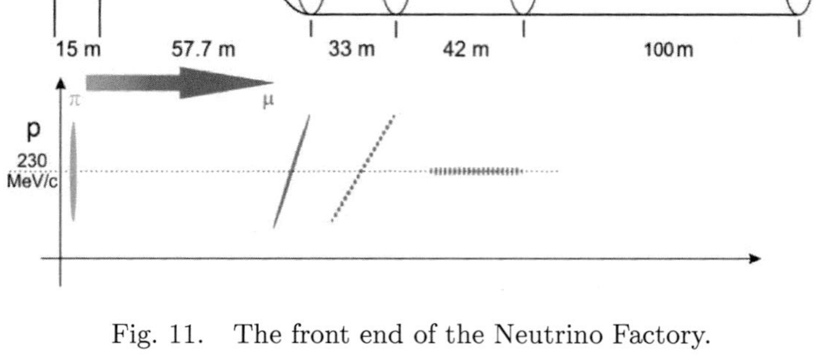


Fig. 11. The front end of the Neutrino Factory.

An ionization cooling channel follows the phase rotation using an absorber to reduce both the longitudinal and transverse components of the muon momentum. The lost longitudinal momentum is then restored using RF cavities, giving a net reduction in transverse momentum (transverse cooling). However, while cooling through energy loss, the absorber also heats up through multiple scattering and the best balance between the two is achieved by using a low atomic number material, such as liquid hydrogen or lithium hydride. In addition, the cooling efficiency is significantly increased if the absorber is in a region in which the beam is highly convergent or divergent, thus requiring a superconducting field around the absorber region. Superconducting magnets are also required around the RF cavities to aid transport. Because of the complexity, an engineering demonstration of the cooling technique is being constructed [Blondel (2012)], and will give a first demonstration of ionization cooling during 2013. In addition, as the RF cavities of the baseline cooling cell will be in a large magnetic field, measurements of the effect on the accelerating gradient are being made by the MuCool project [Bross *et al.* (2010)]. To minimize potential problems, alternative cooling lattices that reduce the magnetic field at the cavities while maintaining the same performance are being studied. An engineering view of the cooling channel can be seen in Figure 12.

The final muon energy is now 10 GeV, following the results of measurements on the oscillation parameter $\theta_{13}$ [Tanaka (2012)]. Two options exist for the muon acceleration system. The first uses a linac to 0.8 GeV, followed by two recirculating linear accelerators (RLAs), one to 2.8 GeV and the second to 10 GeV. The second option uses a linac to 1.2 GeV, an RLA to 5 GeV,

Fig. 12. The Muon Ionization Cooling Experiment (MICE): Two cooling cells can be seen in the middle of the channel. The muons are traced individually through the channel to measure the total cooling effects.

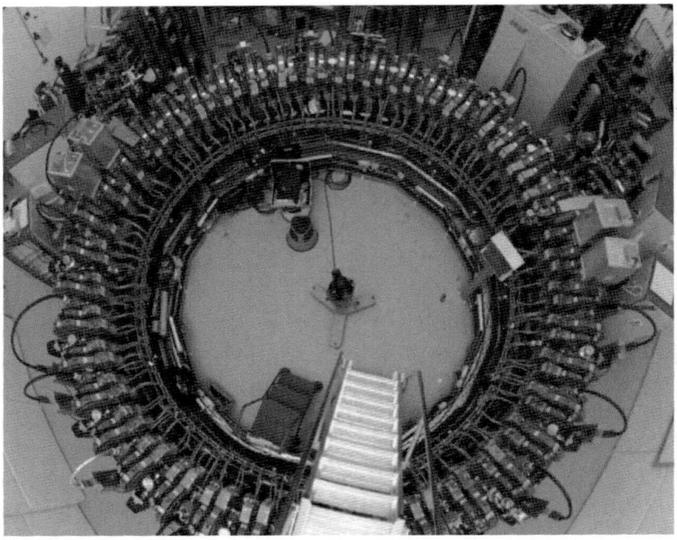

Fig. 13. The EMMA proof-of-principle accelerator at the Daresbury Laboratory.

and a nonscaling fixed field alternating gradient (ns-FFAG) accelerator to 10 GeV. Both options are under study to determine which would be best based on performance and cost (Figure 9).

An ns-FFAG is proposed as its properties of fixed magnetic fields and pseudo-isochronous operation mean that muon acceleration will be very fast, plus it has the large acceptance required for the high emittance muon beam, even after cooling. However, it is an entirely novel type of accelerator, so a proof-of-principle machine called EMMA [Barlow *et al.* (2010)] has been constructed at the STFC Daresbury Laboratory, see Figure 13. Many of the novel features of the muon accelerator, in particular serpentine acceleration and multiple resonance crossings work [Machida *et al.* (2010)] have been demonstrated in EMMA. Some parameters of the EMMA prototype and the full scale machine proposed for the NF are shown in Table 3.

The Neutrino Factory Decay Ring long straight sections are where the useful decays take place. It is planned to produce and accelerate bunches of both $\mu^+$ and $\mu^-$ at the same time. These will arrive in three bunches each, of 250 ns length, separated by 120 $\mu$s. The Decay Ring will have a total circumference of 1286 m, of which 470 m will form a production straight section for neutrinos in the direction of the far detector for both muon charges. The ring will be tilted at an angle of about 10 degrees for the 2000 km long baseline. An outline injection system design that will inject all of the bunches

Table 3. A selection of parameters of the ns-FFAG for the Neutrino Factory compared to the EMMA prototype.

| Parameter | Unit | Muon FFAG | EMMA | Ratio |
|---|---|---|---|---|
| Momentum | MeV/c | 12.6–25 × $10^3$ | 10–20 | 1:0.001 |
| RF voltage | MV | 1214 | 2.28 | 1:0.002 |
| Number of cells | — | 64 | 42 | 1:0.66 |
| Circumference | m | 667 | 16.6 | 1:0.025 |
| QD/QF length | m | 2.251/1.087 | 0.0777/0.0588 | 1:0.035/0.054 |
| Straight section | m | 5 | 0.2 | 1:0.04 |
| Aperture | mm | 300 | 30 | 1:0.1 |

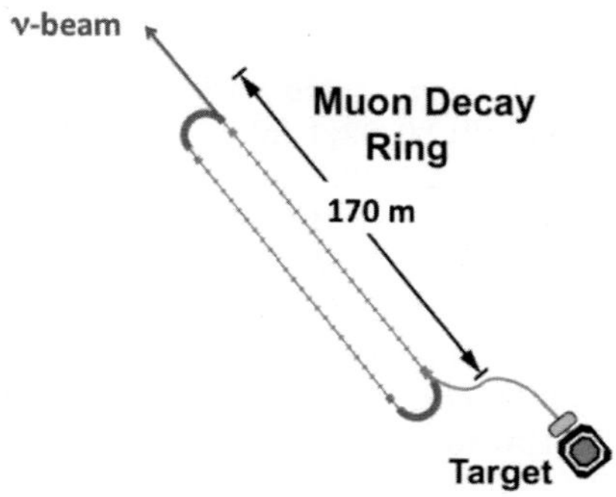


Fig. 14. $\nu$STORM.

into the ring has been made. A minimum separation of at least 100 ns is required between bunches to distinguish which bunch the detected neutrinos come from. With the expected 2% energy spread of the muon beam, this will exist for four muon lifetimes, allowing the vast majority of muons to decay.

Amongst the recommendation in the EUROnu project conclusion was to start building a small test facility. $\nu$STORM [Kyberd *et al.* (2012)] is a facility to test several aspects of a Neutrino Factory, and to give precise information on cross-sections for neutrinos in the detectors, see Figure 14. The principle is to build a small muon decay ring with long straight sections of about 150 m. Pions from the target are guided and injected into the storage ring, where they decay essentially on the first straight section. The produced muons will circulate for some 100 turns before they decay. Challenges are to build a short pion transport channel from the target, making possible also to build a proton absorber in the vicinity, a stochastic injection system where large aperture magnets are needed and a lattice that can house the two species, pions and muons, at the same time on the first straight section. Pions extracted after the first straight section can be used for muon cooling experiments. Proposals have been delivered at CERN and Fermilab for a collaborative approach to this very important start of a "next generation" neutrino program. $\nu$STORM can also give important contribution related to the existence of sterile neutrinos.

## References

R. J. Abrams *et al.* (2012). Interim Design Report No. CERN-ATS-2011-216; arXiv:1112.2853.
L. Agostino *et al.* (2012). arXiv:1206.6665.
R. Barlow *et al.* (2010). *Nucl. Instrum. Methods Phys. Res., Sect. A* **624**, 1.
E. Baussan *et al.* (2012a). arXiv:1212.0732v1 [physics.acc-ph].
E. Baussan *et al.* (2012b). arXiv:1212.5048.
M. Benedikt *et al.* (2011). *Eur. Phys. J. A* **47**, 24.
A. Blondel (2012). *3rd International Particle Accelerator Conference*, New Orleans, 2012 (IEEE, Piscataway, NJ), MOEPPB002.
Y. Blumenfeld *et al.* (2007). *AIP Conf. Proc.* **891**, 147.
A. Bross *et al.* (2010). In *1st International Particle Accelerator Conference*, Kyoto, Japan, 2010 (ICR, Kyoto), THPEA046.
O. Brunner *et al.* (2009). IDS-NF-20. *Phys. Rev. ST Accel. Beams* **12**, 070402.
A. Chancé (2012). `https://espace.cern.ch/betanu/MeetingsWP4/WP4Meeting9/2011_10_20_Chance_DecayRing.pdf`.
T. R. Edgecock *et al.* (2013). *Phys. Rev. ST Accel. Beams* **16**, 021002.
ESS (2012). `europeanspallationsource.se/`.
R. Garoby *et al.* (2012). CERN-NEUTRINO-FACTORY-NOTE-157.
F. Gerigk (2007). CERN, Geneva, Switzerland, 01–05 Oct 2007, pp. 143–146.
D. C. Heinrich *et al.* (2011). *AIP Conf. Proc.* **1382**, 211.
S. D. Holmes *et al.* (2012). IPAC2012, p. 3945.
JPARC (2012). `http://indico.cern.ch/getFile.py/access?contribId=2&sessionId=5&resId=1&materialId=slides&confId=124959`.
B. Kyberd *et al.* (2012). arXiv:1206.0294.
LAGUNA (2012). FP7 Research Infrastructure Design Studies, LAGUNA (Grant Agreement No. 212343, FP7-INFRA-2007-1).
L. Latrasse *et al.* (2012). *Rev. Sci. Instrum.* **81**, 02A324.
S. Machida *et al.* (2010). *Nat. Phys.* **8**, 243.
D. Neuffer. (2013). `https://hepconf.physics.ucla.edu/higgs2013/talks/neuffer.pdf`.
C. Orme (2010). arXiv:1004.0939.
C. Rubbia *et al.* (2012). *Nucl. Instrum. Methods Phys. Res., Sect. A* **568**, 475.
T. Stora *et al.* (2012). Low-Q ion production, $^{6}$He and $^{18}$Ne measurements and perspectives, in *4th EUROnu Annual Meeting*, Paris, France, 2012.
H. A. Tanaka (2012). *Pramana* **79**, 941–952.
E. Vardaci *et al.* (2012). Cross-section measurements of $^{8}$Li and $^{8}$B, in *4th EUROnu Annual Meeting*, Paris, France, 2012.

M. Vretenar (2012). Status of Linac4 construction at CERN, in *Proceedings of Linear Accelerator Conference LINAC2010*, KEK, Tsukuba, Japan, 2010.

E. Wildner *et al.* (2012). *Conf. Proc. C* **1205201**, 3939–3941.

E. Wildner *et al.* (2013). *Phys. Rev. ST Accel. Beams* **17**, 071002.

Y. Zhang (2010). In *1st International Particle Accelerator Conference*, Kyoto, Japan, 2010 (ICR, Kyoto), MOZMH01.

P. Zucchelli (2002). *Phys. Lett. B* **532**, 166.

# Part 3

# Particle Accelerators for the XXI Century

## Chapter 38

# Plasma acceleration

Ralph Assmann (CERN)

A dramatic improvement in energy gain per unit cost is mandatory for the future of very high-energy accelerators. Plasma acceleration opens a possible path to more compact and ultimately more cost-effective accelerators for the 21st century.

## 1 Introduction

For more than three decades the collision energy in particle colliders has increased exponentially in time. This remarkable success of accelerator science is described by the so-called Livingston curve. A schematic version of the Livingston curve is shown in Fig. 1, including conventional and novel technologies of various kinds. It is seen that up to around 1985 the maximum beam energy in accelerators followed the Livingston curve. This rapid increase in energy reach provided a thriving particle physics community with multiple experimental opportunities. In parallel accelerators for completely new fields of science emerged at lower energy, developing further the technologies that had been established for high-energy machines. Prime examples for this spin-off are the accelerators for photon science, including superconducting free-electron lasers.

The success story of high-energy accelerators has slowed down in the recent decades. Though future machines significantly extend the available energy reach in particle physics, they arrive late, such that they fall below the Livingston curve. In addition, fewer high-energy projects are built in a given time period. M. Tigner [2001] discussed the situation in 2001, opening the question whether accelerator-based particle physics does have a future.

The signs of saturation in beam energy indicate that practical and technical limits are indeed being approached for high-energy accelerators. The most important practical limitation that high-energy accelerators face today is the cost per GeV of beam energy. The goal of accelerator research is a

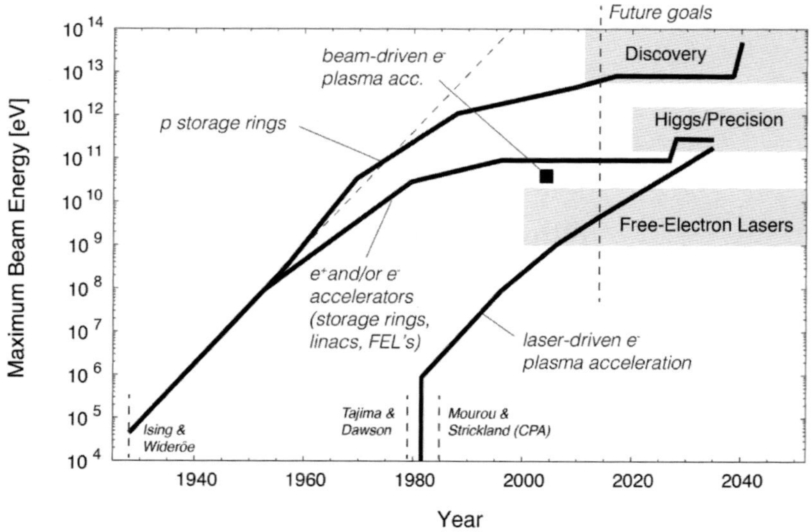


Fig. 1. Livingston plot for conventional and advanced technology, showing the maximum reach in beam energy for different technologies versus year. To increase clarity, the various results on beam-driven electron acceleration in plasmas are represented by the result with highest energy gain. At the bottom a few major breakthrough inventions are indicated. Grey bands visualize accelerator applications of interest. The curves represent achievements up to 2014. Beyond 2014 they indicate the goals for the various technologies.

low-cost technology and an accelerator design that provides many more GeV per cost unit. One possible solution is a technology that provides a higher accelerating gradient and therefore reduces the size of accelerators. The concept for such a novel technology emerged in a canonical paper by Tajima and Dawson [1979], proposing a plasma accelerator based on laser-driven wakefields. They showed that the accelerating gradients in their scheme would be three to four orders of magnitude higher than in conventional accelerators. However, in 1979 the required lasers did not yet exist. This changed when Mourou and Strickland invented the compression of amplified chirped optical pulses (CPA) [1985]. The required lasers became available and a new research field on plasma accelerators emerged.

Experiments were performed in various competing laboratories. The achieved beam energy in laser-driven plasma accelerators has experienced a steep increase over the last three decades, forming the Livingston curve of advanced accelerators as shown in Fig. 1. Multi-GeV electron beams have been produced inside a few cm of plasma with accelerating gradients of 100–200 GV/m. The novel technology of plasma accelerators has now entered the region of interest for modern free-electron lasers. The prospect of an ultra-compact, table-top FEL has become much closer to the real axis and worldwide R&D is presently strongly expanding towards achieving this goal.

Plasma accelerators can also be driven by short pulses of particle beams. Experiments on this alternative path of work were first proposed and approved in 1997 for the SLAC linac beam [Assmann *et al.* 1997]. Follow-up experiments achieved electron acceleration from 42 GeV to 85 GeV from the wakefields in a 0.85 m long plasma channel [Blumenfeld *et al.* 2007]. This result indicates that plasma accelerators are adequate for producing also very high-energy electron beams. Beam-driven plasma accelerators have disadvantages (long and expensive accelerators as drivers) but also very important advantages (higher energy stored in driver pulse), when being compared to the laser-driven option. They open an alternative path for high-energy physics. Recently the AWAKE experiment at CERN has been approved for studying proton beams as drivers for plasma accelerators [AWAKE collaboration 2014].

It is noted that another thriving field of development is the generation of proton and ion beams from plasmas. Here, a high-power laser impacts on a solid target. A plasma is generated and strong longitudinal fields extract and accelerate protons or ions to many MeV of beam energy. Relevant literature can be found in [Maksimchuk 2000] and [Cowan 2004]. This topic is not included here.

## 2 Path to compact accelerators and high frequency RF

The synchronicity between the accelerating electromagnetic wave and the accelerator beam (most often at light velocity) is achieved through the use of oscillating fields (RF fields). Specific frequency bands are defined and are used to label the accelerator technologies, as listed in Table 1. It is noted that the longitudinal size of the accelerator scales with the inverse of the RF frequency. The path to compact accelerators is therefore also a path to high frequency acceleration. Accelerator technology based on metallic cavities has presently explored frequencies up to 30 GHz, the former frequency of the CLIC project ("Compact Linear Collider") at CERN. The 30 GHz technology had to be abandoned due to damage problems at the metallic walls, as shown in Fig. 2. Metallic cavities at 12 GHz (X band) for CLIC allow accelerating gradients above 100 MV/m without damage [Corsini 2012].

Plasma accelerators operate at much higher frequencies in the W band. Accelerating gradients are above 1 GV/m and have reached in experiments values of up to 200 GV/m. The size of a plasma cell is equal to half the wavelength of the accelerating voltage and therefore is quite small (around

Table 1. Summary of frequency bands that are used and often referred to in RF acceleration. Accelerating gradients are listed.

| Band | Frequency [GHz] | Gradient [MV/m] | Comments |
|---|---|---|---|
| L band | 1–2 | 24 | This band is used by superconducting RF technology. The cell dimensions are large (around 10 cm) and disturbing wakefields are weak. |
| S band | 2–4 | 21 | Technology of the SLAC linac, completed in 1966. This is still the technology behind many accelerators. |
| C band | 4–8 | 35 | New technology developed in Japan and used for the construction of the SACLA linac in Japan. |
| X band | 8–12 | 70–100 | Technology developed from the 1990's onwards for linear collider designs, like NLC and CLIC. The cell length is up to a factor 10 shorter than in L band. |
| Ku band | 12–18 | | — |
| K band | 18–27 | | — |
| Ka band | 27–40 | 70 | Investigated for a CLIC linear collider technology at 30 GHz. Abandoned after damage problems. |
| V band | 40–75 | | — |
| W band | 75–110 | $10^3$–$10^5$ | Advanced acceleration schemes with ultra high gradients and very short cell lengths (<1 mm). |

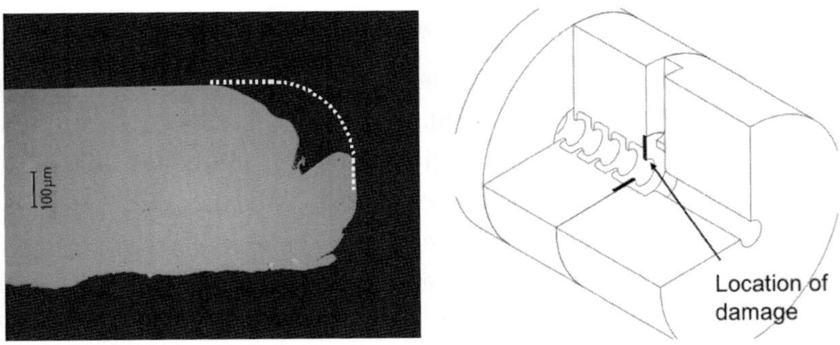


Fig. 2. Photograph of the damage to a single feed power coupler of a 30 GHz cavity, after being operated with 16 ns long RF pulses and a local accelerating gradient of 66 MV/m. Courtesy W. Wuensch (2002).

100 $\mu$m). The electron beam must be placed into a well-controlled position within this length. Longitudinal tolerances are evidently much tighter and more difficult to achieve than, for example, in L band technology where dimensions are 1000 times larger.

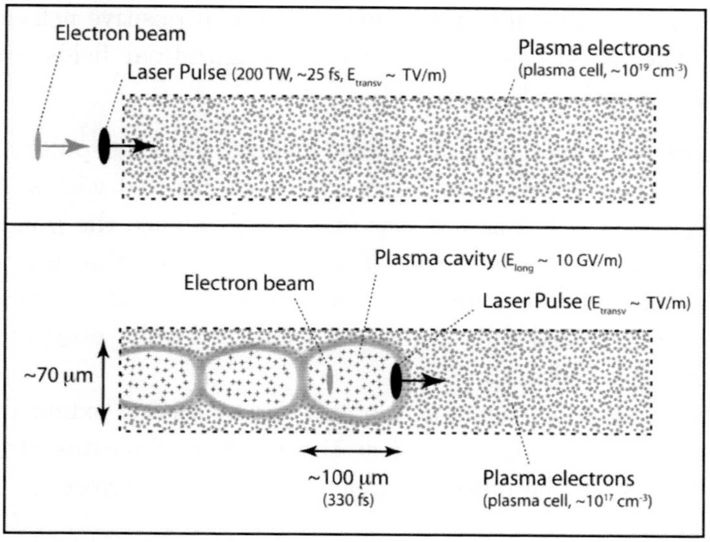


Fig. 3. Illustration of laser-induced plasma wakefield acceleration with injection of an external electron beam into the plasma cavity. The red dots represent mobile plasma electrons, the black crosses the stationary ions after expulsion of the electrons. Ions are not shown in the regions of unperturbed plasma electrons.

## 3 The physics of plasma acceleration for electrons

Tajima and Dawson described how plasmas can be used to transform transverse fields into longitudinal fields [1979]. Transverse fields in the plasma can be excited by short pulses of laser light (high transverse electrical fields, ponderomotive force) or by pulses of charged particles (space charge force). The principle is illustrated in Fig. 3. The plasma response to a short laser pulse is as follows:

- The laser pulse enters the plasma and expels the free electrons that are transversely accelerated (transverse driving force). The plasma ions move a negligible amount due to their higher mass.
- Along the path of the laser pulse a positively charged ion channel is formed. The ponderomotive force has pushed out plasma electrons transversely.
- Once the laser pulse has passed, the plasma electrons rush back in, attracted by the ion channel (transverse restoring force).
- Due to their speed the electrons over-shoot the center of the ion channel, rush back out and are attracted back by the ion channel. A space charge driven oscillation has formed.

- Alternating regions of negative net charge and positive net charge form behind the driver laser pulse. Strong longitudinal fields are induced ("plasma wakefields").

If a short electron bunch is injected behind the laser pulse at the correct distance, then this electron beam will be accelerated with an ultra-high gradient. For a particle beam driver one would replace the ponderomotive force by the repelling space charge of an electron beam that is injected into the plasma channel. For a proton beam driver, the plasma electrons are transversely accelerated due to the attraction from the positively charged proton beam ("sucked in").

The processes in a plasma accelerator are to a large extent determined by the plasma density $n_0$, defined as the number of plasma electrons per volume. This defines the plasma wavelength $\lambda_p$ and therefore the size of the plasma bubble:

$$\lambda_p \approx 1\,\text{mm} \cdot \sqrt{\frac{10^{15}\,\text{cm}^{-3}}{n_0}}$$

The disturbing pulse (either laser or beam) induces oscillations of the plasma electrons. These oscillations generate strong longitudinal and transverse fields. These fields can be accurately evaluated by state-of-the-art theory [Easrey *et al.* 2009] and computer simulation codes. The fields can also be estimated by simple formulae. The maximum accelerating field $\mathrm{W}_z$ in a plasma accelerator is given as follows:

$$W_z\ (\text{V/m}) \approx 100 \cdot \sqrt{\frac{n_0}{(\text{cm}^{-3})}}$$

For a plasma density of $10^{18}$ cm$^{-3}$ we estimate a gradient of 100 GV/m, three orders of magnitude above the gradients achievable in conventional accelerators. On the other side, conventional accelerators have, to good approximation, no transverse fields that affect the beams. For plasma accelerators, however, strong transverse fields are generated. The focusing gradient $g = W_r/r$ in radial direction $r$ is estimated as follows in the ion channel (assuming immobile ions):

$$g\left(\frac{\text{T}}{\text{m}}\right) \approx 960\pi \cdot \left(\frac{n_0}{10^{14}\,\text{cm}^{-3}}\right)$$

The transverse focusing gradients reach 30 MT/m for our example of a plasma density of $10^{18}$ cm$^{-3}$. It is easily seen why plasma accelerators must operate in the regime of transverse focusing fields: similarly strong defocusing fields would immediately defocus and destroy the accelerated beam. This has interesting consequences:

- A plasma accelerator cannot operate at the maximum accelerating gradient (crest), as this would place the electron beam close to the defocusing regime (close to the point where the plasma electrons rush back through the beam axis).
- The longitudinal accelerating field has a large dependence on the longitudinal position at the location where the electron bunch can be placed (large slope). The head of the accelerated bunch is therefore always accelerated less than its tail. Every plasma accelerator therefore induces a correlated energy spread that is of the order of a few percent, depending on the bunch length and counter-acting beam loading effects.
- There is no region where the bunch is focused and at the same time the accelerating field has the opposite slope. The correlated energy spread is therefore always induced in the same direction and cannot be directly compensated.

The limited useful phase of plasma accelerators has been shown by Ruth and Chen [1986]. Figure 4 illustrates the various regimes and the useful phase for a linear plasma oscillation. The transverse beam dynamics of a plasma accelerator has been described by Assmann and Yokoya [1998] and Ferrario [2011]. Here it is shown that the beta-functions (as defined in accelerator physics) that correspond to typical plasma channels range from a couple of cm for low plasma densities ($2 \times 10^{14}$ cm$^{-3}$) to a few mm for high plasma

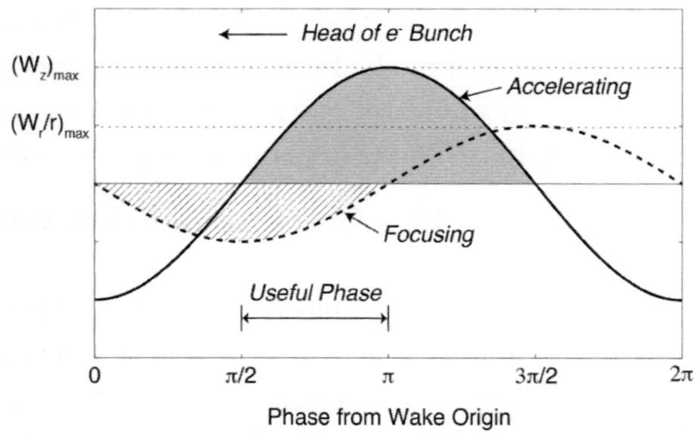


Fig. 4. Illustration of the accelerating and focusing properties of a plasma accelerator and its "useful" phase over one plasma wavelength (linear wakefield regime). Any accelerated electron bunch must be placed into the useful phase with its head pointing towards zero phase. The head of a bunch is therefore accelerated less than its tail. The figure is based on a graph in the paper of Ruth and Chen [1986].

densities ($10^{17}$ cm$^{-3}$). Matching in and out of these plasmas becomes a major issue. Transverse tolerances are in the range of a few $\mu$m and below, becoming highly demanding.

We note that the case of external injection (Fig. 3) can be staged, i.e. allows placing several of these plasma accelerators behind each other for acceleration of the same electron beam. It can therefore be extended to high beam energy.

## 4 Plasma accelerators with internally captured e$^-$ beam

Pukhov and Meyer-ter-Vehn described a possibility to capture plasma electrons in the plasma wakefields of the highly non-linear broken-wave regime [2002]. They showed that a well-defined bunch of electrons can be generated and accelerated. This new idea was based on theoretical considerations and computer simulations. In the following years the proposed "bubble regime" was investigated in various experiments. Electron beams captured from plasma electrons were successfully established by Geddes *et al.* [2004], Mangles *et al.* [2004] and Faure *et al.* [2004]. This success of theory and simulation illustrates the excellent understanding of the physics processes in plasmas and the high predictive power of the present simulation programs.

The principle of a plasma accelerator with "internal injection" and operating in the bubble regime is illustrated in Fig. 5. The originally proposed regime uses a plasma density of $10^{19}$ cm$^{-3}$. The strong transverse electrical fields of an intense laser pulse generate a ponderomotive force that drives the plasma electrons outwards and creates a "bubble" of ions, surrounded by an over-density of electrons. A plasma cavity with strong accelerating fields is formed. The highly non-linear fields trap some of the plasma electrons from the back of the plasma cavity, bunch them, focus them and accelerate them coherently. This process can generate multi-GeV electron beams with energy spreads in the percent regime. The process is limited by several factors:

- Depletion of laser power. This favors lasers pulses with higher pulse energy.
- Dephasing between the relativistic electron bunch and the wakefield with phase velocity below light velocity. This favors low plasma densities with faster phase velocity of the plasma wakefield.
- Rayleigh length of the laser beam. The limited Rayleigh length is counteracted by self-guiding or external guiding of the laser in the plasma channel (see for example [Lu *et al.* 2011]).

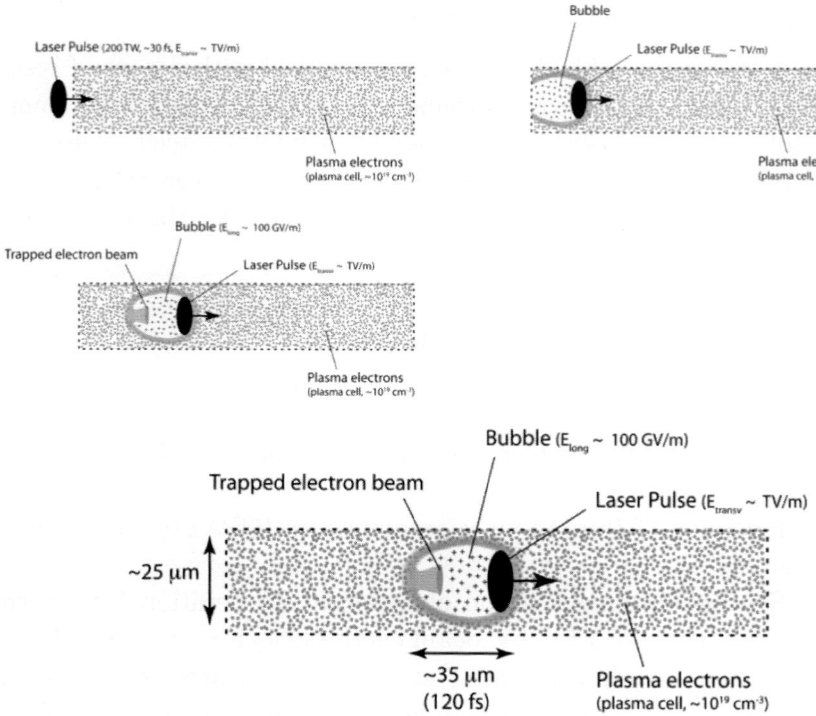


Fig. 5. Illustration of laser-induced plasma wakefields that trap, bunch and accelerate plasma electrons ("internal injection"). Top left: laser enters undisturbed plasma. Top right: ponderomotive force (transverse electrical fields) of laser expels plasma electrons and the plasma cavity starts forming. Middle: plasma electrons have rushed back to axis, plasma cavity has formed, some plasma electrons are trapped. Bottom: plasma cavity propagates through plasma channel and accelerates the trapped electron bunch.

It is noted that all these limitations apply both to laser wakefield accelerators with an externally injected beam and with internally injected, captured electrons. Internal injection is a highly complex system where many parameters must be controlled at the same time in order to generate high quality beams. Presently attempts are ongoing to improve control of the internal injection process, for example by a second laser pulse that triggers the injection into the plasma cavity [Faure *et al.* 2006, Geddes *et al.* 2008]. Plasma accelerators with internal injection have the advantages that neither external beam infrastructure, optics matching nor injection of an external beam is required. One can even hope to use the strong transverse fields for undulating the generated electron beam and generating synchrotron photons for users.

## 5 Selected highlights for plasma accelerators

Plasma acceleration is a highly active research field with dozens of research groups vigorously contributing around the globe. A fair and complete overview of important results would require a book in itself. Here we can only provide a selection of highlight results that have been achieved in the various research directions. The results are grouped by the type of driver used to excite the plasma wakefield.

### 5.1 *Plasma acceleration driven by laser pulses*

Important milestones include the following results:

- 1979: Proposal of laser-driven plasma accelerators from Tajima and Dawson [1979].
- 1985: Invention of chirped-pulse amplification (CPA) by Strickland and Mourou [1985].
- 1993: Successful demonstration of wakefield acceleration for externally injected electrons of 1.5 MeV beam energy [Clayton *et al.* 1993].
- Around 1995: Work progressed at various laboratories on laser self-injected electrons in the self-modulation regime of laser wakefield acceleration. Beams with energies above 100 MeV were produced over an acceleration length of about 1 mm. The energy spread of beams was on the order of 100%.
- 2002: Description of the highly non-linear broken-wave regime for laser wakefield acceleration by Pukhov and Meyer-ter-Vehn [2002]. Accelerating gradients reach 200 GV/m [Malka *et al.* 2002].
- 2004: Laser-driven plasma wakefield accelerators produce beams with small emittances of $\gamma\varepsilon \sim 1$ mm·mrad [Fritzler *et al.* 2004]. The predicted regime of the highly non-linear broken-wave is confirmed experimentally by Geddes *et al.* [2004], Mangles *et al.* [2004] and Faure *et al.* [2004]. The produced electron beams have energies around 100 MeV, charges of about 100 pC and a record small energy spread of a few percent.
- 2006–2010: For the first time an electron beam with more than 1 GeV is produced in a laser wakefield accelerator [Leemans *et al.* 2006]. This result was achieved by channeling of the wake-driving laser pulses. Another group showed that the control of the internal injection process is improved by counter-propagating laser beams [Faure *et al.* 2006]. The stability of the accelerated electron beam is improved by better control of laser parameters [Mangles *et al.* 2007, Hafz *et al.* 2008], the use of new internal injection schemes [Faure *et al.* 2006, Geddes *et al.* 2008], and

the implementation of so-called "steady-state-flow gas cells" [Osterhoff *et al.* 2009]. For the first time undulator radiation with photon energies above 100 eV is produced with an electron beam from a laser wakefield accelerator [Fuchs *et al.* 2009], opening the path to a table-top FEL based on plasma acceleration [Grüner *et al.* 2007].

- 2013: Wang *et al.* [2013] generate a 2 GeV electron beam with a divergence of 0.5 mrad, a pointing stability of 1.4 mrad (r.m.s.) and an energy spread of 10% (FWHM). Independently, Kim *et al.* [2013] report acceleration by over 3 GeV using a dual-stage laser wakefield acceleration setup. Lasers in the petawatt regime are now becoming available and drive the progress.
- 2014: Leemans *et al.* [2014] report that they have generated an electron bunch with beam energy of 4.25 GeV at the new BELLA facility.

## 5.2 *Plasma acceleration driven by electron beams*

Important milestones include the following results:

- 1985: Developing further the ideas on plasma accelerators, Chen *et al.* [1985] propose the use of an electron bunch for driving a plasma wakefield.
- 1997: The E-157 experiment is approved at SLAC, aiming at establishing 1 GeV beam acceleration in a 1.4 m long plasma wakefield that is driven by the 30 GeV SLAC electron beam [Assmann *et al.* 1997]. This research line is extended in various follow up experiments (E-162, E-164, E-167) and recently entered into the setup of the FACET facility at the SLAC linac.
- 2002: Wang *et al.* [2002] measure the X-rays produced by betatron motion in the strong transverse focusing fields of the beam-driven plasma wakefield. This demonstrates the potential of plasmas for ultra-compact wigglers and undulators. Measurements of the transverse beam dynamics in plasma accelerators are published in [Clayton *et al.* 2002], testing important concepts of plasma acceleration.
- 2003: Positrons are accelerated by 79 MeV in beam-driven plasma wakefields, providing unique data on the transport and acceleration of positrons in plasma channels [Blue *et al.* 2003]. The accelerating gradient reaches 56 MV/m.
- 2004: Electrons are accelerated by up to 280 MeV in a plasma wakefield driven by the SLAC electron beam [Muggli *et al.* 2004]. Muggli *et al.* used a plasma density of $1.9 \times 10^{14}$ cm$^{-3}$ resulting in a much longer plasma

wavelength than for laser-driven plasma accelerators. The injected electron beam was matched to the plasma channel and its transverse focusing field.

- 2007: A beam driven plasma acceleration experiment at SLAC achieves an absolute acceleration of 42 GeV with accelerating gradients of up to 53 GV/m over a plasma length of 85 cm [Blumenfeld *et al.* 2007]. Ideas emerge on the creation of micro-bunches that can resonantly drive plasma wakefields [Ferrario *et al.* 2007, Muggli *et al.* 2010].
- 2012: A new method is published for producing an ultracold electron bunch with so-called plasma photocathode emission and acceleration in the beam-driven plasma blowout regime [Hidding *et al.* 2012]. The electron beam emittance can be orders of magnitude smaller than achieved in conventional photocathodes, assuming that the electron bunches can be maintained and extracted from the plasma. These plasma-generated beams would be superior to presently available electron bunches.
- 2013: The FACET facility [Muggli and Hogan 2009] at SLAC has been commissioned and further experiments on beam-driven plasma experiments are being started.

### 5.3 *Plasma accelerators driven by proton beams*

Important milestones include the following results:

- 2009: Caldwell *et al.* [2009] describe in detail the possibility to use proton bunches for driving plasma wakefields. They show that the large energy stored in the proton beams would allow driving electron beams to higher beam energies in a single plasma stage than it is possible with electron or laser drivers. The problem of energy depletion in the driver pulse is relaxed by orders of magnitude.
- 2011: Caldwell *et al.* [2011] describe possibilities to use the PS or SPS proton beams at CERN to perform experiments on proton-driven plasma acceleration. Early accelerator studies are being performed at CERN [Assmann *et al.* 2011].
- 2013: The AWAKE experiment is approved at CERN for experiments on proton-driven plasma acceleration [AWAKE collaboration 2014]. In view of the long proton bunches available at CERN, a technique of plasma wakefield induced self-modulation and microbunching of the proton bunches will be used [Lotov 2010, Kumar *et al.* 2010, Caldwell *et al.* 2011]. Electron bunches will be injected for measuring accelerating wakefields with gradients of up to 1 GV/m.

## 6 Future research directions and outlook

Plasma accelerators have experienced a rapid development over the recent decades. The beam energy achieved in plasma accelerators grows roughly exponentially with time (see Fig. 1) and there is no known fundamental limitation that is expected to prevent further progress. A rich and diverse research landscape has developed, bringing together universities, national and international research laboratories and industry (mainly on the rapidly evolving laser front).

The progress is accelerated by the fact that plasma acceleration has now reached the beam energy that is required to build modern free-electron lasers (as indicated in Fig. 1). Intense studies are ongoing on solving the last feasibility issues for plasma-based FELs and on developing their technical design. The technical problems that are addressed are summarized in a word cloud in Fig. 6. It is seen that the topic considered most urgent for plasma accelerators is "stability and beam quality". As explained earlier, the large energy spread induced by plasma accelerators is a particular and unavoidable worry. Longitudinal effects couple into the transverse plane and can easily destroy beam quality.

Plasma accelerators with high beam quality and stability will require a compensation or active correction of the induced energy spread. The internal compensation of the slope of the accelerating field with beam loading effects is an intense field of work. Such compensation works in principle [Katsouleas *et al.* 1987, Rosenzweig *et al.* 2004] but induces even more constraints on

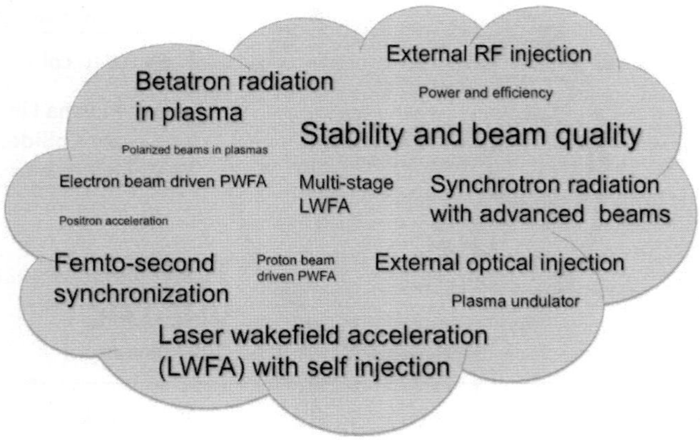


Fig. 6. Graphical visualisation of research topics in advanced acceleration R&D that are pursued in various facilities. The font size reflects the number of facilities that are addressing the specific research topic and the priority attached to this by the research teams. This word cloud is based on data collected by the European Network for Novel Accelerators from a number of universities and laboratories.

the plasma acceleration process. An alternative path will require the development of active correction mechanisms, as they are used in conventional accelerators.

Plasma accelerators provide a path to high beam energies in a much reduced length. They are therefore of great interest for particle physics. Beam energies and beam quality are not yet at the level required for applications in particle physics. However, as soon as a plasma-based FEL can be operated, many of the issues will have been solved. Several publications have explored the possibilities for a plasma based linear collider at 500 GeV CM and more recently at the Higgs energy [Schroeder *et al.* 2010, Hillenbrand *et al.* 2013, Adli *et al.* 2013].

The striking feature of plasma-based linear colliders is their small size, about a factor 10 smaller than a conventional linear collider. This would open an alternative future for particle physics. In case that the next generation of colliders with a 100 km footprint cannot be built, an alternative path to much more compact accelerators is available, as illustrated in Fig. 7.

There are a number of challenges that must be addressed before a plasma linear collider becomes a reality. For example, the efficiency of the plasma acceleration process and the drivers (e.g. lasers) must be improved, beam

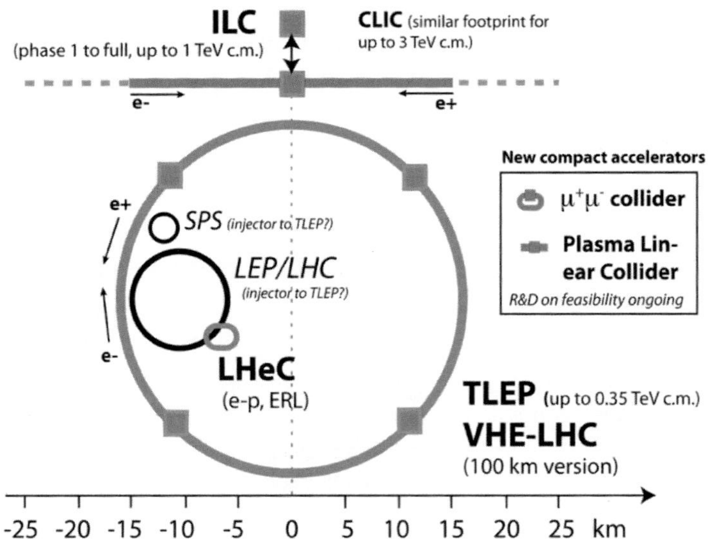


Fig. 7. Comparison of the various proposed paths to a future $e^+e^-$ or $\mu^+\mu^-$ Higgs factory, as indicated by blue lines. The SPS and the LEP/LHC tunnels are shown for comparison (black lines). Red boxes indicate future detectors, the ILC having two push-pull detectors. The dimensions are to scale. The 100 km TLEP/VHE-LHC concepts have recently been renamed as the FCC (Future Circular Collider) project at CERN.

quality and stability goals must be achieved, positron acceleration must be fully mastered and, last but not least, it must be proven that plasma accelerators are more cost-effective than conventional accelerators. Work is ongoing on these topics. Mourou *et al.* have described the idea and successful initial tests on coherent combination of many highly efficient fibre lasers into a single driving pulse for a plasma accelerator [Mourou *et al.* 2013]. This could become a breakthrough result towards high efficiency driver technology. The outlook to exawatt and zetawatt lasers [Mourou *et al.* 2012] could provide a possibility for very high-energy plasma accelerators with internal injection and in a single stage.

## References

Adli, E. *et al.* A Beam Driven Plasma-Wakefield Linear Collider: From Higgs Factory to Multi-TeV. SLAC-PUB-15426 (2013).

Assmann, R. *et al.* 1-GeV Beam Acceleration in a One Meter Long Plasma Cell. SLAC-PROPOSAL-E-157 (1997).

Assmann, R. and K. Yokoya. Transverse Beam Dynamics in Plasmas. *NIM A* **410**, 544–548 (1998).

Assmann, R. *et al.* Accelerator Studies on a Possible Experiment on Proton-Driven Plasma Wakefields at CERN. *2nd International Particle Accelerator Conference*, San Sebastian, Spain, 4–9 Sep 2011. EuCARD-CON-2011-055. CERN-ATS-2011-169.

Assmann, R. *et al.* (AWAKE collaboration). Proton-Driven Plasma Wakefield Acceleration: A Path to the Future of High-Energy Particle Physics. *Plasma Physics and Controlled Fusion* **56**, 084013 (2014).

Blue, B. E. *et al.* Plasma-Wakefield Acceleration of an Intense Positron Beam. *Phys. Rev. Lett.* **90**, 214801 (2003).

Blumenfeld, I. *et al.* Energy Doubling of 42 GeV Electrons in a Metre-Scale Plasma Wakefield Accelerator. *Nature* **445**, 741 (2007).

Caldwell, A., K. Lotov, A. Pukhov, and F. Simon. Proton-Driven Plasma-Wakefield Acceleration. *Nature Phys.* **5**, 363 (2009).

Caldwell, A., K. Lotov, A. Pukhov, and G. Xia. Plasma Wakefield Excitation with a 24?GeV Proton Beam. *Plasma Phys. Control. Fusion* **53**, 014003 (2011).

Caldwell, A. and K. Lotov. Plasma Wakefield Acceleration with a Modulated Proton Bunch. *Phys. Plasmas* **18**, 103101 (2011).

Chen, P. *et al.* Acceleration of Electrons by the Interaction of a Bunched Electron Beam with a Plasma. *Phys. Rev. Lett.* **54**, 693–696 (1985).

Clayton, C. E., K. A. Marsh, A. Dyson, M. Everett, A. Lal, W. P. Leemans, R. Williams, and C. Joshi. Ultrahigh-Gradient Acceleration of Injected Electrons by Laser-Excited Relativistic Electron Plasma Waves. *Phys. Rev. Lett.* **70** (1), 37–40 (1993).

Clayton, C. E. *et al.* Transverse Envelope Dynamics of a 28.5-GeV Electron Beam in a Long Plasma. *Phys. Rev. Lett.* **88**, 154801 (2002).

Clayton, C. E. *et al.* Self-Guided Laser Wakefield Acceleration beyond 1 GeV Using Ionization-Induced Injection, *Phys. Rev. Lett.* **105**, 105003 (2010).

Corsini, R. *et al.* Experimental Verification of the CLIC Two-Beam Scheme, Status and Outlook. *Proceedings IPAC12*. CERN-ATS-2012-180 (2012).

Cowan, T. E. *et al.* Ultralow Emittance, Multi-MeV Proton Beams from a Laser Virtual-Cathode Plasma Accelerator. *Phys. Rev. Lett.* **92**, 204801 (2004).

Esarey, E., C. B. Schroeder and W. P. Leemans. Physics of Laser-Driven Plasma-Based Electron Accelerators. *Rev. Mod. Phys.* **81**, 1229 (2009).

Fritzler, S., E. Lefebvre, V. Malka, F. Burgy, A. E. Dangor, K. Krushelnick, S. P. D. Mangles, Z. Najmudin, J.-P. Rousseau, and B. Walton. Emittance Measurements of a Laser-Wakefield-Accelerated Electron Beam. *Phys. Rev. Lett.* **92** (16), 165006 (2004).

Faure, J., Y. Glinec, A. Pukhov, S. Kiselev, S. Gordienko, E. Lefebvre, J. P. Rousseau, F. Burgy, and V. Malka. A Laser-Plasma Accelerator Producing Monoenergetic Electron Beams. *Nature* **431** (7008), 541–544 (2004).

Faure, J., C. Rechatin, A. Norlin, A. Lifschitz, Y. Glinec, and V. Malka. Controlled Injection and Acceleration of Electrons in Plasma Wakefields by Colliding Laser Pulses. *Nature* **444** (7120), 737–739 (2006).

Ferrario, M., M. Boscolo *et al.* A Train of Micro-Bunches for PWFA Experiments Produced by RF Photoinjectors. *Int. J. Mod. Phys. B* **21**, 415 (2007).

Ferrario, M. Accelerator Physics: Basic Principles on Beam Focusing and Transport. *Proc. of International School of Physics "Enrico Fermi — Laser-Plasma Acceleration"*, Varenna, 20–25 June 2011, SPARC-BD-12/01.

Fuchs, M., R. Weingartner, A. Popp, Zs. Major, S. Becker, J. Osterhoff, R. Hörlein, G. D. Tsakiris, U. Schramm, T. P. Rowlands-Rees, S. M. Hooker, D. Habs, F. Krausz, S. Karsch, and F. Grüner. Laser-Driven Soft-X-Ray Undulator Source. *Nature Physics* **5**, 826 (2009).

Geddes, C. G. R., C. Tóth, J. van Tilborg, E. Esarey, C. B. Schroeder, D. Bruhwiler, C. Nieter, J. Cary, and W. P. Leemans. High-Quality Electron Beams from a Laser Wakefield Accelerator Using Plasma-Channel Guiding. *Nature* **431** (7008), 538–541 (2004).

Geddes, C. G. R., K. Nakamura, G. R. Plateau, C. Tóth, E. Cormier-Michel, E. Esarey, C. B. Schroeder, J. R. Cary, and W. P. Leemans. Plasma-Density Gradient Injection of Low Absolute-Momentum-Spread Electron Bunches. *Phys. Rev. Lett.* **100** (21), 215004 (2008).

Grüner, F., S. Becker, U. Schramm, T. Eichner, M. Fuchs, R. Weingartner, D. Habs, J. Meyer-ter-Vehn, M. Geissler, M. Ferrario, L. Serafini, B. van der Geer, H. Backe, W. Lauth, and S. Reiche. Design Considerations for Table-Top, Laser-Based VUV and X-Ray Free Electron Lasers. *Applied Physics B: Lasers and Optics* **86** (3), 431–435 (2007).

Hafz, N. A. M., T. M. Jeong, I. W. Choi, S. K. Lee, K. H. Pae, V. V. Kulagin, J. H. Sung, T. J. Yu, K.-H. Hong, T. Hosokai, J. R. Cary, D.-K. Ko, and J. Lee. Stable Generation of GeV-Class Electron Beams from Self-Guided Laser-Plasma Channels. *Nature Photonics* **2**, 571–577 (2008).

Hidding, B., G. Pretzler, J. B. Rosenzweig, T. Königstein, D. Schiller, D. L. Bruhwiler *et al.* Ultracold Electron Bunch Generation via Plasma Photocathode Emission and Acceleration in a Beam-Driven Plasma Blowout. *Phys. Rev. Lett.* **108**, 035001 (2012).

Hillenbrand, S., A. S. Müller, R. Assmann, D. Schulte. Considerations for a Higgs Facility Based on Laser Wakefield Acceleration. *Proceedings IPAC13.*

Katsouleas, T. *et al.* Beam Loading in Plasma Accelerators. *Particle Accelerators* **22**, 81–99 (1987).

Kim, H. T. *et al.* Enhancement of Electron Energy to the Multi-GeV Regime by a Dual-Stage Laser-Wakefield Accelerator Pumped by Petawatt Laser Pulses. *Phys. Rev. Lett.* **111**, 165002 (2013).

Kumar, N., A. Pukhov, and K. Lotov. Self-Modulation Instability of a Long Proton Bunch in Plasmas. *Phys. Rev. Lett.* **104**, 255003 (2010).

Leemans, W. P. *et al.* GeV Electron Beams from a Centimetre-Scale Accelerator. *Nature Physics* **2**, 699 (2006).

Leemans, W. P. *et al.* Multi-GeV Electron Beams from Capillary-Discharge Guided Subpetawatt Laser Pulses in the Self-Trapping Regime. *Phys. Rev. Lett.* **113**, 245002 (2014).

Lotov, K. Simulation of Proton Driven Plasma Wakefield Acceleration. *Phys. Rev. ST Accel. Beams* **13**, 041301 (2010).

Lu, H. *et al.* Laser Wake?eld Acceleration of Electron Beams Beyond 1 GeV from an Ablative Capillary Discharge Waveguide. *Applied Physics Letters* **99** (9), 091502 (2011).

Maksimchuk, A., Gu, S., Flippo, K., Umstadter, D., and Bychenkov, V. Forward Ion Acceleration in Thin Films Driven by a High-Intensity Laser. *Phys. Rev. Lett.* **84**, 4108–4111 (2000).

Malka, V., S. Fritzler, E. Lefebvre, M.-M. Aleonard, F. Burgy, J.-P. Chambaret, J.-F. Chemin, K. Krushelnick, G. Malka, S. P. D. Mangles, Z. Najmudin, M. Pittman, J.-P. Rousseau, J.-N. Scheurer, B. Walton, and A. E. Dangor. Electron Acceleration by a Wakefield Forced by an Intense Ultrashort Laser Pulse. *Science* **298** (5598), 1596–1600 (2002).

Mangles, S. P. D., C. D. Murphy, Z. Najmudin, A. G. R. Thomas, J. L. Collier, A. E. Dangor, E. J. Divall, P. S. Foster, J. G. Gallacher, C. J. Hooker, D. A. Jaroszynski, A. J. Langley, W. B. Mori, P. A. Norreys, F. S. Tsung, R. Viskup, B. R. Walton, and K. Krushelnick. Monoenergetic Beams of Relativistic Electrons from Intense Laser-Plasma Interactions. *Nature* **431** (7008), 535–538 (2004).

Mangles, S. P. D., A. G. R. Thomas, O. Lundh, F. Lindau, M. C. Kaluza, A. Persson, C.-G. Wahlström, K. Krushelnick, and Z. Najmudin. On the Stability of Laser Wakefield Electron Accelerators in the Monoenergetic Regime. *Physics of Plasmas* **14** (5), 056702 (2007).

Mourou, G. *et al.* Exawatt-Zettawatt Pulse Generation and Applications. *Opt. Comm.* **285**, 720 (2012).

Mourou, G., B. Brocklesby, T. Tajima and J. Limpert. The Future is Fibre Accelerators. *Nature Photonics* **7**, 258–261 (2013).

Muggli, P. *et al.* Meter-Scale Plasma-Wakefield Accelerator Driven by a Matched Electron Beam. *Phys. Rev. Lett.* **93**, 014802 (2004).

Muggli, P., M. J. Hogan. Review of High-Energy Plasma Wakefield Experiments, *C. R. Physique* **10**, 116–129 (2009).

Muggli, P. *et al.* Simple Method for Generating Adjustable Trains of Picosecond Electron Bunches, *Phys. Rev. ST Accel. Beams* **13**, 052803 (2010).

Osterhoff, J., A. Popp, Zs. Major, B. Marx, T. P. Rowlands-Rees, M. Fuchs, R. Hörlein, F. Grüner, D. Habs, F. Krausz, S. M. Hooker, and S. Karsch. Stable Laser-Driven Electron Beams from a Steady-State-Flow Gas Cell. In *Advanced Accelerator Concepts* (edited by C. Schroeder, E. Esarey, and W. P. Leemans). *AIP Conference Proceedings*, accepted for publication (2009).

Pukhov, A. and J. Meyer-ter-Vehn. Laser Wakefield Acceleration: The Highly Non-Linear Broken-Wave Regime. *Applied Physics B: Lasers and Optics* **74** (4–5), 355–361 (2002).

Rosenzweig, J. B., N. Barov, M. C. Thompson, and R. B. Yoder. Energy Loss of a High Charge Bunched Electron Beam in Plasma, *Phys. Rev. ST Accel. Beams* **7**, 061302 (2004).

Ruth, R. D. and P. Chen. Plasma Accelerators. SLAC-PUB-3906 (1986).

Schroeder, C. B. *et al.* Physics Considerations for Laser-Plasma Linear Colliders. *Phys. Rev. ST Accel. Beams* **13**, 101301 (2010).
Strickland, D. and G. Mourou, Compression of Amplified Chirped Optical Pulses. *Opt. Comm.* **56** (3), 219–221 (1985).
Tajima, T. and J. M. Dawson. Laser Electron Accelerator. *Phys. Rev. Lett.* **43**, 267 (1979).
Tajima, T. and G. Mourou. Zettawatt-Exawatt Lasers and Their Applications in Ultrastrong-Field Physics. *Phys. Rev. ST Accel. Beams* **5**, 031301 (2002).
Tigner, M. Does Accelerator-Based Particle Physics have a Future? *Physics Today* **54** (1) Jan 2001.
Wang, S. *et al.* X-Ray Emission from Betatron Motion in a Plasma Wiggler. *Phys. Rev. Lett.* **88**, 135004 (2002).
Wang, X. Physics of Electron Acceleration Beyond 1 GeV at the Texas Petawatt Laser, Talk at the *Advanced Accelerator Concepts Workshop 2012*, Austin, TX, United States, 12 June 2012.
Wang, X. *et al.* Quasi-Monoenergetic Laser-Plasma Acceleration of Electrons to 2 GeV. *Nat. Commun.* **4**, 1988 (2013).

# Chapter 39

# Energy recovery linacs

Christopher Tennant (Jefferson Laboratory)

## 1 Introduction

An increasing number of scientific applications require intense electron beams of superior quality (extremely dense 6-dimensional phase space), for the investigation of fundamental processes as well as the generation of highly coherent, high average brightness photon beams. These applications include high average power free electron lasers (FELs), synchrotron light sources, terahertz and Compton sources as well as electron cooling devices and electron–ion colliders for nuclear and particle physics research.

Traditionally the demands for beams with these characteristics have been met by storage rings. Synchrotrons (and more generally, storage rings) pass beam through a modest RF accelerating system, incrementally gaining energy over many turns. While this allows rings to generate beams with high power, it also makes the beam susceptible to degradation from error sources and beam dynamical effects, such as synchrotron radiation. Furthermore, the ultimate performance of such systems is limited by the fact that electrons are stored for many turns in an equilibrium state. The equilibrium between radiation damping and quantum excitation sets a fundamental limit on the minimum emittance and bunch length that can be achieved. Linear accelerators (linacs), on the other hand, are single pass devices in which the beam is transported through multiple RF accelerating structures. Consequently, beam quality can be preserved from the source. However in terms of cost effectiveness, linacs are inefficient in translating wall plug power to beam power.

### 1.1 *Recirculating linac*

One way to efficiently generate and preserve high brightness beams is to use the accelerating structure of a linac multiple times to accelerate the beam. After each pass through the linac, the beam is transported back to

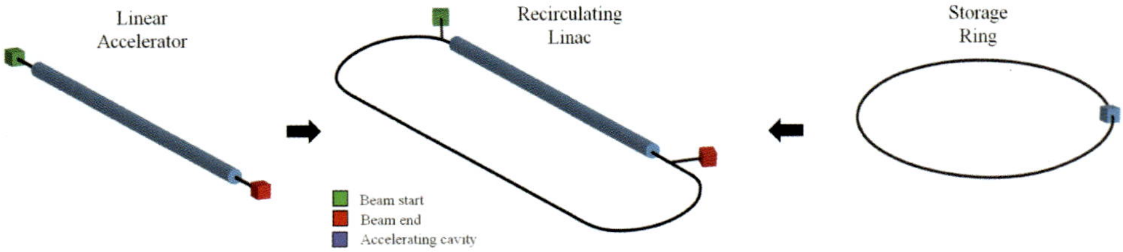


Fig. 1. A recirculating linac (center) borrows from traditional accelerator architectures — the linear accelerator (left) and storage ring (right) — to efficiently generate and preserve high brightness beams.

the entrance and re-injected — in phase with the radio-frequency fields — for further acceleration. The result is a more cost effective scheme than a linac and which generates better beam quality than a storage ring. The various architectures are illustrated in Fig. 1. Using superconducting radio-frequency (SRF) technology further enhances the capabilities of a recirculating linac by supporting high gradient continuous-wave (CW) operation, allowing for higher beam currents and minimizing harmful short range wakefields that exist in normal conducting structures. However, for applications which require high average beam current — with corresponding amounts of RF power — a recirculating linac begins to lose its cost effectiveness.

### 1.2 *Energy recovery linac*

In instances where high average current is required, the concept of energy recovery presents an attractive solution. Energy recovering linacs (ERLs) are a class of novel accelerators which are uniquely qualified to meet the demands for a wide variety of applications by borrowing features from traditional architectures to generate linac quality beams with near storage ring efficiency [1]. After acceleration through a linac section, the electrons in an ERL are returned 180° out of phase with respect to the RF accelerating field for energy recovery. The beam deposits energy into cavity fields, which can then accelerate newly injected bunches, thereby effectively canceling the beam loading effects of the accelerated beam. The data in Fig. 2 shows this effect. The red trace represents the power requested by the cavity during the transit of a 250 $\mu$s macropulse without energy recovery. With energy recovery, the signal is nulled (blue trace) save for RF transients at the ends of the macropulse. Therefore the only RF power required for linac cavities is that required to establish cavity fields and to have sufficient overhead to deal with RF transients (see Section 2.2.1). Therefore ERLs can accelerate very high average currents with only modest amounts of RF power. Further, since

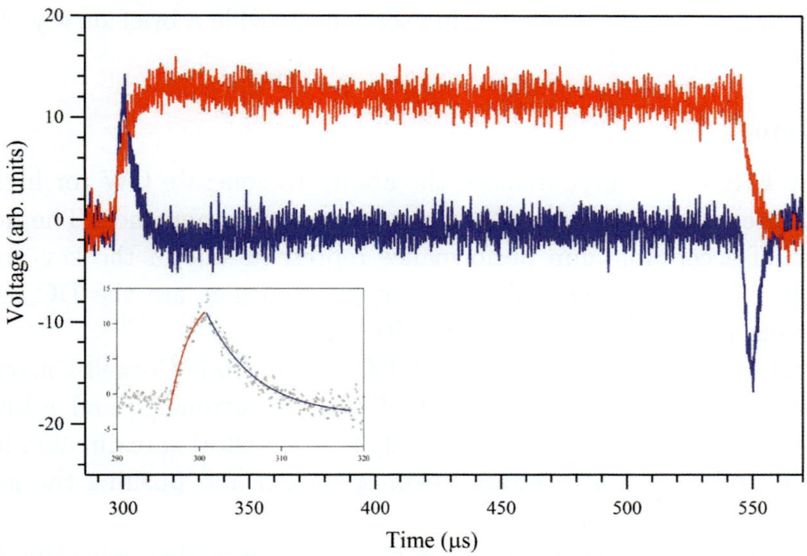


Fig. 2. Data from the CEBAF with energy recovery experiment showing the requested cavity power with (blue) and without (red) energy recovery. With the initial passage of the leading edge of the pulse, power is drawn for 4.3 $\mu$s which corresponds to the recirculation time of the machine (inset).

the energy of the decelerated beam is approximately equal to the injection energy, the dump design becomes considerably easier.

Because in an ERL the beam is constantly being renewed, it never reaches an equilibrium state. Consequently, the phase space can be manipulated to tailor the beam properties for a specific application. For instance, in the Jefferson Laboratory IR FEL Upgrade Driver, the injected longitudinal phase space (temporally long with a low momentum spread) is rotated upright (short bunch with modest momentum spread) at the wiggler using nonlinear bunch compression. During the transport from the wiggler to the beam dump, the process is reversed and the bunch is energy compressed (see Section 4.4). Meanwhile the transverse phase space is reflected in a series of skew quadrupoles for beam breakup (BBU) mitigation (see Section 2.2.2). As this example illustrates, ERL dynamics are fully 6-dimensional, providing an opportunity for novel phase space manipulations, as well as potentially creating adverse effects to the beam if care is not exercised in the design of the machine.

## 2 Challenges

Due to their very nature — namely, the generation and preservation of high power, high brightness beams — ERLs are subject to myriad technical and operational challenges. The following sections provide a brief survey of some of the most important.

### 2.1 *Guns*

Because ERLs are source limited, the ability to generate CW (or high repetition rate), high brightness beams is of utmost importance. The demonstration of acceptable gun performance represents one of the key areas of research and development. The primary technologies are the DC, normal conducting RF, and superconducting RF guns.

Recent notable achievements with DC guns include Cornell University's prototype injector generating 65 mA of average current [2] and KEK high voltage processing their gun to 550 kV [3]. Current efforts are directed toward maintaining high voltage during beam operation and pushing the average current to 100 mA.

Superconducting RF guns are attractive because they allow for higher gradients off the cathode to counteract the deleterious effects of space charge, thereby maintaining beam brightness. Compared to the other gun types, SRF guns are the least mature technology. One of the design challenges is incorporating a normal conducting photocathode (high quantum efficiency) into an SRF environment. The gun at ELBE represents the current state of the art, generating a maximum bunch charge of 400 pC up to 125 kHz [4].

Normal conducting RF guns can generate very high peak gradients, providing the opportunity to generate incredibly bright beams. Historically, NCRF guns have been utilized in systems with low repetition frequency. For high average current applications, issues of thermal loading and maintaining adequate vacuum must be addressed.

### 2.2 *Superconducting radio frequency*

The ability of SRF cavities to support CW, or high duty factor, operation, to efficiently couple energy to the beam, and to support high average current make it a natural choice for ERL applications. With the exception of the THz FEL in Novosibirsk (see Section 5.2), all current and proposed ERLs are based on SRF technology.

An ERL involves two distinct regions that utilize SRF: the injector and the linac. As the injector is not energy recovered, there is an emphasis on developing high-power input couplers and minimizing coupler kicks to the low energy beam. The linac, being energy recovered, must efficiently damp higher-order modes (HOMs), minimize microphonics, and maintain adequate phase and amplitude control of the cavity fields.

### 2.2.1 *RF drive system*

Dynamic loading due to incomplete energy recovery is an issue for all ERLs [5]. In some machines it is due to unintentional errors imposed on the energy recovered beam; for instance, path length errors in large-scale light sources. In other machines, such as high power ERL-based FEL drivers, it is done intentionally. In cases where there is the potential for rapid changes in the relative phase of the energy recovered beam, dynamic loading would be difficult to completely control using fast tuners. In such cases adequate headroom in the RF power must be designed into the system.

Take as an example an ERL-driven FEL. In addition to the increasing the energy spread, the FEL process leads to a decrease of the central energy of the bunch as energy is transferred from the electron beam to the optical beam. This reduction in energy couples to the nonzero momentum compaction ($M_{56}$) of the recirculator lattice to generate a change in the path length (or equivalently, a phase shift). Thus the RF system must deal with a phase shift of several degrees as the laser turns on and off. Because the phase shifts occur on the timescale of the laser turn on/off, even piezo-tuners cannot tune the cavities fast enough. During this time sufficient RF power must be delivered to maintain the gradient in the cavities at a level consistent with the available energy aperture of the machine. The absence of sufficient RF overhead will lead to beam loss and an eventual machine trip.

### 2.2.2 *Beam breakup*

The term beam breakup, as associated with ERLs, refers to a type of regenerative instability that can develop in recirculating linacs (of which ERLs are a specific type). Beam breakup occurs when the electron beam interacts with the HOMs of an RF cavity on the accelerating pass and again on the energy recovering pass. The feedback loop between the beam and the RF cavity is what distinguishes regenerative-type beam instabilities from the so-called cumulative instabilities, and is of particular concern in the design and operation of high average current ERLs utilizing SRF technology — where dipole modes with quality factors several orders of magnitude higher than in normal conducting cavities can exist if not properly damped.

The instability is initiated when a beam bunch passes through an RF cavity off-axis, thereby exciting a dipole HOM. The magnetic field of an excited mode deflects following bunches traveling through the cavity. Depending on the details of the machine optics, the deflection produced by the mode can translate into a transverse displacement at the cavity after recirculation. The recirculated beam induces, in turn, an HOM voltage which depends on the

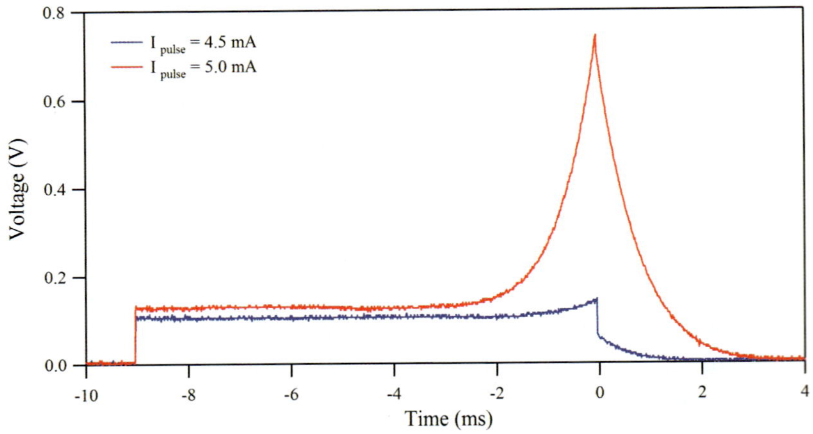


Fig. 3. Measured HOM voltage from the Jefferson Laboratory IR FEL Upgrade showing BBU developing within a sufficiently long macropulse for two different currents.

magnitude and direction of the beam displacement. Thus, the recirculated beam completes a feedback loop which can become unstable if the average beam current exceeds the threshold for stability. Figure 3 displays data showing the onset of BBU, marked by exponential growth in the HOM voltage, in a long macropulse at two different currents.

A thorough suite of measurements to characterize the BBU instability and successfully benchmark data with existing simulation codes has been performed at the Jefferson Lab IR Upgrade Driver [6,7,8]. Using this data, and through two different beam optical suppression techniques, BBU is no longer an operational impediment. The first is a phase trombone, whereby a judicious change in the phase advance can achieve point-to-point focusing ($M_{12}$ or $M_{34} = 0$) at the location of an unstable mode so that an HOM-induced kick on the first pass results in a zero displacement on the second pass. The second, and the default mode for high average current operations, is a five-skew-quadrupole rotator to interchange the horizontal and vertical phase spaces, thereby effectively breaking the feedback loop between the beam and the offending HOM. That is to say, a vertically induced kick on the first pass through a cavity is outcoupled to a horizontal displacement on the second pass.

It should be noted that these techniques were especially effective because a single, well localized, HOM was the cause of BBU. In an ERL utilizing multiple passes, rotating the phase space will not completely break the feedback loop, as the beam will still interact with the HOM on higher passes. Likewise, in ERLs comprised of long linac sections, the possibility exists for many harmful HOMs — of different polarizations — to be present. In these instances, beam optical suppression techniques may not be particularly effective, but active damping on targeted HOMs may. Measurements confirmed that additional damping of a specific HOM could be achieved through the use of an external feedback circuit; by coupling voltage from the cavity, selecting the frequency of interest, and returning it 180° out of phase, the quality factor of the mode could be decreased by factors of a few with a commensurate increase in the threshold current [8].

## 2.3 *Beam dynamics*

The chief goal of any machine design is to deliver an appropriately configured phase space for the designated application, whether it is a wiggler, cooling channel or interaction point. To that end, one must maintain beam quality throughout the system, from the low energy regime of the gun and injector, through the merger, during acceleration, through the multiple bends of the recirculator and during recovery. Despite their difference in architecture, ERLs face many of the challenges common to other accelerators. Those of principal concern are briefly outlined below.

### 2.3.1 *Halo*

Halo is defined as the relatively diffuse and potentially irregularly distributed components of beam phase that can reach large amplitudes. It is of concern because ERL beams are manifestly non-Gaussian and can have beam components of significant intensity beyond the beam core [9]. Through sampling large amplitudes, halo responds to the external focusing of the accelerator transport system in a predictable manner. It is therefore not always at large spatial amplitude, but will at some locations instead be small in size but strongly divergent. Halo can therefore present itself as "hot spots" in a beam distribution, and thus may be thought of as a lower intensity, co-propagating beam that is mismatched to the core beam focusing, timing, and energy.

Because they are high power, CW systems, losses in ERLs must be limited to a few parts per million to avoid undue heating and potential damage to beamline components. In addition to machine safety, halo also complicates efforts to match the beam to the lattice using conventional diagnostics. Energy recovery linac beams are not Gaussian, and as a consequence, even a straightforward measurement of the beam size requires high-dynamic range imaging techniques to see the core as well as the diffuse, large amplitude components of the distribution. Recent progress in imaging techniques have resulted in beam size measurements with a dynamic range of $10^5$ (see Fig. 4) [10]. Utilizing this type of imaging in conjunction with phase space

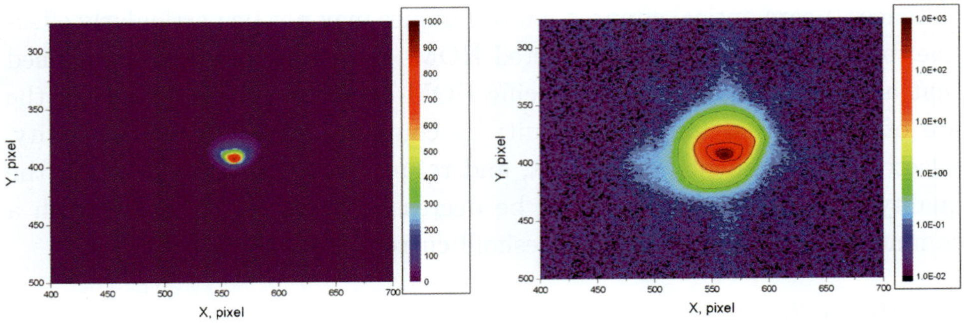


Fig. 4. An example of a beam spot imaged using high-dynamic techniques displayed on a linear (left) and logarithmic (right) scale.

tomography techniques may offer unique insights into (core) beam and halo evolution.

Numerous sources contribute to the halo. Operational experience at various laboratories suggest that the biggest culprits are: stray light striking the photocathode, photocathode emission effects, field emission/dark current from the gun, beam dynamics during beam formation and evolution, and field emission/dark current in SRF cavities.

### 2.3.2 *Collective effects*

Two of the most significant means by which the beam can interact with itself, and so become degraded, are coherent synchrotron radiation (CSR) and space charge (SC). While many ERLs achieve high beam power through modest bunch charge and high repetition rate, the role of space charge forces (both transverse and longitudinal) often dictate many operational aspects of the machine, such as injected bunch length and charge. Maintaining beam brightness during the sub-relativistic regime at the exit of the gun and then after the booster through the merger is vitally important. Common practice is to inject a long(er) bunch to reduce the charge density, take advantage of emittance compensation schemes to preserve beam quality, accelerate the beam as quickly as possible and compress the bunch at high energy. However, there is a competing desire to cut costs and improve system efficiency by decreasing the injected beam power (energy) which aggravates space charge effects on the beam. Therefore the choice of the injected beam energy requires a careful optimization of all the relevant parameters.

Once injected into the linac, the beam energy at the front end is often low enough that space charge forces cannot be completely neglected. While the effect of the transverse space charge force is predominantly in the injector, longitudinal space charge (LSC) can be an operational impediment in the linac. Longitudinal space charge is a force which manifests itself by an energy spread asymmetry about the linac on-crest phase [11]. The LSC wakes act to accelerate the head of the bunch while decelerating the tail. Operating on the rising part of the waveform leads to a decrease in the correlated energy spread, while accelerating on the falling side leads to an increase. These observations inform where acceleration, and how the longitudinal match, is performed.

Like linac-driven FELs, ERL-based light sources also suffer from the effects of CSR. This is not surprising since both system architectures require transporting a short bunch through a dipole, giving rise to coherent radiation and its attendant effects on the beam (i.e. phase space distortion, emittance growth, beam mismatch to the downstream lattice). However, while CSR is customarily associated with light sources, low energy and low energy spread beams typical of ERL-based electron cooler designs, are also susceptible [12].

Thus far CSR has not hindered operational performance of ERLs. In fact it is often used as a diagnostic to tune up the longitudinal match. The bunch length is properly compressed for the FEL when CSR begins to "turn on" [11,13]. At the Jefferson Laboratory FEL systems beam filamentation was, and is, evident when the bunch becomes strongly compressed (see Fig. 5). Initial beam-based measurements to characterize CSR have been taken, however due to the complexity of the longitudinal phase space it becomes difficult to distinguish the contributions from LSC, CSR and other environmental wakes.

In addition to the beam interacting with itself, the beam can also interact with its environment in unwanted ways. As with other system architectures intended to handle high-brightness beams, ERLs can be performance limited by wakefield effects. Not only can beam quality be compromised by interaction of the beam with environmental impedances, there is also significant potential for localized power deposition in beamline components. Resistive wall and RF heating have proven problematic during the operation of the Jefferson Laboratory IR FEL Upgrade [14]. Extrapolation of this experience to higher bunch charges and beam powers leads to serious concern regarding heating effects. Careful analysis and management of system component impedances is required.

### 2.3.3 *Magnet field quality*

An often overlooked aspect of ERL design, and one with significant implications for system performance, is magnetic field quality. The necessary transverse–longitudinal coupling required for energy compression in high

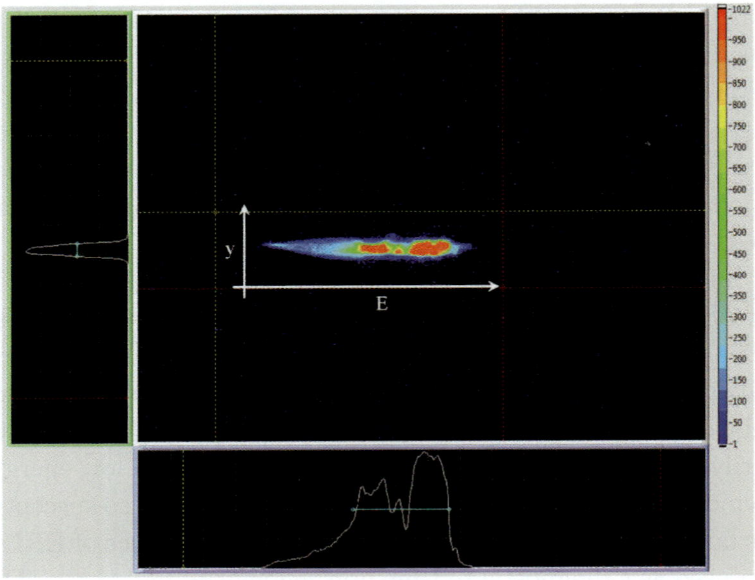


Fig. 5. Screenshot of the synchrotron light monitor in the second arc of the Jefferson Laboratory FEL Upgrade driver showing beam filamentation due to strong bunch compression upstream.

power FEL drivers also creates the means by which magnetic field errors can generate energy errors. Poor field quality lead to transverse steering errors, which due to the non-zero $M_{52}$ of the recirculator, lead to path length errors (or equivalently, phase shifts). Such phase shifts, in turn, increase the energy spread of the bunch and can lead to an unmanageably large energy spread at the dump [15].

## 3 Historical overview

It is important to note that ERLs are not a mature technology in the same way as, for example, storage rings are. Rings have been successfully operated at laboratories throughout the world, reliably over the course of many decades. Conversely, the cumulative beam time of all the ERLs that have ever been in operation pales in comparison to that of rings. Therefore in order to provide proper context for the discussion of currently operating ERLs, a brief historical overview of their development is given; many of which were simply "demonstration" experiments.

The first demonstration of energy recovery occurred at Chalk River Nuclear Laboratories in 1977 using a two-pass reflexotron which passes the beam through an accelerating structure and is returned through the structure in the opposite direction by a 180° reflecting magnet [16]. By changing the distance of the reflecting magnet from the accelerating structure, the phase of the beam relative to the accelerating field can be made to generate either energy doubling or energy deceleration and recovery. Using this method, output energies between 5 MeV (with energy recovery) and 25 MeV (with energy doubling) were achieved. In 1985 a 400 MeV electron beam was energy recovered to 23 MeV at the MIT-Bates Linac as part of an experiment to operate the recirculation system under a variety of conditions [17]. In 1986, Stanford University's Superconducting Accelerator (SCA) energy recovered 150 $\mu$A of average beam current from 55 MeV to 5 MeV [18]. This experiment was significant in that it marked the first time energy recovery had been demonstrated in a superconducting RF environment. At about the same time, the free electron laser at Los Alamos National Laboratory demonstrated energy recovery in a unique configuration where the decelerated beam deposited energy in a different cavity from which it was accelerated [19]. This scheme represents a departure from the previous examples of "same-cell" energy recovery. Using this setup, they successfully energy recovered 21 MeV to 5 MeV. Despite its success, this method of energy recovery has not been used since. More recently, in 2002 the JAEA (formerly JAERI) ERL-driven FEL achieved first light [20]. This prototype machine successfully recovered 17 MeV to 2.5 MeV.

## 4 Jefferson laboratory

For nearly two decades, the implementation of energy recovery has been most active at Jefferson Laboratory. Over the course of 16 years, from 1993 to 2009, same-cell energy recovery was successfully demonstrated in five different accelerators. Combining the principle of energy recovery with SRF cavities leads to an accelerator capable of generating an intense beam with excellent beam qualities in an efficient and economical manner. Initial experience with SRF cavities, however, presented formidable challenges. In the early 1970s, when Stanford University began operation of the SCA, multipactoring in the SRF cavities severely limited the gradients and consequently the final beam energy. To overcome this obstacle, transport elements were installed to recirculate the beam multiple times through the linac. When the beam was recirculated, insufficiently damped HOMs caused beam breakup, thereby limiting the achievable average beam current. Thus, despite the great potential of SRF cavities, the first ERL to implement SRF technology was limited in beam energy (due to multipactoring) and average beam current (due to BBU).

### 4.1 *CEBAF front end test*

When in 1985 it was proposed to build a 4 GeV electron accelerator for nuclear physics based on SRF technology at Jefferson Laboratory, a great effort was made to address the issues of implementing SRF technology on a large scale [21]. By this time Cornell University had designed a cavity using an elliptical cell shape which all but eliminated multipactoring. And while the Cornell cavity exhibited greater HOM damping than the cavities used in the SCA, much was done to address the potential problem of beam breakup. During the initial construction of the Continuous Electron Beam Accelerator Facility (CEBAF), the injector linac was used in conjunction with a single recirculation line to experimentally investigate the problem of BBU [22,23]. The injector was capable of providing over 200 $\mu$A of average beam current. Beam was injected into the linac at 5.5 MeV and accelerated to 43 MeV by two cryomodules. Next, the beam was recirculated and sent through the linac for a second pass where it could either be accelerated to 80 MeV or the recirculator could be configured for energy recovery in which the beam was decelerated to 5.5 MeV. In neither operating scenario were there indications of BBU developing.

### 4.2 *Jefferson laboratory IR demo*

Even before the construction of CEBAF was complete, a proposal was put forward to use it as a driver for an FEL [24]. In addition to the ability of an SRF linac to maintain superior beam quality, the ability for CW operation opened up the possibility of achieving high average output power while using bunches of modest charge. It had been recognized that invoking energy recovery would increase the system efficiency while at the same time reducing the need for expensive, high power RF sources. An initial design for an ERL-based driver for an FEL at Jefferson Laboratory was developed in 1991 [25]. This design was significant in that it marked the first time energy recovery was implemented as the nominal mode of operation. By 1998 the Jefferson Laboratory IR FEL Demo successfully energy recovered 5 mA of average beam current through a single cryomodule from 48 MeV to the injection energy of 10 MeV [26]. By the end of 2001, as the IR Demo was being decommissioned to prepare for an upgrade, the machine had operated at, or exceeded, design parameters. As a result of the IR FEL Demo's demonstrated success, the attractive features of an SRF linac with energy recovery became apparent. Applications of ERLs were extended to synchrotron radiation sources, electron cooling and electron–ion colliders. Many of these applications require a significant extrapolation of the operating parameters achieved at the FEL, such as beam energy and current.

### 4.3 *CEBAF with energy recovery*

In 2001, a proposal was put forward to non-invasively test energy recovery on a large scale using CEBAF [27]. Because it is a recirculating linac, operating CEBAF with energy recovery requires only minor modifications, namely the installation of a magnetic chicane — to provide a half-RF wavelength delay — and a beam dump. In 2003, 80 $\mu$A of average beam current was successfully energy recovered from 1056 MeV to the injection energy of 56 MeV [28]. The experiment demonstrated that large scale energy recovery — through 312 SRF cavities and transported through 1.3 km of beamline — is feasible. One of the important issues that the CEBAF-ER experiment addressed is that the beam quality could be preserved in a common transport channel (in the presence of steering and focusing errors) over a large dynamic range of energy. During the experiment, maximum-to-injector energy ratios ($E_{\rm max}/E_{\rm inj}$) of 19:1 and 51:1 were demonstrated by operating with two different injector energies. For comparison, in the IR FEL Demo this ratio was 5:1. The CEBAF-ER experiment was, and is, important because it represents the first attempt to bridge the gap between the existing lower energy (order 100 MeV), compact (up to 3 cryomodules), SRF-based ERLs and the proposed large-scale ERL drivers. While many test facilities are being designed and constructed to address the issues of high average current, it becomes prohibitively expensive to construct an ERL test facility that addresses issues at high energy. For that reason, this experiment provides an important data point in parameter space.

### 4.4 *IR upgrade FEL*

The most mature ERL, in terms of operational experience, is the Jefferson Lab IR Upgrade which began beam operations in 2003. The facility has served as an invaluable testbed to study, among other items, each of the challenges outlined in Section 2 [11,14].

The FEL Upgrade Driver is an energy recovery based linear accelerator used to condition an electron beam for high average power lasing in the infrared. Electrons are generated in a DC photocathode gun, accelerated to 9 MeV and injected into the linac where they are further accelerated up to 135 MeV through three cryomodules (each containing 8 SRF cavities). The beam is transported to an undulator where in excess of 14 kW of laser power has been generated. Because the SRF linac supports CW beam, high

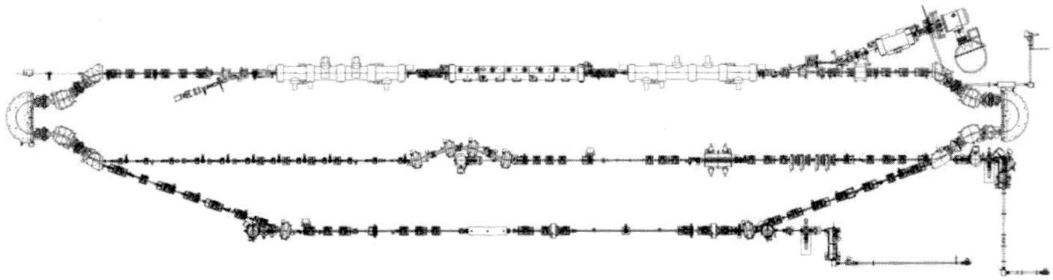

Fig. 6. The Jefferson Lab IR and UV FEL. The SRF linac (top), IR FEL transport line (middle) and UV FEL transport line (bottom) are shown. Note the absence of a compressor chicane in the UV line.

average laser power can be achieved with a high bunch repetition rate and only modest single bunch charge. The spent electron beam is recirculated and phased in such a way that the beam is decelerated through the linac on the second pass. A schematic of the Driver is shown in Fig. 6.

Reduced to its primary objective, the ERL driver must generate a short bunch (high peak current) at the undulator and energy compress and energy recover the large longitudinal phase space of the spent electron beam following the undulator [15,29]. The injector is designed to generate a long bunch with low momentum spread. Acceleration through the linac occurs off-crest at $-10°$ so as to impart a phase–energy correlation across the bunch. The first- and second-order momentum compactions of the first Bates-style recirculation arc are set to rotate the bunch upright at the wiggler and to eliminate phase space curvature, producing a short bunch and high peak current. Following the undulator, the longitudinal phase space must be rotated back by 90° to energy compress the beam which acquires a large momentum spread from the FEL interaction. The energy recovery transport consists of a second Bates-style endloop. Trim quadrupoles, sextupoles, and octupoles in the arc adjust momentum compactions through third-order to longitudinally rotate the short, very large momentum spread bunch and adjust its curvature (second-order) and torsion (third-order). Because energy recovery occurs off-trough, the imposed phase–energy correlations are selected to generate energy compression during energy recovery, yielding a long (of order 30° at 1497 MHz), low momentum spread bunch at the dump. All apertures in the energy recovery loop are chosen to allow lossless transmission of large energy spreads to the beam dump. Operational experience with the IR Upgrade shows the Bates endloop to be a very robust design; at least 12% (full) energy spread beam has been transported cleanly to the end-of-line dump.

As a consequence of the large energy spread, care must be taken to ensure that the entire bunch precedes the trough of the RF waveform during energy recovery. Failure to do so leads to an unmanageable high energy tail after deceleration. To first-order, to prevent the beam from falling into the trough of the RF waveform, the deceleration phase must exceed

$$\phi_{\text{dec}} = \cos^{-1}\left(1 - \frac{1}{2}\frac{\Delta E}{E}\right) \tag{1}$$

where the fractional energy spread, $\Delta E/E$, is approximately six times the FEL extraction efficiency. Thus an extraction efficiency of 2% requires operation with the bunch centroid 20° before trough.

By shifting the bunch further up the waveform to accommodate the full energy spread, the beam central energy is no longer 180° out of phase from the accelerated beam, a condition referred to as incomplete energy recovery. If one tries to restore the condition of perfect energy recovery by accelerating further off-crest, the resulting energy spread will exceed the acceptance of the undulator. To circumvent this one might consider injecting a shorter bunch; however this results in phase space dilution due to LSC. In short, the longitudinal operating point lives in a highly constrained parameter space where a number of trade-offs must be considered.

A recent nuclear physics experiment utilizing the Upgrade Driver demonstrated the machine's flexibility and the wide ranging applications for high brightness, high power, CW beams. The goal of the experiment was to simulate high power ERL operation with an internal gas-jet target by controlling power deposition from beam loss and impedance/wake effects from both the beam core and halo components through a 10 cm long, small aperture (6, 4, and 2 mm diameter). By independently controlling the betatron match of the primary beam and halo, low beam loss transmission of CW high power sub-100 $\mu$m beam at 100 MeV was established. Measured beam interception was as low as 3.2 parts-per-million through the 2 mm aperture while running 450 kW of beam power (100 MeV, 60 pC at 74.85 MHz) [30].

### 4.5 *UV upgrade FEL*

The UV driver ERL shares the linac and the recirculator endloops with the IR Upgrade driver. It is, however, a distinct system with respect to operating parameters and beam handling configuration. For UV operation, the corner dipoles of the IR transport operate at half their IR field, halving the bend angle at the end (beginning) of the IR delivery (recovery) arc. The reduction in angle directs beam toward the UV wiggler; the bend onto (off of) the axis of the optical cavity is completed achromatically through use of a

FODO-focusing transport managing dispersion, controlling beam envelopes, and allowing chromatic correction with sextupoles.

As the UV system shares Bates arcs with the IR, the longitudinal match is both robust and flexible. This process has three unique features. First, compression is performed using arc momentum compactions; there is no compressor chicane. This allows, secondly, full compression with acceleration on either side of crest of the RF waveform; operation is not restricted to the rising side. Thirdly, linearization of RF curvature effects is performed with the transport system sextupoles (and, for energy compression required for lossless recovery, using octupoles as well); harmonic RF is not used or needed [31].

A recent experiment demonstrated that equally good lasing performance could be achieved while operating on the falling side of the RF waveform. This feat is possible only because the UV does not have a compressor chicane. Furthermore, the experiment has shown that it is not only possible, but even desirable from a beam physics standpoint, to accelerate on the falling part of the RF waveform and compress using a positive momentum compaction ($M_{56}$) [32].

Collective effects differ in character from those in the IR machine. Lower bunch charge (60 pC compared to 135 pC for IR operation) alleviates space charge effects — improving beam brightness and reduces average current — mitigating instabilities and interaction of the beam with the environment. Thus, for example, adequate control of BBU is maintained by choice of pass-to-pass phase advance, and does not require the use of a skew-quadrupole rotator.

## 5 Other currently operating ERLs

In addition to Jefferson Laboratory, there are two other ERLs that are currently in operation. Table 1 summarizes relevant beam parameters of the ERLs discussed in Sections 3 and 4 and represents a complete listing of ERL operations to date.

### 5.1 *ALICE at Daresbury Laboratory*

The ALICE (Accelerators and Lasers In Combined Experiments) facility, based at Daresbury Laboratory, is the first ERL in Europe. Initially conceived as a prototype for the 4GLS project, ALICE has evolved into a robust and multifunctional facility servicing a wide range of projects. In addition to serving as a valuable testbed for accelerator physics, ALICE is an IR FEL, a THz radiation source with application to the life sciences, and is the

Table 1. Parameters of ERLs, past (italicized) and present.

| | $E$ (MeV) | $I_{\text{ave}}$ (mA) | $Q_b$ (pC) | $\varepsilon_N$ ($\mu$m) | Rep. (MHz) | Duty (%) |
|---|---|---|---|---|---|---|
| *Chalk River* | 25 | 30 | 10 | 50 | 3000 | 0.1 |
| *MIT Bates* | 400 | 10 | 3.5 | 10 | 2856 | 1 |
| *HEPL* | 48 | 0.6 | 50 | 10 | 11.8 | pulsed |
| *LANL* | 21 | 0.2 | 8000 | 50 | 1300 | pulsed |
| *CEBAF-FET* | 45 | 0.3 | 0.2 | 5 | 1497 | 100 |
| *JLab IR Demo* | 20–50 | 5 | 60 | 10 | 75 | 100 |
| *CEBAF-ER* | 1050 | 0.08 | 0.2 | 1 | 500 | 100 |
| *JAEA* | 17 | 8 | 400 | 40 | 20.8 | pulsed |
| BINP | 22 | 30 | 2000 | 30 | 22.5 | 100 |
| JLab IR Upgrade | 165 | 9 | 135 | 10 | 75 | 100 |
| ALICE | 27.5 | 8.125 | 100 | 1.2 | 81.25 | 0.1 |
| JLab UV Demo | 135 | 2.5 | 60 | 5 | 37.5 | 100 |

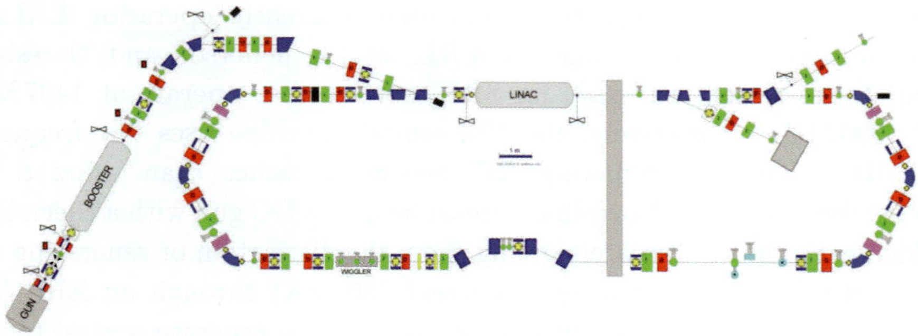


Fig. 7. Schematic of the ALICE facility at Daresbury.

injector for EMMA, a non-scaling FFAG (fixed-field alternating gradient) accelerator [13].

A schematic of the ALICE facility is shown in Fig. 7. Electrons are generated from a DC photocathode gun, accelerated to 6.5 MeV in a booster and injected into the SRF linac where they are further accelerated to 26 MeV. The recirculation arcs are triple bend achromats. Mounting the arcs on translation stages provides a means of path length control. Embedded sextupoles are used to linearize the bunch longitudinally and counteract the curvature imposed by the RF waveform during acceleration. Following the first arc — which is tuned to be isochronous — the beam enters a 4-dipole chicane which compresses the bunch for delivery to the undulator.

For IR FEL operation, the driver must generate a low momentum spread bunch with high peak current at the undulator. To that end, careful control and tuning of the longitudinal dynamics must be maintained. Due to site constraints, ALICE has a long injection line which exacerbates the effects of space charge and velocity bunching. These effects on the evolution of the beam dynamics have been the subject of recent studies [33]. With the recent installation of a new HV gun ceramic, the operating gun voltage has been increased from 230 kV to 325 kV. To achieve the required longitudinal manipulations, the bunch is accelerated 10° off-crest through the linac to induce a phase–energy correlation. No bunching occurs in the first arc, however the phase space is linearized using sextupoles thereby avoiding the need for a harmonic linearizer. The bunch compressor provides the required momentum compaction to rotate the longitudinal phase space upright at the undulator entrance.

### 5.2 *Budker institute of nuclear physics*

The Novosibirsk ERL driven FEL, at the Budker Institute of Nuclear Physics, represents a departure from other currently operating ERLs in many respects. Firstly, unlike the ERLs at the Jefferson and Daresbury Laboratories which are based on SRF systems that operate at 1497 MHz and 1300 MHz, respectively, the Novosibirsk machine uses low frequency (180 MHz), normal conducting RF. Secondly, rather than using a DC photocathode gun, the Novosibirsk machine uses a DC gun with a thermionic gridded cathode — ultimately giving them the distinction of generating and transporting the highest average current (30 mA) through an ERL [34]. Thirdly, limited to low repetition rates, in order to generate several tens of milliamperes of current, Novosibirsk operates with 1.5 nC bunch charges. This is in contrast to the approach of the Jefferson Laboratory FELs, for instance, where high average current is obtained by using modest bunch charge (135 pC) and taking advantage of high repetition rates afforded by the use of an SRF system. Like the facility at Jefferson Laboratory, the Novosibirsk facility has multiple FELs which share a common linac. What makes the facility unique, however, is that in addition to one of the FELs being built out of plane of the other, it is the only operating multi-turn ERL (see Fig. 8). Recent highlights include 4-pass up (acceleration) and 4-pass down (deceleration) operation of the facility [35].

The Novosibirsk FEL is a THz radiation source with seven user workstations. The injector provides 2 MeV electron bunches to the linac which are accelerated to 11 MeV. One might expect that with 1.5 nC bunches at 2 MeV space charge would destroy the beam quality, however, the bunch length is kept relatively long (1.1 ns from the gun, 100 ps at the FEL) which reduces the charge density thereby mitigating its effects.

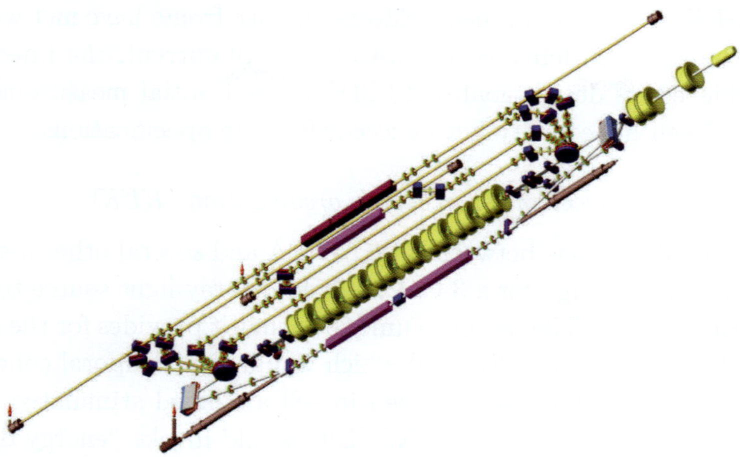

Fig. 8. Layout of the Novosibirsk facility, showing the multi-orbit configuration of the machine.

## 6 ERLs on the horizon

In addition to the four ERLs currently in operation, there exist a number of ERL-based machines in the design stage, under construction, or in some instances, beginning commissioning. A brief survey of the proposed machines demonstrates the robustness of ERLs for meeting the needs of a wide range of applications.

### 6.1 *Synchrotron light source*

Storage rings have been used effectively as sources of synchrotron radiation by virtue of the high average currents they can produce and maintain. One way to overcome the fundamental processes that constrain the achievable beam quality in rings is to use a high brightness injector and high energy SRF linac. To manage the enormous energy consumption of such a machine requires the use of energy recovery. The result is a machine capable of generating smaller emittances, smaller energy spread and shorter beams than a storage ring based system, ultimately leading to better spectral brightness and higher coherence.

#### 6.1.1 *Cornell University*

Cornell University has produced a detailed design report for a 5 GeV, 100 mA, hard X-ray light source [36]. The design leverages the existing infrastructure of the CESR storage ring in the design of the ERL. In order to utilize the ring as well as to seamlessly integrate into the surrounding university campus, the machine features a unique split linac topology. The current focus is on demonstrating key technologies, namely, the DC photocathode gun and SRF cavity development. Efforts on both fronts have met with much success; the gun has delivered 65 mA of average current, the injector SRF cavity achieved its design goal of 13 MV/m, and initial measurements of a prototype 7-cell linac cavity have exceeded design specifications.

### 6.1.2 *High energy acceleration research organization (KEK)*

In Japan, collaborations between KEK, JAEA and several other institutions have resulted in a design for a 3 GeV, 100 mA, X-ray light source to be built on the KEK site [37]. The design is unique in that it provides for the option of an X-ray FEL oscillator (XFEL-O) which would give temporal coherence — a property which cannot be obtained in self-amplified stimulated emission (SASE) type FELs. However, the XFEL-O would invoke "energy doubling" (two passes of acceleration) of the machine rather than operate as an ERL.

Like Cornell, current efforts are focused on research and development of the key technologies, namely, the gun and SRF cavities. Notable achievements include a high voltage test of the KEK gun design to 550 kV and prototyping of both the injector and linac SRF cavities. Before making the jump to the full-scale ERL, the collaboration has decided to build a test facility to successfully demonstrate the technologies working together (see Section 6.4.2).

## 6.2 *Electron–ion colliders*

Energy recovering linacs have been invoked in several machine designs to replace the electron storage ring for high luminosity, electron–ion colliders. This application requires an extrapolation of more than an order of magnitude in energy from what has been demonstrated thus far. Brookhaven National Laboratory and CERN have both proposed electron–ion colliders based on a linac–ring configuration. Unlike a storage ring, electrons from an ERL are stored only a few turns, and can therefore tolerate increased disruption due to beam–beam effects before they are dumped. As a result, the number of ions per bunch can be increased yielding a commensurate increase in the luminosity.

### 6.2.1 *eRHIC at BNL*

Based on the existing Relativistic Heavy Ion Collider (RHIC) facility, Brookhaven National Laboratory's (BNL) design for an electron–ion collider utilizes an ERL to provide 5–30 GeV, polarized electrons [38]. To reduce civil construction costs, the ERL shares the same tunnel as RHIC. A multi-staged approach is taken whereby the electron energy is increased incrementally by adding SRF cavities to each of the two linacs, located in straight sections of the RHIC tunnel. The final stage will have electrons from the gun entering a pre-accelerator ERL where they reach 600 MeV and are then accelerated to a maximum energy of 30 GeV by passing six times through the main ERL. Due to the multi-turn nature of the machine, special care is required in the design of splitters and recombiners.

### 6.2.2 *LHeC at CERN*

Leveraging 7 TeV protons from the Large Hadron Collider (LHC) and colliding with 60 GeV electrons would provide the opportunity to explore lepton–proton collisions with a center of mass in the TeV regime. Several options for the Large Hadron electron Collider (LHeC) have been put forward, including ring–ring and linac–ring options [39]. Pushing the luminosity higher requires CW operation, which in turn, leads to the choice of an SRF-based linac. Its length and cost can be reduced by reusing a shorter linac multiple times with beam recirculation. By invoking energy recovery the costs are reduced further by increasing the wall plug efficiency of the system. The current ERL design is a racetrack topology with 10 GeV linacs in each of the two straight sections. A 500 MeV beam from the injector is accelerated through three passes to the final energy of 60 GeV. The average beam current is a modest 6.4 mA. An important aspect of the design is the choice of arc parameters to minimize energy loss due to synchrotron radiation. The current design uses an arc radius of 1 km which leads to a total circumference of 8.9 km.

## 6.3 *Electron cooler*

Electron cooling is achieved when a relatively low energy electron beam co-propagates with a relatively high energy ion beam through a dedicated channel and removes thermal energy from the ion beam. This process allows for collisions at higher luminosities.

### 6.3.1 *MEIC at Jefferson Laboratory*

Over the last decade, Jefferson Laboratory has been investigating the science and various machine feasibility studies for an electron–ion collider. The result has been a design for a polarized Medium Energy Ion Collider (MEIC), a ring–ring collider with three interaction points, which covers beam energies up to 11 GeV for electrons, 100 GeV for protons and 40 GeV/u for heavy ions [40]. A critical component to this design is the ambitious medium energy electron cooler, a 55 MeV, 1500 mA machine which necessitates the use of energy recovery. Unlike the FELs (past and present) at Jefferson Laboratory which achieve high currents using modest bunch charges at high repetition frequencies, the cooler design represents a challenge in that it uses high repetition rates (750 MHz) in conjunction with high bunch charge (2 nC).

## 6.4 *Test facilities*

From the survey of previously, and currently, operational ERLs, one can see that most of the operational experience is at lower energy (∼100 MeV) and with modest currents (few mA). Achieving the parameters outlined above requires a leap of an order of magnitude in energy and/or beam current. To that end, several test facilities have been proposed which seek to demonstrate more fully the technologies required for the next generation of ERLs, that high brightness, CW beam can be generated (gun, injector) and preserved during acceleration (SRF). Each one has been approved by their respective funding agencies and they are in various stages of construction (except for MESA, which only just recently received financial support). In terms of the near future, these machines will be the next to populate the ERL landscape.

### 6.4.1 *BERLinPro at Helmholtz-Zentrum Berlin*

At Helmholtz-Zentrum Berlin (HZB) plans are moving forward with a 50 MeV, 100 mA facility named BERLinPro [41]. One of the emphases of the program is not only to minimize beam loss but develop techniques and diagnostics to detect loss. Further, there is an emphasis on providing flexible beam optics: to control BBU via phase advance, minimize impact of CSR driven degradation, to provide large acceptance transport, and to use tunable momentum compactions — first and second order — to compress and linearize the longitudinal phase space. Additionally, work is proceeding on the development of a 1.5–2.0 MeV SRF photoinjector (77 pC) aimed at generating normalized emittances of 1 mm-mrad.

### 6.4.2 *cERL at KEK*

As an intermediate step before building the 3 GeV X-ray light source at KEK, a collaboration comprised of laboratories and universities is making progress in the design and construction of a Compact ERL (cERL) facility. The primary purpose is to demonstrate proficiency in the two key technologies: the gun and the SRF accelerator. The cERL machine is taking a staged approach and will deliver 35 MeV and 10 mA (7.7 pC) of beam in the first phase, before ramping up to 245 MeV and 100 mA (77 pC) — through a 2-pass/up down configuration — in the second phase [42].

### 6.4.3 *Mainz Energy-Recovering Superconducting Accelerator (MESA)*

Building on its experience with CW, multi-turn accelerators, the University of Mainz is moving forward with a 105 MeV, 1 mA multi-turn machine design called MESA [43]. The facility can be operated as an ERL or as a recirculating linac to increase the beam energy. Unlike other test facilities which are generally intermediate steps to light sources, the aim of MESA is to serve as a user facility for particle physics experiments.

### 6.4.4 *Brookhaven National Laboratory*

Like many of the test facilities described, the ERL at BNL will investigate high current phenomena, halo generation and mitigation, high beam brightness generation and preservation, and SRF issues associated with high power beams, however it will do so with ampere-class beam currents [44]. The SRF photocathode gun is designed to deliver 500 mA (0.7 nC at 703.5 MHz) at 2 MeV. Construction is nearly complete and commissioning with beam is expected in early 2013.

### 6.4.5 *Institute of high energy physics*

In order to demonstrate proficiency with a DC photocathode gun (500 keV, 77 pC) and SRF cavities, the Institute of High Energy Physics (IHEP) in Beijing is designing a 35 MeV, 10 mA ERL test facility [45]. Using nonlinear bunch compression, the facility is designed to generate radiation in an oscillator-type FEL as well as terahertz from CSR. The motivation for the facility is to gain operational and technical experience with an ERL as they consider incorporating a large-scale ERL as the last phase of the proposed Beijing Advanced Photon Source (BAPS) [46]. The unique design calls for a single SRF linac that will drive a 6–8 GeV XFEL-O (without energy recovery) as well as a 5 GeV X-ray source (utilizing energy recovery).

### 6.4.6 *Peking University*

Peking University is in the midst of constructing a 30 MeV ERL-based IR FEL which will serve as a user facility [47]. The machine will utilize a DC photocathode gun (60 pC) and nonlinear bunch compression to achieve the required peak current at the wiggler. Limited by cryogenic capacity, the machine will operate in pulsed mode [48].

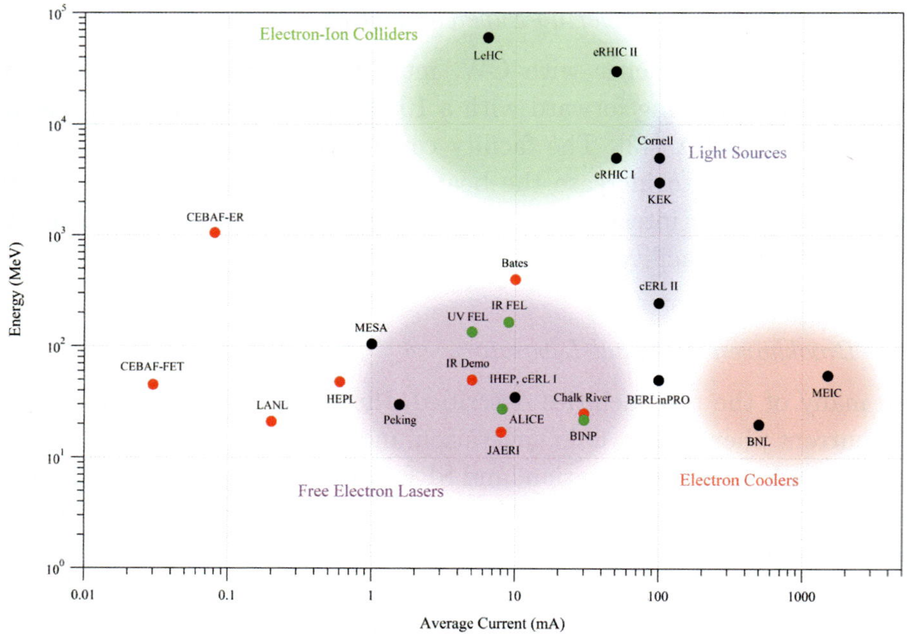


Fig. 9. ERL landscape. Red markers denote previous demonstrations of energy recovery, green markers indicate ERLs currently in operation and black markers represent future ERLs. (If the machine was, or will be, operated in pulsed mode, the value of the average current represents the current in the macropulse.)

## 7 Summary

A graphical representation of the current "ERL landscape" is depicted in Fig. 9, showing past, present, and proposed ERLs and where they lie in parameter space (energy and average current). The successes of early demonstrations of energy recovery have led to a proliferation of proposals for ERLs to meet a wide variety of demands — a testament to their versatility and robustness. In recent years advances in developing key technologies for the next generation of machines have been made, from high average current guns, to a better understanding of collective effects, beam instabilities and mitigation techniques, to high performance SRF systems. Several ERL test facilities are expected to be operational within a few years and will be able to provide valuable insight as they probe previously unreached regions of parameter space.

## Acknowledgements

The author would like to thank the Jefferson Laboratory FEL Team for their fruitful discussions and assistance in assembling this material. This work was supported by the Office of Naval Research, the Joint Technology Office, the Commonwealth of Virginia, the Air Force Research Laboratory, and by the DOE Contract DEAC05-84ER40150.

## References

[1] L. Merminga, D. Douglas, G. Krafft, *Annu. Rev. Nucl. Part. Sci.* **53**, 387 (2003).
[2] R. Eichhorn, Talk presented at *TESLA Technology Collaboration Meeting*, Newport News, VA (2012).
[3] R. Nagai *et al.*, *Rev. Sci. Instrum.* **81**, 033304 (2010).
[4] J. Teichert *et al.*, in *Proceedings of the Free Electron Laser Conference*, Malmö, Sweden, pp. 443–456 (2010).
[5] T. Powers and C. Tennant, in *Proceedings of the 41st Advanced ICFA Beam Dynamics Workshop on ERLs*, Daresbury Laboratory, UK, pp. 75–79 (2007).
[6] C. Tennant, Ph.D. thesis, College of William and Mary (2006).
[7] C. Tennant *et al.*, *Phys. Rev. ST Accel. Beams* **8**, 074403 (2005).
[8] D. Douglas *et al.*, *Phys. Rev. ST Accel. Beams* **9**, 064403 (2006).
[9] D. Douglas *et al.*, Technical Note 12-017, Jefferson Laboratory (2012).
[10] P. Evtushenko *et al.*, in *Proceedings of the Free Electron Laser Conference*, Nara, Japan, pp. 670–673 (2012).
[11] C. Tennant, in *Proceedings of the Particle Accelerator Conference*, Vancouver, BC, 2009 (IEEE, Piscataway, NJ, 2009), pp. 3125–3129.
[12] C. Tennant and D. Douglas, Technical Note 12-027, Jefferson Laboratory (2012).
[13] F. Jackson *et al.*, in *Proceedings of the Particle Accelerator Conference*, New Orleans, LA, (IEEE, Piscataway, NJ, 2012), pp. 610–612.
[14] S. Benson *et al.*, in *Proceedings of the Particle Accelerator Conference*, Albuquerque, NM, 2007 (IEEE, Piscataway, NJ, 2007), pp. 79–81.
[15] D. Douglas, in *Proceedings of the 14th Beam Instrumentation Workshop*, Santa Fe, NM, pp. 506–515 (2010).
[16] S. Schriber *et al.*, in *Proceedings of the Particle Accelerator Conference*, Chicago, IL, 1977 (IEEE, Piscataway, NJ, 1977), pp. 1061–1063.
[17] J. Flanz and C. Sargent, in *Proceedings of the Particle Accelerator Conference*, Vancouver, British Columbia, 1985 (IEEE, Piscataway, NJ, 1985), pp. 3213–3215.
[18] T. Smith *et al.*, *Nucl. Instrum. Methods A* **259**, 1 (1987).
[19] D. Feldman *et al.*, in *Proceedings of the Particle Accelerator Conference*, Washington, DC, 1987 (IEEE, Piscataway, NJ, 1987), pp. 221–223.
[20] N. Nishimori *et al.*, in *Proceedings of the Free Electron Laser Conference*, Berlin, Germany, pp. 265–272 (2006).
[21] Report Project 87-R-203, Continuous Electron Beam Accelerator Facility (1985).
[22] N. Sereno *et al.*, in *Proceedings of the Particle Accelerator Conference*, Washington, DC, 1993 (IEEE, Piscataway, NJ, 1993), pp. 3246–3248.
[23] N. Sereno, Ph.D. thesis, University of Illinois at Urbana-Champaign (1994).
[24] G. Krafft and J. Bisognano, in *Proceedings of the Particle Accelerator Conference*, Chicago, IL, 1989 (IEEE, Piscataway, NJ, 1989), pp. 1256–1258.
[25] D. Douglas, Technical Note 91-017, Jefferson Laboratory (1991).
[26] G. Neil *et al.*, *Phys. Rev. Lett.* **84** (2000).
[27] D. Douglas, Technical Note 01-018, Jefferson Laboratory (2001).
[28] C. Tennant *et al.*, in *Proceedings of the 11th Workshop on RF Superconductivity*, Travemunde, Germany (2003).

[29] P. Piot *et al.*, *Phys. Rev. ST Accel. Beams* **6**, 030702 (2003).
[30] R. Alarcon *et al.*, *Phys. Rev. Lett.* **111**, 164801 (2013).
[31] D. Douglas *et al.*, in *Proceedings of the International Particle Accelerator Conference*, New Orleans, LA, (IEEE, Piscataway, NJ, 2012), pp. 2111–2115.
[32] D. Douglas *et al.*, Invention Disclosure (2012).
[33] Y. Saveliev *et al.*, in *Proceedings of the International Particle Accelerator Conference*, New Orleans, LA, (IEEE, Piscataway, NJ, 2012), pp. 616–618.
[34] N. Vinokurov *et al.*, in *Proceedings of the Russian Particle Accelerator Conference*, Protvino, Russia, pp. 133–135 (2010).
[35] N. Vinokurov, Private communication (2012).
[36] G. Hoffstaetter, S. Gruner, and M. Tigner, eds, "Cornell ERL Project Definition Design Report," http://erl.chess.cornell.edu/PDDR (2011).
[37] N. Nakamura, in *Proceedings of the International Particle Accelerator Conference*, New Orleans, LA, (IEEE, Piscataway, NJ, 2012), pp. 1040–1044.
[38] V. Ptitsyn *et al.*, in *Proceedings of the International Particle Accelerator Conference*, San Sebastian, Spain, (IEEE, Piscataway, NJ, 2011), pp. 3726–3728.
[39] C. Adolphsen *et al.*, CERN Technical Note LHeC-Note-2011-003 GEN (2011).
[40] Y. Zhang *et al.*, in *Proceedings of the International Particle Accelerator Conference*, New Orleans, LA, (IEEE, Piscataway, NJ, 2012), pp. 2014–2016.
[41] J. Knobloch *et al.*, in *Proceedings of the International Particle Accelerator Conference*, New Orleans, LA, (IEEE, Piscataway, NJ, 2012), pp. 601–603.
[42] S. Sakanaka *et al.*, *Proceedings of the International Particle Accelerator Conference*, New Orleans, LA, (IEEE, Piscataway, NJ, 2012), pp. 607–609.
[43] R. Heine *et al.*, *Proceedings of the International Particle Accelerator Conference*, New Orleans, LA, (IEEE, Piscataway, NJ, 2012), pp. 1993–1995.
[44] I. Ben-Zvi *et al.*, *ICFA Beam Dynamics Newsletter* **58**, 151–182 (2012).
[45] J. Zhai, Talk presented at *50$^{th}$ Advanced ICFA Beam Dynamics Workshop on ERLs*, Tsukuba, Japan (2011).
[46] S. Wang *et al.*, *Chinese Physics C* **36**(5), 469–474 (2012).
[47] S. Huang, Talk presented at *50$^{th}$ Advanced ICFA Beam Dynamics Workshop on ERLs*, Tsukuba, Japan (2011).
[48] K. Liu, Private communication (2012).

## Chapter 40

# FFAGs: Front-end for neutrino factories and medical accelerators

Yoshiharu Mori (Kyoto University)

The idea of Fixed Field Alternating Gradient (FFAG) accelerator was originated by different people and groups in the early 1950s. It was independently introduced by Ohkawa [Ohkawa (1953)], Symon *et al.* [Symon *et al.* (1956)], and Kolomensky [Kolomensky and Lebedev (1966)] when the strong Alternate Gradient (AG) focusing and the phase stability schemes were applied to particle acceleration. The first FFAG electron model was developed in the MURA accelerator project led by Kerst and Cole in the late 1950s. Since then, they have fabricated several electron models in the early 1960s [Symon *et al.* (1956)]. However, the studies did not lead to a single practical FFAG accelerator for the following 50 years. Because of the difficulties of treating non-linear magnetic field and RF acceleration for non-relativistic particles, the proton FFAG, especially, was not accomplished until recently. In 2000, the FFAG concept was revived with the world's first proton FFAG (POP) which was developed at KEK [Aiba (2000); Mori (1999)]. Since then, in many places [Berg (2004); Johnstone *et al.* (2004); Mori (2011); Ruggiero (2004); Trbojevic (2004)], FFAGs have been developed and constructed.

An FFAG accelerator has various advantages: strong focusing in 3D space and fast beam acceleration. Strong focusing in 3D space is achieved by the transverse strong (AG) focusing and the longitudinal phase focusing in RF acceleration. A static magnetic field allows for the rapid acceleration and also large repetition rate in operation, where the acceleration of short-lived particles such as muons and also intense averaged beam current could be achieved.

The FFAG accelerator features a static magnetic field and strong (AG) focusing. In order to coordinate these two issues in beam dynamics, the beam loss caused by betatron resonance crossing during beam acceleration has to be overcome. There are two possible ways to conduct this difficulty;

one is to arrange a special beam optics keeping the operating betatron tunes constant, thus featuring a zero-chromatic optics. The other is to cross the betatron resonances as quickly as possible to minimize the beam losses during acceleration. The former type is called "scaling FFAG" and the latter "non-scaling FFAG".

In the scaling FFAG where the betatron tunes are always constant during acceleration, a very large momentum acceptance of more than $\pm 100\%$ could be possible. On the other hand, the non-scaling FFAG in which all optical elements are essentially linear could have large dynamic apertures when the fast resonance-crossing, that is, fast acceleration becomes possible. Since the orbit excursion of non-scaling FFAGs is rather small compared to that of scaling FFAG, small aperture magnets become an option.

## 1 Zero-chromatic beam optics in scaling FFAG

The zero-chromatic optics in scaling FFAG, where the betatron tunes in transverse plane are constant for different beam momentum in the circular orbit ($s = r\theta$), can be achieved with specific magnetic field configurations to satisfy the orbit similarity and constant geometrical field index:

$$\frac{d(r^2/\rho^2)}{dp} = 0, \tag{1}$$

and

$$\frac{d(K\rho^2)}{dp} = 0. \tag{2}$$

Here, $K$ is defined as a form with magnetic field gradient:

$$K = -\frac{1}{B\rho}\frac{\partial B_z}{\partial r}. \tag{3}$$

The concept of FFAGs was originally limited to ring accelerators where the orbit excursion evolves horizontally just as in an ordinary cyclotron and synchrotron. However, Okawa proposed later a vertical excursion FFAG in 1955 [Ohkawa (1955)] and he named it as "electron cyclotron" because the optics satisfies an isochronous condition for relativistic particles such as electrons. Also even in vertical direction, the zero-chromaticity is realized. This idea was revisited recently by Brooks [Brooks (2004)].

In case of the horizontal orbit excursion, the orbit similarity and constant geometrical field index required for the zero-chromaticity shown in Eqs. (1)

and (2) lead to the following magnetic field configuration:

$$B_z = B_0 \left(\frac{r}{r_0}\right)^k f\left(\theta - \zeta \ln \frac{r}{r_0}\right), \tag{4}$$

where

$$\zeta = \tan \xi. \tag{5}$$

Here, $\xi$ is a spiral angle and $k$ is a geometrical field index shown as

$$k = \frac{r}{B_z}\left(\frac{\partial B_z}{\partial r}\right). \tag{6}$$

Using this field configuration, two types of beam optics can be derived that accomplish a zero-chromatic scaling FFAG. One is called the radial sector lattice and the other the spiral sector lattice, respectively. In the radial sector lattice, the AG focusing takes a FODO structure (Focusing Bend Defocusing Bend sequence) with a negative bend gradient magnet. On the other hand, in the spiral sector lattice, the alternating focusing and defocusing can be realized with the edge effect. Figures 1 and 2 show the orbit configurations for these two optics, respectively, in a circular scaling FFAG.

In case of the vertical orbit excursion, the zero-chromatic condition requires the following magnetic field configurations [Brooks (2004); Ohkawa (1955)].

$$B_z = B_0 \exp\left(\frac{n}{\rho} z\right). \tag{7}$$

Here, $n$ is an ordinary field index and the beam optics is expressed with horizontal and vertical coupled betatron motions.

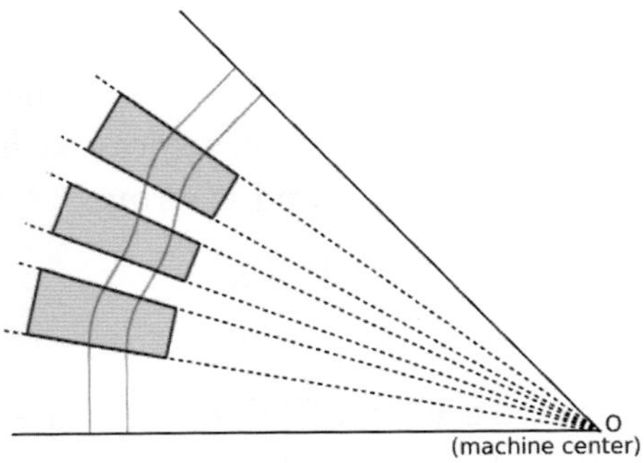


Fig. 1. Radial sector type.

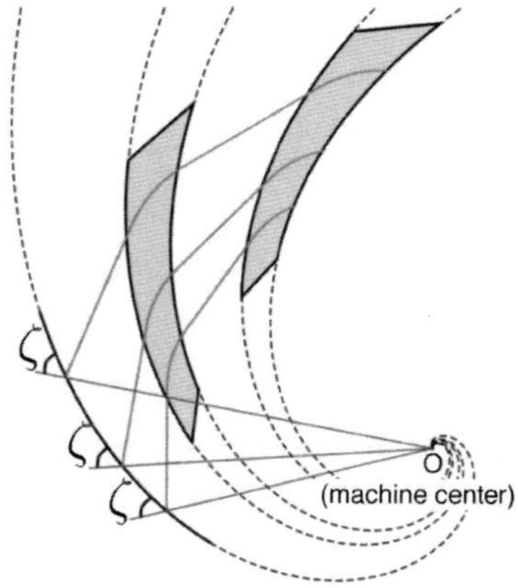


Fig. 2. Spiral sector type.

In the ordinary circular scaling FFAG lattice assuming an azimuthal symmetry, there are some disadvantages. Large dispersion and orbit excursion require large horizontal apertures of the magnet and the RF cavity, and the space of the magnet-free straight section is rather small for placing the injection/extraction devices and RF cavities for beam acceleration. Having a long straight line to keep zero-chromaticity and match to the scaling arc can overcome these difficulties.

The orbit curvature and the field index must be constant to satisfy the zero-chromaticity in this frame, which leads the magnetic field configuration to be an exponential form shown as [Lagrange *et al.* (2009)]:

$$B_z = B_0 \exp\left[\left(\frac{n}{\rho}\right) x\right]. \tag{8}$$

Using the scaling FFAG straight lattice, we could realize a dispersion suppressor and also matching insertion with the curved scaling FFAG lattice. For the dispersion suppressor, successive $\pi$-cells in the horizontal plane can suppress the dispersion.

In order to match the straight line with the circular ring FFAG lattice, the 1st order (linear) matching condition expressed in Eq. (9) has to be satisfied between the straight section and the ring.

$$\frac{k+1}{r} = \frac{n}{\rho} \tag{9}$$

Using a newly discovered scaling FFAG straight line, the design of scaling FFAG becomes more flexible and capable for various applications and the scaling FFAG opens a new advanced stage as shown in Fig. 3.

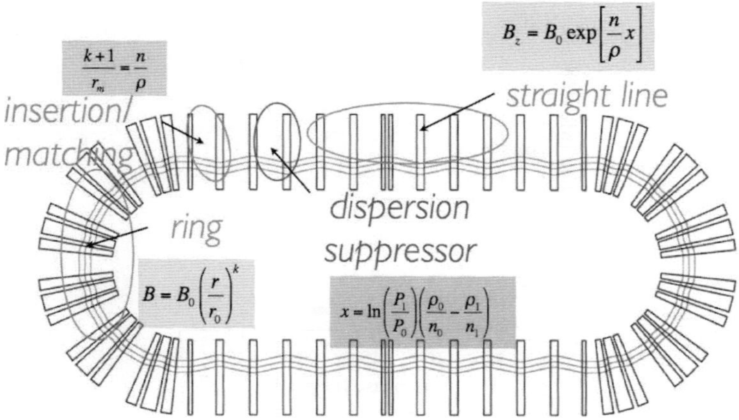


Fig. 3. A new advanced stage of scaling FFAG optics.

## 2 Beam acceleration in FFAGs

In RF acceleration, we also had an advancement in the scaling FFAG design. The beam acceleration in the scaling FFAG has some varieties because the momentum compaction is always strictly constant for different beam energies and has no higher orders. This situation takes either variable frequency of fixed frequency RF in beam acceleration. For the variable frequency RF acceleration, a broad-band RF cavity using magnetic alloys becomes feasible, which has actually been used for the world's first proton FFAG (POP-FFAG) at KEK [Aiba (2000); Mori (1999)]. And, for the fixed frequency RF acceleration, the stationary bucket acceleration scheme can be useful for ultra-relativistic high energy particle such as muons and electrons.

There was also a new advancement in the fixed frequency RF acceleration. In the strong focusing machine, two RF buckets below and above the transition energy are interfered with some conditions, which were analyzed by Symon and Sessler in 1960s [Symon and Sessler (1956)], and a serpentine acceleration path channel between two buckets existed. The serpentine acceleration path shown in Fig. 4 was devoted to accelerate ultra-relativistic particles in the non-scaling FFAG [Machida *et al.* (2012)].

In the scaling FFAG, Hamiltonian describing the longitudinal particle motion can be analytically derived [Yamakawa *et al.* (2009)] as presented in Eq. (10),

$$H = 2\pi m_0 c^2 \left[\frac{(\gamma_s^2 - 1)^\lambda}{2\gamma_s}\frac{(\gamma^2 - 1)^{-\lambda+1}}{(1-\lambda)} + \gamma\right] + \frac{eV_{\mathrm{RF}} f_0}{h}\cos\phi, \qquad (10)$$

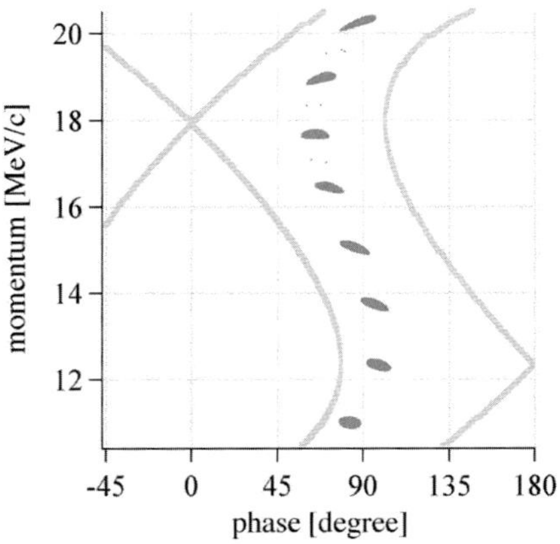


Fig. 4. Serpentine channel acceleration path [Machida *et al.* (2012)].

where

$$\lambda = \frac{k}{2(k+1)}. \tag{11}$$

The Hamiltonian shows that a serpentine path for acceleration exists for both non-relativistic and ultra-relativistic particles. Thus scaling FFAGs could use the serpentine acceleration scheme either for lepton or hadron beams.

## 3 Hadron FFAG accelerator and medical applications

After the success of the world's first proton FFAG (POP) shown in Fig. 5, a higher energy proton FFAG was constructed at KEK in 2004. The maximum energy of this machine was 150 MeV and a very high repetition rate of 100 Hz was demonstrated in operation. In 2007, this proton FFAG was moved to Kyusyu University and is used as a multi-purpose machine for various application fields, especially as a tool to accelerate secondary particles such as isomers [Yonemura (2008)].

In Kyoto University, Research Reactor Institute (KURRI), the experimental studies of ADSR (Accelerator Driven Sub-critical Reactor) have been implemented combining a FFAG proton accelerator with a KUCA (Kyoto University Critical Assembly) reactor.

The 150 MeV proton FFAG accelerator at KURRI is composed of Injector, Booster and Main Ring, and they are all FFAG rings as shown in Fig. 6. The beam is transported from the FFAG to KUCA through the long beam transport line.

In March 2009, the first beam from the FFAG was successfully injected into the KUCA reactor starting the ADSR experimental studies [Pyeon

Fig. 5. Photograph of the world's first proton FFAG (POP) [Aiba (2000); Mori (1999)].

Fig. 6. FFAG complex at KURRI for ADSR experiment [Aiba (2000); Mori (1999); Pyeon (2009)].

(2009)]. As the FFAG operates with 10–100 Hz, prompt neutrons are created every 33 msec, then, the delayed neutrons amplified by nuclear fission reactions came out depending on the reactor sub-criticality. Recently, the beam intensity has been successfully increased almost 100 times by charge-exchanged $H^-$ ion beam injection with a newly built injector of 11 MeV $H^-$ linac [Ishi *et al.* (2010); Uesugi *et al.* (2008)].

Applications of the FFAG accelerators for medical use have been proposed in two different fields: hadron beam therapy and boron neutron capture therapy. The FFAG accelerators are seen as good candidates for hadron therapy applications, with various potential advantages such as conformal spot scanning treatment with high repetition pulsed mode operation compared to cyclotrons or pulsed synchrotrons.

In France, the RACCAM project [Antoine *et al.* (2009)] has been initiated, which aims at producing a preliminary design study of a variable energy proton installation based on a variable energy, 5 to 15 MeV, $H^-$ injector cyclotron followed by a spiral lattice FFAG ring with 70 to 180 MeV extraction energy. A schematic layout of scaling proton FFAG for proposed RACCAM cancer therapy project is shown in Fig. 7. In the UK, the PAMELA project [Yokoi (2011)] for designing a hadron therapy accelerator has been funded to invoke the achievements of the non-scaling FFAG accelerator, EMMA.

A cancer therapy hadron accelerator composed of three concentric non-scaling FFAGs has been proposed and the largest ring of the setup is shown in Fig. 8 [Keil *et al.* (2007)]. The smaller pair of FFAGs would accelerate protons to 250 MeV and the larger pair carbon ions to 400 MeV/u. Each ring is composed of 48 doublet cells and the circumference of the largest ring is $C = 52$ m. Modest RF voltages of less than 220 kV are sufficient to keep good beam quality while crossing many betatron resonances in acceleration.

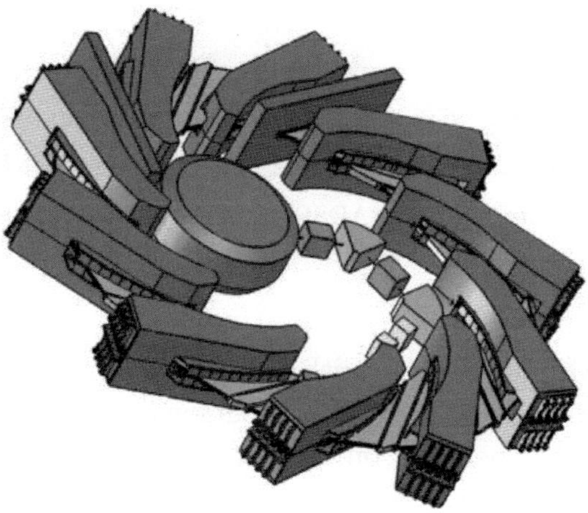

Fig. 7. A schematic layout of scaling proton FFAG for proposed RACCAM cancer therapy project [Antoine *et al.* (2009)].

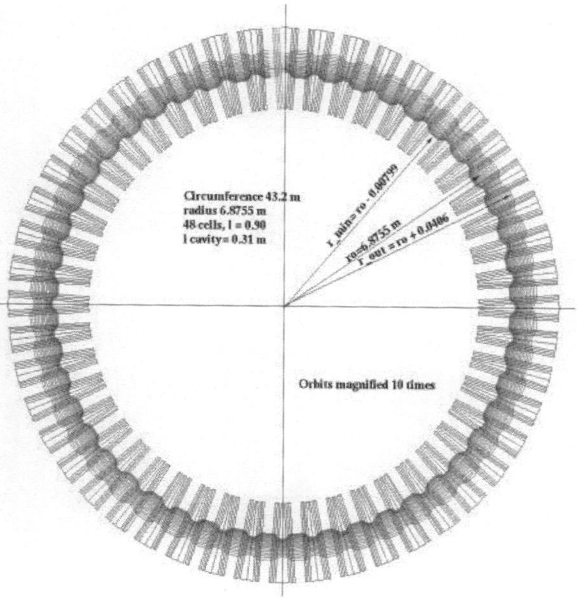


Fig. 8. Non-scaling FFAG ring of carbon accelerator for cancer therapy [Keil *et al.* (2007)].

A new type of compact neutron source called FFAG-ERIT (Emittance Recovery Internal Target) has been developed for the boron neutrino capture therapy at KURRI [Mori (2006)]. Figure 9 shows a photograph of FFAG-ERIT developed at KURRI. Neutrons are generated at the Be target placed internally into the proton FFAG storage ring. To suppress the emittance growth caused by Rutherford scattering, ionization cooling with energy recovery was adopted. The FFAG-ERIT ring has worked successfully as expected and a neutron yield of more than $10^{13}$ n/sec was obtained [Mori (2009); Okabe (2010)].

## 4 Muon accelerator for neutrino factory front-end

A muon phase rotation ring with FFAG optics for reducing the energy spread of muon beams called PRISM (Phase rotation Ring for Intense Slow Muons) has been developed at Osaka University. The ring has been developed for the experiment with rare $\mu$–e conversion events where the lower limit of the branching ratio should be less than $10^{-18}$ [LOI-PRISM (2003)]. Before constructing a full model, they have carried out a demonstration test of phase rotation for a beam with such large momentum spread with $\alpha$-particles (see Fig. 10) [Sato (2004)]. Although the number of RF cavities was just one for the test (five to six RF cavities are needed for the real experiment), the demonstration test could still clarify the phase rotation in principle, and

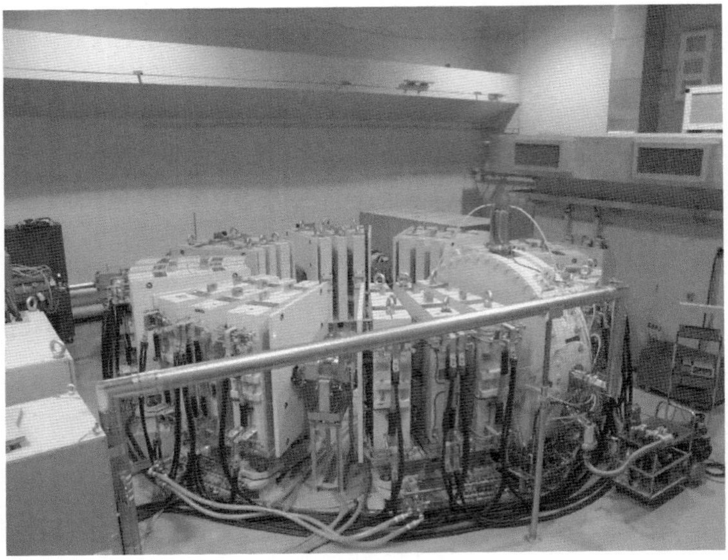

Fig. 9. FFAG-ERIT for intense neutron production.

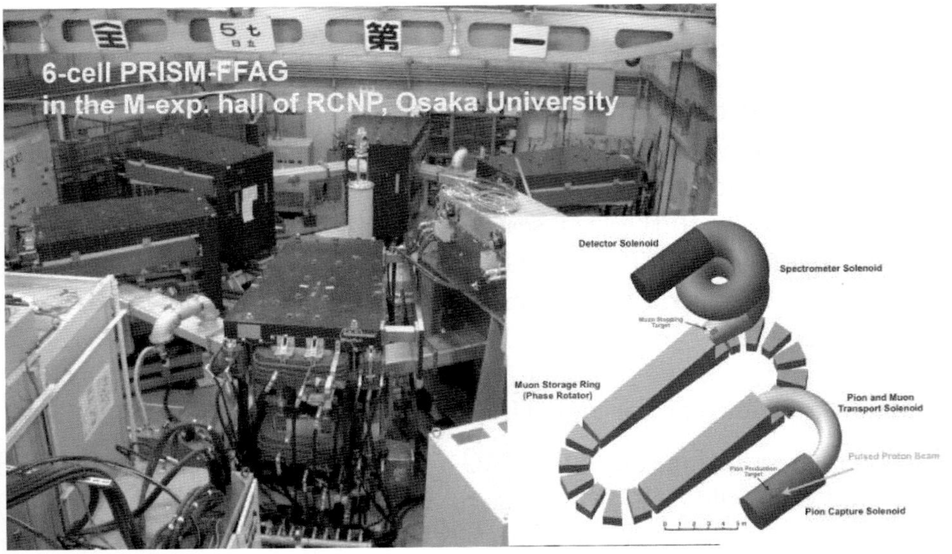


Fig. 10. Muon phase rotation ring with scaling FFAG for PRISM project [LOI-PRISM (2003); Sato (2004)].

the experimental results showed a good agreement with the results predicted by the beam simulation.

The world's first non-scaling FFAG electron model of muon accelerator (EMMA) for future neutrino factory has been successfully developed

Fig. 11. Photograph of EMMA: World's first non-stacking FFAG [Machida *et al.* (2012)].

recently in the UK [Machida *et al.* (2012)]. Figure 11 shows a photograph of EMMA. One of the unique features of this machine is the beam acceleration using fixed frequency RF described above. The neutrino factory (NuFact) which devotes the lepton flavor international collaboration experiment with high energy neutrino beams is based on the muon accelerator complex with non-scaling FFAG [NuFact (2011)]. As an injector of non-scaling FFAG, a recirculating linear accelerator (RLA) is a candidate in the present design of NuFact. However, the RLA is a cost-driving accelerator in the NuFact complex, and a scaling FFAG using stationary bucket or harmonic-number jump acceleration is also under consideration as an alternative [Planche *et al.* (2009)].

Advancements in the beam optics and dynamics based on the studies of zero-chromaticity and flexible designs in 3D phase space become possible only now. A very-low-energy neutrino factory (VLENF) project has also been discussed as a front-end for a neutrino factory. The facility is composed of a pion production system and a muon decay race-track ring with a straight length of between 50–75 m, where a race-track scaling FFAG consisting of zero-chromatic straight sections shown in Eq. (8) with a center momentum of 3.8 GeV/c, as shown in Fig. 12, is promising [Lagrange *et al.* (2013)]. The ring has large transverse and momentum acceptances of more than 0.01 m.rad and 20%, respectively.

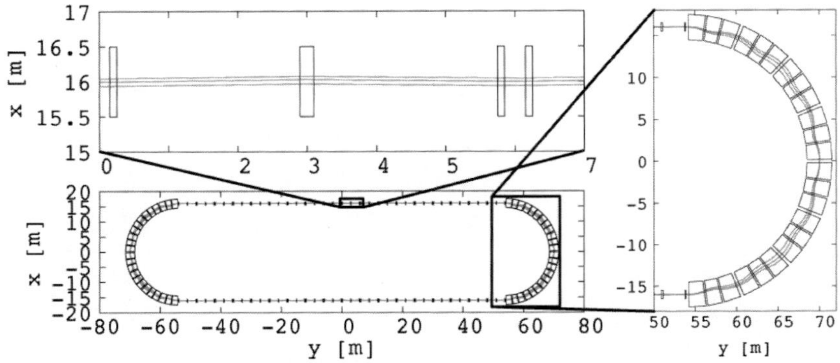


Fig. 12. Design of a race-track scaling FFAG for 3.8 GeV/c muon decay ring for nuSTORM project [Lagrange and Mori (2012)].

## 5 Summary

The unique features of FFAG compared with the ordinary accelerators are that it allows a strong beam focusing in 3D space (AG focusing and phase stability) and fast beam acceleration. These are both beneficial for high intensity beam acceleration and also for rapid acceleration of very short lived particles. The scaling zero-chromatic FFAG, in particular, has fairly large momentum acceptance, which could also be very useful for large emittance secondary particle beam acceleration. Actually, these features are not provided by ordinary ring accelerators such as cyclotrons or synchrotrons and recent studies and developments on scaling and non-scaling FFAGs show that FFAGs have very large capabilities in the fields of high intensity proton accelerator for ADS and muon accelerators for a neutrino factory. As for beam acceleration in FFAGs, fixed frequency RF acceleration such as serpentine path, stationary bucket and harmonic-number jump accelerations become feasible to allow very fast acceleration. Moreover, various RF gymnastics such as stacking, coalescing and multi-bunch acceleration can be possible. Advancements of the beam optics and dynamics of FFAGs based on the studies of zero chromaticity present more flexibility of beam optics designs in 3D phase space such as zero-chromatic straight line, race-track FFAG and vertical FFAG.

## 6 Acknowledgements

The author would like to express his sincere appreciation to all the people who gave him the various useful informations and results on their FFAG developments.

## References


M. Aiba *et al.*, *Proc. of EPAC2000*, Vienna, Austria, 2000, pp. 581–583.

S. Antoine *et al.*, *Nucl. Instr. Meth. A* **602**(3) (2009) 293–305.

T. Baba *et al.*, *Proc. of EPAC08*, Genoa, Italy, 2008, pp. 3371–3373.

S. Berg, *Proc. of Int. Workshop on FFAG Accelerators* (FFAG04), KEK, Tsukuba, 2004, pp. 1–9.

S. Brooks, *Proc. of IPAC'11*, 2011.

Y. Ishi *et al.*, *Proc. of IPAC'10*, Kyoto, Japan, 2010, pp. 1327–1329.

C. Johnstone *et al.*, *Int. J. Mod. Phys. A*, **26** (2011) 1690.

E. Keil, A. M. Sessler and D. Trbojevic, *Phys. Rev. ST-AB*, **10** (2007) 054701.

A.A. Kolomensky and A.N. Lebedev, *Theory of Cyclic Accelerators*, North-Holland, Amsterdam (1966), p. 340.

J.B. Lagrange and Y. Mori, internal report, acc-kurri-0508-02-2012, KURRI, 2012.

J.B. Lagrange, J. Pasternak and Y. Mori, *Proc. of NuFact'13*, IHEP Beijing, China, 2013.

J.B. Lagrange *et al.*, *Proc. of PAC'09*, Vancouver, Canada, 2009.

LOI (The PRISM Project) for Nuclear and Particle Physics Experiments at the J-PARC (2003).

S. Machida *et al.*, *Nature Physics*, **8** (2012) 243.

Y. Mori, *Nucl. Instr. Meth. PRS A* **563** (2006) 591–595.

Y. Mori, *Proc. of PAC'09*, Vancouver, Canada, 2009.

Y. Mori, *Int. J. Mod. Phys. A*, **26** (2011) 1744.

Y. Mori *et al.*, *Proc. of 12th Symp. on Accelerator Science and Technology*, Wako, Japan, 1999.

Neutrino Factory Design Group, IDS-NF, 2011.

T. Ohkawa, Workshop at annual meeting of JPS, 1953.

T. Ohkawa, *Phys. Rev.*, **100** (1955) 1247.

K. Okabe *et al.*, *Proc. of IPAC'10*, Kyoto, Japan, 2010, pp. 3897–3899.

T. Planche *et al.*, *Proc. of PAC'09*, Vancouver, Canada, 2009.

C.H. Pyeon *et al.*, *Journal of Nuclear Science and Technology*, **46**(12) (2009) 1091–1093.

A. Ruggiero, *Proc. of Int. Workshop on FFAG Accelerators* (FFAG04), KEK, Tsukuba, 2004, pp. 53–61.

A. Sato *et al.*, *Proc. of EPAC2004*, Edingburgh, Scotland, 2004, pp. 2508–2510.

K.R. Symon, D.W. Kerst *et al.*, *Phys. Rev.*, **103** (1956) 1837.

K.R. Symon and A.M. Sessler, *Proc. Int. Conf. High Energy Acc.* (1956), p. 44.

H. Tanaka *et al.*, *Proc. of Cyclotorons 2004*, Tokyo, Japan, 2004.

D. Trbojevic, *Proc. of Int. Workshop on FFAG Accelerators* (FFAG04), KEK, Tsukuba, 2004, pp. 9–10.

T. Uesugi *et al.*, *Proc. of EPAC08*, Genoa, Italy, 2008, pp. 1013–1015.

E. Yamakawa *et al.*, *Proc. of FFAG09*, FNAL, Chicago, USA, 2009.

T. Yokoi, *Int. J. Mod. Phys. A*, **26** (2011) 1887.

Y. Yonemura *et al.*, *Proc. of EPAC08*, Genoa, Italy, 2008, pp. 3523–3525.

# Chapter 41

# Fast cooling, muon acceleration and the prospect of muon colliders

Mark Palmer (Fermilab)

Facilities based on stored muons offer unique potential for future high-energy physics capabilities. Three key characteristics of the muon make this possible:

- The muon is a lepton;
- The muon is roughly 200 times as massive as the electron;
- The muon decays to an electron and two neutrinos.

As the next heavier members of the lepton family with respect to the electron and positron, $\mu^+$ and $\mu^-$ beams can be collided to provide a precision lepton probe of the electroweak couplings. This makes a muon collider a suitable option for a lepton collider companion to a hadron collider discovery machine.

The large mass of the muon has several implications. First it means that relatively little synchrotron radiation is emitted by these leptons when their trajectory is bent in a magnetic field. Thus circular colliders, which can support multiple interaction regions and their detectors, and multi-pass acceleration systems at high energies can be utilized. Similarly, the large muon mass means that beamstrahlung effects at the interaction point of a TeV-scale collider are minimized and a larger fraction of the interactions will take place near the nominal center-of-mass energy. Another aspect of muon collider interactions is that direct s-channel production of species such as the Higgs, with a much larger cross-section than that for associated production, is possible. These features offer unique physics potential for muon colliders, particularly in the several TeV center-of-mass energy regime.

The fact that muons are unstable particles means that a well-defined muon beam can serve as a precision source of electron and muon neutrinos through the decay process:

$$\mu^- \rightarrow e^- \bar{\nu}_e \nu_\mu$$
$$\mu^+ \rightarrow e^+ \nu_e \bar{\nu}_\mu$$

The use of a stored muon beam for neutrino oscillation experiments was first proposed in 1980 [1] and the physics potential of the "neutrino factory" (NF) was developed in detail in 1998 [2]. Particular advantages of having a precision source of this type are that it provides superb sensitivity to new physics, minimizes background issues, and provides a well-characterized beam, which aids in the control of systematic effects in the measurements. Thus NF capabilities are well-suited for discovery and precision measurement efforts in the neutrino sector.

Deployment of a muon accelerator complex could support our quest for both a world leading neutrino source as well as a precision lepton collider at center-of-mass energies up to several TeV. Of course, the unstable nature of muon beams, which helps provide this promising breadth of capability, is the source of some of the most significant technology challenges in effectively utilizing these beams. Thus, in order to form beams of useful intensities, the muons must be produced in copious quantities by tertiary production ($p \rightarrow \pi \rightarrow \mu$) where a suitable target is bombarded with a high-power proton beam. The large phase space volume occupied by the muons generated in this way leads to the necessity of cooling the emittance of the resulting muon beams. The short muon lifetime also means that all beam manipulations, including acceleration to the desired final beam energy, must be completed on timescales far shorter than those required for any other high-energy physics accelerators. An active R&D program is presently underway to design and evaluate the required technologies. Cost effective solutions for these technology challenges would pave the way to unique future high-energy physics capabilities.

## 1 Key challenges for muon-based HEP accelerators

Figure 1 shows block diagrams of the accelerator systems required for a long-baseline neutrino factory (NF) and a muon collider (MC). These two applications require the highest intensity muon sources under consideration for any physics application. Each design includes:

- A high-power *proton driver* capable of providing MW-class proton beams to a pion production target — a summary of key proton driver concepts has been provided in Chapter 34;

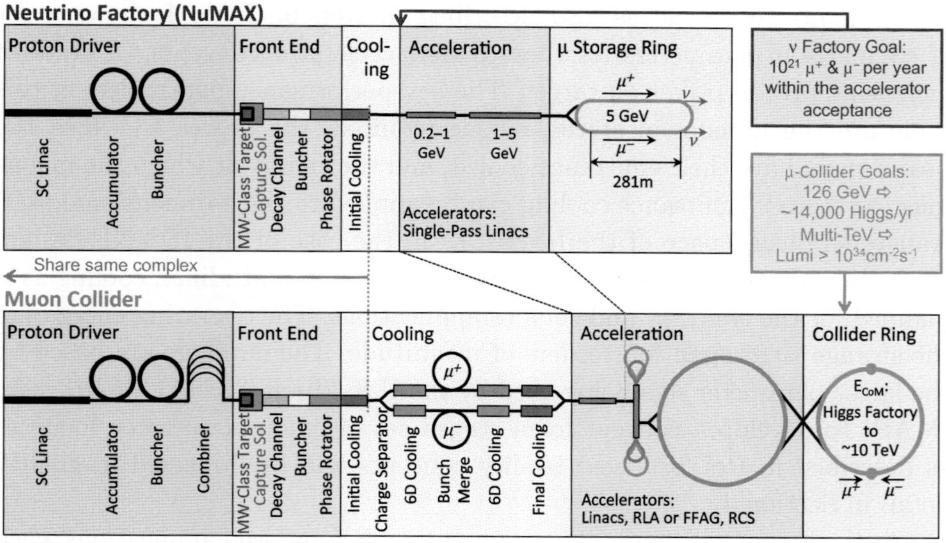


Fig. 1. The two block diagrams show the major muon accelerator sub-systems required for a long-baseline neutrino factory and a muon collider. As discussed in Section 4, the significant overlap of the sub-systems leads to a logical staging plan for deploying these capabilities.

- A *front end* which includes a high-power target for pion production, a capture solenoid capable of simultaneously capturing both positive and negative pions, a decay channel followed by phase rotator and buncher sections to properly prepare the resulting muon beams for further manipulation;
- A *muon cooling* section capable of significant reduction in the muon beam emittance and which can provide muon beams with suitable bunch structure for NF or MC use;
- An *acceleration system* which can take muon beams to the roughly 5–10 GeV scale for the NF and to the 0.1–10 TeV-scale for the MC;
- Muon *storage rings* to serve as the decay ring for the NF and as the collider ring for the MC.

The challenges for high-energy physics (HEP) muon applications begin with the production of muons in sufficiently large quantities to enable useful measurements. In order to set the scale, we can consider the number of muons per year required for three classes of proposed applications: a short-baseline NF as envisioned in the nuSTORM concept [3, 4], a long-baseline NF such as the design developed by the International Design Study for a Neutrino Factory (IDS-NF) [5] and for the NuMAX NF concept [6], and finally for a collider at center-of-mass energies from the Higgs resonance to the multi-TeV scale as also described in Ref. 6. Table 1 shows a set of comparison parameters for these muon accelerator concepts, in order of increasing proton power on target. The key performance parameters in this table are the number of stored muons required, the degree to which the beams must have their emittance cooled, and the energy at which the muons must be stored. Emittance cooling can be applied to the 4-dimensional (4D) transverse phase space of the beams, as in the case of the IDS-NF design, or can include 6-dimensional (6D), transverse plus longitudinal, cooling as is planned for the NuMAX and collider applications. The operating energies of the storage rings span four orders of magnitude. The numbers are based on the latest parameter sets being used by the U.S. Muon Accelerator Program (MAP) to develop a concept for a staged set of muon accelerator facilities as described in Ref. 6. These facility concepts will be discussed in greater detail in Section 4.

At the intensities required for high-energy NF and MC applications, each of the above systems has unique challenges. Radiation issues are particularly relevant to the front end, which must be able to handle the spent beam and undesirable side products coming from the high-power target. All of the systems beyond the target must be designed to operate in the radiation load of muon decays. Furthermore, all of the beam manipulations must be carried out quickly enough to avoid unacceptable loss of muon intensity.

The significant overlap between the NF and MC systems as shown in Figure 1 leads immediately to consideration of a facility concept aimed at the support of both capabilities in a staged fashion. This concept will be discussed in greater detail in Section 4. The next two sections will discuss two of the unique technical challenges for muon accelerator capabilities — how to produce low emittance muon beams of high intensity and how to accelerate the muon beams to very high energies within the very short lifetime of the muon.

## 2 Muon cooling concepts

With the mass of the muon being too large for synchrotron radiation damping to be effective, an alternative method to cool the emittance of the muon beams is required. Furthermore, given the short lifetime of the muon, active cooling methods that have been used with hadron beams do not offer high enough damping rates to be effective. Thus, the process of choice for cooling muon beams is to use $dE/dx$ energy loss as the damping mechanism — the process commonly known as "ionization cooling."

Table 1. Nominal muon and proton requirements for proposed HEP muon applications. Key performance drivers are the number of stored muons required (which is tightly coupled to the required proton power on target), whether the muon emittance cooling process is employed, and the final energy of the beams.

| | nuSTORM* | NuMAX (commission) | NuMAX | NuMAX+ | IDS-NF | Collider |
|---|---|---|---|---|---|---|
| Stored muons/yr (per species) | $8 \times 10^{17}$ | $1.3 \times 10^{20}$ | $4.7 \times 10^{20}$ | $1.3 \times 10^{21}$ | $1 \times 10^{21}$ | $3 \times 10^{20}$ |
| Storage ring energy (GeV) | 3.8 | 5 | 5 | 5 | 10 | $10^2$–$10^4$ |
| Muon cooling | No | No | 6D | 6D | 4D | 6D |
| Nom. protons on target/yr | $1 \times 10^{20}$ | $9.2 \times 10^{21}$ | $9.2 \times 10^{21}$ | $2.5 \times 10^{22}$ | $3 \times 10^{22}$ | $3.7 \times 10^{22}$ |
| Proton beam energy (GeV) | 120 | 6.75 | 6.75 | 6.75 | 10 | 6.75 |
| Proton beam power (MW) | 0.2 | 1 | 1 | 2.75 | 4 | 4 |

*nuSTORM stores a single muon species at a time, whereas the other facilities store both species simultaneously.

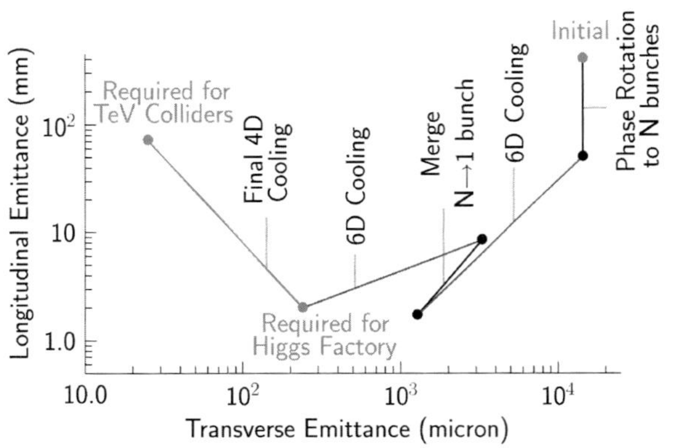


Fig. 2. Plot of the required evolution of the muon beam emittance for application as a Higgs factory, where very low beam energy spread is desirable, and for TeV-scale colliders, where the transverse emittance must be minimized in order to achieve the desired luminosity performance. In the present designs being considered by MAP, the initial phase rotation would result in approximately 20 bunches in the initial muon bunch train.

Figure 2 shows the required evolution of the muon beam emittance in an ionization cooling channel in order to support MC applications. It should be noted that the target emittance for a Higgs factory (MCHF) is quite different from that for a multi-TeV collider. For the MCHF, the goal is to provide the collider ring with a beam having very small longitudinal emittance to minimize the energy spread in the collider ring. Thus MCHF lattice designs are able to exploit energy spreads of a few parts in $10^5$ to directly probe the width of the Higgs. On the other hand, multi-TeV collider designs require luminosities of $>10^{34}\,\mathrm{cm}^{-2}\mathrm{s}^{-1}$ in order to provide sufficient event rates to adequately probe the physics processes of the terascale.

Following Neuffer [7,8], the rate of reduction of the normalized transverse beam emittance, $\epsilon_n$, in the absorber medium in a cooling channel can be written approximately as:

$$\frac{d\epsilon_n}{ds} = -\frac{1}{\beta^2}\left\langle\frac{dE_\mu}{ds}\right\rangle\frac{\epsilon_n}{E_\mu} + \frac{1}{2\beta^3}\frac{\beta_\perp(0.0136\,\mathrm{GeV})^2}{m_\mu E_\mu X_0}$$

where $s$ is the path length, $\beta$ is the muon velocity, $E_\mu$ is the muon energy, $\beta_\perp$ is the transverse beta function (which is the same for both transverse dimensions in a cylindrically symmetric channel), $m_\mu$ is the muon mass, and $X_0$ is the radiation length of the absorber material. The first term is the cooling term while the second gives the growth due to multiple scattering. The corresponding equilibrium emittance can be written as:

$$\epsilon_{n,\mathrm{eq}} = \frac{\beta_\perp(0.0136\,\mathrm{GeV})^2}{2\beta m_\mu X_0\left|\frac{dE_\mu}{ds}\right|}.$$

The equilibrium emittance is minimized by designing a lattice with a small $\beta_\perp$ and utilizing an absorber with a large radiation length, $X_0$. The energy lost by the muons while passing through this medium is restored by RF cavities that are incorporated into the cooling channel lattice.

In order to obtain effective cooling in the longitudinal dimension, absorbers of varying density or geometric extent can be used in regions of dispersion to induce emittance exchange between the longitudinal and transverse degrees of freedom, thus resulting in full 6D cooling lattices.

Figure 3 shows the familiar $dE/dx$ curve for a variety of materials. The optimal energy for operating a 6D ionization cooling channel is driven by two considerations:

(1) If the slope of the $dE/dx$ curve is negative in the channel, then particles with smaller energy will experience greater energy loss leading to an increase in longitudinal emittance. This is the case for energies that are less than the minimum ionizing energy, and is generally to be avoided.
(2) For energies above minimum ionizing, the slope of the $dE/dx$ curve is slightly positive, which could aid in the 6D cooling. However, since the damping rate scales as the fractional loss in energy, the required RF to replace the energy losses quickly becomes prohibitive.

Thus, the optimum energy for operating an ionization cooling channel is generally quite close to the minimum ionizing point on the $dE/dx$ curve, i.e. near a momentum of 200 MeV/c for muons.

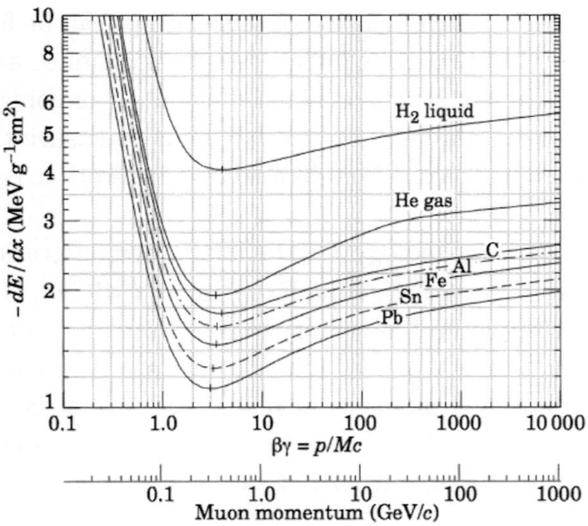


Fig. 3. $dE/dx$ curve for the interaction of muons with various materials.

Recent design efforts have focused primarily on two cooling channel technology concepts:

- Cooling channels utilizing RF cavities operating in vacuum with discrete absorbers for energy loss [9] — the "vacuum cooling channel" (VCC) concept;
- Cooling channels filled with high-pressure hydrogen gas to help suppress cavity breakdown in high magnetic fields and serve as a distributed absorber material [10] — the "helical cooling channel" (HCC) concept.

Detailed simulations of 6D cooling lattices based on these two technologies demonstrate performance consistent with achieving the target MCHF emittance as shown in Figure 2 while preserving sufficient muon transmission to provide the required bunch charges to the storage ring.

The final step in achieving the required transverse emittances for multi-TeV collider applications is to transit the red line shown in Figure 2. This "Final Cooling" stage describes a trade-off between the relatively large transverse and small longitudinal emittance of the Higgs factory operating point to an operating point with significantly smaller transverse and larger longitudinal emittance. In order to move between these two points, a process of emittance exchange at low beam energies is envisioned. Thus this final stage of cooling now explicitly takes advantage of the negative slope of the $dE/dx$ curve to allow the transverse emittance to cool at the expense of the longitudinal emittance. A description of this concept can be found in Ref. 11.

A key issue for the 6D cooling lattice designs mentioned above is the need to operate RF cavities in strong magnetic fields. An ongoing R&D effort in the MuCool Test Area (MTA) at Fermilab seeks to ensure that RF designs capable of supporting the cooling channel re-acceleration requirements are feasible [12, 13].

In addition to this technology R&D effort, a demonstration of ionization cooling, the Muon Ionization Cooling Experiment (MICE), is being deployed at the Rutherford Appleton Laboratory in the United Kingdom [14–17]. This experiment offers a unique opportunity to validate muon interactions with absorbers in the relevant momentum regime as part of its Step IV experimental program scheduled to begin in 2015 and to subsequently provide a demonstration of cooling with RF re-acceleration as part of its Step V experimental program.

## 3 Muon acceleration

Once cold muon beams have been obtained, they must be rapidly accelerated to their desired energy to avoid unacceptable decay losses. This requires highly efficient acceleration systems, particularly at low energies. For neutrino factory applications, three accelerator types have been considered for use in the early phase acceleration: single-pass linacs; multi-pass recirculating linear accelerators (RLA); and circular fixed-field alternating gradient (FFAG) machines. The International Design Study for a Neutrino Factory (IDS-NF) [18] has explored various combinations of these machines. In the present IDS-NF baseline configuration, acceleration to 10 GeV is based on an initial linac followed by two RLAs. The accelerator systems utilize 200 MHz superconducting cavities in order to have sufficient aperture to accelerate beams that have been cooled by roughly a factor of 2 transversely in a 4D ionization cooling channel.

A key issue for the design of the acceleration system is the degree to which the muon beam emittance has been cooled before entering the acceleration stage of the machine. Thus the NuMAX design, which is being developed by MAP, proposes to use a 6D Initial Cooling section to provide smaller beams which can fit within an acceleration system that begins with 325 MHz superconducting cavities, but which rapidly moves to 650 MHz structures to minimize overall facility costs. The NuMAX concept is discussed in greater detail in the following section as part of a staged facility concept that could ultimately provide muon collider capabilities.

For acceleration to collider energies, the principal issue is to optimize the acceleration efficiency as one moves to progressively higher energies. For beam energies up to roughly 63 GeV (the Higgs factory target), the use of RLAs with multi-pass arcs provides an efficient concept. The use of the multi-pass arcs is novel for this technology, and a scaled test employing electron beams has been proposed [19]. Beyond the energies required for a Higgs factory, the use of pulsed rapid cycling synchrotrons (RCS) appears to be the most effective solution [20]. Figure 4 shows a quarter-cell section for such an RCS. The warm dipoles, which are rapid cycling, would have to operate at $\geq$400 Hz to satisfy the rapid cycling lattice requirements.

A final acceleration issue for multi-TeV colliders is the need for a dedicated low energy acceleration system capable of accepting the significantly increased bunch lengths and very low beam energy resulting from the Final Cooling stage described in the preceding section. A low energy inductive linac would likely be the only solution for the initial stages of acceleration in this scenario (see, for example, Ref. 21).

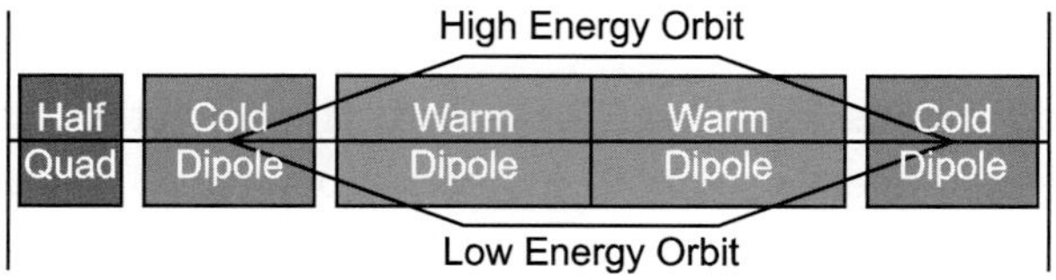


Fig. 4. A quarter-cell schematic for a pulsed muon RCS.

## 4 The prospect for muon colliders

As described in Ref. 6, one of the most important practical considerations when comparing the cost-effectiveness of multi-TeV colliders is the luminosity that can be provided per unit of wall power supplied to the facility. Figure 5 shows a comparison between linear collider options and a muon collider option using this metric. For center-of-mass energies above ~2 TeV, the muon collider option clearly offers better performance, by this measure, than the linear collider options.

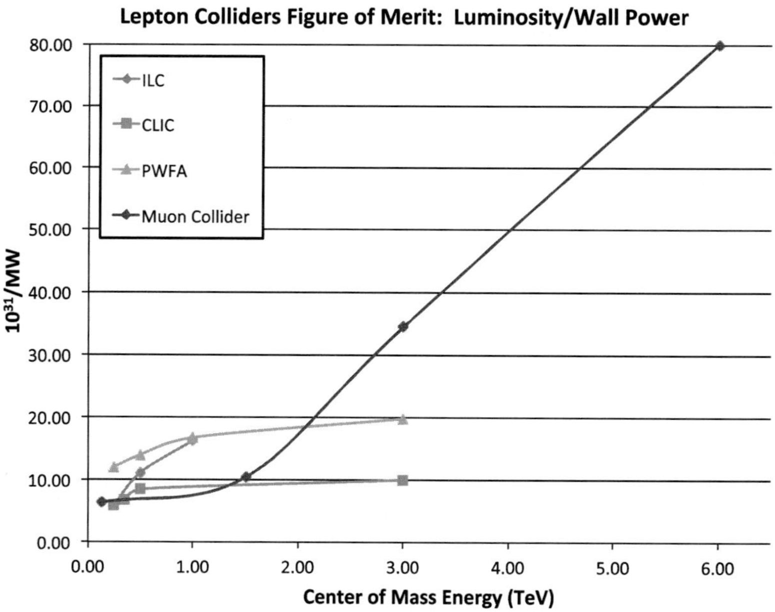


Fig. 5. The luminosity per unit wall power provided by a lepton collider represents a useful figure of merit when comparing the cost effectiveness of such facilities (see Ref. 6). For a muon collider, a significant "entry cost" exists due to the power required for the MW-class proton driver. As one moves to higher energies, however, the power requirements for the facility grow at a relatively modest rate, since synchrotron radiation is not a dominant issue, thus making a muon collider a very appealing multi-TeV facility option.

## 5 The muon accelerator staging study

In late 2012, MAP launched the Muon Accelerator Staging Study (MASS) to explore potential options for a staged set of muon capabilities that would enable both demonstration of the novel technologies required to produce and manipulate high intensity muon beams as well as provide a broad and continuous range of physics output. The staging concept was detailed in the white paper delivered by the Muon Accelerator Program to the 2013 U.S. Community Summer Study on high-energy physics [22]. Five potential stages of muon accelerator capabilities were identified. These stages included both short- and long-baseline neutrino factory options and extended to colliders that could reach the multi-TeV scale. Each stage was shown to provide unique physics reach:

**Stage 1** — nuSTORM [23] (**neu**trinos from **STOR**ed **M**uons):

nuSTORM is a proposed short-baseline NF for which no novel technology developments are required, thus representing a capability that could be deployed in the near term. The nuSTORM proposal describes a muon storage ring and detector capable of delivering $3 \times 10^{17}$ each of electron and muon neutrinos per year to a short-baseline detector. The integrated flux from storing $8 \times 10^{17}$ muons per year in the ring, which are produced by providing an integrated flux of $\sim 10^{20}$ protons per year (at an energy of 120 GeV) on target, would enable a test of the LSND (the Liquid Scintillation Neutrino Detector) short-baseline anomaly with a sensitivity $>10\sigma$ within five years. In addition, such a facility would provide an ideal vehicle for precision characterization of neutrino cross sections, particularly for $\nu_e$, which would significantly enhance the sensitivity of long-baseline neutrino oscillation studies.

**Stage 2** — NuMAX (**Neu**trinos from a **M**uon **A**ccelerator Comple**X**) [6]:

NuMAX is a long-baseline NF, which would serve as a precise and well-characterized neutrino source that exceeds the performance of conventional superbeam technology, particularly in terms of providing precisely specified flux determinations for each of the neutrino species in the beam. The NuMAX concept is based on a site-specific assumption that it would be located at Fermilab and would send beams to the Sanford Underground Research Facility in South Dakota. An operating energy for the muon storage ring in the range 4–6 GeV is optimal for this baseline and the nominal NuMAX energy has been chosen to be 5 GeV. However, flexibility in the operating energy is envisioned if motivated by the physics needs. It is assumed that the Fermilab Proton Improvement Plan (PIP [24], PIP-II

[25]) upgrade path will ultimately be extended to provide a 3 GeV $H^-$ linac, which could then be upgraded with accelerator systems as specified in the NuMAX plan. These systems include a dual-use $H^-$ and muon linac, accumulator and buncher rings to properly format the proton beams, the NF target and front-end systems, a muon pre-accelerator, and the muon storage ring. A diagram summarizing the major elements of the NuMAX accelerator complex is shown in Figure 6.

In its commissioning configuration, NuMAX would require no muon cooling and would operate with 1 MW of proton power on target. Addition of a 6D Initial Cooling stage, which would provide an increase in neutrino flux by a factor of $\sim$4, would establish the nominal operating point of the facility.

As with the IDS-NF detector concept (see Ref. 5), NuMAX would require a magnetized far detector (see Ref. 22), which, as with the accelerator, could be deployed in stages. This could be a variant of the MIND detector [26]

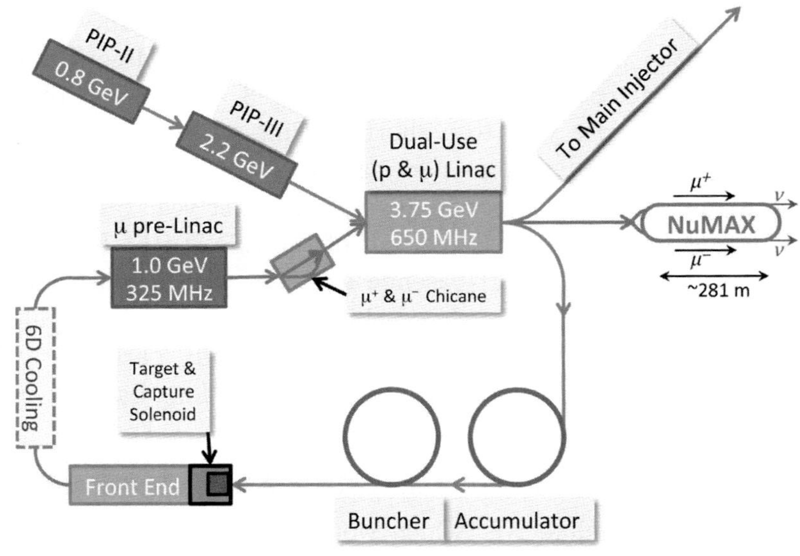


Fig. 6. Block diagram showing the principal elements of the NuMAX NF layout. Blue lines/arrows show the paths for the $H^-$/p beam and red lines/arrows show the paths for muons. NuMAX deployment would begin with the 3.75 GeV dual-use linac. The beam from this linac could be used to feed beam to the Main Injector while the remainder of the NuMAX accelerator complex was installed and commissioned. These additional elements include: the accumulator and buncher rings to format the proton beam before sending it to the pion production target; the production target, capture solenoid and front end systems, the muon pre-linac; a timing chicane to properly synchronize the $\mu^+$ and $\mu^-$ beams at the transition between the 325 MHz pre-linac and the 650 MHz dual-use linac; and the NuMAX muon decay ring, which would be pointed towards SURF in South Dakota. The 6D cooling section is envisioned to be added after initial commissioning of the other systems is complete.

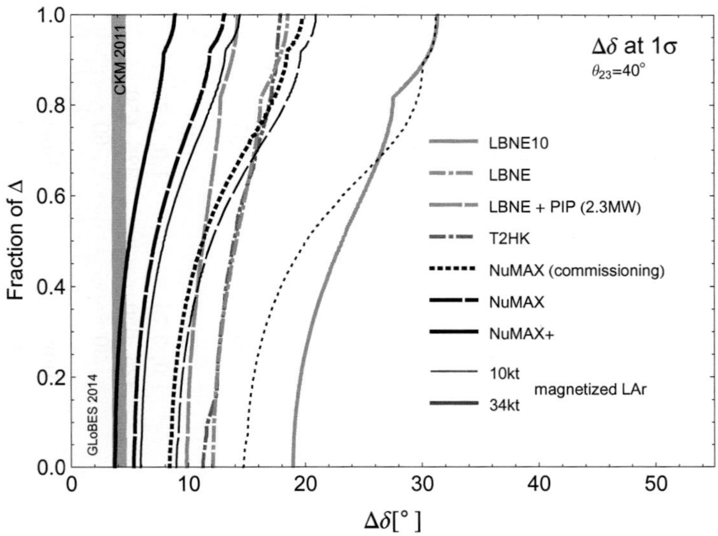


Fig. 7. Comparison of NuMAX(+) sensitivity for measuring the CP violating phase with other proposed neutrino oscillation experiments (from Ref. 6). The detector masses are those corresponding to the use of a magnetized LAr detector, if that concept can be successfully implemented for detectors of this mass scale.

optimized for operation at lower neutrino energies, a fully active magnetized plastic scintillator detector, or possibly a magnetized liquid argon TPC.

**Stage 3** — NuMAX+:

The ultimate NF performance would be obtained by upgrading the front end and target to handle 2.75 MW of proton power. The resulting sensitivity to CP violation in the neutrino sector approaches that of the sensitivities presently achieved in the quark sector. Comparisons with selected other long-baseline experimental options are shown in Figure 7.

Parameters for nuSTORM and the various NuMAX stages are shown in Table 2.

**Stage 4** — A muon collider Higgs factory (MCHF) [6, 27, 28]:

Muon collider capabilities could be deployed incrementally by augmenting the facilities previously deployed for NuMAX. The MCHF would require the addition of more aggressive 6D cooling for the muon beams, a further upgrade in proton beam power, the addition of an RLA to accelerate the muon beams to 63 GeV, and the addition of a collider ring and detector. Figure 8 shows the MAP concept for how a NF complex and MCHF might be deployed on the Fermilab site.

Table 2. Key parameters for the nuSTORM short-baseline NF and the NuMAX long-baseline neutrino factory concepts.

| System | Parameters | Unit | nuSTORM | NuMAX Commissioning | NuMAX | NuMAX+ |
|---|---|---|---|---|---|---|
| Performance | $v_e$ or $v_\mu$ to detectors/year | — | $3 \times 10^{17}$ | $4.9 \times 10^{19}$ | $1.8 \times 10^{20}$ | $5.0 \times 10^{20}$ |
| | Stored $\mu^+$ or $\mu^-$/year | — | $8 \times 10^{17}$ | $1.25 \times 10^{20}$ | $4.65 \times 10^{20}$ | $1.3 \times 10^{21}$ |
| Detector | *Far Detector:* | | SuperBIND | MIND/Mag LAr | MIND/Mag LAr | MIND/Mag LAr |
| | Distance from ring | km | 1.9 | 1300 | 1300 | 1300 |
| | Mass | kT | 1.3 | 100/30 | 100/30 | 100/30 |
| | Magnetic field | T | 2 | 0.5–2 | 0.5–2 | 0.5–2 |
| | *Near Detector:* | | SuperBIND | Suite | Suite | Suite |
| | Distance from ring | m | 50 | 100 | 100 | 100 |
| | Mass | kT | 0.1 | 1 | 1 | 2.7 |
| | Magnetic field | T | Yes | Yes | Yes | Yes |
| Neutrino Ring | Ring momentum ($P_\mu$) | GeV/c | 3.8 | 5 | 5 | 5 |
| | Circumference ($C$) | m | 480 | 737 | 737 | 737 |
| | Straight section length | m | 184 | 281 | 281 | 281 |
| | Number of bunches | — | | 60 | 60 | 60 |
| | Charge per bunch | $1 \times 10^9$ | | 6.9 | 26 | 35 |
| Acceleration | Initial momentum | GeV/c | — | 0.25 | 0.25 | 0.25 |
| | Single-pass linacs | GeV/c | — | 1.0, 3.75 | 1.0, 3.75 | 1.0, 3.75 |
| | | MHz | — | 325, 650 | 325, 650 | 325, 650 |
| | Repetition frequency | Hz | — | 30 | 30 | 60 |
| Cooling | | | No | No | Initial | Initial |
| Proton Driver | Proton beam power | MW | 0.2 | 1 | 1 | 2.75 |
| | Proton beam energy | GeV | 120 | 6.75 | 6.75 | 6.75 |
| | Protons/year | $1 \times 10^{21}$ | 0.1 | 9.2 | 9.2 | 25.4 |
| | Repetition frequency | Hz | 0.75 | 15 | 15 | 15 |

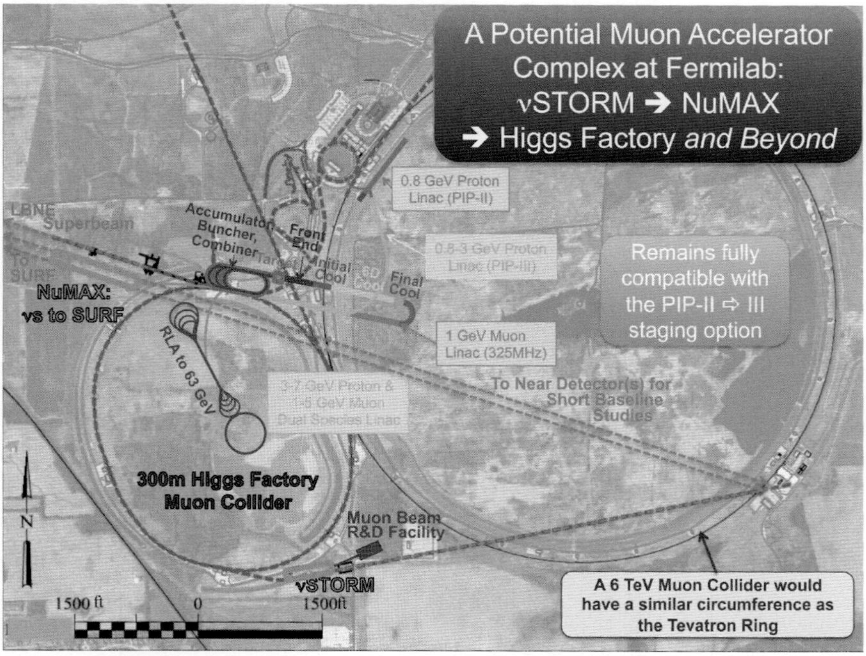


Fig. 8. The MAP concept for the layout of nuSTORM, NuMAX and a 126 GeV CoM muon collider Higgs factory on the Fermilab site. A proposed PIP-II and PIPIII linac layout, which is compatible with the deployment of muon accelerator capabilities, is also shown. The RLA and collider ring for the MCHF are shown on the infield of the Fermilab Main Injector. If extended beyond the MCHF facility, a multi-TeV muon collider facility would readily fit within the Fermilab site boundaries.

Table 3 summarizes the parameters of various MC options including the MCHF. The key features of the MCHF are its exquisite energy resolution, which would enable direct measurement of the Higgs width. The s-channel coupling of muons to the Higgs, which is $4 \times 10^4$ greater than for an electron–positron collider, allows the MCHF to be competitive at relatively low luminosity with the baseline parameters allowing $>10^4$ Higgs to be produced per Snowmass year ($10^7$ s). As described in Ref. 22, preliminary detector studies indicate that the detector backgrounds due to the muon decays in an MC can be satisfactorily managed by incorporating the necessary shielding in the machine/detector design and having a suitably pixelated detector with good timing resolution.

**Stage 5** — A multi-TeV muon collider:

A multi-TeV muon collider requires updates to the cooling channel (i.e., the addition of Final Cooling), the addition of acceleration capabilities, and deployment of a high-energy collider ring and detector. As with the MCHF,

Table 3. Parameters for a range of potential muon collider rings ranging from Higgs factory center-of-mass energies to the several TeV scale.

| Parameter | Units | Higgs factory: Startup operation | Higgs factory: Production operation | Top threshold options: High resolution | Top threshold options: High luminosity | Multi-TeV baselines | Multi-TeV baselines | Accounts for site radiation mitigation |
|---|---|---|---|---|---|---|---|---|
| CoM Energy | TeV | 0.126 | 0.126 | 0.35 | 0.35 | 1.5 | 3.0 | 6.0 |
| Avg. luminosity | $10^{34}\,cm^{-2}s^{-1}$ | 0.0017 | 0.008 | 0.07 | 0.6 | 1.25 | 4.4 | 12 |
| Beam energy spread | % | 0.003 | 0.004 | 0.01 | 0.1 | 0.1 | 0.1 | 0.1 |
| Higgs* or top$^\dagger$ production/$10^7$ sec | | 3,500* | 13,500* | 7,000$^\dagger$ | 60,000$^\dagger$ | 37,500* | 200,000* | 820,000* |
| Circumference | km | 0.3 | 0.3 | 0.7 | 0.7 | 2.5 | 4.5 | 6 |
| No. of IPs | | 1 | 1 | 1 | 1 | 2 | 2 | 2 |
| Repetition rate | Hz | 30 | 15 | 15 | 15 | 15 | 12 | 6 |
| $\beta^*$ | cm | 3.3 | 1.7 | 1.5 | 0.5 | 1 (0.5–2) | 0.5 (0.3–3) | 0.25 |
| No. muons/bunch | $10^{12}$ | 2 | 4 | 4 | 3 | 2 | 2 | 2 |
| No. bunches/beam | | 1 | 1 | 1 | 1 | 1 | 1 | 1 |
| Norm. trans. emittance, $\varepsilon_{TN}$ | $\pi$ mm-rad | 0.4 | 0.2 | 0.2 | 0.05 | 0.025 | 0.025 | 0.025 |
| Norm. long. emittance, $\varepsilon_{LN}$ | $\pi$ mm-rad | 1 | 1.5 | 1.5 | 10 | 70 | 70 | 70 |
| Bunch length, $\sigma_s$ | cm | 5.6 | 6.3 | 0.9 | 0.5 | 1 | 0.5 | 0.2 |
| Proton driver power | MW | 4 | 4 | 4 | 4 | 4 | 4 | 1.6 |

the preliminary detector studies suggest that a high-energy muon collider can be effective in its measurements.

As described in Ref. 22, the interactions at a MC operating at several TeV CoM energy are dominated by the fusion process — thus a multi-TeV MC is effectively an electroweak boson collider. In this sense, a MC will have as much discovery potential in the electroweak sector as a proton–proton collider of roughly 7 times the CoM energy. Given that the MC also offers superb energy resolution, this makes it a unique machine to explore the electroweak sector, including the Higgs, and if the LHC shows evidence for a supersymmetric spectrum, characterize physics processes beyond the Standard Model.

For the international collaboration working on muon accelerator technologies for HEP applications, the goal is to assess the feasibility of these technologies during the current decade. As described above, these machines have tremendous potential for high-energy physic research ranging from neutrino oscillation experiments to collider physics. If this assessment is positive, we could be in a position to consider deployment of such capabilities through the 2020's and 2030's.

## References

[1] D. Cline, D. Neuffer, "A muon storage ring for neutrino oscillations experiments," *AIP Conf. Proc.* **68**, 846–847 (1980).
[2] S. Geer, "Neutrino beams from muon storage rings: Characteristics and physics potential," *Phys. Rev. D* **57**, 6989 (1998).
[3] D. Adey *et al.*, "nuSTORM — Neutrinos from STORed Muons," arXiv:1308.6228.
[4] D. Adey *et al.*, "Light sterile neutrino sensitivity at the nuSTORM facility," *Phys. Rev. D* **89**, 071301(R) (2014).
[5] R. J. Abrams *et al.*, "Interim design report," arXiv:1112.2853.
[6] J.-P. Delahaye *et al.*, "A staged muon accelerator facility for neutrino and collider physics," *Proc. of IPAC'14*, Dresden, Germany, 2014, WEZAO2.
[7] D. Neuffer, in *Advanced Accelerator Concepts*, F. E. Mills, ed., *AIP Conf. Proc.* **156**, 201 (1987).
[8] R. C. Fernow, J. C. Gallardo, *Phys. Rev. E* **52**, 1039 (1995).
[9] D. Stratakis *et al.*, "Complete 6-dimensional cooling channel for a muon collider," *Proc. of IPAC'14*, Dresden, Germany, 2014, TUPME020, and references therein.
[10] K. Yonehara *et al.*, "Development of six-dimensional helical muon beam cooling channel for muon colliders," *Proc. of IPAC'14*, Dresden, Germany, 2014, TUPME014 and references therein.
[11] H. Sayed *et al.*, "Design and simulation of a high field-low energy muon ionization cooling channel," *Proc. of IPAC'14*, Dresden, Germany 2014, TUPME019.
[12] D. Bowring *et al.*, "RF design and operation of a modular cavity for muon ionization cooling R&D," *Proc. of IPAC'14*, Dresden, Germany 2014, MOOCA02.
[13] M. Chung *et al.*, "Pressurized H2 RF cavities in ionizing beams and magnetic fields," *Phys. Rev. Lett.* **111**, 184802 (2013).

[14] MICE, http://mice.iit.edu/, and references therein.
[15] J. C. Nugent, "Physics programme of next MICE Step IV," *Proc. of IPAC'14*, Dresden, Germany, 2014, TUPME010.
[16] P. Snopok *et al.*, "The status of the construction of MICE Step IV," *Proc. of IPAC'14*, Dresden, Germany, 2014, TUPME011.
[17] D. Kaplan *et al.*, "Progress towards completion of the MICE demonstration of muon ionization cooling," *Proc. of IPAC'14*, Dresden, Germany 2014, THPRI030.
[18] IDS-NF, https://www.ids-nf.org/wiki/FrontPage, and references therein.
[19] S. A. Bogacz *et al.*, "JEMMRLA — Electron model of a muon RLA with multi-pass arcs," *Proc. of IPAC'13*, Shanghai, China, 2013, WEOAB202.
[20] J. S. Berga and A. A. Garren, "Hybrid fast-ramping accelerator to 750 GeV/c: Refinement and parameters over full energy range," BNL-98171-2012-IR, MAP-doc-4335, and references therein.
[21] R. B. Palmer, "An overview of muon colliders," *ICFA Beam Dynamics Newsletter*, No. 55, M. A. Palmer (ed.), August 2011, Section 2.2.12.
[22] J.-P. Delahaye *et al.*, "Enabling intensity and energy frontier science with a muon accelerator facility in the U.S.: A white paper submitted to the 2013 U.S. Community Summer Study of the Division of Particles and Fields of the American Physical Society," arXiv:1308.0494.
[23] P. Kyberd *et al.*, "nuSTORM: Neutrinos from STORed Muons," arXiv:1206.0294.
[24] Proton Improvement Plan, http://www-ad.fnal.gov/proton/PIP/PIP_index.html.
[25] Proton Improvement Plan-II, https://pip2.fnal.gov/.
[26] A. Bross *et al.*, "A toroidal magnetised iron neutrino detector (MIND) for a neutrino factory," arXiv:1306.5327.
[27] Y. Alexahin *et al.*, "Preliminary design of a Higgs factory $\mu^+\mu^-$ storage ring," *Proc. of IPAC2013*, Shanghai, China, 2013, TUPFI061.
[28] D. Neuffer *et al.*, "A muon collider as a Higgs factory," *Proc. of IPAC2013*, Shanghai, China, 2013, TUPFI056.

# Part 3

# Particle Accelerators for the XXI Century

## Chapter 42

# The accelerator facility of the Heidelberg Ion-Beam Therapy Centre (HIT)

Andreas Peters (Universität Heidelberg)

## 1 Introduction

In 1946, Robert Wilson at FermiLab laid the foundation for hadron therapy with his famous article in *Radiology* about the therapeutic interest of protons for treating tumors [1]. He combined the discovery of the Bragg peak by William Henry Bragg in 1904 [2] with the fast inventions and development in accelerator technology in nineteen thirties and forties [3]. Only eight years later, the first patient treatments were started in 1954 at the Radiation Laboratory in Berkeley with proton, deuteron and helium ion beams from the 184 inch synchrocyclotron. From the sixties to the end of the eighties of the 20th century particle radiotherapy was based exclusively on accelerator facilities developed for nuclear physics, with beam-lines and treatment rooms adapted to the needs of radiotherapy. Then the first hospital based installations occurred: at first, the MC60 62.5 MeV proton cyclotron, delivered by Scanditronix, operating at the Clatterbridge Oncology Centre (UK) since 1989 and then from 1990 a dedicated 250 MeV proton synchrotron, developed by FermiLab at Loma Linda University (California, USA), the first dedicated clinical facility equipped with three rotating gantries [4].

Inspired by the Berkeley experiences biophysics studies were started at GSI in Germany in the late seventies, using the beam of the heavy ion linac UNILAC (<20 MeV/u). From 1989 high energetic beams were available from the SIS18 expanding the possibilities for in-depth experiments — 30 cm penetration depth in water demands a $^{12}C^{6+}$ beam of 430 MeV/u. In parallel, technical developments took place to create 3D conformal therapy options. While at PSI, Switzerland, a spot scanning proton gantry was built up [5], using a 1D magnetic pencil beam scanning plus passive range stacking (digital range shifter), in parallel at GSI a 2D magnetic pencil beam raster scanning system plus active range stacking (energy, spot size, intensity) in

Fig. 1. The HIT building with the cupreous front embedded in the Heidelberg hospital ring (left: National Centre for tumor diseases (NCT); Head Clinics and Medical Hospital in the background; right: Children's Clinics).

the accelerator was developed [6]. From 1993 a pilot project at GSI was set-up using this technology for treating around 450 patients with carbon beam from 1997–2008 [7, 8].

Based on the experiences from this installation and a first proposal for a clinical facility in 1998, the technical design study was worked out and after evaluation the HIT project started in 2002 [9, 10]. The building was constructed from 2003–2007, the accelerator installation started in 2006.

## 2 Beam parameters

The HIT accelerator facility and its beam parameters were driven by the raster scan technique of delivering an optimized 3D dose distribution into a predefined treatment volume, which is converted into a series of N $\times$ 2D fluence distributions, i.e. the number of stopping particles per $cm^2$. Namely, the treatment volume is virtually dissected into a series of iso-energy slices

Table 1. Requested accelerator parameters for the HIT facility.

| | |
|---|---|
| Ion species | p, $^{4}$He, $^{12}$C, $^{16}$O |
| Penetration depth (in water) | 20–300 mm |
| Energies | p: 48–221 MeV,<br>He: 51–221 MeV/u,<br>C: 88–430 MeV/u,<br>O: 103–430 MeV/u |
| Max. beam intensities (particles per spill) | p: $4 \times 10^{10}$,<br>He: $1 \times 10^{10}$,<br>C: $1 \times 10^{9}$,<br>O: $5 \times 10^{8}$ |
| Intensity variation | $(10^{-3}–1) \times N_{\max}$ |
| Beam spot sizes | 4–10 mm FWHM (at full energy) |
| Max. irradiation field | $20 \times 20\,\text{cm}^2$ |

(IES). Each IES will then be irradiated using, wherever possible, a single synchrotron cycle to scan the focused beam along a precalculated pattern of beam positions. Following these requirements, together with biological demands (like the necessary penetration depth), the requested beam parameters were deduced, which are shown in Table 1. From the beginning, the treatment and/or research both with low LET (p, He) and high LET (C, O) ions were foreseen.

## 3 General layout of the HIT facility

The Heidelberg Ion-Beam Therapy (HIT) Centre has two patient treatment rooms with fixed horizontal beams and one with the worldwide first ion beam gantry providing 360° angular freedom. The accelerator facility is comprised of the following subsystems, see Fig. 2:

- Two ECR ion sources for the routine operation of proton and carbon beams at 8 keV/u; in the meantime a third ion source was added to produce especially helium beams [11].
- A compact 216.8 MHz linac consisting of an RFQ and an IH-DTL with end energy of 7 MeV/u for all ions; a foil stripper directly located behind these cavities is designed to produce fully stripped ions.
- A synchrotron of 65 m circumference, capable to accelerate protons, helium, carbon and oxygen to predefined end energies, e.g. for carbon ions from 88 to 430 MeV/u in 255 steps.

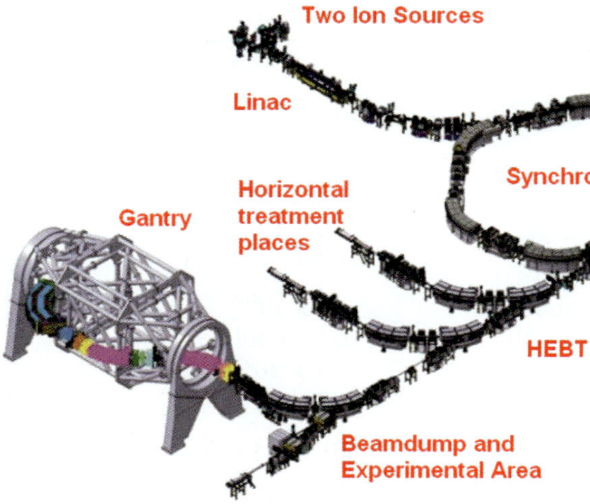


Fig. 2. Layout of the HIT accelerator facility (status 2007), in the meantime a third ion source for parallel helium production was added.

- Besides the three patient treatment rooms the high energy transport lines serve two additional destinations, an experimental area for research activities and a beam dump equipped with dedicated beam diagnostics, which can be operated independently.

In order to guarantee a high degree of reliability the design of most of the components was rather conservative and no operation up to the technical limits was foreseen. Most of the proposed operation modes at the different sections are well established. The special requirements of the raster scan method (beam properties, safety aspects), evaluated and successfully tested at GSI, were fully embedded in the layout.

## 4 The accelerator chain in detail

### 4.1 *Injector linac*

The injector consists of three 14.5 GHz permanent magnet ECR ion sources from PANTECHNIK with a modified extraction system by HIT [11]; the requested beam intensities are listed in Table 2. After selecting the desired ion species with the slits following the 90° analyzing magnet, a cylindrical beam shape along the straight section downstream of the switching magnet is produced by a quadrupole triplet. After adjusting the beam center along the whole LEBT with the steerer magnets, a transmission of typically 80–100% is achieved up to the RFQ entrance.

Table 2. Specified ions and intensities behind the 90° analyzing magnet.

| Ion | $I/\mu A$ | $U_{source}/kV$ |
|---|---|---|
| $^{3}H^{1+}$ | 700 | 24 |
| $^{4}He^{2+}$ | 500 | 24 |
| $^{12}C^{4+}$ | 200 | 24 |
| $^{16}O^{6+}$ | 150 | 21.3 |

The 216.8 MHz injector linac comprises a 400 keV/u radio frequency quadrupole accelerator (RFQ) of 1.4 meter length and a 20 MV IH-type drift tube linac (IH-DTL), built as a 3.8 m long milled steel tank [12]. The cavities are fed by a 200 kW amplifier from THOMSON for the RFQ and a 1.5 MW amplifier from BERTRONIX using a THALES tetrode TH 526 B for the IH-DTL tank.

### 4.2 *Synchrotron*

The synchrotron is the key to the enormous variety of beam parameters provided by the HIT accelerator [13]. Its maximum magnetic rigidity is 6.6 Tm, corresponding to carbon ions at 430 MeV/u and 30 cm range in water. Three bumper magnets allow for multi-turn injection; six dipoles, twelve quadrupoles and two sextupoles are used for the ion beam optics and two redundant cavity gaps are fed by HITACHI amplifiers in the frequency range from 1–7 MHz for acceleration. The beam is extracted horizontally using 3rd order resonance extraction above $Q_h = 5/3$ by RF noise excitation (so-called “knock-out” extraction). The synchrotron RF is kept active during extraction to smoothen the spill microstructure. The extraction time is fixed to 5 s, but extraction may be aborted earlier by the treatment system. The spill may be interrupted asynchronously up to three times per second, thus enabling the irradiation of disjoint iso-energy slices of the target volume in a single acceleration cycle. In the first years of operation, care was taken to avoid magnetic memory effects: in the synchrotron, this is done by ending each cycle with a conditioning ramp; in the beam lines, all bending magnets are field controlled using hall probes. In this way, ion type and energy can be requested from the accelerator in arbitrary sequences. Meanwhile, the synchrotron dipoles are also driven using magnetic field control, the four quadrupole groups will follow soon [14].

Fig. 3. The HIT 360° ion beam gantry.

### 4.3 *Gantry*

Besides the two horizontal treatment rooms (see Fig. 2), where the clinical operation started in November 2009 and September 2010, the HIT facility houses in addition the worldwide first 360° rotatable ion beam gantry — Figure 3 gives an impression of the mechanical structure of the HIT gantry, more details on the beam transport assembly can be found in [15]. In order to keep the gantry design compact, the two scanner magnets are located upstream to the 90° dipole so ion optical characteristics of this dipole have to be strongly considered.

The general requirements on the beam demand: (a) full transmission (beta functions within aperture limits even with beam scanning along the requested irradiation field of $20 \times 20\,\text{cm}^2$ in the iso-center); (b) an optical setting keeping dispersion in iso-center within small limits, and keeping beam focus and position independent of gantry angle yet compensating the coupling of horizontal and vertical phase space. This includes the optimisation of the ion beam at the gantry entrance point, i.e. size and divergence (equalised at the gantry entrance) for a single energy. Since the vertical emittance depends on the beam energy (due to adiabatic damping) the matching at the entrance point depends on the energy as well. (c) A phase advance which is a multiple integer of 180° (minimum position dependence on the gantry angle).

For a single energy, the control data for the accelerator components is calculated from physical input parameters. The data supply model (part of the control system) accounts for scaling of the process data with the magnetic and electric rigidity for different energies. Other energy-dependent effects have to be compensated by semi-automated adjustment of components, i.e. quadrupoles, dipoles, steerers in the HEBT and gantry. The final beam focus is adjusted by means of the last quadrupole doublet, the final beam position with dipoles and steerer magnets. Beam size and position in the iso-center are measured either with a viewing target and camera mounted on the rotating nozzle of the gantry or a multi-wire proportional chamber (MWPC) fixed on a rotatable mechanics installed on the patient positioning robot.

In order to minimize the commissioning effort, only a small subset of input parameters is determined as interpolation points; i.e. for a few energy steps (about 10–15), foci (4), and gantry angles (about 8–12) making up for about 1–3% of the overall library of beam characteristics. The achieved procedures ensure that the beam size ranges within a ±25% limit around the values needed for patient treatment within the whole library of desired beam properties for the carbon and proton beam with only a few exceptions. The beam position of the "centered" beam ranges so far within ±2 mm with only a slight dependence on the gantry angle. This deviation can be easily compensated by the beam scanning system.

After two longer commissioning phases [16] the first patients were treated at the gantry with protons from October 2012 and with carbon beams from November 2012, see Fig. 4 for an illustration of the patient positioning procedure. The completion of the beam line settings for all angles of proton and carbon beams, which have to pass all necessary medical qualification steps, took until the beginning of 2014.

## 5 Operational aspects of a particle therapy facility

Besides designing, manufacturing, installation and commissioning of the accelerator and medical systems a lot of further aspects have to be taken into account for a successful management of such a particle therapy facility:

- Economics: In contrast to a research institute, a public (university) hospital has to operate cost-covering, whereas private clinics aim at generating substantial yields. Therefore the investment costs of such a particle

therapy facility (about 100–150 M€) as well as the running annual costs of about 20–25 M€ for staff, capital charges, energy and maintenance costs, etc., need to be financed through refunds of costs by the health insurances. With a treatment capability of about 1000 patients per year after running up, a reimbursement of about 20–25 k€ per patient is thus necessary — in the case of HIT the numbers are a little bit more relaxed, because of a government aid of 50% for the investment costs. (All data are valid for Europe, the conditions in the US and Asia may differ.)
- Operations: For the most part, the particle therapy facility belongs to the radio-oncological section of a hospital. The daily therapy workflow with 50–80 patients demands a high degree of availability of the dose delivery system, including the whole accelerator. Only a well-trained technical team can guarantee the reliable operation and short reaction times in case of failures; see Section 6 for a detailed discussion.
- Safety and regulatory aspects: A particle therapy facility is a medical product and has to comply with comprehensive legal requirements, which also influence the operation and the maintenance procedures of the accelerator part, see Section 7 for more details.

## 6 24/7 accelerator operation at 335 days per year

The HIT accelerator core team hired in 2004 and 2005 only comprised of six employees. The team grew by about a factor of three in the period from mid-2006 to mid-2008 in order to operate the facility, at first in two shifts from Monday to Friday, stepwise expanding to a 24/7 mode. Continuous training by the GSI commissioning team on one side as well as intense briefing on the supplied devices by the contracted companies was organized besides "training on the job". This was necessary because only 1/3 of the employed persons worked in the accelerator field before or had to be retrained as most of the HIT equipment was newly designed or has undergone substantial changes in contrast to the original GSI designs. Within one year it was possible to operate the whole accelerator chain by the newly formed HIT team [17].

The accelerator shifts consist of two operators coming from the three technical teams. The physicists among the HIT accelerator crew are also responsible for retuning the optic parameters in case the performance deviates from the predefined limits for intensity, beam position and foci at the treatment location. These conditions are checked every morning. Although the intensity adjustment is done regularly in time intervals of some days, retuning of sections of the linac, synchrotron or the beam lines is only

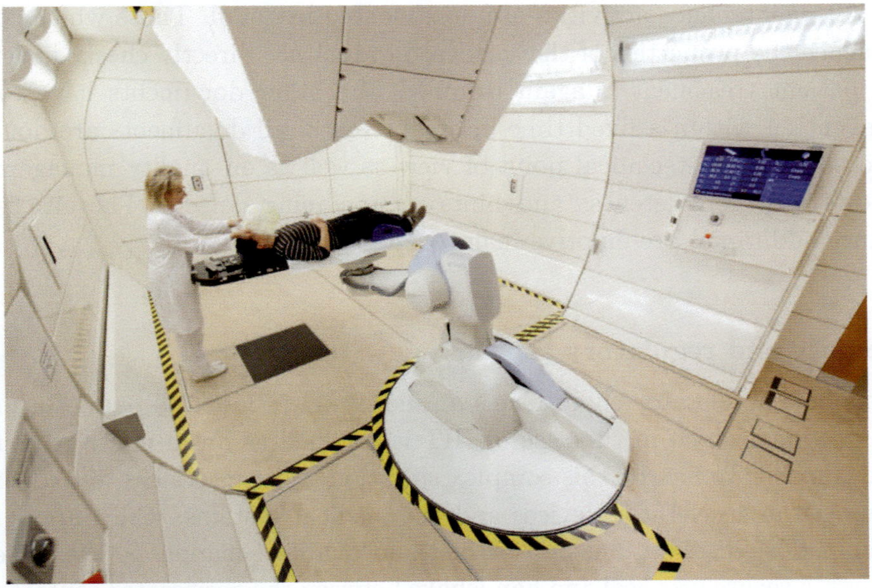

Fig. 4. Patient positioning procedure at the HIT gantry treatment place — for rotation of the gantry the closed folding ground floor is bent down.

necessary every two to six months. The stability is mainly influenced by stable temperatures of ambient air and cooling water.

Up until the end of 2011, two longer shutdowns of 2–4 weeks per year were used for maintenance and new installations as well as single device service shifts every 2–3 weeks. During these relatively long maintenance intervals the patients have to be phased out, resulting in "ramping down" the patient numbers before the shutdown and "ramping up" afterwards; this reduces the available time for patient treatment. Together with the medical management of HIT the following maintenance slots were defined from 2012.

(a) Six maintenance blocks with four days length (from Thursday to Sunday), two of the four days reserved for service only, followed by the restart of the accelerator, retuning of ion source and beam optics, if necessary, and the comprehensive quality assurance of the medical treatment systems.
(b) Maintenance shifts on Monday morning once every three weeks between the service blocks for shorter (visual) inspections, smaller repairs and update works.

Within these boundary conditions all regular maintenance tasks and deferrable repairs must be carried out to guarantee around 335 operation days of the HIT accelerator per year.

All maintenance activities are documented in detail. Those which could influence the properties of the medical product are directly checked, before the patient treatment is again resumed; the other documents have to be transmitted within a fixed time interval. Nearly all documents are double-checked from the technical point of view as well as on quality management aspects, e.g. completeness and hints to upcoming defects. The aim is to deduce additional pre-emptive maintenance measures.

## 7 Safety and regulatory aspects

The whole accelerator chain forms an industrial product as part of a medical product, which in the case of HIT is certified as an in-house manufactured device. To handle this complex situation, clear interfaces were defined between the accelerator devices and the medical systems, which apply the planned dose distributions in a correct and qualified manner to the patient. At first, a secure protocol and handshake between the accelerator control system (ACS) and the therapy control system (TCS) — part of the IONTRIS system from SIEMENS — was necessary to link both worlds by dedicated and safe interconnects. In this way it is ensured that the beam characteristics requested by the TCS are safely produced by the accelerator. In addition, the personnel safety system (PSS) has a dedicated interface to the TCS. Major components of this interconnection are secure interruptions of magnet power supplies, e.g. for switching dipoles leading the beam to the treatment rooms. Furthermore the TCS reads out all position information of vacuum valves and beam diagnostic devices in the HEBT to guarantee that the correct energy is delivered to the patient treatment place, unaffected by any matter in the beam line. Interfaces like these ensure that the beam can be redundantly stopped in case of emergencies. The functions of all these safety systems have to be checked regularly, especially after maintenance procedures [18]. All these measures are part of a comprehensive quality management system to comply with all legal requirements.

## 8 Status and perspectives

In March 2014 the two-thousandth patient was treated at HIT and routine operation is fully established with up to 60 patients per day. This is based on more than 98% availability of the accelerator throughout the planned operation periods, short maintenance times and, first of all, an excellent cooperation of highly-motivated technical and medical teams at HIT.

Increasing numbers of treated patients per year — 2012: 560, 2013: 640, 2014: 670, 2015: 700 planned — are the result of a continuous improvement process in addition to the daily operation. Further upgrades like spill feedback and magnetic field control in the synchrotron have been implemented, which have led to significant shortened dose delivery times [19, 20]. Treatment-plan-specific spill feedback was implemented in 2014 [21]. A big potential for faster treatment is offered by the multiple-energy operation within one synchrotron cycle, presently studied at HIMAC [22]. HIT has seized this suggestion and plans a similar prototype installation using its very flexible control system in the near future. However a full implementation will take some years.

Altogether it was a long, but very successful route from the first HIT proposal in 1998 at GSI to the full routine operation in Heidelberg about 15 years later. But there are still a lot of improvements possible in the whole chain from the accelerator facility to the dose delivery systems. HIT as the first clinically based installation in Europe producing proton and ion beams will sustain the necessary progress in the field of particle therapy to give more patients the chance to be treated with this highly effective method. A next step is already done, the University Clinics in Heidelberg took over the responsibility for the Marburg Ion-Beam Theraphy Center [23], where patient treatment started in October 2015.

## References

[1] R.R. Wilson, Radiological use of fast protons, *Radiology* **47**, 487–491 (1946).

[2] W.H. Bragg, *Studies in Radioactivity*, Macmillan, London, pp. 29–38 (1912).

[3] H. Wiedemann, *Particle Accelerator Physics*, Springer-Verlag, Berlin, pp. 1–4 (1993).

[4] http://ptcog.web.psi.ch/ptcentres.html.

[5] E. Pedroni *et al.*, The 200-MeV proton therapy project at the Paul Scherrer Institute: Conceptual design and practical realization, *Med. Phys.* **22**, 37–53 (1995).

[6] Th. Haberer, W. Becher, D. Schardt, and G. Kraft, Magnetic scanning system for heavy ion therapy, *Nucl. Instr. Meth. A* **330**, 296–305 (1993).

[7] H. Eickhoff, Th. Haberer, G. Kraft. U. Krause, K. Poppensieker, M. Richter, and R. Steiner, The GSI cancer therapy project, *Proc. PAC'97*, Vancouver, Canada, see http://www.jacow.org/.

[8] D. Schardt, Th. Elsässer, and D. Schulz-Ertner, Heavy-ion tumor therapy: Physical and radiobiological benefits, *Rev. Mod. Phys.* **82**, 383–425 (2010).

[9] R. Bär, A. Dolinskii, H. Eickhoff, Th. Haberer, A. Peters, M. Rau, B. Schlitt, and P. Spiller, HICAT — Heavy Ion Cancer Therapy accelerator facility for the clinic in Heidelberg, GSI (2000).

[10] Th. Haberer, J. Debus, H. Eickhoff, O. Jäkel, D. Schulz-Ertner, and U. Weber, The Heidelberg Ion Therapy Center, *Rad. Onc.* **73**(2), 186–190 (2004).

[11] T. Winkelmann, R. Cee, T. Haberer, B. Naas, and A. Peters, Test bench to commission a third ion source beam line and a newly designed extraction system, *Rev. Sci. Instrum.* **83**, 02B904 (2012).

[12] B. Schlitt *et al.*, Commissioning of the 7 MeV/u, 217 MHz injector linac for the Heavy Ion Cancer Therapy Facility at the University Clinics in Heidelberg, *Proc. LINAC 2006*, Knoxville, Tennessee, USA, see http://www.jacow.org.

[13] D. Ondreka and U. Weinrich, The Heidelberg Ion Therapy (HIT) accelerator coming into operation, *Proc. EPAC'08*, Genoa, Italy, see http://www.jacow.org.

[14] E. Feldmeier, T. Haberer, M. Galonska, R. Cee, S. Scheloske, and A. Peters, The first magnetic field control (b-train) to optimize the duty cycle of a synchrotron in clinical operation, *Proc. IPAC2012*, New Orleans, Louisiana, USA, see http://www.jacow.org.

[15] M. Galonska, S. Scheloske, R. Cee, K. Höppner, T. Winkelmann, A. Peters, and T. Haberer, Commissioning of the ion beam gantry at HIT, *Proc. IPAC2011*, San Sebastián, Spain, see http://www.jacow.org.

[16] M. Galonska, S. Scheloske, S. Brons, R. Cee, K. Höppner, J. Mosthaf, A. Peters, and T. Haberer, The HIT gantry: From commissioning to operation, *Proc. IPAC2013*, Shanghai, China, see http://www.jacow.org.

[17] A. Peters, R. Cee, E. Feldmeier, M. Galonska, T. Haberer, and K. Höppner, Five years of operation experience at HIT, *Proc. IPAC2012*, New Orleans, Louisiana, USA, see http://www.jacow.org.

[18] A. Peters, R. Cee, T. Haberer, and T. Winkelmann, The HIT accelerator as part of a medical product: Impacts on the maintenance strategy, *Proc. IPAC2013*, Shanghai, China, see http://www.jacow.org.

[19] Th. Haberer, S. Brons, R. Cee, E. Feldmeier, K. Höppner, J. Naumann, R. Panse, A. Peters, S. Scheloske, C. Schömers, and T. Winkelmann, Improving the synchrotron performance of the Heidelberg Ion Beam Therapy Centre, *Proc. IPAC2012*, New Orleans, Louisiana, USA, see http://www.jacow.org.

[20] Th. Haberer, E. Feldmeier, M. Galonska, A. Peters, and C. Schömers, Novel techniques and challenges in hadron therapy, *Proc. IPAC2013*, Shanghai, China, see http://www.jacow.org.

[21] C. Schoemers, E. Feldmeier, J. Naumann, R. Panse, A. Peters, and T. Haberer, The intensity feedback system at Heidelberg Ion-Beam Therapy Center, *Nucl. Instr. Meth. A* **795**, 92–99 (2015).

[22] Y. Iwata *et al.*, Multiple-energy operation with quasi-DC extension of flattops at HIMAC, *Proc. IPAC2010*, Kyoto, Japan, see http://www.jacow.org.

[23] http://www.marburger-ionenstrahl-therapiezentrum.de/.

# Chapter 43

# MedAustron: The Austrian ion therapy facility

Michael Benedikt (CERN)

MedAustron is a synchrotron-based light-ion beam therapy center for cancer treatment as well as for clinical and non-clinical research in Wiener Neustadt, Austria. The center is designed for the treatment of up to 1200 patients per year and for non-clinical research in the areas of radiobiology and medical radiation physics, as well as in experimental physics. MedAustron is an interdisciplinary project, benefiting from close cooperation and knowledge transfer with medical, scientific and research institutes on the national and international level. Three medical irradiation rooms will allow quasi-permanent patient treatment during two shifts on working days. The remaining beam time will be used for non-clinical research applications in a dedicated fourth irradiation room. The expected start of operation of MedAustron is mid-2016.

The realization strategy for MedAustron was based on the establishment of strong international collaborations. The heart of the MedAustron center is the accelerator complex, consisting of several ion sources, a linear accelerator and a synchrotron as the main accelerator. The accelerator design is mainly based on the CERN-PIMMS study [1] and its technical implementation by the Italian CNAO center [2]. In an agreement with CNAO, all relevant technical documentation for the injector, the synchrotron and the high-energy transfer line has been made available to the EBG MedAustron company [3] that is charged with the construction in Austria. The procurement of accelerator components and the construction of the accelerator facility as well as training of future operation personnel were done in close collaboration between CERN and EBG MedAustron, since no domain specific know-how existed in Austria. A collaboration with the Swiss Paul Scherrer Institute (PSI) was concluded, on work relating to developments in the field of ion therapy, in particular for the construction of the proton gantry for MedAustron which is based on the design of the newly-developed PSI Gantry-2 [4].

## 1 Introduction

Due to their physical and radiobiological selectivity, light-ion beams offer advantages over classical photon therapy for certain tumor indications. While protons and carbon ions exhibit an enhanced geometrical selectivity by conforming the dose to the tumor and sparing normal tissues, carbon ions have in addition an increased radiobiological effect, due to their high ionization density at the end of their track, in the so-called Bragg-peak region. Using other light ions like helium or oxygen is also being considered for the future, calling for a flexible design of the accelerator complex.

The layout of the medical part of the facility is defined mainly by the number of patients to be treated. Once fully operational, about 24,000 fractions should be delivered in medical operation per year, corresponding to around 1200 patients. Consequently three treatment rooms are needed to efficiently use the accelerator and to achieve the desired patient numbers, since treatment preparation (i.e. patient positioning and position verification) takes about three times longer than the actual irradiation process. The choice was made for two fixed-beam rooms for proton and carbon-ion treatments and one gantry room for proton treatment only. One of the fixed-beam rooms is equipped with a horizontal beam line and the other with a horizontal and a vertical beam line with a common iso-center.

In addition MedAustron is equipped with labs and workshops for research in radiobiology, medical radiation physics and experimental physics. These disciplines will profit from a dedicated irradiation room with a horizontal beam port, equipped with a beam-delivery system identical to the medical treatment rooms.

## 2 Basic choices and parameters

The choice of a synchrotron as main accelerator [5] is a natural consequence of two major medical requirements. Due to its operational flexibility the synchrotron can provide different particle types, i.e. protons and carbon ions, and it also satisfies the requirements of active scanning (relatively low beam intensities and an active variation of beam energy on a pulse-to-pulse basis). This means that all beam characteristics (energy, focus, intensity, extraction duration, ion type) are fully electronically selectable; there are no mechanical beam-intercepting components involved.

The required penetration depth in tissue defines the ion-beam energies and the dynamic range of the synchrotron. To reach 27.5 cm, carbon ions have to be accelerated to 400 MeV/u and protons to 250 MeV. The corresponding magnetic rigidity of the carbon ions at 400 MeV/u is

Table 1. Medical requirements and accelerator parameters.

| Medical requirements | | Units |
|---|---|---|
| Typical dose per fraction | 2 | Gy |
| Max. treatment volume | 2 | liter |
| Penetration depth | 3.5–27.5 | cm |
| Intensity variation | 1–625 | |
| Protons | | |
| Kinetic energy (min–max) | 60–800 | MeV |
| Beam intensity (max) | $1 \times 10^{10}$ | per pulse |
| Extraction time (min–max) | 0.1–10 | s |
| Repetition rate (max) | 1 | Hz |
| Carbon ions | | |
| Kinetic energy (min–max) | 120–400 | MeV/u |
| Beam intensity (max) | $4 \times 10^{8}$ | per pulse |
| Extraction time (min–max) | 0.1–10 | s |
| Repetition rate (max) | 1 | Hz |

significantly higher than the one of protons, meaning that protons with energies up to 1.1 GeV could still be guided by the magnet systems. MedAustron will make partial use of this feature and provide proton beams of up to 800 MeV for non-clinical research, mainly in the areas of proton scattering and detector research [6].

Fully active beam delivery is foreseen for all treatment rooms. The target volume is subdivided in layers of a few-mm thickness, each at a constant penetration depth, corresponding to a specific particle energy. Every spot of such an iso-energy layer is then irradiated by a particle beam with a 4–10 mm cross-section. When the required dose for the spot is reached, the beam is directed to the next spot.

The above considerations led to the main parameters [7] as listed in Table 1.

## 3 Civil engineering aspects

The MedAustron project was subject to an environmental impact assessment, with radiation protection being one of the main aspects. Therefore specific attention was paid to that subject and in particular the optimization of the shielding to allow unlimited access to the areas outside the accelerator bunker and the treatment rooms. Figure 1 shows the result of Monte Carlo calculations for shielding dimensioning and a yearly dose of less than 0.1 mSv (i.e. a factor 10 below the legal limit for public areas) outside this shielding.

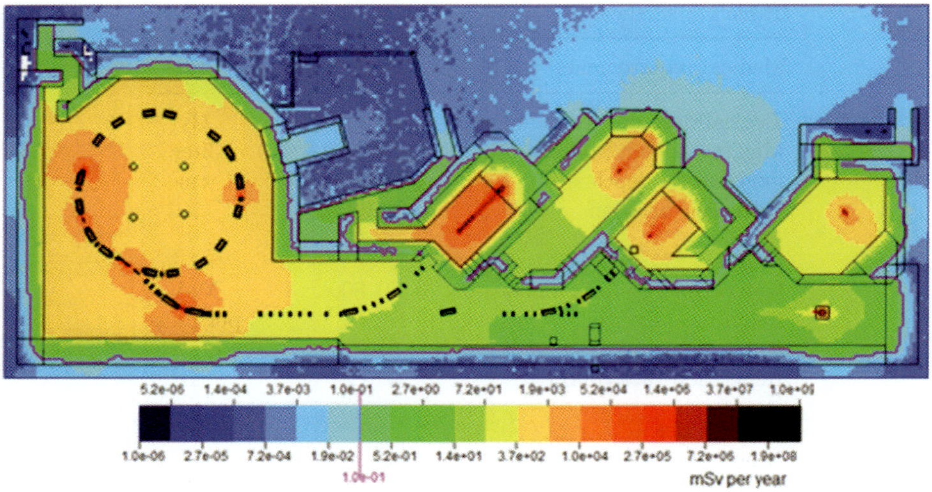


Fig. 1. Simulation of the yearly effective dose in the accelerator area based on nominal operation conditions. The magenta line inside the shielding walls indicates a yearly dose level of 0.1 mSv.

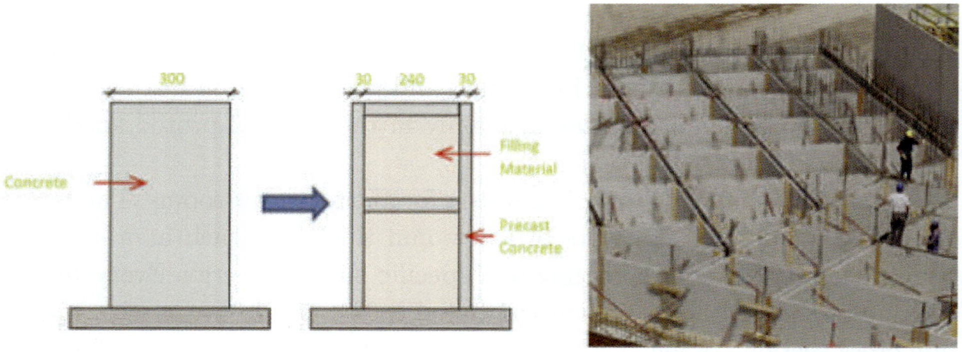


Fig. 2. Schematics of replacement of solid concrete by a sandwich structure with lateral pre-fabricated walls and material fill-in, and photo of the real structures and filling process.

For the construction of shielding walls the sandwich construction [8] technique was applied. Floors, walls and ceilings are made of a honeycomb structure of precast concrete and are then filled with excavation material that is compressed, as indicated in Figure 2.

This approach brings important economic and ecologic advantages. In the case of MedAustron about 25,000 $m^3$ of concrete were spared by using excavation material and also traffic from and to the site was correspondingly limited.

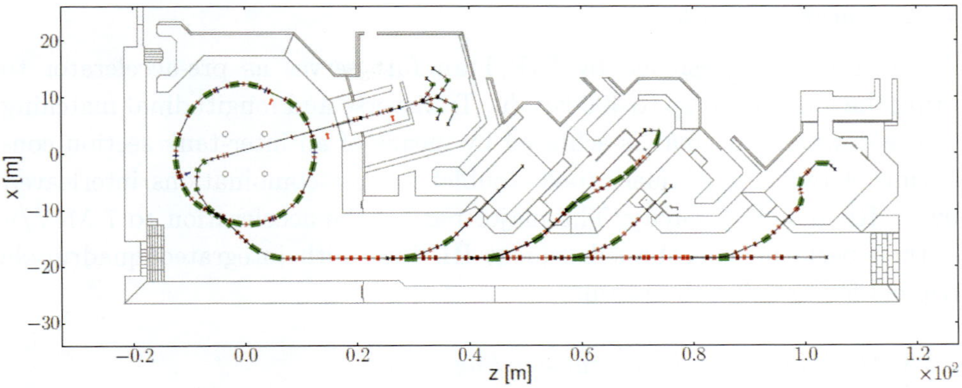


Fig. 3. Accelerator layout with injection chain, extraction and transport lines to the treatment rooms.

## 4 Accelerator layout

### 4.1 *Overview*

The layout of the accelerator complex and the high-energy beam transport lines is shown in Figure 3. The ion sources are situated in a separate hall from injector and synchrotron, allowing access during operation for maintenance purposes. The injector is composed of Electron Cyclotron Resonance (ECR) ion sources, a Radio Frequency Quadrupole (RFQ) and an Interdigital H-mode linear accelerator (IH linac).

The synchrotron design is closely following the CERN-PIMMS study [2] with a particular emphasis on the slow-resonant extraction scheme to allow for extraction durations between 0.1 and 10 seconds. The third-order integer extraction is performed by means of a betatron core that accelerates the beam into a sextupole-driven resonance via induction principle [9].

### 4.2 *Injector complex*

#### 4.2.1 *Sources and low-energy beam transport*

The MedAustron injector is designed for ions with a charge-to-mass ratio $q/m \geq 1/3$. Three identical commercially available Electron Cyclotron Resonance sources are used to feed the injector for medical operation with $H_3^+$ (for proton) and C-ions, with the third source serving as a back-up. A fourth ion source will be added later to provide further flexibility in the particle type for research. The ions are extracted at an energy of 8 keV/u and transported to the downstream RFQ for pre-acceleration.

### 4.2.2 *Linear accelerator*

A four-rod RFQ designed by IAP/Frankfurt serves as pre-accelerator to bring the ion energy up to 400 keV/u. Transverse and longitudinal matching to the downstream linac is achieved by means of an inter-tank section consisting of two sets of quadrupole-doublet/steerer combinations interleaved by an RF bunching cavity. The major part of the acceleration to 7 MeV/u is then performed in the subsequent IH linac with integrated quadrupole triplets for focusing the beam.

### 4.2.3 *Medium-energy beam transfer line*

At the exit of the IH linac, the beam is focused onto a thin 100 $\mu$g/cm$^2$ carbon foil in order to strip the remaining electrons of the ions. Thereby the charge-to-mass ratios of the ions are changed to $q/m = 1$ for protons and $q/m = 1/2$ for carbon ions, which also changes the magnetic rigidities. The medium-energy beam transfer line also houses a debuncher cavity to reduce the energy spread for injection into the synchrotron.

## 4.3 *Synchrotron*

### 4.3.1 *Magnet lattice*

The synchrotron lattice has a super-symmetry of two, with a mirror symmetry within each of the two super-periods. The linear lattice is of the split-FODO type with 16 dipoles, 16 focusing quadrupoles in two circuits and eight defocusing quadrupoles in a single circuit. Four sextupoles are connected in pairs of two to adjust the natural chromaticity to the optimum value for the slow extraction process. One additional elements, the resonance sextupole shares a dispersion-free region with the RF cavity and is used to drive the slow-resonance extraction process. The injection (from the inside) and extraction (to the outside) are combined in the other dispersion-free straight section. The lattice functions of the synchrotron and an indication of the layout is given in Figure 4.

### 4.3.2 *Multi-turn injection*

The injection into the synchrotron features a classical multi-turn scheme, using two thin magnetic and an electrostatic septum in combination with a fast synchrotron orbit variation, controlled with ferrite bumpers, to inject over 15–20 turns. The electrostatic septum with a 0.1 mm thick molybdenum foil as septum sheet is used to minimize injection losses.

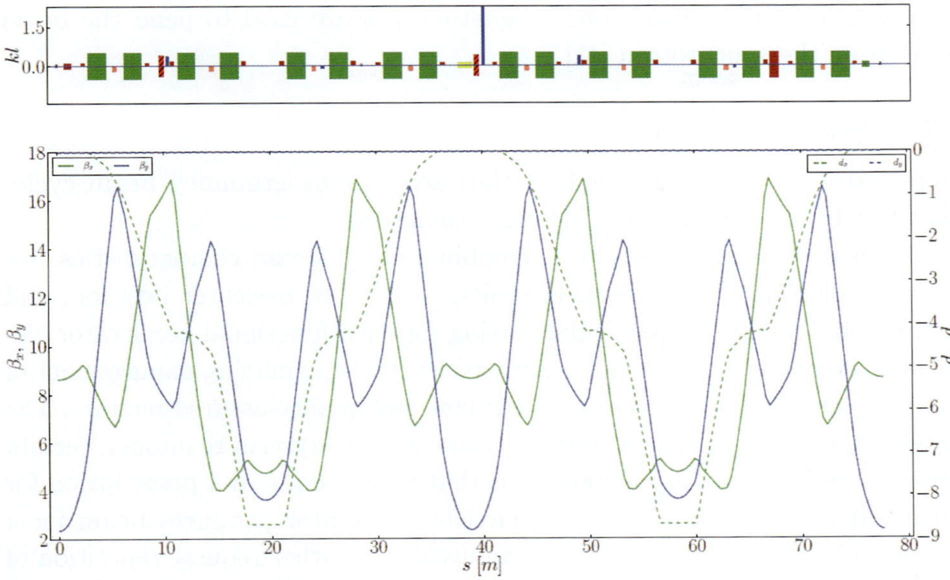


Fig. 4. Optics functions of the synchrotron with indication of geometrical sequence.

### 4.3.3 *Acceleration and RF system*

After injection into the synchrotron the coasting beam is adiabatically trapped into a single bunch by slowly switching on the RF at harmonics $h = 1$ within about 50 ms. The further frequency program for acceleration is derived from an online magnetic field measurement in a reference dipole magnet, connected in series with the synchrotron main dipoles. The high-power part of the accelerating RF system consists of two wide-band, six-gap cavities loaded with FINEMET® nano-crystalline metal alloy rings, powered by 12 identical 1 kW semiconductor amplifiers. The low-level beam control system is entirely digital. Both high- and low-level systems were developed in close collaboration with CERN, exploiting synergies with the CERN Booster RF system upgrade project [10].

### 4.3.4 *Extraction*

The slow-resonance extraction scheme has to provide the required variable pulse length between 0.1 s and 10 s. The third-integer resonance is excited by first powering the resonance sextupole, accelerating the beam into the resonance with the betatron core that is slowly ramped. Beam particles that have become unstable increase their horizontal amplitude under the action of the resonance, until they jump the blade of the electrostatic extraction septum and are deflected towards the extraction channel. Downstream one thin and three thick laminated magnetic septa are used to bend the beam away from the synchrotron [11].

#### 4.3.5 *Operation principle*

Accelerator operation is based on the use of preprogrammed beam cycles that last from a fraction of a second to minutes.

Each cycle corresponds to a combination of beam characteristics, for instance particle type, energy, intensity, beam size, isocenter position, and extraction duration. A particular setting for each functional accelerator element is assigned to one or more beam cycles, thus permitting management of the accelerator configuration in an efficient and quality-assured manner. The accelerator is at rest until a medical-beam delivery system requests a certain beam cycle from a list of beam cycles that corresponds to a prescription for an irradiation session. While the particle accelerator produces beam for a beam cycle, the beam delivery system decides to either request repetition of the same beam cycle in case the intensity is not sufficient to complete the depth-dose layer or to request the next cycle in the prescription list. In such a way, dead-times between cycles are avoided. The chosen mode of operation decouples the medical systems in charge of delivering a prescribed dose to a defined volume from the beam production systems [12].

### 4.4 *High-energy beam transport*

The High Energy Beam Transport line (HEBT) transfers the beam from the synchrotron to the four irradiation rooms or the beam dump.

The transverse beam size is adjusted according to the medical requirements in a specific lattice module downstream of the chopper in the common part of the HEBT. The module has six independent quadrupoles which control five variables: the two beta functions and their slopes as well as the horizontal phase advance [13]. The further HEBT lines feature optics modules with 1:1 optical properties that are then used to transfer the beam to the different irradiation rooms [14]. In order to match the beam to the rotated coordinate system of the proton gantry, a so-called rotator is included in the main extraction line. This is a 1:1 (horizontal phase space) and 1:−1 (vertical phase space) transfer module consisting of seven quadrupoles which are physically rotated by half of the gantry rotation angle [15].

Each of the individual irradiation lines is equipped with a fail-safe beam stopper, which is located between the branch-off dipoles, in order to protect the room from unintended irradiation. Figure 5 shows a drawing of the accelerator with the transport lines and treatment rooms.

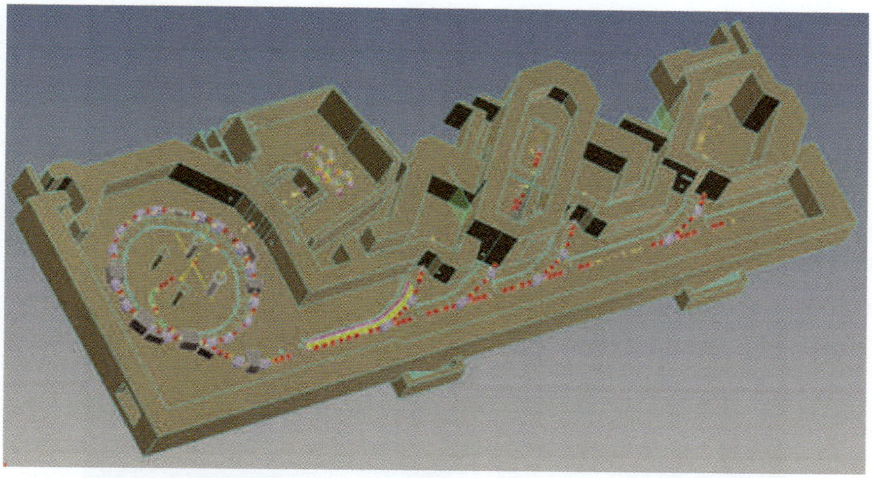

Fig. 5. CAD drawing of the MedAustron accelerator complex and treatment rooms.

## 4.5 *Irradiation rooms*

### 4.5.1 *Non-clinical irradiation room*

The non-clinical irradiation room features a fixed horizontal beam line with a transverse scanning system. The beam can be focused over a length of about 5 meters to allow for the installation of several experiments. Although non-clinical research will be mainly in the areas of radiobiology and medical radiation physics, experimental physics experiments are also envisaged such as the development and test of particle counters, proton scattering and computerized proton tomography at 800 MeV [6, 16].

### 4.5.2 *Medical treatment rooms*

The three medical treatment rooms are all equipped with active scanning systems for dose application that are provided by CNAO [2]. A high conformity of the deposited dose to the target shape is achieved by adapting the energy and the position of the beam. Laterally, the beam is fast scanned by magnetic deflectors upstream of the patient. The modulation in depth is achieved by dynamically changing the energy of the ion beams from cycle to cycle. The major advantages of the active scanning technique compared to passive scattering are the additional dose sparing due to the variable modulation of the range, the dose delivery without patient-specific hardware and the capability to deliver intensity-modulated ion therapy without additional system modifications and any beam-intercepting devices.

Fig. 6. The photo on the left shows the PSI Gantry-2 which has been reproduced for MedAustron (courtesy of PSI).

In the horizontal beam line room, the patient is positioned relative to the fixed beam to achieve the required delivery angle. The combination of horizontal and vertical beam lines in one treatment room gives additional degrees of freedom for the treatment delivery.

MedAustron will also be equipped with a proton gantry to allow full flexibility in the choice of beam entrance angle. The proton gantry for MedAustron is based on the new PSI Gantry-2 (Figure 6) [4], with some revisions and adaptations of the design.

The gantry is designed as a dynamic beam delivery device, in order to speed up pencil-beam scanning, based on a fast double parallel magnetic scanning. It is characterized by an isocentric compact gantry layout. By scanning the beam upstream of the last 90° bending magnet, the diameter of the rotating structure is limited to less than 8 m. The space in the treatment room is comfortably large and the access around the patient table is on a fixed floor. There is enough space to use a sliding gantry Computer Tomograph (CT) within reach of the treatment table for positioning verification as well as for image-guided radiotherapy purposes. The rotation of the gantry is limited to $-10°$ to $+180°$ in the vertical direction in order to have a flat, fixed floor around the patient table. The most critical component of the gantry beam line is the last 90° bending magnet. The gap height of the magnet has been chosen as 15 cm. This permits double parallel scanning over a scanning area of 12 cm $\times$ 20 cm [17].

## 5 Project status and outlook

The MedAustron project has entered the transition from development, which was performed under the guidance of CERN, to installation and commissioning carried out by EBG MedAustron at the facility site in Wiener Neustadt, Austria. After the delivery of the buildings and technical infrastructure in autumn 2012, accelerator installation started with the injector. Beam commissioning of the injector was performed in 2013; in parallel the installation of the synchrotron was taking place [18]. Most of 2014 was used to take the synchrotron and HEBT into operation, leaving 2015 for medical commissioning with the goal of starting patient treatments by mid-2016.

The realization of MedAustron via international collaborations with experienced partners is an excellent example for large-scale technology transfer from basic research to applications valuable for the general public at large.

## References

[1] M. Crescenti, P. Holy, A. Maier, M. Pullia, S. Reimoser, S. Rossi, and P. Knaus, in *PIMMS, Proton-Ion Medical Machine Study*, L. Badano, M. Benedikt, and P. Bryant (eds.) (CERN 2000-006, Geneva, 2000), ISBN: 92-9083-166-9.

[2] S. Rossi, The status of CNAO, *Eur. Phys. J. Plus* **126** (2011) 78.

[3] MedAustron, http://www.ebgmedaustron.at.

[4] E. Pedroni, The PSI-Gantry 2: A second generation proton scanning gantry, *Z. Med. Phys.* **14** (2004) 25–34.

[5] M. Benedikt, J. Gutleber, M. Palm, W. Pirkl, U. Dorda, and A. Fabich, Overview of the MedAustron: Design and technology choices, in *Proceedings of First International Particle Accelerator Conference (IPAC'10)*, Kyoto, Japan, 23–28 May 2010.

[6] M. Benedikt, H. Leeb, and T. Schreiner (eds.), *Physics Opportunities at MedAustron — White Book* (Vienna University of Technology, 2009), ISBN: 978-3-901-1671-4-0.

[7] M. Benedikt and A. Wrulich, MedAustron — Project overview and status, *Eur. Phys. J. Plus* **126** (2011) 69.

[8] J. Forster, Forster Bau GmbH Ingolstadt, www.forster-bau.de.

[9] G. Feldbauer, M. Benedikt, and U. Dorda, Simulations of various driving mechanisms for the 3rd order resonant extraction from the MedAustron medical synchrotron, in *Proceedings of International Particle Accelerator Conference (IPAC'11)*, San Sebastian, Spain, 4–9 Sep 2011.

[10] M. Angoletta, CERN's PS Booster LLRF Renovation: Plans and initial beam tests, in *Proceedings of First International Particle Accelerator Conference (IPAC'10)*, Kyoto, Japan, 23–28 May 2010.

[11] J. Borburgh *et al.*, Construction and initial tests of the electrostatic septa for MedAustron, in *Proceedings of International Particle Accelerator Conference (IPAC'13)*, Shanghai, China, 12–17 May 2013.

[12] J. Gutleber *et al.*, The MedAustron accelerator control system, in *Proceedings of 13th International Conference on Accelerator and Large Experimental Physics Control Systems*, Grenoble, France, Oct 2011, pp. 9–12.

[13] M. Benedikt, P. Bryant, and M. Pullia, A new concept for the control of a slow-extracted beam in a line with rotational optics, *Nuclear Instruments and Methods in Physics Research A* **430** (1999) 523–533.

[14] U. Dorda, M. Benedikt, and P. J. Bryant, Layout and optics of the MedAustron high energy beam transfer line, in *Proceedings of International Particle Accelerator Conference (IPAC'11)*, San Sebastian, Spain, 4–9 Sep 2011.

[15] M. Benedikt and C. Carli, Optical design of a beam delivery system using a rotator, CERN/PS 96-41 (DI), CERN (1996).

[16] D. Georg *et al.*, *MedAustron — Research Opportunities for Medical Radiation Physics and Radiation Biology* (MedAustron Research, 2010), ISBN: 978-3-200-01792-4.

[17] A. Koschik, U. Dorda, D. Meer, and E. Pedroni, The MedAustron proton gantry, in *Proceedings of International Particle Accelerator Conference (IPAC'12)*, New Orleans, USA, 20–25 May 2012.

[18] A. Koschik, F. Osmic, P. Urschütz, and M. Benedikt, Status of MedAustron the Austrian Ion Therapy and Research Centre, in *Proceedings of International Particle Accelerator Conference (IPAC'14)*, Dresden, Germany, 15–20 June 2014.

# Chapter 44

# Industrial projects from an accelerator-based point of view: Siemens and IBA

Stephen Myers (CERN)

## 1 Overview

At present, there are numerous hadron therapy projects in operation or being designed around the world and there are several reasons for this new enthusiasm. It has been clearly established that if the cost of such a facility could be set at around the same level as that for the modern day X-ray radiotherapy centres, then, hadron therapy could become a very favourable treatment for many of the tumors presently treated with conventional radiotherapy. For equivalent prices, it is likely that hadron therapy could become a standard procedure for cancer treatments where radiation is being indicated.

With such a technical challenge and with a potentially large market, it is clear that many different groups have already started to think about improvements for the accelerators but, even more importantly on new approaches for the design of the gantries. Currently, these new developments are being studied by dedicated commercial companies and by scientific laboratories, but ultimately these new facilities will be operated by hospitals. As a consequence, it is likely that these machines will be purchased by hospitals as turnkey solutions directly from just a few industrial suppliers.

This brief summary will focus on the industrial landscape in Europe, although it should be acknowledged that there is also considerable activity in the USA and in Japan. A few introductory comments on some of the basic principles are provided below.

## 2 Introduction

The primary benefit of hadron therapy over radiotherapy is the manner in which the radiation dose is deposited along its path inside the human body. With a well defined particle energy, it is possible for a beam of hadrons to penetrate through the body with a minimal dose deposition along its path until it reaches a critical energy where most of the energy is deposited at a well defined location — the well known Bragg peak effect. In addition, the ions will have lost all their energy at the position of the tumor and there will be almost no additional dose deposition behind the tumor. This particular situation is very different from X-ray radiotherapy where the majority of the radiation dose deposition occurs shortly after the photons (or leptons) have entered the body and the dose deposition follows an exponential decay along its path through the body. Radiotherapy therefore exhibits two fundamental differences with respect to hadron therapy: firstly, the maximum of the dose deposition never occurs at the location of the tumor and secondly, all the organs located behind the tumor are likely to be irradiated and this cannot be avoided. There are other important parameters to take into account when comparing the two methods, but the basic differences mentioned above are significant enough to explain why hadron therapy may be an interesting candidate for the future applications.

Another significant difference between hadron therapy and X-ray radiotherapy is that with hadron therapy tumor irradiation, it is necessary to compute the energy of the ions required to reach the expected penetration depth in the body. This rather simple condition has two serious operational implications:

1. For a given penetration depth, the required energy will depend on the type of ions being used;
2. Since the tumor occupies a certain volume, the energy of the ions has to be modified during the treatment in order to penetrate the full depth to be treated.

As will be shown later, the question arises of whether the clinical objective will be to use a single species of ion or different types of ions; this will have a major impact on the choice of the accelerator being used in the facility.

## 3 The industrial aspects

Although the industrial landscape may evolve in the near future, the present situation involves two well defined types of facilities:

1. Centres which are using exclusively protons (proton therapy, PT);
2. Centres which are using (or planning to use) at least two types of ions (generally protons and carbon ions).

This leads to two main classes of industrially produced hadron therapy centres, namely those limited to one type of ions and provided by cyclotrons, and those allowing for different types of ions which (presently) need synchrotrons to produce the ion beams. The differences between these two types of machines are fundamental and will be briefly described below.

### 3.1 *Cyclotron-based centres*

The cyclotron is a circular accelerator composed of one single large dipole magnet working at a constant magnetic field. The particles coming out of the source are accelerated by the radio-frequency system and, because of the constant magnetic field, are spiraling outwards until they reach the outer radius of the machine where they are extracted at a given, pre-defined, energy.

The advantage of the cyclotron is that the availability of particles at the exit of the machine is practically continuous. In addition, the mandatory presence of an energy degrader to vary the energy as required for the treatment of the depth of the tumor and of collimators allow for an optimisation of the shape of the beam which will be very useful for the design optimisation of a rotating gantry in the treatment room. When working with low energy ions (e.g. protons), the size of the accelerator can be extremely compact.

The major drawback of this type of accelerator is that it produces particles at a single pre-defined energy and thus imposes the need for a degrader section which leads to increased levels of induced radioactivity in the area.

### 3.2 *Synchrotron-based centres*

The conventional synchrotron is also a circular machine but equipped with many small dipole magnets and a magnetic field which can be varied. It thus follows that the particles are accelerated while remaining on the same circular trajectory and the extraction energy can be easily varied as long as the available radio-frequency power and magnetic field are sufficient. The advantage of this type of machine is that the extraction energy of the ions can be varied and thus removes the need for an energy degrader.

Main drawbacks of the synchrotrons are the size of the machine (50–70 metres circumference), which is extremely large as compared to that of a cyclotron and, even more importantly, once the ions have been extracted, the magnets have to be ramped down back to the injection energy, where new particles are being injected into the synchrotron for the next acceleration to take place. The whole process is relatively time consuming, so the beam for the treatment might be available only every two seconds compared with the continuous beam available from the cyclotron.

As explained above, the accelerator components related to the two approaches are very different so it is not too surprising that the industrial suppliers fall into two separate camps.

It is correct to mention that both types of accelerators are relatively widely used with possibly a recent resurgence of interest in synchrotrons since treatment with carbon ions have been demonstrated to be efficacious (i.e. use of both protons and carbon ions). Japan is presently leading the field of research with carbon ions, so there is a clear preference in Japan for synchrotrons supplied by the companies — Hitachi and Mitsubishi.

## 4 Hadron therapy in Europe

In Europe, the situation has been relatively quiet in the sense that the two accelerator approaches have developed in parallel without impacting on each other too much. The situation is now changing since the medical requirements have become essential parameters with significant implications on the design of the medical facilities.

Until the year 2000, hadron therapy was considered to be a field of research where the main objectives had been to demonstrate the feasibility of the treatment procedures and try to establish adequate certification procedures.

Even before this, the IBA company (Ion Beam Applications, Louvain-la-Neuve, Belgium) had clearly established itself as the world leader in producing cyclotrons for medical isotope production mainly for industrial customers. On the basis of their experience in designing and building commercial cyclotrons, it is not surprising that they were able to establish themselves rapidly as leaders in the development of cyclotrons for proton therapy centres and IBA could offer total turnkey solutions for these facilities.

For synchrotrons, there were almost no commercial offerings since proton therapy was tested and operated mainly in scientific laboratories which had existing synchrotrons. The real stimulus for an industrial interest took place around 2000 with the publication of the CERN report on Proton–Ion Medical Machine Study (PIMMS) [1]. This report combined with the clinical expertise accumulated at GSI (Darmstadt, Germany) coincided with the real start of hadron therapy centres in Europe. This new area of treatment in Europe began with the successful treatment of a few hundred patients at GSI with proton therapy. This experience motivated Siemens to start a new business initiative using GSI expertise. The HIT facility at Heidelberg, which is operating today, is a good example of a collaboration between an accelerator laboratory such as GSI, which lead the design and construction of the accelerator part of the facility, and industrial partners such as Siemens which design and build the medical part surrounding the patient. A new department for "turnkey" hadron therapy centres was established at Siemens. Based on the same philosophy of partnership between accelerator laboratories and industry, two other projects were launched, namely the CNAO facility in Pavia, Italy (typically a PIMMS machine) which started to treat patients by the end of 2011 and the Austrian MedAustron project (see Chapter 43) which is in its final phase of construction.[1]

At this time, the main commercial objectives of these industrial facility projects were the certification procedures, the equipment reliability and cost containment of the facility building and operations. Although expertise for successful clinical operations existed, high capital costs were being foreseen and consequently there was little appetite to develop new but potentially higher risk technologies. This is probably the reason why all operating centres, independent of the accelerator type, seemed to adopt the so-called scattering beam delivery system rather than the new pencil beam scanning method [3]. It is worth recalling that the scattering delivery approach had a few inherent drawbacks in the sense that it required the construction of dedicated equipment such as masks, filters, collimators etc. customised differently for each individual patient which resulted in considerably longer treatment times.

The new competitor technology, i.e. the pencil beam scanning, is conceptually much simpler since it avoids the manufacturing and assembly of most of this patient dependent equipment.

For pencil beam scanning, irradiation of the tumor target volume is made with the help of two fast scanning magnets deflecting the beam in the $x$ and $y$ directions (i.e. a "slice" of the tumor at a given depth), while the third dimension (depth) is scanned by varying the energy of the incoming beam. Until now, this third parameter largely favoured the use of a cyclotron, since it delivers an almost continuous beam whose energy can be relatively rapidly changed via the energy degrader, while in the case of synchrotrons, there is a delay of around 1–2 seconds between two different energies of the beam.

It is important to acknowledge the major contribution made by the Paul Scherrer Institute (PSI, Villigen, Switzerland) since their innovative ideas and realisations gave the impetus to a totally new set of specifications for the treatment protocols which then impacted on the design, build and operation of hadron therapy centres supplied by industry [2,3,7].

[1]Stand as of 2014.

The cyclotron at PSI was built by the VARIAN company with additional modifications implemented by the PSI people to meet their particular needs. PSI developed a new beam delivery scheme (PROSCAN [2,3]) which represented a major step forward which is likely to influence most of the new accelerator designs, since PSI were able to perform 3-D irradiation scans at speeds which looked simply not be previously achieved.

It now looks like the technical specifications for a modern hadron therapy centre have significantly changed and can be summarised as:

- The total external dimensions of the centre (i.e. the accelerator, the transfer/preparation lines and the treatment rooms etc.) must be small enough to be accommodated easily at a hospital location.
- The method for the treatment should definitively be scanning rather than scattering.
- The treatment rooms should have small dimensions with compact beam delivery systems, and include all the required imaging systems needed to check the performance of the irradiation session *in situ* and if possible online.
- For a centre with more than one treatment room, one of the treatment rooms should be equipped with a gantry rotating azimuthally around the patient.
- The capital and operating costs of such a centre and the number of patients to be treated each year should be planned such that the centre operations would be financially viable.
- Ideally, the centre should be able to provide patient treatments with different types of light ions. As will be shown later in this report, this requirement might finally prove impossible for certain centres.

With the basic parameters having been explained, let us now turn to the existing industrial situation in Europe. PSI is performing extremely well, but PSI cannot be considered as an industrial contributor, because it remains a national scientific laboratory treating only a relatively small number of patients each year. For CNAO in Pavia, the centre is still in its initial phase working with a small number of patients to obtain the corresponding certification procedures. One is therefore basically left with two industrial projects, namely IBA (Belgium) and Siemens (Germany), which will be considered separately in the following.

### 4.1 *Siemens*

Siemens acquired the techniques and knowhow developed at GSI (see Chapter 42) and adopted the synchrotron approach. The Heidelberg Ion-Beam

Therapy Center (HIT) works well and has the advantage of having a rotating gantry. Similar to all machines based on a synchrotron, HIT has the capability to operate with both protons and carbon ions. The number of patients treated in Heidelberg was around 640 patients in 2013. A target of 750 patient treatments has been set for 2014, and there is confidence that the goal of 1000 patients per year with proton beam treatment will be achieved in the following years.

However this specific aspect raises certain problems. Based on the success at Heidelberg, Siemens moved forward to sign contracts for the construction of three new centres: one in Marburg (Germany), one in Kiel (Germany) and a third one in Shanghai (China). The main reasoning behind these commercial decisions was clear — based on the operational performance obtained at Heidelberg plus the incorporation of new developments by Siemens, supported by DANFYSIK, it was estimated that about 2500 patients per year could be treated. Assuming a treatment cost at the facility of 16,000 Euros per treatment and considering that some insurance companies had already agreed to reimbursement levels of 20,000 Euros per treatment which seemed similar to the conditions expected at CNAO, it seemed to Siemens and its partners that it could develop this into a profitable business.

Unfortunately, after factoring in all the time needed to set up correctly all the fine tuning of the beam and the beam energy, as well as complying with the rigorous certification procedures, there appeared to be a significant drop in the projected patient throughput numbers. Following careful re-evaluation of these different operational parameters, Siemens came to the conclusion that the numbers recorded at Heidelberg would be at best the maximum achievable, i.e. around 1000 patients per year.

By the time this was realized in mid-2011, the Marburg centre was fully operational and the Kiel centre was well advanced with an anticipated completion date near the end of 2011. At this very late stage with hundreds of millions of Euros already invested, Siemens and its partners made a business decision to stop with immediate effect the two projects based in Marburg and Kiel and Siemens announced that they would definitively abandon any commitment to the construction of any further hadron therapy centre, apart from the already launched project in Shanghai, and that they would focus on imaging procedures which will, by definition, become essential tools for future treatment centres.

At this stage, it might be interesting to add a personal and non-economical consideration which might or might not be relevant to reinforce the decision to abandon: the centres proposed by Siemens were indeed based on synchrotrons (i.e. different particles and variable energy), but with little possibility to increase significantly the repetition rate of the beam delivery. In this respect, it implies that the 3-D painting technique which seems to have established itself as a basic future requirement could not have been achieved with these machines. This specific aspect will be briefly addressed at the end of the report.

As a conclusion, it can be said that Siemens is no longer a 'player' in the industrial segment of hadron therapy centre supply. However, this indicates that Siemens will put all the required efforts on the development of the associated imaging systems which will become an essential ingredient in the future treatments with hadron therapy.

### 4.2 *IBA*

The situation and the evolution of IBA (Ion Beam Applications) are very different from that of Siemens, since IBA has followed a very successful evolution in the development of the accelerator system. As indicated earlier, IBA has been involved in the production of cyclotrons from its very beginning with the successful production of different machines for radioisotopes production. From this initial phase onwards, IBA has always adapted to the new demands and related developments arising in the market and the medical environment. For hadron therapy IBA has focused on the steadily evolving requirements of the medical community.

When it was established that proton therapy centres should provide protons with an energy range between 100 and 250 MeV, IBA came on the market with a very conventional solution referred to as the "C235" cyclotron. This conventional cyclotron was still large, using the conventional method of energy variation by scattering.

With the imperative requirement to reduce as much as possible the overall size and concomitant cost of a centre and possibly more importantly, to focus on the pencil scanning approach, IBA developed a new system referred to as the "Proteus One®" which provided an ultra compact system, including two new components — a superconducting synchro-cyclotron and a new gantry system allowing for full pencil beam scanning capabilities.

This new system is compact, relatively cheap, and targeted to meet customer needs in new markets whilst still employing the latest technology in the accelerator field. This design might well fulfill the requirements of the numerous medical centres interested in proton therapy.

In parallel to the move to superconducting cyclotron for the "Proteus One®", IBA decided to also try to cover the field of carbon ions therapy.

Here the challenge is more difficult since in order to penetrate the body to the same depth as protons, the energy of the carbon ions has to be increased to around 400 MeV/u. To build a cyclotron reaching this energy with dimensions remaining within "reasonable limits" has imposed the need to study a much more innovative system and in addition, the associated gantry is far from being straightforward. Despite all of these technical challenges, IBA and partners have proposed the "C400" cyclotron concept which will consist of a complete treatment centre with carbon ions (including a cyclotron, transfer lines and gantries) with the cyclotron and gantries using superconducting magnet technologies.

It would appear that the "Proteus One®" system is mature enough to become the major 'player' on the proton therapy market, while the "C400" project still needs to be demonstrated but offers promising solution for hadron therapy.

For more than 20 years, IBA has demonstrated a remarkable ability to identify and recognize customer needs, to adapt itself and to propose new technical solutions in this fascinating field which holds undoubtedly a bright future.

## 5 A personal point of view from the author

CERN is clearly a scientific laboratory and not an industrial operator, nevertheless CERN has been examining possible future trends in the field of hadron therapy centres. A detailed review is in preparation [4], but there are a few very basic considerations which should be mentioned at this early stage.

A hadron therapy centre comprises three distinct main components:

- The accelerator itself (including the ion sources), where the ions themselves are accelerated to the required energy. In using protons or alpha particles, the solution of a cyclotron seems to be the most appropriate accelerator solution, since the cyclotron provides an almost continuous beam and presents rather favourable conditions for the 3-D pencil scanning method. Also activation of the zone containing the energy degrader and the collimators seems to be acceptable.
  For accelerating more than one type of ion, the need to vary the energy of the ion beam over a large range becomes essential. This makes the synchrotron the most appropriate accelerator solution (actually, there exist also other acceleration mechanisms fulfilling this condition, as will be discussed in the more detailed report in preparation).

For the sake of completeness, it should be mentioned that CERN has completed a study demonstrating the feasibility of a rapid cycling synchrotron (30 Hz) [5] which could represent an interesting alternative which circumvents the apparent limitation of synchrotrons for the depth scanning via fast energy variation. Also this type of accelerator would have the capability for intensity modulation of the beam which is required in most patient dose plans. Intensity modulation is already used in many machines (or projects), but it is certainly a positive aspect that it could be included in the rapid cycling option.

- Transfer lines are required to transport the beam coming out from the accelerator to the treatment room (gantry) and to perform the required optical manipulations. In the case of a synchrotron, the transfer line is the place where beam optical functions are manipulated so that the beam entering the gantry has the correct beam optical parameters, which have to be independent of the angle of the rotating gantry. In a cyclotron, this is the part of the machine where the energy degrader is located as well as the collimators to shape the beam to conform with the requirements of the gantry (e.g. to create a round beam). It is also often the place where the intensity modulation takes place.
- The treatment room is mainly composed of a rotating gantry and a key component in the design of the gantry is the final dipole magnet whose role is to deflect the beam in such a way that the beam arrives perpendicular to the table where the patient is positioned. This last magnet is probably the most important part of the machine since the magnetic field it can produce will directly influence the size (volume) of the treatment room. As mentioned in the beginning, one of the objectives for a new machine is its compactness, and in this respect the size of the treatment room is a major issue for a machine designed to work with carbon ions. As an illustration, with a conventional resistive magnet the required radius above and below the patient for the rotating gantry will be of the order of 10 metres (i.e. a minimum height of about 20 metres for the room) [6].

  Should this dimension be considered excessive, the size of the room can obviously be reduced by simply increasing the magnetic field of the last magnet, i.e. by using a superconducting magnet. The downside of choosing a superconducting magnet is that it might prevent a rapid longitudinal scanning variation of the energy for depth penetration.

  In fact, when the energy of the beam is modified, it is necessary to vary the magnetic field in the magnet, so that the dimensions and the position

of the beam remain unchanged. This very basic consideration might be problematic for a superconducting magnet, because its field cannot be changed sufficiently rapidly to match with the expected speed of the energy scanning.
At present, it seems as if this specific requirement has possibly been slightly underestimated. However, the only solution would be to select a conventional magnet for the gantry which will obviously work against the requirement of a small, compact treatment room.

In conclusion, the main goal of this report has been to evaluate the status of industrial projects in Europe for hadron therapy. The main issues which have been identified can be summarised as follows:

- Industrial projects have developed in two distinct directions. Both have been strongly influenced by the pioneering work performed since 2005 at the Paul Scherrer Institute (PSI) in Villigen, as far as the basic requirements for the treatment procedure itself are concerned. New developments and innovative solutions continue to be developed at PSI and elsewhere, but since this work is mainly originating from scientific laboratories rather than from industry, it is not included in the present discussion.
- The first widely distributed industrial option was followed by IBA (Louvain-la-Neuve, Belgium) who opted for a cyclotron solution, because of their huge experience accumulated over the last 20 years, starting with dedicated cyclotron for radioisotope production, and contributed to the evolution of equipment and new clinical requirements in the field of proton therapy. They are commercializing cyclotron-based multi-room systems including a variety of options and treatment modalities, called "Proteus Plus®". Complementarily, they are now proposing a new superconducting synchro-cyclotron in the single-room, compact and affordable "Proteus One®" system, which fulfils most of the latest requirements of the active scanning treatment method. In parallel, IBA has proposed the design for a new large facility dedicated to therapy with carbon ions. The whole system will be superconducting and the accelerator part will be composed of a large superconducting cyclotron, the "C400" which will be able to accelerate carbon ions to 400 MeV/u and protons up to 230 MeV (the machine will also be able to accelerate any ion species having a charge to mass ratio of 1/2 up to carbon). Installation of the first prototype of this accelerator at the Archade centre in Caen, France, is being discussed.

- The second industrial option is that retained by Siemens, which is based on a synchrotron solution. This choice is justified for accelerating protons and some heavier ions (e.g. carbon) in the same structure. Synchrotrons are well understood machines and are presently used in all centres using carbon ions (two centres in Japan, Heidelberg, CNAO and later at MedAustron). Siemens' decision to opt for this solution relied on the successful experimentation and positive clinical results achieved at GSI. Siemens therefore committed itself to build three centres for hadron therapy on top of the Heidelberg facility, namely in Marburg, Kiel and Shanghai. As already mentioned, the technology of synchrotrons is well under control but it might be that the difficulties were underestimated for the delivery of the beam to the patients and the compliance with the necessary certification procedures. Siemens' stated position is that the expected number of patients to be treated each year would have been much smaller than initially estimated and that, as a consequence, the idea of operating such centres could not be financially justified and therefore the centres should be simply abandoned. It is worth recalling that, by the time the decision to give-up was taken, Marburg was totally completed and Kiel was almost finished. Despite of the availability of these new tools, the decision was taken to dismantle the two centres. It remains true that it was also decided to continue with the Shanghai project, although it is not yet clear whether this centre will be used as a research centre or dedicated to treat patients.
- It should be recalled that for carbon ions a synchrotron is a simpler option but not fulfilling some basic boundary conditions such as a compact design and allowing for a "fast" 3-D scanning. There is still some expectation that this latter limitation can be removed either by moving to a fast cycling synchrotron or by using a different type of accelerator (e.g. linac, FFAG or DWA). Many development studies are being actively pursued at many laboratories given the obvious commercial interest and the potential market for this exciting radiotherapy treatment field.

## References

[1] CERN Yellow Report, Proton-Ion Medical Machine Study (PIMMS), CERN-2000-006 (2000).

[2] J.M. Schippers *et al.*, "The SC cyclotron and beam lines of PSI's new proton therapy facility PROSCAN", *NIM in Physics Research B* **261** (2007) 773–776.

[3] J.M. Schippers and A.J. Lomax, "Emerging technologies in proton therapy", *Acta Oncologica* **50** (2011) 838–850.

[4] D. Brandt, Report in preparation, CERN-DG-DI-DAT (2013).

[5] K. Hanke *et al.*, "Study of a rapid cycling synchrotron to replace the CERN PS Booster", CERN-ATS-2011-054 (2011).
[6] E. Keil, "Thoughts about gantries", CERN-ATS-2012-063 (2012).
[7] T. Haberer *et al.*, "Beam Delivery Techniques: Scanning Beams", in *PTCOG 48*, Heidelberg, Germany, September 28–October 3, 2009.

## Chapter 45

# OPENMED: A facility for biomedical experiments based on the CERN Low Energy Ion Ring (LEIR)

Christian Carli (CERN)

At present protons and carbon ions are in clinical use for hadron therapy at a growing number of treatment centers all over the world. Nevertheless, only limited direct clinical evidence of their superiority over other forms of radiotherapy is available [1]. Furthermore fundamental studies on biological effects of hadron beams have been carried out at different times (some a long time ago) in different laboratories and under different conditions. Despite an increased availability of ion beams for hadron therapy, beam time for pre-clinical studies is expected to remain insufficient as the priority for therapy centers is to treat the maximum number of patients. Most of the remaining beam time is expected to be required for setting up and measurements to guarantee appropriate good quality beams for treatments. The proposed facility for biomedical research [2] in support of hadron therapy centers would provide ion beams for interested research groups and allow them to carry out basic studies under well defined conditions. Typical studies would include radiobiological phenomena like relative biological effectiveness with different energies, ion species, and intensities. Furthermore possible studies include the development of advanced dosimetry in heterogeneous materials that resemble the human body, imaging techniques and, at a later stage, when the maximum energy with the LEIR magnets can be reached, fragmentation.

## 1 LEIR facility and usage for biomedical experiments

The Low Energy Ion Ring (LEIR) [3, 4], reusing to a large extent the hardware of the former Low Energy Antiproton Ring (LEAR) described in Chapter 19, has been constructed in order to generate dense heavy ion bunches for the LHC by accumulation with electron cooling. In addition, a fixed target ion program using the SPS is based on beams coming from

Linac3 and LEIR. The LEIR synchrotron has a size and an energy range very similar to typical machines designed for hadron therapy with carbon ions. Furthermore, the machine is used only for limited periods when beam is required for physics experiments and for setting up beforehand. LEIR will be maintained in the long term for LHC ion operation, so with limited upgrades and little additional efforts, beams can also be provided for biomedical experiments. Even though usage of LEIR for physics program tends to increase and scheduling is likely to become an issue, beam time could be available for a biomedical program provided sufficient priority is given.

If such a facility is used as is (without an upgrade of its injector and uses Linac3 as installed at present), biomedical experiments and setting up for them will be possible only during dedicated runs and will not allow the production of "medical ions" simultaneously with beams for high energy physics experiments. Even though this scenario is reasonable for a first exploitation period of the facility, studies on an upgrade of the LEIR injector are on-going in parallel. One option is to equip the present LEIR injector Linac3 with a second dedicated source and RFQ optimized for lighter ions of interest for radiobiological studies. Another option is the construction of a dedicated injector replacing Linac3. With such an upgrade, operation of a biomedical facility becomes possible in time-sharing mode during LHC ion runs between fills. Switching between heavy ion operation for LHC and light ions could only be done within about a minute due to the time constants of the massive magnets in the beam line.

Heavy ion beams are extracted towards the PS ring by a standard fast extraction using kicker magnets installed in the straight section upstream from the ejection and a septum magnet. For biomedical experiments, a new extraction as indicated in Fig. 1 has to be created.

## 2 Implementation of a slow extraction in LEIR

At present, the ion beams accumulated in LEIR can only be extracted by "fast extraction" towards the PS ring. It would be non-trivial to send the beam from this fast extraction channel to a location suitable for biomedical experiments. However, a slow extraction through another straight section could be implemented with a limited number of hardware modifications such that the beam leaves the machine in a direction compatible with the installation of experiments in the "south hall", where sufficient space can be made available. The implementation of a slow extraction not only allows more flexibility for experiments, but is also technically simpler, as it is not possible

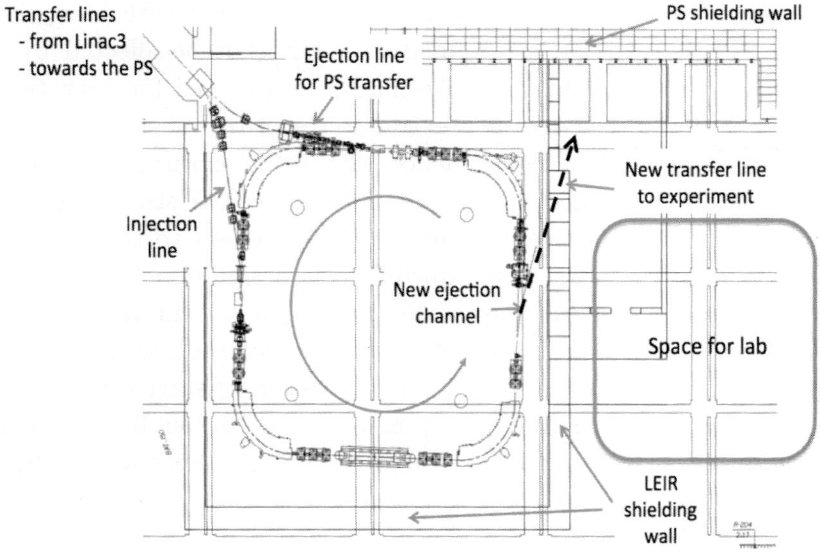


Fig. 1. A sketch of the new facility.

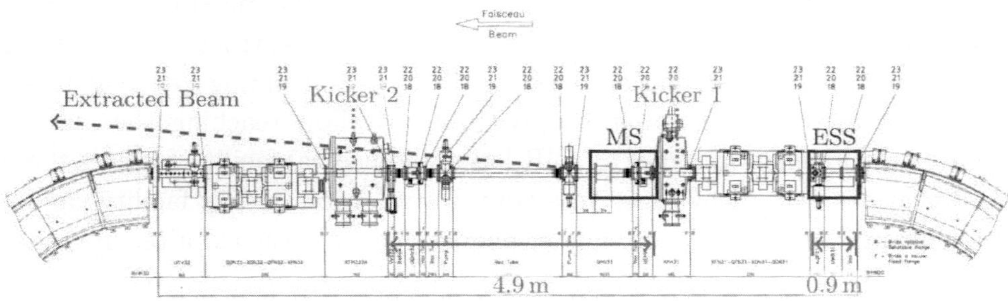


Fig. 2. Sketch of the new extraction section to implement a new slow extraction.

to find space for the installation of new additional kickers for extraction of the medical beam.

The extraction section for the implementation of a new slow extraction scheme [5] is shown in Fig. 2. New hardware installed must be compatible with the two kicker magnets required for the fast extraction of heavy ion beams towards the PS. An electrostatic septum is installed in the first part of the straight section between the bending magnet and a quadrupole doublet. For the extraction of beams with limited energy, one magnetic septum installed further downstream is sufficient. Extraction of beams with higher energy corresponding to the maximum possible magnetic field of the bending magnets and possibly implemented as a second step will require a second electrostatic and magnetic septum.

The large acceptances required for LEIR operation with heavy ions for LHC and SPS have to be maintained and must not be affected by a new slow extraction scheme. Thus, unlike machines designed for operation with slow extraction, the electrostatic extraction septum must not reduce the machine acceptance. As a consequence, either the septum must be moved between operation for heavy ions and for biomedical studies or a closed orbit bump bringing the beam during extraction closer to the septum must be implemented. The latter is assumed for the design done at present.

Unlike typical synchrotrons of hadron therapy centers, LEIR is not shielded from above; this is permitted because of the low intensities and the low maximum energy for heavy ions. However, operation for a biomedical facility at the maximum energy possible with the bending magnets will require additional shielding of LEIR and the transfer lines to be added.

## 3 Status and outlook

Both the implementation of a slow extraction and a transfer line to a vertical and a horizontal target station are studied at present.

Various technical components presently limit the maximum possible beam energy. Some of the magnets required for slow extraction reach their upper limit, ambient radiation without additional shielding becomes unacceptable and the main power supply does not allow reaching the magnet limit. The implementation of a LEIR based biomedical facility foresees these modifications and is envisaged in a staged way. A low energy facility can be implemented more easily and higher energies can be reached after upgrades later.

## References


[1] M. Dosanjh, B. Jones and S. Myers, A possible biomedical facility at the European Organization for Nuclear Research (CERN), *The British Journal of Radiology* **86** (1025), 20120660 (2013).

[2] D. Abler *et al.*, Feasibility study for a biomedical experimental facility based on LEIR at CERN, *J. Radiat. Res.* **54**, 162–167 (2013).

[3] P. Belochitskii *et al.*, LEIR commissioning, in *Proceedings of EPAC 2006*.

[4] D. Manglunki *et al.*, Performance of the CERN heavy ion production complex, in *Proceedings of IPAC 2012*.

[5] A. Garonna *et al.*, Proposal for a new LEIR slow extraction scheme dedicated to biomedical research, CERN-ACC-2014-0061.